Cellular Functions in Immunity and Inflammation

Editors-in-Chief

Joost J. Oppenheim, M.D.
Chief of the Cellular Immunology Section
National Institute of Dental Research
National Institutes of Health
Bethesda, Maryland

David L. Rosenstreich, M.D.
Associate Professor of Microbiology and Medicine
Department of Medicine and Department of Microbiology and Immunology
Albert Einstein College of Medicine
Bronx, New York

Michael Potter, M.D.
Chief of Immunochemistry Section of the Laboratory of Cell Biology
National Cancer Institute
National Institutes of Health
Bethesda, Maryland

Elsevier/North-Holland
New York • Amsterdam

Elsevier North Holland, Inc.
52 Vanderbilt Avenue, New York, New York 10017

Sole distributors outside the United States and Canada:

Edward Arnold (Publishers) Ltd.
41 Bedford Square, London WC1B 3DQ

© 1981; by Elsevier North Holland, Inc.

Library of Congress Cataloging in Publication Data

Main entry under title:

Cellular functions in immunity and inflammation.

    Bibliography: p.
    Includes index.
    1. Immunity. 2. Inflammation. 3. Immunocompetent cells.
    1. Oppenheim, Joost J. II. Rosenstreich, David L. III. Potter, Michael.
QR185.3.C44 1981   616'.0473   81-4678
ISBN 0-444-00554-4   AACR2

Cover photo provided by B. K. Wetzel and S. M. Wahl: Scanning
electron micrograph of a migrating guinea pig peritoneal macrophage
on a 5μ polycarbonate filter.
Desk Editor Danielle Ponsolle
Design Edmée Froment
Design Editor Virginia Kudlak
Cover Design Paul Agule Design
Production Manager Joanne Jay
Compositor Waldman Graphics
Printer Halliday Lithograph

Manufactured in the United States of America

To the many investigators
from Louis Pasteur, Robert Koch, Elie Metchnikoff, Paul Ehrlich on,
who discovered the cells and factors responsible
for immunological and inflammatory reactions,
that have made this book possible.

Contents

Inflammation and Wound Healing   453
Sharon M. Wahl, Ph.D.

Index   467

Preface

The study of the processes that underly host defense against foreign substances has become ever more compartmentalized as our knowledge has increased. Immunologists study the specific immune reactions of the host to an antigenic stimulus, which primarily involve lymphocytes and immunoglobulins. In contrast, students of inflammatory reactions investigate the non-specific inflammatory reactions that are mediated by polymorphonuclear leukocytes, macrophages, complement enzymes, prostaglandins, and kinins. Moreover, even within these broad categories, specialization is the rule therefore many investigators study only a specific cell type or factor.

Although such specialization is an extremely efficient means of producing rapid progress in a small area, it is clear to most of us that these processes to not occur in isolation, but are mechanistically interrelated and also occur in a close temporal relationship, either simultaneously or sequentially. The more we know about these systems, the more we realize how they interact with each other. For example histamine and serotonin which are released in acute inflammatory reactions influence lymphocyte-dependent delayed hypersensitivity whereas some lymphocyte products result in "nonspecific inflammation." Thus cells and their products participate both in immunologically mediated and nonspecific inflammatory processes.

For these reasons, we felt that it was important that the processes of immunity and inflammation be considered as an integral unit rather than as isolated subjects. We felt that this approach would give the working scientist some new insights, and the beginning student a less fragmented, better structured framework on which to build his knowledge.

With this in mind, we organized a post-graduate course focusing on the mechanisms of host defense at the National Institutes of Health that included lectures from a number of related disciplines not usually considered together. Although each area was taught by specialized investigators the common theme of interaction between diverse biological systems was stressed. This course has proved to be popular and informative for student and instructors alike. However, in the course of organizing and giving the course, one noted a conspicuous absence of a single book that used this approach or included the diversity of subjects under one cover that we considered to be essential. Because of this, we decided such a text would be useful, and this book is the result.

The contents of the book are a combination of classical subjects covered in text books on immunology and inflammation. Thus, there are chapters on T- and B-lymphocytes, macrophages, immunoglobulins, complement, and on inflammatory cells such as neutrophils, eosinophils, and basophils, as well as chapters concerning important molecules such as acute phase proteins, kinins, prostaglandins, interferon and lymphokines modulating inflammation. However, we have also included chapters on other important interactions that are often over looked, or studied in isolation such as the effects of hormones on inflammation and the effect of inflammatory processes in tissue repair. In addition to a thorough discussion of each topic, we have tried to emphasize the interrelationships and interactions of cells and factors that serve to culminate in immunity and inflammation. In order to keep this volume to a manageable size some topics, such as platelets, that did not warrant a separate chapter were briefly considered in a number of the chapters.

We would like to thank all the contributors for their work and cooperation, our colleagues for their many helpful suggestions, Drs. Stanley Cohen and Gerald Weissman for their constructive critiques of this volume and the secretarial and editorial assistance that made this book possible.

Joost J. Oppenheim, M.D.
David L. Rosenstreich, M.D.
Michael Potter, M.D.

List of Contributors

Arthur O. Anderson, M.D.
Assistant Professor of Pathology and Biology, The University of Pennsylvania, Philadelphia, Pennsylvania

Norman D. Anderson, M.D.
Associate Professor of Medicine and Surgery, The Johns Hopkins Medical Institutions, Baltimore, Maryland

Sheldon G. Cohen, M.D.
Director of Immunology, Allergic and Immunologic Disease Program, National Institute of Allergy and Infectious Diseases, Bethesda, Maryland

John M. Davis, M.D.
Assistant Professor of Surgery, Laboratory of Clinical Investigation, National Institute of Allergy and Infectious Diseases, National Institutes of Health, Bethesda, Maryland

Anthony S. Fauci, M.D.
Chief of Clinical Physiology Section, Laboratory of Clinical Investigation, National Institute of Allergy and Infectious Diseases, National Institutes of Health, Bethesda, Maryland

Robert M. Friedman, M.D.
Chief of Laboratory of Experimental Pathology, National Institutes of Arthritis, Metabolism, and Digestive Diseases, National Institutes of Health, Bethesda, Maryland

John L. Gallin, M.D.
Head of the Bacterial Diseases Section, Laboratory of Clinical Investigation, National Institute of Allergy and Infectious Diseases, National Institutes of Health, Bethesda, Maryland

Michael Kaliner, M.D.
Head of the Allergic Diseases Section, Laboratory of Clinical Investigation, National Institute of Allergy and Infectious Diseases, National Institutes of Health, Bethesda, Maryland

Allen P. Kaplan, M.D.
Professor of Medicine and Division Head, Division of Allergy, Rheumatology and Clinical Immunology, Department of Medicine, SUNY, Stony Brook, Health Sciences Center, Stony Brook, New York

Dean D. Metcalfe, M.D.
Senior Clinical Investigator, Laboratory of Clincial Investigation, National Institute of Allergy and Infectious Diseases, National Institutes of Health, Bethesda, Maryland

David L. Nelson, M.D.
Senior Investigator, Metabolism Branch, National Center Institute, National Institutes of Health, Bethesda, Maryland

Joost J. Oppenheim, M.D.
Chief of the Cellular Immunology Section, National Institute of Dental Research, National Institutes of Health, Bethesda, Maryland

Eric A. Ottesen, M.D.
Senior Investigator of the Labor of Parasitic Disease and Laboratory of Clinical Investigation, National Institute of Allergy and Infectious Diseases, Immunology, Allergic and Immunologic Diseases Program, Bethesda, Maryland

Michael Potter, M.D.
Chief of Immunochemistry Section of the Laboratory of Cell Biology, National Cancer Institute, National Institutes of Heath, Bethesda, Maryland

David L. Rosenstreich, M.D.
Associate Professor of Microbiology and Medicine, Department of Medicine and Department of Microbiology and Immunology, Albert Einstein College of Medicine, Bronx, New York

Ann Sandberg, Ph.D.
Chief of Humoral Immunity Section, Laboratory of Microbiology and Immunology, National Institute of Dental Research, National Institutes of Health, Bethesda, Maryland

Irwin Scher, M.D.
Chairman, Department of Immunology, Naval Medical Research Institute and Uniformed Services University of the Health Sciences, Bethesda, Maryland

Ethan M. Shevach, M.D.
Senior Investigator of the Laboratory of Immunology, National Institute of Allergy and Infectious Diseases, National Institutes of Health, Bethesda, Maryland

Jean D. Sipe, Ph.D.
Associate Research Professor of Medicine and Biochemistry–Arthritis Section, Boston University Medical Center, Boston, Massachusetts

Reuben P. Siraganian, M.D.
Chief of Clinical Immunology Section, Laboratory of Microbiology and Immunology, National Institute of Dental Research, National Institutes of Health, Bethesda, Maryland

Henry C. Stevenson, M.D.
Chief, Basic Mechanisms Section, Biologic Response Modifiers Program, National Cancer Institute, National Institutes of Health, Frederick, Maryland

Sharon M. Wahl, Ph.D.
Senior Investigator, Laboratory of Microbiology and Immunology, National Institute of Dental Research, National Institutes of Health, Bethesda, Maryland

Immunity and Inflammation

Joost J. Oppenheim, M.D. and Michael Potter, M.D.

The cardinal signs of inflammation, rubor, calor, dolor, and tumor, represent the manifestations of the basic vascular and cellular responses of the mammalian organism to infection, tissue injury, and intrusion of foreign materials. These inflammatory responses serve to maintain the integrity of the host by eliminating dead tissue, microbes, toxins, and inert foreign substances (Figure 1). It is the purpose of this chapter to delineate the role of immunity in the various types of inflammatory reactions. We will classify inflammatory responses as either nonimmunological, when they only involve vascular changes and nonlymphocytic inflammatory cells, or as immunological, when they involve lymphocyte dependent reactions. Inflammation and immunity are closely interrelated, and in vertebrate species most inflammatory responses have an immunological component. Furthermore, the effector mechanisms of inflammation and immunity are virtually identical. Immunity provides a mechanism for focusing inflammatory reactions on a specific target and thus has enabled inflammatory reactions to evolve into a more efficient, effective, and rapid host defense response. Immunity can either preempt or suppress the inflammatory response to a foreign substance or, alternatively, mobilize an inflammatory reaction that at times may be excessive and even self destructive. Conversely, nonimmunological stimulants can have adjuvant effects and markedly augment immunologically mediated host reactions.

Historical Overview

Historically, the concepts that inflammation and immunity participate in host defense were developed independently of one another. The close interrelationship of these two processes has been discovered in a piecemeal fashion during the past century.

Rudolph Virchow, the father of cellular pathology, recognized in 1858 that inflammation represents the host response to an exogenous insult.

Whilst until quite recently it was the custom to look upon inflammation as a real entity, as a process everywhere identical in its essence, after I made my investigations no alternative remained but to divest the notion of inflammation of all that was ontological

From the National Institute of Dental Research, National Institutes of Health, Bethesda, Maryland, and the National Cancer Institute, National Institutes of Health, Bethesda, Maryland.

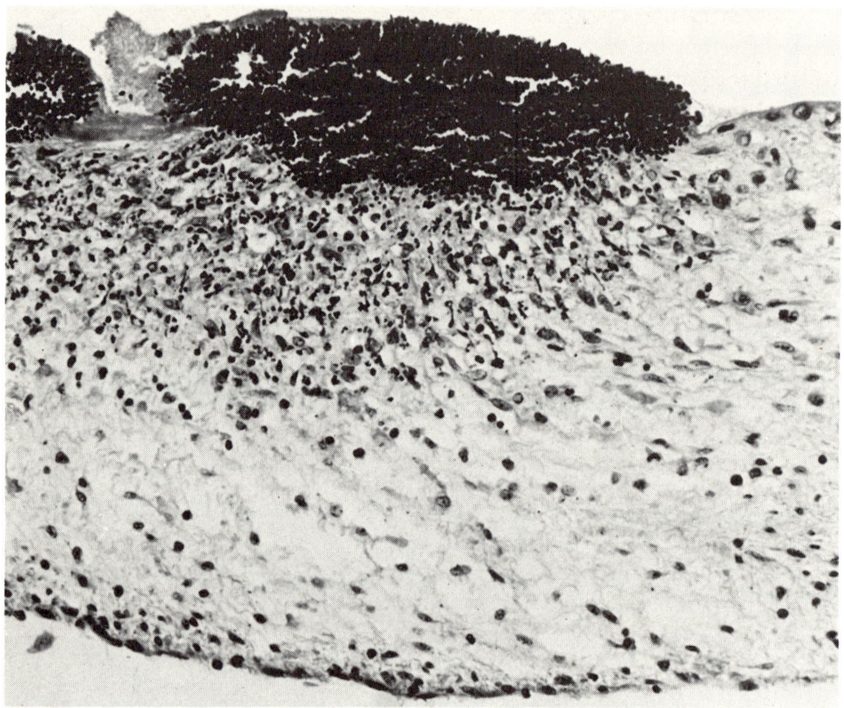

Figure 1. Yeast vegetation on wall of right atrium following central venous catheter related septic episode. Note chronic mononuclear inflammatory cell response to the organisms. (Courtesy of A. Anderson.)

in it, and no longer to look upon the processes but only to regard it as one differing in its form and course. We cannot imagine inflammation to take place without an irritating stimulus (irritament) and the first question is, what conception we are to form of such a stimulus.

Virchow explains the term "irritament":

The term irritament (Reiz which however, sometimes means irritant; stimulus) is intended to express the change (mechanical or chemical, palpable anatomical or molecular) which takes place in a tissue in consequence of the action of an irritant—a change therefore, which is of a fairly passive nature (lesion) and which (subsequently) provokes changes in the neighboring parts not directly altered by the irritant. The consequence of which is their action or reaction.

Virchow then thought of inflammation as host reactions to irritants but not as a nosological entity. Inflammation was a process that could vary in form and course.

Julius Cohnheim in 1867 described the pathophysiology of inflammation as being induced by touching the cornea of a frog's eye with a stick of lunar caustic, and he correlated the changes with histologic sections. He found that the edge of the necrotic site became infiltrated with white blood cells. He developed a vital stain for phagocytic cells and was able to trace the origin of the cells from lymph to blood to inflammatory site. Cohnheim's research focused attention on vascular changes in inflammation and their consequences: the transudation of fluids and the migration of white blood cells into inflammatory sites. What remained to be described was the kinds of cells that occupied an inflammatory site and their function.

The first natural forms of inflammation to be studied in great depth were associated with infections by microorganisms. Robert Koch in 1877 and Louis Pasteur in 1879 isolated the pathogens of ovine anthrax and fowl cholera. These were exciting findings that led quickly to one of the most momentous discoveries in all of medicine: a rediscovery of Edward Jenner's (1798) principle of immunization by Louis Pasteur. For the first time, an isolated propagatable organism that induced immunity was available for detailed study. No finding better illustrates Pasteur's famous saying "chance favors the prepared mind" as did the study of fowl cholera.

The causative agent of fowl cholera was a small gram-negative coccobacillus, now in the *Pasteurella* group. Pasteur, Chamberland, and Roux succeeded in isolating and culturing the organism in a chicken broth medium. They were able to transmit the disease with cultured organisms, and all chickens infected with the cultured organisms administered even in the minutest doses died within 2 to 4 days. In late summer of 1879, the work on fowl cholera came to a temporary halt for the summer vacation. The cultures apparently were not passaged during this period but had remained viable. When work resumed, organisms from the old cultures now failed to transmit the virulent disease. So attempts were quickly made to reestablish virulent organisms. During this period of testing, normal chickens were used, but some "economy minded" person probably threw in a few of the birds that had been inoculated with the old non-virulent cultures. For the first time some chickens survived the fatal disease. Pasteur questioned the source of these survivors and discovered they had been inoculated previously with the nonvirulent cultures. He then recognized he had rediscovered a new means to induce immunity that was related to Jenner's vaccination reported 81 years earlier. A series of interesting experiments followed.

Three quotes from Pasteur's original article on fowl cholera give the facts succinctly.

Par certain changement dans le mode de culture on peut faire que le microbe infectieux soit diminué dans sa virulence.

La diminution dans la virulence se traduit dans les cultures par un faible retard dans le développement du microbe; mais au fond il y a identité de nature entre les deux variétés du virus. Sous le premier de ses états, l'état très infectieux, le microbe inoculé peut tuer vingt fois sur vingt. Sous le second de ses états il provoque vingt fois sur vingt la maladie et non la mort. . . .

Il me paraitrait superflu de signaler les principales conséquences des faits que je viens d'avoir l'honneur d'exposer devant l'Académie. Il en est deux cependant qu'il n'est peut-être pas sans utilité de mentionner: c'est d'une part, l'espoir d'obtenir des cultures artificielles de tous les virus, de l'autre une idée de recherche des virus vaccins des maladies virulentes qui ont désolé à tant de reprises et désolent encore tous les jours l'humanité et qui sont une des grandes plaies de l'Agriculture dans l'élevage des animaux domestiques (Pasteur 1880).[1]

---

[1]"By a certain change in the mode of culture one can bring about a diminution in the virulence of the infectious microbe. The diminution in virulence is reflected in the cultures by a slight delay in the development of the microbe but basically the nature of the two varieties of virus are identical. In the first of these states, the very infectious state, the inoculated microbe can kill twenty times out of twenty. In the second state it provokes an illness twenty times out of twenty but no death.

It would seem to me to be superfluous to point out the principle consequences of the facts that I have the honor to present to the Academy. There are two however, that seem to me to be worth mentioning. One is the hope of obtaining artificial cultures of all viruses, the other the idea of seeking viral vaccines for the virulent diseases which have so grievously desolated and still desolate humanity day after day and which are one of the great scourges of agriculture in breeding domestic animals." (Pasteur 1880)

And with this historic note published in February 1880, modern immunology began. This experiment captivated the minds of many talented investigators and launched a whole new field. The science of medicine would never be the same again.

Elie Metchnikoff, a Russian zoologist, made the first observations that dealt with the relationship of immunity and inflammation. His primary interest was the cellular basis of inflammation. Metchnikoff spent the summer of 1880 with his family in Messina, Italy, where he conducted experiments on provoking inflammations in transparent starfish larvae. These immature invertebrates lacked both a vascular and a nervous system. Metchnikoff placed rose thorns under the skin of the larvae and observed to his delight that wandering phagocytes migrated and attached to these foreign bodies in an attempt to destroy them. He had been intrigued with extraordinary digestive capabilities of phagocytes and their role in metamorphosis. On reading of the exciting work of Pasteur on the role of microorganisms in diseases, rose thorns were replaced by bacteria, and starfish by vertebrates. As in invertebrate inflammation, phagocytic cells migrated to the site of an infection in mammals.

From this followed the conclusions that the diapedesis and accumulation of white corpuscles in inflammatory dieases must be regarded as modes of defense of the organism against microorganisms. The leukocytes in this struggle devouring and destroying the parasites (Metchnikoff 1901).

In 1887 Metchnikoff decided to join Pasteur and left this poignant description:

On arriving at the laboratory destined for antirabic vaccinations I saw an old man rather undersized with a left hemiplegia, very piercing gray eyes, a short beard and moustache and slightly gray hair covered with a black skull cap. His pale and silky complexion and tired look betokened a man who was not likely to live many more years. He received me very kindly and immediately spoke to me of the question which interested me most, the struggle of the organism against microbes. "I at once placed myself on your side" he told me, "for I have for many years been struck by the struggle between the diverse microorganisms which I have had occasion to observe. I believe you are on the right road."

Metchnikoff set to work. His wife Olga began seeing a transformation in him as revealed in this interesting comment that probably typifies the excitement of the era:

As long as Metchnikoff was but a zoologist the scientific atmosphere around him remained calm and serene. But everything changed suddenly when he entered the Pathology with his theory of phagocytes and phagocytosis. Here was the realm of secular traditions deeply rooted and theories generally admitted but resting on no biological basis.

The experiments on inflammation provoked Metchnikoff to propose that phagocytes were also the mediators of immunity.

I thought as already mentioned above that the observation on absorption and leukocytes which had been accumulating for years in pathological histology had sufficiently paved the way for a favorable reception to the idea that the amoeboid cells are defensive elements of the body capable of guaranteeing to it immunity and care (Metchnikoff 1905).

Metchnikoff always argued that the only way an organism could rid itself of a microbial infection was by phagocytosis and the intracellular destruction of organisms. He equated this with immunity.

He explained the mechanisms in vaccination by proposing that natural or artificial vaccination by attenuated microbes allows phagocytes to become gradually accustomed to digest more virulent ones, and this confers the immune state. The notion was simplistic and compelling and Metchnikoff defended it vigorously and ingeniously to the end, even though considerable evidence indicating the role of serum factors became available.

VonBehring and Kitasato in 1891 showed they could transfer antitoxic immunity to diphtheria and tetanus toxins by cell-free serum. Metchnikoff (1905) considered it a special case not a general phenomenon of immunity. He took Pfeiffer's experiments on cholera (1894) as a greater challenge. Pfeiffer showed that the body fluids of guinea pigs immunized to cholera possess substances that destroy cholera vibrios. Metchnikoff repeated these experiments immediately and argued they took place only under special conditions and, further, that some of the active substances (cytases) were actually produced by phagocytes and liberated by them. Pfeiffer had demonstrated that cell-free materials could induce bacteriolysis. Bordet (1895) in Metchnikoff's laboratory extended this work and actually showed two substances were involved in the Pfeiffer phenomenon, a heat-labile substance called alexin (later complement) and a heat-stable sensitizing substance later called antibody. Bordet's experiments indicated the heat-stable substance had specificity for certain targets, thus beginning an appreciation of antibody specificity.

Metchnikoff, ever defending his phagocytic theory, believed at least until 1901 that acquired immunity, as he defined it, arose in phagocytes as a consequence of the digestive process (phagocytosis). In this he was basically wrong: immune bodies, as Ehrlich suggested but by no means demonstrated, had their "origin in special cells of the body" (i.e., not phagocytes). Although Metchnikoff and his contemporaries failed to understand the underlying lymphocytic basis of immunity, his arguments fore-shadowed a relationship of immune and inflammatory processes.

It was not until 1903–1905 that Almoth Wright found that antibody-coated particles (microbes), served to prepare them for phagocytosis and that this is an essential requirement for the phagocytosis of many pathogens. This established the cooperation of the immune system with phagocytosis and led to an explanation of how immunity and inflammation interacted. The macrophage was established as the cornerstone of the inflammatory response. Further work established the details of the phagocytic process and elucidated those situations in which microorganisms subverted this function by dividing intracellularly or inhibiting some aspect of phagolysosome activity. From this research emerged the concept of chronic inflammation.

Pasteur's experiments of 1879 on the induction of immunity to fowl cholera triggered intensive study of immunity to the organisms that caused anthrax, tuberculosis, diphtheria, tetanus, cholera, typhoid fever, and many others. The plan, at first straightforward, was to identify and culture the organism to prepare a vaccine. Fowl cholera unfortunately proved to be an exceptional model, but other diseases soon presented new and unforeseen problems.

In 1882 Robert Koch isolated the tubercule bacillus. The race was on to try to find a means for generating immunity to these organisms. Koch (1891) injected living tubercule bacilli subcutaneously in guinea pigs, which caused the formation of a nodule that formed in about 10 days and finally ulcerated but never healed. When he injected living bacilli into the skin of a guinea pig that had been previously infected with tubercule bacille, he found that an inflammatory reaction developed at the injection site in 1 to 2 days. In contrast to the former primary infection this lesion ulcerated but quickly healed over. Koch also showed he could induce the phenomenon with killed organisms or even a protein isolated from these organisms called tuberculin. The Koch phenomenon is an example of *delayed hypersensitivity*.

Although Pasteur attempted to induce immunity to fowl cholera with nonliving materials, he was not successful. Emil Roux (1887), however, succeeded in inducing immunity to a septicemia caused by a bacillus that was isolated from soil and the intestines of sheep. This laboratory-induced disease was 100% fatal in guinea pigs, usually within 24 hours. Roux showed he could immunize guinea pigs with 80 ml of sterile culture filtrates, and protection from death would result.

Roux, Yersin, and Loeffler discovered that cultures of *Corynebacterium diphtheriae* contained a toxin. The lethal effects of diphtheria infection in human beings were caused by the obstructive effects of a local noninvasive pharyngeal and tracheal exudate or by the diphtheria toxin. VonBehring showed that antibodies to the toxin had neutralizing properties and very soon thereafter he and Kitasato (1891) were able to transfer the antitoxin by injecting serum. Collectively, these and other observations established that chemical substances could induce immunity and that such substances (antigens) were the targets of specific immune bodies "antikorpers." Thus, the concepts of antibodies and antigens began to be appreciated.

The transfer of antitoxins by serum triggered extensive investigations into the nature of the active materials (1891–1901). Paul Ehrlich studied "immune bodies" in serum in great detail. By 1900, in his Croonian lecture, he was able to speak of antibodies as follows:

Much more complex than in the cases hitherto discussed are the conditions when instead of the relatively simple metabolic products of microbes, the living microorganisms themselves come to be considered, as in immunization against cholera, typhoid, anthrax, swine fever, and many other infectious diseases. There then come into existence alongside of the antitoxins, produced as a result of the action of the toxins, manifold other reaction products. This is because the bacterium is a highly complicated living cell, of which the dissolution in the organism yields a great number of bodies of different nature, in consequence of which a multitude of "Antikorper" are called into existence. Thus we see, as a result of the injection of bacterial cultures, that there arise alongside of the specific bacteriolysins, which dissolve the bacteria, other products, as, for example, "coagulins" (Kraus, Bordet) i.e., substances which are able to cause the precipitation of certain albuminous bodies contained in the culture fluid injected; also the so much discussed "agglutinins" (Durham, Gruber, Pfeiffer), the antiferments (VonDungern), and no doubt many other bodies which we have not yet recognized.

It is by no means unlikely that each of these reaction products finds its origin in special cells of the body: on the other hand, it is quite likely that the formation of any single one of these bodies is not of itself sufficient to confer immunity. Thus in case of the introduction of bacteria into the body we have to do with a many-sided production of different forms of "Antikorper" each of which is directed only against one definite quality or metabolic product of the bacteria cell.

At the turn of the century many studies were made on antigens and antibodies. Two new kinds of inflammation were discovered. In 1903 Maurice Arthus found that horse serum injected into normal rabbits was not toxic nor did it produce any lesions. If, however, he sensitized rabbits by giving multiple injections of horse serum several days apart, they developed a striking reaction to a subcutaneous injection. In sensitized rabbits a localized sterile inflammatory lesion developed at the injection site. These lesions usually were characterized by considerable swelling and induration; some even became necrotic. This phenomenon is known as the Arthus reaction. The capability to develop an Arthus reaction can be passively transferred by injecting serum from a sensitized animal into a normal animal and then injecting antigen into a local site. The Arthus reaction therefore results from an antigen–antibody interaction *in vivo* and is the basis for immune-complex disease.

Other experiments involving multiple injection of antigens in animals produced other kinds of acute inflammatory responses, such as anaphylactic shock, that were attributed to hypersensitivity to antigens. Richet and Portier (1902) first encountered anaphylaxis in dogs sensitized with a variety of highly antigenic foreign substances such as eel serum. The term allergy, meaning altered reactivity, was introduced in 1906 by VonPirquet. Other allergic inflammatory responses in man, such as urticaria, hay fever, and asthma, were recognized.

By 1905, three new types of inflammatory processes that stem from immune responses, e.g., delayed hypersensitivity, immune complex disease, and immediate hypersensitivity (allergy), had been discovered. However, the relationship of immune and inflammatory processes was not well understood and awaited the development of much more precise information regarding the cellular basis of the immune response. This information evolved slowly over the next 70 years, and often key advances did not rest on a single discovery but on numerous findings that happened seemingly simultaneously. The crucial missing information about the identity of the cellular system that produced immune substances was revealed by two distinct lines of investigation.

Antibodies were first shown to be produced by plasma cells in 1948 by Astrid Fagraeus. The identification of the precursors of plasma cells, however, would have to await an understanding of the functions of lymphocytes. Two remarkable advances that established much of the direction of immunology in recent years were embodied in the clonal selection theory of antibody formation and the demonstration of the recirculation of lymphocytes. Niels Jerne in 1955 opened the era with his natural selection theory of antibody formation, and Sir Macfarlane Burnet modified it with the clonal selection theory. The clonal selection theory hypothesizes that the precursors of antibody-forming cells must be selected by antigen and that this involved an interaction of antigen and a specialized cell. Shortly thereafter, Nossal and Lederberg provided the first evidence to support Burnet's theory when it was realized that individual plasma cells, which were known to be end-stage cells, were restricted to making only a single kind of homogeneous immunoglobulin (antibody). A small number of genetically determined precursor cells are activated by antigen to proliferate and mature to the plasma cell stage (i.e., immunoglobulin-secreting cells) thus providing a specific response to an antigen.

The experiments of James Gowans on lymphocyte recirculation provided the evidence that lymphocytes were precursors of antibody-secreting cells by demonstrating that some circulating lymphocytes entered tissue sites (e.g., the homing of large lymphocytes to the lamina propria of the gut), where they became plasma cells. Tumors of plasma cells, like their normal counterparts, were also specialized and provided homogeneous immunoglobulins, further evidence that supported the clonal theory. Specific fluorescent antibody probes for immunoglobulins permitted further identification of immunoglobulin-producing lymphocytes, as well as evidence of specialization in immunoglobulin formation as exemplified by the different heavy chain classes.

In 1943 Karl Landsteiner and Merrill Chase demonstrated that delayed hypersensitivity to a specific antigen could be transmitted to a naïve host by cell populations rich in lymphocytes but not by serum. This experiment established the fact that there was a second form of immunity associated with cells rather than with cell-free molecules (antibodies). Over the next few decades lymphocytes were implicated as the cells mediating delayed hypersensitivity. Understanding the function of lymphocytes depended on developing suitable assays and means for isolating specific cell populations. The first clue came from Jacques Miller's experiments on the effect of neonatal thymectomy in mice. Neonatally thymectomized mice were shown to have deficiencies in cell-mediated immunological functions, implicating thymus-derived lymphocytes. Shortly, a comparison of lethally irradiated mice reconstituted with thymus-derived versus non-thymus-derived lymphocytes led to the demonstration by Claman and by Miller and Mitchell that two distinct populations of lymphocytes were essential for cell-mediated and humoral immune responses (later to be called T- and B-lymphocytes, respectively). Three immunological functions attributable to thymus derived lymphocytes were (1) cytotoxicity (in graft rejections), (2) cooperation with B-lymphocytes, and (3) suppressor effects. During the past two decades the development of tissue culture technology has enabled investigators to dissect the cellular interactions involved in immunity and inflammation. Instead of merely describing the host response to an irritant, the identity of the cells or their products or both that are responsible for the host reactions is becoming clearer.

The elucidation of the cellular basis of immune responses, coupled with increasing knowledge on the special functions of inflammatory cells (neutrophil, eosinophil, basophil, macrophage, platelet), has provided the facts and information missing in earlier studies. Inflammatory sites contain both inflammatory cells and immunocytes, and these cells interact in a variety of ways and communicate with each other by liberating various molecular signals that trigger various other effects, just as Virchow pictured it (see above). In this chapter we describe the characteristics of the various types of inflammatory reactions and their relationship to immunological processes. The subsequent chapters in this book discuss in detail the manner in which various cell types and their products contribute to inflammation and immunity.

Phylogenetic Considerations

One of the essential components of a multicellular animal system is a mechanism for protecting and policing the organism's intracellular spaces from the intrusion of foreign materials. In invertebrates, e.g., the coelenterate, wandering phagocytes serve this purpose. As multicellular organisms have increased in size and complexity, the intracellular spaces have also expanded and become compartmentalized, creating potential cavities, e.g., the coelom, and vascular systems. In higher invertebrates, phagocytic cells associated with these spaces continued to be major effectors in removing foreign particulates. Concomitantly, new demands and functions, including the mechanism for handling traumatic tissue destruction and infections (intrusions by replicating agents), required expansion and refinement of phagocytic processes, e.g., circulating phagocytic cells that could migrate into tissues, a system of fixed phagocytic tissue cells to aid in clearing particles from the circulation, the evolution of specialized granular phagocytes, and the appearance of circulating substances that could aid the process of phagocytosis (opsonins). Nonphagocytic (lymphocyte-like) cells with killer or cytotoxic properties have also evolved in invertebrates.

The invertebrate systems involved in the defense of the intracellular spaces, though effective for the organisms, are relatively primitive in form when compared with their counterparts in the vertebrates. In vertebrates, remarkable specialization has evolved. For example, the mobile phagocytic system comprises three specialized cell types: the neutrophil, the monocyte–macrophage, and the eosinophil. An extensive system of tissue-associated phagocytic cells includes the fixed macrophage of the liver sinusoids (Küpffer cells), alveolar macrophages of the lung, peritoneal macrophages, and non-phagocytic Langerhans cells of the skin. Other specialized cell types developed, such as basophils and mast cells. These cells and platelets can be stimulated to release vasoactive substances.

Moreover, vertebrates differ from other animals by having developed a novel group of inflammatory cells: the lymphocytes and plasma cells that are able to interact specifically with foreign intruders. The lymphocytic system is the organ equivalent for the immune system. Although considerable effort has been directed toward finding the counterparts of the immune system in invertebrates, the precise relationships have not been established. The association has been even more puzzling because of the extensive evolution of immune systems that has occurred in vertebrates. This process has taken place both at the cellular level, with the appearance of new cell types, and at the organ level, with the emergence of specialized tissues for generating inflammatory and immune cells as well as organs that permit these cells to interact and function more efficiently, e.g., blood- and lymph-filtering organs such as the spleen and lymph nodes.

The novel feature of the immune system is its ability to focus on specific targets. This is handled by special components of the immune system: thymic-derived (T) lymphocytes with specific antigen-binding receptors and bone marrow–derived (B) lymphocytes that produce antibody molecules. The genome of the organism, embodied in its germ cells and stem cells, has the potential for generating a great variety of antigen-binding and immunoglobulin-producing lymphocytes. Through a complex differentiation process, a lymphocyte bearing antigen-binding receptors or producing immunoglobulin is limited to producing receptors or antibodies with only a single binding specificity. Both the antigen-binding receptors bearing T-cells and the B-cells secreting immunoglobulins fulfill a number of highly specialized immune functions. The basis of the immune response is the selective proliferative expansion of the clone of specifically reactive lymphocytes, which bear receptors that recognize (bind) an irritant and are thus activated to respond to the irritant.

Early Events in Inflammation

Before discussing the unique aspects of nonimmunological and immunologically mediated inflammation, ubiquitous early changes in inflammation in general should be discussed. It must be emphasized that the initial events in inflammation occur largely independently of the nature of the noxious agent (Hurley 1978).

Chemicals, such as toxins and irritants, or physical injuries owing to trauma, infarcts, irradiation, and thermal burns cause cell damage and induce local inflammation in a multifactorial fashion. Direct damage to the vascular walls of capillaries, arterioles, or venules results in leakage and edema, fibrin formation, platelet aggregation, and leukocyte accumulation. This can begin immediately or, if milder, several hours may elapse before the onset of inflammation, and it may persist for up to 24 hours.

Vascular Changes

There are four general vascular manifestations of inflammation, (1) changes in the caliber and blood flow of small vessels in the region of inflammation, (2) changes in venule permeability, (3) adherence of leukocytes and/or platelets to vascular walls, and (4) triggering components of the clotting cascade.

Following any type of injury there is usually an immediate but transient constriction of arterioles followed by dilation of arterioles and venules. Blood flow through an injured area may increase up to 10-fold for minutes to several hours and is followed by a period of stasis prior to normalization of the flow. There is a concomitant immediate increase in vascular permeability with leakage of protein-rich exudate and development of local edema that begins within a few minutes and persists for up to several hours or longer in the case of chronic inflammation. This leakage occurs through transient gaps (0.1 to 0.4 μ) that form at the junctions of endothelial cells of venules and veins. Direct injury will result in leakage from all types of blood vessels.

Pavementing, Emigration, and Chemotaxis of Phagocytic Cells

After several hours in most types of inflammation, neutrophils, some eosinophils, and monocytes begin to adhere to the walls of inflamed vessels, resulting in pavementing. The mechanism of adhesion is unclear. In contrast, lymphocytes do not participate in pavementing. They adhere selectively only to postcapillary high endothelial venules (HEV) in normal lymphoid tissues (see Chapter 2). In the case of vessel injury, platelet adherence to vascular walls is more prominent, and release of vasoactive substances from platelets in turn triggers the clotting process. Erythrocytes escape passively into the tissues in proportion to the degree of injury.

Pavementing is followed by emigration of the inflammatory cells into the injured tissues. The immediate vascular changes that occur nonspecifically during tissue injury or intrusion by foreign materials are concomitantly associated with the active emigration of large numbers of phagocytic cells into the involved site. This occurs predominantly across locally involved vascular structures via the intercellular junctions between endothelial cells (Figure 2). The first cells to arrive in large numbers are the neutrophils and eosinophils. The blood monocytes enter as well; these cells undergo further maturation in target sites to become macrophages, activated macrophages, or histiocytes. Later in chronic inflammation the macrophages can transform into multinucleated giant cells or epithelioid cells. Mediators such as histamine increase permeability and allow the leukocytes to cross the basement membrane into the tissues. A wide variety of chemotactic signals, such as bacterial products, products of the complement or clotting cascades, certain lymphokines, cytophilic antibody, and products of damaged cells, can induce the leukocyte to migrate into the inflamed tissues.

Nonimmunological Inflammatory Reactions

Local Reaction to Tissue Injury

The initial vascular changes and inflammatory response to an injury lead to a secondary sequence of reactions mediated by inflammatory cell products. Injured and dying cells release lysozomal enzymes, oxygen radicals (see Chapter 5), prostaglandin (see Chapter

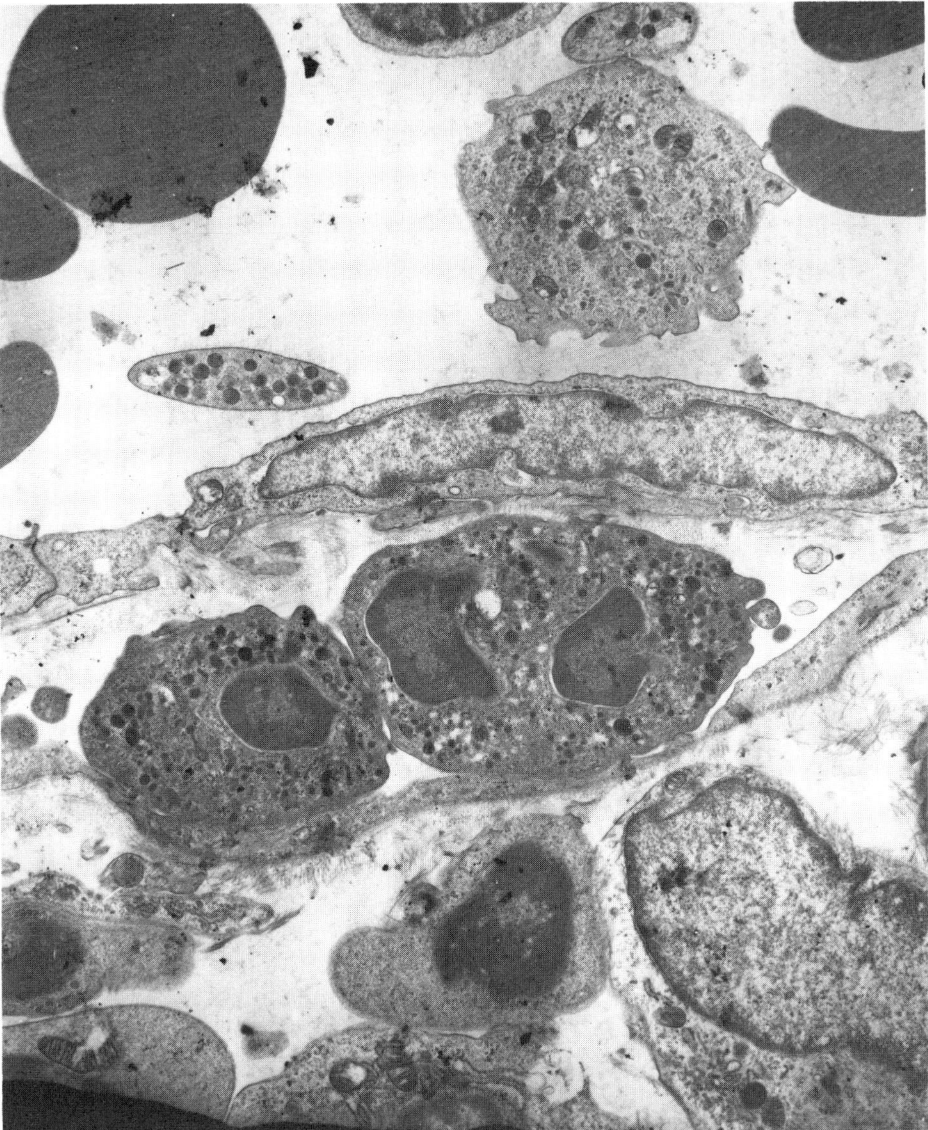

Figure 2. Diapedesis in early acute inflammation: neutrophils that have penetrated blood vessel wall are located between endothelial and perivascular cells. (Courtesy of A. Anderson.)

14), chemotactic factors and histamine (see Chapter 13). These and other factors in turn initiate the clotting sequence, kinin (see Chapter 16), and complement cascades (see Chapter 15), all of which promote inflammation. However, immunological mechanisms may also participate in the early host response to injured cells and debris. Scattered evidence suggests the existence of opsonic and cytotoxic antibodies to altered self-constituents. Such antibodies may play a very important role in facilitating the removal of damaged cells and their products. As such they would serve to reduce inflammatory reactions in tissue injury. The rapidity with which the noxious agent,

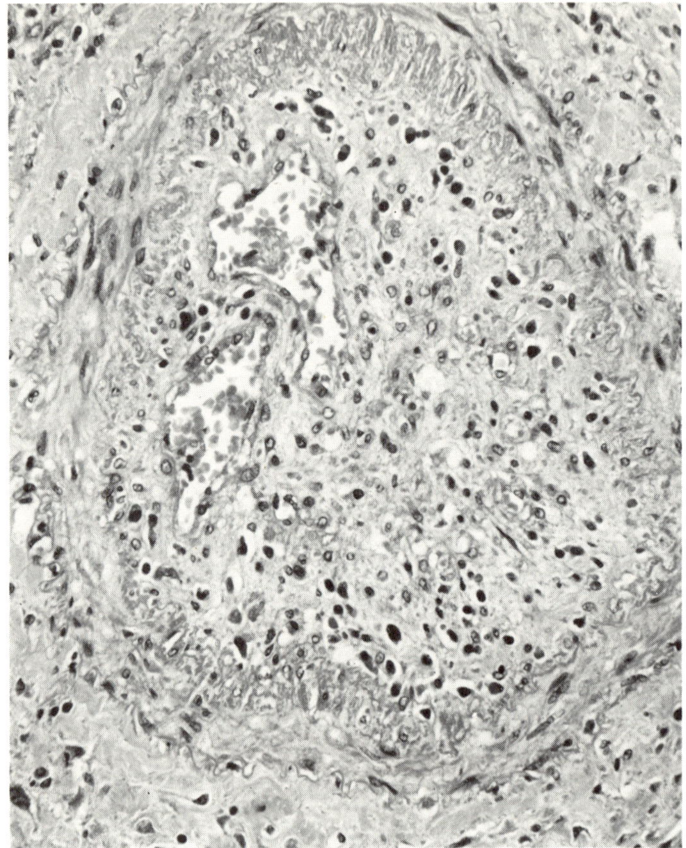

Figure 3. Recannulating artery after thrombosis. Organizing connective tissue includes many chronic mononuclear inflammatory cells. (Courtesy of A. Anderson.)

fibrin, leukocytes, and tissue debris can be removed determines the subsequent course of inflammation (Figure 3). It can either resolve with or without fibrotic scar formation or suppuration or become a site of chronic granulomatous inflammation.

Systemic Reactions to Irritants

The localized inflammatory reactions involving cell injury and death, if sufficiently severe, can result in systemic reactions. Thus, mediators may be produced in sufficient amounts to result in systemic changes such as fever and the elaboration of acute phase reactants (see Chapter 17). The route of administration of irritants also influences the nature of the reaction. For example, when carrageenan, the nonimmunogenic algal polysaccharide, is injected intraperitoneally, it results in an acute reaction consisting of edema, erythema, fever, and the mobilization of inflammatory cells (Vinegar et al. 1976). These effects of carrageenan are multifactorial and include the activation of the kinin and alternate complement pathways. Macrophages that are sublethally injured by carrageenan are stimulated to secrete or release their lysozomal enzymes and a variety of mediators, including lymphocyte-activating factor (LAF) (Stamenkovic et al. 1979), endogenous pyrogen (EP), which induces fever, and a monokine, which stimulates hepatocytes to produce acute phase protein such as "C reactive" protein (CRP), serum amyloid A (SAA), and ceruloplasm. Recent evidence suggests that EP and the hepatocytic-stimulating factors may be identical to LAF. Thus, macrophage mediators play a crucial role in causing the systemic manifestations of inflammation. In addition, the generation of LAF potentially provides a pathway by which a non-immunogenic irritant can function as an adjuvant and augment concomitant immunological reactions to antigens (see Chapter 5). Other types of toxic irritants, such as turpentine, induce acute predominantly neutrophilic reactions. This is presumably mediated by a variety of chemotactic signals either released by the damaged tissues and by enzymatically activated kinin or complement pathways. Nonimmunogenic irritants and toxins can thus damage local tissues and induce deleterious systemic reactions. However, if antibodies, such as antitoxins, are present due to prior immunization they can intercept and neutralize the toxins and prevent both direct tissue damage and self-destructive inflammatory responses. Even in the absence of antibodies, these reactions are self-limiting and subside after the removal of the irritants and cell debris.

Foreign Body Granulomas

Nonantigenic irritants that resist degradation produce foreign body granulomas (Boros 1978). They are formed in response to noxious agents such as carrageenan, silica, bentonite, talc, and nondigestible mineral oil. The foreign body granulomas are characterized by the presence of only a few neutrophils and a predominant number of macrophages and fibroblasts but few lymphocytes. Epithelioid and Langhans giant multinucleated cells may develop (Figure 4). The irritants are phagocytosed by macrophages, which are activated to produce lysozomal enzymes, neutral proteases, and monokines. Activated macrophages produce monokines such as lymphocyte-activating factor (LAF), which may augment concomitant T-cell activation, and chemotactic factors that attract more inflammatory leukocytes. The amount of irritant within the macrophages and their degree of toxicity determines the rate of macrophage turnover. The more injurious agents, such as silica and cord factor, stimulate considerable macrophage lysozomal enzyme release, cell damage, and death, as well as reactive fibrosis and scarring which can destroy organ functions (see Chapter 18).

Role of Immunity in Inflammation

Lymphocyte Functions

The immune component of the mobile body defense system is embodied in the lymphocytes and their products; antibodies (immunoglobulins from B-lymphocytes) and lymphokines. The T-derived lymphocytes possess specific cell-membrane–associated antigen-binding receptors. Direct binding of T-cell receptors to target antigens can result in two different types of effector actions: either cytotoxic killing of the target (see Chapter 7) or the release of lymphokines that regulate the migration and functional capabilities of other inflammatory cells, as well as other T- and B-lymphocytes (see Chapter 10).

The second group of lymphocytes, the B-lymphocytes and plasma cells, produce and secrete immunoglobulins with a wide variety of antibody-combining sites that interact with the target. Complexes of the antibodies with antigens bind preferentially

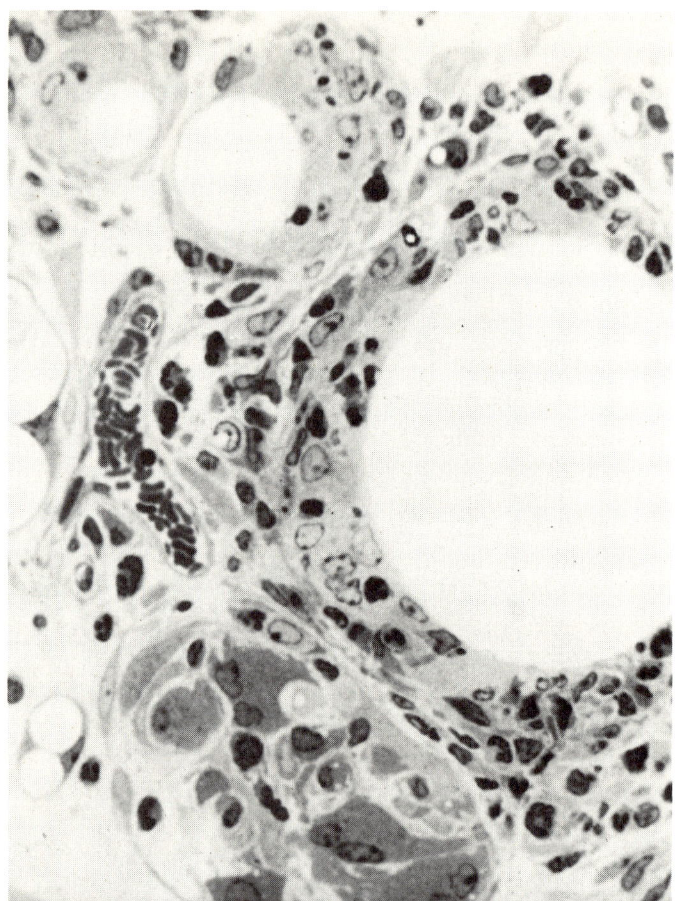

Figure 4. Foreign body (pristane oil) granuloma. Note neutrophils and monocytes surrounding clear dropletlike spaces and accumulation of epithelioid and multinucleated cells at lower left. (Courtesy of A. Anderson and M. Potter.)

to inflammatory cells of the phagocytic system by the constant (C) region sites of the Ig molecule, and they may activate the humoral complement system (see Chapter 15). The generated complement components have chemotactic, anaphylatoxic, virus-neutralizing, cytolytic, and other biological effects. Thus, lymphocytes not only marshal specific inflammatory reactions to the antigenic stimulus but also focus nonspecific inflammatory responses on the target. This provides vertebrates with the capacity to adapt and magnify reactions designed to eliminate deleterious agents rapidly and efficiently.

Characteristics of Immunogens and Antigens

An immunogen is a substance that activates the immune system. The capacity of a substance to act as an immunogen is a function of its general characteristics and form, the route of administration, the dose, and the genetic makeup and age of the recipient (Gill 1972). The outcome may be tolerance or specific humoral and cell-mediated immunity. By contrast, an antigen is a macromolecular substance with which a component of the immune system reacts. These components are either antibody molecules or antigen-binding receptors on lymphocytes. As a rule, antibodies react only with some of the chemical determinants on antigenic molecules called haptens (Leskowitz 1972). Haptens include simple small aromatic chemicals, peptides, oligosaccharides, steroids, drugs, nucleic acids, and lipids. When coupled to macromolecular "carrier" molecules, haptens become immunogenic. All macromolecules in the appropriate doses and schedule are probably capable of inducing immune reactions, but some smaller molecules (molecular weight: 500 to 10,000) are also immunogenic. Polypeptides and carbohydrates generally are more immunogenic than polynucleotides, whereas lipids are least stimulatory. In the case of polypeptides, a minimum chain length of seven amino-acids plus a hapten is required to induce an immune reaction. The presence of aromatic amino acids enhances immunogenicity, whereas the D-configuration and highly charged molecules are inhibitory. Although all immunogens are antigens, most smaller antigens are not immunogenic. In addition, particulate forms of antigens are more immunogenic than soluble forms. In fact, many adjuvants augment the immune response to weak immunogens merely by promoting their aggregation.

Immunogenicity is determined by the manner in which a substance interacts with the cells of the recipient. Aggregated molecules are more immunogenic than small soluble molecules, because they are more readily taken up and presented by macrophages to lymphocytes (see Chapter 5). Furthermore, even though small soluble antigens can readily bind antigen-binding receptors on lymphocytes, they fail to stimulate the lymphocytes. Immunogenic substances also bind the lymphocyte receptors, but immunogens also have the capacity to cross-link lymphocyte-membrane receptors, which is required to activate the lymphocytes (see Chapter 6). Therefore, macromolecules bearing multiple haptenic determinants are usually more immunogenic than molecules too small to cross-link lymphocyte receptors.

Operationally, immunogens do not usually elicit an inflammatory response until a second exposure. The initial exposure to these antigens serves to induce an expansion of the clone of lymphocytes bearing receptors specific for the antigen. These lymphocytes are called memory cells. Thus, the interaction of antigen with the subpopulation of receptor-bearing lymphocytes culminates in the selective activation of relevant antigen-reactive lymphocytes. A subsequent exposure to the antigen (provided sufficient time has elapsed to expand the clone of initially reactive cells) will then result in a magnified secondary response by the increased number of reactive lymphocytes. The type of inflammatory response that results from such a secondary response is a function of the nature, dose, and route of administration of the antigen (Table 1).

Types of Immunologically Mediated Inflammation

Humoral and cell-mediated responses to antigens result in several characteristic types of inflammatory reactions (Gell and Coombs 1975). Immunologically induced inflammation is categorized most readily by evaluating the cutaneous inflammatory responses of sensitized subjects to antigens injected intradermally. Sensitized subjects manifest three distinct types of inflammatory response to antigens, depending on the nature of their immune status (Table 2).

Immediate Hypersensitivity (Allergic) Reactions. Immediate hypersensitivity reactions can be caused only by cell-associated IgE or IgG_1 types of antibodies. Allergic

Table 1. Conditions for Inducing Different Types of Immunologically Mediated Inflammatory Reactions

	Humoral immunity		Cell-mediated immunity
	Allergic Immediate Hypersensitivity	Immune Complex Arthus Reactions	Delayed hypersensitivity
Composition of Inducers	Polypeptides (helminthic)	Polypeptides Carbohydrates Polynucleotides	Polypeptides Glycopeptides
Doses	Small doses with alum adjuvant	Higher doses of soluble antigens with or without adjuvants	Smaller doses of aggregated antigen with adjuvants
Optimal route	Nasopharyngeal Bronchial	Intravenous Intraperitoneal	Intradermal
Reactive component of antigen	Bivalent haptens	Multivalent haptens	Hapten and carrier

subjects that have these antibodies develop immediate pruritic wheal-and-flare reactions in response to minute doses of antigens administered intracutaneously. Exposure of their nasal or bronchial passages to these antigens also results in immediate hypersensitivity reactions characterized by edema, basophilic and eosinophilic infiltrates, and bronchospasm (asthma). These reactions are mediated by histamine release from antigen-activated IgE- or IgG_1-coated basophils and mast cells. The mechanism and ramifications of this type of inflammatory response are discussed in detail in Chapter 13.

Systemic administration of antigen to sensitized animals possessing IgE or IgG antibodies can result in an anaphylactic reaction, which is manifested in guinea pigs by scratching, sneezing, coughing, respiratory distress, incontinence, convulsions, prostration, and death within 2 to 5 minutes (Waksman 1970). The pathophysiology of systemic anaphylaxis is typified by degranulation of mast cells, basophils, and neutrophils. These cells release pharmacologically active mediators including histamine, heparin, serotonin, slow-reacting substance of anaphylaxis (SRS-A), kinins, and lysozomal enzymes. These in turn produce increased permeability of venules and veins, smooth muscle contraction, hypocoagulability of blood, and hypersecretion of mucous glands. Anaphylactic reactions are characterized by vascular constriction and bronchial constriction, shock, hypothermia, anoxia, thrombocytopenia, embolization, leukopenia, hypocomplementemia, necrosis, and hemorrhage. Obviously the consequences of systemic antigen–antibody reactions can be devastating.

Antigen–Antibody Complex Induced Arthus Reaction. The reaction of circulating antibodies of the IgM and IgG type with antigens to form antigen–antibody complexes, activates complement, and results in a variety of inflammatory reactions. Intracutaneous injections of antigen into sensitized animals or of antigen–antibody complexes into unsensitized recipients results in Arthus reactions that are characterized by hemorrhagic wheal and erythema within 1 to 24 hours. These reactions are usually maximal in 2 to 5 hours, and may, if severe, produce tissue necrosis. Preformed

Table 2. Characteristics of Three Types of Immunologically Mediated Inflammation

	Humoral immunity		Cell-mediated immunity
	Allergic reactions	Immune complex Arthus reactions	Delayed hypersensitivity
Cutaneous manifestations	Wheal and Erythema	Erythema hemorrhage and edema	Erythema and induration
Onset	5–15 minutes	1–4 hours	8–24 hours
Duration	1 hour	24 hours	$\geq$ 72 hours
Infiltrating leukocytes	Basophils and eosinophils	Neutrophils and eosinophils	Basophils followed by mononuclear cells
Cellular basis	B-lymphocytes	B-lymphocytes	T-lymphocytes
Transfer	Serum IgE or IgG antibody	Serum IgG antibody	T-lymphocytes Transfer factor
Predominant mediators	Histamine and SRS-A	Complement	Lymphokines
Disease manifestations	Anaphylactic shock Asthma Rhinitis	Anaphylaxis Serum sickness Arthritis Glomerulonephritis Polyarteritis	Hypersensitivity granulomas Contact hypersensitivity Graft rejection GVH reactions

antibodies present in the sensitized subject react with the injected antigen, activate the complement system, and result in the generation of complement components including chemotactic signals to the inflammatory cells (see Chapter 15). The deposition of the antigen–antibody complexes and complement fragments produce inflammatory reactions ranging from edema to severe hemorrhage and tissue necrosis. Histologic examination reveals vascular fibrinoid necrosis, edema, platelet aggregation, and thrombosis (Figure 5). Arthus reactions are also characterized predominantly by neutrophilic and some eosinophilic infiltration (Waksman 1970). The phagocytes internalize immune complexes and exocytose their lysosomal contents. These released acid hydrolases (cathepsins D and E) and cationic proteins can cause vascular necrosis and rupture basement membranes. Heparinization, reduction in platelets, and depletion of polymorphonuclear leukocytes and complement in animal models inhibit Arthus reactions. After the cells phagocytize and degrade the antigen–antibody complexes, the reaction subsides within a few days. Thus, the inflammatory reaction is self-limiting and subsides when the host response succeeds in eliminating or sequestering the antigen.

Systemic reactions of antigens with IgM or IgG-2, IgG-3, or IgG-4 can also result in anaphylactic reactions similar to those of the immediate type. These reactions activate complement and generate anaphylatoxins. However, IgM and IgG antibodies play little or no role in the induction of anaphylactic shock in primates.

Serum sickness can be attributed to systemic inflammation of the Arthus type. It results from antigen persistence and results in chronic immune complex formation with IgG or IgM antibodies. Serum sickness is characterized by urticarial and Arthuslike rashes, leukopenia, fever, lymphadenopathy, polyarthritis, edema, and albuminuria. Passive entrapment of circulating immune complexes and complement

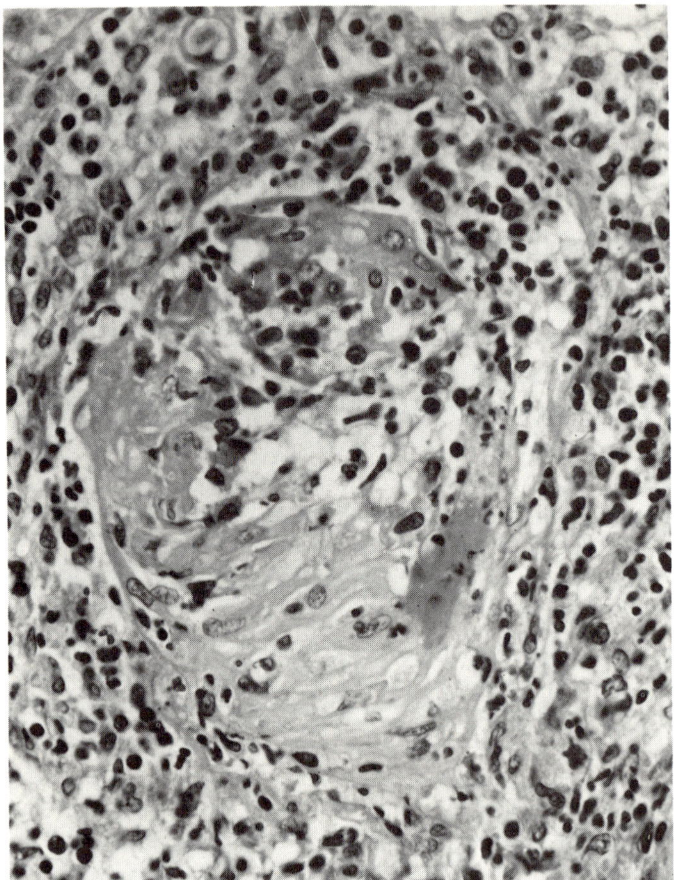

Figure 5. Acute Arthus reaction with necrosis of blood vessel wall and thrombotic obliteration of the lumen. Note extensive perivascular neutrophilic and monocytic infiltration. (Courtesy of A. Anderson.)

components in vascular structures can result in polyarteritis, endocarditis, glomerulonephritis, arthritis, and hyperplasia of the lymphoid tissues. Histologically, the lesions resemble those of the Arthus reaction. The lesions seen in autoimmune syndromes often involve an immune-complex-mediated serum sickness type of reaction based on the persistence of autologous antigen and chronic immune-complex formation.

Humoral antibodies also cause tissue injury or malfunction by direct interaction with antigens. For example, cytotoxic antibodies directed against thyroid cell membranes destroy the target cells, as in Hashimoto's thyroiditis and in autoimmune hemolytic anemias. Antibodies-to-basement-membrane antigens are involved in the pathogenesis of some forms of nephritis (Cohen et al. 1979). Antibodies-to-cell products such as intrinsic factor are involved in pernicious anemia, and antibodies directed against acetyl choline receptors are pathognomonic of myasthenia gravis. However, the mechanisms of tissue injury in these autoimmune processes is similar to those of Arthus reactions and is also mediated by immune complex formation and complement activation.

Delayed Hypersensitivity (Cell-mediated) Reactions. Delayed-type hypersensitivity (DTH) reactions usually take from 24 to 72 hours to develop and are characterized by erythema and induration (nonpitting edema) caused by vascular dilation and increased permeability of the postcapillary venules. This reactivity can be transferred adoptively only with lymphocytes, but not serum, from sensitized donors and is therefore said to be cell, rather than antibody, mediated. Immunological reactions based on such cell-mediated processes include transplant rejection, graft-versus-host reactions (GVHR), some forms of autoimmunity, tumor immunity, chronic granuloma formation, and contact hypersensitivity.

The cellular basis of these reactions depends on the interaction of antigen with specific clones of thymus-derived (T) lymphocytes. The intradermal route or administration of antigen in adjuvant favors the induction of DTH (Table 1). The antigen is presented in association with a major histocompatibility product (Ia) present on some macrophages, Langerhans cells of the skin, and dendritic cells in the spleen. Antigens present on the macrophage surface therefore is thought to be recognized by T-lymphocytes bearing complementary receptor(s) as a change in the configuration of "self ", i.e., a modification of the appearance of the major histocompatibility components (MHC) on the cell surface. The activation of the T-lymphocytes results in the production of a great variety of hormone-like signals (lymphokines) that rapidly attract and promote the differentiation and proliferation of neutrophils, macrophages, lymphocytes, and eosinophils that become involved in the inflammatory reaction. Hence the histology of DTH is characterized by an initial neutrophil infiltration coincident with the increased vascular permeability. Increased basophilia also appears within.8 hours in perivascular locations. By 24 to 72 hours the perivascular infiltrate becomes predominantly mononuclear (80 to 90% monocytic and 10 to 20% lymphocytic) T-cells predominate (Figures 6 and 7). The inflammatory cells emigrate between the endothelial cells of the postcapillary venules into the tissue. The reaction usually subsides in several days after degradative elimination of the antigen.

Hypersensitivity Granuloma Formation. The inability of macrophages to degrade intracellular antigens or to kill and digest microbes results in focal chronic inflammatory responses. These responses are characterized by the accumulation and proliferation of leukocytes, predominantly of the mononuclear type, called granulomas (Boros 1978). Thus, hypersensitivity granulomas are evoked by nondegradable particulate or slowly released soluble antigens that also evoke the characteristic delayed hypersensitivity cell-mediated immune reactions. In contrast with foreign body granulomas, the immunologically induced granulomas are T-lymphocyte and lymphokine mediated. The hypersensitivity granulomas characteristically contain macrophages, both T- and B-lymphocytes, eosinophils, few neutrophils, occasional basophils, and tend to become encapsulated by fibroblasts. Macrophages under the influence of lymphokines metamorphose into epithelioid cells and multinucleate giant cells. Agents that typically cause hypersensitivity granulomas include those intracellular replicating microbes that cause persistent infections such as histoplasmosis, tuberculosis, leprosy, syphilis, brucellosis, and salmonellosis. Infectious organisms such as schistosomes and filaria that are too large to be phagocytosed also cause granulomas. Metals such as beryllium and zirconium salts can act as haptens and bind proteins, inducing delayed hypersensitivity and hypersensitivity granulomas. These granulomas wall off, ingest, and eventually destroy (within 3 weeks) or, alternatively, sequester the nondegradable antigenic pathogens or sensitizing agents.

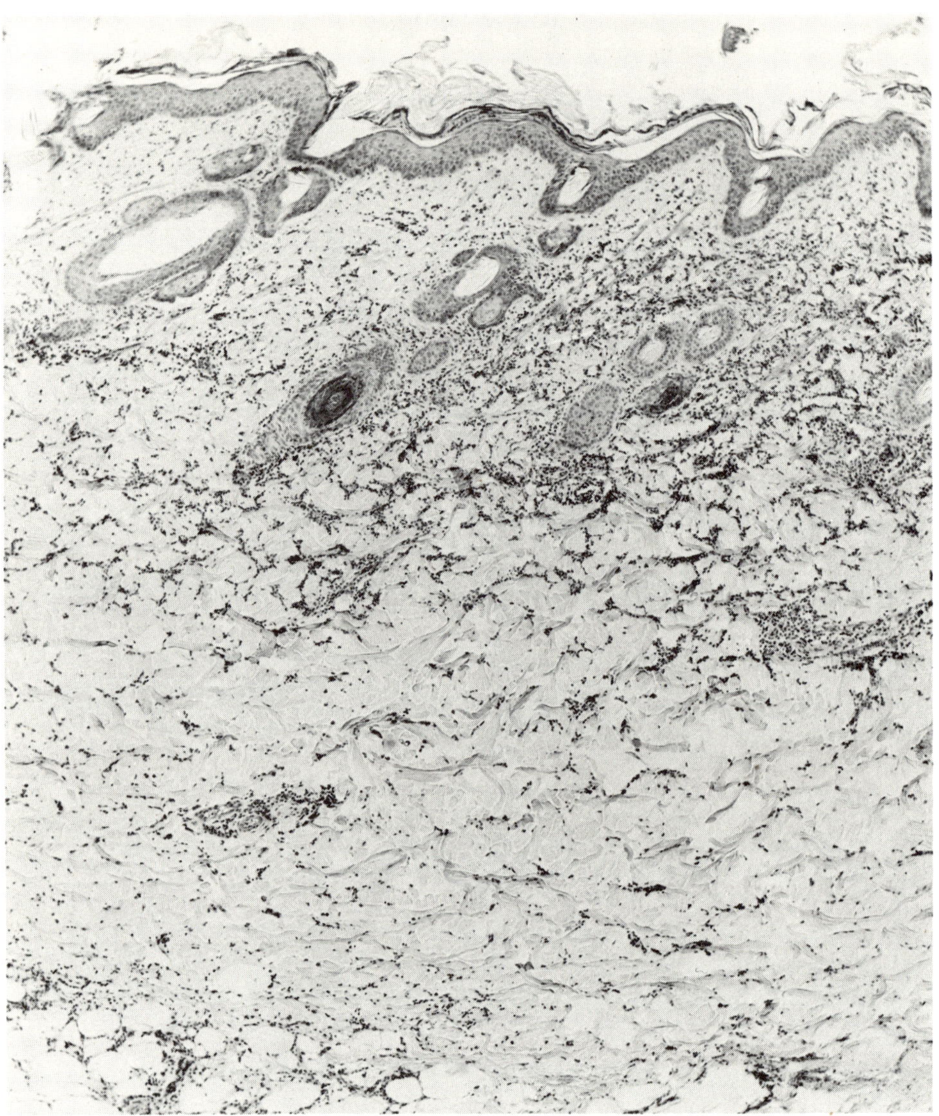

Figure 6. Q-fever–induced delayed cutaneous hypersensitivity response in guinea pig (24 hours). Note heavy mononuclear cell infiltrate in the dermis, particularly in perivascular locations and in connective tissue spaces between fat cells. (Courtesy of A. Anderson and M. Ascher.)

The granulomatous responses represent the ultimate cell-mediated inflammatory response and illustrate the self-destructive potential of cell-mediated immune responses. When the granulomatous response fails to eliminate the antigen, the resulting prolonged inflammation with its associated release by macrophages of lysozomal enzymes, collagenase, elastase, etc., can cause considerable progressive destruction of previously noninvolved tissues. Excessive fibrosis and necrosis may result in severe organ derangement. Thus, the host's immune response can result in excessive deleterious inflammatory reactions (see Chapter 19).

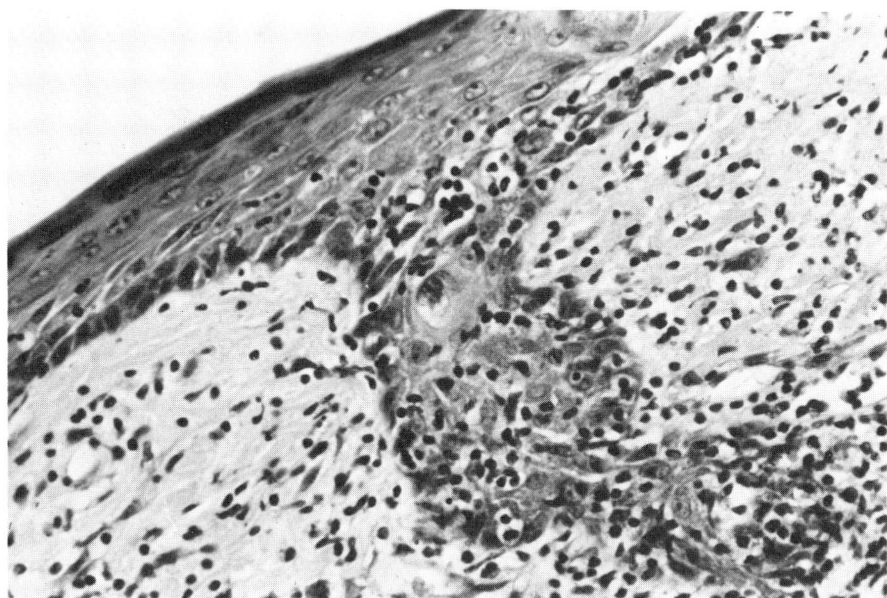

Figure 7. Contact hypersensitivity cutaneous reactions in rat ear. Note subcutaneous and intradermal penetration by lymphocytes and monocytes and the intraepithelial edema. (Courtesy of A. Anderson.)

Immunologically Mediated Suppression of Inflammation

The immune system has also developed a number of pathways to actively prevent or contain inflammatory reactions. The immune system has developed the means of discriminating between self and nonself antigens. This nonreactivity to self is called tolerance and is an integral part of the immune response. A failure to prevent reactions to self may be the basis of autoimmune diseases and reactivity to one's own tissues may have detrimental consequences.

Tolerance to soluble antigens can be obtained more readily in immature than adult animals. It is possible to suppress humoral or cell-mediated immune reactions or both selectively. Various forms or modifications of antigens or certain immunization regimens can favor the establishment of a tolerant instead of an immune state. High doses of antigens and intravenous administration in particular promote tolerance. Removal of the aggregated component of protein immunogens results in residual soluble antigens that preferentially induce tolerance rather than reactive immunity. This can be accomplished experimentally by ultracentrifugation of the antigen to separate the particulate and more immunogenic fraction from the more soluble, tolerogenic supernatant components. Administration of such deaggregated soluble antigens intravenously, subcutaneously, or intramuscularly in large doses without adjuvant favors the induction of tolerance, e.g., specifically blocks active immune responses to subsequent injections of the particulate, more immunogenic fraction. Maintenance of tolerance requires the continuous presence of the tolerogen. Tolerance may also be induced by the injection of either nonimmunogenic fragments of antigen or nonmetabolizable D-isomers of the antigens. Such molecules persist for a long time in the circulation, as do higher doses of antigen, and thus inhibit the immune response preferentially.

The more immunogenic-aggregated component of immunogens is taken up preferentially by macrophages, which favors the activation of T-helper lymphocytes. Conversely, the use of soluble antigens that circumvent macrophage "processing" favors the induction of immunological tolerance.

The mechanism of induction of tolerance is far from clear. It is known that antigens can either activate inhibitory T-suppressor lymphocytes or alternatively can directly paralyze or delete immunoreactive T- or B-lymphocytes.

T-cell Suppression. Considerable evidence suggests that in a normal functioning immune system, immune responses to autologous tissue antigens are continually depressed. This implies that tolerance is an active process in which some lymphocytes can be activated to suppress the antigen-reactive lymphocytes that mount inflammatory responses. Subpopulations of T-lymphocytes have been isolated that function to suppress immune responses. They are distinguishable and separable from other T-cells by distinct surface markers (antigens). They regulate immune reactions in conjunction with other distinct subsets of reactive helper T-lymphocytes. Immunological responses thus represent the net result of augmenting (helper) and inhibitory (suppressor) activities of opposing regulatory subpopulations of T-lymphocytes (see Chapter 6).

The presence of the thymus is necessary for the development of T-suppressor cells. In younger animals, T-suppressor cell reactivity predominates. High doses of some antigens (e.g., KLH) activate T-suppressor cells preferentially. Suppression of T-cells that can be transferred with T-lymphocytes is called infectious tolerance. This can be achieved even when nude (athymic) mice are used as recipients. Removal of T-cells should restore any immune response inhibited by T-suppressor cells.

The effector mechanism of T-cell suppression is unclear (see Chapter 6). The suppression of T-cells can interfere either at the afferent level, with the induction of T-helper functions, or with the efferent expression of immune responses by B-lymphocytes. Antigen-specific suppression probably requires cell contact between lymphocytes and mediation by lymphokines. A number of mediators produced by lymphocytes that suppress immunological reactions—such as interferon, an inhibitor of DNA synthesis (IDS), soluble immune response suppressor (SIRS), and MIF inhibitory factor (MIFIF)—have been described (see Chapter 10). However, these lymphokines all suppress immunological reactions nonspecifically, independent of antigens. They, therefore, can not account for specific suppression in which only responses to the tolerizing antigen are absent or diminished. This requires cell contact or is mediated by "specific" suppressive factors (see Chapter 6).

Immune Deviation (T-Cell Tolerance). It is possible to selectively suppress CMI while leaving humoral immunity intact. The mechanism for such selectivity is not clearly understood but may be related to differing susceptibility of T- and B-lymphocytes to tolerization. It is easier to induce long-lasting T- than B-cell tolerance. It is generally more difficult to suppress induction of antibody formation than cell-mediated reactions. The administration of an antigen intravenously or subcutaneously without an adjuvant will result in antibody production but block the subsequent induction of delayed hypersensitivity in response to a dose of antigen in complete Freunds adjuvant. This phenomenon is called "immune deviation" and may be caused by selective suppression of T-cells involved in DTH but not of T-helper subpopulations. In all likelihood, commonly used immunization procedures that induce antibody production and Arthus reactions without an accompanying DTH reaction are probably a net result of simultaneous suppression of DTH and activation of T-helper cells.

Desensitization. In addition to the type of T-cell suppression discussed previously, large intravenous doses of antigen can transiently suppress the development of an already established delayed hypersensitivity state. The latter process is called desensitization. It may result initially in generalized anergy (skin-test unresponsiveness to many antigens) for up to 1 to 2 weeks, followed by inhibited cutaneous delayed hypersensitivity only to the specific injected antigen. This may persist for an additional week or so, depending on the antigen. The mechanism of this desensitization and its relationship to tolerance is unclear. However, it is possible to transfer DTH using lymphocytes from a desensitized donor, but it is not possible to overcome the desensitized state by the administration of reactive cells. It has been suggested that the presence of both circulating paralytic doses of antigen and lymphokines may contribute to this anergic effect of the environment of the desensitized animal (Cohen et al. 1979).

Clonal Deletion Hypothesis

Tissues are immunogenic, as demonstrated by graft rejection of allogeneic transplants. However, autologous tissues normally are not rejected. Burnet (1957) has suggested that naturally occurring tolerance to self is induced *in utero* and in the early neonatal period to the developing tissues present at that time. In fact, tolerance to an antigen is much easier to induce in immature than in adult animals. Allografts of closely related strains are not rejected by newborn mice and can induce a life-long specific tolerance to all allografts from the same donor strain. In contrast, when adult recipients are used, the same allografts are sensitizing and become rejected. The immunological maturity of a recipient's lymphoid system determines whether the predominant response to an allograft is graft survival because of development of tolerance, as if it is a self-antigen, or whether it is rejected by CMI reactions. Burnet has proposed that "clones" of self-reactive lymphocytes are deleted or paralyzed more effectively in the neonatal thymus. Alternatively, rather than being deleted the clones may be actively suppressed, since it has also been observed that T-suppressor lymphocyte functions are predominant in younger animals, whereas the activity of T-helper lymphocytes increases progressively with age. This latter view predicts that removal of suppressor T-cells should break tolerance to self. Indeed, there is some evidence for this, since the decline with age in host T-cell function is associated with an increased incidence of autoimmune diseases. However, the clonal deletion hypothesis can not be excluded. Rather it has been expanded to include the phenomenon of clonal abortion, in which exposure of immature lymphocytes to antigen results in antigen-specific unresponsiveness, regardless of age.

B-Cell Tolerance. Evidence suggests antigens can also have direct suppressive effects on B-cell antibody production. For example, although athymic nude mice are almost completely devoid of T-lymphocytes, they can be rendered specifically tolerant for a long time by high doses of pneumococcal polysaccharide (Type III) antigen. Furthermore, purified neonatal B-lymphocytes and, to a lesser extent, adult B-lymphocytes can be rendered tolerant by a relatively short exposure to antigen (see Chapter 8). A number of hypotheses have been proposed to explain B-cell tolerance, including "clonal abortion," "receptor blockade," a deficiency of T-helper cells relative to T-suppressor cells in the neonatal environment, "central failure," and the generation of B-cell suppressive factors or antiidiotype antibodies. This type of tolerance can not be transferred to normal lymphocytes and is not affected by removal of T-lymphocytes.

Suppression of Immune Responses by Antibodies. Injection of a large amount of specific antibody against an immunizing antigen can suppress both humoral- and CMI-type immune responses to that antigen. The mechanism is believed to be based on feedback inhibition of antigen recognition by the immune system. Blocking of antigenic determinants by the attached antibodies interferes with antigen binding and lymphocyte activation. The relevant antigenic sites are covered by antibodies that inhibit interaction of antigen with cell receptors and subsequent lymphocyte activation. The inhibitory effect of antibodies is augmented by the increased opsonic phagocytosis of immune complexes by neutrophils and macrophages. The complexed antigens are destroyed subsequently by intracellular enzymatic cleavage. Thus the immune process has come full circle from initial presentation of antigen on "processing" macrophages to final degradation of the antibody-coated antigen by scavenger phagocytic macrophages and neutrophils. However, it should be added that the passive transfer of antibody can conceivably result in the formation of antiidiotype antibody, which may reduce immune responses by another mechanism.

Effects of Polyclonal Stimulants

Another group of potent inflammatory agents belongs in a distinct category. Polyclonal activators are not inherently cytotoxic, but they have the capacity to activate macrophages and lymphocytes nonspecifically. Although they exhibit little immunogenicity, they still depend on the immune system for their effects. Polyclonal activators include ubiquitous bacterial fractions, such as purified protein derivative (PPD) of mycobacteria, peptidoglycan, and lipopolysaccharide endotoxin (LPS). They may contain a few antigenic determinants, but their basic macromolecular core, although nonimmunogenic, can stimulate polyclonal lymphocyte proliferation and antibody production.

The classical polyclonal activator is the LPS moiety from the cell wall of many gram-negative bacteria. LPS is composed of antigenic epitopes (haptenic side chains) on a backbone composed of lipid A. The lipid A is the polyclonal activating component and generates chemoattractants (by activating complement and or B-lymphocytes) for leukocytes. Lipid A is also a potent nonspecific activator of macrophages, a nonspecific mitogen for B-lymphocytes, and a polyclonal activator of B-cell antibody production.

As a consequence of its capacity to stimulate a considerable proportion of lymphocytes and macrophages when the polyclonal activator, LPS, is administered locally or systemically, it induces unique and severe inflammatory responses (Waksman 1970). The local Shwartzman reaction is a hemorrhagic necrotizing cutaneous lesion that occurs in normal animals given an intradermal dose of LPS, followed 24 hours later by a provocative intravenous dose of LPS. Provocative injections by any agent that cause clumping of neutrophils and platelets, including immune complexes, starch, dextran, etc., are effective. Generalized Shwartzman reactions are produced by two successive intravenous doses of LPS, resulting in generalized intravascular clotting, infarctions, and hemorrhagic necrosis, especially of the kidneys. The serum of mice and rabbits undergoing Shwartzman reactions contain the monokines, LAF, and tumor necrosis factor (TNF). Leukocyte depletion blocks these reactions. These types of inflammatory reactions are seen in patients with gram-negative infections.

Effects of Nonimmunogenic Stimulants on Immunological Reactions

Another consequence of the polyclonal stimulation of macrophages and B-lymphocytes by an activating agent such as LPS is the immunoenhancing (adjuvant) effects on the immune responses to its attached polysaccharide antigens or to antigens that are administered along with it. Adjuvants such as LPS, muramyl dipeptide (Oppenheim et al. 1980), peptidoglycan, or mycobacterial organisms emulsified in oil (complete Freund's adjuvant) by attracting and activating macrophages and T- or B-lymphocytes can thus facilitate the interaction of concomitantly administered antigens with greater numbers and more active macrophages and lymphocytes. This accelerates and magnifies both specific cell-mediated as well as humoral antibody responses.

Some substances with adjuvant activity are neither immunogenic nor polyclonal activators, including irritants such as mineral oil, aluminum hydroxide, carrageenan, and bentonite. The adjuvant effects of these irritants presumably are based on their ability to attract and or activate macrophages or, alternatively, to aggregate antigens. Both these effects serve to increase the antigen-presenting capabilities of macrophages. In contrast, polyclonal activators are potent adjuvants because they activate lymphocytes as well as macrophages.

Evidence suggests that microbial polyclonal-nonspecific stimulants play a very important role in promoting the normal development and functional capabilities of the immune system. The vital role of nonspecific environmental stimulants, and in particular of LPS to the integrity of host immune defenses, is best illustrated by studies of C3H/HeJ mice that have a genetically defective response to LPS as well as a number of other defects. These mice survive as well as LPS reactive mice in a conventional environment. However, C3H/HeJ mice show a drastically impaired survival when infected by a number of widely different pathogens including *Klebsiella pneumoniae* (Chedid et al. 1975), *Salmonella typhimurium* (O'Brien et al. 1980), and *Rickettsia akari* (Nacy and Meltzer 1980). This decreased resistance to some pathogenic organisms may be attributable to the failure of C3H/HeJ macrophages to be activated by ubiquitous adjuvant effects of LPS. In the absence of nonspecific stimulation by LPS, and despite numerous compensatory mechanisms, these C3H/HeJ mice show a number of other deficiencies, including diminished capacity of their macrophages to be induced by lymphokines to become tumoricidal (Ruco et al. 1978). C3H/HeJ inflammatory cells also appear to be functionally immature and relatively undifferentiated, since their macrophages bear fewer Fc receptors and phagocytose less well than normal macrophages (Vogel and Rosenstreich 1979). The capacity of C3H/HeJ precursor cells to form granulocyte–macrophage colonies also appears to be defective (MacVittie and Weinberg 1980). These observations all suggest that the capability of normal animal cells to react to the nonspecific stimulating effects of LPS is very important for achieving the necessary development and differentiation of inflammatory cell functions and for obtaining optimal inflammatory reactions to specific or nonspecific immunological challenges.

Another line of evidence indicative of the importance of exogenous stimulants in the development of inflammatory reactions is derived from studies of germ-free animals (Mandel and Asofsky 1969). The lymphoid tissues of germ-free animals are hypoplastic, and their specific humoral and cell-mediated immunological reactions are abnormally low. Germ-free mice are also hyporeactive to LPS, suggesting that bacterial organisms present in the normal environment provide the necessary signals for the development of normal lymphoid tissues and immunological reactions. It is difficult to establish whether specific or nonspecific stimulants play a more prominent role in restoring the lymphoid tissues of the germ-free animal. However, the facts that the lymphoid tissue develops so rapidly and that the ability of a conventionalized germ-free mouse to respond normally to a wide spectrum of stimuli is rapidly established, suggests that polyclonal activators of macrophages and T- and B-lymphocytes must contribute to the restoration of the immune response.

Normally we are sensitized repeatedly by a multiplicity of microbial antigens as well as by nonspecific polyclonal activating agents present in our conventional "leaky" gastrointestinal tract, e.g., by transient bacteremia that frequently accompanies tooth-brushing, mastication, and defecation. Our phagocytic and immune system enables us to cope with the vast majority of such challenges without any evident inflammatory response. Most invasive organisms are eliminated rapidly and efficiently by antibody-mediated immune responses. By and large, only those organisms that are, for one reason or another, difficult to eliminate are actually pathogenic and elicit a considerable inflammatory cell-mediated response. These include organisms that have protective coats, secrete immunosuppressive signals, mimic self-antigens, or contain poorly digestible constituents. These organisms prove troublesome and induce detectable at times deleterious host inflammatory responses. Thus, the immune response can be considered a modulator of inflammatory reactions. The most effective immune responses are accompanied by minimal inflammation. The more aggregated, indigestible, and persistent stimulants induce more cellular immunity, resulting in considerable and potentially self-destructive inflammatory reactions.

Conclusion

All inflammatory reactions, whether immunologically or nonimmunologically mediated, have common features. They all constitute the host response to a potentially injurious agent. The unifying concept held since the time of Virchow, that cell injury sets in motion a series of reactions designed to eliminate the irritant, is still applicable. This view is complicated but not contradicted by the observation that an exuberant inflammatory response can in turn be injurious to previously noninvolved tissues of the host.

Our current understanding of inflammatory reaction is to a considerable extent based on information provided by immunological studies. Immunological studies have been complemented by biochemical, pharmacological, and physiological investigations, as well as by the development of sophisticated tissue culture techniques that have revealed the central role of interactions of macrophages, lymphocytes, and other leukocytes and their products, e.g., antibodies, lymphokines, monokines, cell receptors, and enzymes, in regulating inflammatory responses. These cells and their products will be discussed in detail in subsequent chapters in an effort to delineate the multifactorial nature of inflammation.

We would like to acknowledge the critical discussion of this chapter by Ann Sandberg, Reuben Siraganian, Sharon Wahl, David Wray, Stefanie Vogel, Terry Hoffeld, Philip Baker, Sam Broder, and David Rosenstreich.

References

Arthus, M. (1903) Injections répétées de serum de cheval chez le lapin. Comptes Rendu Soc. Biol. 55,817–820.

Bordet, J. (1898) Sur l'égglutination et la dissolution des globules rouges par le serum d'animaux injectés de sang de 'fibrine'. Ann. Inst. Pasteur 12,688–695.

Boros, D.L. (1978) Granulomatous inflammation. Prog. in Allergy 24,184–267.

Burnet, F.M. (1957) A modification of Jerne's theory of antibody production using the concept of clonal selection. Austral. J. Sci. 20,67–69.

Chedid, L., Parent, M., Damais, C., Parant, D., Juy, D., Galelli, A. (1975) Failure of endotoxin to increase nonspecific resistance to infection of LPS low responder mice. Infect. Immunol. 13,722.

Claman, H.N., Chaperon, E.A., Triplett, R.F. (1966) Immunocompetence of transferred thymus marrow cell combinations. J. Immunol. 97,828–832.

Cohen, S., Ward, P.A., McCluskey, R.T. (1979) *Mechanisms of Immunopathology*. New York, Wiley and Sons.

Ehrlich, P. (1900) On immunity with special reference to cell life. Croonian Lecture. Proc. Roy. Soc.

Fagraeus, A. (1948) Antibody production in relation to the development of plasma cells. Acta Med. Scand. (suppl.) 204,11–122.

Gell, P.G.H., Coombs, R.R.A. (1975) *Clinical Aspects of Immunology,* 3rd ed. Oxford, Blackwell Scientific.

Gill, T.J., III (1972) The chemistry of antigens and its influence on immunogenicity. In: *Immunogenicity* Borek, F., (ed.) Amsterdam, North Holland, p. 544.

Gowans, J.L., Knight, E.J., (1963) The route of re-circulation of lymphocytes in the rat. Proc. Royal Soc. London, Series B, 159,257–282.

Hurley, J.V. (1978) The sequence of early events in inflammation. In: *Handbook of Experimental Pharmacology*. Vane, J.R., Ferreira, S.M., (eds.) New York, Springer-Verlag, pp. 26–67.

Jerne, N.K. (1955) The natural selection theory of antibody formation. Proc. Natl. Acad. Sci. U.S.A. 41,849–857.

Landsteiner, K., Chase, M.W. (1942) Experiments on transfer of cutaneous sensitivity to simple compounds. Proc. Soc. Exp. Biol. Med. 49,688–690.

Leskowitz, S. (1972) The immune response to haptens. In: *Immunogenicity*. Borek, F., (ed.) Amsterdam, North Holland, pp. 131–151.

MacVittie, T.J., Weinberg, S.R. (1980) Murine macrophage colony-forming cells (M-CFC). Their response to endotoxin in C3Heb/FeJ and C3H/HeJ mice. In: *Genetic Control of Natural Resistance to Infection and Malignancy*. Skamene, E., Kongshavn, P.A.L., and Landy, M., (eds.) New York, Academic Press, pp. 511–518.

Mandel, M.A., Asofsky, R. (1969) The effect of heterologous antithymocyte sera in mice. III. High susceptibility of germ-free mice to suppression effects of IgG from rabbit anti-mouse thymocyte serum. J. Exp. Med. 129,1203–1216.

Metchnikoff, E. (1905) *Immunity in Infectious Diseases*. Translated by Francis G. Binnie. Cambridge, University Press. (French edition (1901).

Miller, J.F.A.P. (1961) The immunological function of the thymus. Lancet 2, 748–749.

Miller, J.F.A.P., Mitchell, G.F. (1968) Cell to cell interaction in the immune response. I. Hemolysin forming cells in neonatally thymectomized mice reconstituted with thymus or thoracic duct lymphocytes. J. Exp. Med. 128,801–820.

Movat, H.Z. (1976) Pathways to allergic inflammation: The sequelae of antigen antibody complex formation. Fed. Proc. 35,2435–2440.

Nacy, C.A., Radlick, G., and Meltzer, M.S. (1980) Role of activated macrophages in natural resistance to Rickettsia akari. In: *Genetic Control of Natural Resistance to Infection and Malignancy*. Skamene, E., Kongshavn, P.A.L., and Landy, M., (eds.) New York, Academic Press, pp. 555–564.

Nossal, G.J.V., Lederberg, J. (1958) Antibody production by single cells. Nature (London) 181, 1419–1420.

O'Brien, A.D., Rosenstreich, D.L., Scher, I., Campbell, G.H., MacDermott, R.P., Formal, S.B. (1980) Genetic control of susceptibility to *Salmonella typhimurimum* in mice: Role of the LPS gene. J. Immunol. 124,20.

Oppenheim, J.J., Togawa, A., Chedid, L., Mizel, S. (1980) Components of mycobacteria and muramyl dipeptide with adjuvant activity induce LAF. Cell. Immunol. 50,71–81.

Pasteur, L. (1880) Sur les maladies virulentes, et en particulier sur la maladie appelée vulgairement cholera des poules. C.R. Acad. Sci. Paris 40, 239–248.

Roux, E., Chamberland, M. (1887) Immunité contre la septicemie conferee par des substances solubles. Ann. Inst. Pasteur 1,561–572.

Ruco, L.P., Meltzer, M.S., Rosenstreich, D.L. (1978) Macrophage activation for tumor cytotoxicity. Control of macrophage tumoricidal capacity by the LPS gene. J. Immunol. 121,543–548.

Spector, W.G. (1969) The granulomatous inflammatory exudate. Int. Rev. Exp. Path, 8,1–51.

Stamenkovic, M., Stosic-Grujicic, S., Simic, M.M. (1979) Carrageenan-induced macrophage soluble factors with costimulatory effects in the proliferative response of T lymphocytes to mitogens. Periodicum Biologorum 81,171–173.

Vinegar, R., Truax, J.F., Selph, J.L. (1976) Quantitative studies of the pathway to acute carrageenan inflammation. Fed. Proc. 35,2447–2454.

Virchow, R. (1860) *Cellular Pathology as Based Upon Physiological and Pathological Histology*. Translated by F. Chance. London, John Churchill.

Vogel, S.N., Rosenstreich, D.L. (1979) Defective Fc-receptor mediated phagocytosis by C3H/HeJ macrophages. I. Correction by lymphokine induced stimulation. J. Immunol. 123,2842.

Waksman, B.H. (1970) *Atlas of Experimental Immunobiology and Immunopathology*. New Haven, Yale University Press.

Structure and Physiology of Lymphatic Tissues

Arthur O. Anderson, M.D. and Norman D. Anderson, M.D.

Immunity is a property of living organisms that results in recognition and elimination of foreign antigens through secretory and cytotoxic mechanisms. In single-celled and other primitive organisms this discriminatory function is mediated by cell-surface molecules defined by the genetic background of each cell, and recognition of foreign cells or products leads directly to protective behavior such as phagocytosis, increased secretion, or cytotoxicity. In larger animals, such as mice or men, discrimination of self from nonself is also a property of cell-surface molecules, but the size and structural complexity of the host dictates there be division of labor and regional specialization for defense. In an organism whose structural complexity precludes direct interaction of all its cells with the external environment, lymphocytes, macrophages, and other leukocytes in blood and connective tissue have differentiated to provide protective functions we know as inflammation and immunity. Immunological diversity enables the mammalian host to respond to a vast number of different antigenic determinants (Cold Spring Harbor Symposia 1977), but each lymphocyte is limited by the structure of its antigen-binding surface receptors to recognizing only one antigenic epitope. Since selective cellular interactions with other lymphocytes and accessory cells must occur before immunity is initiated, an immune response in mammals requires the help of an efficiently organized cell- and antigen-sorting system.

In mice and human beings the cardiovascular, respiratory, nervous, musculoskeletal and digestive systems have taken the place of assorted behaviors exhibited by primitive organisms. The cells, tissues, and organs of the lymphatic system have likewise developed to assume both recognition and protection functions of host defense. The lymphatic system provides a dynamic environment for the generation and maturation of lymphocytes that are tolerant to self-antigens but recognize nonself when it is presented to them within antigen-concentrating sites located downstream from physical barriers at the skin, blood-tissue interface, and mucous membranes. While it is possible that antigens may be encountered anywhere in the body, compelling argu-

From the United States Army Medical Research Institute of Infectious Diseases, Frederick, Maryland, and The Johns Hopkins Medical Institutions, Baltimore, Maryland.

The views of the authors do not purport to reflect the positions of the Department of the Army or the Department of Defense.

ments suggest that the peripheral lymphatic tissues provide the most efficient loci for selection of antigen-specific clonal precursors, clonal expansion, and dissemination of immune effectors and immunological memory to the rest of the body.

All connective tissues of the body contain afferent lymphatic vessels that drain centripetally toward lymph nodes serving as lymph filters and temporary lodging sites for billions of lymphocytes, macrophages, and other accessory cells; the spleen provides filtration for the blood, residence for lymphocytes, and trapping of antigens that may have escaped capture by lymph nodes; and specialized surface epithelium of mucosal-associated lymphatic tissues endocytose environmental antigens present in the respiratory or digestive system lumens and deposit them onto dense concentrations of mature and proliferating lymphoid cells. There is no structural continuity among the anatomically dispersed lymphatic organs except for their connections with blood and lymphatic vessels, yet nearly all cells of the immune system are capable of entering and leaving each organ. Lymphatic tissues are populated and replenished by circulating immunocompetent cells that emigrate repeatedly from the blood into lymph nodes, spleen, and mucosal lymphatic tissues where they may remain to divide, differentiate, secrete, or leave. Exiting cells migrate into efferent lymph, where they will be returned to the blood to begin additional cycles of recirculation through the various organs of the lymphatic system (Gowans 1959). Lymphocytes emigrate into nonlymphoid tissues at a much slower rate and follow circuitous paths through connective tissues past nearly all cells of the body until they ultimately enter afferent lymphatics draining toward lymph nodes. Persistence of antigen in peripheral depots results in chronic inflammation, increased local emigration of lymphocytes and macrophages, and formation of nonencapsulated lymphatic tissue that now contributes some of the immune function of lymph nodes.

Since it is true that an immune response can be induced *in vitro* using only lymphocytes and adherent accessory cells, then why is such a complex organ system necessary? The answer to this question is the subject of this chapter on the structure and physiology of the tissues and cells of the immune system. The fine structure of lymphatic microenvironments presented here will also illustrate how many of the selective cellular interactions revealed through *in vitro* methods may actually occur in lymphatic tissues.

Tissue Components of the Immune System

The lymphatic tissues in mammals can be grouped into two major types. The central lymphoid organs consist of bone marrow and thymus, where stem cells give rise to a progeny of proliferating and differentiating lymphocytes through processes completely independent of antigen stimulation. The peripheral lymphatic tissues include the lymph nodes, spleen, and lymphoid nodules dispersed beneath the mucous membranes of the respiratory and gastrointestinal systems, where lymphoid development is antigen-dependent.

Central Lymphoid Organs

Bone Marrow. The bone marrow is the site of origin of precursors of all T- and B-cells, platelets, erythrocytes, and other leukocytes. While its major role in adults is to replenish the red and white cells and platelets in the blood, the bone marrow serves as a protected environment in which immune cells undergo the antigen-independent proliferative steps and genetic recombinations necessary for the continued generation of diverse repertoires of antigen-sensitive cells. The aggregate mass of bone marrow tissue in most mammals is about equal to that of the liver.

Structure. The bone marrow is composed of central and radiating venous sinuses that surround and separate wedge-shaped interstitial compartments containing hematopoietic cells (Figure 1). The hematopoietic compartments are separated from the vascular spaces by a thin endothelium that forms an irregular sandwich with the adventitial cells of the interstitium. Bone marrow sinus endothelial cells contain numerous pinocytotic vesicles and few-to-moderate numbers of lysosomes. This endothelium has been described as being phagocytic, but this function is more often performed by macrophages within the interstitium that insinuate their endocytic processes between the sinus endothelial cells. The irregular granular basement mem-

Figure 1. The hematopoietic tissue in bone marrow is separated by radially oriented venous sinuses that communicate with a large central vein. The large multinucleated cell is a megakaryocyte and the other small cells are myeloid, lymphoid, and erythroid cell precursors.

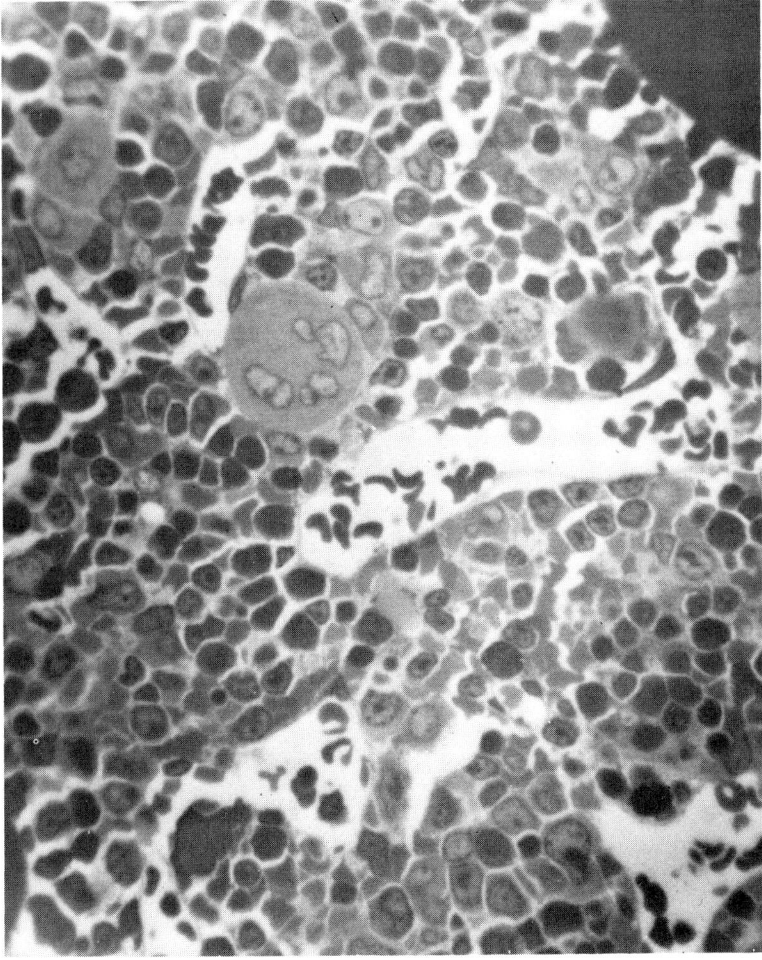

brane that separates endothelial and adventitial cells shares many morphological characteristics with the reticular fibers of peripheral lymphatic tissues (Weiss 1972). The adventitial cells that face the hematopoietic areas extend cytoplasmic processes into the cellular compartment, where they make contact with numerous different cells. The white cell components generated within the interstitium traverse the endothelium by actively migrating between endothelial and adventitial cells. Whether or not intravasating leukocytes pass directly through sinus endothelium is still a subject of controversy. The endothelial/adventitial cell complex is thought to have some regulatory influence on circulating leukocyte levels by controlling which cells enter the blood (Weiss 1972). Only cells of sufficient maturity appear capable of crossing this vascular barrier.

The hematopoietic tissue of bone marrow contains undifferentiated stem cells, intermediate precursor cells, and mature erythrocytes, granulocytes, lymphocytes, megakaryocytes, monocytes, macrophages, and mast cells. Erythroblasts and immature erythrocytes lie against the sinus wall, while granulocytic precursors tend to be positioned further away. Megakaryocytes also lie close to the sinus wall and discharge platelets directly into the lumen from their extensively folded cytoplasm. Macrophages are found frequently within clusters of erythrocytic precursors where they ingest the extruded nuclei of maturing erythrocytes. Since these macrophages also contain large numbers of ferritin-filled granules in their cytoplasm, they may also participate in the scavenging and recycling of iron for use by erythroid cells.

Lymphocytes and their precursors are found scattered among the other hematopoietic cells or are present in discrete nodules. Up to 20% of the nucleated cells in marrow may be lymphoid cells, but in hyperimmunized animals or animals subjected to corticosteroid treatment, the concentration of the cells in marrow is much higher. Macrophages frequently are seen associated with clusters of rapidly dividing lymphocytes and may serve a "nurse cell" function for these cells by releasing nucleic acids and other factors that favor continued proliferation. The marrow is a site of rapid lymphocyte production with estimated turnover time of 3 days, but this proliferation does not appear to be altered by depleting the peripheral lymphocyte stores or depression of serum immunoglobin levels.

Thymus. The thymus develops from endoderm derived from the third and a portion of the fourth branchial pouch, which migrates caudally into the base of the neck and upper thorax. During early fetal development this epithelial structure is infiltrated by lymphocytes and other mesodermal elements derived from bone marrow (Clark 1973). Precursor T-cells move through the bloodstream and pass through the walls of blood vessels in the outer cortex of the thymus to infiltrate the epithelial reticulum. These cells proliferate rapidly within the gland, exhibiting mitotic indices 10 times those seen in lymph nodes and have a population turnover time of 24 to 36 hours. Many thymocytes appear to die *in situ,* but evidence of large-scale cell death is not morphologically obvious. The surviving small lymphocytes acquire new surface markers and exit from the gland by moving back into postcapillary venules and efferent lymphatics. These migrant cells pass through the blood and seed peripheral tissues, where they begin to function as immunocompetent T-cells. Although this antigen-independent proliferation and differentiation of lymphocytes is greatest in the thymuses of young animals, T-cell differentiation inevitably continues at reduced rates in the involuted thymuses of adults.

In most mammalian species the thymus is developed almost completely before birth; its maturation precedes the development of the other peripheral lymphoid tissues, such as lymph nodes, spleen, and mucosal lymphoid tissues. Extirpation of the thymus from those strains of newborn rats and mice that require 1 or 2 days more after birth to complete thymic maturation, results in severe depletion of "thymic-dependent" lymphoid tissue in the lymph node paracortex, periarteriolar lymphatic sheath of the spleen, and the perifollicular tissue of Peyer's patches (Parrott et al. 1966). This cellular deficiency is accompanied by a diminution in the ability to produce antibodies to "thymic-dependent" antigens, reduced delayed-cutaneous hypersensitivity, failure to reject skin allografts, and numerous other phenomena requiring immunocompetent T-cell interactions, as discussed in Chapters 5 through 8 and Chapter 10.

Structure. The thymus is a pale, lobulated organ that is largest, relative to total body weight, at birth; it gradually increases in size until puberty, after which it involutes slowly to form a fatty ill-defined structure. The thymic cortex forms the periphery of each lobule; the medulla branches to form the central portions (Figure 2). This demarcation is due to differences in the relative proportions of lymphocytes to thymic epithelial cells in the cortex and medulla. Within the cortex, large numbers of rapidly dividing lymphocytes form multicellular nests that stretch and distort the desmosome-linked epithelial cells into a honey-combed reticulum. These epithelial cells display major histocompatibility antigens, including immune-response-related subgroups coded as I-A, H-2K, or H-2D in the mouse (Rouse et al. 1979). At the corticomedullary junction are a thin basal lamina and specialized mesenchymal cells. There are

Figure 2. Lobular architecture of the thymus. The branching medullary tissue contains relatively fewer lymphocytes and more epithelial cells, in addition to the characteristic Hassal's corpuscles. The cortex, which partially surrounds each prong of medullary tissue, is more densely populated by lymphocytes that stretch and distort the desmosome-linked epithelium (modified after Weiss 1972).

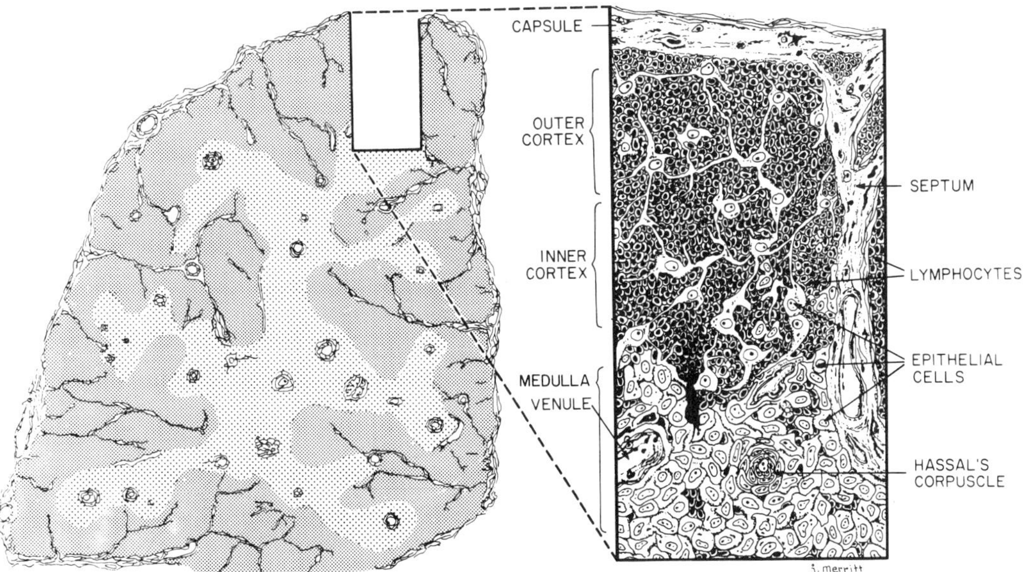

fewer lymphoid cells in the medulla, and the thymic epithelial cells are more polygonal in shape. In addition, the medulla contains specialized epithelial structures formed of concentric whorls of spindled and keratinizing epithelium. These structures are called Hassall's corpuscles; their function still remains obscure, despite their distinctive appearance.

Macrophages. Macrophages and cells that resemble Langerhans cells are found within the epithelial stroma of the thymus. The macrophages contain phagocytosed nuclear debris and periodic acid–Schiff-positive membrane glycoproteins in vacuoles within their cytoplasm. It has been suggested that these macrophages, which are usually found in association with clusters of proliferating cortical T-cells, also serve a "nurse cell" role, like those of the bone marrow. They might release substances, such as lymphocyte-activating factor, that induce mitosis and also provide recycled proteins and nucleic acids derived from degradation of engulfed cells. At the corticomedullary junction, a "special cell" is found that has distinctive electron-lucent cytoplasm, lacks phagocytic granules, and possesses extensively branched cytoplasmic processes (Oláh et al. 1975). This cell bears a close similarity to the novel dendritic cell found in the lymph node paracortex and to Langerhans cells. The similarity to Langerhans cells is further supported by the presence in their cytoplasm of numerous striated particles that are virtually identical to Birbeck granules. Since Langerhans cells are found normally among the epithelial cells of the skin, their possible appearance in a lymphoepithelial organ, such as the thymus, is not surprising. Despite our ability to identify these macrophage-like cells in tissue sites associated with thymus-dependent cell functions, we still know very little about the actual functions of these "special" cells.

The blood–thymus barrier. A blood–thymus barrier was proposed initially because of differences in immune responsiveness to antigens injected directly into the thymus versus those injected intravenously (Marshall and White 1961). Since intrathymic injection of antigen resulted in local development of germinal follicles and other structures associated with B-cell immune responses, the differences in immune responsiveness observed after immunization by these routes could be attributed to changes in the thymic microenvironment rather than a blood–thymus barrier. Horiuchi and Waksman (1968) used a different antigen but a similar experimental design and found that intrathymic inoculation with antigen resulted in the induction of tolerance. This apparently contradictory result led to reexamination of the blood–thymus barrier, since systemic antigens may produce subtle effects on thymus lymphoid populations. A number of anatomical studies, which specifically addressed the question of blood vessel permeability within the thymus, showed that cortical thymocytes, in contrast with those of the medulla, are protected from circulating macromolecules (Chapman and Bopp 1970; Raviola and Karnovsky 1972). Blood vessels in the thymic cortex were not permeable to the physiologic tracers, but the postcapillary venules of the thymic medulla permitted tracers to leak along the clefts between migrating lymphocytes and endothelial cells. The tracers, which crossed medullary venule walls, had limited distribution in the thymic parenchyma because macrophages in the perivascular space ingest and retain much of the leaked tracer. The tissue distribution of the small amounts of antigenic tracer that escaped endocytosis was further limited to the medulla by the phagocytic cells and basal lamina at the corticomedullary border.

Peripheral Lymphoid Organs

The peripheral lymphoid organs are encapsulated and nonencapsulated reticular structures that provide sites for the interaction of immune cells with antigen by way of their afferent lymph and blood vascular connections and their constant access to circulating immunocompetent cells. Variations in the structure and location of these organized lymphatic tissues appear to represent specific adaptations that assist the host in mobilizing immune responses against pathogens invading distant tissue sites, the bloodstream, and mucous membranes.

Lymph Nodes. As Virchow pointed out in 1858, lymph nodes are complex filters that interrupt the current of afferent lymph draining specific regions of the body (Table 1). The lymph is forced to percolate through a reticulum containing lymphoid cells before gaining access to more recognizable efferent lymphatic channels.

Passage of lymph through lymph nodes. Afferent lymphatic vessels originate in delicate plexuses beneath the epithelium of the skin, gut, and urogenital tract and within the connective tissues of all organs. These capillaries merge into larger lymphatics that drain into lymph nodes (Yoffey and Courtice 1956). Efferent lymph from regional nodes may drain into one or two additional nodes before flowing into the large lymphatics that join the thoracic duct (from the lower part of the body) or the right lymphatic ducts (from the upper thorax and head). These major efferent ducts return lymph to the blood by emptying into the great veins at the base of the neck. The lymph nodes, which are interspersed along this network of channels, serve as filtering chambers and are constructed as a meshwork of reticular cells and fibers, macrophages, and specialized blood vessels enclosed in a tough connective tissue sack. Since all lymph passes through at least one lymph node before returning to the blood, any environmental antigens, such as bacteria, viruses, or virus proteins expressed on host cell membranes, protein antigens, and tumor cells will be trapped; under normal conditions, these materials would not enter the blood. Any antigens that might escape lymph node entrapment ultimately will be removed from the blood by macrophages in the spleen, liver, or bone marrow with the exception of living microbes with specific mechanisms for avoiding phagocytosis.

Table 1. Distribution of Antigen to Lymphatic Tissues after Inoculation in Various Sites

Site of inoculation	Route	Location of primary immune response
Ear	Subcutaneous	Postauricular lymph node
Food pad	Subcutaneous	Popliteal lymph node
Lateral thorax	Subcutaneous	Axillary lymph node
	Intramuscular	Subscapular lymph node
Lateral abdomen	Subcutaneous	Inguinal lymph node
	Intramuscular	Inguinal and periaortic lymph nodes
Tongue and buccal membranes	Subcutaneous	Submandibular lymph node
Intestinal mucosa	Pinocytosis by M cells	Mesenteric lymph node
Femoral vein	Intravenous	Spleen
Abdominal cavity	Intraperitoneal	Perithymic lymph nodes and spleen

The argument in Virchow's time was whether or not lymphatic channels were continuous throughout lymph nodes. Sophisticated electron microscopic studies were needed to show that both viewpoints were correct. Subcapsular sinuses are connected to medullary sinuses by delicate plexuses of intermediary sinuses (Figure 3a), and by channels created by vertically oriented reticular meshworks (Figure 3b) that communicate with the floor of the subcapsular sinus through pores (Figure 4). Macrophages are found clinging to reticular cell processes inside lymph sinusoidal spaces and between intranodal lymphatics and the lymphocyte-filled interstitium. Antigens entering a lymph node for the first time will be bound to the surfaces of portal-guarding macrophages that endocytose most of the antigen but leave some remaining on their surfaces (Nossal and Ada 1971). Reticular cells also bind antigen without phagocytosing. After 3 or 4 days, cytophilic antibody and complement are thought to facilitate binding of antigen by the dendritic cells within germinal follicles. Antigen remains bound to follicular dendritic cells for a fairly long time, and the same antigen will bind to these cells with greater avidity on secondary exposure (Hanna and Szakal 1968). The molecular mechanisms of antigen-binding during primary antigenic exposure may be related to natural antibody or to nonspecific opsonins, such as fibronectin, displayed on macrophage and reticular cell surfaces, or to absorption of inflammatory products, such as complement components, present in the antigen-containing afferent lymph (Pryjma and Humphrey 1975). Antigen-binding after a secondary immune response is much more efficient, due to specific recognition provided by cytophilic antibodies bound to macrophage membrane at Fc receptors. The phagocytosis of antigens within the node may provide the antigen-processing and

Figure 3A. Reticular cell networks divide lymph nodes into lymphatic, interstitial, and vascular compartments. Motile lymphocytes, which enter lymph nodes via migration across HEV and to a lesser extent via afferent lymph, pass among these compartments and accumulate in cortical lobules and medullary cords. Germinal follicles created by proliferating B-cells occupy a space created by displacement of the surrounding reticulum.

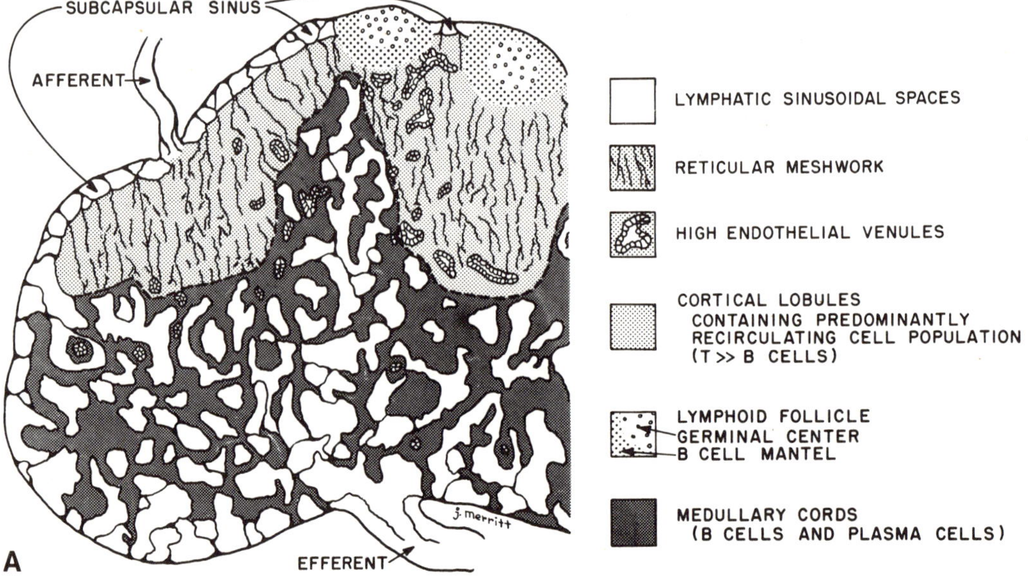

facilitate the macrophage helper functions for immune responses to thymus-dependent antigens.

Lymph node structure. Lymph nodes can be divided into cortex and medulla, based on the relative density of small lymphocytes. In large lymph nodes, cortical tissue is distributed as rounded lobules of densely packed lymphocytes that are delimited on one side by the floor of the subcapsular sinus and on the other by the terminations of vertically oriented reticular fibers at the origins of medullary cords (Figure 3b). Germinal follicles in the superficial cortex partially interrupt the subcapsular sinus. These structures are generated by proliferating lymphocytes that displace the cortical reticulum to form a basket-like enclosure. Inside the enclosure are lymphoblasts, intermediary lymphocytes, dendritic cells, and debris-filled macrophages. On the outside is a mantel composed almost entirely of small B-lymphocytes. Immunofluorescence (Weissman 1975) and antibody-coated erythrocytes (Dukor et al. 1970) delineate the follicular mantel as a B-cell zone. The lymphocytes within the internodular areas and the deeper cortex are largely T-cells (the thymus-dependent zone). Since both B- and T-cells enter the lymph node parenchyma by passing through the

Figure 3B. Antigens enter lymph nodes via afferent lymphatics and flow into cortical parenchyma (CP) through pores in the subcapsular sinus (SCS) and via plexuses of intermediary sinuses (IS). Medullary sinuses (MS) serve as collecting chambers for lymph, which leaves the node via an efferent lymphatic in the hilum. Macrophages that bind antigen are located strategically in lymphocyte channels and at interfaces between vascular, interstitial, and lymphatic spaces. This arrangement of lymph node structures facilitates antigen-presentation by macrophages and cellular cooperation.

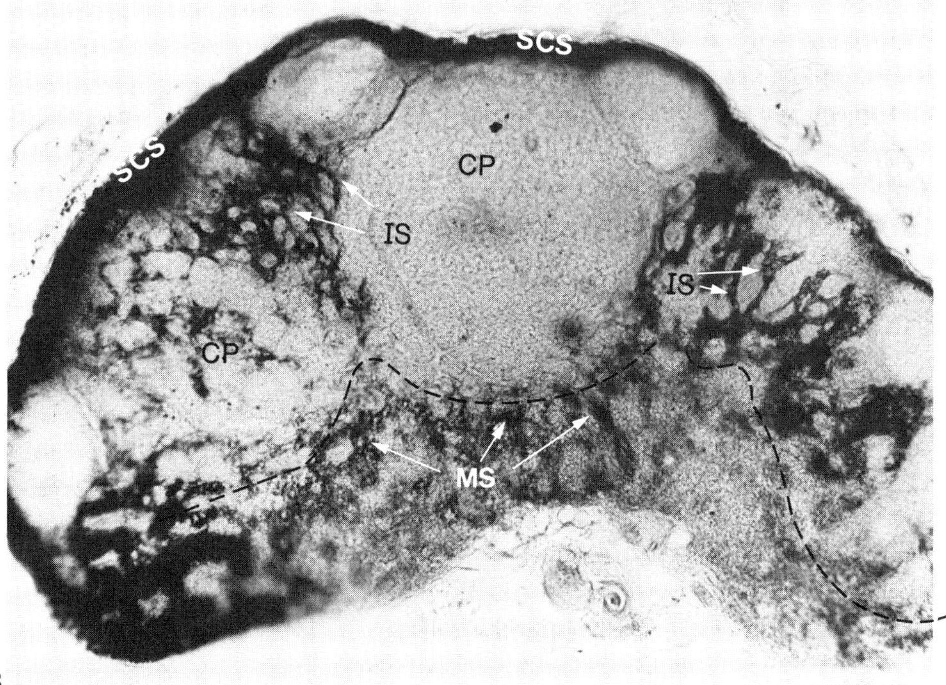

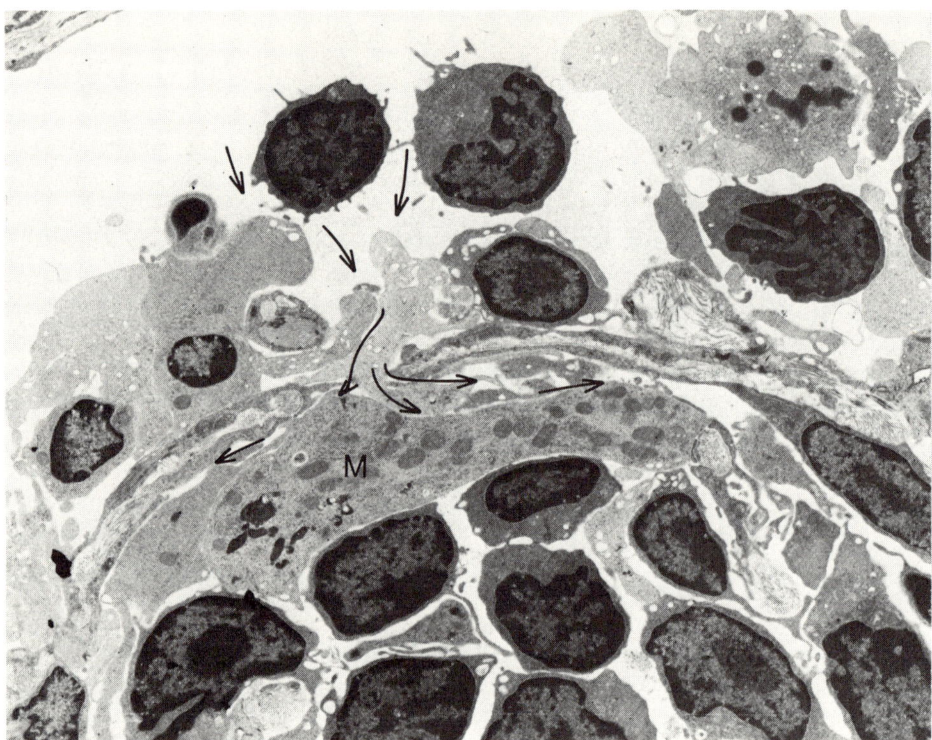

Figure 4. Pores in floor of the subcapsular sinus permit antigen-containing lymph (arrows) to diffuse into the superficial cortex. Macrophages (M) are usually found "guarding" these sites.

walls of specialized venules in the deep cortex, the thymus-dependent zone should be thought of as a transient-cell zone containing about the same proportions of T- and B-cells presented to it by the blood. In most cases, this is 75% T-cells and 25% B-cells, but antigenic stimulation (Gery et al. 1977) or treatment with adjuvants (Anderson and Reynolds 1979) can alter these ratios considerably by supplementing the normal cell traffic with additional T- or B-cells from the blood or from the injection site.

Beneath the cortex and filling all the spaces between cortical lobules of larger lymph nodes is the medullary tissue composed of wide sinuses that surround cords of reticulum filled with antibody-forming B-lymphocytes and plasma cells.

Lymph node vasculature. The circulatory system of the lymph node serves the special physiologic requirements of a tissue that is populated by transient cells with widely fluctuating metabolic activities and capable of rapid cellular fluxes after antigenic stimulation.

Arterial vessels enter lymph nodes at the hilus and branch once or twice as they cross into cortical lobules. These vessels give off capillary twigs and arteriovenous communications at arcades of capillaries in the superficial cortex and in medullary cords. Arteriovenous communications and short venules connect with vessels lined by large polygonal endothelial cells (Figure 5) called high endothelial venules (HEV). The distal segments of HEV open into capacitance vessels with wide lumens and thin

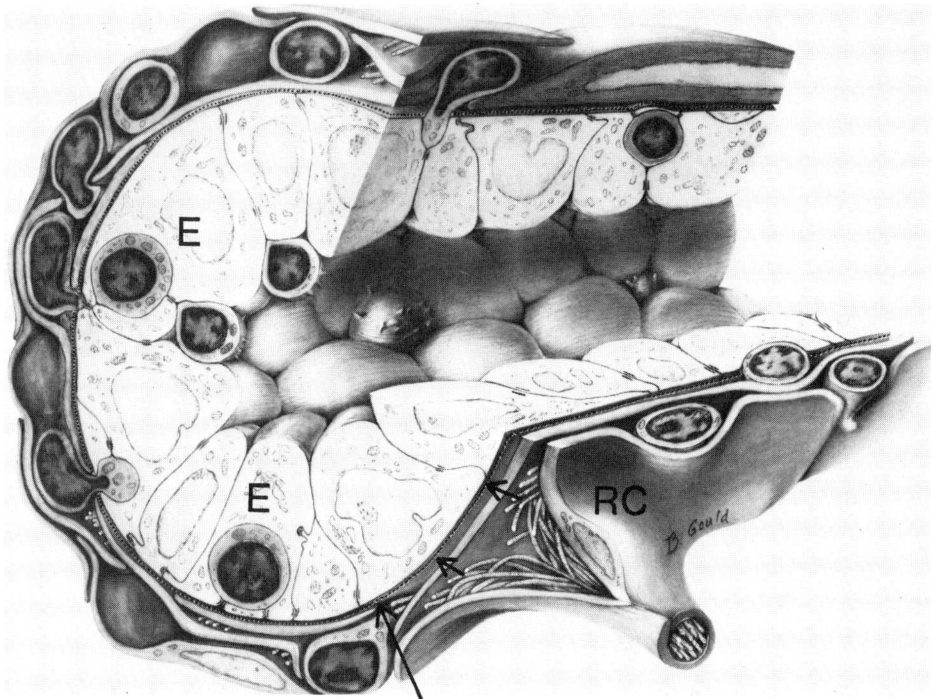

Figure 5. High endothelial venules in the deep cortex facilitate large-scale emigration of lymphocytes from the blood into lymph nodes. The underlapping polygonal endothelial cells (E), delicate basal lamina (arrows), and multilaminar reticular cell sheath (RC) allow transmural migration of lymphocytes with minimal blood extravasation.

walls that drain segments of cortical lobules and merge to form lobular veins. At sites where these veins anastomose with each other and with the systemic circulation, muscular contractile structures are found that may serve to prevent dramatic intranodal circulatory changes from influencing the systemic blood flow.

Studies of circulation changes in stimulated lymph nodes have demonstrated a biphasic shift in blood flow following antigen inoculation. An early increase in blood flow, up to 30 times normal, is associated with hyperemic effect induced by antigen (Hay and Hobbs 1977). The gradual return to increased blood flow that occurs later in the response is thought to result from increased vascularity of the node. In morphological studies, this second phase of increased flow is associated with elongation and increased branching of the network of HEV; autoradiography following infusion of [^{3}H] thymidine revealed that the endothelial cells of HEV located nearest to primary follicles became labeled between 3 and 6 days after alloantigenic stimulation (Anderson et al. 1975). It is likely that these alterations in the microcirculation directly affect transvascular lymphocyte migration into antigen-stimulated lymph nodes initially by increasing the rate of migration across existing HEV and subsequently by enlarging the area of endothelium available for migration.

The endothelial cells of HEV are unlike those of normal venules anywhere in the body. These cells characteristically exhibit diffuse nonspecific esterase activity in their abundant cytoplasm, almost reaching the staining intensity of monocytes. The presence of a prominent Golgi apparatus, multivesicular bodies, and phagolysosomes in the cytoplasm of HEV cells (which also appear to lack typical Weibel–Palade bodies) suggests that this endothelium may be composed of monocyte-like cells (Anderson et al. 1976). Indeed, the analogy of HEV cells to monocytes can be carried further, because the binding of lymphocytes to HEV endothelial cells is similar to that of cultivated mononuclear cells (Lipsky and Rosenthal 1973). However, the observation that HEV cells form a continuous lining linked by junctional complexes that lie on a basal lamina secreted by these cells is more consistent with a uniquely specialized endothelium than a lining formed by modified monocytes.

Lymphocyte emigration into lymph nodes.   High endothelial venules were shown by Gowans and Knight (1964) to be the site of large-scale blood-to-lymph emigration of recirculating lymphocytes (Figure 5). Other investigators proposed direct entry of nodal lymphocytes into the blood at these sites, because HEV lumens frequently contain high concentrations of lymphocytes. Subsequent studies indicated that cellular accumulation of HEV lumens was due to selective adhesion of lymphocytes to endothelial surfaces (Figure 6a) presumably via receptor molecules located on the tips of microvilli (Anderson and Anderson 1976). A specific "homing" receptor has not been isolated and characterized, but various studies suggest it may be a sialoglycoprotein, which requires divalent cations and transmembrane connections to a functional cytoskeleton to mediate recognition and adhesion of lymphocytes to endothelial surfaces (Woodruff and Gesner 1968; Ford et al. 1978; Anderson et al. 1979). *In vitro* models of lymphocyte homing have been developed (Figure 6a–c) that may help in the chemical and biological characterization of the receptor (Stamper and Woodruff 1976; de Bono 1976; Butcher et al. 1979).

Transvascular migration in normal lymph nodes is not restricted entirely to lymphocytes, since monocytes and cells resembling novel dendritic cells are found infrequently among the lymphoid cells crossing HEV walls (Figure 7). Lymphocytes and monocytes attach to luminal surfaces of HEV, squeeze between the endothelial cells, and enter the perivascular sheath composed of multiple laminations of sheath-like reticular cell processes (Figure 5). After gaining access to the paracortical interstitium, these cells may either crawl along reticular fibers toward reactive foci in T- or B-cell areas or they may cross lymphatic endothelium and enter intermediate sinuses, which ultimately delivers them into the efferent lymph.

Elegant studies by Sprent (1972) and Ford (1975) showed that the mean transit time for migration of normal lymphocytes across lymph nodes was 18 to 20 hours for T-cells and 28 to 30 hours for B-cells. This observation that B-cells transit lymph nodes more slowly than T-cells probably is related to the motile behavior of the respective cells. B-cells locomote less frequently and more slowly than T-cells (see Intrinsic Motility of Lymphocytes). T-cells appear to be attracted by short-range factors secreted by B-cells, since T-cells turn and migrate toward B-cells whenever they pass within a 30 to 40 μ radius (Anderson and Warren, 1980). This may obviate the need for equal locomotory activity by B-cells in establishing T-B interactions within lymphatic tissues, since T-cells can overtake B-cells during their journey from blood to lymph.

The controversy about whether lymphocytes pass through or between HEV endothelial cells (Marchesi and Gowans 1964) has been resolved to some extent by numerous recent studies using extracellular tracers, serial sections, and scanning electron microscopy, which show that lymphocytes enter and cross HEV by migrating between the endothelial cells (Schoefl 1972; Wenk et al. 1974; van Ewijk et al. 1975;

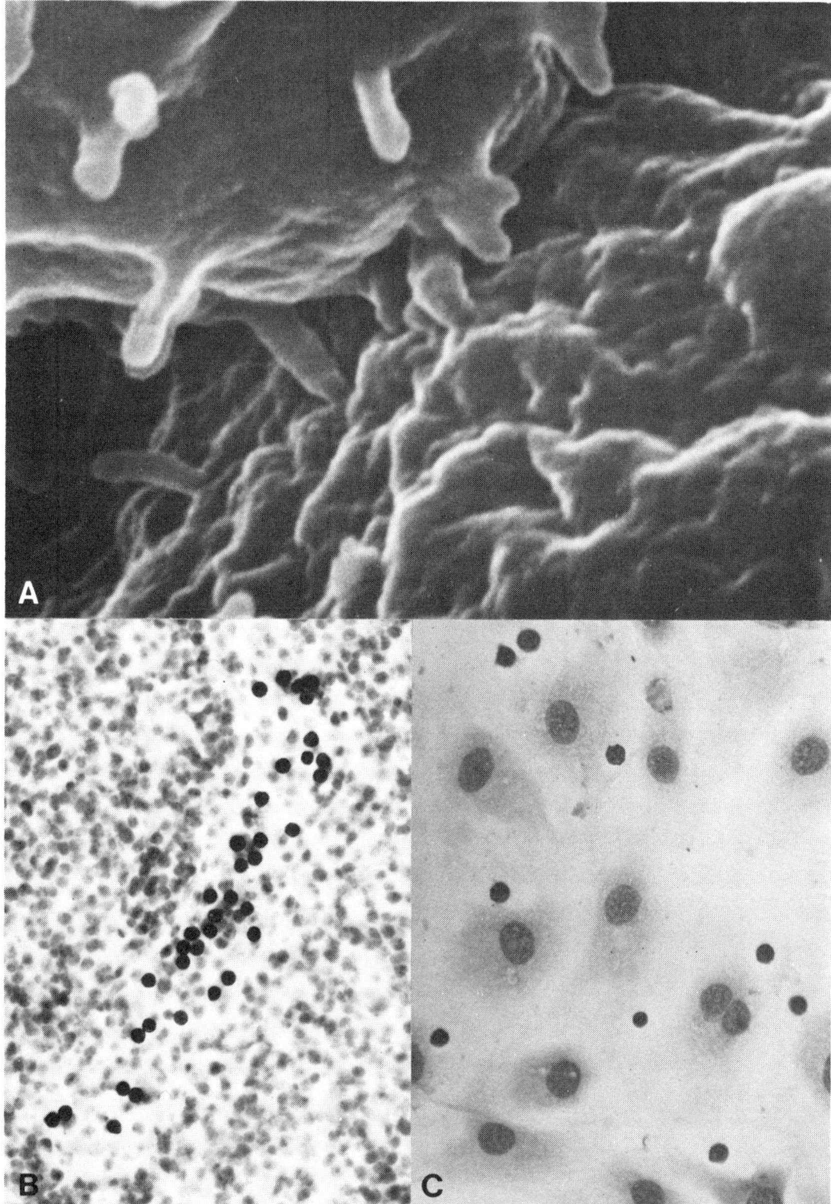

Figure 6 A–C. Methods of studying lymphocyte-endothelial interactions. (A) *In situ* lymphocytes that have attracted to HEV luminal surfaces via microvilli-pit interactions resist hydrodynamic shearing stresses generated by perfusion-flushing and perfusion-fixation. (B) The homing receptor for lymphocyte binding to HEV endothelium is highly conserved in nature since lymphocytes from higher vertebrates bind to sectioned HEV *in vitro* according to their phylogenetic similarity. (Courtesy of E. Butcher). (C) Similar binding of rat, guinea pig, pig and human lymphocytes to cultivated porcine endothelial cells can be demonstrated (Anderson, A.O. and J.L. Middlebrook, unpublished observations).

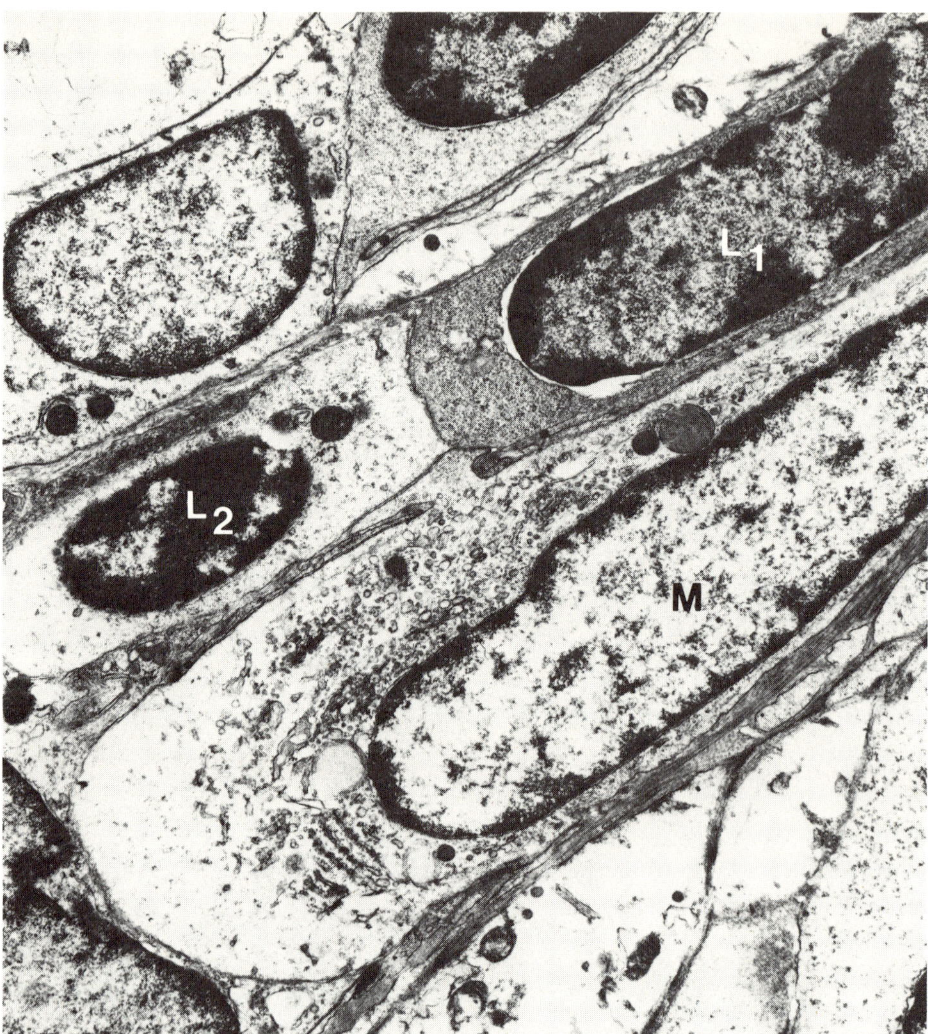

Figure 7. The cells that enter the lymph node paracortex via HEV remain in the reticular sheath for variable periods of time after crossing the endothelium. Here, two lymphocytes (L_1 and L_2) with differing ribosomal density (presumably T- and B-cells) are interacting simultaneously with each other and with a process from the macrophage (M) in the next lamination. The HEV sheath is one of many sites where immune cell cooperation might occur.

Anderson and Anderson 1976). The controversy had some basis, since lymphocytes deeply indent the endothelial cells as they cross HEV without breaking the macular interendothelial junctions. In acute inflammation these endothelial junctions separate so that all varieties of blood leukocytes may pass through.

Special permeability characteristics of the HEV wall permit limited intravasation of macromolecules from within the lymph node parenchyma without allowing significant extravasation, even in the presence of transmigrating lymphocytes. This unidirectional permeability derives from the "flap-guarded valve" structure of the HEV wall composed of overlapping endothelial and reticular cells. Diffusion of materials from the lymphatic tissue into the interendothelial spaces may provide a mechanism by which chemotactic gradients are generated. This is especially true under circumstances of antigen stimulation where there is altered blood flow and increased tissue fluid pressures.

The Spleen. The spleen is the largest lymphoid organ in the body and acts as a large discriminatory filter set across the bloodstream. The spleen is largely a meshwork of reticular cells and their fibers containing entrapped macrophages that can remove and destroy foreign pathogens in the blood and also provide for the selective monitoring and removal of aged and damaged erythrocytes, leukocytes, and platelets passing through its interstices.

In immune reactions, the spleen provides for the same processes of antigen trapping, cellular collaboration, lymphocyte proliferation, and antibody production seen in peripheral lymph nodes. As a rule, the immune responses to antigens and pathogens invading the blood stream are concentrated within the spleen. In addition, some of the antigen administered locally at distant tissue sites can pass through the regional node and be sequestered within the spleen. The immature lymphocytes mobilized into efferent lymph during immune responses in peripheral lymph nodes pass through the blood and frequently establish secondary residence in splenic cords and sinuses, where the B-cells mature into antibody-secreting plasma cells.

Structure. The spleen is composed of a dense fibrous capsule with trabeculae extending inward that subdivide the organ into lobules and merge with the reticular meshwork supporting the splenic pulp (Figure 8). The spleen has no cortex and medulla, but its interior is grossly divisible into white pulp, which forms cylindrical collections of lymphocytes about the arteries, and red pulp, which contains erythrocyte-rich blood.

Figure 8. Vascular, reticular, and lymphoid compartments of the spleen (modified after Weiss 1972).

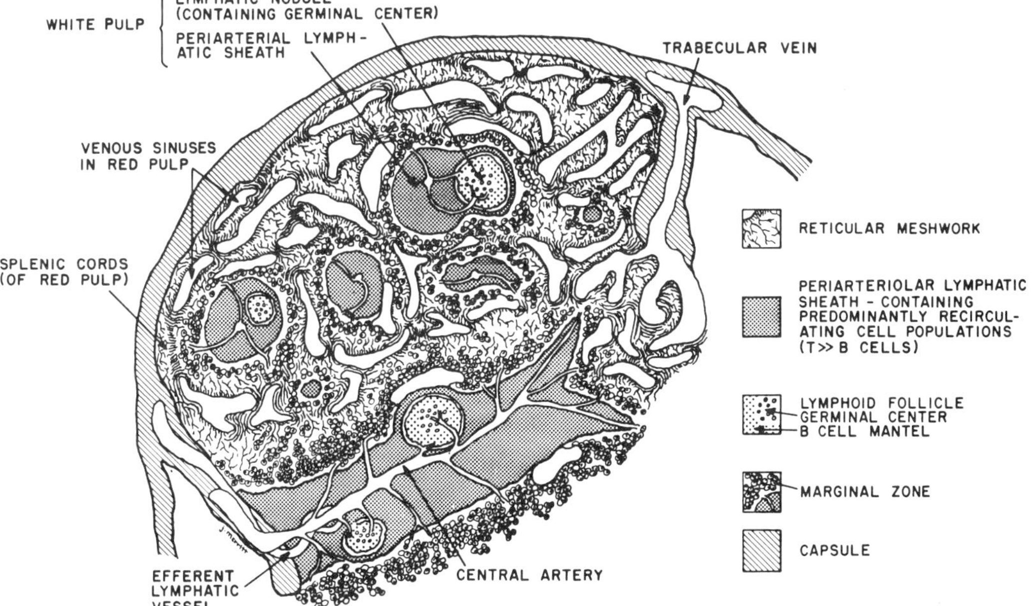

The red pulp contains the splenic cords that are continuous partitions of tissue crisscrossed by reticulum lying between the splenic sinuses. The cords contain erythrocytes, lymphocytes, macrophages, granulocytes, and numerous antibody-secreting plasma cells. The splenic sinuses are vascular channels between the cords that are lined by endothelial (stave) cells and hoop-like arrangements of basement membrane (Figure 9). The erythrocytes liberated into the splenic cords must squeeze through slits between these endothelial cells to enter the sinuses. The spleen monitors for abnormal erythrocytes by this mechanical seiving, and cells that have lost their plasticity are destroyed as they pass from pulp to sinus (Chen and Weiss 1973).

The spleen is supplied by the splenic artery, which passes through the hilus and branches to pass through trabeculae into medium-sized arteries surrounded by a sheath of lymphatic tissue called the periarteriolar lymphatic sheath (PALS). These vessels terminate in small arterioles that empty into the pulp cords, pulp sinuses, or the white pulp and its surrounding marginal zone, which is rich in macrophages. Blood leaves the spleen through a sinusoidal system that connects by gradual transition to pulp veins, trabecular veins, and finally the splenic veins. The spleen has no afferent lymphatic vessels but does possess efferent lymphatics that arise within the white pulp, pass through trabeculae, and exit at the hilus.

Lymphocyte emigration into the spleen. Lymphocyte traffic through the splenic red and white pulp is not regulated by the same surface-recognition mechanisms that determine selective lymphocyte homing into lymph nodes. Many of the lymphocytes entering the spleen in the arterial blood bypass the white pulp and flow along with other blood elements into the reticulum of the red pulp cords and sinuses to exit via the splenic vein. These cells would be returned to the blood within 2 to 3 hours after entering the spleen. However, some small arterioles terminate in sinuses adjacent to the white pulp; others empty directly into the marginal zone. Lymphocytes arriving at these sites move across an antigen-binding interface provided by macrophages in the marginal zone and enter the PALS. Cells entering here may remain within the spleen for up to 12 hours. There is no clear-cut segregation of T- and B-cells into discrete zones within the splenic white pulp, but follicles and their mantels contain B-cells, and recirculating T- and B-cells are found in the PALS. After variable periods of residence here, some recirculating cells exit through the efferent lymphatics, while others appear to move back out via bridging zones into the red pulp sinuses (Mitchell 1973). The magnitude of lymphocyte recirculation via splenic efferent lymph or venous return remains to be determined. However, if these immunocompetent cells encounter an appropriate antigenic stimulus, they proliferate in the same manner as that seen in peripheral nodes; the B-cells can form large germinal centers in this process. A few antibody-secreting cells can be found within the periarteriolar lymphatic sheath, but most of the mature plasma cells appear within the splenic cords following B-cell migration through the blood or their direct movement out from the white pulp.

The Mucosal Immune System. The mucosal immune system consists of noncapsulated lymphoid nodules and diffuse lymphocytic infiltrates closely associated with the epithelium lining the intestinal and respiratory tracts. These structures provide the initial site of antigen priming for the local synthesis and selective transport of secretory IgA antibodies into external secretions, promoting local immunity to pathogens and foreign proteins that invade the body through mucosal surfaces. This

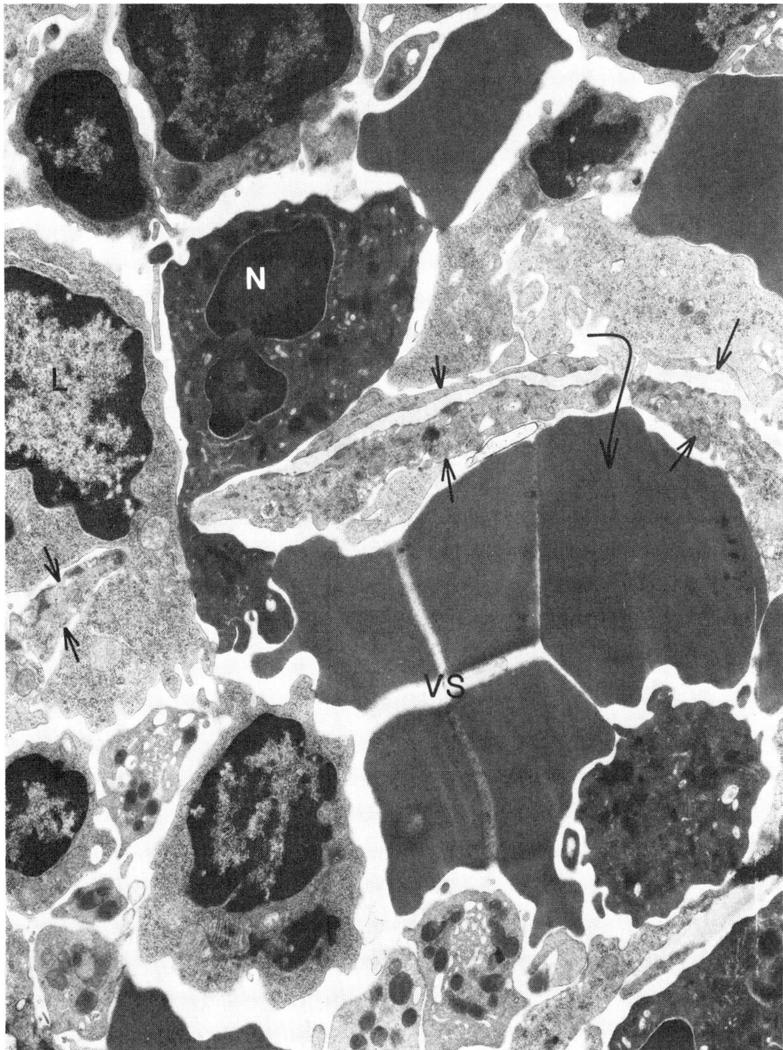

Figure 9. The intricate "barrel stave and hoop" structure of splenic venous sinusoids (VS) are shown. The barrel and the staves, composed of endothelial–reticular cell sandwiches encompassing thick basement membrane material (arrows), can be seen "mechanically screening" a neutrophil (N) and a lymphocyte (L) passing from the red pulp to the blood. Large particulate antigens, nonmotile cells, and red cells that have lost their plasticity will be retained in the red pulp where they may interact with resident macrophages.

modified immunoglobin consists of two IgA molecules linked together by a J chain (molecular weight: 15,000) and contains a molecule of secretory component coiled about the Fc portions of the joined immunoglobin molecules. This structure stabilizes secretory IgA against proteolysis and facilitates its transport across the mucosal epithelium.

Secretory IgA plays an important role in host resistance against many viral diseases by impairing colonization of the mucous membranes and preventing reinfection by the same virus. The efficacy of this process has been well established by the use of oral immunization with live, attenuated viruses, such as the Sabin poliovaccine, which prevents both reinfection and the establishment of the carrier state. Absence of adequate mucosal immunity can result in severe local hypersensitivity reactions when the pulmonary or intestinal epithelium is reinfected by the same viral agent, despite evidence of potent systemic immunity (i.e., the hypersensitivity pneumonitis seen in children immunized parenterally with killed measles vaccine and infected subsequently with live virus). Although secretory IgA probably does not promote direct lysis or phagocytosis of bacteria, it does interfere with bacterial adherence to mucous membranes. When this effect is combined with the normal cleansing actions provided by mucous secretion, cilial beating, cellular desquamation, intestinal motility, etc., secretory IgA can contribute in preventing bacterial colonization within the lung and gut. In addition, evidence suggests that secretory IgA can combine with toxins and macromolecules within the gut lumen, inhibiting their binding and absorption by intestinal epithelium.

Structure. The organized lymphatic tissue components of the mucosal immune system consist of nodular collections of lymphocytes within a reticular meshwork, which are located in the lamina propria and submucosa of the small intestine, appendix, colon, bronchi, tear ducts, salivary glands, and mammary glands. Their structure differs from that seen in lymph nodes, because they lack a well-developed capsule, afferent lymphatic connections, and a discrete medulla containing cords of plasma cells (Parrott 1976).

The correlation between structure and function of these organs is best established in the Peyer's patches (Figure 10a–c). These patches are discrete, subepithelial collections of lymphoid tissue found along the antimesenteric wall of the intestines. Individual patches are found from the duodenum to the ileum but are more numerous and larger in the distal intestinal tract. Each patch contains multiple individual follicles containing dendritic cells, macrophages, and proliferating B-cells and is surrounded by a mantel of small B-cells (Figure 10a). The interfollicular spaces are infiltrated diffusely with small- and medium-sized lymphocytes, usually of T-cell origin. The patches contain specialized HEV and efferent lymphatics, which provide for lymphocyte entry and egress (Figure 10b,c). Although Peyer's patches lack true afferent lymphatics, about 30% of the efferent intestinal lymph from the lamina propria lymphatic plexus diffuses through them before emptying into lymphatic trunks that originate beneath the follicles and drain into mesenteric lymphatics. Regardless of how many follicles a patch contains, each is only one follicle thick, providing an intimate association between the follicle and the overlying epithelium. The specialized intestinal epithelium covering each follicle is heavily infiltrated by lymphocytes. These epithelial cells, called microfold or M-cells, are stretched by their burden of interepithelial lymphocytes, lack prominent microvilli, and their cytoplasm contains numerous transport vesicles that appear to be adapted specifically to sample intestinal contents (Owen 1977). These cells endocytose and transport antigens across their cytoplasm in vesicles and deposit them among the lymphoid cells that lie above germinal follicles. The direct transport of intact antigens into the patch probably explains how these structures provide immune surveillance in the absence of afferent lymphatics (Bockman and Cooper 1973). Recent studies have demonstrated remarkably similar transcellular transport of antigen and structural organization in the bronchial-associated lymphatic tissues, so these features may be common to all components of the mucosal immune system (Rácz et al. 1977).

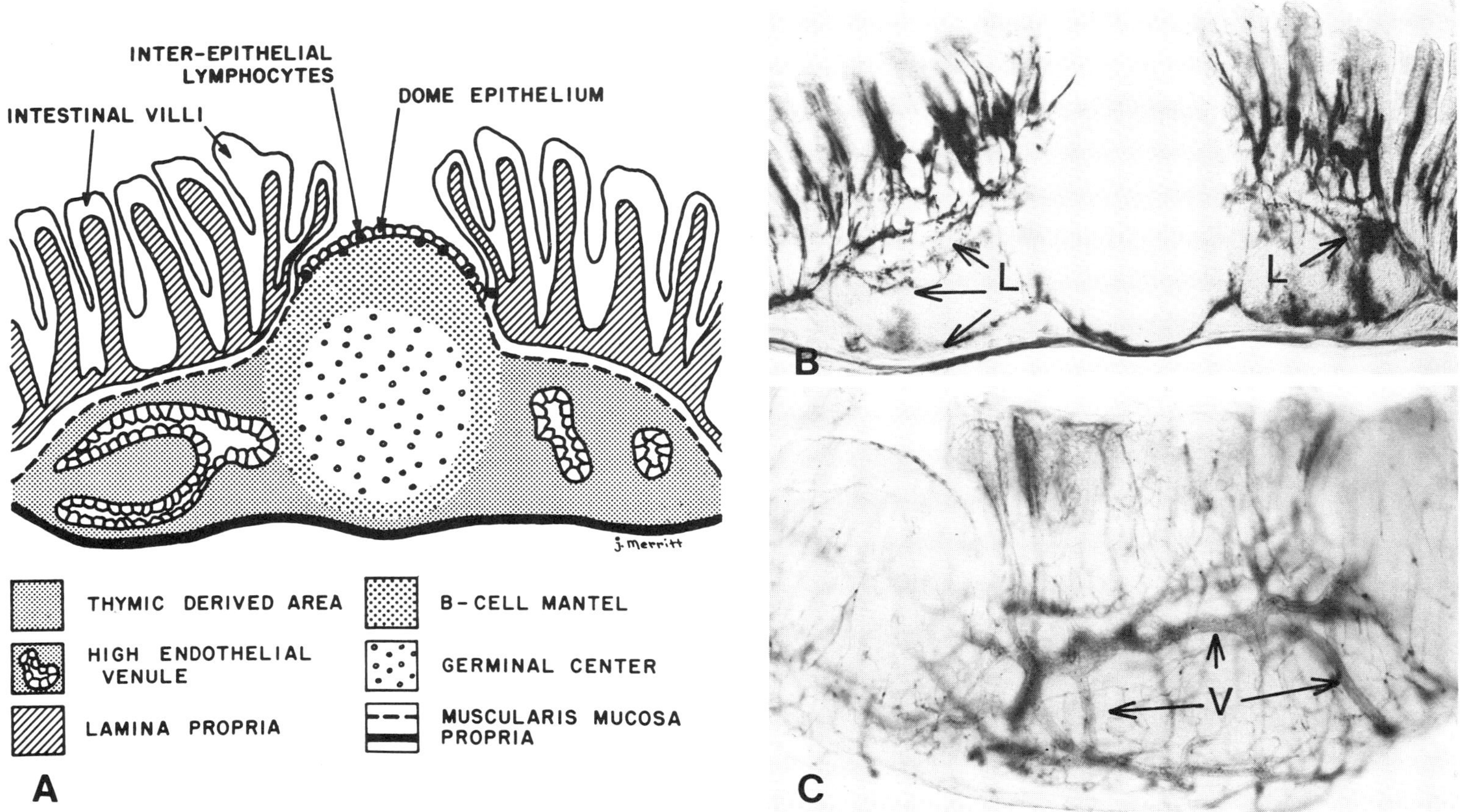

Figure 10 A–C. Structure of the Peyer's patch (PP). (A) Distribution of functional classes of lymphocytes in a simple PP. The cleared sections (B) and (C) show the lymphatic (L) and vascular (V) channels in similar PP. Dye-stained efferent lymph from the lamina propria lymphatic plexus passes around HEV in the thymic-derived area before connecting with subserosal lymphatics and efferent lymph trunks exiting the patch at the base of the follicle. Note also that the follicle and dome area do not appear to be penetrated by the stained lymph.

Peyer's patches are heavily populated with B-lymphocytes, but virtually no IgA secretion occurs in them, despite evidence that many of the B-cells in the patch are precommitted for IgA synthesis (Cebra et al. 1974). When antigen is delivered into the patch, these cells, with the help of macrophages and T-lymphocytes, undergo a clonal burst of proliferation. Maturation to antibody-secreting cells does not occur in the patch, however. The immature cells leave through efferent lymphatics and move through the mesenteric nodes and thoracic duct back into the bloodstream. After completing various differentiation steps in the mesenteric node or the spleen, these cells preferentially establish residence at submucosal sites, such as the lamina propria of the intestine (Cebra et al. 1974; Husband and Gowans 1978). Although still unproved, this distinctive lymphocyte traffic pattern has been variously attributed to the presence of environmental antigens within intestines or the appearance of IgA dimers on the B-cell surface, causing them to localize beneath epithelial cells, synthesizing and releasing the secretory component of IgA. Once they have reached these submucosal sites, the B-cells mature into plasma cells.

Virtually all the IgA found in external body secretions is produced by local synthesis within plasma cells. After the IgA molecules are synthesized and assembled, the J piece is added during secretion and release of the dimers into submucosal tissue. These dimers diffuse across the basement membrane where they are coupled to secretory components synthesized and bound to the surface of epithelial cells. The resulting complex is endocytosed and transported through the epithelial cell cytoplasm to be secreted into the intestinal or bronchial lumens. In the lumens, it can bind with the appropriate antigen or be lost from the body during normal passage of the intestinal contents and mucous to the exterior. In addition, some IgA may be reabsorbed into the blood, removed by the liver, and secreted back into the intestine with the bile, providing a form of enterohepatic circulation that may greatly augment intestinal immunity (Reynolds et al. 1978).

Peripheral Sites of Chronic Inflammation

Antigens presented in a form that is not easily removed by cells of the mononuclear phagocyte system, e.g., emulsified in a nondegradable oily base, result in the formation of chronic inflammatory foci. These lesions possess follicular and diffuse lymphocytic infiltrates (Figure 11), in addition to collections of foam, epithelioid and giant cells, and other monocyte components of granulomata (Spector and Willoughby 1963). Similar collections of diffuse lymphoid tissue may be seen in (1) lesions resulting from infections by facultative intracellular organisms of viral, bacterial, and parasitic origin; (2) tissues exposed chronically to microbes present in the fluids that bathe them, such as the bladder or renal pelves of patients with chronic bacilluria due to pyelonephritis; (3) organs undergoing chronic rejection following transplantation; and (4) organs attacked by autoallergic phenomena, such as Hashimoto's thyroiditis, allergic orchitis, and allergic encephalomyelitis.

All these chronic inflammatory lesions contain structural elements similar to those of nonencapsulated lymphatic tissue. The inflammatory infiltrate that succeeds neutrophils in these sites is composed predominantly of monocytes, macrophages, and lymphoid cells. While monocytes and lymphocytes usually arrive at the site simultaneously, large-scale lymphoid cell emigration occurs only after additional microenvironmental structures are generated. Chronic inflammation typically develops in connective tissue sites that are well supplied by lymphatic channels. The inductive

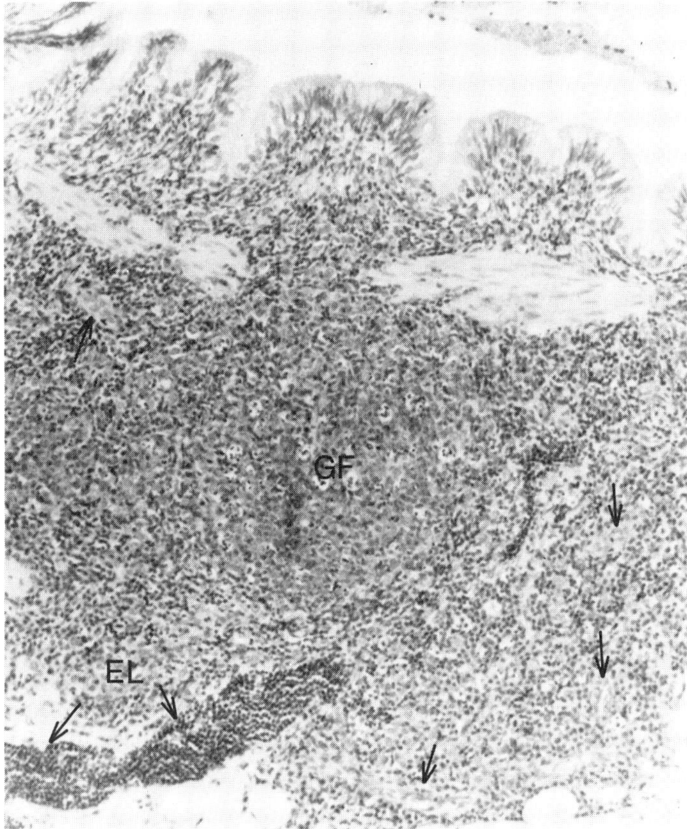

Figure 11. Laboratory rats frequently have respiratory mycoplasmal infections that induce the formation of submucosal lymphoid aggregates. These lesions resemble normal bronchial-associated lymphatic tissue, except they are hyperplastic, more numerous, ectopically located, and appear to lack specialized dome epithelium. These chronic inflammatory lesions have germinal follicles (GF), efferent lymphatics (EL), HEV (arrows), and regions of diffuse lymphoid tissue populated by recirculating cells.

stimulus, possibly provided by a number of factors released by monocytes and lymphoblastic cells, results in proliferation of fibroblasts and lymphatic and blood vascular endothelium (Figure 12). Specialized vascular segments develop by mitotic division, that morphologically, histochemically, and functionally resemble HEV of lymph nodes (Astrom et al. 1968; Smith et al. 1970; Graham and Shannon 1972). These new vessels lined by plump, esterase-positive endothelial cells arise in chronic inflammatory foci prior to the appearance of dense aggregates of small lymphocytes (Graham and Shannon 1972). Circulating lymphocytes home to these vessels, emigrate from the blood, and accumulate temporarily in the lesion before passing into afferent lymphatics and regional lymph nodes. Lymphoblastic cells also enter the lesion at these sites and differentiate into plasma cells or participate in the formation of germinal follicles. The antibody specificities of these cells are frequently unrelated to the antigens present at the site.

This large-scale lymphoid cell traffic through sites of chronic inflammation has been documented in studies where the outputs of cannulated afferent lymphatics

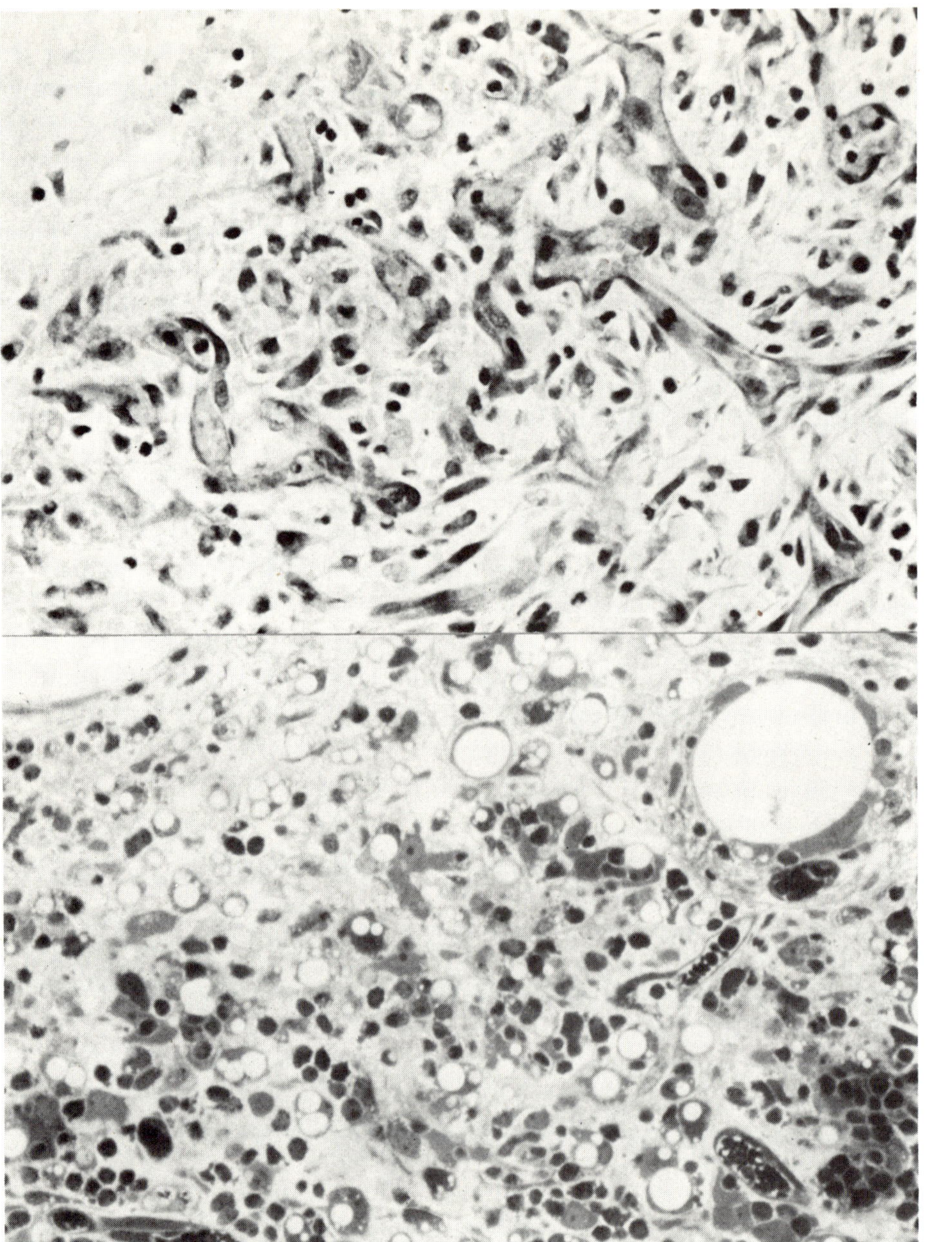

Figure 12. In connective tissue sites, chronic inflammation evolves as granulation tissue composed of proliferating vascular twigs (arrows) penetrates necrotic debris left by dying acute inflammatory cells. Subsequently, the leukocyte traffic across venular segments causes these lesions to be populated with monocytes (M) and lymphocytes (L). Secretion of factors by monocytes unable to degrade ingested materials increases both vascular proliferation and the mononuclear cell emigration.

draining adjuvant-induced granulomata were measured (Smith et al. 1970), or where the distributions of transfused [^{3}H]uridine-labeled lymphocytes were traced in tissues by radioautography (Anderson and Reynolds 1979). In the former studies, the afferent lymph output in sheep inoculated with Freund's adjuvant reached 3.1 $\times$ 10^7 lymphocytes per hour, which is nearly equal to the output of a normal lymph node, and is at least 10 times that of normal afferent lymph. Increased lymphocyte traffic via the granuloma site persisted through the 70 days of the study and probably would have continued as long as the lesion was present. In the latter study the lymphocytes present in the afferent lymph were identified as predominantly B-cells. This augmentation of B-cell traffic to the regional lymph node via the granuloma site correlated with the pronounced increase in antibody production and with the altered ratios of T- to B-lymphocytes found in these nodes. The enriched B-cell populations present in the afferent lymph may have been recruited from the circulation, but the formation in germinal follicles is also possible.

Macrophages in the lesion, which are unable to completely remove the antigen-containing material, become activated and secrete mediators that affect the local microenvironment and that of the regional lymph nodes by stimulating lymphocyte division, chemotaxis of monocytes and lymphocytes, vascular permeability, blood flow, and cell proliferation (Unanue et al. 1976). These peripheral sites of chronic inflammation resolve when all the antigen is neutralized and eliminated by phagocytes; they may, therefore, serve a beneficial purpose by establishing new reticular microenvironments that serve the same antigen- and cell-sorting functions as peripheral lymphatic tissues. However, these foci also cause discomfort and disease when the antigen is (1) a normal constituent of the host; (2) not degradable or neutralizable; or (3) capable of changing its antigenic determinants in response to immunologic pressure. In addition, hydrolytic enzymes that may be released in the course of phagocytosis, may damage other neighboring structures, such as the articular surfaces of joints in rheumatoid arthritis and the arthritis associated with autoallergic diseases.

Cellular Components of the Immune System

Although specific *in vitro* interactions, metabolism, and functions of individual cellular components of the inflammatory and immune system will be discussed in greater detail in subsequent chapters, some morphologic descriptions and biologic characterizations of these cells as they appear within intact organisms seems both instructive and appropriate.

Lymphocytes are the major cell type in lymphatic tissues. Reticular cells and macrophages found intercollated among the lymphocytes are nearly as prevalent. Mast cells, fibroblasts, vascular endothelial cells, Langerhans cells, basophils, eosinophils, and neutrophils are also present in progressively decreasing numbers, respectively. Mast cells are well represented in all lymphatic tissues, while neutrophils are so rare that their presence suggests the existence of acute inflammation. Only the reticular cells, fibroblasts, and cells of neurovascular tissue remain as relatively permanent structures after all the mobile cells are removed. The various leukocytes pass constantly into and out of lymphatic tissues at different rates and by various routes.

Lymphocytes traverse peripheral lymphatic tissues so rapidly that a significant portion of the small lymphocytes in each lymphatic organ will be replaced every 24 hours. During the same time interval the circulating pool of lymphocytes, which comprises 20 to 40% of the peripheral blood leukocytes, will be replaced 10 times (Gowans 1959).

Macrophage populations in lymphatic tissues turn over more slowly, since some become fixed to the reticular meshwork, some die *in situ,* others undergo proliferation, and a number exit in the efferent lymph in a form morphologically indistinguishable from lymphoid cells. Macrophages are replenished in lymphatic tissues indirectly through migration of tissue monocytes along afferent lymphatics draining peripheral connective tissue and directly through migration of bone marrow-derived monocytes across the vascular endothelium. The fate of macrophages after they enter lymphatic tissues remains a mystery.

Little information is available about the origin and life cycle of mast cells in lymphatic tissues, although some investigators have theorized that they develop from cells within the thymus (Ishizaka et al. 1976), whereas others consider mast cells to be bone marrow-derived (Kitamura et al. 1979). It is assumed that polymorphonuclear inflammatory cells enter and leave lymphatic tissues by the same mechanisms that control their passage through sites of acute inflammation.

Lymphocytes

Morphology and Biology. Each of the body's 10^{12} lymphocytes belongs to one of two functional classes. Thymus-dependent lymphocytes (T-cells) are best identified by serological labeling of class-specific surface antigens. These theta-antigen-positive cells mediate delayed hypersensitivity, graft rejection, and cellular cytotoxicity. B-cells, which are said to be derived from the bursa of Fabricius in chickens and bone marrow of mammals, are recognized by their surface expression of immunoglobulins and produce antibodies after exposure to antigens. Morphological characteristics cannot be used to distinguish between these functional classes of lymphocytes, but categorizing these cells according to size and relative organelle content often provides useful information, especially when the class of lymphocyte is known (Figure 13). Size correlates with the level of "activity" of the cell, and the cytoplasmic organization, histochemical reactivity, and other morphological markers yield additional information, which can be interpreted in the light of specialized functions.

Small lymphocytes. Small lymphocytes typically display a dense, heterochromatic nucleus containing some euchromatin (Figure 13). Their scant cytoplasm contains free and clustered ribosomes. Lysosomes and vesicles are present in some cells, but not in others. All small lymphocytes have a Golgi zone, which is located near the centriole and nuclear indentation; two to six mitochondria may be seen clustered nearby. T- and B-cells are virtually indistinguishable morphologically, but the former cells incorporate radiolabeled RNA precursors (^{3}H-uridine) more rapidly than B-cells, which facilitates their identification in tissue, blood, or lymph (Howard et al. 1972). In addition, Knowles and Holck (1978) have shown that prolonged incubation in substrates for nonspecific esterase activity results in punctate staining of the Golgi zones of T-cells but not B-cells. These are useful nonimmunologic techniques for enumerating T- and B-cell populations in tissues.

The surfaces of normal circulating lymphocytes, examined by scanning electron microscopy, are usually studded with numerous microvilli (Figure 14), whereas some lymphocytes may exhibit relatively smooth surfaces. Such differences appear to reflect variation in the biologic activities of individual cells or changes introduced during

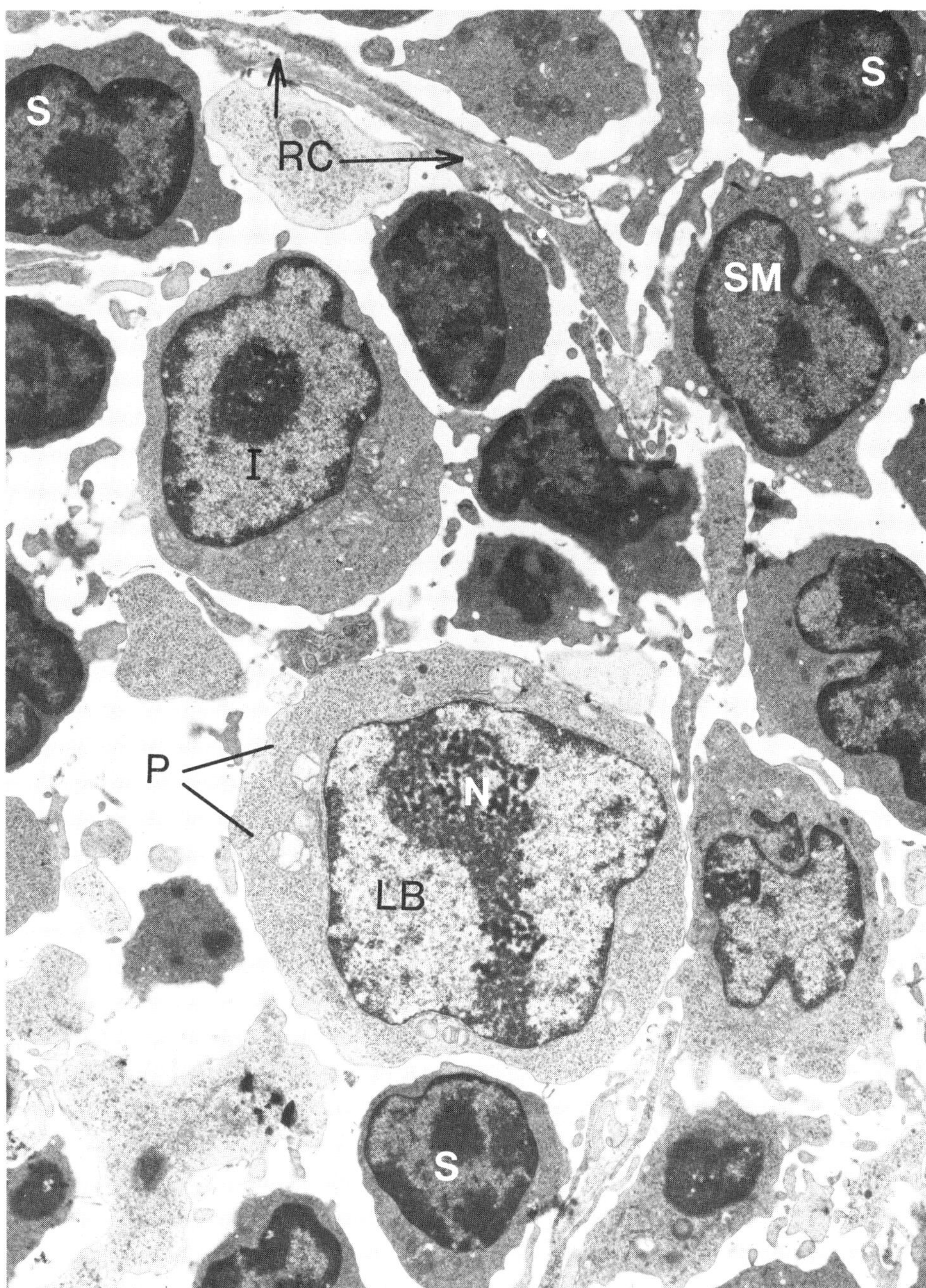

Figure 13. This electron micrograph of lymph node paracortex includes an "immunoreactive" lymphoblast (LB) containing a giant nucleolus (n) and numerous polyribosomes (P); abundant cytoplasm and a prominent nucleolus distinguish intermediate-sized lymphocytes (I) from small lymphocytes (S), which have little cytoplasm and a dense heterochromatic nucleus. Small macrophages (SM) and reticular cell processes (RC) can also be seen.

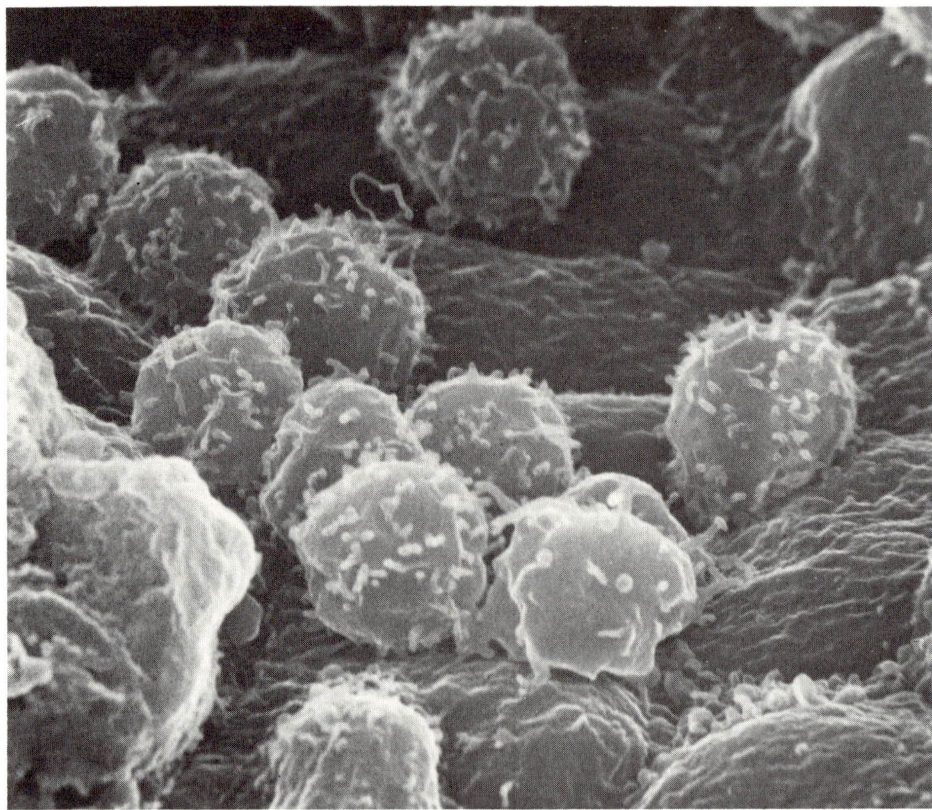

Figure 14. Scanning electron micrograph of the luminal surface of a high endothelial venule demonstrating that all lymphocytes exhibit microvilli on their surfaces when they have settled onto the specialized endothelial cells of these vessels.

processing. For example, the microvilli move toward one pole of the lymphocyte during capping or induction of locomotion, leaving behind a relatively smooth or ruffled surface. Treatments that affect the integrity of the cytoskeleton, such as use of metabolic inhibitors or exposure to cold, also result in smoothing of microvilli. Therefore, the presence or absence of surface microvilli cannot be used to discriminate between T- and B-cells, as had been suggested.

Lymphoblasts. When small lymphocytes are stimulated by appropriate antigens or mitogens, they undergo a series of metabolic changes preparatory for division and form large blast cells. These cells range in diameter from 10 to 30 μ and have a large vesicular nucleus, prominent nucleolus, and abundant cytoplasm containing numerous polyribosomes and a few rings of rough endoplasmic reticulum. Heavy staining of the increased ribonucleoproteins in histochemical preparations results in their characteristic pyroninophilia. These cells also display insulin receptors, glycosyl transferase activities, and secrete lymphokines.

Transitional lymphocytes. A small number of cells with size and morphologic characteristics intermediate to those described for small lymphocytes and lymphoblasts are present in blood and thoracic duct lymph (TDL); these cells are thought to be the progeny of proliferating blast cells that are in varying stages of maturation. All these cells display a prominent cytoplasm with a variable content of polyribosomes, endoplasmic reticulum, vesicles, and Golgi development. However, only the cells with a well-developed endoplasmic reticulum exhibit surface immunoglobulin markers, and antibody synthesis consistent with their ultimate maturation into plasma cells.

Plasma cells. Plasma cells represent the end stage of B-cell differentiation. Although these cells are found rarely in blood and lymph, they are one of the dominant populations present in spleen and lymph nodes. These oval cells range between 8 and 20 μ in diameter and exhibit an eccentric nucleus with condensed chromatin. These cells can be distinguished readily in electron micrographs by their extensively developed rough endoplasmic reticulum (Figure 15). Although mature plasma cells are a prime source of antibody synthesis and secretion (Leduc et al. 1955), they lack the surface immunoglobulin markers and motility of precursor B-cells.

Intrinsic Motility of Lymphocytes. It has been known for nearly 50 years that lymphocytes exhibit cycles of motility alternating with resting stages of variable

Figure 15. Intermediate lymphocytes of B-cell origin differentiates into plasma cells (PC) after completing antigen-induced proliferation. Antibody secretion is a function of the rough endoplasmic reticulum (RER) cysternae found in their prominent cytoplasm.

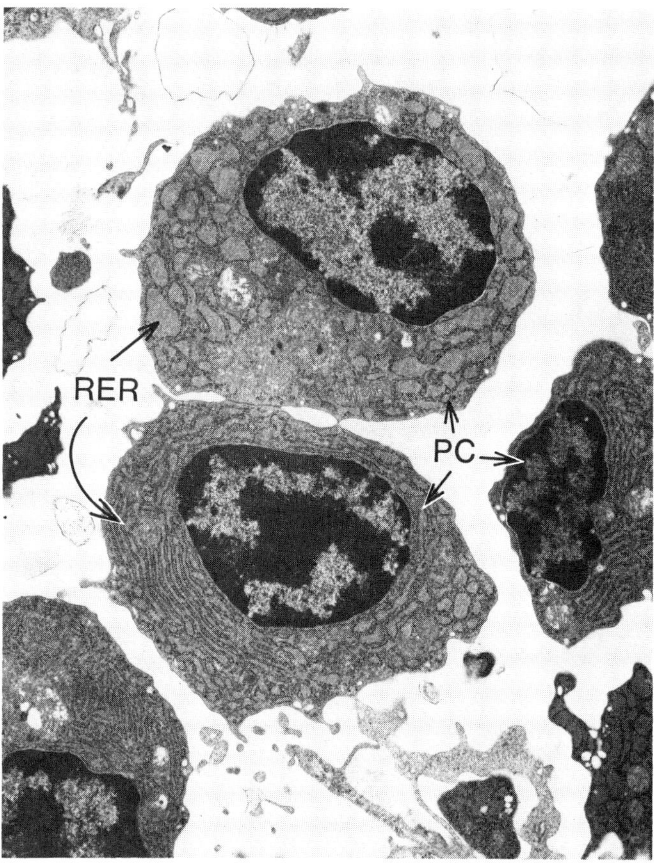

length (Lewis 1931). Resting lymphocytes assume a rounded shape with uniformly distributed microvilli studding their surfaces. As the motile phase begins, the lymphocyte probes its surroundings by intermittently extending and retracting villous projections and lamellipodia. When a lamellipodium "sticks" to the substratum, the cell elongates into the characteristic "hand mirror" shape with anterior nucleus and posterior uropod (Figure 16). As it translocates, the lymphocyte continuously extends lamellipodia anteriorly. Surface contact triggers a wave of cytoskeletal contraction which constricts the nucleus as the wave passes from front to rear. By transmission electron microscopy, the zone of constriction contains circumferential thickened sub-plasmalemmal filament mat. The surface overlying the constriction zone is devoid of microvilli, as seen in scanning electron microscopic preparations (Anderson et al. 1979). Both T- and B-lymphocytes exhibit spontaneous motility, although rates of locomotion vary. Small T-lymphocytes from TDL move relatively rapidly (7 to 12 μ/minute), while B-cells move slowly (3 to 5 μ/minute), if at all. Blast cells of either class are faster than small lymphocytes and migrate nearly as rapidly as neutrophils and monocytes.

Capping of ligands bound to surface receptors is another manifestation of lymphocyte motility (Schreiner and Unanue 1976). Most capping phenomena appear to involve participation of the cytoskeleton, glycolysis, and cellular respiration, since drugs that affect these activities inhibit capping; in rare instances, capping may result from passive diffusion and aggregation of cross-linked receptors within the fluid membrane. Antibody-induced cross-linking and capping of immunoglobulin receptors on B-cells is apparently a signal that precipitates locomotion, since it is frequently followed by translational movement. Regardless whether the B-cell caps or locomotes first, the anti-Ig cap is usually found at the end of the uropod of migrating cells. This suggests that oriented movement of surface receptors occurs from front to rear during locomotion. Since the centriole is always found in the uropod end of locomoting and capping lymphocytes, this structure is probably responsible for the cytoplasmic and cytoskeletal polarization necessary for these motile functions. A primary cilium generated by the centriole appears to serve this orienting function in other types of eukaryotic cells.

The organization of the lymphocyte cytoskeleton is still uncertain, but an umbrella-like network of microtubules may be seen radiating from the centriole around the nucleus toward the lamellipodia of lymphocytes that have become polarized before locomoting. A network of 5 nm filaments exists as a continuous mat beneath the plasmalemma and is arranged within the core of microvilli as longitudinally oriented fibers. Ten-nanometer filaments appear to extend anteriorly beside the microtubules and can be seen forming loops in the lamellipodia of motile cells. The complex interrelationships of these fibers during various biological activities have yet to be determined.

Chemotaxis. The directed migration of motile cells along a chemical gradient is generally accepted as playing an important *in vivo* role in regulating cellular traffic and promoting the accumulation of inflammatory cells at sites of tissue injury. However, the chemotactic responsiveness of lymphocytes is still controversial. Recent evidence indicates that mitogen-stimulated T-lymphoblasts and some lines of transformed lymphocytes display true chemotactic responses to casein hydrolysates, denatured albumin, endotoxin-treated serum, submitogenic doses of plant lectins, and staphylococcal protein (Russell et al. 1975; O'Neill and Parrott 1977; Wilkinson et

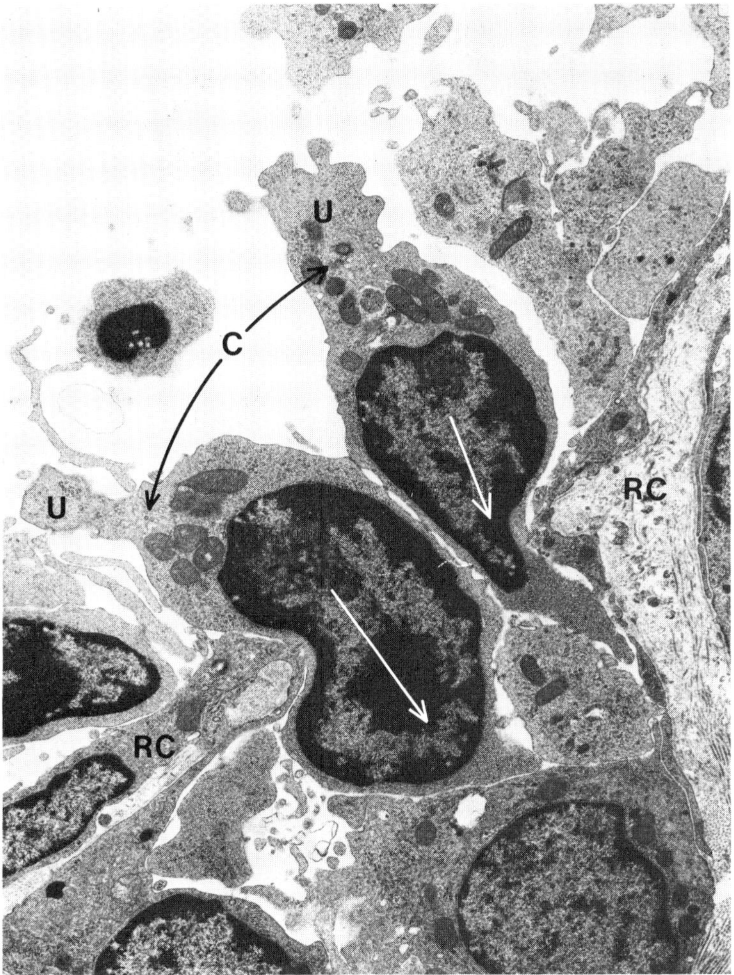

Figure 16. Lymphocytes caught in locomotion at the time of fixation exhibit cytoplasmic polarization. The centriole (c), mitochondria, and other organelles are typically concentrated in the trailing end, or uropod (U), while the nucleus and organelle-poor lamellipodia are in front. The arrows indicate the probable course these migrating lymphocytes were taking while passing between reticular cell (RC) processes.

al. 1977). In addition, supernatants isolated from mixed lymphocyte cultures, mitogen-stimulated lymphocytes, and activated macrophages evoke chemotactic responses by spleen and lymph node cells (Ward et al. 1971, 1977). Such findings suggest that some lymphocyte subpopulations exhibit directed migration toward selected stimuli under appropriate *in vitro* conditions. As each of these chemoattractants exhibits equally potent chemotactic effects for other leukocytes, the only selective lymphocyte chemoattractants identified to date are low doses of anti-Ig, which appear to attract B-cells (Higuchi et al. 1975; Ward et al. 1977). Further studies are needed to determine whether chemotaxis plays any role in regulating the characteristic traffic patterns displayed by different lymphocyte subpopulations *in vivo*.

Life-Span. Because of the discrepancy between the high rate of production of lymphocytes in lymphatic tissues and efferent lymph and the relatively low numbers of lymphocytes in the blood, early investigators assumed that all lymphocytes lived short lives. Ottesen (1954) provided the first clear indication that some small lymphocytes live for long periods of time. Administration of ^{32}P-labeled lymphocytes *in vivo* and DNA extraction of peripheral blood cells indicated that only 20% of blood lymphocytes had an average life-span of at least 2 to 3 days, but less than 14 days. In contrast, 80% of the lymphocytes remained in interphase for longer than 6 months. These findings were confirmed subsequently by other investigators who studied (1) [^{3}H]-thymidine uptake and turnover in rodent models; (2) thoracic duct drainage following nuclear labeling; and (3) lymphocytes bearing irradiation-induced chromosomal damage present in the blood from patients with accurate histories of therapeutic irradiation. From these investigations, it is now known that some small lymphocytes can live longer than 10 years in man and for nearly a year in rodents (reviewed by Sprent 1977). Thus, there are clearly two populations of lymphocytes with respect to longevity: long-lived cells and short-lived cells that survived in a detectable state less than 14 days. Long-lived lymphocytes form the bulk of the small lymphocytes in lymph nodes, recirculate continually from blood to lymphatic tissues and back again via the major efferent lymphatics, and possess immunological memory and immunoregulatory functions. The short-lived lymphocytes are more difficult to characterize because they may either be derived directly from the antigen-independent proliferation of lymphocytes in the bone marrow and thymus, or they may result from antigen-dependent proliferation of cells that may originally have been part of the long-lived population before meeting antigen. Regardless of their origin, short-lived lymphocytes do not recirculate; they lodge in the lamina propria of the intestine, in the spleen, at sites of inflammation, and (in smaller numbers) in other lymphatic tissues (Gowans and Knight 1964).

While it is clear that most long-lived lymphocytes are T-cells, careful labeling studies have shown that some B-cells are also long-lived. Since TDL contains approximately 80% T-cells and 20% B-cells, and TDL is composed of 90% long-lived cells, part of the long-lived lymphocyte population should include B-cells. In the absence of thymic helper cells, which affect the antigen-induced turnover of B-cells, nearly 100% of lymph node lymphocytes in nude mice appear to be long-lived cells. Because of the inability to discriminate between long-lived and short-lived B-cells, their average life-span must be considered 5 to 7 weeks in mice, although some of these cells, which are reactive to antigenic determinants that are rare in the environment, may remain in interphase as long as T-cells (Sprent 1977).

Reticular Cells

Reticular cells are the principal fixed cells of the immune microenvironment. These cells take many conformations within the encapsulated and diffuse lymphatic tissue. Reticular cells form the internal scaffolding of encapsulated lymphatic tissues, and produce the lacy networks of reticulum which are found within the connective tissue of all organs.

Morphology. It is difficult to appreciate the complex structural organization of lymphatic tissue reticular cells without use of histochemical methods, such as silver impregnation or the periodic acid–Schiff reaction to stain glycosaminoglycans. By electron microscopy, reticular cells have elongated branching cytoplasmic processes of moderate electron density. Multiple small deposits of heterochromatin and up to four nucleoli may be present in the irregular nucleus (Figure 17). Reticular cells characteristically lie on and invest specialized fibers composed of Types I and III collagen bundles and amorphous granular material resembling basement membrane. These fibers usually are covered by reticular cell cytoplasm throughout their entire length. A few lysosomes may be present in the cytoplasm of reticular cells, but they are not a prominent feature. The numerous small vesicles in the peripheral cytoplasm

Figure 17. The same reticular (RC) shown in Figure 23. The nucleus of this cell is usually located near the junctions of two or more reticular fibers (arrows), which are ensheathed by cytoplasmic processes of the reticular cell.

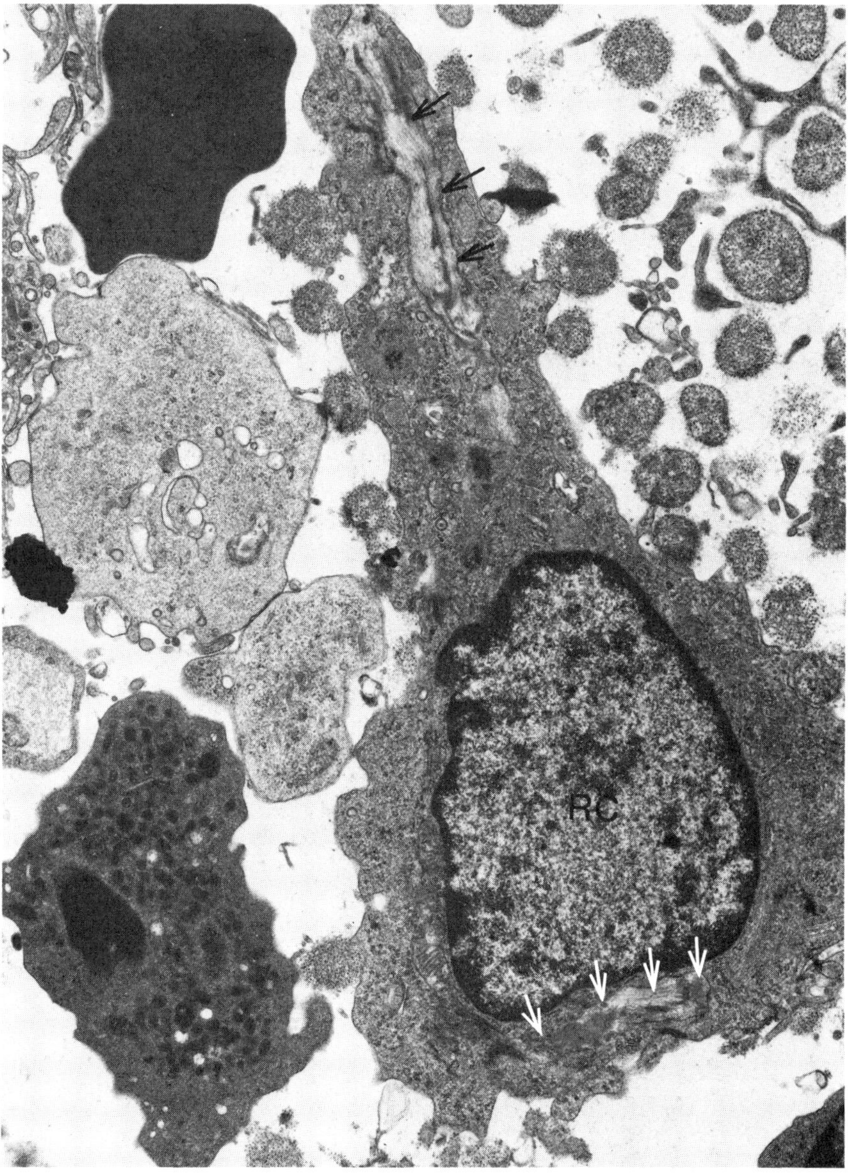

and the Golgi apparatus and rough endoplasmic reticulum of the central cytoplasm often contain granular basement membrane-like material, which suggests that reticular cells secrete and maintain reticular fibers. These fibers have also been shown to conduct molecular tracers along their lengths in a wick-like manner (Anderson and Anderson 1975), which may aid in the rapid distribution of protein antigens to specialized lymph node structures.

Functions of Reticular Cells. In their simplest form, reticular cells are arranged to produce a spongework or meshwork that fills a three-dimensional space, but the flattened endothelium of lymphatics, lymph, and vascular sinuses in lymph nodes and bone marrow, and the complex sheath of high endothelial venules in lymph nodes and venous sinusoids in the spleen are also composed of reticular cells. Since lymphatic and blood vascular reticular fibers and sheets are all continuous with the structural meshwork of encapsulated lymphatic tissue, it is possible to see how these cells physiologically regulate vessel diameter and permeability. This is especially true of terminal afferent lymphatics in peripheral connective tissue, which dilate rather than collapse when tissue fluid pressure exceeds lymph pressure. Paradoxical dilatation occurs because the lymphatic endothelial cells are guy-wired to the surrounding connective tissue by anchoring filaments (Leak and Burke 1968).

Reticular cell meshworks provide a microenvironment conducive to (1) antigen trapping; (2) adhesion and locomotion of lymphocytes and monocytes; and (3) dynamic sorting and functional grouping of cells that participate in immune responses. At one time, reticular cells were thought to become the immune cells through proliferation, but sensitive tissue culture techniques failed to demonstrate lymphoid differentiation by these cells.

Reticular cells display fibronectin on their surfaces, which may contribute to establishing a lymphatic microenvironment (Stenman and Vaheri 1978). Fibronectin is an important cell-surface glycoprotein that participates in such diverse processes as coagulation (Mosher 1975), particle opsonization (Blumenstock et al. 1978), and contact inhibition of proliferation and locomotion (Vaheri et al. 1976).

Reticular cells are said to be phagocytic, but these cells primarily provide anchoring sites for macrophages, which perform the bulk of the phagocytosis. These cells may participate in "selective" endocytosis and may recycle complex molecules secreted by neighboring cells, since reticular cells engulf the heparin-rich stroma of secreted mast-cell granules. The antigen-trapping function of these cells is related to the ability of meshworks of these cells to cause turbulence in lymph flow, which impacts particles against the surfaces of reticular cells and sinus macrophages. In addition, immunofluorescence studies indicate that immunoglobulins and antigens bind nonspecifically to reticular cell surfaces; possibly by utilizing intermediary molecules, such as complement components and fibronectin.

Macrophages

The components of the mononuclear phagocyte system (van Furth et al. 1972) exhibit a remarkable functional and morphological heterogeneity (Unanue 1972; Carr 1973; van Furth et al. 1975). Within the bone marrow a committed stem cell transforms into a monoblast which produces promonocytes and subsequently matures to form differentiated monocytes. Monocytes enter the blood circulation and either remain there or enter tissues where they become fixed or wandering macrophages. Free macrophages in loose connective tissue are called histiocytes; macrophages that are fixed between endothelial cells of liver sinusoids are called Kupffer cells. Bone contains osteoclasts; the central nervous system has microglia and "gitter" cells. Both free and fixed macrophages are found in lung alveoli, lymph nodes, spleen, and serous cavities. Specialized macrophages, which serve specific functions in immunity and inflammation, include the dendritic cells of germinal follicles and the epithelioid, foam, and giant cells found in granulomas. Of particular interest is a recently described class of monocytes that is smaller than the circulating monocyte (5 to 10 μ), looks surprisingly like a lymphocyte, but provides highly efficient accessory cell function in immune responses initiated *in vitro* (Arenson et al. 1978; Lee et al. 1979, Norris et al. 1979). There is reason to suspect that this kind of small macrophage is also found in TDL from antigen-stimulated animals, since 18-hour incubation of TDL in complement-containing media induces some of the lymphoid cells to transform into macrophages (Anderson et al. 1980).

In addition, there are macrophage-like cells whose origins are presently obscure but which often assume the functions of macrophages in specialized organs. These cells include nonphagocytic Langerhans cells found in the epidermis, dendritic cells in the lymphoid organs, dome epithelial cells in Peyer's patches and bronchial-associated lymphoid tissue, and special phagocytic epithelial cells interspersed among the renal tubular epithelial cells. These cells may serve to sample and transport antigens or altered proteins that would not make direct contact with macrophages unless the epithelial barrier were breached. These cells may also recover secreted products or complexes that might otherwise be lost to the external environment.

Morphology and Biology. *Peripheral blood monocytes.* Blood monocytes originate in the bone marrow and remain in the blood circulation for approximately 40 hours before transforming into tissue macrophages. Peripheral blood monocytes appear to be present in a bimodal distribution with respect to size. "Small" phagocytic, esterase- and peroxidase-positive cells that resemble lymphocytes are 5 to 10 μ in diameter and contain a mildly lobulated nucleus with densely clumped heterochromatin. The cytoplasm contains rough endoplasmic reticulum and a relatively prominent Golgi zone. This cell, called an M-1 cell by Arenson (1978), usually exhibits Ia and C_3 receptors. A larger cell (or M-2 cell) is 12 to 15 μ in diameter, has a large lobulated nucleus that is ringed by a thin band of heterochromatin, and is filled with light euchromatin. The cytoplasm contains a prominent Golgi apparatus and multiple small vesicles. Phagolysosomes are not often seen in this cell, unless bacteremia, viremia, or circulating cell debris has triggered its phagocytic function. The cytoplasm stains diffusely for nonspecific esterase activity, while vesicles and rough endoplasmic reticulum contain endogenous peroxidase activity, which accounts for the ability of this cell to exhibit chemiluminescence on exposure to particles and opsonins. The cell surface has Fc and C_3b receptors in addition to Ia antigens. Macrophages secrete alpha-2-macroglobulin (Hovi et al. 1977) and the nonspecific opsonin fibronectin has been demonstrated on macrophage membranes (Colvin et al. 1979). *In vitro,* these cells either spread on surfaces, giving them a fried-egg appearance on scanning electron microscopy, or they assume a rounded appearance and display multiple folds and cup-like processes.

Tissue macrophages. Wandering and fixed macrophages are virtually indistinguishable, except for the presence of "desmosome-like" cytoplasmic densities at sites where fixed macrophages make contact with reticular or endothelial cells (Figure 18). The fully differentiated macrophage, called an M-3 cell by Arenson et al. (1978), is a large cell measuring 15 to 20 μ in diameter. Its nucleus is often deeply indented and rimmed with irregularly dispersed heterochromatin. The cytoplasm characteristically contains abundant endocytic vacuoles, lysosomes, and phagolysosomes. Mitochondria and bundles of microtubules and microfilaments are also present. Time-lapse cinematography combined with fluorescent antibody studies have illustrated that these cytoskeletal constituents, which are arranged beneath the plasma membrane, participate in adhesion, endocytosis, vesicle movement, and fusion of lysosomes with phagosomes and the plasma membrane. A prominent Golgi apparatus is located within a cavity formed by the large indented nucleus. Smooth and rough endoplasmic reticulum

Figure 18. Macrophages (M) attach to reticular fibers (RF) crossing the lumens of intranodal lymphatic sinuses. During immune responses lymphocytes adhere to these macrophages surfaces and form lymphocyte–macrophage clusters similar to those observed *in vitro* by Lipsky and Rosenthal (1973).

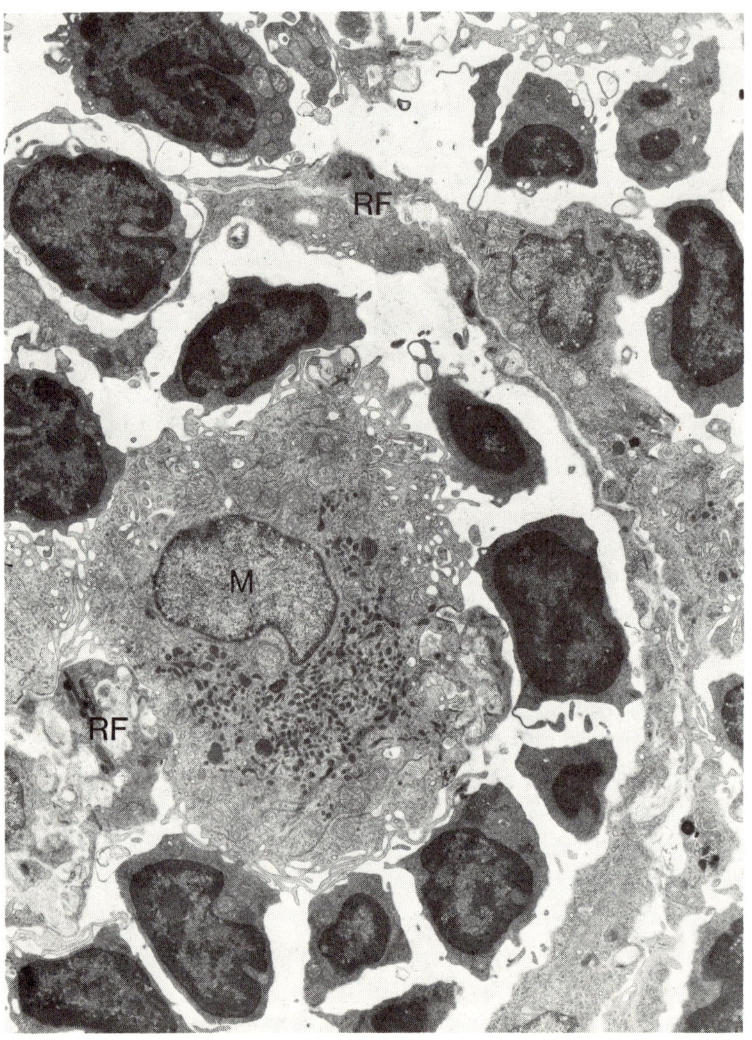

located nearby is the site of synthesis of lysosomal enzymes that ultimately pass into the Golgi for packaging. There is a great deal of variation in the morphology of tissue macrophages, depending on their site of activation, which in turn depends on local microenvironmental influences exerted by complement components, immune complexes, and lymphokines.

Novel dendritic cells of lymphoid tissue. A subclass of wandering macrophage located in the paracortex of lymph nodes has been called the cortical interdigitating cell by Hoefsmit et al. (1979) or the novel dendritic cell of peripheral lymphoid organs by Steinman et al. (1975). Ultrastructural and *in vitro* studies indicate both to be nonphagocytic cells, with branching, narrow cytoplasmic processes. In tissue, these processes can be seen interdigitating between small lymphocytes and lymphoblasts within the lymph node paracortex or splenic periarteriolar lymphatic sheath (Figure 19). *In vitro,* the same characteristic cytoplasmic branching and propensity to link together small and large lymphocytes are demonstrable. These cells possess no distinguishing cytoplasmic features, but tubulovesicular structures and Langerhans-like granules may be infrequently found. When these cells are seen among lymphocytes, their cytoplasm is typically electron-lucent, whereas the cytoplasm of traditional macrophages is denser than that of lymphocytes. Dendritic cells frequently are found juxtaposed to reticular fibers and clustered near small blood vessels.

Follicular dendritic cells. Nossal et al. (1968), and Hanna and Szakal (1968) described antigen-binding dendritic reticular cells (macrophages) located within germinal follicles of the spleen (Figure 20). This cell type has a bland, rounded nucleus containing delicate clumps of peripheral heterochromatin. Its extensively branching cytoplasm may contain small electron-dense granules, but normally this cell exhibits very little phagocytic activity. Classical macrophages are present in germinal follicles as well. They are easily recognized by clusters of large debris-filled phagolysosomes. When these cells are found to contain nuclei of dead lymphocytes they are called "tingible" body macrophages. The follicular dendritic cells described by Nossal et al. (1968) are not likely to be identical to the "novel dendritic cells" described by Steinman et al. (1975), because Steinman's cell effects T-cell behavior *in vitro* while Nossal's cell seems to influence only B-cells. The real significance of these unusual cells may lie in the observation that they begin to bind antigen within a few days following immunization. These dendritic cells persistently display antigen on their surfaces for months and possibly years. For this reason, these cells are thought to be related to the process of generating and maintaining B-cell memory within the germinal center (Humphrey 1976). In this case, relatively short-lived B-cells must continue to be induced to divide by antigen to maintain the level of clonal precursors needed for secondary immune responses.

Location of macrophages in lymphoid tissue. In central lymphatic tissues, macrophages are found near clusters of rapidly dividing cells and have been attributed functions as "nurse" cells that remove effete or aberrant cells and products of proliferation. In the thymus, these cells are characteristically periodic acid–Schiff-positive because they contain large amounts of membrane-associated glycoproteins. Within peripheral lymphatic tissues, macrophages guard every portal of entry. Lymph flowing from peripheral tissues must pass by macrophages that are found attached to reticular cell processes bridging lymphatic channels in lymph nodes. Similarly, the location of macrophages

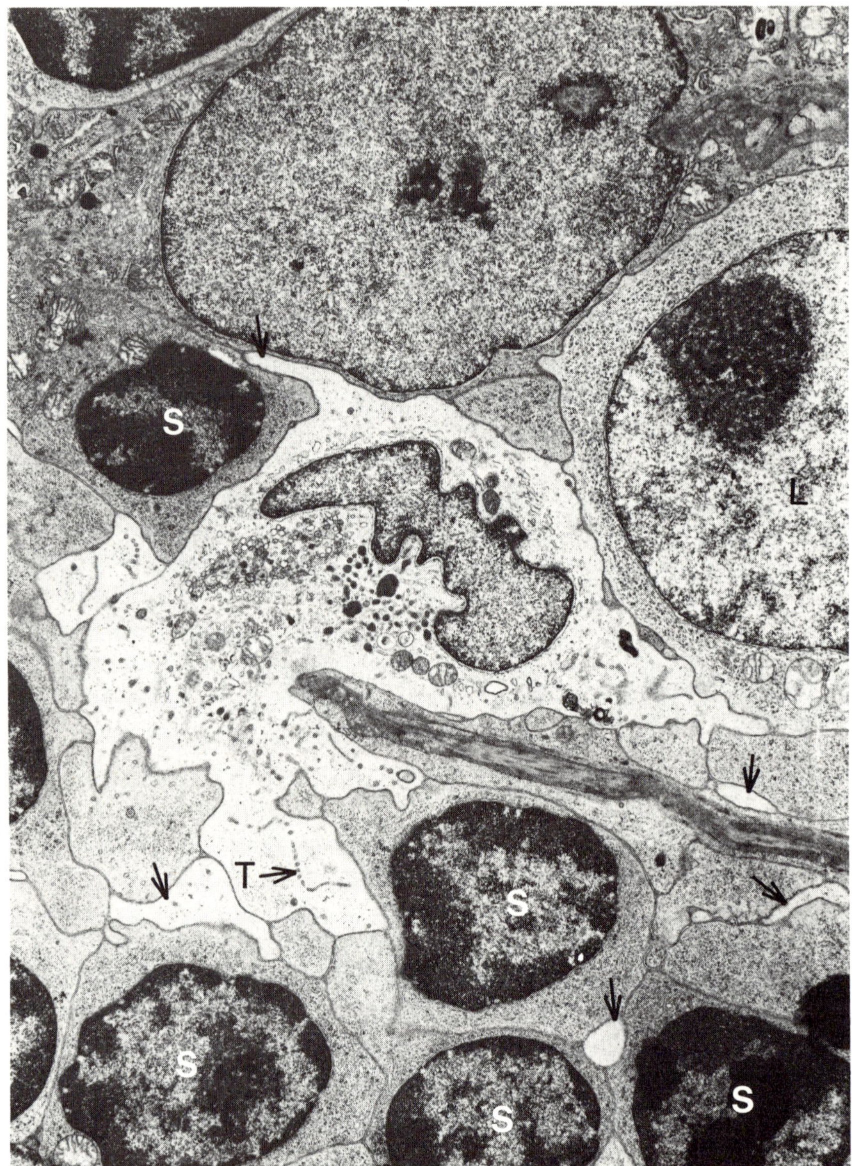

Figure 19. This macrophage-like cell found in the lymph node paracortex is not phagocytic; it contains tubulovesicular structures (T) in its characteristically electronlucent cytoplasm; and its branching cytoplasmic processes (arrows) interdigitate among the small (S) and large (L) lymphocytes of the thymic-dependent regions of lymphatic tissues. These cells have been said to resemble Langerhans cells and the "special" cells found at the corticomedullary border of the thymus.

among the reticulum of cords of red pulp, in the lining of venous sinuses, and in the marginal zone surrounding the white pulp affords a strategic placement with respect to the flow of blood through the spleen. In both organs, macrophages are found at breaks in the littoral cell barrier that separates the lymphatic or vascular spaces of lymphatic tissues from the interstitial spaces diffusely populated by lymphocytes.

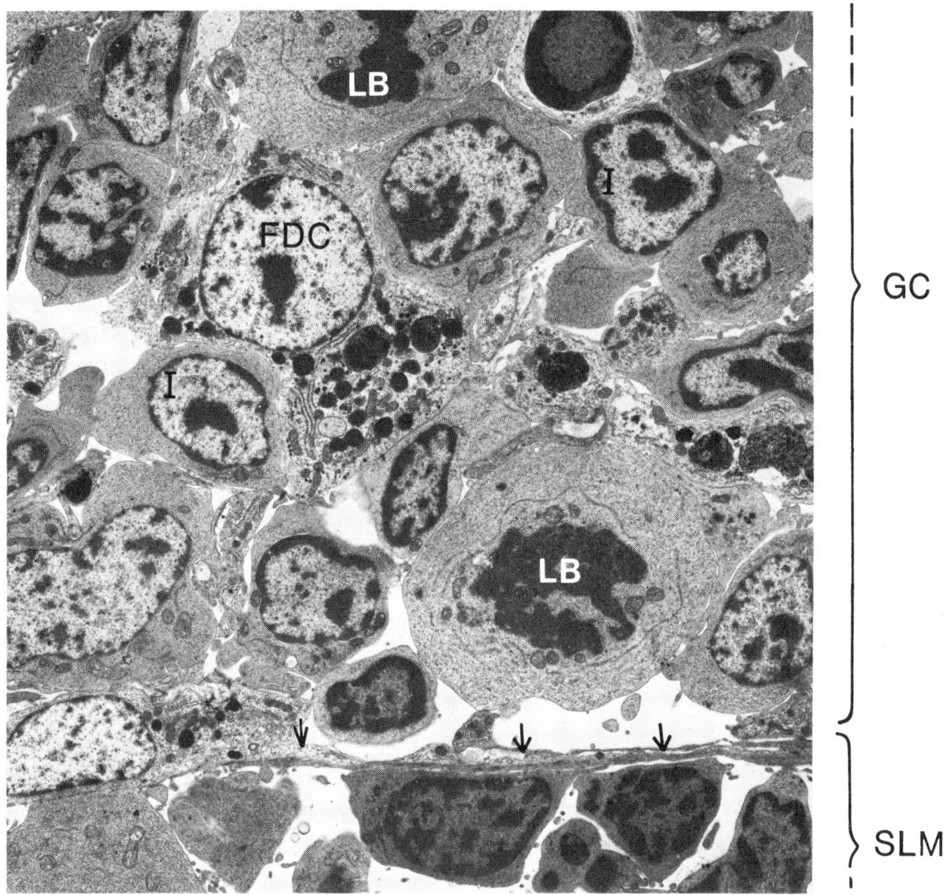

Figure 20. Germinal follicle. The border between the germinal center (GC) and the small lymphocyte mantel (SLM) is formed by a basket of centripetally displaced reticular cell processes (arrows). Within the germinal center lymphoblasts (LB), intermediate-sized lymphocytes (I), and follicular dendritic cells (FDC) are found. Lymphocytes make contact with antigen-binding dendritic cell processes during all phases of division, differentiation, and locomotion within the follicle.

Macrophage motility. Monocytes and macrophages exhibit spontaneous locomotion essential to enable them to enter or leave tissue. Their rate of migration *in vitro* is midway between that of neutrophils and lymphocytes. Macrophages respond chemotactically to various components of the complement cascade, products of fibrin breakdown, certain bacterial products, many denatured proteins, and factors released by activated lymphocytes. It is interesting to note that higher concentrations of chemotactic factors cause macrophages to stop migrating and become more firmly adherent to the substratum in the new environment. This phenomenon of macrophage migration-inhibition serves a useful purpose in the immunological schema, since it retains antigen-binding macrophages near to the portals of entry of other inflammatory cells, such as lymphocytes and immunoblasts.

Immune functions. Monocytes and macrophages have the important immune functions of processing antigen and presenting it to T-cells in association with surface Ia molecules and of secreting mediators that modify immune responses. The details of these functions will be discussed in other chapters of this book.

Mast Cells

Mast cells, first described by Paul Ehrlich (1879), measure 20 to 30 μ in diameter and contain characteristic metachromatic, electron-dense granules. They have a rounded central nucleus, which is smaller than that of a monocyte but larger than the nucleus of a lymphocyte. The cytoplasm contains mitochondria, a Golgi apparatus, endoplasmic reticulum, microtubules, and microfilaments, in addition to the more obvious granules. Mast cells apparently travel in the blood in an unrecognizable form (possibly a lymphoid cell) and differentiate after they seed the tissues of the recipient animal, where they remain as long-lived cells. The origin of mast cells from a lymphoid precursor was suggested by the *in vitro* observations of Ishizaka et al. (1976) where mast cells differentiated from thymic monolayers. However, recent *in vivo* studies suggest mast cells originate in the bone marrow (Kitamura et al. 1979). Ishizaka suggests there may be two forms of mast cells depending upon the site of origin (Ishizaka, personal communication). They are physiologically important cells because they release pharmacologically active mediators that affect blood flow and permeability (histamine), coagulability (heparin), muscle tone (slow reactive substance of anaphylaxis), and perception of pain (prostaglandins). Mast cells also release factors that attract eosinophils (eosinophil chemotactic factor) and modify the behavior of other inflammatory cells (Becker and Henson 1973). Their surfaces display receptors that

Figure 21. Cross-section of a rat lymph node removed and cleared immediately after vascular perfusion with the polycationic dye, alcian blue. The luminal and abluminal surfaces of high endothelial venules (HEV) stain strongly because they are covered with a thick sialoglycoprotein surface coat. Extravasation of this dye has stained the strongly anionic granules of mast cells (MC) located in the medullary cord region. The demarcation between cortex and medulla formed by convergence of reticular fibers at the corticomedullary border is indicated by arrows.

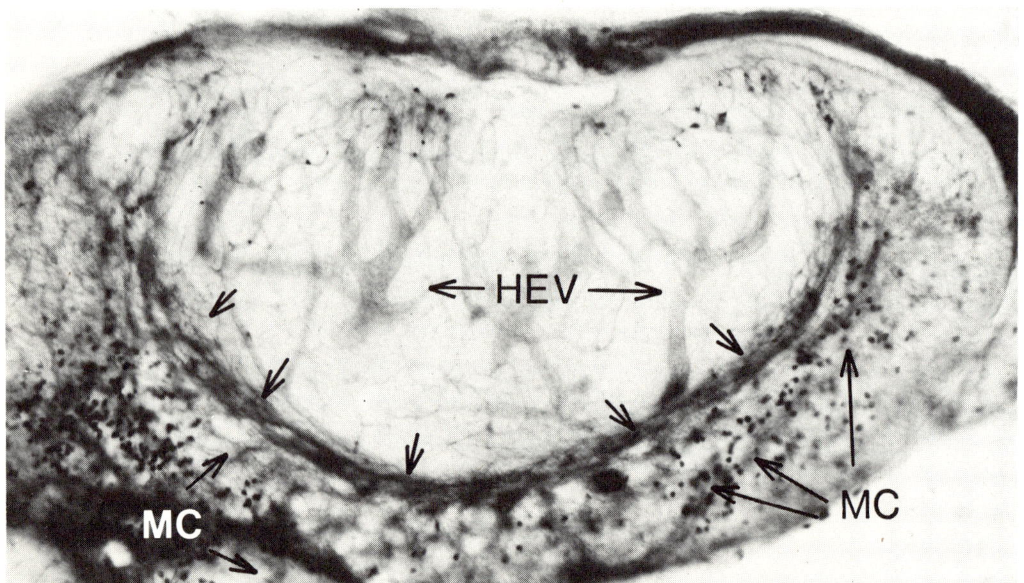

specifically bind the Fc portion of the IgE molecule, whose binding in conjunction with antigen signals this cell to degranulate.

Location in Lymphatic Tissue.   Mast cells are distributed in perivascular connective tissue. Concentrations of them are found in the thymus, bone marrow, and capsule and medullary sinuses of lymph nodes (Figures 21 and 22) and within the connective tissue stroma of the spleen. Mast cells located in the dermis of the skin, lamina propria of the gut, and submucosa of the bronchial epithelium may affect lymphatic tissues indirectly through release of their mediators into afferent lymphatics. Mast cells in lymphatic tissues may regulate blood flow and permeability in lymph nodes and spleens subjected to antigenic stimulation. Increased numbers of mast cells are found

Figure 22.   Most intranodal mast cells (MC) are located among the macrophages, lymphocytes, and reticular fibers in medullary lymphatic sinuses, but they are also present in the capsule and along neurovascular connective tissue. In these locations, mast cells are often exposed to materials present in the afferent lymph flowing from the peripheral tissues into the lymphatic spaces of the node.

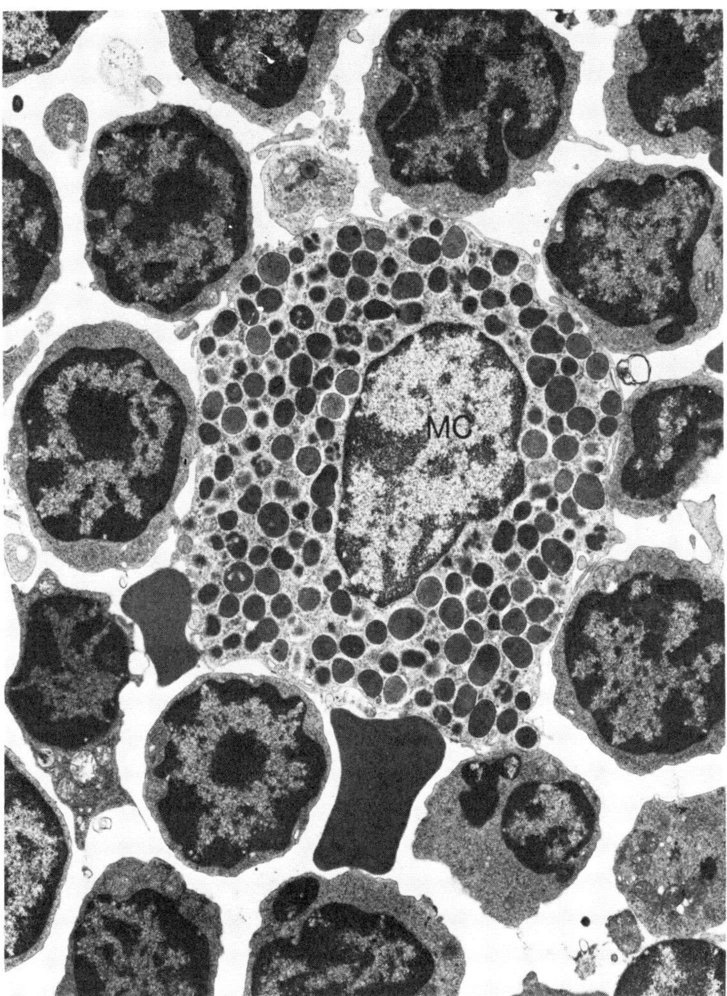

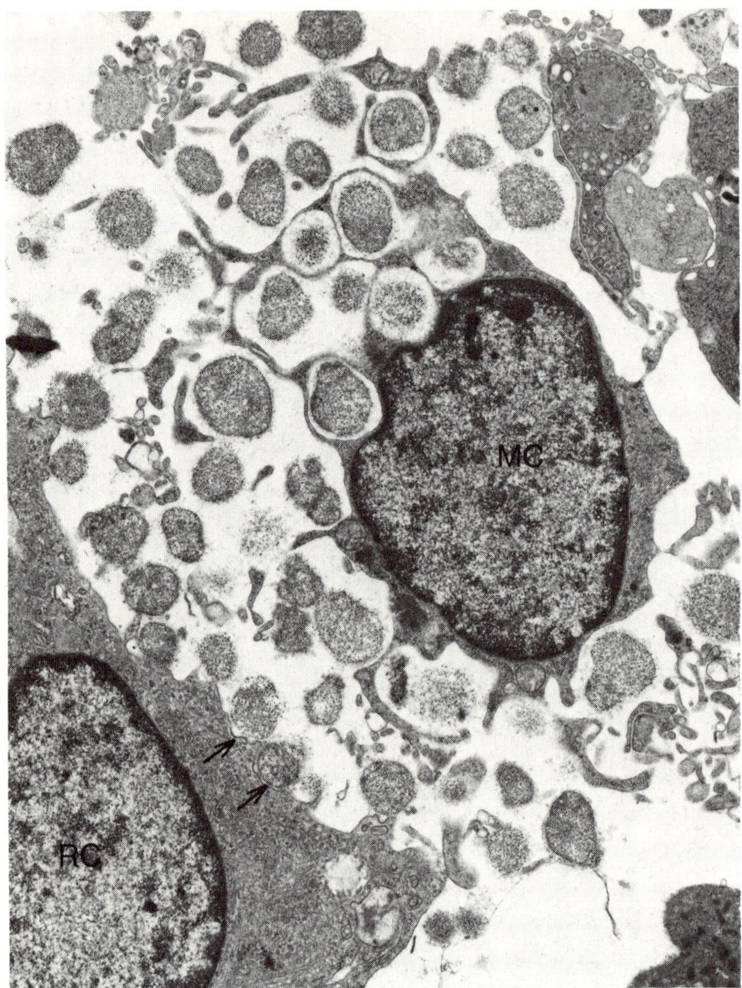

Figure 23. Forty to fifty percent of the mast cells (MC) in popliteal lymph nodes were degranulated 24 hours after foot-pad injection of 10^5 allogeneic lymphocytes. Surprisingly, many of these granules were endocytosed by "coated pits" (arrows) in reticular cell (RC) membranes. Endocytosis of mast cell granules by reticular cells has not been described previously.

in lymph nodes draining tumors (Bowers et al. 1979); mast cell degranulation has been described in lymph nodes during the initial 48-hour period following application of allogeneic skin grafts (Anderson et al. 1975) and within hours following intralymphatic instillation of the water-soluble adjuvant, muramyl dipeptide. This degranulation may influence the flux of migrating lymphoid populations by affecting blood flow and permeability of lymph node vasculature. It is interesting to note that the secreted granules are rapidly ingested by reticular cells, which use coated pits to endocytose the granule stroma (Figure 23). It is possible that the heparin-rich granule stroma may be metabolized and either secreted or used by the reticular cell. Redistribution of anionic molecules derived from mast cell granules by reticular fibers may serve to decrease the adhesion of lymphocytes to reticulum and thus clear out extraneous lymphocyte cell populations. Alternately, release of linear anionic polymers may play a role in detoxifying agents present in the lymph, by decreasing the positive charge on toxic proteins and lipids.

Circulation of Lymphocytes During an Immune Response

Successful initiation of a specific immune response requires that the lymphocytes must engage and bind the appropriate antigen with its surface receptor. Studies in non-immune animals have shown that very few of the lymphocytes present within a single lymph node are capable of reacting with a given antigenic determinant. If these cells were static or their movements were sorted randomly throughout the body, the likelihood of chance collisions between reactive lymphocytes and the appropriate antigen displayed on an accessory cell would be very remote. However, this is clearly not the case *in vivo,* where immunocompetent T- and B-cells recirculate continually between blood and lymphatic tissues (Sprent 1972). Long-lived lymphocytes flowing in the blood show a unique homing instinct for lymph node HEV. These cells cross HEV and emigrate into the reticular meshwork, where they crawl along reticular cell surfaces, collide with macrophages and other lymphocytes, and either stay to initiate proliferation or leave.

Lymphocytes recirculate through other lymph nodes and mucosal lymphatic tissues via HEV, providing a constant form of surveillance by immunocompetent lymphocytes that move through antigen-binding meshworks. This phenomenon probably enables a small depot of antigen to recruit a large number of antigen-specific cells from the body's lymphocyte pool (Figure 24).

When an individual lymphocyte encounters an appropriate antigenic stimulus within the node, it binds the antigen, interacts with other cell types, and is trapped in the node, where it gives rise to a clonal burst of proliferation and differentiation of lymphocytes that mediate immune responses. Many of the immature lymphocyte progeny leave the stimulated node 48 to 100 hours after antigen exposure and disseminate to distant nodes, the spleen, and other tissues, where they can mature into specific effector cells. The B-cells activated in this response move into the medullary cords of regional and distant nodes, where they mature into antibody-secreting plasma cells (Frost et al. 1976).

Although T- and B-lymphocytes emigrate from the same segments of HEV, they are sorted by unknown mechanisms within the nodal parenchyma. The T-lymphocytes establish residence in the deep cortex for relatively short time intervals before moving out into the efferent sinuses. B-cells emigrate into the superficial cortex and probably remain there for longer periods before exiting via the sinusoidal pathways. This migration pattern appears to permit the T- and B-cells to interact with antigen-binding macrophages and engage in cellular collaboration before they redistribute into their respective zones within the nodal cortex.

Other factors also influence lymphocyte traffic patterns in the body. The ability of T-cell subpopulations to recirculate appears to depend upon their state of maturation (i.e., at least some effector T-cells leave the circulation at random to enter sites of inflammation). B-lymphocytes display similar variations in their emigration patterns. The immature B-cells appearing shortly after antigenic challenge may leave the node, but they frequently lodge in the spleen and do not recirculate through the thoracic duct. However, "memory B-cells" arising at the late stages of an immune response do recirculate in a typical manner between blood and lymph (Ponzio et al. 1977). The

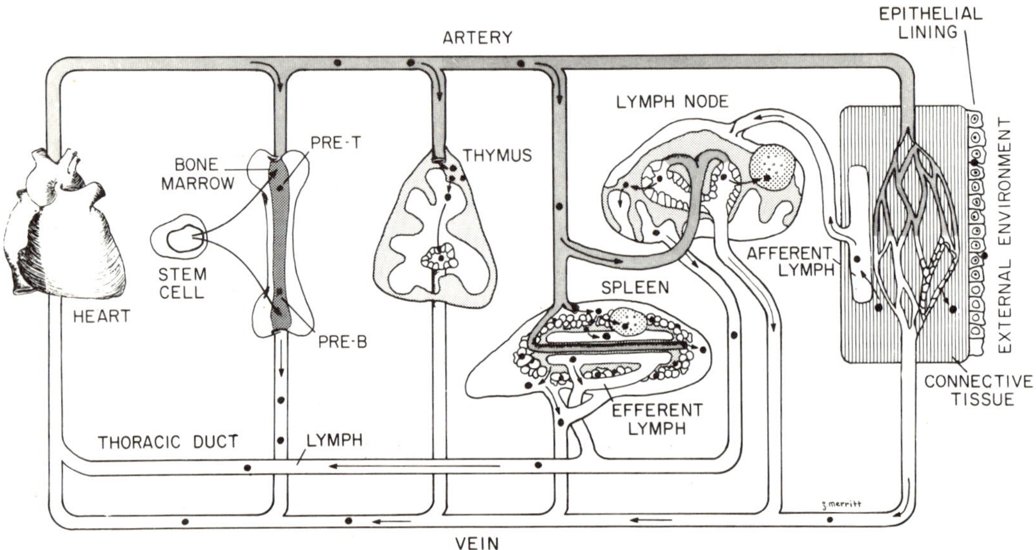

Figure 24. Circuits taken by recirculating and lodging lymphocytes. During maturation and differentiation, stem cells differentiate into pre-T- and pre-B-cells. Pre-T-cells home to the thymic cortex, multiply, and migrate past medullary epithelium to postcapillary venule (PCV) or directly via efferent lymphatics. These T-cells may be fully mature or may require further differentiation steps in the spleen and lymph nodes. Mature recirculating $Ly^{(1+)}$ T-cells never return to the thymus after exiting at PCV, but $Ly^{(2+,3+)}$ T-cells, which exit via lymphatics, may return under conditions associated with the induction of tolerance. Recirculating T- and B-cells pass from blood to lymphatic tissue by emigrating across HEV in lymph nodes and diffuse lymphatic tissues and by crossing the marginal zone in the spleen. The migration patterns of recirculating cells do not appear to be restricted to specific organs, and these cells randomly seek the periarteriolar lymphatic sheath of the spleen and the paracortex of lymph nodes. Of the lymphocytes leaving a lymph node via efferent lymphatics, only 4% arrive via afferent lymph, 10% are generated by proliferation inside the node, and 86% enter via migration across HEV. Nonrecirculating cells and short-lived cells generated by antigenic stimulation seek out the spleen, submucosa, and inflammatory lesions.

B-cell precursors of IgA secretion also possess distinct and quite different migration pathways. This circuit includes generation of precursor progeny in Peyer's patch follicles, which complete various maturational steps in the mesenteric node and spleen before finally lodging in the lamina propria of the gut as IgA-secreting plasma cells.

In nonstimulated animals, lymphocyte recirculation is characterized by a balanced flux of cellular traffic across afferent and efferent terminals of lymphatic tissues. This kinetic equilibrium is distorted rapidly in the regional nodes draining sites of inflammation, infection, or antigenic challenge. This is accompanied by a rapid increase in lymphocyte accumulation within the lymph node cortex, which is not associated with cellular replication. Such nodal enlargement probably reflects the combined result of increased lymphocyte traffic into the node and decreased egress, or markedly increased entry with normal rates of exiting. These early changes are believed to be produced, in part, by changes in blood flow and by release of secretory factors and lysosomal enzymes from activated macrophages that alter lymphocyte surface adhesiveness and transit times within the node (Unanue et al. 1976). As this early sequestration of recirculating lymphocytes within the stimulated node subsides, blast cell transformation and mitotic activity appear in the T- and B-cell zones of the cortex, reflecting the antigen-dependent cellular proliferation.

The specificities of the lymphocytes entering lymphatic tissues in the immediate 6- to 48-hour period following stimulation are largely unrelated to the antigen that initiated the response. In fact, similarly nonspecific lymphocytic accumulations can be induced by adjuvants that are themselves poorly immunogenic. However, as the sequestered lymphocytes are released into the efferent lymph, they are depleted of cells specifically reactive to the priming antigen (Rowley et al. 1972). The retention of antigen-specific lymphocytes in regional lymph nodes has been termed "specific recruitment" and was documented using double-label techniques. Specific memory cells generated *in vivo* in the presence of $[^{14}C]$thymidine were mixed with equal numbers of $[^3H]$thymidine-labeled control memory cells and subsequently transfused to immunized syngeneic recipients. The relative accumulation of specific versus nonspecific cells was measured in regional and contralateral nodes by scintillation spectroscopy of extracted DNA. These kinds of studies clearly demonstrated increased traffic of specifically reactive cells in antigen-draining nodes that withstood reciprocal specificity controls (Sprent et al. 1971; Thursh and Emeson 1972; Parrott and Ferguson 1974). The numbers of labeled immunospecific cells available to lodge selectively were small enough to tax the resolving power of the assays, but studies of efferent lymph indicated that specific cell traffic remained elevated throughout the immune response. Specifically reactive lymphocytes enter the efferent lymph in maximal numbers between 72 and 100 hours after antigen exposure. Peak accumulations of antigen-reactive lymphocytes correlated chronologically with early formation of reactive centers in the cortex and proliferation of new segments of HEV (Anderson et al. 1975), suggesting that these vascular structures may have some function in selective recruitment, by lymphocyte chemotactic factors or modified antigens secreted by macrophages may also favor the emigration of specific cells at HEV.

The increased migration of lymphocytes into antigen-stimulated lymphatic tissues is not restricted to long-lived recirculating lymphocytes. Lymphocytes in the cortex of the thymus undergo enhanced proliferation within the first 3 days following peripheral antigen inoculation. The systemic stimulus that causes thymocyte proliferation is presently unknown, but these newly formed cells leave the thymus and accumulate in the marginal zone of the spleen, Peyer's patches, and some lymph nodes (Durkin et al. 1978). These cells, composed predominantly of $Ly^{(1+,2+,3+)}$ cells, may be precursors of antigen-specific suppressor cells that require T-cell help to complete their differentiation. Adult thymectomy, which removes the short-lived cortical thymocytes but not the long-lived recirculating cells, prevents the formation of Freund's adjuvant-induced T-suppressor cell populations, further supporting the suggestion that the cortical thymocytes that lodge in the spleen are precursors of suppressor cells. As early as 6 days after immunization, the numbers of antigen-specific suppressor cells begin to increase until the immune response is terminated.

These changes in lymphocyte traffic combined with local proliferation of immunoreactive cells are responsible for the biphasic 2- to 6-fold enlargement of regional lymph nodes draining sites of antigen inoculation or infection. In the absence of additional free antigen, the suppressive effects of antiidiotypic antibodies or suppressor cells or both eventually result in gradual reduction in traffic and proliferation. It may take a month or more for the regional lymph node to return to its normal size and physiologic activity. Following secondary exposure to antigen, the kinetics of lymphocyte traffic and immunocyte proliferation in the regional and distant lymph nodes are greatly accelerated, possibly because of the dissemination of foci of immunological memory to nearly all the lymphatic tissues of the body.

Conclusion

Immunity is a process by which an organism distinguishes self from nonself resulting in the recruitment of cellular and molecular effectors which destroy and remove the foreign invader. This oversimplified definition of immunity belies the incredible complexities of both the afferent and efferent aspects of the process. This chapter has attempted to place the immune response within the context of the anatomical structures which are responsible for the generation and "education" of immune cell precursors, and through which mature immune cells migrate and encounter antigen trapped by blood or lymph filtration or by transport of antigen across mucosal surfaces. The anatomically dispersed organs of the immune system are linked together by the cardiovascular and lymphatic systems. The cells of the immune system spend part of their life cycles circulating in the blood from whence they enter and leave lymphatic tissues through recognition and migration processes analogous to embryological development. Since development stops when the cells stop moving, the organs of the immune system can be regarded as tissues which never stop developing. Continuous motion of subsets of T- and B-lymphocytes and mononuclear cells through and within lymphatic tissues are vitally important for the remarkably selective inductive and regulatory cellular interactions which must take place during the initiation and course of an immune response. Regardless of the presence or absence of antigen, lymphocyte migration and recirculation proceed at a relentless pace regulated by mechanisms which continue to remain a mystery despite our knowledge of the anatomy and physiology of the lymphatic system.

The authors are grateful to Jonathan W. Warren for technical assistance, Jenny Merrit for artwork, Diane Hill and Valerie Ives for typing, and Phebe W. Summers for editorial assistance.

References

Anderson, A.O., Anderson, N.D. (1975) Studies on the structure and permeability of the microvasculature in normal rat lymph nodes. Am. J. Pathol. 80,387–418.

Anderson, A.O., Anderson, N.D. (1976) Lymphocyte emigration from high endothelial venules in rat lymph nodes. Immunology 31,731–748.

Anderson, A.O., Reynolds, J.A. (1979) Adjuvant effects of the lipid amine CP-20,961 on lymphoid cell traffic and antiviral immunity. J. Reticuloendothel. Soc. 26,667–680.

Anderson, A.O., Warren, J.T. (1980) Lymphocyte chemotaxis under agarose, cell interactions. Fed. Proc. 39,698.

Anderson, A.O. et al. (1979) Basic mechanisms of lymphocyte recirculation in Lewis rats. In: *Function and Structure of the Immune System*. Müller-Ruchholtz, W., Müller-Hermelink, H.K., (eds.) New York, Plenum Press, pp. 73–83.

Anderson, N.D. et al. (1975) Microvascular changes in lymph nodes draining skin allographs. Am. J. Pathol. 81,131–160.

Anderson, N.D. et al. (1976) Specialized structure and metabolic activities of high endothelial venules in rat lymphatic tissues. Immunology 31,455–473.

Arenson, E. et al. (1978) Subsets of monocytes in human blood. Blood 52(Suppl. 1),130 (abstr.).

Astrom, K.E. et al. (1968) The initial lesion in experimental allergic neuritis. A phase and electron microscopic study. J. Exp. Med. 128,469–495.

Becker, E.L., Henson, P.M. (1973) *In vitro* studies of immunologically induced secretion of mediators from cells and related phenomena. Advan. Immunol. 17,93–193.

Blumenstock, F.A. et al. (1978) Biochemical and immunological characterization of human opsonic α_2SB glycoprotein: its identity with cold-insoluble globulin. J. Biol. Chem. 253,4287–4291.

Bockman, D.E., Cooper, M.D. (1973) Pinocytosis by epithelium associated with lymphoid follicles in the bursa of Fabricius, appendix, and Peyer's patches. An electron microscopic study. Am. J. Anat. 136,455–477.

Bowers, H.M. et al (1979) Numbers of mast cells in the axillary lymph nodes of breast cancer patients. Cancer 43,568–573.

Burnet, F.M. (1959) *The Clonal Selection Theory of Acquired Immunity*. Nashville, Tenn., Vanderbilt University Press.

Butcher, E. et al. (1979) Lymphocyte-high endothelial venule interactions: examination of species specificity. In: *Function and Structure of the Immune System*. Müller-Ruchholtz, W., Müller-Hermelink, H.K., (eds.) New York, Plenum Press, pp. 65–72.

Cahill, R.N.P. et al. (1976) The effects of antigen on the migration of recirculating lymphocytes through single lymph nodes. J. Exp. Med. 143,870–888.

Cantor, H., Boyse, E.A. (1977) Regulation of the immune response by T-cell subclasses. Contemp. Top. Immunol. 7:47–67.

Carr, I.A. (1973) *The Macrophage: A Review of Ultrastructure and Function*. New York, Academic Press.

Cebra, J.J. et al. (1974) Cell types contributing to the biosynthesis of IgA. In: *The Immunoglobulin A System*. Mestecky, J., Lawton, A.R., (eds.) New York, Plenum Press, pp. 23–33.

Chapman, A.L., Bopp, W.J. (1970) Electron microscopy of vascular barrier in thymus, tonsil, and lymph node of Beagle pups. Am. J. Vet. Res. 31,1255–1268.

Chen, L.-T., Weiss, L. (1973) The role of the sinus wall in the passage of erythrocytes through the spleen. Blood 41,529–537.

Clark, S.L., Jr. (1973) The intrathymic environment. Contemp. Top. Immunobiol. 2:77–99.

Cold Spring Harbor Symposia on Quantitative Biology (1977) *The Origins of Lymphocyte Diversity*, Vol. XLI. Cold Spring Harbor, N.Y., Cold Spring Harbor Laboratory.

Colvin, R.B. et al. (1979) Macrophage fibronectin (cold-insoluble globulin, LETS protein). Fed. Proc. 38,1408 (abstr.).

de Bono, D. (1976) Endothelial-lymphocyte interactions *in vitro*. 1. Adherence of nonallergized lymphocytes. Cell Immunol. 26,78–88.

Dukor, P. et al. (1970) Tissue localization of lymphocytes bearing a membrane receptor for antigen-antibody-complement complexes. Proc. Natl. Acad. Sci. U.S.A. 67,991–997.

Durkin, H.G. et al. (1978) Antigen-induced increase in migration of large cortical thymocytes (regulatory cells?) to the marginal zone and red pulp of the spleen. J. Immunol. 121,1075–1081.

Ehrlich, P. (1879) Beitrage zür Kenntnis der granulierten Bindegewebszellen und der eosinophilen Leukocyten. Arch. Anat. Physiol. 3,166–169.

Everett, N.B., Tyler, R.W. (1967) Lymphopoiesis in the thymus and other tissues: functional implications. Int. Rev. Cytol. 22,205–237.

Ford, W.L. (1975) Lymphocyte migration and immune responses. Progr. Allergy 19,1–59.

Ford, W.L. et al. (1978) Possible clues to the mechanism underlying the selective migration of lymphocytes from the blood. In: *Cell-Cell Recognition, Symposia of the Society for Experimental Biology*, Vol. XXXII. Cambridge, Cambridge University Press, pp. 359–392.

Friedman, H., (ed.), (1975) Thymus factors in immunity. Ann. N.Y. Acad. Sci. 249,1–547.

Frost, H. et al. (1976) Antipolysaccharide antibodies of restricted heterogeneity secreted by a single lymph node. J. Exp. Med. 143,707–711.

Gery, I. et al. (1977) Selective accumulation of cells with 'B' properties in stimulated lymph nodes. Immunology 33,727–731.

Ginsburg, H., Sachs, L. (1963) Formation of pure suspensions of mast cells in tissue culture by differentiation of lymphoid cells from the mouse thymus. J. Natl. Cancer Inst. 31,1–39.

Gowans, J.L. (1959) The recirculation of lymphocytes from blood to lymph in the rat. J. Physiol. 146,54–69.

Gowans, J.L., Knight, E.J. (1964) The route of recirculation of lymphocytes in the rat. Proc. R. Soc. B. 159,257–282.

Graham, R.C., Jr., Shannon, S.L. (1972) Peroxidase arthritis. II. Lymphoid cell-endothelial interactions during a developing immunologic inflammatory response. Am. J. Pathol. 69,7–24.

Hall, J.G., Morris, B. (1965) The immediate effect of antigens on the cell output of a lymph node. Br. J. Exp. Pathol. 46,450–454.

Hanna, M.G., Jr., Szakal, A.K. (1968) Localization of ^{125}I-labeled antigen in germinal centers of mouse spleen: histologic and ultrastructural autoradiographic studies of the secondary immune reaction. J. Immunol. 101,949–962.

Hay, J.B., Hobbs, B.B. (1977) The flow of blood to lymph nodes and its relation to lymphocyte traffic and the immune response. J. Exp. Med. 145,31–44.

Higuchi, Y. et al. (1975) Production of chemotactic factor for lymphocytes by neutral SH-dependent protease of rabbit PMN leukocytes from immunoglobulins, especially IgM. Cell Immunol. 15,100–108.

Hoefsmit, E.C.M. et al. (1979) Cells containing Birbeck granules in the lymph and the lymph node. In: *Function and Structure of the Immune System.* Müller-Ruchholtz, W., Müller-Hermelink, H.K., (eds.) New York, Plenum Press, pp. 389–394.

Horiuchi, A., Waksman, B.H. (1968) Role of the thymus in tolerance. VIII. Relative effectiveness of nonaggregated and heat-aggregated bovine γ globulin, injected directly into lymphoid organs of normal rats, in suppressing immune responsiveness. J. Immunol. 101,1322–1332.

Hovi, T. et al. (1977) Cultured human monocytes synthesize and secrete α_2-macroglobulin. J. Exp. Med. 145,1580–1589.

Howard, J.C. et al. (1972) Identification of marrow-derived and thymus-derived small lymphocytes in the lymphoid tissue and thoracic duct lymph of normal rats. J. Exp. Med. 135,200–219.

Humphrey, J.H. (1976) The still unsolved germinal center mystery. In: *Immune Reactivities of Lymphocytes. Development, Expression and Control.* Feldman, M., and Globerson, A., (eds.) New York, Plenum Press, pp. 711–723.

Husband, A.J., Gowans, J.L. (1978) The origin and antigen-dependent distributions of IgA-containing cells in the intestine. J. Exp. Med. 148,1146–1160.

Ishizaka, T. et al. (1976) Development of rat mast cells *in vitro.* I. Differentiation of mast cells from thymus cells. J. Immunol. 166,747–754.

Kitamura, Y. et al. (1979) Distribution of mast-cell precursors in hematopoietic and lymphopoietic tissues of mice. J. Exp. Med. 150,482–490.

Knowles, D.M., Holck, S. (1978) Tissue localization of T-lymphocytes by the histochemical demonstration of acid α-naphthyl acetate esterase. Lab. Invest. 39,70–76.

Leak, L.V., Burke, J.F. (1968) Ultrastructural studies on the lymphatic anchoring filaments. J. Cell Biol. 36,129–149.

Leduc, E.H. et al. (1955) Studies on antibody production. II. The primary and secondary responses in the popliteal lymph node of the rabbit. J. Exp. Med. 102,61–72.

Lee, K.-C. et al. (1979) Antigen-specific murine T-cell proliferation: role of macrophage surface Ia and factors. Cell Immunol. 48,79–90.

Lewis, W.H. (1931) Locomotion of lymphocytes. Bull. Johns Hopkins Hosp. 49,29–36.

Lipsky, P.E., Rosenthal, A.S. (1973) Macrophage-lymphocyte-interaction. I. Characteristics of the antigen-dependent binding of guinea pig thymocytes and lymphocytes to syngeneic macrophages. J. Exp. Med. 138,900–924.

Marchesi, V.T., Gowans, J.L. (1964) The migration of lymphocytes through the endothelium of venules in lymph nodes: an electron microscope study. Proc. Roy. Soc. B. 159,283–290.

Marshall, A.H.E., White, R.G. (1961) The immunological reactivity of the thymus. Br. J. Exp. Pathol. 42,374–385.

Miller, J.F.A.P. (1975) T-cell regulation of immune responsiveness. Ann. N.Y. Acad. Sci. 249,9–26.

Miller, R.G., Phillips, R.A. (1975) Development of B lymphocytes. Fed. Proc. 34,145–150.

Mitchell, J. (1973) Lymphocyte circulation in the spleen. Marginal zone bridging channels and their possible role in cell traffic. Immunology, 24,93–107.

Mosher, D.F. (1975) Cross-linking of cold-insoluble globulin by fibrin-stabilizing factor. J. Biol. Chem. 250,6614–6621.

Mosier, D.E., Cohen, P.L. (1975) Ontogeny of mouse T-lymphocyte function. Fed. Proc. 34,137–144.

Norris, D.A. et al. (1979) Isolation of functional subsets of human peripheral blood monocytes. J. Immunol. 123,166–172.

Nossal, G.J.V. (1979) Differentiation of B cells and the clonal abortion theory. In: *Function and Structure of the Immune System.* Müller-Ruchholtz, W., Müller-Hermelink, H.K., (eds.) New York, Plenum Press, pp. 253–258.

Nossal, G.J.V., Ada, G.L. (1971) *Antigens, Lymphoid Cells and the Immune Response.* New York, Academic Press.

Nossal, G.J.V. et al. (1968). Antigens in immunity. XV. Ultrastructural features of antigen capture in primary and secondary lymphoid follicles. J. Exp. Med. 127,277–290.

Oláh, I. et al. (1975) *Ultrastructure of Lymphoid Organs: An Electron-Microscopic Atlas.* Philadelphia, J.B. Lippincott.

O'Neill, G.J., Parrott, D.M.V. (1977) Locomotion of human lymphoid cells. 1. Effect of culture and Con A on T and non-T lymphocytes. Cell Immunol. 33,257–267.

Osoba, D., Miller, J.F.A.P. (1964) The lymphoid tissues and immune responses of neonatally thymectomized mice bearing thymus tissues in Millipore diffusion chambers. J. Exp. Med. 119,177–194.

Ottesen, J. (1954) On the age of human white cells in peripheral blood. Acta Physiol. Scand. 32,75–93.

Owen, R.L. (1977) Sequential uptake of horseradish peroxidase by lymphoid follicle epithelium of Peyer's patches in the normal unobstructed mouse intestine: an ultrastructural study. Gastroenterology 72,440–451.

Parrott, D.M.V. (1976) The gut as a lymphoid organ. Clin. Gastroenterol. 5,211–228.

Parrott, D.M.V., Ferguson, A. (1974) Selective migration of lymphocytes within the mouse small intestine. Immunology 26,571–588.

Parrott, D.M.V. et al. (1966) Thymus-dependent areas in the lymphoid organs of neonatally thymectomized mice. J. Exp. Med. 123,191–203.

Ponzio, N.M. et al. (1977) Transfer of memory cells into antigen-pretreated hosts. I. Functional detection of migration sites for antigen-specific B cells. Cell Immunol. 34,79–92.

Pryjma, J., Humphrey, J.H. (1975) Prolonged C3 depletion by cobra venom factor in thymus-deprived mice and its implication for the role of C3 as an essential second signal for B-cell triggering. Immunology 28,569–576.

Rácz, P. et al. (1977) Functional architecture of bronchial associated lymphoid tissue and lymphoepithelium in pulmonary cell-mediated reactions in the rabbit. J. Reticulendothel. Soc. 22,59–83.

Raviola, E., Karnovsky, M.J. (1972) Evidence for a blood-thymus barrier using electron-opaque tracers. J. Exp. Med. 136,466–498.

Reynolds, J. et al. (1978) Delivery of IgA-antibody to the intestine in bile. Fed. Proc. 37,838 (abstr.).

Rouse, R.V. et al. (1979) Expression of MHC antigens by mouse thymic dendritic cells. J. Immunol. 122,2508–2515.

Rowley, D.A. et al. (1972) The specific selection of recirculating lymphocytes by antigen in normal and preimmunized rats. J. Exp. Med. 136,499–513.

Russell, R.J. et al. (1975) Chemotaxis of lymphoblasts. Nature (London) 256,646–648.

Schoefl, G.I. (1972) The migration of lymphocytes across the vascular endothelium in lymphoid tissues. J. Exp. Med. 136,568–588.

Schreiner, G.F., Unanue, E.R. (1976) Membrane and cytoplasmic changes in B lymphocytes induced by ligand-surface immunoglobulin interaction. Advan. Immunol. 24,37–165.

Smith, J.B. et al. (1970) The migration of cells through chronically inflamed tissues. J. Pathol. 100,21–29.

Spector, W.G., Willoughby, D.A. (1963) The inflammatory response. Bacteriol. Rev. 27,117–154.

Sprent, J. (1972) Circulating T and B lymphocytes of the mouse. I. Migratory properties. Cell Immunol. 7,10–39.

Sprent, J. (1977) Migration and lifespan of lymphocytes. In: *B and T Cells in Immune Recognition*. Loor, F., Roelants, G.E. (eds.) New York, John Wiley and Sons, pp. 59–82.

Sprent, J., Basten, A. (1973) Circulating T and B lymphocytes of the mouse. II. Lifespan. Cell Immunol. 7,40–59.

Sprent, J., Miller, J.F.A.P. (1976) Effect of recent antigen priming on adoptive immune responses. III. Antigen-induced selective recruitment of subsets of recirculating lymphocytes relative to H-2 determinants. J. Exp. Med. 143,585–600.

Sprent, J. et al. (1971) Antigen-induced selective recruitment of circulating lymphocytes. Cell Immunol. 2,171–181.

Stamper, H.B., Jr., Woodruff, J.J. (1976) Lymphocyte homing into lymph nodes: *in vitro* demonstration of the selective affinity of recirculating lymphocytes for high-endothelial venules. J. Exp. Med. 144,828–833.

Steinman, R.M. et al. (1975) Identification of a novel cell type in peripheral lymphoid organs of mice. IV. Identification and distribution in mouse spleen. J. Exp. Med. 141,804–820.

Stenman, S., Vaheri, A. (1978) Distribution of a major connective tissue protein, fibronectin, in normal human tissues. J. Exp. Med. 147,1054–1064.

Thursh, D.R., Emeson, E.E. (1972) The immunologically specific retention of recirculating long-lived lymphocytes in lymph nodes stimulated by xenogeneic erythrocytes. J. Exp. Med. 135,754–763.

Unanue, E.R. (1972) The regulatory role of macrophages in antigenic stimulation. Advan. Immunol. 15,95–165.

Unanue, E.R. et al. (1976) Regulation of immunity and inflammation by mediators from macrophages. Am. J. Pathol. 85,465–478.

Vaheri, A. et al. (1976) Fibroblast surface antigen (SF): molecular properties, distribution *in vitro* and *in vivo* and altered expression in transformed cells. J. Supramol. Struct. 4,63–70.

van Ewijk, W. et al. (1975) Scanning electron microscopy of homing and recirculating lymphocyte populations. Cell Immunol. 19,245–261.

van Furth, R., (ed), (1975) *Mononuclear Phagocytes in Immunity, Infection and Pathology*. Oxford, Blackwell Scientific Publications.

van Furth, R. et al. (1972) The mononuclear phagocyte system: a new classification of macrophages, monocytes, and their precursor cells. Bull. W.H.O. 46,845–852.

Virchow, R. (1858) *Cellular Pathology. As Based Upon Physiological and Pathological Histology*. (reprint) Birmingham, Ala., Classics of Medicine Library, 1978, pp. 172–176.

Ward, P.A. et al. (1971) Chemoattractants of leukocytes, with special reference to lymphocytes. Fed. Proc. 30,1721–1724.

Ward, P.A. et al. (1977) Chemotaxis of rat lymphocytes. J. Immunol. 119,416–421.

Weiss, L. (1972) *The Cells and Tissues of the Immune System. Structure, Functions, Interactions*. Englewood Cliffs, N.J., Prentice Hall, Inc.

Weissman, I.L. (1975) Development and distribution of immunoglobulin-bearing cells in mice. Transplant. Rev. 24,159–176.

Wenk, E.J. et al (1974) The ultrastructure of mouse lymph node venules and the passage of lymphocytes across their walls. J. Ultrastruct. Res. 47,214–241.

Wilkinson, P.C. et al. (1977) Antigen-induced locomotor responses in lymphocytes. J. Exp. Med. 145,1158–1168.

Woodruff, J., Gesner, B.M. (1968) Lymphocytes, circulation altered by trypsin. Science 161,176–178.

Yoffey, J., Courtice, F. (1956) *Lymphatics, Lymph and Lymphoid Tissue*. Cambridge, Mass., Harvard University Press.

The Neutrophil

John M. Davis, M.D. and John I. Gallin, M.D.

The neutrophil is the most abundant cell in the circulation that deals with host resistance to infection. It shares certain structural and developmental similarities with the monocyte and the other polymorphonuclear leukocytes, the eosinophil, and the basophil. Its importance was appreciated by Elie Metchnikoff in the 1880s, whose profound appreciation of zoology and careful observations in animals led him to formulate a foundation for our modern concepts of the role of the phagocytic cell in host resistance (Metchnikoff 1907). The 1960s was the period during which major technological advances opened investigation into mechanisms involved in leukocyte physiology. As a result, disorders of neutrophil function have been defined in numerous diseases where infection is a common complicating event.

In this chapter, the neutrophil is followed from its maturation in the bone marrow to its ultimate fate in the tissues, as schematically outlined in Figure 1. This topic has been the subject of recent comprehensive books (see Cline 1975; Klebanoff and Clark 1978; Weissmann 1980). Accordingly, this chapter is limited in scope to selected advances in the field with the anticipation that the references provided serve as a guide to further reading.

Neutrophil Development

The only site of neutrophil production in normal human beings is the bone marrow. Recent evidence suggests the neutrophil evolves from a single pluripotent stem cell, which also gives rise to the other cells of the granulocyte series, the eosinophil, the basophil, and the monocyte (Cline and Golde 1979).

Figure 2 is a schematic diagram of neutrophil development in the bone marrow; it has been reviewed by Cline (1975). The myeloblast is the first recognizable precursor cell of the polymorphonuclear leukocyte (PMNL) series. This cell is approximately twice as large as the mature neutrophil and contains numerous primitive cellular organelles including ribosomes and a Golgi apparatus. Following cell division, a cell approximately two and one-half times larger than the mature neutrophil appears,

From the Bacterial Diseases Section, Laboratory of Clinical Investigation, National Institute of Allergy and Infectious Diseases, National Institutes of Health, Bethesda, Maryland.

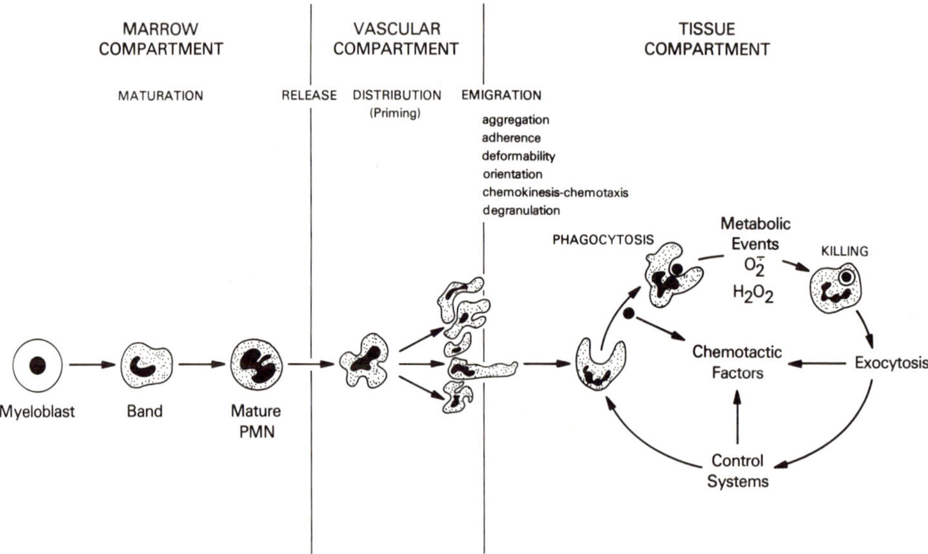

Figure 1. Accumulation of neutrophils at inflammatory sites (from Gallin, J.I., Revs. Inf. Dis., 1981).

called the promyelocyte. Its cytoplasm contains rough endoplasmic reticulum, mitochondria, and a prominent Golgi apparatus. Developing from this Golgi apparatus are the azurophilic or primary granules. These granules are called azurophilic, because of their blue appearance when stained with Wright's stain, and primary, because they are the first granules to appear during neutrophil maturation. With subsequent cell division, production of primary granules ceases and they become distributed in the cytoplasm of the daughter cells. Thus, primary granules are more abundant in immature cells. The promyelocyte undergoes cell division to produce the myelocyte, a cell similar in size to the myeloblast and responsible for the development of the specific or secondary granules (Bainton and Farquhar 1966; Bainton et al. 1971). These are called specific granules because they are only found in neutrophils and secondary granules, since they appear after the development of primary granules. Some of the distinguishing features of primary and secondary granules are shown in Table 1. During the myelocyte stage the specific granules increase in number, and in mature cells are the most prevalent granule type. The identification of different markers for each granule type (e.g., lactoferrin and B-12 binding protein for specific granules and β-glucuronidase and myeloperoxidase for the azurophil granules) has provided valuable mechanisms for monitoring the different granule classes.

Table 1. Neutrophil Granule Contents

Azurophil (1°)	Specific (2°)
Lysozyme	Lysozyme
Acid hydrolases	Lactoferrin
Neutral proteases (elastase)	B-12 binding protein
Cationic proteins	Neutral proteases (collagenase)
Myeloperoxidase	
Acid mucopolysaccharide	
β-Glucuronidase	

NEUTROPHIL MATURATION

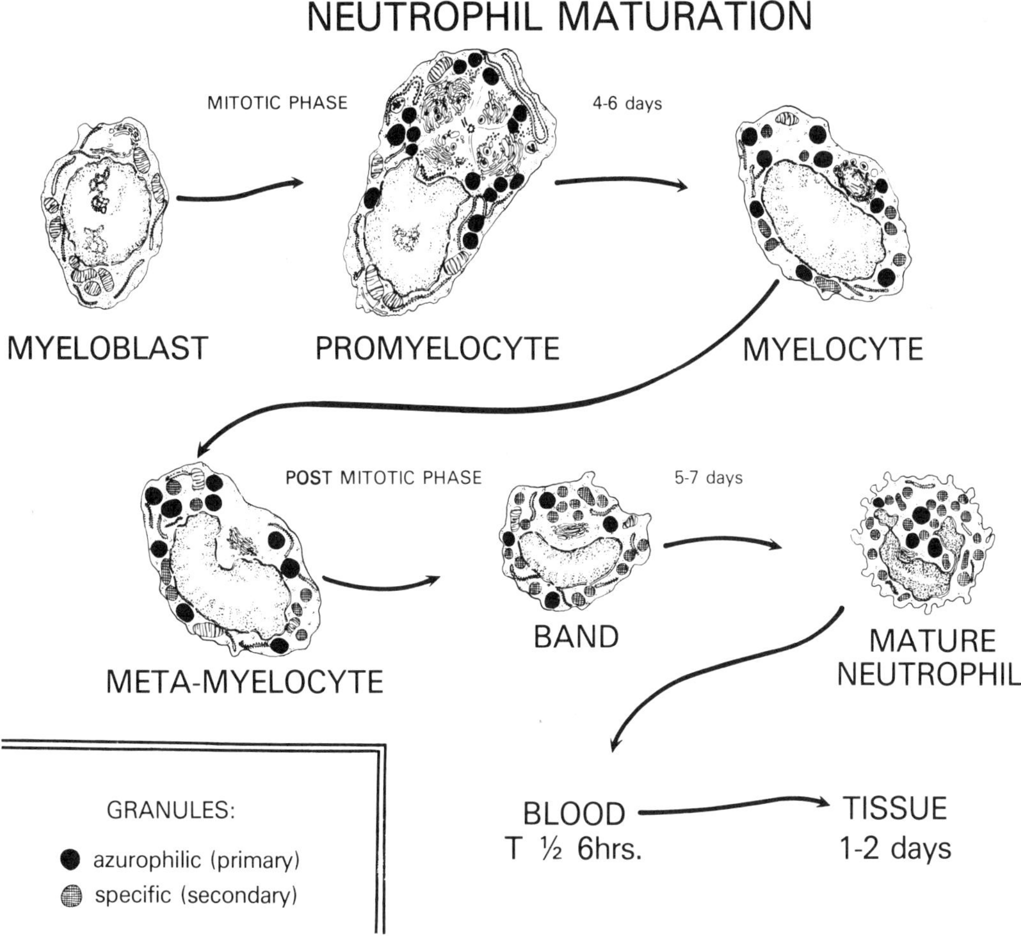

Figure 2. The maturation of neutrophils in the bone marrow (adapted from Bainton and Farquhar [1966]).

The period of cell development leading to the myelocyte stage is 4 to 6 days. The maturation of the myelocyte to the mature neutrophil, with acquisition of the characteristic morphologic and functional features, takes 5 to 7 days. During this period, there is no cell division but several cell stages have been defined. The first stage in this phase is the metamyelocyte, which may occasionally be found in the peripheral blood. The nucleus of this cell has a characteristic kidney-bean shape, and the cytoplasm is slightly more abundant than that of the preceding cells. The azurophilic granules are now less prevalent and the specific granules become predominant. The metamyelocyte, which matures into the band neutrophil, is smaller than all its precursors and approaches the size of the mature PMNL of approximately 10 to 12 μ. The metamyelocyte is incapable of cell division and has some phagocytic ability, although not as much as the band or mature neutrophil. The band neutrophil has a sausage-shaped nucleus. As the cell matures, the nucleus becomes lobulated. Metchnikoff speculated that the multilobed nucleus may enable the cell to squeeze between endothelial cells as it migrates to tissues (Hirsch 1975).

The mature neutrophil is surrounded by a trilaminar-structured plasma membrane that is able to extend pseudopods toward a chemotactic gradient (Berlin et al. 1975). Adjacent to the plasma membrane are the microfilaments (actin and myosin) that generate the contractile forces required for locomotion (Stossel 1974). In the mature cell the specific granules outnumber the azurophilic granules approximately 3 to 1. Histochemical, ultrastructural, and biochemical studies of isolated granules have indicated that there are additional subclasses of granules (Baggiolini et al. 1974; Baggiolini 1980). The segmented neutrophil is incapable of cell division, and the nucleus consists of clumped chromatin with no nucleoli. The number of lobulations characteristic of the neutrophil are not related to cell age. The cytoplasm also contains distinct glycogen particles, which first appear in the myelocyte and increase in number as the cell matures; these are thought to contribute to the energy source for the cell. The microtubule system, which organizes about the centriole, is thought to be important in stabilizing the cell during locomotion and organizing intracellular granules during degranulation (Malech et al. 1977; Hoffstein 1980).

Bone Marrow and Peripheral Blood Compartments

During the final period of neutrophil maturation changes in the cell membrane occur with decreases in cell negative surface charge and associated acquisition of increased membrane deformability (Lichtman 1970; Lichtman and Weed 1972). After cell maturation the bone marrow serves as a storage area for large numbers of neutrophils. This pool is estimated to contain several times the number of circulating neutrophils as demonstrated in Figure 3 (Boggs 1975). It is assumed that release from the marrow is preferential for segmented neutrophils over bands, since there are proportionally many more segmented PMNLs in the peripheral blood than in the marrow.

The cell, which has matured from a primitive stem cell to a complex phagocytic cell, is capable of sustained movement, engulfment of foreign material, and microbial killing. Recent work assessing the availability of immunoglobulin Fc receptors on circulating neutrophils suggests that there are two subclasses of neutrophils distinguished by their ability to rosette immunoglobulin-coated red cells (Klempner and Gallin 1978). The rosetting, or Fc$^+$, cells comprise about 80% of normal circulating neutrophils and have superior locomotion and bactericidal capacity compared with nonrosetting cells. Whether such distinctions in Fc receptor availability represent subclasses of neutrophils or different states of maturation of a single cell line is currently not known.

The number of circulating neutrophils is dependent on their adequate production and maturation, storage, release, and orderly destruction. The concept of a marginated pool of leukocytes that adheres to the surface of blood vessels was first suggested in the 1890s but was not established until the availability of appropriate technology in the 1960s. In 1960, using radiolabeled cells from normal volunteers, Athens et al. (1961) were able to calculate the total number of cells in the circulating and marginated pool. They estimated that the circulating granulocyte pool is approximately 0.3×10^9/kg of blood with a marginated pool of 0.4×10^9/kg of blood. The marrow supplies to this total blood granulocyte pool 1.5×10^9 cells/kg of blood per day and has in reserve 8.8×10^9 cells/kg of blood stored in the marrow (Figure 3). Additional precursor neutrophils in the bone marrow with inefficient host defense capability are estimated at 2.8×10^9 cells/kg of blood.

NEUTROPHIL KINETICS

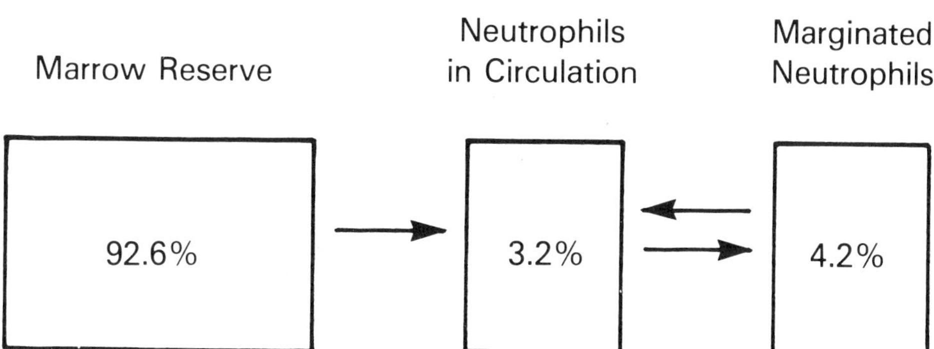

Figure 3. The bone marrow serves as a site of maturation and of storage of the neutrophil. Upon appropriate stimulation, the neutrophil leaves the bone marrow and enters the circulation where it equilibrates with the marginated pool.

The regulatory mechanism for the stimulation of maturation and release of cells from the bone marrow has been credited to several factors. Diffusible granulopoietic substance (Rothstein et al. 1971) has been shown to stimulate ectopic hematopoiesis in neutropenic laboratory animals. Colony stimulating factor (Cline and Golde 1979) is necessary to maintain bone marrow cultures *in vitro* and is also found in the sera of neutropenic human subjects. This substance has been found recently to be a class of glycoproteins that may be produced by lymphocytes and macrophages. Endotoxin and etiocholanolone promote the release of granulocytes from the marrow probably via a mediating substance. This latter factor may be neutrophil releasing factor (Boggs et al. 1966), which has been found in the sera of laboratory animals rendered neutropenic by myelotoxic agents and in the sera of human subjects treated with endotoxin. Evidence suggests leukocyte endogenous pyrogen (Kampschmidt et al. 1972), as well as fragments from the third component of complement (C3e), can also stimulate leukocyte release from the bone marrow.

The half-life of the circulating neutrophil is relatively short, only 6 hours. The neutrophils in the intravascular compartment are thought currently to be primed by substances, probably chemotactic factors, that stimulate neutrophil aggregation, adhere to endothelial surfaces, then finally migrate out of the circulation. Migration of neutrophils out of the circulation is referred to as diapedesis. In the healthy adult there is evidence that most neutrophils leave the circulation by migration through the mucous membrane of the gastrointestinal tract (Fliedner et al. 1964).

The integrity of the vascular bed is critical for delivery of leukocytes to tissues. In disease states in which the integrity of the vascular bed is compromised, such as diabetes and peripheral vascular disease, distribution of the neutrophil to sites of tissue injury is impaired, contributing to the decreased resistance to infection and increased local tissue destruction seen in these conditions.

Margination

Recent studies of the peripheral leukocyte counts in patients undergoing dialysis, who have a transient neutropenia attributable to margination, provide convenient models for the study of margination. In these patients margination has been thought to be related to activation of the alternate pathway of complement with generation of the chemotactically active cleavage fragment of C5 (Craddock et al. 1977). It has been suggested that the chemotactic factors initiate leukocyte aggregation and subsequent adherence in capillary beds (for a review of granulocyte adherence, see MacGregor 1980). Support for this concept comes from *in vitro* studies of neutrophil and endothelial cell interaction, which showed increased PMNL adherence to cultured endothelial cells after pretreatment of the leukocytes with chemotactic agents (Lackie and DeBono 1977). Recent data suggest the mechanism for cell aggregation and margination may be related to neutralization of the negative charge on the surface of the neutrophil by limited secretion of intracellular cationic granule contents, increasing the adhesive properties of the cell by electrostatic mechanisms (Gallin 1980b).

Chemotactic Factors

There is a growing list of chemotactic stimuli that include products of invading microorganisms as well as products produced in response to noxious stimuli (Figure 4). Among the best-studied mediators are the products of complement activation, which have been reviewed elsewhere (see Ward et al. 1979 and Chapter 15). The complement system is a group of proteins found in the serum that are activated in a sequential manner. The end products of complement activation have many functions. For example, a cleavage product of the third complement component, C3b, opsonizes particles, rendering them more likely to adhere to neutrophil C3b receptors. C3a, which has anaphylatoxic activity, was also thought to have chemotactic properties, but recent evidence suggests this product of the complement cascade has relatively feeble chemotactic activity (Fernandez et al. 1978). Another product of complement activation is the cleavage product of the fifth component (C5a), which has low molecular weight (approximately 15,000). C5a has the most potent anaphylatoxic and chemoattractant activities and appears to have a short life *in vivo*. One degradation product of C5a is $C5_{a\ des\ arg}$, which is inactive as an anaphylatoxin, retaining potent chemotactic activity (Fernandez et al. 1978) but only in the presence of whole serum. Since C5a is rapidly broken down to $C5_{a\ des\ arg}$, some investigators believe it is the most important chemoattractant *in vivo*.

Activation of the kinin-generating pathway by Hageman factor activation has been shown to generate plasminogen activator and kallikrein, both of which are reported to have chemotactic activity for neutrophils (Kaplan et al. 1972; 1973). Kallikrein promotes the formation of bradykinin, a potent vasodilator that enhances capillary permeability facilitating neutrophil emigration (see Chapter 16). Fibrinopeptide B and fibrin degradation products are other substances produced by the coagulation system that have reported chemotactic activity for the neutrophil (Kay et al. 1974).

Many chemotactic factors have been found to be products of cell metabolism. Recently, cultured fibroblasts have been shown to secrete chemotactic substances with preferential chemotactic activity for neutrophils (Sobel and Gallin 1979). These fibroblast products are distinct from collagen, which also has chemotactic activity for

NEUTROPHIL CHEMOTACTIC FACTORS

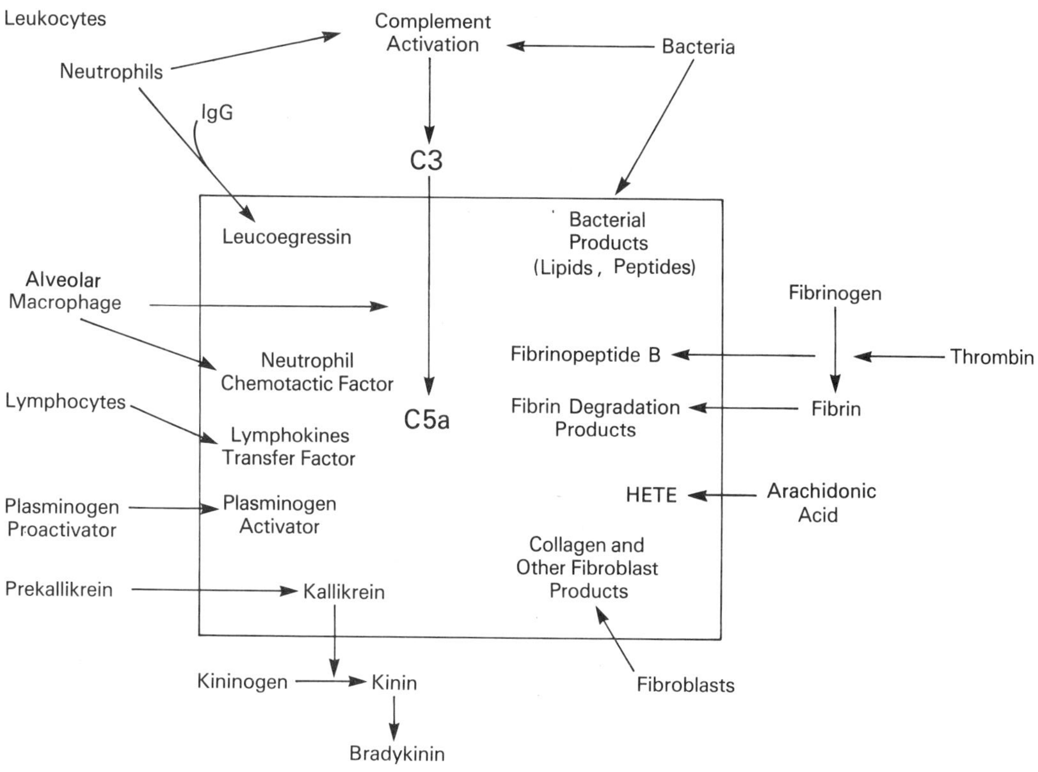

Figure 4. Naturally occurring chemotactic factors.

the neutrophil (Chang and Houck 1970). Chemotactic lymphokines have been described (Altman 1978), and transfer factor, a dialyzable product of lymphocytes, contains chemotactic activity that has been demonstrated both *in vitro* and *in vivo* (Gallin and Kirkpatrick 1974). Kazmierowski et al. (1977) have demonstrated that the pulmonary macrophage of monkeys produces a C5a-like molecule as well as a low molecular weight peptide capable of attracting neutrophils *in vitro*. A similar alveolar macrophage product has been shown in human beings (Gadek et al. 1978).

Neutrophil degranulation appears to have an important modulatory role in inflammation (Weissmann et al. 1969; Gallin et al. 1978b; Weissmann et al. 1980). For example leukoegressin, a chemoattractant derived from the sulfhydryl-dependent proteolysis of IgG (Hayashi 1975), can be generated by proteolytic enzymes located in neutrophil granules. Neutrophil specific granules, which are particularly accessible to extracellular release both *in vitro* and *in vivo* (Wright and Gallin 1979), contain materials that activate the alternate pathway of complement to generate C5a (Wright and Gallin 1977). This activation of complement by neutrophil products, like the generation of leukoegressin, may provide an important amplification system of the inflammatory response. Inactivation of C5a and other chemoattractants by azurophil granule products, which are released late in the phagocytic process, may limit the extent of inflammation and thereby provide another modulatory system.

In 1888, Leber showed that bacterial products possess chemotactic activity. The distinction between chemotactic factors derived in the serum through the activation of complement by bacterial products and the chemotactic activity of bacteria products was first made by Keller and Sorkin (1967). Subsequently, Schiffmann and others (1975) synthesized a group of biochemical compounds, such as f-methionine-leucine-phenylalanine, that are related structurally to chemotactically active products in bacterial culture filtrates. These synthetic peptides were shown to have a range of activity with strict structure-activity relationships; a review of their activity appeared recently (Schiffmann and Gallin 1979).

Prostaglandins and related compounds are also thought to be important humoral components of the inflammatory response (Ferreira 1979). Arachidonic acid, the parent compound of the prostaglandins and the thromboxanes, is metabolized by platelet lipoxygenase to a substance known as HETE (12L-OH-5,8,10,14-eicosatetraenoic acid). HETE has been shown to have chemotactic properties and share certain chemical similarities with the chemotactic peptides and bacterial chemotactic factors.

The prostaglandins also effect the inflammatory response by modulating neutrophil responsiveness by altering levels of cyclic nucleotides. Cyclic GMP has been associated with enhancement and cyclic AMP with inhibition of chemotaxis. However, evidence for the role of cyclic nucleotides is confusing and more data are required before any conclusions can be reached (Schiffmann and Gallin 1979).

It should be obvious from these comments that many chemotactically active substances have been demonstrated (Figure 4). It is probable that this multitude of chemoattractants provides the host with numerous "backup" systems to establish an inflammatory response. Even if a host lacks specific mediators there are usually enough alternative ways to recruit neutrophils to inflammatory foci so that survival is possible.

Diapedesis

Following margination, neutrophils migrate between endothelial cells to tissues. This locomotory response of neutrophils has been studied extensively and has been recently summarized in two monographs (Wilkinson 1974; Gallin and Quie 1978). Unlike bacteria, leukocytes crawl; they do not swim. The locomotory response has been characterized *in vitro*. Spontaneous nondirected (random) migration is the pattern of migration of cells in buffer. Chemokinesis or stimulated nondirected migration is the enhanced migration response of cells suspended in buffers containing a chemical stimulus (chemoattractants). Chemotaxis is the directed migration of cells exposed to a gradient of chemoattractant. The cell biology and biochemistry of chemotaxis is a complex field being actively studied and is the subject of recent reviews (Zigmond 1978; Gallin 1980a; Schiffmann and Gallin 1979).

When considering the mechanism of chemotaxis, it is convenient to consider an orderly sequence of events that presumably occur (Gallin et al. 1978c, 1979b). There appears to be a mechanism for sensing the signal and for transduction of the signal across the plasma membrane. In addition, there are mechanisms for activation of the machinery for locomotion and maintenance of cell orientation and stability. A complex system to modulate cell migration also exists. Studies by Zigmond (1974) of the sensing mechanism suggest that the neutrophil can detect a gradient as small as 1% over its membrane surface enabling it to move toward a stimulus in a relatively direct path.

Chemoattractant Receptors

The availability of synthetic N-formylmethionyl peptide chemoattractants through the work of Schiffmann and his collaborators (Schiffmann et al. 1978) has made it possible to assess neutrophil chemoattractant receptors. Aswanikumar et al. (1977) and Williams et al. (1977) demonstrated the presence of chemoattractant receptors on human neutrophils. Evidence for the presence of chemotactic factor receptors included the following: (1) saturable binding of a labeled chemoattractant was time dependent; (2) binding was enhanced about twofold in the presence of proteolytic inhibitors; (3) a high affinity dissociation constant of 1.5 nM was found for the binding site with an estimated 10^5 receptors per cell; (4) concentrations of chemotactic factors optimal for chemotaxis were close to their dissociation constant (Kd) for receptor binding (concentration when about 50% of the receptors are bound); and (5) peptides competed for the binding site in the same rank order as their chemotactic effectiveness with a very close correlation between the concentration of each peptide causing half-maximal chemotaxis and half-maximal inhibition of binding. Recently, receptors for C5a (Chenoweth and Hugli 1978) and a cell-derived chemotactic (Spilberg and Mehta 1979) have also been shown in neutrophils.

Transduction of Chemoattractant Signal

Little is known about the mechanism of transduction of the signal. Studies on macrophages have suggested that alteration of membrane potential may play some role in initiation of the chemotactic response. E.K. Gallin et al. (1975), using standard intracellular recording techniques, showed that peritoneal resident or induced macrophages from several species, as well as human macrophages obtained from cultured peripheral blood monocytes, exhibit spontaneous fluctuations in membrane potential characterized as hyperpolarizations lasting 4 to 8 seconds in duration and 10 to 50 mV in amplitude. Similar observations were reported subsequently in primary cultures of mouse liver macrophages (Okado et al. 1977a,b) and dextran and oil-induced peritoneal exudate cells (Dos Reis and Oliveira-Castro 1977). The macrophage hyperpolarizations could be induced by mechanical stimulation, application of hyperpolarizing currents of 2 to 8 nA, and the calcium ionophore A23187. Based on studies of their reversal potential, varying extracellular K^+, utilizing tetraethyl ammonium ions and ethylene(bis)glycoltetraacetic acid, the hyperpolarizations were attributable to increased membrane permeability to K^+ that were hypothesized to be regulated by fluctuations of intracellular ionized calcium (Gallin E.K. et al. 1975; 1980; Dos Reis and Oliveira-Castro 1977).

The possible relationship of the macrophage spontaneous hyperpolarizations to cell motility prompted studies of the effect of chemotactic factors on macrophage membrane potential (Gallin and Gallin 1977; Gallin, E.K. et al. 1980). A variety of chemotactic factors were employed including endotoxin-activated serum, partially purified C5a, formyl–methionyl peptides, and kallikrein. The chemotactic stimuli were applied to cells by allowing the test stimulus to diffuse from a blunted microelectrode placed adjacent to the cell. The chemoattractant induced a brief depolarization followed by more prolonged hyperpolarization. Repetitive stimulation caused desensitization of the response. In some cells an oscillation of membrane potential accompanied the hyperpolarization (Gallin, E.K. et al. 1980). These membrane potential changes often preceded detectable morphologic changes such as pseudopod formation. Furthermore, in studies with a limited number of the synthetic formyl–methionyl peptides there was a good correlation between the chemotactic activity of the peptides and their ability to elicit potential changes (Gallin, E.K. et al. 1980).

It has not been possible to obtain data on membrane potential changes in small and highly motile cells, such as neutrophils or monocytes, using microelectrodes. Therefore, indirect techniques have been developed including radiolabeled triphenyl methyl phosphonium ions (TPMP$^+$) (Schuldiner and Kaback 1976) and fluorescent cyanine dyes (Waggoner 1976). These membrane potential sensitive probes have been used to study potential changes in numerous small cells. Korchak and Weissmann (1978), using TPMP$^+$, have reported evidence of membrane potential changes in human granulocytes during phagocytosis. Studies using the fluorescent cyanine dye di-O-C5(3) have provided evidence for chemoattractant-induced membrane hyperpolarizations in neutrophils (Seligmann et al. 1980) and monocytes (Gallin, E.K. et al. 1980). High extracellular potassium ablated the response, implying that the response is a hyperpolarization related to potassium efflux. These and other experiments, using the calcium ionophore A23187 and Mg-EGTA, have provided evidence that the chemoattractant-induced alterations in membrane potential in neutrophils and monocytes, assessed with cyanine dyes, have a similar ionic basis to those seen in macrophages using direct intracellular recording techniques.

It is not known whether these potential changes are intimately associated with cell movement. For example, in the macrophage studies using intracellular microelectrodes, morphologic changes in some cells were noted in the absence of changes in membrane potential whereas in other cells, changes in membrane potential were not accompanied with pseudopod formation (Gallin and Gallin 1977). Thus, proof that membrane potential changes transduce the chemotactic signal awaits additional study. As will be discussed below, the membrane potential changes may relate to other functions, such as secretion and superoxide generation which chemoattractants initiate.

Other data indicate a role for cations in neutrophil chemotaxis. For example, Ward and Becker (1970) showed that chemotaxis is inhibited by the Na–K ATPase blocker ouabain. In addition Ca^{++} and Mg^{++} are required for neutrophil chemotaxis through micropore filters (Becker and Showell 1972; Gallin and Rosenthal 1974), and submembranous deposits of calcium appear to occur at the leading edge of neutrophils oriented in a gradient of chemoattractant (Gallin et al. 1978c). The possible relationship of cation flux across the plasma membrane or the redistribution of intracellular calcium to membrane deformability, to the actin and myosin system that generates the contractile forces of locomotion, or to the microtubule network that may be important in cell orientation and maintenance of cell stability during chemotaxis is intriguing and has been discussed elsewhere (see Lichtman 1970; Stossel 1978; Gallin and Rosenthal 1974; Malech et al. 1977; Hoffstein 1980).

Tissue Function

Once the neutrophil arrives at the site of tissue injury a complex series of events occur that result in destruction of invading bacteria. Particles opsonized with IgG antibodies and complement components adhere to neutrophils primarily by interaction with the Fc and C3b receptors (Lay and Nussenzweig 1968). Interaction with the neutrophil

Fc receptors is thought to trigger phagocytosis (Henson and Oudes 1975). In addition tuftsin or leukokinin, a tetrapeptide fragment of IgG distinct from leukoegressin, has been found to enhance phagocytosis two- to threefold (Najjar 1974, 1975).

The various experimental requirements for phagocytosis *in vitro* are quite similar to those conditions required for chemotaxis. Calcium and magnesium are required (Stossel 1973), as are physiologic tonicity and a physiologic temperature range (Peterson et al. 1977). Although the content of an abscess cavity may have a low pH, evidence suggests that lowered pH reduces phagocytic ability of the PMNL so that most *in vitro* studies are conducted at a pH of 7.4 (Klebanoff and Clark 1978).

The Respiratory Burst

Ingestion of foreign particles is associated with a marked increase in the uptake of oxygen by the phagocytosing neutrophil (Sbarra and Karnovsky 1959). This change in metabolism is referred to as the metabolic or respiratory burst because normally the cell is rather dormant metabolically. The enzymic mechanisms responsible for the respiratory burst have centered largely on (1) the nature of the primary oxidase responsible for the increase in oxygen consumption, superoxide generation, and H_2O_2 formation; (2) the link between this oxidase activity and stimulation of the hexose-monophosphate shunt; and (3) the nature of the event that triggers the processes. It is generally accepted that the primary oxidase responsible for the respiratory burst is located on the outer surface of the neutrophil plasma membrane (Briggs et al. 1975; Goldstein et al. 1977; Dewald et al. 1979). The nature of the primary oxidase of phagocytosis remains controversial, but the primary candidates at the moment include NADH oxidase (Karnovsky et al. 1966) and NAD(P)H oxidase (DeChatelet et al. 1975); a role for other oxidases is still being considered.

Activation of the hexose-monophosphate shunt, the other major component of the respiratory burst aside from the primary oxidase of phagocytosis, has been thought to be linked to NADPH oxidase, an NADPH-linked lactic dehydrogenase (Evans and Karnovsky 1962), the glutathione cycle (Karnovsky et al. 1971), the pyridine nucleotide transhydrogenase (Baehner et al. 1972), and an NAD kinase. The details of the possible complex interrelationships of these pathways is summarized in Klebanoff and Clark (1978).

Initiation of the respiratory burst occurs upon membrane stimulation, prior to ingestion and degranulation (Rossi et al. 1972). Studies by Korchak and Weissmann (1978) have provided evidence that alteration of neutrophil membrane potential may be the trigger that transduces the signal for activation of the respiratory burst. Changes in membrane potential antecede the respiratory burst.

The products of the respiratory burst include formation of the superoxide anions (O_2^-), hydrogen peroxide (H_2O_2), singlet oxygen ($^1O_2$), and hydroxyl radicals (OH^-), as reviewed in Klebanoff and Clark (1978). These toxic oxygen derivatives not only kill microorganisms but also attack tumors and parasites (Klebanoff and Clark 1978). They can also damage normal host cells such as synovial tissues in joints (Babior 1978). There are enzymes, such as superoxide dismutase, that converts superoxide anion to hydrogen peroxide, which is then converted to water by catalase. In addition, ceruloplasmin, a copper-containing serum protein shown by Goldstein et al. (1979) to be an effective scavenger of superoxide anion, may be elaborated during acute inflammation and may help to protect the host from toxic effects of excessive oxygen derivatives.

The reactive oxygen species emit light as the electrons in excited states return to more stable orbits, a phenomenon called chemiluminescence. Because of the wide range of events that occur in the neutrophil with the initiation of phagocytosis, a host of assays reflect various parameters of the phagocytic process, the details of which can be obtained in recent reviews (Babior 1978; Klebanoff and Clark 1978; Stossel 1974). It is important to emphasize that assays that measure phagocytosis are independent of bactericidal properties of the neutrophil. The ability of cells to ingest bacteria is an important precursor to killing, but killing of bacteria does not always follow, as will be discussed later.

Degranulation

In association with the oxygen burst, degranulation occurs. The studies of Cohn and Hirsch (1960) demonstrated that lysosomal enzymes sequestered in lysosomes are discharged from granules into phagocytic vacuoles or secondary lysosomes following ingestion of bacteria. Subsequently, it was shown that lysosomes of neutrophils contain neutral proteases capable of attacking extracellular substrates (Davis et al. 1970; Weissmann et al. 1971) and other investigators showed the release (regurgitation) of lysosomal enzymes when neutrophils are in contact with foreign materials (Weissmann et al. 1971), appropriately coated surfaces, or during phagocytosis (Henson 1971). The specific granules appear to fuse with the phagocytic vacuole before the azurophil granules (Bainton 1973). As will be discussed below, fusion of these granules results in release of a variety of enzymes important in bacterial killing and subsequent digestion.

Bacterial Killing

The systems within the neutrophils that are responsible for antimicrobial action may be divided into oxygen dependent and oxygen independent, as summarized in Table 2. The importance of oxygen has been underscored by work by Mandell (1974) who showed that certain bacteria (including *Staphylococcus aureus, Escherichia coli, Serratia marcescens, Klebsiella pneumoniae, Proteus vulgaris,* and *Salmonella typhimurium*), which could be killed quite effectively aerobically were not as effectively killed in an anaerobic environment. On the other hand, Mandell's study emphasized that alternate anaerobic systems of killing also occurred. For example, in the absence of oxygen, *Staph. epidermidis, Staph. enterococci, Streptococcus veridans, Pseudomonas aeruginosa,* and *Peptostreptococcus magnus* were readily killed.

A complex array of killing mechanisms have been described. The myeloperoxidase system requires halide, hydrogen peroxide, and myeloperoxidase (Rosen and Klebanoff

Table 2. Antimicrobial Systems in the Neutrophil

Oxygen dependent	Oxygen independent
Myeloperoxidase dependent	Sequestration of phagocytic vacuole
MPO plus halide plus H_2O	Acid environment
Myeloperoxidase independent	Lysozyme
Superoxide anion (O_2^-)	Lactoferrin
Hydroxyl radical (OH^-)	Cationic proteins
Singlet oxygen ($^1O_2$)	Lysozomal hydrolases
Hydrogen peroxide (H_2O_2)	Neutral proteases

1979). This system has been studied extensively *in vitro* and has been found to be less effective when any of the three components are missing. In addition, enzymes that may be produced by ingested bacteria, such as catalase, can limit the capacity of the system by decreasing available hydrogen peroxide.

The oxygen-dependent but myeloperoxidase-independent systems are activated during the metabolic burst associated with the phagocytic process (Table 2). The four by-products of the metabolic burst (hydrogen peroxide, superoxide anion, hydroxyl radicals, and singlet oxygen) appear to have antimicrobial activity augmented by the presence of myeloperoxidase and halide. Acting alone, these antimicrobials work more slowly and without as broad a spectrum of effectiveness.

The oxygen independent systems are more limited in their range of activity. Lysozyme, one of the most abundant enzymes in the neutrophil, is highly specific in its function and is very effective for sensitive microbes. However, the majority of gram-negative bacteria are protected from this enzyme's action by their mucopolysac-charide cell wall. Other antimicrobial mechanisms that may be less specific include sequestration in the phagocytic vacuole, which may in certain circumstances result in bacterial death by removing the bacterium from its source of nutrition and exposure of the sensitive bacterium to the acid environment created inside the vacuole (Klebanoff and Clark 1978). Lactoferrin is thought to inhibit growth of certain bacteria by chelating iron (Schade and Caroline 1946), an essential component of growth of several microorganisms. Cationic proteins and hydrolases isolated from granules also have antimicrobial activity *in vitro* (Hirsch 1965).

Microorganisms are killed by intact neutrophils without major structural alteration, and macromolecular synthesis by the organisms may persist or increase during the degradation process (Elsbach et al. 1974). It is generally assumed that bacterial digestion results from action of acid hydrolases released from neutrophil granules (Ginsberg and Sela 1976), although evidence for the participation of the neutral protease, elastase, in the digestion of bacterial proteins has been presented (Janoff and Blondin 1974; Blondin and Janoff 1976). The degradation of microbial lipids by granulocyte enzymes would appear to be more closely related to bacterial death than effects of neutrophil products on bacterial protein synthesis (Elsbach 1972).

Clinical Considerations

The host may be rendered susceptible to infection due to either decreased production or impaired performance of the neutrophil. These problems may arise from a variety of reasons including inherited traits, acquired, or iatrogenic causes. These "experiments of nature" serve as important tools for understanding the biology of normal neutrophil function in inflammation and immunity.

Disorders of Chemotaxis

Disorders of chemotaxis are due to functional defects associated with abnormal cell adherence, orientation, and deformability, as well as random and directed locomotion. A list of the various disorders is shown in Table 3 (see reviews by Clark 1978; Quie and Cates 1978; Gallin et al. 1980). One of the most thoroughly studied patients with abnormal cell adherence and chemotaxis was an infant with recurrent pyogenic infections described by Boxer et al. (1974). This patient's neutrophils spread poorly on glass surfaces and did not accumulate into superficial skin abrasions. Biochemical

Table 3. Disorders of Chemotaxis

Inherited	
Chediak–Higashi syndrome	Clark and Kimball (1971)
Hyperimmunoglobulin E syndrome	Quie and Cates (1978)
Abnormal spreading	Crowley et al. (1980)
Actin dysfunction	Boxer et al. (1974)
Microtubule dysfunction	Gallin et al. (1978a)
Chronic granulomatous disease	Ward and Schlegel (1969); Clark and Klebanoff (1978)
Congenital ichthyosis	Clark (1978)
α-Mannosidase deficiency	Desnick et al. (1976); Gallin et al. (1976)
Diabetes mellitus	Mowat and Baum (1971a)
Acquired	
Neonates	Miller (1971)
Active infection	McCall et al. (1971)
Thermal injury	Warden et al. (1974); Davis et al. (1980)
Bone marrow transplantation	Clark et al. (1976)
Idiopathic juvenile peridontitis	Clark (1978); Shurin et al. (1979)
Rheumatoid arthritis	Mowat and Baum (1971b)
Hypogammaglobulinemia	Gallin (1975)
Hypophosphatemia	Craddock et al. (1974)

studies of this patient's neutrophils demonstrated abnormal actin polymerization, and the data presented provided evidence for a close association between actin proteins, leukocyte locomotion, and host defense.

An inherited abnormality of neutrophil spreading was described by Crowley et al. (1980). Neutrophils from a 5-year-old boy with recurrent bacterial infection failed to spread on surfaces, leading to a severe defect in chemotaxis and a mild impairment of phagocytosis. Studies of family members suggested that this was an X-linked congenital disease. Gel electrophoresis revealed that a protein with a molecular weight of 110,000, present in a particulate fraction of normal neutrophils was absent from the patient's cells. The findings suggested that 110,000 molecular weight granule particulate protein is necessary for the spreading of neutrophils onto surfaces and that the functional abnormality was caused by its absence.

Two other clinical settings have provided evidence for an important role of microtubules in chemotaxis. The first was a patient with recurrent pyogenic and viral infections associated with markedly diminished neutrophil chemotaxis (Gallin et al. 1978a). The nucleus and granules in this patient's cells did not orient properly in a gradient of chemoattractant. Ultrastructural studies revealed marked increases in microtubule assembly, and biochemical studies provided evidence for elevated leukocyte cyclic GMP. (An example of the abnormal cells is shown in Figure 5.) The abnormalities in this patient's cells were compatible with the concept that proper microtubule assembly is required for normal neutrophil migration and that cyclic nucleotides may modulate microtubule function. Unfortunately, the patient died before additional studies could be performed.

Further evidence for a role of microtubules in neutrophil chemotaxis has been provided by a group of patients with a somewhat opposite cellular defect (i.e.,

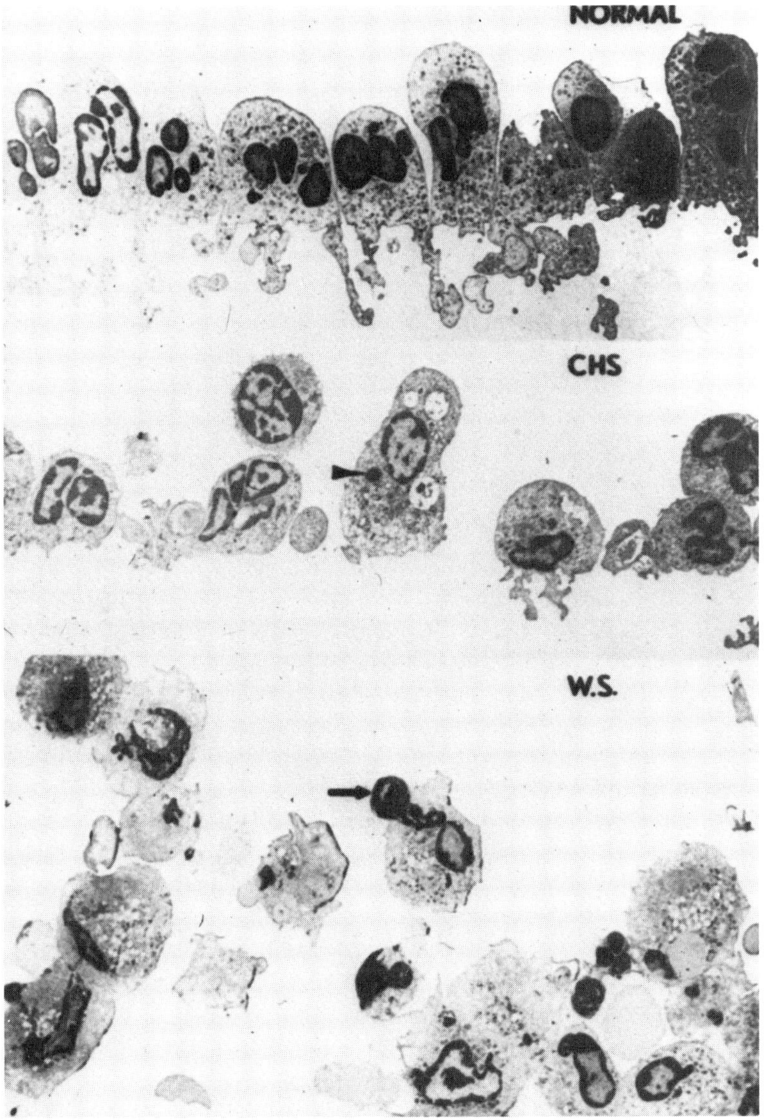

Figure 5. Electron micrographs of neutrophils at the surface of 0.45-μ pore cellulose nitrate filters after 45-minute incubation under conditions of directed migration (chemotaxis) with *Escherichia coli* endotoxin-activated serum as the chemoattractant. Upper panel: neutrophils from a normal subject (× 2400); middle panel: from a patient with the Chediak–Higashi syndrome (× 2400); lower panel: from a patient with recurrent infections (W.S.) (× 2400). From Gallin et al. (1978c).

decreased microtubule polymerization and elevated cyclic AMP). These patients have a rare disease called the Chediak–Higashi syndrome, which is characterized by recurrent pyogenic infections and characteristic giant lysosomal granules in all lysozomal-containing cells (Blume and Wolff 1972). Neutrophils from these patients have defective chemotaxis (Clark and Kimball 1971) and a delay of degranulation of lysozomal enzymes (Root et al. 1972). These neutrophil abnormalities have been associated with decreased microtubule polymerization and possibly abnormal cyclic nucleotide metabolism (Oliver 1978). An abnormality of orientation of intracellular structure has been suggested (Figure 5) with the Chediak–Higashi neutrophils having characteristics of normal cells treated with low concentrations of the microtubule-disrupting agent colchicine. Additional support for these conclusions has been provided by Boxer et al. (1976), who treated a few patients with Chediak–Higashi syndrome with ascorbic acid, a stimulant of cyclic GMP (Sandler et al. 1975). The treated patients showed improved neutrophil function as well as clinical improvement (Boxer et al. 1976). However, this effect of ascorbic acid was not seen in all patients with this disease (Gallin et al. 1979a), and the role of microtubules in the pathophysiology of this disease remains controversial (Frankel et al. 1978). Recently Nath et al. (1980) have provided data, using a sensitive biochemical assay and monitoring the tyrosylation of the α chain of tubulin, that indicate excessive tyrosylation of tubulin α chains in two brothers with Chediak–Higashi syndrome, compatible with increased depolymerized tubulin in the patient's cells. Furthermore the defect was corrected after *in vitro* or *in vivo* ascorbic acid. Resolution of the relevance of microtubules in the pathogenesis of this disease will await additional biochemical studies.

Patients with abnormal chemotaxis have been described in whom there is no reason to suspect abnormalities of actin polymerization and in whom no defect of microtubule assembly has been detected. The best studied individuals of this group of patients have a disease referred to as Job's syndrome, or the syndrome of hyperimmunoglobulin E and recurrent pyogenic (mostly staphylococcal) infections, particularly of the skin and lungs. It has been suggested recently by Schopfer et al. (1979) that the underlying cause of this disease is related to an abnormal immunologic response manifested by the formation of IgE antistaphylococcal antibodies. These patients have depressed neutrophil chemotaxis (Quie and Cates 1978) and have, in addition to recurrent infections, an allergic history consisting of eczema and mild eosinophilia. In addition, some of the patients have mucocutaneous candidiasis. These latter patients are skin test-negative for candida antigen, and have a T-cell abnormality. In patients with this spectrum of disease, assays for microbicidal activity and oxidative functions are normal, as are levels of neutrophil lysozomal enzymes. A potentially exciting observation is that the antihelminthic and immunostimulatory agent, levamisole, improves the chemotactic responsiveness of neutrophils from patients with this disease, *in vitro* as well as *in vivo* (Wright et al. 1977).

Neutrophil functional defects have also been described in less obscure diseases. For example, defects in neutrophil adherence, chemotaxis, phagocytosis, and microbicidal activity have been described in diabetes mellitus. The inflammatory response *in vivo,* as estimated by the Rebuck skin window, was normal in a stable diabetic state but abnormal in patients who were in a ketoacidotic state (Perillek et al. 1962). However, in another study the chemotactic *in vitro* response (Boyden chamber) was abnormal in 31 patients with diabetes, none of whom were in a ketoacidotic state (Mowat and Baum 1971a). Restoration of defective neutrophil chemotaxis has been reported by incubating patient's cells with glucose and insulin. Thus, it has been suggested that intracellular glycogen and potassium may be important in a normal chemotactic response.

Other clinical situations in which abnormal host responses are associated with depressed neutrophil chemotaxis include: (1) thermal injury (Davis et al. 1980; Warden et al. 1974), (2) malnutrition (Susskind 1977), (3) uremia (Baum et al. 1975), (4) periodontal disease (Clark 1978; Shurin et al. 1979), (5) the neonate (Miller 1971), and (6) the Wiskott–Aldrich syndrome (Altman et al. 1974).

Disorders of Killing

Disorders of killing can be considered in terms of disordered opsonization, ingestion of bacteria, abnormal granule formation and degranulation, as well as abnormal bactericidal systems (Table 4).

Disordered opsonization is seen in patients deficient in C3 and, therefore, unable to produce the opsonic fragment C3b (Alper et al. 1969). In sickle cell anemia, there is also reported deficient opsonization related to abnormal C3 metabolism (Johnston 1974). Another disorder of opsonization is thought to exist in patients with Leiner's disease, a disorder of C5 metabolism (Miller and Nilsson 1970). A C5 product seems to be an important opsonic factor for certain microorganisms. Leiner's disease, which occurs in neonates, is characterized by exfoliative dermatitis and is important to recognize, since patients have been managed successfully by exchange transfusions; after several months, they may "outgrow" the disease.

Other acquired disorders of ingestion include immature cells, as in patients with agranulocytic anemia and certain leukemias (Goldman and Catovsky 1972). Defective ingestion has been reported in a patient with a deficiency of tuftsin, the tetrapeptide that stimulates phagocytosis. In the setting of acute viral influenza, adherence of virus particles to the cell membrane is reported to interfere with ingestion (Najjar 1975).

There have been two important descriptions of patients with disorders of granule formation and degranulation. The first group includes patients with Chediak–Higashi syndrome in whom the large characteristic neutrophil granules have delayed fusion with the phagocytic vacuole following ingestion (Root et al. 1972). The second group of patients includes those patients with myeloperoxidase deficiency. Patients with this disease have increased risk for candida infection (Lehrer and Cline 1969). In myeloperoxidase deficiency the metabolic burst, as measured by oxygen consumption and superoxide generation, is greater than normal, presumably as a compensatory reaction by the cell. In addition the clinical manifestations may be minimal and, in fact, in

Table 4. Disorders of Killing

Defect	Disease	Reference
Opsonization	C3 deficiency	Alper et al. (1969)
	Leiner's disease	Miller and Nilsson (1970)
	Sickle cell disease	Johnston (1974)
Ingestion	Immature cells	
	Leukemia	Goldman and Catovsky (1972)
	Aplastic anemia	
	Other	
	Actin dysfunction	Boxer et al. (1974)
	Tuftsin deficiency	Najjar (1975)
Granule formation and degranulation	Chediak–Higashi syndrome	Blume and Wolff (1972)
	Myeloperoxidase deficiency	Grignaschi et al. (1963)
Interleukocytic metabolism	Glucose-6-phosphate dehydrogenase deficiency	Cooper et al. (1972)
	Chronic granulomatous disease of childhood	Quie and Cates (1978)
	Thermal injury	Heck et al. (1973)
	Postradiation	Baehner et al. (1973)

many patients the only clinical finding may be dermal abnormalities (Rosen and Klebanoff 1976).

Some patients have a severe defect in intraleukocytic metabolism. Perhaps the best example of this is chronic granulomatous disease (CGD). Quie et al. (1967) outlined the metabolic defect associated with these patients' susceptibility to infection. This inherited disease, which usually has an X-linked pattern of inheritance, may also be inherited as an autosomal recessive trait (15% of patients). Other parameters of the CGD cell that are normal include maturation, morphology, and phagocytosis. Because the humoral systems are not affected by this disease, the ability to opsonize microbes is not changed. Normal levels of complement have been found as well as normal delayed cutaneous hypersensitivity.

Neutrophils from patients with CGD have defective oxygen metabolism during the respiratory or metabolic burst associated with phagocytosis. An important feature of this disease is that bacteria that lack catalase and do not destroy their own hydrogen peroxide are only rarely found to be pathogenic to these patients (e.g., *Streptococcus pneumonia* and other streptococci). In addition, *in vitro* studies have shown that cells from CGD patients readily kill catalase-negative bacteria, which provide a rich source of H_2O_2. In contrast, CGD patients have increased incidence of infections by catalase-positive microorganisms such as *Staphylococcus aureus* and the enterobacteriaceae (Johnston and Baehner 1971), and CGD neutrophils have impaired killing of these bacteria *in vitro*. It is thought that the bacteria lacking catalase do not destroy their H_2O_2, and this important component of the bactericidal system can correct the defective cell function.

A number of enzyme defects have been related to decreased oxidative metabolism associated with CGD. In addition to the nicotinamide adenine dinucleotide (NADH) oxidase deficiency, nicotinamide adenine dinucleotide phosphate (NADPH) oxidase deficiency has been described (Hohn and Lehrer 1975). Glutathione peroxidase deficiency has been reported in patients with CGD (Holmes et al. 1970). A glucose-6-phosphate dehydrogenase (G6PD) deficiency has also been described in some (Bellanti et al. 1970) but not all patients with CGD (Baehner et al. 1970).

Recent data suggest that the abnormality in chronic granulomatous disease may relate to an abnormal activation of an oxidase rather than an enzyme deficiency. Studies using indirect probes of membrane potential indicated an abnormality of elicited changes of membrane potential in response to various stimuli of superoxide generation (Seligmann and Gallin 1980; Whitin et al. 1980). The possibility that CGD is related to an abnormal activation mechanism is intriguing and will await studies of ion permeability.

Clinically, patients with CGD often present with complications of recurrent draining lymph nodes as well as with abscesses of the liver. Although most patients present with infection in early childhood, some patients have their first infection later in life (Klebanoff and Clark 1978). Treatment of this disease is by aggressive antibiotic therapy and surgical intervention when required. In addition, there is increasing evidence that chronic sulfisoxazole therapy may be beneficial (Johnston et al. 1975), although this approach is controversial. The current understanding of this disease is more extensive than most of the other neutrophil defects, but the prognosis of these patients remains poor.

Patients with deficiency of G6PD, the first enzyme of the hexose monophosphate shunt, lack the metabolic pathway that leads to generation of O_2^-, $^1O_2$, OH^-, and H_2O_2 (Cooper et al. 1972). Patients with this deficiency have many of the same clinical and laboratory manifestations as patients with CGD, including failure to reduce nitroblue tetrazolium (Spirer et al. 1971). The similarity between CGD and G6PD deficiency is so close as to suggest that G6PD is the enzyme missing in CGD (Baehner et al. 1970; Bellanti et al. 1970). However, methylene blue, an agent that can regenerate NADP from NADPH in neutrophils from patients with CGD, has no effect on G6PD cells (Klebanoff and Clark 1978). Moreover, the level of G6PD in the cell must be less than 20% of normal to produce symptoms, which is not the case in patients with CGD (Baehner et al. 1970). Consequently, most investigators believe that G6PD is a distinct entity from CGD.

Conclusion

Although the importance of the neutrophil has been known for over 100 years, it is only recently that technology has become sophisticated enough to become the biochemical basis for the physiological process to be understood. With these advances, it is hoped that rational therapeutic modalities will evolve. Present manipulation of the cellular response to excessive inflammation consists of inhibition with agents such as steroids and colchicine. It is hoped that agents capable of stimulating the cellular response in settings of depressed inflammation can be used in the future, and two drugs have shown promise. One of these substances, ascorbate, has been shown to enhance chemotaxis in a few patients with Chediak–Higashi syndrome (Boxer et al. 1976) and hyperimmunoglobulin E syndrome (Friedenberg et al. 1979). Another agent, levamisole, has been used in patients with hyperimmunoglobulin E syndrome and has been associated with improved chemotaxis with improvement of clinical symptoms (Wright et al. 1977; Gallin et al. 1980). Both agents are thought to affect the intracellular levels of cyclic nucleotides; however, this is controversial. The results of clinical trials currently being carried out are greatly anticipated.

References

Alper, C.A., Propp, R.P., Klemperer, M.R., Rosen, F.S. (1969) Inherited deficiency of the third component of human complement (C'3). J. Clin. Invest. 48,553–557.

Altman, L.C. (1978) Chemotactic lymphokines: A review. In: *Leukocyte Chemotaxis: Methods, Physiology and Clinical Implications*. Gallin, J.I., Quie P.G. (eds.), New York: Raven Press, pp. 267–287.

Altman, L.C., Snyderman, R., Blaese, R.M. (1974) Abnormalities of chemotactic lymphokine synthesis and mononuclear leukocyte chemotaxis in Wiscott-Aldrich syndrome. J. Clin. Invest. 53,486–493.

Aswanikumar, S.B., Corcoran, B., Schiffmann, E., Day, A.R., Freer, R.J., Showell, H.J., Becker, E.L. (1977) Demonstration of a receptor on rabbit neutrophils for chemotactic peptides. Biochem. Biophys. Res. Commun. 74,210–217.

Athens, J.W., Haab, O.P., Raab, S.O., Mauer, A.M., Ahsenbrucker, H., Cartwright, G.E., Wintrobe, M.M. (1961) Leukokinetic studies. IV. The total blood, circulating and the granulocyte turnover rate in normal subjects. J. Clin. Invest. 40,989–995.

Babior, B.M. (1978) Oxygen-dependent microbial killing by phagocytes. N. Engl. J. Med. 298,659–668, 721–725.

Baehner, R.L., Johnston, R.B., Jr., Nathan, D.G. (1972) Comparative study of the metabolic and bactericidal characteristics of severely glucose-6-phosphate dehydrogenase-deficient polymorphonuclear leukocytes and leukocytes from children with chronic granulomatous disease. J. Reticuloendothel. Soc. 12,150–169.

Baehner, R.L., Nathan, D.G., Karnovsky, M.L. (1970) Correction of metabolic deficiencies in the leukocytes of patients with chronic granulomatosis disease. J. Clin. Invest. 49,865–870.

Baehner, R.L., Nathan, D.G., Castle, W.B. (1971) Oxidant injury of Caucasian glucose-G-phosphate dehydrogenase-deficient red blood cells by phagocytosing leukocytes during infection. J. Clin. Invest. 50,2466–2473.

Baehner, R.L., Neiburger, R.G., Johnson, D.E., Murrmann, S.M. (1973) Transient bactericidal defect of peripheral blood phagocytes from children with acute lymphoblastic leukemia receiving craniospinal irradiation. N. Engl. J. Med. 289,1209–1213.

Baggiolini, M. (1980) The neutrophil. In: *The Cell Biology of Inflammation. Handbook of Inflammation 2.* Glynn, L.E., Houck, J.C., Weissmann, G. (eds.), Amsterdam: Elsevier North Holland, pp. 163–187.

Baggiolini, M., Bretz, U., Gusus, B. (1974) Biochemical characterization of azurophilic and specific granules from human and rabbit polymorphonuclear leukocytes. Schweiz Med. Wochenschr. 104, 129–132.

Bainton, D.F. (1973) Sequential degranulation of the two types of polymorphonuclear leukocyte granules during phagocytosis of microorganisms. J. Cell. Biol. 58:249–264.

Bainton, D.F., Farquhar, M.G. (1966) Origin of granules in polymorphonuclear leukocytes: Two types derived from opposite faces of the golgi complex in developing granulocytes. J. Cell. Biol. 28,277.

Bainton, D.F., Ullyot, J.L., Farquhar, M.G. (1971) The development of neutrophilic polymorphonuclear leukocytes in human bone marrow: Origin and content of azurophil and specific granules. J. Exp. Med. 134,907–934.

Baum, J., Cestero, R.V.M., Freeman, R.B. (1975) Chemotaxis of the polymorphonuclear leukocyte and delayed hypersensitivity in uremia. Kidney Int. (Suppl.) 7,147–153.

Becker, E.L., Showell, H.J. (1972) The effect of Ca^{++} and Mg^{++} on the chemotactic responsiveness and spontaneous motility of rabbit polymorphonuclear leukocytes. Immun. Fortschr. 143,466–476.

Bellanti, J.A., Cantz, B.E., Schlegel, R.J. (1970) Accelerated decay of glucose-6-phosphate dehydrogenase activity in chronic granulomatosis disease. Pediatr. Res. 4,405–411.

Berlin, R.D., Oliver, J.M., Ukena, T.E., Yin, H.H. (1975) The cell surface. N. Engl. J. Med. 292,515–520.

Blondin, J., Janoff, A. (1976) The role of lysosomal elastase in the digestion of *Escherichia coli* proteins by human polymorphonuclear leukocytes. *J. Clin. Invest.* 58,971–979.

Blume, R.S., Wolff, S.M. (1972) The Chediak–Higashi syndrome: studies in four patients and a review of the literature. Medicine 51,247–280.

Boggs, D.R. (1975) Physiology of neutrophil proliferation maturation and circulation. Clin. Haematol. 4,535–551.

Boggs, D.R., Cartwright, G.E., Wintrobe, M.M. (1966) Neutrophilia inducing activity in plasma of dogs recovering from drug induced myelotoxicity. Am. J. Physiol. 211,51–60.

Boxer, L.A., Watanabe, A.M., Rister, M., Besch, H.R., Allen, J., Baehner, R.L. (1976) Correction of leukocyte function in Chediak–Higashi syndrome by ascorbate. N. Engl. J. Med. 295,1041.

Boxer, L.A., Hedley-Whyte, E.T., Stossel, T.P. (1974) Neutrophil actin dysfunction and abnormal neutrophil behavior. N. Engl. J. Med. 291,1093–1099.

Briggs, R.T., Drath, D.B., Karnovsky, M.L., Karnovsky, M.J. (1975) Localization of NADH oxidase on the surface of human polymorphonuclear leukocytes by a new cytochemical method. J. Cell. Biol. 67,566–586.

Chang, C., Houck, J.C. (1970) Demonstration of the chemotactic properties of collagen. Proc. Soc. Exp. Biol. Med. 134,22–26.

Chenoweth, D.E., Hugli, T.E. (1978) Demonstration of specific C5a receptor on intact human polymorphonuclear leukocytes. Proc. Natl. Acad. Sci. U.S.A. 75,3943–3947.

Clark, R.A. (1978) Disorders of granulocyte chemotaxis. In: *Leukocyte Chemotaxis: Methods, Physiology and Clinical Implications.* Gallin, J.I., Quie, P.G. (eds.), New York: Raven Press, pp. 329–356.

Clark, R.A., Kimball, H.R. (1971) Defective granulocyte chemotaxis in the Chediak–Higashi syndrome. J. Clin. Invest. 50,2645–2652.

Clark, R.A., Klebanoff, S.S. (1978) Chronic granulomatous disease, studies in a family with impaired neutrophil chemotactic, metabolic and bactericidal function. Am. J. Med. 65,941–948.

Clark, R.A., Johnson, F.L., Klebanoff, S.J., Thomas, E.D. (1976) Defective neutrophil chemotaxis in bone marrow transplant patients. J. Clin. Invest. 58,22–31.

Cline, M.J. (1975) *The White Cell*. Cambridge, Mass.: Harvard University Press.

Cline, M.J., Golde, D.W. (1979) Controlling the production of blood cells. Blood 53,157–165.

Cohn, Z.A., Hirsch, J.G. (1960) The influence of phagocytosis on the intracellular distribution of granule-associated components of polymorphonuclear leukocytes. J. Exp. Med. 112,1015–1022.

Cooper, M.R., DeChatelet, L.R., McCall, C.E., LaVia, M.F., Spurr, L.L., Baehner, R.L. (1972) Complete deficiency of leukocyte glucose-6-phosphate dehydrogenase with defective bactericidal activity. J. Clin. Invest. 51,769–778.

Craddock, P.R., Yawata, Y., Van Santen, L., Gilberstadt, S., Silvis, S., Jacob, H.S. (1974) Acquired phagocyte dysfunction: A complication of the hypophosphatemia of parenteral hyperalimentation. N. Engl. J. Med. 290,1403–1407.

Craddock, P.R., Fehr, J., Dalmasso, A.P., Brigham, K.L., Jacob, H.S. (1977) Hemodialysis leuko-penia: Pulmonary vascular leukostasis resulting from complement activation by dialyzer cellophane membrane. J. Clin. Invest. 59,879.

Crowley, C.A., Curnutte, J.T., Rosin, R.E., André-Schwartz, J., Gallin, J.I., Klempner, M., Sny-derman, R., Southwick, R.S., Stossel, T.P., Babior, B.M. (1980) An inherited abnormality of neutrophil adhesion. Its genetic transmission and association with a missing protein. N. Engl. J. Med. 302,1163–1168.

Davis, J.M., Dineen, P., Gallin, J.I. (1980) Neutrophil degranulation and abnormal chemotaxis following thermal injury. J. Immunol. 124,1467–1471.

Davis, P., Krakauer, K., Weissmann, G. (1970) Subcellular distribution of neutral proteases and peptidases in rabbit polymorphonuclear leukocytes. Nature 228,761–762.

DeChatelet, L.R., McPhail, L.C., Mullikin, D., McCall, C.E. (1975) An isotypic assay for NADPH oxidase activity and some characteristics of the enzyme from human polymorphonuclear leukocytes. J. Clin. Invest. 55,714–721.

Desnick, R.J., Sharp, H.L., Grabowski, G.A., Brunning, R.D., Quie, P.G., Sung, J.H., Gorlin, R.J., Ikonne, J.U. (1976) Mannosidosis: clincal, morphologic, immunologic, and biochemical stud-ies. Pediatr. Res. 10,985–996.

Dewald, B., Baggiolini, M., Curnutte, J.T., Babior, B.M. (1979) Subcellular localization of superoxide forming enzyme in human neutrophils. J. Clin. Invest. 63,21–29.

Dos Reis, G., Oliveira-Castro, G. (1977) Electrophysiology of phagocytic membranes. I. Potassium-dependent slow membrane hyperpolarizations in mice macrophages. Biochim. Biophys. Acta 469, 257–263.

Elsbach, P. (1972) Lipid metabolism by phagocytes. Semin. Hematol. 9,227–239.

Elsbach, P., Beckerdite, S., Pettis, P., Franson, R. (1974) Persistence of regulation of macromolecular synthesis by *Escherichia coli* during killing by disrupted granulocytes. Infect. Immun. 9,663–668.

Evans, W.H., Karnovsky, M.L. (1962) The biochemical basis of phagocytosis. IV. Some aspects of carbohydrate metabolism during phagocytosis. Biochemistry 1,159–166.

Fernandez, H.N., Henson, P.M., Otani, A., Hugle, T.E. (1978) Chemotactic response to human C3a and C5a anaphylatoxins. I. Evaluation of C3a and C5a leukotaxis *in vitro* and under simulated *in vivo* conditions. J. Immunol. 120,109–115.

Ferreira, S.H. (1979) Prostaglandins. In: *Chemical Messengers of the Inflammatory Process. Handbook of Inflammation 1*. Houck, J.C. (ed.), Amsterdam: Elsevier North Holland, pp. 113–152.

Fliedner, T.M., Cronkite, E.P., Robertston, J.S. (1964) Granulopoiesis. I. Senescence and random loss of neutrophilic granulocytes in human beings. Blood 24,402–414.

Frankel, F.R., Tucker, R.W., Bruce, J., Stenberg, R. (1978) Fibroblasts and macrophages of mice with the Chediak–Higashi-like syndrome have microtubules and actin cables. J. Cell. Biol. 79,401–408.

Freidenberg, W.R., Marx, J.J., Jr., Hansen, R.L., Haselby, R.C. (1979) Hyperimmunoglobulin E syndrome: response to transfer factor and ascorbic acid therapy. Clin. Immunol. Immunopathol. 12,132–142.

Gadek, J., Hunninghake, G., Lawley, G., Kelman, J., Fulmer, J., Crystel, R. (1978) Role of immune complexes in amplifying the alveolitis of idiopathic pulmonary fibrosis. Clin. Res. 76,446A.

Gallin, E.K., Gallin, J.I. (1977) Interaction of chemotactic factors with human macrophages: induction of transmembrane potential changes. J. Cell. Biol. 75,277–289.

Gallin, E.K., Seligmann, B.E., Gallin, J.I. (1980) Alteration of macrophage and monocyte membrane potential by chemotactic factors. In: *Mononuclear Phagocytes—Functional Aspects.* VanFurth, R. (ed.), The Hague: Martinus Nijhoff, pp. 505–523.

Gallin, E.K., Wiederhold, M.L., Lipsky, P.E., Rosenthal, A.S. (1975) Spontaneous and induced membrane hyperpolarizations in macrophages. J. Cell. Physiol. 86,653–662.

Gallin, J.I. (1975) Abnormal chemotaxis: Cellular and humoral components. In: *The Phagocytic Cell in Host Resistance.* Bellanti, J.A., Dayton, D.H. (eds.), New York: Raven Press, pp. 227–248.

Gallin, J.I. (1980a) The cell biology of leukocyte chemotaxis. In: *The Cell Biology of Inflammation. Handbook of Inflammation 2.* Glynn, L.E., Houck, J.C., and Weissmann, G. (eds.), Amsterdam: Elsevier North Holland, pp. 299–335.

Gallin, J.I. (1980b) Degranulating stimuli decrease the negative surface charge and increase the adhesiveness of human neutrophils. J. Clin. Invest. 65,298–306.

Gallin, J.I., Kirkpatrick, C.H. (1974) Chemotactic activity in dialyzable transfer factor. Proc. Natl. Acad. Sci. U.S.A. 71,498–502.

Gallin, J.I., Quie, P.G. (1978) *Leukocyte Chemotaxis: Methods, Physiology and Clinical Implications.* New York: Raven Press.

Gallin, J.I., Rosenthal, A.S. (1974) The regulatory role of divalent cations in human granulocyte chemotaxis: Evidence for an association between calcium exchanges and microtubule assembly. J. Cell. Biol. 62,594–609.

Gallin, J.I., Wright, D.G., Fauci, A.S., Rosenwasser, L.J., Chusid, M.J., Taylor, H.A., Thomas, G., Libaers, I., Shapiro, L.J., Neufeld, E.F. (1976) Defective leukocyte chemotaxis in mannosidosis. Clin. Res. 24,344A.

Gallin, J.I., Wright, D.G., Malech, H.L., Davis, J.M., Klempner, M.S., Kirkpatrick, C.H. (1980) Disorders of phagocyte chemotaxis. Ann. Intern. Med. 92,520–538.

Gallin, J.I., Elin, R.J., Hubert, R.T., Fauci, A.S., Kaliner, M.A., Wolff, S.M. (1979a) Efficacy of ascorbic acid in Chediak–Higashi syndrome: Studies in humans and mice. Blood 53,226–234.

Gallin, J.I., Gallin, E.K., Schiffmann, E. (1979b) Mechanisms of chemotaxis. Adv. Inflamm. Res. 1,123–138.

Gallin, J.I., Malech, H.L., Wright, D.G., Whisnant, J.K., Kirkpatrick, C.H. (1978a) Recurrent severe infections in a child with abnormal leukocyte function: possible relationship to increased microtubule assembly. Blood 51,919–933.

Gallin, J.I., Wright, D.G., Schiffmann, E. (1978b) Role of secretory events in modulating human neutrophil chemotaxis. J. Clin. Invest. 62,1364–1374.

Gallin, J.I., Gallin, E.K., Malech, H.L., Cramer, E.B. (1978c) Structural and ionic events during leukocyte chemotaxis. In: *Leukocyte Chemotaxis: Methods, Physiology and Clinical Implications.* Gallin, J.I., and Quie, P.G. (eds.), New York: Raven Press, pp. 195–228.

Ginsburg, I., Sela, M.N. (1976) The role of leukocytes and their hydrolases in the persistence, degradation and transport of bacterial constituents in tissues: Relation to chronic inflammatory processes in staphylococcal, streptococcal and mycobacterial infections and in chronic periodontal disease. C.R.C. Crit. Rev. Microbiol. 4,249–332.

Goldman, J.M., Catovsky, D. (1972) The function of the phagocytic leucocytes in leukemia. Br. J. Haematol. 23,223–230.

Goldstein, I.M., Kaplan, H.B., Edelson, H.S., Weissmann, G. (1979) Ceruloplasmin: A scavenger of superoxide anion radicals. J. Biol. Chem. 254,4040–4045.

Goldstein, I.M., Cerqueira, M., Lind, S., Kaplan, H.B. (1977) Evidence that the superoxide-generating system for human leukocytes is associated with the cell surface. J. Clin. Invest. 59,249–254.

Grignaschi, V.I., Sperperato, A.M., Etcheverry, M.J., Macario, A.J.L. (1963) Un nuevo caudro citoquimico: Negativiadad espontanea de las reacciones de peroxidasas oxidasas y lipido en la progenie neutrofila y en los monocitos de dos hermanos. Rev. Asoc. Med. Argent. 77,218–221.

Hayashi, H. (1975) Intracellular neutral SH-dependent proteases associated with inflammatory reactions. Int. Rev. Cytol. 40,101–151.

Heck, E.L., Browne, L., Curreri, P.W., Baxter, C.R. (1973) Evaluation of leukocyte function in burned individuals by *in vitro* oxygen consumption. J. Trauma 15,486–489.

Henson, P.M. (1971) The immunologic release of constituents from neutrophil leukocytes. I. The role of antibody and complement on nonphagocytosable surfaces or phagocytosable particles. J. Immunol. 107,1535–1546.

Henson, P.M., Oudes, Z. (1975) Stimulation of human neutrophils by soluble and insoluble immunoglobulin aggregates. Secretion of granule constituents and increase oxidation of glucose. J. Clin. Invest. 53,1053–1061.

Hirsch, J.G. (1975) The phagocytic cell in host resistance: a perspective summation. In: *The Phagocytic Cell in Host Resistance.* Bellanti, J.A., Dayton, D.H. (eds.), New York: Raven Press, pp. 333–340.

Hirsch, J.G. (1965) Phagocytin: A bactericidal substance from polymorphonuclear leukocytes. J. Exp. Med. 103,589–611.

Hoffstein, S.T. (1980) Intra- and extracellular secretion from polymorphonuclear leukocytes. In: *The Cell Biology of Inflammation.* Weissmann, G. (ed.), Amsterdam: Elsevier North Holland, pp. 387–430.

Hohn, D.C., Lehrer, R.I. (1975) NADPH oxidase deficiency in X-linked chronic granulomatous disease. J. Clin. Invest. 55,707–713.

Holmes, B., Park, B.H., Malawista, S.W., Quie, P.G., Nelson, D.G., Good, R.A. (1970) Chronic granulomatous disease in females. A deficiency of leukocyte glutathionine peroxidase. N. Engl. J. Med. 283,217–221.

Janoff, A., Blondin, J. (1974) The role of elastase in the digestion of *E. coli* proteins by human polymorphonuclear leukocytes. I. Experiments *in vitro.* Proc. Soc. Exp. Biol. Med. 145,1427–1430.

Johnston, R.B., Jr. (1974) Increased susceptibility to infection in sickle cell disease: Review of its occurrence and possible causes. 5th Med. J. 67,1342–1348.

Johnston, R.B., Jr., Baehner, R.L. (1971) Chronic granulomatosis disease: Correlation between pathogenesis and clinical findings. Pediatrics 48,730–739.

Johnston, R.B., Jr., Wilfert, C.M., Buckley, R.H., Webb, L.S., DeChatelet, L.R., McCall, C.E. (1975) Enhanced bactericidal activity of phagocytes from patients with chronic granulomatous disease in the presence of sulphisoxazole. Lancet 1,824–827.

Kampschmidt, R.F., Long, R.D., Upchurch, H.F. (1972) Neutrophil releasing activity in rats injected with endogenous pyrogen. Proc. Soc. Exp. Biol. Med. 139,1224–1226.

Kaplan, A.P., Goetzl, E.J., Austen, K.F. (1973) The fibrinolytic pathway of human plasma. II. The generation of chemotactic activity by activation of plasminogen proactivator. J. Clin. Invest. 52, 2591–2595.

Kaplan, A.P., Kay, A.B., Austen, K.F. (1972) A prealbumin activator of prekallikrein. III. Appearance of chemotactic activity for human neutrophils by the conversion of human prekallikrein to Kallikrein. J. Exp. Med. 135,81–97.

Karnovsky, M.L., Baehner, R.L., Githens, S., III, Simmons, S., Glass, E.A. (1971) Correlation of metabolism and function in various phagocytes. In: *Immunopathology of Inflammation.* Forscher, H., Houck, J.C. (eds.), Amsterdam: Excerpta Medica, pp. 121–132.

Karnovsky, M.L., Shafer, A.W., Cagan, R.H., Graham, R.C., Karnovsky, M.J., Glass, E.A., Saito, K. (1966) Membrane function and metabolism in phagocytic cells. Trans. N.Y. Acad. Sci. 28,778–787.

Kay, A.B., Pepper, D.S., McKenzee, R. (1974) The identification of fibrinopeptide B as a chemotactic agent derived from human fibrinogen. Br. J. Haematol. 27,669–677.

Kazmierowski, S.A., Gallin, J.I., Reynolds, H.Y. (1977) Mechanism for the inflammatory response in primate lungs. J. Clin. Invest. 59,273–281.

Keller, H.U., Sorkin, E. (1967) Studies on chemotaxis. V. On the chemotactic effects of bacteria. Lut. Arch. Allergy 31,505–517.

Klebanoff, S.J., Clark, R.A. (1978) *The Neutrophil: Function in Clinical Disorders.* New York: Elsevier North Holland.

Klempner, M.S., Gallin, J.I. (1978) Separation and functional characterization of human neutrophil subpopulations. Blood 51,659–669.

Korchak, H.M., Weissmann, G. (1978) Changes in membrane potential of human granulocytes antecede the metabolic responses of surface stimulation. Proc. Natl. Acad. Sci. U.S.A. 75,3818–3822.

Lackie, J.M., DeBono, D. (1977) Interaction of neutrophil granulocytes and endothelium *in vitro.* Microvas. Res. 13,107–112.

Lay, W.H., Nussenzweig, V. (1968) Receptors for complement on leukocytes. J. Exp. Med. 51,45.

Leber, T. (1888) Uber die Entstchung der Entalindung und die Wirkung der entzüdungserregenden Schüdlichkeiten. Fortschr. Med. 4,460–464.

Lehrer, R.I., Cline, M.J. (1969) Interaction of candida albicans with human leukocytes and serum. J. Bacteriol. 98,966–1004.

Lichtman, M.A. (1970) Cellular deformability during maturation of the myeloblast. Possible role in marrow egress. N. Engl. J. Med. 283,943.

Lichtman, M.A., Weed, R.I. (1972) Alteration of the cell periphery during granulocyte maturation: Relationship to cell function. Blood 39,301–315.

McCall, C.E., Caves, J., Cooper, R., DeChatelet, L. (1971) Functional characteristics of human toxic neutrophils. J. Infect. Dis. 124,68–75.

MacGregor, R.R. (1980) Granulocyte adherence. In: *The Cell Biology of Inflammation. Handbook of Inflammation 2*. Glynn, L.E., Houck, J.C., Weissmann G. (eds.), Amsterdam: Elsevier North Holland, pp. 267–298.

Malech, H.L., Root, R.K., Gallin, J.I. (1977) Structural analysis of human neutrophil migration: Centriole, microtubule, and microfilament orientation and function during chemotaxis. J. Cell. Biol. 75,666–693.

Mandell, G.L. (1974) Bactericidal activity of anerobic and anerobic polymorphonuclear neutrophils. Infect. Immun. 9,337–341.

Metchnikoff, E. (1907) *Immunity in Infective Diseases*. Translated by Starling, F.A., and Starling, E.H. London: Paul Trench, Trubner.

Miller, M.E. (1971) Chemotactic function in the neonate: humoral and cellular aspects. Pediatr. Res. 5,487–492.

Miller, M.E., Nilsson, U.R. (1970) A familial deficiency of the phagocytosis-enhancing activity of serum related to a dysfunction of the fifth component of complement (C5). N. Engl. J. Med. 282,354–358.

Mowat, A.G., and Baum, J. (1971a) Chemotaxis of polymorphonuclear leukocytes from patients with diabetes mellitus. N. Engl. J. Med. 284,621–627.

Mowat, A.G., Baum, J. (1971b) Chemotaxis of polymorphonuclear leukocytes from patients with rheumatoid arthritis. J. Clin. Invest. 50,2541–2549.

Najjar, V.A. (1975) Defective phagocytosis due to deficiencies involving tetrapeptide tuftsin. J. Pediatr. 87,1121–1124.

Najjar, V.A. (1974) The physiologic role of gamma globulin. Adv. Enzymol. 4,129–178.

Nath, J., Flavin, M., Gallin, J.I. (1980) Tubulin tyrosylation in normal and Chediak–Higashi syndrome neutrophils. J. Cell. Biol. (Abstr.) 87,1952.

Okado, Y., Roya, G., Tsuchiya, W., Inouye, K., Inouye, A. (1977a) Oscillation of membrane potential in L cells. I. Basic properties. J. Membr. Biol. 35,319–335.

Okado, Y., Roya, G., Tsuchiya, W., Doida, Y., Inouye, A. (1977b) Oscillation of membrane potential in L cells. II. Effect of monovalent ion concentrations and conductance changes associated with oscillations. J. Membr. Biol. 35,337–350.

Oliver, J.M. (1978) Cell biology of leukocyte abnormalities. Am. J. Pathol. 93,221–259.

Perillek, P.E., Nolan, J.P., Finch, S.C. (1962) Studies of the resistance to infection in diabetes mellitus: Local exudative cellular response. J. Lab. Clin. Med. 59,1008–1015.

Peterson, P.K., Verhoef, J., Quie, P.G. (1977) Influence of temperature of opsonization and phagocytosis of staphylococci. Infect. Immun. 15,175–179.

Quie, P.G., Cates, K.L. (1978) Clinical manifestations of disorders of neutrophil chemotaxis in leukocyte chemotaxis. In: *Leukocyte Chemotaxis: Methods, Physiology and Clinical Implications*. Gallin, J.I., and Quie, P.G. (eds.), New York: Raven Press, pp. 307–328.

Quie, P.G., White, J.G., Holmes, B., Good, R.A. (1967) *In vitro* bactericidal capacity of human polymorphonuclear leukocytes: diminished activity in chronic granulomatous disease in childhood. J. Clin. Invest. 46,668–679.

Root, R.K., Rosenthal, A.S., Balestra, D.J. (1972) Abnormal bactericidal, metabolic, and lysozomal function of Chediak–Higashi syndrome leukocytes. J. Clin. Invest. 51,649–665.

Rosen, H., Klebanoff, S.J. (1979) Bactericidal activity of a superoxide anion-generating system. A model for the polymorphonuclear leukocyte. J. Exp. Med. 149,27–39.

Rosen, H., Klebanoff, S.J. (1976) Chemiluminescence and superoxide production by myeloperoxidase-deficient leukocytes. J. Clin. Invest. 58,50–60.

Rossi, F., Romeo, D., Patriarca, P. (1972) Mechanism of phagocytosis associated oxidative metabolism in polymorphonuclear leukocytes and macrophages. J. Reticuloendothel. Soc. 12,127–149.

Rothstein, G., Christensen, R.D., Hugl, E.H., Athens, J.W. (1971) Stimulation of granulopoiesis by a diffusable factor *in vivo*. J. Clin. Invest. 50,2004.

Sandler, J.A., Gallin, J.I., Vaughan, M. (1975) Effects of serotonin, carboxylcholine and ascorbic acid on leukocyte cyclic GMP and chemotaxis. J. Cell. Biol. 67,480–484.

Sbarra, A.J., Karnovsky, M.L. (1959) The biochemical basis of phagocytosis. I. Metabolic changes during the ingestion of particles by polymorphonuclear leukocytes. J. Biol. Chem. 234,1355–1362.

Schade, A.L., Caroline, L. (1946) The iron binding component in human blood plasma. Science 104,340–341.

Schiffmann, E., Gallin, J.I. (1979) Biochemistry of phagocyte chemotaxis. In: *Current Topics in Cellular Regulation,* vol. 15. Studtman, R. (ed.), New York: Academic Press, pp. 203–261.

Schiffmann, E., Showell, H.J., Corcoran, B.A., Ward, P.A., Smith, E., Becker, E.L. (1975) The isolation and partial characterization of neutrophil chemotactic factors from *E. coli*. J. Immunol. 114,1831–1837.

Schiffmann, E., Corcoran, B.A., Aswanikumar, S. (1978) Molecular events in the response of neutrophils to synthetic N-f-met chemotactic peptides: demonstration of a specific receptor. In: *Leukocyte Chemotaxis: Methods, Physiology and Clinical Implications.* Gallin, J.I. (ed.), New York: Raven Press, pp. 97–111.

Schopfer, K., Baerlocher, K., Price, P., Krech, U., Quie, P.G., Douglas, S.D. (1979) Staphylococcal IgE antibodies, hyperimmunoglobulinemia E and *Staphylococcus aureus* infections. N. Engl. J. Med. 300,835–838.

Schuldiner, S., Kaback, H.R. (1976) Membrane potential and active transport in membrane vesicles from Escherichia coli. Biochemistry 14,5451–5461.

Seligmann, B.E., Gallin, J.I. (1980) Use of lipophilic probes of membrane potential to assess human neutrophil activation: Abnormality in chronic granulomatous disease. J. Clin. Invest. 66,493–503.

Seligmann, B.E., Gallin, E.K., Martin, W., Shain, W., Gallin, J.I. (1980) Interaction of chemotactic factors with human polymorphonuclear leukocytes: Studies using a membrane potential sensitive cyanine dye. J. Membr. Biol. 52,257–272.

Shurin, S.B., Socransky, S.S., Sweeney, E., Stossel, P. (1979) A neutrophil disorder induced by capnocytophaga, a dental micro-organism. N. Engl. J. Med. 301,849–854.

Sobel, J.E., Gallin, J.I. (1979) Polymorphonuclear leukocyte and monocyte chemoattractants produced by human fibroblasts. J. Clin. Invest. 63,609–618.

Spilberg, I., Mehta, J. (1979) Demonstration of a specific neutrophil receptor for a cell-derived chemotactic factor. J. Clin. Invest. 63,85–88.

Spirer, Z., Binor, Z., Bogair, N. (1971) G6PD and childhood infection. Lancet 2,661–662.

Stossel, T.P. (1978) The mechanism of leukocyte locomotion. In: *Leukocyte Chemotaxis: Methods, Physiology and Clinical Implications.* Gallin, J.I., Quie, P.G. (eds.), New York: Raven Press, pp. 143–160.

Stossel, T.P. (1974) Phagocytosis. N. Engl. J. Med. 290,717–723, 774–780, 833–839.

Stossel, T.P. (1973) Quantitative studies of phagocytosis. Kinetic effects of cations and heat labile opsonin. J. Cell. Biol. 58,346–356.

Susskind, R.M. (1977) *Malnutrition and the Immune Response.* Kroc Foundation Series, vol. 7. New York: Raven Press.

Waggoner, A. (1976) Optical probes of membrane potential. J. Membr. Biol. 27,317–334.

Ward, P.A., Becker, E.L. (1970) Potassium reversible inhibition of leukotaxis by ouabain. Life Sci. 9,355–360.

Ward, P.A., Schlegel, R.A. (1969) Impaired leucotactic responsiveness in a child with recurrent infections. Lancet 2,344–347.

Ward, P.A., Hugli, T.E., Chenoweth, D.E. (1979) Complement and chemotaxis. In: *Chemical Messengers of the Inflammatory Process.* Handbook of Inflammation 1. Houck, J.C. (ed.), Amsterdam: Elsevier North Holland, pp. 153–178.

Warden, G.D., Mason, A.D., Pruitt, B.A. (1974) Evaluation of leukocyte chemotaxis *in vitro* in thermally injured patients. J. Clin. Invest. 54,1001–1004.

Weissmann, G. (1980) *The Cell Biology of Inflammation*. Amsterdam: Elsevier North Holland.

Weissmann, G., Spilberg, I., Krankauer, K. (1969) Arthritis induced in rabbits by lysates of granulocyte lysosomes. Arthritis Rheum. 12,103–116.

Weissmann, G., Zurier, R.B., Spieler, P.J., Goldstein, I.M. (1971) Mechanisms of lysosomal enzyme release from leukocytes exposed to immune complexes and other particles. J. Exp. Med. 134,149s–165s.

Weissmann, G., Snollen, J.E., Korchak, H.M. (1980) Release of inflammatory mediators from stimulated neutrophils. N. Engl. J. Med. 303,27–34.

Whitin, J.C., Chapman, C.E., Simmons, M.E., Chovaneic, M.E., Cohen, H.J. (1980) Correlation between membrane potential changes and superoxide production in human granulocytes stimulated by phorbol myristate acetate. Evidence for defective activation in chronic granulomatous disease. J. Biol. Chem. 255,1874–1878.

Wilkinson, P.C. (1974) *Chemotaxis and Inflammation*. Edinburgh: Churchill Livingston.

Williams, L.T., Snyderman, R., Pike, M.C., Lefkowitz, R.J. (1977) Specific receptor sites for chemotactic peptides on human polymorphonuclear leukocytes. Proc. Natl. Acad. Sci. U.S.A. 74,1204–1208.

Wright, D.G., Gallin, J.I. (1977) Functional differentiation of human neutrophil granules: Generation of C5a by a specific (secondary) granule product and inactivation of C5a by azurophil (primary) granule products. J. Immunol. 119,1068–1076.

Wright, D.G., Gallin, J.I. (1979) Secretory responses of human neutrophils: Exocytosis of specific (secondary) granules by human neutrophils during adherence *in vitro* and during exudation *in vivo*. J. Immunol. 123,285–294.

Wright, D.G., Kirkpatrick, C.H., Gallin, J.I. (1977) Effects of levamisole on normal and abnormal leukocyte locomotion. J. Clin. Invest. 59,941–950.

Zigmond, S. (1978) Chemotaxis by polymorphonuclear leukocytes. J. Cell. Biol. 77,269–287.

Zigmond, S. (1974) Mechanisms of sensing chemical gradients by polymorphonuclear leukocytes. Nature 249,450–452.

Eosinophils in Immune Function

Sheldon G. Cohen, M.D. and Eric A. Ottesen, M.D.

Paul Ehrlich was the first to describe a specific leukocyte type with unique protoplasmic granules of basic pH that could be identified feasibly by acidic dye markers. After the introduction of eosin into conventional laboratory use as a cytohistochemical reagent, the term "eosinophil" became synonymous with Ehrlich's "acidophil."

As cellular participants in both immune and inflammatory processes, eosinophils have been noted to be especially prominent in situations primarily associated with manifestations of allergy and parasitic helminth infection. In both these instances, it is likely that the eosinophil's appearance in circulating blood and involved tissue may at least partially depend on a common mechanism related to immediate hypersensitivity. Thus, an understanding of relevant relationships to the mast cell and IgE appears to be especially important for the interpretation of eosinophil production and migration. However, appreciation of certain features of eosinophil structure, function, kinetics, and relationship to cellular immune mechanisms may help to explain other instances of eosinophilia where immediate hypersensitivity phenomena are not evident,[1] e.g., neoplastic and myeloproliferative diseases, immunodeficiency disorders, adrenal hypofunction.

Morphology and Structure

The living, motile eosinophil is polymorphous in shape with a diameter of 10 to 15 μm with a multilobed nucleus (average, 2.3 lobes) and features characteristic of nonreplicating, end-stage cells, i.e., condensed heterochromatin and no nucleolus. Cytoplasmic organelles, such as the Golgi apparatus, endoplasmic reticulum, and mitochondria, have been variously described as being more or less abundant than those found in neutrophils, but clearly these organelles are less prominent in the eosinophil than they are in the more metabolically active mononuclear leukocytes, i.e., the lymphocytes and monocytes.

From the Immunology, Allergic, and Immunologic Diseases Program and Laboratory of Parasitic Diseases, National Institute of Allergy and Infectious Diseases, Bethesda, Maryland.

[1]For detailed discussion and more extensive reference citations on material presented in this chapter the reader is referred to a more exhaustive review by the authors (Ottesen and Cohen 1978) and to other general bibliographic sources (Beeson and Bass 1977; Butterworth 1977; Zucker-Franklin 1974).

Cell Surface

The surface membrane of the eosinophil appears to differ from those of other leukocytes in a number of important ways. First, it is antigenically distinct. Antisera raised in rabbits against mouse, guinea pig, or human eosinophils can be shown to have specificities for determinants not found on other leukocyte types (Mahmoud et al. 1973). Second, the receptors for immunoglobulin and complement on the eosinophil membrane appear to be both quantitatively and qualitatively different from those on other leukocytes. While it is generally agreed that the eosinophil lacks receptors for IgA, IgM, and IgD, IgG receptors have been variably reported on between 5 and 25% of normal human eosinophils. Furthermore, there is some suggestion that this percentage is increased in certain disease states, including parasitic helminth infections (Ottesen et al. 1977). In no situation, however, does either the percentage of IgG receptor-positive eosinophils or the receptor density on individual cells approach that found for neutrophils. Because of conflicting data, it is not clear whether or not eosinophils have IgE receptors. Complement receptors (for the C4, C3 fragments) have been described on 30 to 50% of normal human eosinophils, and again there is the suggestion that this number is increased in disease states including helminth infections. Additionally, it has been reported that histamine and at least one of the eosinophil chemotactic factors (ECF-A, described in the section Chemotactic Factors) can increase the percentage of eosinophils displaying complement receptors (Anwar and Kay 1978). Thus, it is possible that these receptor bearing cells represent a subpopulation of "activated" eosinophils ready to carry out whatever functional roles are required of them. Other "pharmacologic" receptors for substances such as estrogen, ECF-A, and histamine have been described on the eosinophil, but both their uniqueness to this cell type and their precise physiologic roles remain to be fully defined.

Granules

Though the number, size, and even staining properties of the specific granules of the eosinophil vary among different species, it is these granules that best distinguish the eosinophil from other cell types (Figures 1 and 2). The mature human eosinophil contains approximately 200 such granules per cell, generally oval in shape with diameters of 0.9 to 1.3 μm in length. The granules are membrane bound vacuoles with a characteristic electron dense, crystalloid core surrounded by a less dense, enzyme-rich homogeneous matrix (Figure 1). The core is a lamellate structure with repeating units every 40 Å (Miller et al. 1966) and because of its unique biochemical makeup is responsible for the marked acidophilia (eosin uptake) of these cells. This core is almost entirely composed of a major basic protein (MBP) with molecular weight of about 9,000. Major basic protein, first recovered from guinea pig eosinophils, was described as extremely basic (pH 10) with 13 arginine residues, 6 half-cystine residues, and 2 free sulfhydryl groups per molecule. Subsequent purification of MBP from human eosinophils has yielded very similar biochemical findings as well as distinct antigenic cross-reactivity with guinea pig MBP (Gleich et al. 1976). Unfortunately, specific functional capabilities for this protein, other than those characteristic of any basic protein molecule with multiple reactive sulfhydryl groups, have yet to be defined despite the fact that MBP constitutes fully 25% of the total cell protein of the eosinophil.

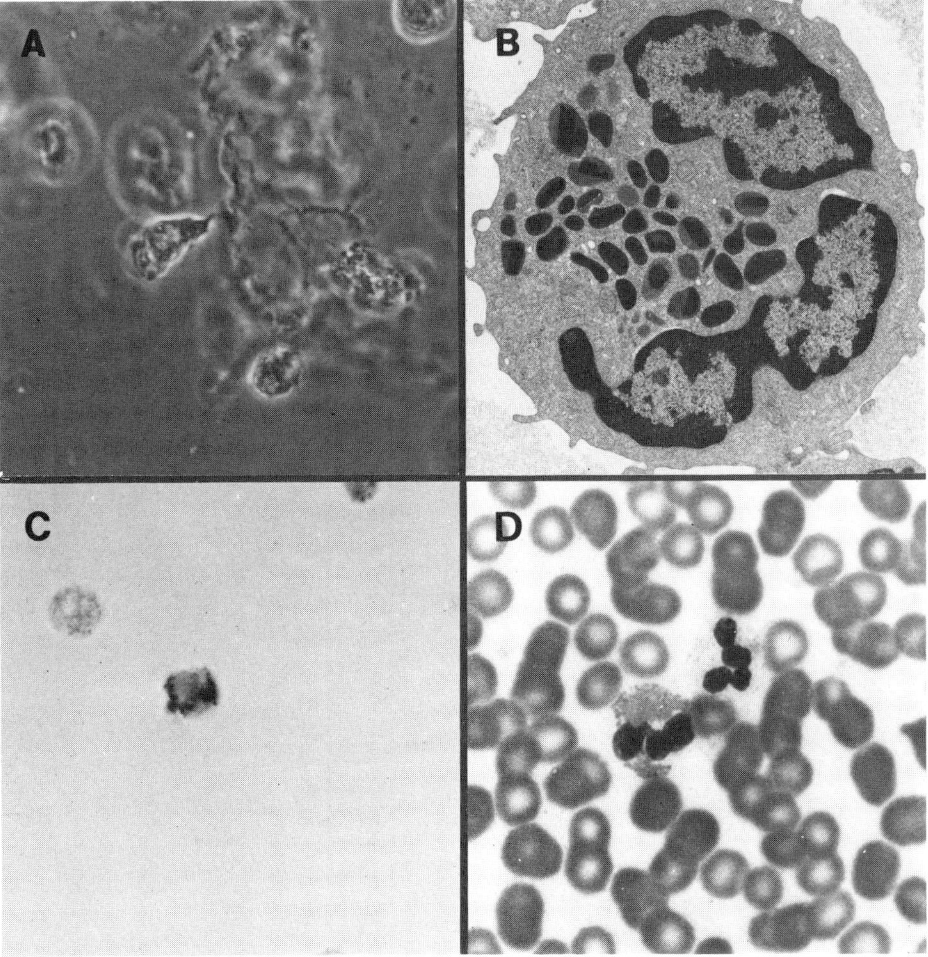

Figure 1. Varied appearance of human eosinophil. (**A**) Pilot stain. Motile eosinophil-extending psuedopod; phase optics (original magnification × 400). (**B**) Electron micrograph of human eosinophil with its characteristic granules composed of dense crystalline core and less dense homogenous matrix (original magnification × 10,000). (Courtesy A.S. Rosenthal, M.D. and K.U. Cehrs.) (**C**) Bromcresol purple vital stain, fluid preparation. More darkly stained eosinophils can be distinguished from lightly stained neutrophil (original magnification × 400). (**D**) Wright stain. Dark granules and bilobed nucleus distinguish eosinophil (left) from polymorphonuclear neutrophil (right). (original magnification × 630). (Reproduced with permission from Ottesen, Eric A., Cohen, Sheldon G. (1978) The eosinophil, eosinophilia, and eosinophil-related disorders. In: Middleton, Elliott, Jr., Reed, Charles E., Ellis, Elliot F. (eds.) Allergy: Principles and Practice. St. Louis: C.V. Mosby, p. 586.)

A second type of granule recently described is smaller, less dense, and lacks a crystalloid core (Parmley and Spicer 1974). Although numerous in tissue eosinophils, these smaller granules are found only occasionally in normal blood eosinophils and are rarely seen in cells within the bone marrow. They are characterized by high concentrations of the enzymes arylsulfatase and acid phosphatase.

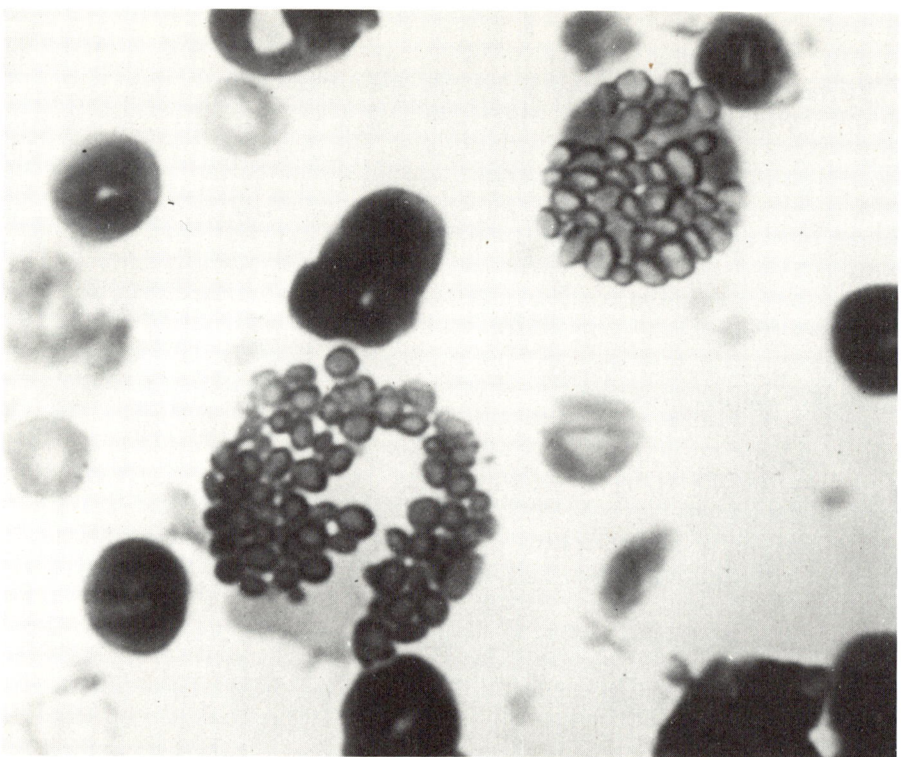

Figure 2. Eosinophil leukocytes in peripheral blood of the horse, Giemsa stain. Granules are larger than those of human eosinophils (original magnification × 1000)

Intracellular Enzymes

Eosinophils have abundant enzymatic activities that may be associated with the granules, cytoplasm, or membranes (Table 1). Many of these activities are similar to those found in the neutrophil (Table 1) but some are distinctly different. Peroxidase, for example, is much more abundant in the eosinophil than it is in the neutrophil. It is present in all of the large specific granules in contrast to the neutrophil in which peroxidase is found only in primary granules, which constitute less than one-third of the total number. Furthermore, not only are there quantitative differences between eosinophil and neutrophil peroxidases but there are also qualitative differences in their specific chemical properties, their antigenic makeups, and even their genetic controls. Other major differences between eosinophil and neutrophil content of enzymes, which can be seen in Table 1, are absence from the eosinophil of lysozyme, phagocytin, and neutrophil basic protein (NBP), all of which play important roles in the antibacterial functions of the neutrophilic leukocytes. Furthermore, it is seen that the eosinophil is relatively rich in enzymes that are antiinflammatory (histaminase, kininase, aryl-sulfatase, and phospholipase D) by virtue of the potential to inactivate many of the important mediators of inflammation (Chapter 13). The possible funtional meanings of these enzyme differences are discussed below.

The salient points of the eosinophil's properties are summarized in Table 2.

Table 1. Comparison of Chemical Constituents of Eosinophilic and
Neutrophilic Polymorphonuclear Leukocytes.

Enzymes	Eosinophils	Neutrophils	Function
Phosphatases			
Acid	GC	G	Metabolic; breakdown of phagocytosed material
Alkaline	?	G	
Oxidoreductases			
Peroxidase	G[b]	G	Antibacterial; antitoxin[a]
Catalase	C	C	Antibacterial; antitoxin[a]
Succinic dehydrogenase	C	C	Metabolic
Lactic dehydrogenase	C	C	Metabolic
Glucose-6-phosphate dehydrogenase	C	C	Metabolic
Phosphogluconate dehydrogenase	C	C	Metabolic
Coenzyme Q	G	G	Metabolic
Histaminase	G	G	Anti-inflammatory (inactivates histamine)
Proteases and peptidases			
Aminopeptidase	G	G	Substrate defined; no precise function delineated[a]
Collagenase	[c]	G	Substrate defined; no precise function delineated[a]
Cathepsin	G	G	Substrate defined; no precise function delineated[a]
Trypsin	C	[c]	Substrate defined; no precise function delineated[a]
Kininase	G	G	Anti-inflammatory (inactivates kinins)
Other	G	G	Substrate defined; no precise function delineated[a]
Nucleases			
Ribonuclease	G	G	Substrate defined; no precise function delineated[a]
Deoxyribonuclease	G	G	Substrate defined; no precise function delineated[a]
Other	ND	G	Substrate defined; no precise function delineated[a]
Esterases			
Phospholipase B	G	ND	Anti-inflammatory, metabolic
Phospholipase D	C	ND	Anti-inflammatory (inactivates PAF)
Other hydrolases			
Lysozyme	Absent	G	Antibacterial[a]
Beta-glucuronidase	G	G	Substrate defined; no precise function delineated[a]
Amylase	C	G	Substrate defined; no precise function delineated[a]
Arylsulfatase	G	G	Anti-inflammatory (inactivates SRS-A)
Nonenzymatic molecules			
MBP	G	Absent	Unknown
NBP	Absent	G	Antibacterial
Phagocytin	Absent	G	Antibacterial
Pyrogen	[c]	[c]	Induces fever
Plasminogen	?	Absent	Plasmin precursor

From Ottesen, Eric A., Cohen, Sheldon G. (1978) The eosinophil, eosinophilia, and eosinophil-related disorders. In: Middleton, Elliott, Jr., Reed, Charles E., Ellis, Elliot F. (eds.) *Allergy: Principles and Practice*. St. Louis: C.V. Mosby Co.
G, granule associated; C, cytoplasm associated; ND, not defined; PAF, platelet activating factor; SRS-A, slow-reacting substance of anaphylaxis.
[a]Defined for neutrophils only [b]Distinctly different from neutrophil enzyme [c]Enzyme present but location undefined

Table 2. Characteristic Features of Eosinophil Leukocytes

Source	bone marrow
Distribution	blood-tissue ratio (approx. 1:100)
Shape	polymorphous
Size	10–15 μm
Nucleus	bilobed
	eccentricaly placed
Membrane Receptors	
Immunologic	IgG, IgE?, C3b, C3d, C4
Pharmacologic	estrogen, histamine, ECF-A
Granules	
Large	crystaloid core of major basic protein, peroxidase and other enzymes
Small	mainly in tissue eosinophils; arylsulfatase and acid phosphatase
Enzymes[a]	
Antiinflammatory	histaminase, kininase, arylsulfatase, phospholipase D
Antibacterial	*lacks* lysozyme, phagocytin and neutrophil basic protein

[a]See Table 1.

Normal Cell Development and Kinetics

There are three phases in the life cycle of an eosinophil: first, its development in the bone marrow; second, its passage in the blood stream; and third, its presence in tissues (Figure 3). Though usually thought of as a cellular element of blood, the eosinophil numerically is much more well represented in the bone marrow and in other tissues. For example, in the guinea pig it has been estimated that for each eosinophil found in the circulation, 400 more reside in the marrow and another 300 in other tissues. In the rat, there are an estimated 300 marrow and 200 to 300 tissue eosinophils for each one found in the blood, and in man an eosinophil tissue to blood ratio of 100:1 has been proposed.

Studies from *in vitro* culture of marrow support the notion of separate commitment stem cells for neutrophils and eosinophils (Dao et al. 1977). Complete development of eosinophils in the bone marrow requires about 5.5 days (Spry 1971) and consists of maturation from the earliest identifiable committed precursor cells which contain the first distinguishable eosinophilic granules, promyelocytes, to eosinophil myelocytes and metamyelocytes before the cells enter a 40 to 60 hours, nondivisional period of maturation. During that time, they are morphologically indistinguishable from blood eosinophils and constitute the animal's eosinophil "marrow reserve." In some species (e.g., the rat) this maturational period occurs in the spleen, but otherwise the development process of the eosinophil appears very similar among all species studied.

The period of time that eosinophils spend in the blood stream is relatively short but not precisely defined. There is evidence in the rat and guinea pig that removal of radiolabeled eosinophils from the blood is exponential (i.e., random) and that the half-life is 6 to 12 hours, values which are similar for neutrophils in the same species. In man, however, the only similar studies have been done in patients with the hypereosinophilic syndrome (described in a following section), and the kinetics of cell removal from the blood of these individuals was extremely complex. An initial exponential decrease was followed by a subsequent rise or plateau in the number of blood eosinophils and then a second more gradual decline for several days thereafter (mean half-life, 44 ± 2 hours) (Dale et al. 1976). This biphasic kinetic pattern suggests that after initial entry into the bloodstream eosinophils are transiently

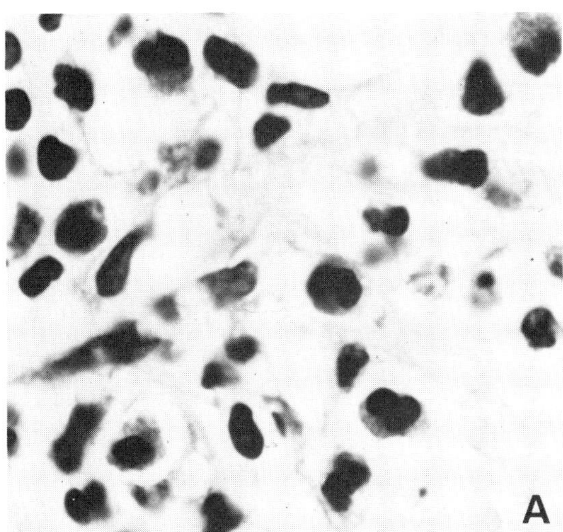

Figure 3. Eosinophils in human disease; tissue sections, hematoxylin and eosin, (original magnification × 125). (A) Nasal polyp; eosinophilic cellular infiltrate within the stroma. (Courtesy M. Samter) (B) Lung in asthma; infiltrate of bronchial mucosa by eosinophils. (Courtesy G.J. Gleich with permission from the Mayo Clinic.) (C) Lung in hypereosinophilic syndrome; eosinophils trafficking in lumen of a pulmonary vessel and infiltrating the alveolar wall. (Courtesy A.S. Fauci)

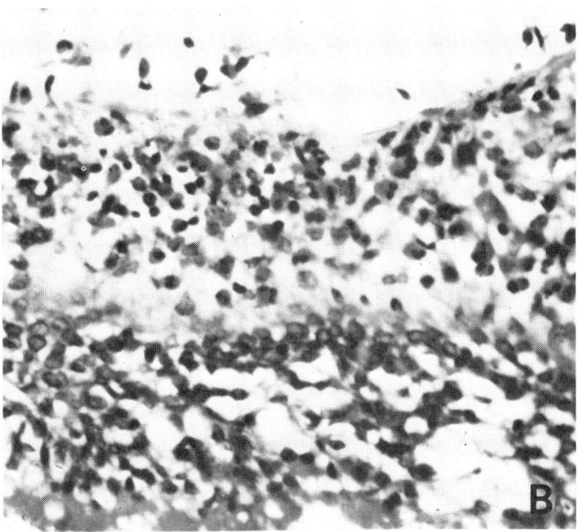

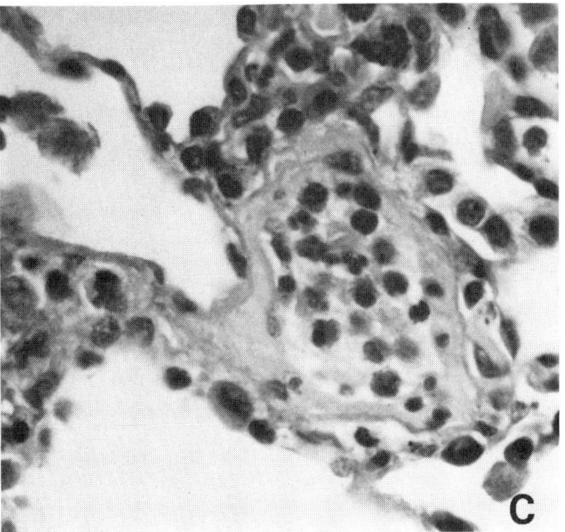

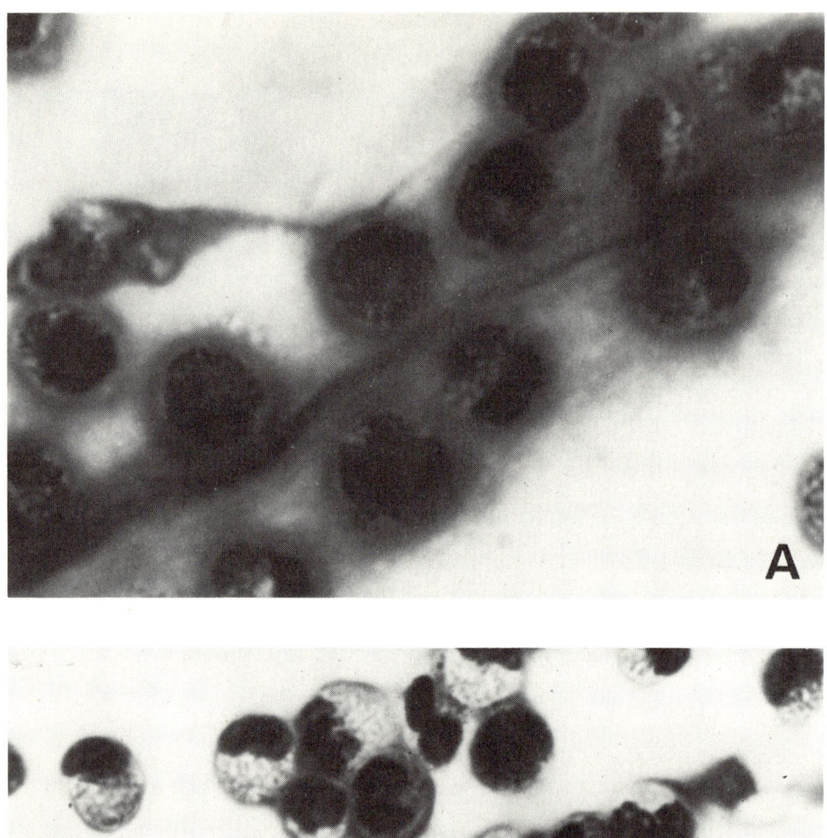
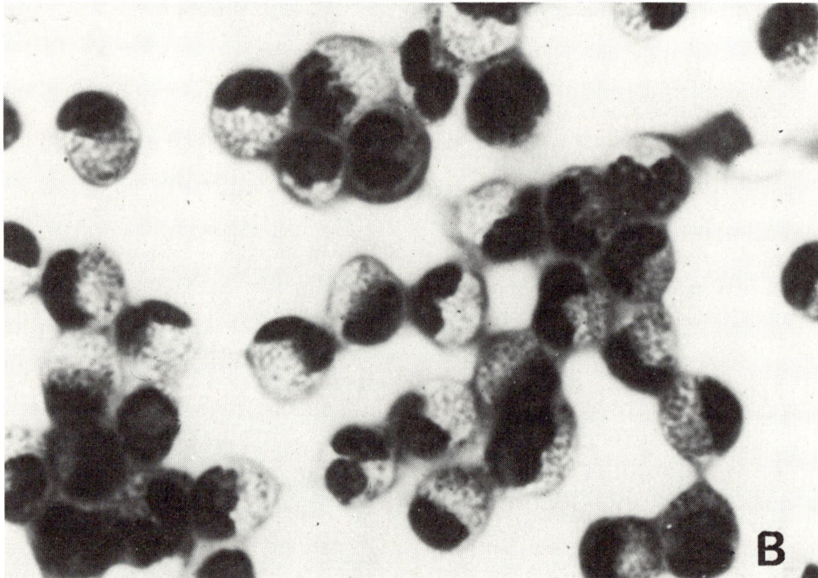

Figure 4. Eosinophils in secretions from human respiratory tract; eosin and methylene blue/ Hansel's stain. (**A**) Sputum in asthma (original magnification × 400). (**B**) Nasal secretion in allergic rhinitis (hay fever) (original magnification × 400).

sequestered, then subsequently released into the circulation again for a longer period before eventual distribution to sites of final tissue localization. Both the mechanisms underlying these observations and their relevance to the normal blood transit time in man at present are not clearly understood.

Once they have left the bloodstream, eosinophils localize in various organs in characteristic patterns, with the gastrointestinal tract, skin, and lungs the most heavily infiltrated tissues (Rytömaa 1960). Their fate thereafter is largely unknown, though it has been speculated that extrusion of intact eosinophils across the mucosal surfaces of the gastrointestinal and respiratory tracts (Figure 4) might serve as a common final step in the completion of the cell's life cycle. Estimates of the length of time that eosinophils spend in tissues have been on the order of several days, but, at present, available data are insufficient for firm conclusions.

Under the influence of numerous homeostatic factors affecting production, development, release from bone marrow, and subsequent tissue distribution, eosinophil levels are maintained in humans at average normal counts of 70 to 450/mm^3 for adults and 34 to 740/mm^3 in children. In addition to these age determinants there is also a sex dependence, as circulating cell counts are usually higher in males than in females. While individually variable fluctuations in eosinophil levels can be demonstrated throughout the day, major diurnal variations are consistent findings. This rhythmic pattern is related to the regulatory role of the adrenal cortex in eosinophil kinetics, i.e., eosinopenia identifiable at the height of adrenocortical activity during the early morning hours and, conversely, maximal counts at the ebb of cortisol secretion in the late afternoon. Also affecting the variability of daily levels of circulating eosinophils is the sympathetic nervous system's influence on blood–tissue distribution through the action of adrenergic products on beta-receptors.

The stimuli for eosinophilia generally originate in the tissues, but the development

Figure 5. Evolution of an eosinophilic infiltrate. Tissue, blood, and marrow eosinophil levels are dissociated temporally after peripheral eosinophilogenic stimulus. Note how need for eosinophils arises in tissue (e.g., lung) in which eosinophilic infiltrate is established at expense of number of cells circulating in blood. Eosinophilia in marrow lags behind that of tissue but persists until tissue infiltrative process is resolved and excessive numbers of eosinophils accumulate in blood. Marrow production of eosinophils then returns to normal level along with numbers of cells in peripheral circulation. (Reproduced with permission from Ottesen, Eric A., and Cohen, Sheldon G. (1978) The eosinophil, eosinophilia, and eosinophil-related disorders. In: Middleton, Elliott, Jr., Reed, Charles E., and Ellis, Elliot F. (eds.) Allergy: Principles and Practice. St. Louis: C.V. Mosby, p. 593.)

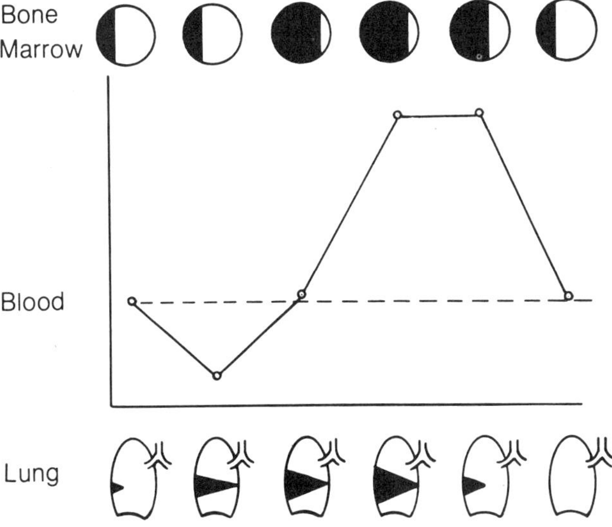

of cellular infiltrates depends on the recruitment of cells, first from the blood and ultimately from the bone marrow. Thus, during the evolution of an eosinophilic infiltrate one would anticipate a temporal dissociation of tissue, blood, and marrow eosinophil levels following a peripheral eosinophilogenic stimulus. This dissociation is depicted graphically in Figure 5, in which a developing infiltrate in the lung is seen to induce a temporary blood eosinopenia that leads to subsequent marrow, then blood, then tissue eosinophilia until the peripheral process is resolved. Much recent investigation has focused on the identification of factors responsible for inducing eosinophils to shift from one compartment to another, i.e., the chemotactic factors that attract eosinophils from the blood to the tissues and the eosinopoietic factors that stimulate the marrow to increase eosinophil production.

Chemotactic Factors

During an inflammatory response, eosinophils migrate from the bloodstream to the tissues by diapedesis through the intercellular junctions of the endothelium. These openings define the vascular permeability. Thus, to form a cellular infiltrate, eosinophils depend on those same vasoactive amines and chemical mediators (e.g., histamine, kinins, prostaglandins) to increase vascular permeability as do other types of inflammatory cells. This fact probably accounts for the following observations made during the first 18 hours of a developing eosinophilic inflammatory response: (1) the infiltrating leukocytes are a mixed population of inflammatory cells; and (2) the number of eosinophils in these infiltrates is proportional to the number of such cells circulating in the blood. Once out of circulation, the inflammatory cells are subject to a variety of activating and chemotactic stimuli with varying degrees of specificity for the eosinophil. Five major types of eosinophil chemotactic factors have been identified: (1) those derived from mast cells; (2) from neutrophils; (3) from lymphocytes; or (4) from serum; and (5) those dependent on the physicochemical nature of the stimulus itself.

The mast cell can be triggered either immunologically through IgE-dependent mechanisms or nonimmunologically by anaphylotoxin (generated from complement), various surface active agents, or trauma. Among releasable contents of this cell are two agents chemotactic for eosinophils, histamine and eosinophil chemotactic factor of anaphylaxis (ECF-A). *In vitro* histamine is chemotactic for eosinophils over a rather restricted range of concentrations (Clark et al. 1975), but both *in vitro* and *in vivo,* it has been shown to inhibit migration of eosinophils and may thus serve as a localizing factor in much the same fashion that migration inhibitory factor (MIF) is thought to serve for macrophages (Chapter 5). Histamine's chemotactic effects appear to be restricted to eosinophils. Eosinophil chemotactic factor of anaphylaxis, on the other hand, is a tetrapeptide released from the mast cells; while preferentially attractive to eosinophils, it is also chemotactic to neutrophils (Kay et al. 1971). Either or both of these mediators could play an important role in determining the eosinophilia associated with immediate hypersensitivity reactions.

Several low molecular weight eosinophil chemotactic factors similar to ECF-A have been described from neutrophil, fibroblast, and various lung tumor sources. Exact chemical relationships among these various low molecular weight chemotactic factors, their cellular specificities, and their *in vivo* importance still remain subjects for investigation.

Lymphocytes have also been shown to produce products (lymphokines) of about

20,000 to 40,000 molecular weight that are chemotactic for eosinophils (Colley 1973). In one experimental system, sensitized guinea pig lymphocytes challenged with homologous antigen produced a factor, eosinophil chemotactic factor precursor (ECF$_p$), which when "activated" with homologous antigen–antibody complexes generated a moiety, ECF, that was specifically chemotactic for eosinophils. A second lymphokine, termed eosinophil stimulation promoter (ESP), has been generated from both mitogen-stimulated and specifically sensitized antigen-stimulated lymphocytes from mice or humans and found to be preferentially attractive for eosinophils. Though precise biologic roles for these eosinophil-attracting lymphokines are still undefined, an extensive series of observations has served to demonstrate the requirement for an intact thymus-dependent (T-cell) immune system for the generation of certain types of eosinophilic response, such as those found in parasitic helminth infections (Basten et al. 1970). Such data point clearly to a potential functional significance for both of the lymphocyte derived chemotactic moieties.

Serum and plasma contain a large number of potent chemotactic factors, most of which preferentially attract neutrophils and exert some effect on the eosinophil. These factors include kallikrein, plasminogen activator, neutrophil proteases, and tissue enzymes, but clearly the most significant of the serum-associated chemotactic factors are those derived from complement. C3a, C5a, and the trimolecular complex $C\overline{567}$ have all been shown to be chemotactic for eosinophils, but C5a (originally described as the eosinophil chemotactic factor derived from complement, ECF-C) is probably the most biologically significant of these complement-derived factors (Kay 1970). Since C5a can be generated by activation of either the classical or the alternative complement pathway (Chapter 15), its role in attracting eosinophils into inflammatory foci can be initiated both by immunologic means (during the union of antigen and antibody or by preformed immune complexes) and by nonimmunologic means (e.g., through molecular aggregates, endotoxin, or tissue injury). Clearly, these complement molecules are of major importance as chemotactic amplifiers in all types of inflammatory response and bear multiple potential relationships to those characterized by the presence of eosinophils.

In addition to these serum-derived chemotactic factors, there are a number of other molecules with demonstrable chemotactic activity for eosinophils. These include the histamine metabolite imadazole acetic acid, the platelet-derived hydroxyheptadecatrienoic acid, the prostaglandin precursors, and extracts of two parasitic worms, *Ascaris suum* and an *Anisakis* species. These factors all have been shown to have *in vitro* chemotactic activity for eosinophils, but their exact biologic role in the generation of eosinophilic inflammatory infiltrates *in vivo* is still uncertain.

Eosinophilopoietic Factors

For an eosinophilic inflammatory response to be sustained, the bone marrow must accelerate its normal production of eosinophils. In experimental models, it has been observed that eosinopoiesis involves enhanced proliferation of eosinophil precursor cells, a decrease in cell cycle time to permit additional divisions of these precursors, and a decrease in the eosinophil emergence time (the time spent in the maturational marrow reserve period of development). However, despite our knowledge of what happens, it is unclear what the factors are that initiate and regulate these accelerated marrow events.

That eosinophilopoietic factors do exist has been documented by *in vivo* experiments in which stimulated mice have been shown capable of generating factors that would induce an eosinophilia in other animals (Mahmoud et al. 1977). Other studies suggest that lymphocytes and monocytes may be the source of these eosinopoietic factors (Ruscetti et al. 1976; Cline and Golde 1979). Further understanding of the mechanisms involved in initiating and regulating eosinophil production by the bone marrow will require additional investigations into this very basic and important area of eosinophil development and kinetics. However, in view of the prominent eosinophilic reactions associated with allergic and parasitic diseases, such chemoattractants presumably play an important *in vivo* role.

Functions

A great many functional capabilities of the eosinophil have been described, but it is still not clear what it is about this cell that makes it uniquely suited for a still largely undefined role in the body's host defense. Current prevailing opinion would ascribe to it a function as modulator of inflammatory reactions (especially allergic) or as a cytotoxic effector important in the immune response to helminth parasites (Figure 6).

Phagocytic Capability

The phagocytic capabilities of the eosinophil have long been recognized in experiments demonstrating the ingestion of bacteria, mycoplasma, yeast, inert particles, and antigen-antibody complexes. However, when the relative efficiencies of eosinophils and neutrophils have been compared, eosinophils have performed consistently less well than neutrophils, both in rate of phagocytosis and in quantity of material ingested. Furthermore, bacterial killing by eosinophils is significantly less effective than that by neutrophils (Cline 1972). Thus, while it can function reasonably well as a phagocyte, it is difficult to imagine that this is the major functional role of the eosinophil.

Modulatory Capabilities

Early observations recording both the accumulation of eosinophils at the margins of inflammatory lesions and their presence in the blood during the healing phase of many infectious processes led to the hypothesis that these cells were, in some manner, modulators of inflammation. Indeed, experiments in ponies showed that prior injection of eosinophils into skin sites challenged subsequently with histamine, serotonin, or bradykinin effectively diminished the inflammatory responses to these mediators (Archer 1959). Furthermore, antigen–antibody complexes, themselves potent initiators of inflammatory responses, were earlier shown to be effectively phagocytosed (and thus detoxified) by eosinophils. Most recently, however, investigational focus has been on the mediator-rich mast cells and basophils important to allergic inflammation and on the properties of the eosinophil that appear to be especially suited for regulating and modulating these cells and their products (Goetzel et al. 1975).

Mast cell and basophil granules contain a number of different inflammatory mediators including histamine, slow reactive substance of anaphylaxis (SRA-A), ECF-A, and heparin. Interestingly, it has recently been shown that eosinophils, which, as discussed above, are attracted to regions of discharged mast cells, are uniquely equipped to inactivate many of the products of mast cell degranulation. The enzymes

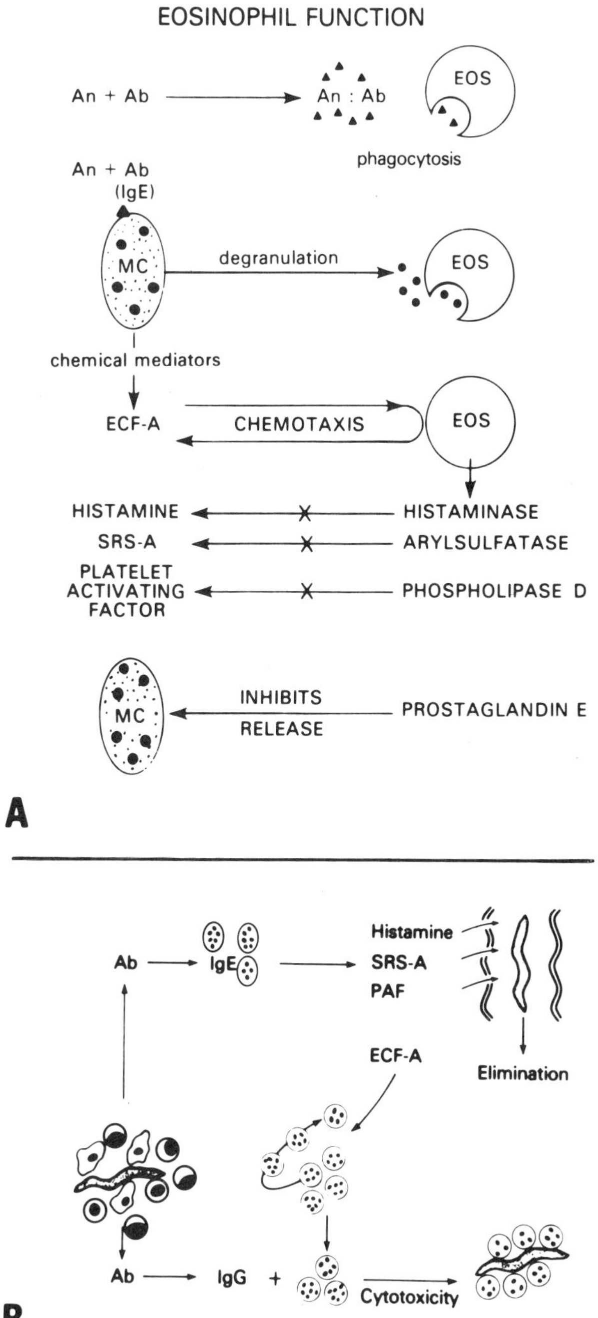

Figure 6. Diagrammatic representation of the eosinophil's functional potentials. (A) Ability to modulate mast cell (MC) derived allergic inflammation by phagocytosis of MC granules, response to and inactivation of MC chemical mediator products and inhibition of further granular release. (B) Host defense against parasite helminth infection. IgE immediate hypersensitivity reaction evokes MC derived vascular permeability factors affecting the helminth and chemotactic migration of eosinophils that participate in IgG antibody and complement dependent cytotoxicity.

histaminase and arylsulfatase (which degrades SRS-A) are, for example, far more abundant in eosinophils than in other types of leukocytes. Furthermore, heparin is rapidly neutralized by the MBP of eosinophils, and ECF-A has been shown to lose its activity after reacting with eosinophils. The latter probably occurs as a result of being degraded by a cell membrane-bound esterase.

Not only can the granule mediators be detoxified, but whole granules can also be ingested and presumably removed from affected sites (Welsh and Geer 1959). Furthermore, even the replenishment of histamine by the discharged mast cells appears to be inhibited by the presence of eosinophils localized in the tissues (Jones and Kay 1976). Finally, the very release of granules by basophils has been shown to be inhibited by prostaglandins (PGE_1 and PGE_2) elaborated by eosinophils (Hubscher 1975) during allergic stimulation. Thus, although precise evidence is still lacking that such is its major biologic function *in vivo,* the eosinophil does seem well equipped to play an important role in modulating allergically mediated inflammatory responses.

Cytotoxic Capabilities

The cytotoxic potential of the eosinophil has only recently been well appreciated, with renewed interest in this subject stimulated largely by observations of *in vitro* interactions between eosinophils and parasitic organisms. The association between eosinophils and parasitic helminth infections has been recognized almost since the first description of this cell type, but the meaning of this association has not been clear. In an interesting series of *in vitro* experiments designed to examine the cytotoxic potential of eosinophils against larval schistosome parasites, it was found that eosinophils were unique among human leukocytes in their ability to kill or damage the parasites (Butterworth et al. 1975). Characteristics of the response were similar to those described for antibody-dependent cellular cytotoxity (ADCC) reactions by other cell types (Chapter 7). Morphologically, eosinophils were seen to bind tightly to parasite surfaces (presumably via the cells' Fc receptors) and undergo degranulation onto the surface of the parasite (Figure 7). The MBP in these granules has been shown to be capable of damaging the parasite surface and thus may be responsible for the observed cytotoxic capacity of the eosinophil (Butterworth et al. 1979). Alternatively, it has been proposed that eosinophil peroxidase in the granules reacts with H_2O_2 and a halide to generate a toxic complex capable of damaging or killing the parasites (or other target cells) and that this mechanism may be primarily responsible for the cell's cytotoxic potential (Jong and Klebanoff 1980).

These observations and others with different experimental systems leave little question about the effectiveness of the eosinophil as a cytotoxic cell *in vitro*. Evidence for the importance of these observations to the host inflammatory response *in vivo,* however, has been more difficult to obtain. There is abundant documentation that eosinophils accumulate in tissues invaded by helminth parasites. However, the most convincing evidence that these cells are important in killing these invading parasites *in vivo* comes from the use of antieosinophil serum (AES) to render mice eosinophil-openic, and then challenging them with schistosome parasites (Mahmoud et al. 1975). After AES, previously established immunity to rechallenge was completely abolished. These observations suggest that the eosinophil, presumably through its cytotoxic capacity, plays an important role in protective immunity against at least this one important helminth parasite. Recently, similar observations using AES have been made with infection of another species (guinea pigs) by a different helminth parasite,

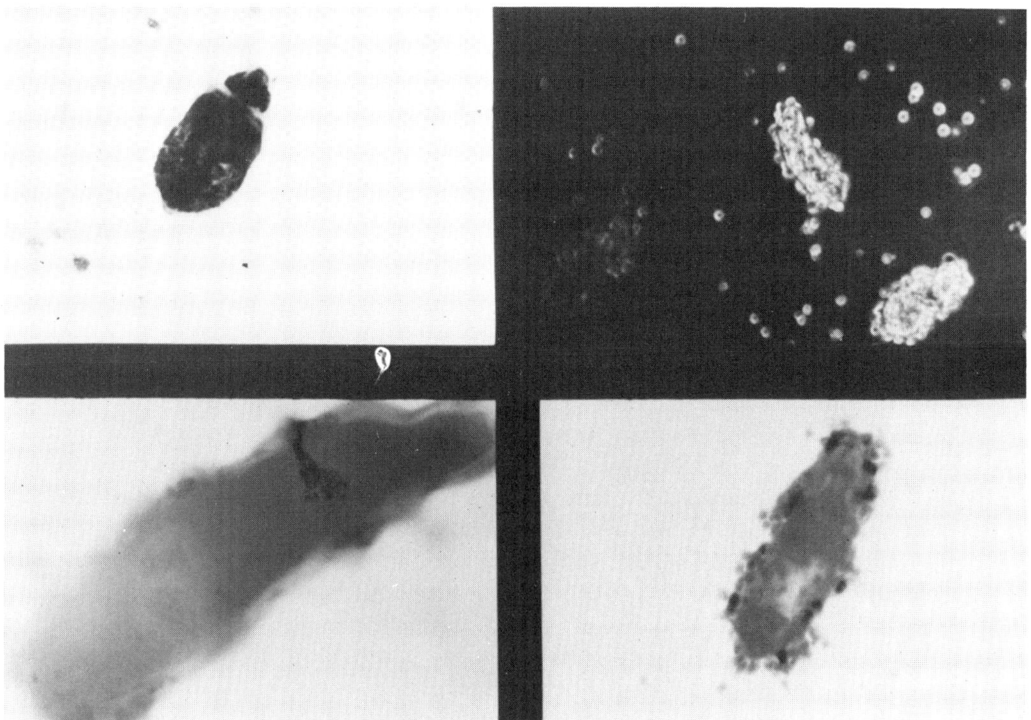

Figure 7. Eosinophil adherence and degranulation on parasite surface (Ottesen et al. 1977). Eosinophils (darkly stained cells) and other leukocytes will not adhere to schistosomules in the presence of normal, heat-inactivated serum (upper left). In the presence of either complement or antibody, however, there is marked cellular adherence (upper right, phase contrast micrograph; lower right, stained permanent preparation). Adherent eosinophils release their granules on the parasite surface (lower left) in a reaction that presumably leads to the recently described eosinophil cytoxicity towards schistosomules (Butterworth et al., 1975).

Trichostrongylus colubriformis, (Gleich et al. in press) thereby confirming the importance of the original observations in defining a cytotoxic role for eosinophils in protection against helminth infections.

Other Immunologic Capabilities

The eosinophil is distinctly different from the other major granulocyte, the neutrophil, in its much closer association with the lymphoid immune system. In fact, an elegant series of experiments on trichinella-infected rats, as well as studies of schistosomiasis and certain other experimental situations, have shown that a functional T-cell immune system is required for the development of eosinophilia. This includes both the primary response following an eosinophilogenic stimulus and the accelerated secondary response developing after reexposure.

A primary role in the efferent limb of the immune response has been suggested by the modulating and cytotoxic capabilities of the eosinophil, as discussed above. However, earlier observations of a close association between eosinophils and macrophages observed morphologically in a number of settings led to speculation that the eosinophil may also play a role in the afferent limb of the immune response. One suggestion was that the eosinophils might be important in the handling of certain types of antigens before they were presented to macrophages for further immunologic processing. Another was that eosinophils might even generate certain types of antibodies. However, in view of the recent evidence emphasizing the importance of the inflammatory modulating and cytotoxic capabilities of the eosinophil, it is probably more likely that this cell's unique functional role lies in an aspect of the efferent limb of the host's immune system, and not in one of these potential afferent functions.

Physiologic and Pathologic Correlates with Eosinophil Properties and Function

As described in the previous section, experimental data on eosinophil biology has defined cellular properties that seem to endow this leukocyte with potential for two diametrically opposed functions. The content of enzymes and basic protein in eosinophil granules capable of cytotoxic action on helminth parasites under experimental conditions suggests that these same materials could have similar destructive effects on autologous tissues. On the other hand, identification of eosinophil products capable of inactivating or inhibiting the chemical mediators of inflammation derived from mast cell release provides a basis for the concept of a modulating function (Figure 6). Illustrative of this dichotomy were two early observations. One, by Paul Ehrlich, noted that the return of eosinophils to the circulation during the course of typhoid fever constituted a favorable prognostic sign (probably indicative of beneficial immune activity). The other relates to the era of arsenical treatment for syphilis when developing eosinophilia served as a premonitory sign to anticipate a severe untoward drug hypersensitivity reaction (also probably caused by an immune phenomenon). Thus, apart from the eosinophil's possible role in host defense against helminth infections, the exact significance of eosinophilia under both physiologic and pathologic conditions requires further definition.

Disorders with Associated Influences on Eosinophil Regulatory Mechanisms

The discussion that follows is intended to relate the earlier section on eosinophil biology to eosinophil expression both under normal conditions and in disordered states. For the reader who may be interested in specific literature citations or in an understanding of the clinical significance of eosinophilia in greater depth than can be covered within the scope of this review, more detailed treatment of the subject can be found in another publication by the authors (Ottesen and Cohen 1978) and in other comprehensive textbook material (Ottesen 1976; Beeson and Bass 1977; Cline 1975).

In view of the effects of myeloproliferative stimuli, inflammatory events, complement generation, and T-lymphocyte regulatory mechanisms on eosinophil processes, it is only to be expected that some degree of peripheral blood and/or tissue eosinophilia may be demonstrable in a variety of pathophysiologic expressions associated with these biologic influences. On the other hand, because of the relatively more specific interrelationships of IgE and eosinophil function, significantly elevated eosinophil levels are most consistently associated manifestations with immediate hypersensitivity states and parasitic helminth tissue invasion.

While it is not known exactly how T-cell function may be involved in eosinopoiesis (Cline and Golde 1979) or related eosinophil regulatory mechanisms, loss of such control may be a responsible factor for the increased eosinophil levels evident in T-lymphocyte disorders (T-cell lymphomas, immunodeficiency syndromes characterized by diminished T-cell function and associated hyperimmunoglobulinemia E), long-term eosinophilia following x-irradiation of the thymus during the neonatal period, and eosinopenia characteristic of certain thymic tumors (thymoma with hypogammaglobulinemia).

In addition to selective eosinopoietic influences on bone marrow precursors, enhanced eosinophil production may result from less specific hematologic disorders. Thus, circulating eosinophilia may be seen during early and presumably general bone marrow stimulating stages of myeloproliferative disease evolving into leukemia of another granular leukocyte type (myelogenous leukemia). Eosinophil production may also be a product of a pancellular myeloproliferative disorder such as polycythemia vera.

The importance of the adrenal's hormonal influence on levels of circulating blood eosinophils is demonstrated in disordered states of this endocrine gland, eosinophilia with loss of function (Addison's disease), and eosinopenia with adrenal hyperfunction. This fact can be applied usefully in tests of adrenal competency designed to look for decreases in circulating levels after stimulatory administration of adrenocorticotropic hormone (Thorn test). Temporary adherence of blood borne eosinophils to vascular endothelial surfaces as a result of the action of released adrenal cortical hormone accounts for this reversible eosinopenic effect.

Effects on blood-tissue distribution of eosinophils via sympathetic nervous system influences can also be demonstrated in pathophysiologic situations. A variant of asthma of unknown cause ("intrinsic asthma") in which beta-adrenergic blockade is believed to play a contributory role is characterized by even higher levels of eosinophilia than are found in patients with asthma in whom allergic factors can be identified. Conversely, administration of a beta-adrenergic-stimulating agent (e.g., epinephrine) can effect a temporary lowering of eosinophil levels.

Until enhanced eosinopoiesis takes place, eosinophils responding to increased tissue demand can only be derived from the circulating pool. Thus, a lag in time between bone marrow receipt of eosinopoietic signals and the availability of above normal cellular blood levels may result in a paradoxical circulating eosinopenia in the face of eosinophil-populated tissue lesions. Conversely, time lags between reversal of hyperproliferation from precursor cells and bone marrow discharge of diminishing numbers of eosinophils can be responsible for the picture of paradoxical blood eosinophilia when local tissue eosinophil lesions are resolving (Figure 5).

Because certain tissues presumably serve as normal filtration beds for elements of the circulating eosinophil leukocyte pool, the finding of small collections of these cells in respiratory, gastrointestinal, and lower urinary tract locations, and in skin and lymphoid tissue (especially the appendix) are not reflections of local disorders. Likewise, a few eosinophils contained in secretory and mucus products of these organs (e.g., nasal mucus, sputum, urine) can be appreciated as only washout effects. There are instances in which exaggeration of this phenomena can be misinterpreted. A striking example is furnished by the picture of eosinophilic infiltrations in the stomach or lung (the latter even large enough to cast shadows on roentgenograms), with cells derived apparently just from increased circulating levels of eosinophils in disorders that do not primarily involve these specific organ sites. In the case of diversion of circulating eosinophils to the uterus during estrus, a physiologic explanation may lie in the estrogen receptors attributed to this cell.

Immune and Hypersensitivity Reactions

Eosinophils may appear among those cellular elements of inflammatory infiltrates primarily involving polymorphonuclear neutrophils (PMNN) and other leukocyte types. As by-products of antigen–antibody activation of the complement cascade, vasoactive histamine and kinins are released. The increases in vascular permeability permit migration of all leukocytes through the affected vessels in numbers that reflect their circulating levels. Thus, a few eosinophils present under such conditions are without specific implication.

Among the granular leukocytes, the primacy of the PMNN for phagocytosis of microorganisms during bacterial invasion and of molecular complexes generated in antigen–antibody reactions is readily demonstrable. However, these processes also lie within the eosinophil's capacity. Thus, eosinophils can often be demonstrated within cell populations responding to bacterial infections and within lesions secondary to immune complex deposition (Figure 8). In some diseases, processes that impair bone marrow production and/or discharge of PMNN (e.g., cyclic neutropenia) lead to increases of eosinophils during neutropenic phases. Though speculative, perhaps eosinophils can effect compensatory host defense functions in such situations.

Since eosinophilia has long been regarded as a unique feature of immune reactions characterized by humoral antibody activity, instances of its occurrence with manifestations of delayed hypersensitivity were confusing. However, with the *in vitro* demonstration of eosinotactic properties of certain lymphokines (ESP, ECF_p), the appearance of small numbers of eosinophils migrating into lesions of delayed hypersensitivity (e.g., allergic contact dermatitis such as poison ivy) along with the mononuclear cells that characterize cell-mediated immune processes may now be better understood.

Of disorders due to immune mechanisms, the highest degrees of eosinophilia are seen with those conditions in which pathology is associated with immediate hypersensitivity mechanisms (e.g., allergic asthma and hay fever) (Figure 4). The identification of IgE as the homocytrophic antibody, its localization on basophil and mast cell surfaces, its role in triggering pathophysiologically responsible (humoral) mediators, and the *in vitro* demonstration that one or more of these mediators has chemotactic properties for eosinophils may now provide leads for understanding these associations. If *in vivo* correlates can be established, a beneficial role may be ascribed to the eosinophil by virtue of its potential for reversing SRS-A-induced bronchospasm in asthma, and reversing the effects that histamine and platelet-activating factors have in inducing the edema of hives and allergic rhinitis. However, the conditions for conversion of the eosinophil's potential for tissue insult (residing in the granule contents) to an actuality is still poorly understood. It is clear that the highest levels of eosinophilia are demonstrated in the most severely affected allergic individuals. Thus, the question arises as to whether the very large numbers of eosinophils are migrating to sites of tissue involvement in a functional attempt to modulate an overwhelming chemically mediated allergic reaction, or whether the severity of the pathologic process is a direct reflection of the eosinophil's harmful inflammatory capability and activity.

Immune reactivity may explain some of the instances of eosinophilia accompanying drug administration. Because some therapeutic agents are proteins derived from biological sources (e.g., penicillin, streptomycin, insulin, antitoxins) and can act as antigens, are chemically structured to act as haptens, or may simulate the effects of

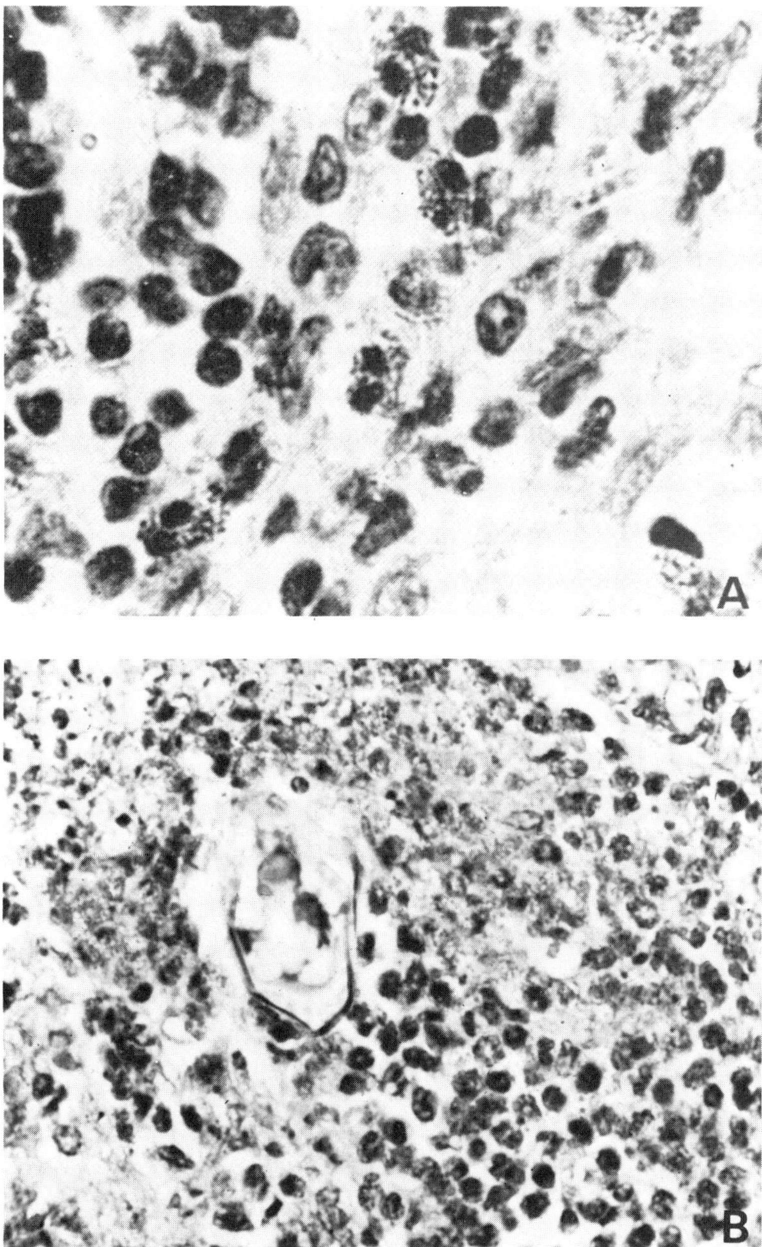

Figure 8. Experimental models of eosinophilia; tissue sections, hematoxylin and eosin (original magnification × 125). (**A**) Guinea pig popliteal lymph node. Eosinophils infiltrating into medullary sinus secondary to deposition of gamma globulin aggregates introduced into foot pad. (Reproduced with permission from Cohen, S.G., Sapp, T.M., Rizzo, A.P., and Kostage, S.T., (1964) J. Allergy 35:349.) (**B**) Mouse liver. Eosinophil participation in granuloma formation surrounding an ovum in experimental *Schistosoma mansoni* infection. (Courtesy S.L. James and D.G. Colley)

antigen–antibody complexes because of high molecular weight, they can play responsible roles in immediate hypersensitivity or immune complex reactions. Even though overt allergic drug reactions may not ensue, development of eosinophilia during the course of therapy that is not attributable to the underlying disease is commonly recognized as indicative of immunologic reactivity involving the therapeutic agent.

Inflammation

Apart from the influences of immune and allergic reactions, eosinophils may participate in tissue inflammation effected by nonspecific evoking agents, for example: (1) complement-activating processes secondary to trauma, chemical and thermal effects, microbial lipopolysaccharides of gram-negative organisms, and high molecular weight compounds and aggregates simulating antigen–antibody complexes in physicochemical character; (2) release of chemical mediators of inflammation with enhancement of vascular permeability by a variety of tissue injurious agents, e.g., histamine, serotonin, prostaglandins, and kinins. In infants, eosinophils can be identified within traumatically induced inflammatory exudates, in which case reduced immune capability of T-lymphocytes and immaturation of immunologic controls might be possible factors.

Because of their differences in phagocytic behavior, there are striking differences in the effects of cellular populations of eosinophils and of PMNN on the tissues infiltrated by each cell type. In contrast to the purulence associated with PMNN autolysis and tissue necrosis resulting from intracellular action with leakage of granule proteolytic enzymes, secretions, and histologically unaltered tissues, structurally intact eosinophils are characteristic of eosinophil infiltrates. There are, however, for reasons not at all understood, some situations where eosinophil-related processes lead to different pathologic consequences. In extreme myeloproliferation (the leukemia-like hypereosinophilic syndrome), extensive space-occupying cell effects impair vital functions of densely infiltrated heart and lungs. Also, in some disorders it is conceivable that once having entered sites of tissue demand, pathogenic mechanisms may exist for triggering the release of granule enzymes and MBP with resultant initiation, perpetuation, or aggravation of irreversible tissue injury. Situations in which this may occur are the following: (1) the purulent and inflammatory tissue changes accompanying hyperplastic sinusitis and asthma of nonallergic causes, both characterized by marked eosinophil infiltrations; (2) complications of focal necrosis and endomyocardial fibroelastosis of the eosinophil-infiltrated heart in the hypereosinophilic syndrome; and (3) the experimental neurotoxicity of eosinophils, which could possibly relate to the central nervous system pathology seen in some patients with the hypereosinophilic syndrome.

Mast Cell–Eosinophil Associations

Functional relationships between mast cells and eosinophils may serve as contributory determinant factors of tissue eosinophil localization. While the interaction of antigen and membrane-bound IgE at the mast cell surface is the most familiar mechanism for release of the eosinophil chemotactic factors ECF-A and histamine, mast cell discharge can occur under nonimmunologic influences. Degranulating properties of a variety of chemical compounds (codeine, morphine, ovomucoid, polymyxin, stilbamadine) and of trauma (e.g., the stroking of skin mast cell tumors) have been recognized and in turn endow these agents with the capacity to effect eosinophil migrations.

Pertinent to putative mast cell-eosinophil association is an examination of some of the situations and circumstances under which eosinophils accumulate in tissues. The organs in which cell deposition occurs as a late phase event in completion of the eosinophil's life cycle are those that are characterized by large mast cell populations (respiratory and gastrointestinal tracts, skin, and uterus). While increases in levels of circulating eosinophils would be expected to be reflected by proportionally increased tissue cell infiltrations at these normal deposition sites, it is noteworthy that the locations of the largest eosinophil accumulations correspond to areas with normally occuring high mast cell density (e.g., pleural surfaces and upper fields of the lung). The phenomenon of mast cell hyperproliferation (or possible basophil accumulations) is a recognized accompaniment of inflammation, which could play an attractant role in the development of the eosinophilia resulting from tumor growth and traumatic processes in mast cell rich organs and at serosal surfaces. Additionally, where hypersensitivity processes and immune complex deposition involve the vasculature, bronchi, kidney, and liver, eosinophils are identified in mast cell-populated areas, i.e., adventitia of small blood vessels, peribronchial sites, renal interstitium, and perilobular hepatic connective tissue, respectively.

Infection

Though there is reason to believe that eosinophils may participate in some aspects of host denfense, there are noteworthy differences in patterns of this cell's response to bacterial and parasitic infections.

Reduction in levels of circulating eosinophils are usually noted to occur during acute phases of bacterial infection. Although exact reasons are not known, explanations advanced have centered on: (1) associated stress-phenomenon effects on the adrenal gland with increased cortical hormone secretion; (2) increased demands on the bone marrow for PMNN production at the expense of other cellular elements; and (3) the existence of an eosinopenic factor of infection of still undefined character. Eosinophils, however, reappear during later or resolving stages of infectious processes (typhoid fever, pneumococcal pneumonia, scarlet fever) and in such situations are presumed to reflect immune phenomena associated with increasing antibody production and immune complex formation and/or hypersensitivity reactions to microbial products. In the case of parasitic infection, eosinophilia as a prominent feature is limited to those infectious processes caused by tissue invasive helminths. That these organisms furnish large persistent depots of complex antigens may be a critical factor in these situations. However, a fact that seems especially relevant is the association of elevated levels of IgE and specific reaginic antibodies with those parasitic helminths with which eosinophilia is identified (e.g., hookworm, ascarids, strongyloides, schistosomes, trichinella). These likely are indicators of the organism's ability to effect immediate hypersensitivity states, a situation not found with bacteria, protozoa, or even other helminths that are not tissue invasive. Other demonstrated potential eosinophil-evoking mechanisms associated with these infections include complement activation and the local mast cell hyperproliferation (or basophil accumulations) that develop at sites of helminth invasion.

Further studies on the role of IgE in helminth infections and the experimental model of antibody-mediated eosinophil cytoxicity should help to clarify a still incompletely understood point. Whether, as suggested, the eosinophil's primary role is one of host defense and its involvement in allergic reactions is as an evolutionary by-product, or whether its primary role is only one of an inflammatory-modulating cell in both allergic disorders and the inflammatory consequences of parasitic infections, requires definition.

It has been postulated that IgE-mediated host defense against certain types of infection represents an early evolutionary development. With the advent of appropriate public health measures and the resultant disappearances of challenge by tissue-invading helminths, the human IgE mechanism was then directed against the challenges of other exposures, e.g., air-borne pollen grains, food and animal products, and insect venoms. Under the influence of the earlier genetic programming, allergic phenomena resulted. Thus, the common denominator of eosinophilia may too represent a product of the evolution of IgE function.

Conclusion

Eosinophils can be readily distinguished from other inflammatory cells on the basis of structure, function, and the situations in which they are associated. Morphologically the eosinophil's most distinctive feature is the characteristic acidophilic granule containing a "major basic protein" and peroxidase among other enzyme components. Both of these appear to be predominating effectors of the antibody-dependent cytotoxicity eosinophils show against helminth parasites, a function which may be one of the most important of this cell. Alternatively, it has been suggested that because of its enzyme contents and products potentially capable of inactivating many of the important chemical mediators of inflammation, the eosinophil's primary functional role is modulating mast cell triggered inflammatory responses, especially those of IgE derivation.

While the eosinophil reacts *in vitro* to most leukocyte chemotactic factors, certain ones derived from mast cells, lymphocytes and parasite worms are relatively specific for this cell. In *in vivo* models and clinical settings eosinophilia is seen most commonly in association with helminth infections and allergic disorders, though eosinophil infiltrates are often found in a variety of diseases which generally share common features of normal function or disorders of the immune system. The reason may be that the eosinophil is unique among granulocytes in its dependence on the T-cell arm of the immune system for initiation and regulation of cellular production by the bone marrow.

References

Anwar, A.R.E., Kay, A.B. (1978) Enhancement of human eosinophil complement receptors by pharmacologic mediators. J. Immunol. 121,1245–1250.

Archer, R.K. (1959) Eosinophil leukocytes and their reactions to histamine and 5-hydroxytryptamine. J. Pathol. Bact. 78,95–103.

Basten, A., Boyer, M.H., Beeson, P.B. (1970) Mechanism of eosinophilia I. Factors affecting the eosinophil response of rats to *Trichinella spiralis*. J. Exp. Med. 131,1271–1287.

Beeson, P.B., Bass, D.A. (1977) *The Eosinophil.* Philadelphia, W.B. Saunders Company.

Butterworth, A.E. (1977) The eosinophil and its role in immunity to helminth infection. Curr. Top. Microbiol. Immunol. 77,127–168.

Butterworth, A.E., Sturrock, R.F., Houba, V., Mahmoud, A.A.F., Sher, A., Rees, P.H. (1975) Eosinophils as mediators of antibody-dependent damage to schistosomula. Nature 256,727–729.

Butterworth, A.E., Wassom, D.W., Gleich, G.J., Loegering, D.A., David, J.R. (1979) Damage to schistosomula of *Schistosoma mansoni* induced directly by eosinophil major basic protein. J. Immunol. 122,221–229.

Clark, R.A.F., Gallin, J.I., Kaplan, A.P. (1975) The selective eosinophil chemotactic activity of histamine. J. Exp. Med. 142,1462–1476.

Cline, M.J. (1972) Microbicidal activity of human eosinophils. RES. J. Reticuloendothel. Soc. 12, 332–339.

Cline, M.J. (1975) *The White Cell* Cambridge, Mass., Harvard University Press, pp. 104–121.

Cline, M.J., Golde, D.W. (1979) Cellular interactions in haematopoiesis. Nature 277,177–181.

Colley, D.G. (1973) Eosinophils and immune mechanisms I. Eosinophil stimulation promoter (ESP): a lymphokine induced by specific antigen or phytohemagglutinin. J. Immunol. 110,1419–1423.

Dale, D.C., Hubert, R.T., Fauci, A. (1976) Eosinophil kinetics in the hypereosinophilic syndrome. J. Lab. Clin. Med. 87,487–495.

Dao, C., Metcalf, D., Bilski-Pasquier, G. (1977) Eosinophil and neutrophil colony-forming cells in culture. Blood 50,833–839.

Gleich, G.J., Loegering, D.A., Mann, K.G., Maldonado, J.E. (1976) Comparative properties of the Charcot-Leyden crystal protein and the major basic protein from human eosinophil granules. J. Clin. Invest. 57,633–640.

Gleich, G.J., Olson, G.M., Herlich, H. The effect of antiserum to eosinophils on susceptibility and resistance of the guinea pig to *Trichostrongylus colubriformis.* Immunology, in press.

Goetzel, E.J., Wasserman, S.I., Austen, K.F., (1975) Eosinophil polymorphonuclear leukocyte function in immediate hypersensitivity. Arch. Pathol. 99,1–4.

Hubscher, T. (1975) Role of the eosinophil in the allergic reactions II. Release of prostaglandins from human eosinophilic leukocytes. J. Immunol. 114,1389–1393.

Jones, D.G., Kay, A.B. (1976) The effect of anti-eosinophil serum on skin histamine replenishment following passive cutaneous anaphylaxis in the guinea pig. Immunology, 31,333–336.

Jong, E.C., Klebanoff, S.J. (1980) Eosinophil-mediated mammalian tumor cell cytotoxicity: Role of the peroxidase system. J. Immunol. 124,1949–1953.

Kay, A.B. (1970) Studies on eosinophil leucocyte migration II. Factors specifically chemotactic for eosinophils and neutrophils generated from guinea-pig serum by antigen-antibody complexes. Clin.Exp.Immunol. 7,723–737.

Kay, A.B., Stechsculte, D.J., Austen, K.F. (1971) An eosinophil leukocyte chemotactic factor of anaphylaxis. J. Exp. Med. 133,602–619.

Mahmoud, A.A.F., Stone, M.K., Kellermeyer, R.W. (1977) Eosinophilopoietin. J. Clin. Invest. 60,675–682.

Mahmoud, A.A.F., Warren, K.S., Boros, D.L. (1973) Production of a rabbit antimouse eosinophil serum with no cross-reactivity to neutrophils. J. Exp. Med. 137,1526–1531.

Mahmoud, A.A.F., Warren, K.S., Peters, P.A. (1975) A role for the eosinophil in acquired resistance to *Schistosoma mansoni* infection as determined by antieosinophil serum. J. Exp. Med. 142,805–813.

Miller, F., de Harven, E., Palade, G.E. (1966) The structure of eosinophil leukocyte granules in rodents and man. J. Cell. Biol. 31,349–362.

Ottesen, E.A. (1976) Eosinophilia and the lung. In: *Immunologic and Infectious Reactions in the Lung.* Kirkpatrick, C.H., Reynolds, H.Y., (eds.) New York, Marcel Dekker, pp. 289–332.

Ottesen, E.A., Cohen, S.G. (1978) The eosinophil, eosinophilia, and eosinophil-related disorders. In: *Allergy: Principles and Practice.* Middleton, E., Ellis, E., Reed, C.E., (eds) St. Louis, C.V. Mosby Company, pp. 584–632.

Ottesen, E.A., Stanley, A.M., Gelfand, J.A., Gadek, J.E., Frank, M.M., Nash, T.E., Cheever, A.W. (1977) Immunoglobulin and complement receptors on human eosinophils and their role in cellular adherence to schistosomules. Am. J. Trop. Med. Hyg. 26,134–141.

Parmley, R.T., Spicer, S.S. (1974) Cytochemical and ultrastructural identification of a small type granule in human late eosinophils. Lab. Invest. 30,557–567.

Ruscetti, F.W., Cypess, R.H., Chervenick, P.A. (1976) Specific release of neutrophilic- and eosino-philic-stimulating factors from sensitized lymphocytes. Blood 47,757–765.

Rytömaa, T. (1960) Organ distribution and histochemical properties of eosinophil granulocytes in the rat. Acta Path. Microbiol. Scand. (Suppl.) 140,1–118.

Spry, C.J.F. (1971) Mechanism of eosinophilia. V. Kinetics of normal and accelerated eosinopoiesis. Cell Tissue Kinet. 4,351–364.

Welsh, R.A., Geer, J.C. (1959) Phagocytosis of mast cell granules by the eosinophilic leukocyte in the rat. Am. J. Path. 35,103–112.

Zucker-Franklin, D. (1974) Eosinophil function and disorders. Adv. Int. Med. 19,1–25.

The Macrophage

David L. Rosenstreich, M.D.

The macrophage is a relatively large, phagocytic cell that belongs to a family of cells which are collectively referred to as "mononuclear phagocytes." The importance of these cells in inflammation was first appreciated by Metchnikoff in the 1880s, and subsequent research has revealed that they play an essential role in many different types of immune and inflammatory reactions. The characteristic that renders macrophages unique is that unlike most cell types, these cells subserve multiple functions. Thus, they are important effector cells, responsible for killing intracellular parasites and tumor cells; they act as scavengers for foreign material and extracellular debris, and, in addition, they act as regulators of immune responsiveness.

A second major characteristic that distinguishes the macrophage from cells to which it is related, such as the neutrophil, is the fact that macrophages undergo a process of differentiation in response to environmental change, that results in an alteration of their morphology, physiology, and functional capabilities. Appreciation of this process, the end result of which is "macrophage activation," is essential to understanding the functional role of macrophages.

A third characteristic of the macrophage, is that this cell is involved in an intimate cycle of interactions with lymphocytes (schematically outlined in Figure 1). Thus, macrophages are essential to the activation of lymphocytes through a variety of mechanisms (to be discussed in later sections). In turn, lymphocytes, via molecular messengers called lymphokines, are the major inducers of "macrophage activation." Activated macrophages reciprocally release increased amounts of lymphostimulatory molecules, but can also act as suppressor cells that serve to regulate this cycle. Like the concept of activation, appreciation of this intimate interaction between macrophages and lymphocytes is helpful in clarifying an otherwise confusing array of seemingly unrelated phenomena.

In this chapter, the subject of macrophages will therefore be divided into four topics. The first will include a description of the morphology, physiology, and functional capabilities of these cells. The second will discuss the nature of "macrophage activation" and the types of stimuli that induce this change. The third section will be

From the Department of Medicine and the Department of Microbiology and Immunology, Albert Einstein College of Medicine, Bronx, New York.

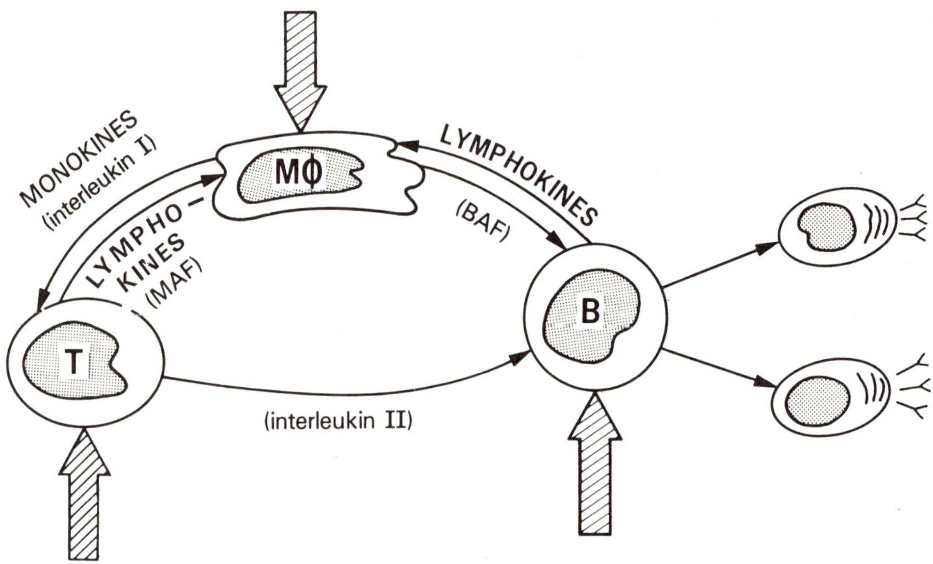

Figure 1. The cellular basis of the immune response. Schematic model for the cellular interactions involved in generating the immune response. Large arrows indicate external stimuli such as foreign proteins, microorganisms, or tumor cells.

a review of macrophage effector functions, and the last section will deal with the regulatory role of macrophages in the immune response. Macrophage function and physiology has been covered in several books and review articles, several of which are listed in the bibliography. The reader is encouraged to consult these whenever more detailed information on any area is required.

Macrophage Properties

Ontogeny

A scheme outlining the development of macrophages is given in Table 1. All macrophages are derived from the bone marrow, from the same precursor stem cell that gives rise to the cells of the granulocyte series (Virolainen 1968). Within the marrow, stem cells differentiate into promonocytes that possess the identifiable macrophage properties of glass adhesiveness, phagocytic ability, and immunoglobulin receptors (Cline and Summer 1972). Promonocytes undergo cell division and further differentiation to become monocytes. Monocytes then leave the marrow and circulate in the blood.

The peripheral tissues contain a number of cells bearing different names, all of which are macrophage-like and derived from the circulating monocyte. Among these are included Kupffer cells in the liver, alveolar macrophages, and connective tissue histiocytes. Together, these cells make up the reticuloendothelial system (RES). In addition, there are some cells, such as osteoclasts and nervous system microglia cells, whose structure and function suggest they are of macrophage origin. Under normal

Table 1. Ontogeny of Macrophages

Cells	Localization
Stem Cell ↓	Bone Marrow
Monoblast ↓	Bone Marrow
Promonocytes ↓	Bone Marrow
Monocytes ↓	Bone Marrow, blood
Macrophages	Connective tissue (histiocytes) Liver (Kupffer cells) Lung (alveolar macrophages) Spleen (free and fixed macrophages) Lymph node (free and fixed macrophages) Bone Marrow (macrophages) Serous Cavity (pleural and peritoneal macrophages) Bone Tissue (osteoclasts?) Nervous System (microglial cells?) Skin (Langerhans cells?) Lymphoid Tissue (Dendritic cells?)

conditions, monocytes circulate to the tissues where their subsequent differentiation occurs. A similar process occurs during inflammation but at a much accelerated rate. In addition, prolonged, chronic inflammatory stimulation results in the development of a "granulomatous reaction" in tissues. The cells that characterize this lesion include epithelioid cells and multinucleated giant cells. These cells are also derived from monocytes (Sutton and Weiss 1966).

Morphological, Histochemical, and Functional Characteristics

The morphology of mononuclear phagocytes varies with their state of differentiation. The peripheral blood monocyte is a relatively large (15μm) cell with a single kidney-shaped nucleus. The cytoplasm is abundant relative to the nucleus and is rich in rough and smooth endoplasmic reticulum, lysosomal granules, and has a well-developed Golgi apparatus, all of which are consistent with a metabolically active cell that is producing and secreting proteins constantly. The cell surface is usually marked by a profusion of ridge-like processes. As monocytes differentiate into macrophages of various types, they tend to enlarge their cytoplasm, with an increase in both lysosomal granules and endoplasmic reticulum (Figure 2).

There are a number of functional and histochemical properties that characterize the macrophage and are useful in its identification, separation, and characterization (Table 2). The distinguishing hallmark of this cell is its phagocytic ability. Cells from the promonocyte stage onward will *phagocytose* particles, through a nonspecific mechanism (i.e., latex beads) and via surface Fc receptors (IgG-coated particles). Details of this complex, but fascinating process are covered elsewhere (Silverstein et al. 1978). In addition, all cells of this series are adherent to glass or plastic surfaces.

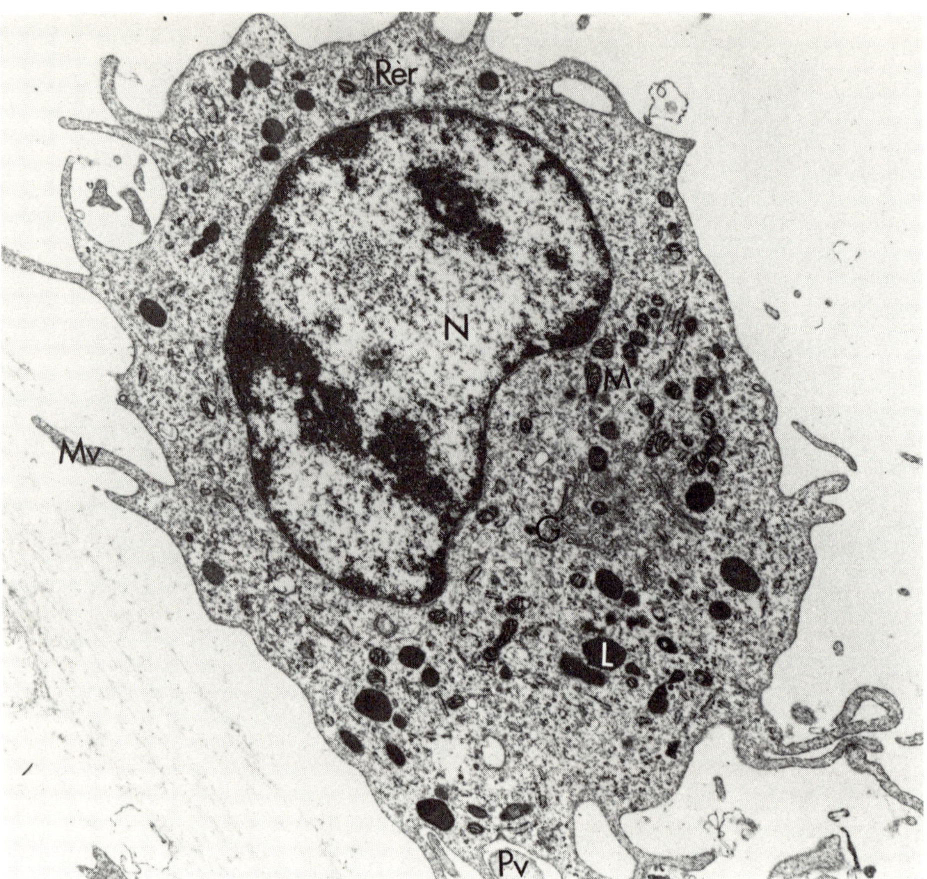

Figure 2. Rabbit alveolar macrophage. Transmission electron micrograph of a normal rabbit alveolar macrophage. N = nucleus; G = Golgi apparatus; M = mitochondria; L = lysosome; PV = phagocytic vacuole; MV = microvillus; RER = rough endoplasmic reticulum.

There are several intracellular or surface enzymes that can be used to characterize macrophages. In general, most macrophages contain a nonspecific cytoplasmic esterase that distinguishes them from neutrophils or lymphocytes. As macrophages differentiate, they undergo alterations in certain enzymes. Peroxidase is absent in promonocytes but develops and is present within the rough endoplasmic reticulum of peripheral blood monocytes. Interestingly, it decreases with further differentiation so that it is usually absent in resident peritoneal macrophages. Because of this property, peroxidase activity can be used as a marker for cells that are newly arrived at an inflammatory site. Although the exact pattern of enzyme activity varies between species (Bodel et al. 1977), 5'-nucleotidase is a macrophage ectoenzyme that increases with differentiation, is present within resident peritoneal macrophages, but is absent from activated macrophages. The loss of 5'-nucleotidase from "activated macrophages" is due to an increased degradation of the enzyme (Edelson and Cohn 1976). Macrophages, but not granulocytes, also have a membrane-bound aminopeptidase, which increases as macrophages differentiate *in vitro* (Wachsmuth and Staber 1977). With the exception of the cytoplasmic esterase that is present in and is a useful marker for all mononuclear phagocytes, the other enzymatic markers may be potentially useful for identifying

Table 2. Distinguishing Functional and Morphological Characteristics of Macrophages

Property	Comment
Functional	
Surface adherence	
Phagocytosis	Through Fc or C3b receptors, or nonspecific
Pinocytosis	mechanisms.
Histochemical	
Peroxidase	Located in rough endoplasmic reticulum. Present in monocytes but diminishes with further differentiation.
Esterase (cytoplasmic)	
5'-nucleotidase (ectoenzyme)	Decreases with activation
Aminopeptidase (ectoenzyme)	Increases with activation
Alkaline phosphodiesterase I (ectoenzyme)	Increases with activation
Antigens	
Macrophage specific antigen	
Activated macrophage antigen	
Ia antigens	There are Ia positive and negative subpopulations

macrophage subpopulations and/or status of differentiation. Unfortunately, these have not as yet been sufficiently well characterized to serve this purpose.

Macrophages also contain several identifiable surface antigens, and they can therefore be distinguished from lymphocytes by using a specific antimacrophage serum. In addition, there appears to be a unique antigen present only on the surface of "activated macrophages," e.g., antibodies specific for activated macrophages can be obtained. Subpopulations of macrophages contain surface Ia antigens as well. This surface component is involved in the immunoregulatory properties of macrophages (described in a following section).

Receptors

In addition to the already discussed surface antigens and enzymes, macrophages also possess a number of surface receptors that have been defined functionally rather than biochemically (Table 3). These receptors all play a role in some aspect of the effector or affector function of the macrophage.

There are at least two, and possibly three distinct Fc receptors on macrophages (Unkeless 1977). The first receptor (FcRI) binds IgG2a, while the second (FcRII) binds immune complexes and IgG2b. These two receptors also differ in trypsin sensitivity. There are several reports that macrophages have a receptor for IgM also, and one report that they possess a receptor for the Fc portion of monomeric IgM. Whether the IgM receptor is similar to one of the other two remains to be established. Fc receptors play an important role in those macrophage functions that require the participation of antibodies, such as antibody-dependent cellular cytotoxicity and opsonic phagocytosis of antibody-coated particles.

Macrophages also possess a receptor for the b fragment of the third component of complement (C3b). This receptor promotes binding of coated particles to the macrophage, but in nonstimulated macrophages it initiates phagocytosis very poorly.

Macrophages are responsive to the effects of a number of lymphokines that induce

Table 3. Macrophage Surface Receptors

Receptor	Specificity
Fc receptor I (FcRI)	Trypsin sensitive, binds IgG2a
Fc receptor II (FcRII)	Trypsin resistant, binds aggregated IgG and IgG2b
IgM receptor	May bind monomeric or complexed IgM
C3b receptor	C3b
Nonspecific lymphocyte receptor	T- and B-lymphocytes
Lymphokine receptors	MIF, CSF, Chemotactic factors
Nonspecific binding sites for protein aggregates	
Fibronectin receptor	
High density lipoprotein receptor	
Lactoferrin receptor	
Asialoglycoprotein receptor	
Insulin receptor	
Fibrinogen receptor	

their growth, differentiation, and activation, and the presence on these cells of receptors for some of these molecules has been documented. Migration inhibitory factor (MIF) is the best studied of these lymphokines, and its receptor appears to be an L-fucose containing glycolipid (Remold 1973; Higgins et al. 1978). Other studies have demonstrated inhibition of MIF activity with other 5-methyl pentose sugars such as L-rhamnose, so that the MIF receptor may contain other sugars as well. Receptor studies on other lymphokines are less well advanced, but preliminary evidence suggests that chemotactic factor activity is also blocked by L-fucose and L-rhamnose while CSF is blocked by another sugar.

Finally, there are two receptors that may also be important in the immunological accessory functions of macrophages. Macrophages trigger lymphocytes by "presenting" them with intact antigen. Since most proteins are rapidly ingested and degraded by macrophages, a small portion of ingested antigens must be handled differently than the rest and remain undegraded. Macrophages have a receptor for aggregated proteins. It is possible that antigens bound via this receptor may not be as rapidly degraded as those that are taken up normally. Recent work has also demonstrated that macrophages will bind fibronectin if it is complexed to denatured collagen (C. Bianco, personal communication.)

Macrophages will also bind both T- and B-lymphocytes nonspecifically (Figure 3). This lymphocyte receptor requires metabolically active macrophages and is species specific (Lipsky and Rosenthal 1973). Thymocytes bound to macrophages lose the ability to rebind, suggesting that they have undergone a maturational step. It is likely that this nonspecific lymphocyte receptor is involved in some aspects of macrophage lymphocyte cooperation.

Biologically Active Secretory Products

The macrophage is primarily a secretory cell, and it rivals the hepatocyte in its ability to produce a variety of different molecules. It is this array of secretory products that, more than anything else, explains the multipotency of the macrophage. Table 4 lists the known secretory products of macrophages.

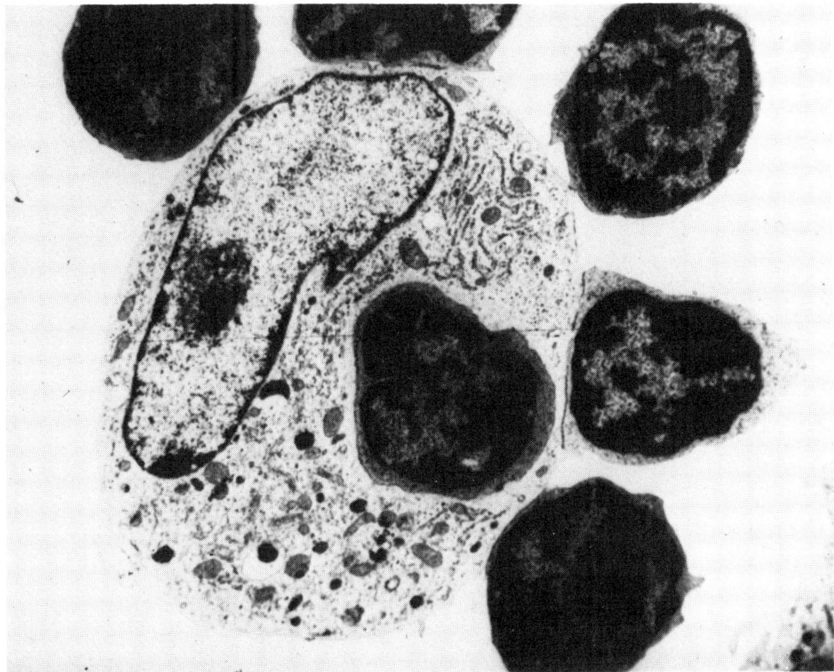

Figure 3. Nonspecific binding of lymphocytes by macrophage. Central cell is a guinea pig peritoneal exudate macrophage. Surrounding cells are guinea pig thymocytes. (From Lipsky and Rosenthal 1973.)

Although some of these products, such as lysozyme, are secreted continually, most are produced in only small amounts or not at all, unless the macrophage is stimulated or activated in some fashion. Mechanisms of stimulation will be covered in a following section.

Hydrolytic Enzymes. Macrophages secrete a number of different hydrolytic enzymes. The acid hydrolases (i.e., cathepsins, hyaluronidase, β-glucuronidase and other glycosidases) are preformed lysosomal enzymes that are primarily involved in the intracellular degradative functions of macrophages. However, upon stimulation, they are also rapidly released into the extracellular environment (Schnyder and Baggiolini 1978). Their extracellular function is unknown although they may play a role in producing the tissue destruction in chronic inflammatory diseases such as rheumatoid arthritis.

Lysozyme is a 14,000 molecular weight protein enzyme that hydrolyses the peptidoglycan of bacterial cell walls. It is a major product of macrophages, accounting for 0.5 to 2.5% of all macrophage protein. Macrophages are the major source of this protein in the body, and diseases where monocytes are increased such as monocytic leukemia are characterized by extremely high urinary levels of lysozyme. This enzyme is constantly secreted by the macrophage and its production is increased relatively little by most stimulatory events. It may be an important antibacterial agent especially in bodily secretions such as tears.

Table 4. Biologically active secretory products of macrophages

Hydrolytic enzymes	
Lysomal hydrolases	Cathepsin B, hyaluronidase, acid phosphatase, β-glucuronidase
Neutral proteinases	Collagenase, elastase, plasminogen activator
Lysozyme	
Arginase	
Cell stimulatory proteins	
Colony stimulating factor (CSF)	
Interferon	
Immunostimulatory factors	B-cell activating factor (BAF)
	Lymphocyte activating factor (LAF)
	T-cell activating factor (TAF)
	Transferrin
Inhibitory factors	Tumor necrosis factor (TNF)
	Arginase
Factors acting on hepatocytes	Glucocorticoid antagonizing factor (GAF)
	Serum amyloid associated (SAA) inducing protein
Endogenous pyrogen	May be identical to LAF
Miscellaneous products	
Prostaglandins	
Complement components	C2, C3, C4, C5, factor B (properdin)
Oxygen intermediates	Hydrogen peroxide, superoxide anion, and hydroxyl radical
Nucleosides	Thymidine
Procoagulant	
α_2Macroglobulin	

The last group of hydrolytic enzymes secreted by macrophages include the neutral proteinases, elastase, collagenase, and plasminogen activator. Unlike the hydrolases, these are not preformed lysosomal enzymes but instead are synthesized *de novo* following a stimulatory event. As such, they are released much later than the acid hydrolases and their production can be blocked by inhibitors of protein systhesis. Collagenase is an enzyme that is highly specific for a single substrate, collagen. Macrophages may help to initiate the process of wound healing (see Chapter 19). Elastase is a related enzyme that will hydrolyze elastin (Werb and Gordon 1975).

Plasminogen activator is a serine esterase that catalyzes the conversion of plasminogen to the proteolytic enzyme, plasmin. Plasmin in turn activates a number of biological pathways including kinin generation, complement consumption, and coagulation and fibrinolysis. Plasminogen activator has received a great deal of study recently because of its documented relationship to cell growth and transformation. Stimulated macrophages release large amounts of this important enzyme (Unkeless et al. 1974).

Cell Stimulatory Agents. The second large group of secretory products have been arbitrarily classified as cell stimulatory agents, since in some way they are all defined by their activity on other cells. This list is rather long, and it may be that future research will reveal that some of these activities are mediated by identical molecules. A number of these products will also be discussed in other sections or chapters, so they will be only briefly described here.

Endogenous pyrogen is a 15,000-molecular weight protein that is responsible for the induction of fever in mammals (see Chapter 17). Macrophages release this substance in response to many types of stimuli, and they seem to be its major source in the body (Bodel and Miller 1977), e.g., in response to phagocytosis of *Staphylococcus aureus*, macrophages release 20 times more pyrogen than do neutrophils. It is detectable in serum and is produced by macrophages and macrophage cell lines *in vitro*. (Bodel 1978).

Colony-stimulating factor (CSF) and interferon are two molecules that are secreted by stimulated macrophages as well as by a number of other cell types. Macrophage interferon is approximately 45,000 daltons. Whether macrophage CSF and interferon are different in any respect from functionally similar products derived from other cell types remains to be established.

In response to challenge with endotoxin, macrophages release two poorly characterized but potentially interesting molecules. Tumor necrosis factor (TNF) will cause necrosis of subcutaneous sarcomas in mice and death of tumor cells *in vitro* (Carswell et al. 1975). It is a glycoprotein of approximately 60,000 molecular weight. In addition, TNF is probably identical to the macrophage-derived, high molecular weight inhibitor of *in vitro* tumor cell growth described by several authors. (Reed and Lucas 1975).

Glucocorticoid antagonizing factor (GAF) is very similar to TNF in its induction, kinetics, and gross physiochemical properties. The action of steroids on hepatocytes is inhibited by GAF, blocking the induction of an enzyme necessary for gluconeogenesis (Moore et al. 1978). This agent is thought to be responsible for the profound hypoglycemia observed in endotoxin-treated animals.

Macrophages release another substance in response to endotoxin that also acts on the liver, serum amyloid-associated protein (SAA) inducing factor (see Chapter 17). Production of this substance by macrophages may be the mechanism by which inflammatory stimuli result in the production by the liver of a number of acute phase reactants (e.g., C-reactive protein, α_2 macroglobulin). In addition, continued production of the SAA inducer and subsequent generation of amyloid protein during chronic inflammatory states may underly the occasional development of amyloidosis in diseases characterized by continuous macrophage stimulation, such as leprosy.

Several macrophage-produced molecules are involved in immune regulation (Rosenstreich and Mizel 1978). The best characterized of these is lymphocyte activating factor (LAF) or interleukin I (IL-1), a 12,000 to 15,000-molecular weight protein that acts on thymocytes and T-lymphocytes. It is probably identical to a similar molecule that stimulates antibody production by B-lymphocytes, B-cell activating factor (BAF). In addition, macrophages secrete another less well characterized immunostimulating molecule, T-cell activating factor (TAF), which has a molecular weight of > 50,000 and acts on peripheral T-cells rather than thymocytes. It appears to be distinct from IL-1 based on target cell and physicochemical differences, but may be a high molecular weight form of LAF.

Recent evidence indicates that IL-1 possesses the same chromatographic properties as endogenous pyrogen, so these two molecules may be identical. In addition semipurified IL-1 preparations will stimulate cultured synovial cells to secrete large amounts of prostaglandins and collagenase. These findings suggest that the molecule IL-1 mediates or regulates several phenomena associated with inflammation or immunity in addition to its effects on T-cells (see Chapter 6).

Macrophages also act as regulators of the immune response by secreting substances that are suppressive. At least five such suppressive products have been identified. The first is a dialyzable low molecular weight molecule that inhibits proliferation of rapidly dividing cells, such as thymocytes or activated lymphocytes. It is likely that the molecule is the nucleoside thymidine, which in the concentrations secreted by stimulated macrophages is capable of blocking several enzyme pathways that are essential to cell division (Stadecker et al. 1977). Macrophages also secrete the enzyme arginase, which catabolizes the essential amino acid arginine. Arginase has been shown to inhibit tumor cell proliferation *in vitro* and may have a similar effect on dividing lymphocytes. Macrophages also produce a high molecular weight suppressor of immune reactions *in vitro*. Similarities of induction and physicochemical properties suggest that this high molecular weight inhibitor is similar or identical to TNF. Macrophage-produced prostaglandins and oxygen intermediates also possess inhibitory properties (see below).

Miscellaneous Products. Prostaglandins are a large group of cyclic fatty acids with a vast array of biological properties. Although these molecules can be shown to be active systemically, it is currently felt their most important function is to act at short distances as regulators of cellular reactions. In response to a number of stimuli, macrophages release several different types of prostaglandins (PGE_1, PGE_2, $PGE_{2\alpha}$, and 6-Keto $PGE_{1\alpha}$) (Kurland and Bockman 1978). Evidence for the regulatory role of these types of prostaglandins is the finding that PGE_1 will greatly enhance collagenase production by endotoxin-stimulated macrophages. Conversely, prostaglandins have antiproliferative, immunosuppressive effects (see Chapter 14). More recently it has been shown that macrophages produce derivatives of arachidonic acid, such as hydroxyeicosatetraenoic acids and slow-reacting substance of anaphylaxis (SRS-A) that are generated by the lipoxygenase pathway.

The complement pathway is involved in a number of immune and inflammatory events. It is therefore of great interest that macrophages are the source of several of the early (and most essential) complement components including C2, C3, C4, and C5 (Littman and Ruddy 1977). In addition macrophages produce factor B of the properdin pathway, which is probably important in early nonspecific host defense against infection (Bently et al. 1978).

The last group of macrophage-produced substances to be discussed are the oxygen intermediates superoxide and hydrogen peroxide, and later metabolites such as hydroxyl radical. Stimulated or activated macrophages undergo a rapid and marked increase in their metabolism of oxygen, the so-called "respiratory burst." The phenomenon of chemiluminescence in macrophages is one way of measuring this respiratory burst. More specifically, increased oxygen metabolism results in the generation of a number of metabolic products of oxygen, the best studied of which are superoxide and hydrogen peroxide. Hydrogen peroxide is known to be a major component of the microbicidal apparatus of neutrophils. More recently, activated macrophages have been shown to produce large amounts of hydrogen peroxide as well (Nathan and Root 1977). Early evidence suggests that some tumoricidal and microbicidal functions of macrophages require hydrogen peroxide. Stimulated macrophages also secrete large amounts of superoxide anion (Johnston et al. 1978) and hydroxyl radical (Weiss et al. 1977). Oxygen intermediates hydroxyl radical and singlet oxygen have been recently implicated in the killing of at least one intracellular parasite, *Toxoplasma gondii*, by macrophages (Murray et al. 1979). Recent work indicates that these oxygen intermediates are also responsible for some of the immunosuppressive activity of stimulated or activated macrophages.

Macrophage Subpopulations, Colonies, and Cell Lines

Mononuclear phagocytes are an extremely heterogeneous population of cells. The previous discussion of their ontogeny and histochemical variability makes it clear that these cells differ markedly in different anatomic locations. It is also important to realize that even within one site, such as the spleen or peritoneal cavity, macrophages vary in their morphological appearance, state of differentiation, and most importantly, in their functional capacities.

The enzymes peroxidase and 5'-nucleotidase define macrophages at different states of differentiation. The functional differences of these subpopulations is not yet clear, although there is a suggestion that the peroxidase-positive macrophage may be the tumoricidal effector cell. Macrophages also vary in their content of surface Ia antigen, and it seems that the Ia-positive group is the subpopulation that is active in antigenic stimulation of lymphocytes. Other macrophage subpopulations are constantly being defined, and it is likely that many of the functions that are now ascribed to "macrophages" as a group will in the near future be more correctly assigned to specific subpopulations.

Other cell types exist that are frequently coisolated with macrophages because of their adherent properties, but that are of an apparently different lineage from macrophages because they are not phagocytic. These cell types are important because of the possibility that they are the true effectors in a number of functions usually ascribed to macrophages. Two such cell types are noteworthy in this regard. A population of nonphagocytic adherent cells (NPAC) exists in peritoneal cell populations that have the morphological appearance of large lymphocytes that were Fc receptor, surface IgM, and esterase positive. The bulk of the evidence suggests that NPAC are a B-lymphocyte subpopulation, and these authors raise the possibility that as a significant contaminant of macrophage monolayers, these cells might be responsible for some activities previously ascribed to macrophages. A second significant population, dendritic cells, has been identified as a minor (1%) population in many lymphoid organs (Steinman and Cohn 1973). The dendritic cell is also nonphagocytic but surface adherent and therefore contaminates many macrophage monolayers. This cell type is of interest because of its high content of surface Ia antigen and its marked stimulatory capacity in the mixed lymphocyte reaction (Steinman and Witmer, 1978). The skin Langerhans cell is functionally similar to this cell. These properties raise the possibility that this cell is responsible for some or all of the immunologically specific accessory functions ascribed to macrophages.

One approach to assigning specific functions to macrophages or to specific subpopulations is to grow these cells *in vitro* from precursor cells. When bone marrow cells are cultured in the presence of macrophage CSF, they produce almost pure colonies of macrophage-like cells. These cells can be obtained in large numbers to test various functional properties. They are extremely useful for demonstrating that a specific function is in fact due to macrophages, since these colonies are virtually free of any other cell type. This approach is also useful for defining the function of specific macrophage subpopulations, since even under these conditions, macrophages are

Table 5. Biological Functions Ascribed to Specific Macrophage Cell Lines

Function	Cell line[a]
Effector	
Antibody-dependent cellular cytotoxicity (ADCC)	RAW 264, PU5-1.8, P388D$_1$, J774, IC-21
Tumoricidal activity	PU5-1.8
Affector	
Accessory cell function in T-cell mitogenesis	P388D$_1$
Accessory cell function in primary antibody responses	IC-21
Production of Biologically Active Products	
Colony stimulating factor (CSF)	PU5-1.8, Wehi-3
Lymphocyte activating factor (LAF)	P388D$_1$
Lysozyme	IC-21, P388D$_1$
Proteinases (elastase, collagenase, and plasminogen activator)	P388D$_1$
Endogenous pyrogen	Human myelomonocytic cell lines

[a]Unless specified all lines are of murine origin.

obtained in all stages of differentiation. Such colony derived macrophages have been used to show that macrophages are tumoricidal *in vitro* and can present antigen to activate immune lymphocytes.

Another approach to overcome the difficulties inherent in studying heterogeneous macrophage populations has been the development and characterization of macrophage cell lines. These lines are extremely valuable, since they can be rendered completely homogeneous and maintained indefinitely *in vitro*. Macrophage cell lines have been utilized to demonstrate that a number of specific functions are in fact mediated by macrophages, to demonstrate the existence and stability of macrophage subpopulations, and to more conveniently investigate aspects of macrophage physiology.

A list of macrophage-like functions found in specific cell lines is given in Table 5. Of most interest is the fact that these lines exhibit some of the important affector and effector functions of macrophages, since the use of cell lines could greatly facilitate studies of these seemingly complex functions. Of equal interest and importance is the presence of specific stable cell lines that are similar to certain macrophages subpopulations. Thus, cell lines have been identified that differ in phagocytic capability, Fc receptors, endotoxin sensitivity, and sensitivity to various lymphokines (Muschel et al. 1977).

Signals Inducing Macrophage Differentiation: Concept of Stimulation and Activation

The life history of the macrophage was outlined in the previous section. Macrophages begin as bone marrow stem cells, circulate in the blood as monocytes, then enter the tissues where they ultimately differentiate into activated effector cells. This process can be divided into a number of discrete steps. For each of these steps an individual group of molecules has been identified that induces the appropriate change. While this process may occur normally, it is greatly accelerated by inflammatory states, and almost all of the relevant molecules are produced by stimulated lymphocytes or are generated from the complement cascade.

Figure 4 presents a hypothetical model for the maturation of macrophages and the relevant inducing molecules. Table 6 lists those molecules that are generated endogenously (i.e., not from external sources such as endotoxin) and sufficiently characterized to enable an approximate molecular weight to be established.

Maturation of Stem Cells to Monocytes: Macrophage Colony Stimulating Factor and Factor Inducing Monocytosis

Two molecules have been identified that act on the earliest stages of macrophage production. Macrophage CSF and a factor that induces monocytosis (FIM).

If bone marrow stems cells are cultured in the presence of CSF, they will differentiate into colonies of either granulocytes or macrophages. Produced by fibroblasts in culture, two distinct fibroblast CSFs have been identified: granulocyte CSF is a 30,000-molecular weight molecule that induces granulocyte colonies; macrophage CSF is a 70,000-molecular weight molecule that induces macrophage colonies (Stanley and Heard 1977). Lymphocytes and macrophages also produce molecules with macrophage CSF activity. These have not been as well characterized as the fibroblast form.

Colony stimulating factor is defined by its activity on bone marrow stem cells. However, there is a molecule, macrophage growth factor (MGF), that induces prolif-

Figure 4. Hypothetical model for the differentiation of macrophages. This scheme of the sequential differentiation of macrophages from their origin as bone marrow stem cells to their development into activated macrophages is based on currently available evidence. Substances that are thought to induce each change are indicated. Multiple arrows between cells indicate a multistep process. See text for details.

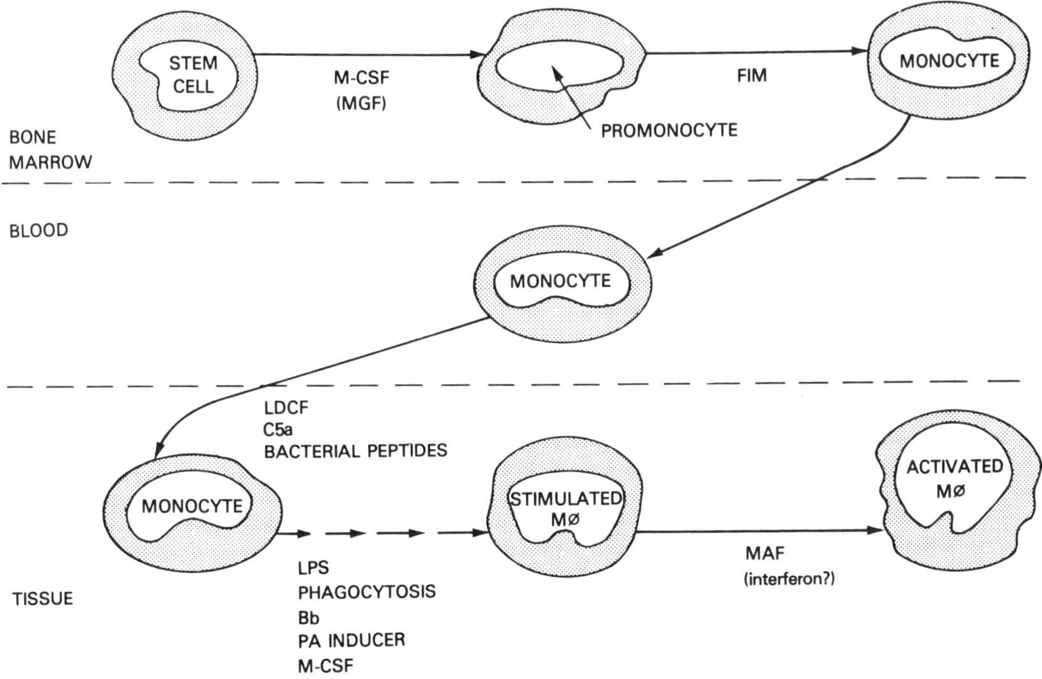

Table 6. Partially Characterized Endogenous Substances Acting on Macrophages

Substance	Physicochemical characteristics (*Mol wt*)	Functional properties
Macrophage colony stimulating factor (CSF)	70,000	Macrophage colony formation from bone marrow cells. Also induces macrophage secretion
Macrophage growth factor (MGF)	Same as CSF	
Factor inducing monocytosis (FIM)	18,000–23,000	Acts on promonocytes
Leukocyte-derived chemotactic factor (LDCF)	12,000	Chemotactic for macrophages
Plasminogen activator inducer (IPA)		Stimulates plasminogen activator secretion
Macrophage stimulatory protein (MSP)	100,000	Increases spreading and phagocytosis
Factor Bb	65,000	Increases spreading and phagocytosis; inhibits migration
Macrophage activating factor (MAF)	50,000–60,000	Increases macrophage tumoricidal and microbicidal function
Migration inhibitory factor (MIF)	25,000–60,000	Inhibits migration
Soluble immune response suppressor (SIRS)	Similar to MIF	
Interferon	Several molecular weight species	Both Types I and II activate macrophages
Macrophage aggregating factor	> 100,000	Distinct from MIF. Causes clumping of macrophages *in vivo* and *in vitro*; may be fibronectin

eration of macrophages in peritoneal exudate populations. Therefore, MGF and macrophage CSF appear to be identical or very closely related molecules based on physicochemical criteria and the interchangeability of molecule and target cell (Stanley et al. 1976). Fibroblasts and activated lymphocytes produce MGF. Thus, in addition to inducing the original production of macrophages in the bone marrow, MGF may also induce the proliferation of macrophages where and when they are most needed, at the actual inflammatory site.

In addition to inducing proliferation of stem cells, recent evidence suggests that macrophage CSF may be a more general macrophage stimulating agent, since it also induces macrophages to secrete prostaglandins (Kurland et al. 1979) and LAF (Moore et al. 1980).

While macrophage CSF and MGF are probably identical, another seemingly distinct molecule has been identified that acts at the bone marrow stage. A factor that induces monocytosis (FIM) has been found in the serum from mice with experimentally induced inflammation. An 18,000 to 23,000-molecular weight protein, FIM is not chemotactic, not derived from a complement component or from the coagulation cascade, and most importantly has no colony stimulating activity *in vitro* (van Waarde et al. 1977). These properties suggest that it is distinct from macrophage CSF. It produces monocytosis by increasing the conversion of promonocytes to monocytes and is specific, since it has no effect on peripheral lymphocytes or granulocytes.

Passage of Monocytes into the Tissues: Chemotaxis

The stimuli that cause monocytes to leave the bone marrow and enter the bloodstream are not known, and this process may occur as a result of maturation. However, there are several substances that induce directional migration of monocytes (chemotaxis) and thereby cause them to leave the circulation and enter tissue spaces. Again, the chemotactic attractants are primarily generated during inflammation, which causes a specific localization of monocytes to where they are most needed.

Agents that are chemotactic for monocytes include bacterial products, C5a, and a lymphocyte-derived chemotactic factor (LDCF). A number of small bacterial peptides are chemotactic for monocytes. In addition, a number of synthetic formyl-methionine-containing tripeptides (i.e., F-met-leu-phe) have been found to be extremely potent monocyte chemotactic agents. Other bacterial products, such as a phospholipid derived from corynebacteria, are also chemotactic. These bacterial chemotactic agents are probably important in causing the migration of monocytes even before inflammatory reaction has developed.

Once an inflammatory reaction has begun, a chemotactic complement C5a is generated (Chapter 15). C5a is a 15,000-molecular weight cleavage fragment of C5. It is generated as a result of the activation of the complement cascade by either the alternative or classical pathways and can also be generated directly from C5 by limited proteolysis with enzymes, such as plasmin or macrophage proteinases. Because of the ease with which it is generated, C5a is probably responsible for the rapid influx of macrophages into inflammatory sites. Experimental evidence suggests that the monocytic influx induced by intraperitoneal agents such as thioglycollate, mineral oil, protease peptone, or endotoxin is largely mediated by the action of C5a (Synderman et al. 1975).

Lymphocyte-derived chemotactic factor is produced by mitogen or antigen-stimulated lymphocytes and probably comes into play during the later stages of inflammatory reactions. The molecular weight of LDCF is 12,500 (human and guinea pig), and it is distinct from C5a (Wahl et al. 1974). It is produced by both T- and B-lymphocytes when these cells are appropriately stimulated. In addition, LDCF has been isolated from inflammatory sites, confirming the relevant role of this molecule *in vivo*.

Differentiation of Monocytes to Macrophages: The Stimulated Macrophage

For reasons of convenience, the process of macrophage maturation has been studied most extensively using cells from the peritoneal cavity. In the absence of any inflammatory stimulation, the newly arrived blood monocyte undergoes some morphological and biochemical modifications and remains as a quiescent cell, the resident macrophage. If, however, an inflammatory stimulus is present, either monocytes or resident macrophages undergo a process of enlargement and metabolic stimulation that results in a cell referred to as an inflammatory or stimulated macrophage. The conversion of the monocyte to a stimulated macrophage probably requires several sequential steps and signals. This is denoted in Figure 4 by the series of broken arrows.

The general properties of stimulated macrophages are listed in Table 7 (Cohn 1978). Overall, this cell is larger, spreads more rapidly, and possesses a number of increased metabolic functions. Several properties are noteworthy. Although many enzymes are increased within these cells, there is a striking decrease in the ectoenzyme, 5'-nucleotidase. Secondly, while their pinocytic and phagocytic rate is increased, there is also a marked increase in their ability to phagocytose particles via the C3b receptor. Third, although increased oxidative metabolism and hexose–monophosphate shunt activity in these cells results in the ability to secrete superoxide anion, they do not secrete proportionally increased amounts of hydrogen peroxide. Greatly increased

Table 7. Properties of Inflammatory Peritoneal Macrophages

	Increase in inflammatory macrophages[a] (Ratio of inflammatory to resident)
Cell protein (μg/10^6 cells)	1.6
Spreading (% cells spread/hr)	24.0
Plasma membrane ectoenzymes	
5'-nucleotidase	(.01) decrease
Alkaline phosphodiesterase I	2.7
Fluid phase pinocytosis (nl/10^6/hr)	5.3
Phagocytosis	
E-IgG (Fc receptor mediated)	2.7
E-IgG-c (C3b receptor mediated)	25.0
Metabolic changes	
ATP (intracellular)	7.5
Superoxide anion (secretion)	12.0
Hydrogen peroxide (secretion)	1.0 (no change)
Secretory enzymes (secreted)	
Lysozyme	0.9 (no change)
Plasminogen activator	800.0
Collagenase	15.0
Elastase	38.0
Lysosomal hydrolases (intracellular)	3.0-14.0

[a]Modified from Cohn, Z. A. 1978.

secretion of hydrogen peroxide appears to be a property of the activated effector macrophage. Finally, as noted previously, while secretion of many enzymes increases, lysozyme production remains comparable to resident macrophage levels.

Signals That Induce Spreading, Secretion, and Other Properties of Stimulated Macrophages

Inflammatory macrophages are generated experimentally *in vivo* by a large array of agents, including endotoxin, thioglycollate, proteose peptone, and mineral oil. Presumably these agents duplicate the changes induced by natural stimuli such as bacteria. Although the mechanisms that induce macrophage differentiation are complex, recent *in vitro* studies have shed some light on this process.

One of the earliest properties developed by stimulated macrophages is their ability to spread rapidly. Rapid spreading of resident macropohages *in vitro* can be induced by components of the complement or coagulation systems (Bianco et al. 1976). Factor Bb, the cleavage product of the B component of the properdin system, has been demonstrated to be responsible for the complement-dependent spreading (Götze et al. 1979). Since many inflammatory agents trigger the complement cascade, it is assumed that fluid phase cleavage of factor B to Bb induces the initial changes in macrophages. In addition, macrophages secrete factor B, so that this form of stimulation could be self-perpetuating as long as there is a means to cleave B to Bb. Cleavage of B to Bb could be induced in the absence of further complement consumption, directly by macrophage proteinases or indirectly by plasmin. Elevated levels of plasmin could be generated by the action of another macrophage product, plasminogen activator. It is, therefore, noteworthy that inflammatory (thioglycollate-induced) macrophages produce very large amounts of plasminogen activator, even without further triggering.

Another stimulatory factor present in serum, which is functionally similar to Bb, has been termed macrophage stimulatory protein (MSP) and is a 100,000-molecular weight protein (Bb, 65,000 molecular weight) that increases macrophage spreading and phagocytosis (Leonard and Skeel 1978).

In certain experimental situations, the inflammatory macrophage constitutively secretes some products without further stimulation, including lysosomal hydrolases, CSF, and plasminogen activator. Production of many other macrophage products usually requires additional stimulatory signals. For many of these products, inflammatory macrophages produce much more than do resident cells after additional stimulation, and can therefore be considered to be metabolically primed. Macrophage products that require stimulation for secretion include collagenase, elastase, endogenous pyrogen, CSF, interferon, LAF, TNF, and GAF. A large number of substances will induce macrophage secretion. Phagocytic stimuli such as antigen-antibody complexes, latex beads, and zymosan will induce secretion, and the presence of undegraded phagocytosed material results in prolonged secretion by macrophages. Endotoxin and related bacterial products are potent inducers of macrophage secretion, have been used extensively *in vitro,* and may be important *in vivo* as well. The compound phorbol myristic acetate (PMA) is also a potent inducer of macrophage secretion.

Two well-defined endogenous molecules have been shown to be involved in the enhancement or induction of macrophage secretion, macrophage CSF and the plasminogen activator inducer (IPA). Purified preparations of macrophage CSF have been shown to induce macrophage secretion of prostaglandin (Kurland et al. 1979) and LAF (Moore et al. 1980). Since macrophage CSF is produced by fibroblasts in response to nonspecific inflammatory stimuli, this molecule may be involved in the induction of macrophage stimulation before the onset of specific immune responses. A product of activated T-cells, IPA induces macrophages to secrete plasminogen activator (Vassalli and Reich 1977), and IPA purified to near homogeneity is also active (Vassalli, personal communication). Unlike macrophage CSF however, IPA is probably not involved until the later stages of inflammatory reactions. As indicated in Table 7, stimulated macrophages exhibit many other altered properties, such as alterations in the number and function of surface receptors, the amount of surface enzymes, and increased metabolism. These changes can be induced by the usual nonspecific inflammatory stimuli. However, the molecular mechanisms that underlie these changes are not defined. It has not yet been reported if factor Bb or macrophage CSF, the two endogenous molecules known to be involved in macrophage stimulation, will induce these other changes.

Induction of Activated Macrophages: Macrophage-Activating Factor

The activated macrophage is defined as a cell with specific effector functions, such as the ability to kill tumor cells or intracellular organisms. Although many types of inflammatory stimuli are capable of stimulating macrophage maturation and secretion, true macrophage activation seems to be induced most effectively by signals generated by activated lymphocytes. The concept of the activated macrophages is based on the pioneering work of Mackaness and his colleagues (Mackaness 1970). Because of this requirement for activated lymphocytes, activated macrophages are generated only late in an inflammatory response and more usually under conditions of chronic inflammation. *In vivo,* activated macrophages are generated experimentally by chronic infection with Bacillus–Calmette–Guerin (BCG), listeria, or corynebacteria. It should be emphasized that macrophage activation is almost entirely nonspecific, so that BCG-activated macrophages will kill tumor cells or other intracellular parasites. Numerous *in vivo* and *in vitro* studies have documented that macrophage activation requires activated lymphocytes, and in most cases activated T-lymphocytes (North 1973). Furthermore, it has been demonstrated that T-cells release a lymphokine(s) that mediates the terminal differentiation of macrophages to the activated state.

The physicochemical nature of this lymphokine has not yet been clearly established, but based on its function it is referred to as macrophage-activating factor (MAF) or macrophage cytotoxicity factor. In the mouse, MAF elutes as a broad peak of approximately 55,000 molecular weight from gel filtration (Sephadex G-100) columns (Leonard et al. 1978). In a series of sequential purification steps, it has not been clearly separable from other lymphokines such as MIF or immune (Type II) interferon. Other studies have demonstrated that purified Type I (fibroblast) mouse interferon would activate macrophages to kill tumor cells, suggesting that MAF and interferon are identical (Schultz et al. 1977). However, this point is controversial, since Type I interferon will not activate macrophages to kill certain intracellular parasites, and it is a much less potent inducer of tumoricidal macrophages than MAF itself.

The metabolic properties that distinguish the activated macrophage from the inflammatory cell are not known, but current knowledge has been summarized concisely by Karnovsky and Lazdins (1978). However, only immunologically activated macrophages are capable of secreting large amounts of hydrogen peroxide when triggered. Hydrogen peroxide is a potent microbicidal agent in granulocytes. In addition, it is involved in macrophage killing of tumor cells (Nathan et al. 1979a), and its secretion correlates with the ability of macrophages to kill an intracellular parasite, *Trypanosoma cruzi* (Nathan et al. 1979). Therefore, it is likely that the ability to secrete hydrogen peroxide may be one step induced by MAF that converts a stimulated macrophage to an activated one.

Other Lymphokine Effects

In addition to activating macrophages, lymphokines will induce in resident cells all those changes associated with the stimulated state that are listed in Table 7. Thus, lymphokines will increase Fc receptors (Vogel and Rosenstreich 1979) and promote phagocytosis via the C3b receptor (Griffin and Griffin, 1979) as well as other macrophage metabolism (Poulter and Turk 1975). The physicochemical nature of the lymphokines mediating these effects is not certain, but most seem to be due to products that have identical chromatographic position with MIF and MAF.

Lymphokines induce two other alterations in macrophages that should be noted: inhibition of migration and aggregation. Migration inhibition was historically the first and most studied lymphokine effect. It is caused by a glycoprotein that exhibits change, weight, and interspecies heterogeneity (see Chapter 10). As noted above, many properties in addition to migration inhibition have been attributed to MIF, but it is not clear whether these multiple effects are due to a single molecule or to the presence of several different molecular species with different biological properties. The biochemical heterogeneity of MIF is consistent with the latter hypothesis.

Some recent work has shed light on the mechanism of migration inhibition. Bianco et al. (1979) have found that factor Bb would inhibit macrophage migration. Furthermore, independent work has shown that IPA chromatographs with and is produced by the same T-cell subsets as MIF (Newman et al. 1978). These two observations raise the interesting possibility that migration inhibition is due to the effect of IPA in generating factor Bb by the pathway discussed in a previous section (i.e., IPA $\rightarrow$

plasminogen activator → plasmin → cleavage of B → Bb → migration inhibition).
Whether the other biological effects of MIF could also be due to IPA requires further
experiments such as determining if IPA acts through a fucose-containing receptor.

The last mediator listed in Table 6 is macrophage aggregating factor. This is a
lymphokine that causes aggregation of macrophages *in vivo* and *in vitro*. Its physiological relevance and place in the scheme outlined in Figure 4 are not known.
However, it is noteworthy that it is clearly physicochemically distinct from MIF
(Postlethwaite and Kang 1976).

Macrophage Functions: General Considerations

Macrophages are involved in four major functional activities. First, they interact with
and degrade all types of foreign material in their environment. Second, they initiate
and enhance the immunological activation of lymphocytes. Third, they are effector
cells capable of destroying tumor cells and certain types of microorganisms, and,
finally, they also act as suppressors of the immune response.

These functions are not exerted by all macrophages at the same time, and not all
macrophages possess each function. Instead, the development of these functional
capabilities occurs in an approximate temporal sequence and concomitantly with
the differentiation process outlined in previous sections. A hypothetical scheme illus-
trating the development of the functional properties of macrophages is shown in
Figure 5.

Figure 5. Hypothetical model for the sequential development of different macrophage functions. This scheme illustrates the temporal relationship between the state of differentiation of
the macrophage and its ability to exert specific functions. Since there is no direct proof that
the activated and suppressor macrophage are identical, the connection between these cells is
indicated with a broken line. See text for details.

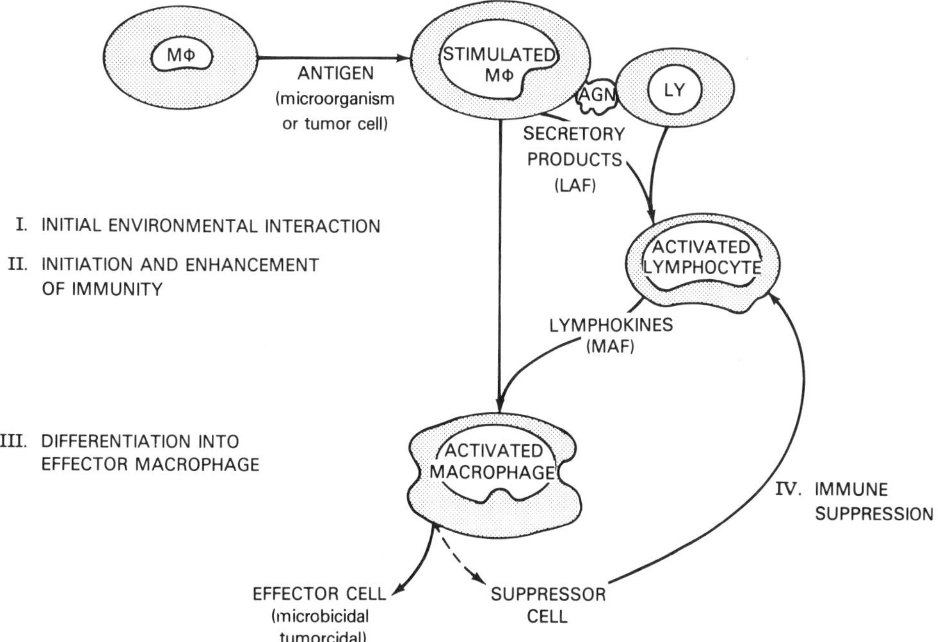

The initial filtering and degradative function of the macrophage is determined by its surface receptors, phagocytic capability, and lysosomal hydrolytic enzymes. This function is one that is probably subserved to some degree by all macrophages. If any foreign material persists in a recognizable form, then the immune system is called into play. Macrophages initiate immune activation by "presenting" antigens to lymphocytes in an immunogenic form. In addition, the initial uptake process may stimulate the macrophage to secrete immunostimulatory molecules such as LAF. The third major macrophage function, the effector role is induced in turn by these activated lymphocytes. Again, reaching this stage probably requires persistence of sufficient antigen to cause continued lymphocyte activation. Finally, intense or prolonged amounts of macrophage activation result in suppression of further lymphocyte activation. This suppression may be mediated by some of the same mechanisms responsible for macrophage effector function. This suppression serves as a feedback regulator of immunity.

Although this scheme is only hypothetical, it helps to clarify individual macrophage functions by orienting them within a larger framework. In addition, this temporal analysis is consistent with the life history of the macrophage, and also explains how a cell can both enhance and suppress immunity. Since the first macrophage function has been discussed in the context of the description of the general properties of macrophages, it will not be discussed further. The last three functions will be discussed in more detail and in the order outlined in Figure 5.

Macrophage Enhancement of Immune Function

The essential role of macrophages in the activation of immune responses is based primarily on *in vitro* studies (see review by Rosenstreich and Oppenheim 1976). Historically, Mosier (1967) was the first to note that removal of macrophages from spleen cell cultures prevented B-cell production of anti-sheep red blood cell (SRBC) antibody. At about the same time, Oppenheim et al. (1968) noted that lymphocyte (probably T-cell) proliferation in response to antigens was also macrophage dependent. Since that time, macrophages have been shown to be essential for the activation of T-lymphocytes in response to antigens, mitogens, and allogeneic cells. Macrophages are required for T-cell proliferation, lymphokine production, and the generation of cytotoxic T-cells. Macrophages have also been shown to be required for the activation of B-lymphocytes by soluble and particulate antigens. Although it has been more difficult to demonstrate, macrophages are also required for activation of B-cells by most mitogens with the notable exception of endotoxin.

The mechanism by which macrophages enhance immune responses can be divided into two categories. The first involves all those functions dealing with antigens and antigen specificity, i.e., antigen uptake, processing, and presentation. The second involves the release of soluble factors that are not for the most part antigen specific.

Antigen Uptake and Storage

Nossal and his coworkers (Nossal et al. 1967) were the first to show that labeled antigens were localized and persisted for long periods of time in the macrophages of the lymph node medulla. *In vitro* studies have confirmed the role of macrophages in the uptake and storage of antigens. Thus, antigens can be fed to macrophages, which in turn can activate lymphocytes. Most studies agree that >90% of the antigens ingested by macrophages are rapidly degraded and rendered nonimmunogenic. However, a small amount of antigen remains within the macrophage in a nondegraded, immunogenic form (Unanue and Cerottini 1970). It is not clear how or where this antigen is stored. Several studies agree that the antigen is not retained on the macrophage surface but that it is slowly released by exocytosis from the cytoplasm (Calderon and Unanue 1974). The mechanism by which antigen can remain unchanged in this cell is currently not understood, despite its importance.

Although most workers feel that relevant antigen remains unchanged within the macrophage, there is a body of literature that deals with the development of "immunogenic RNA." This is supposedly macrophage-derived messenger RNA that will trigger immune lymphocytes. This work is very controversial. Although many investigators agree that RNA has nonspecific immunoenhancing properties, studies that appear to invoke a specific response have usually been attributed to contaminating antigen in the RNA preparation.

Antigen Presentation

After the macrophage has taken up antigen, it is able to "present" it to immune lymphocytes and to induce their activation. Macrophages will present all types of antigens to lymphocytes including red cells, microorganisms and tumor cells (Treves et al. 1976). This is a process that involves the direct contact of lymphocytes and macrophages, and relies on two important macrophage membrane components in addition to processed antigen: surface Ia (immune-associated) antigens and lymphocyte receptors.

The role of macrophage Ia antigens in lymphocyte activation has only been appreciated recently (Immunological Reviews, 1978). Their importance was first noted by Rosenthal and Shevach (1973) who found that T-cells and macrophages had to be histocompatible in order to cooperate effectively. Subsequent studies have shown that it is the macrophage and not the T-cell Ia antigens that are important in this interaction. Only the Ia-positive macrophage subpopulation can act as effective accessory cells for immune responses, (Yamashita and Shevach 1977) and removal of Ia-positive macrophages abrogates immune responses. In contrast, killing of Ia-positive T-cells has no effect (Thomas et al. 1977). Macrophages are extremely potent stimulators of the mixed lymphocyte reaction *in vitro,* which is due to their content of surface Ia and is probably related to their role as antigen-presenting cells.

How the T-cell recognizes macrophage Ia and antigen is still uncertain. One hypothesis is that T-cells possess a dual receptor, with separate recognition units for Ia and for antigens, and that T-cell activation occurs only when both receptors are complexed. A second hypothesis is that a single T-cell receptor recognizes a complex of macrophage Ia with antigen. Recent work by Rosenthal and his co-workers (1977) suggests that macrophages play a very active role in the actual antigen recognition process. Using specific polypeptide chains of insulin, they have found that the ability of a macrophage to "present" a specific antigen is determined by its Ia antigens. The mechanism underlying this "determinant selection" is not understood at this time.

In addition to presenting antigen, macrophages possess a mechanism to passively focus lymphocytes onto their surface so that effective antigen presentation can take place. Macrophages bind lymphocytes by two mechanisms; antigen-independent and antigen-dependent binding steps. When normal macrophages are exposed to lymphocytes *in vitro,* they quickly bind them, forming "rosettes or clusters." This nonspecific or antigen-independent binding is reversible, since with time in culture lymphocytes begin to leave the macrophage. Interestingly, thymocytes appear to differentiate while bound to macrophages, since they will not rebind after they are dissociated (Lipsky and Rosenthal, 1975).

The properties of the nonspecific lymphocyte receptor, described in a previous section, are listed in Table 8. It is important to note that this is a true surface receptor and not merely the "sticking together" of two cells. Presumably, it is the macrophage that bears the receptor, since this cell must be viable to bind lymphocytes and trypsinized macrophages no longer bind lymphocytes. Conversely, even dead or trypsinized lymphocytes will bind to macrophages. Binding is also species specific, although this specificity is lost with dead lymphocytes.

When immune cells come in contact with macrophages containing an antigen for which they carry specificity, a tight bond is formed that ultimately results in a "cluster" of lymphocytes around the macrophage (Lipsky and Rosenthal 1975). This cluster is much longer lived and larger than the one formed nonspecifically and will persist and continue to increase in size with time in culture.

The mechanisms underlying antigen-specific cluster formation are not clear. Studies with antibody suggest that the T-cells do not recognize native antigen on the macrophage surface (Ben-Sasson et al. 1977), but this process is Ia dependent, since antigen specific clusters do not form unless the T-cells and macrophage are Ia identical.

Based on the findings discussed in this section, a hypothetical scheme for macrophage antigen presentation can be constructed. Macrophages take up antigen and reexpress a portion of it in an as yet undetermined fashion. Lymphocytes become exposed to this antigen as they are bound to the surface via the nonspecific receptor. However, with time, the immune lymphocytes remain on the macrophage surface, while the nonimmune lymphocytes dissociate. These immune lymphocytes then become activated due to their exposure to the appropriate combination of processed antigen and macrophage Ia.

Soluble Immunostimulatory Factors

All available evidence suggests that immune lymphocytes cannot be activated without the essential signals generated by macrophage processing and presentation of antigen. However, exposure to processed antigen and macrophage surface Ia alone may not be sufficient to activate lymphocytes. Instead, additional signals in the form of soluble factors produced by macrophages are also required to complete the steps necessary for

Table 8. Antigen Independent Binding of Lymphocytes by Macrophages

Reversible[a]
Species-specific with live lymphocytes
Requires divalent cations
T- and B-lymphocytes bound equally
Requires viable macrophage but not lymphocyte metabolism
Abolished by trypsinization of macrophages but not lymphocytes
Not inhibited by excess immunoglobulin

[a]Modified from Lipsky and Rosenthal 1973.

the initiation of lymphocyte activation. Several growth factors have been described and have been partially characterized.

The best studied of these is the molecule commonly referred to as lymphocyte activating factor (LAF), which was recently renamed IL-1. IL-1 has been described in several mammalian species and ranges from 12,500 to 15,000 molecular weight (reviewed by Mizel 1980). It is a trypsin-sensitive protein that may also be glycosylated, although its activity is not sensitive to periodate treatment. A higher molecular weight intracellular precursor (65,000 molecular weight) dissociates to the low molecular weight form prior to secretion as extracellular IL-1 (Mizel and Rosenstreich 1979). Relatively straightforward purification schemes for human and mouse IL-1 have been developed, and these biological activities can be derived in a homogenous partly purified form from appropriately stimulated macrophage cell lines.

One effect of IL-1, and the one most commonly studied, is to induce proliferation of thymocytes. It also enchances the response of these cells to mitogenic lectins. However, it is much less mitogenic for peripheral T-cells but does increase their capacity to rosette with sheep red blood cells. The biochemical changes induced by IL-1 are not yet completely understood. This is partly because most studies to date have not used purified IL-1. Macrophage supernatants containing IL-1 have been reported to elevate intracellular levels of cAMP in thymocytes and induce expression of Ia,Lyt-1, and H-2 antigens in the thymocyte surface (Beller and Unanue 1977). Among these effects, however, the cAMP elevation seems to be due to prostaglandins rather than IL-1, and the increased H-2 antigen expression may be due to another macrophage mediator. IL-1 is distinct from the similarly sized molecule, T-cell growth factor (TCGF), (IL-2), that promotes the growth of T-cells in culture. However, it is probably identical to the B-cell activating factor (BAF) described by Wood and Cameron (1975) that promotes antibody production *in vitro.*

A second group of molecules produced by macrophages has much more activity on peripheral T-cells and is inactive on thymocytes. In the guinea pig, this molecule(s) has been termed T-cell activating factor (TAF). It is >50,000 molecular weight in size and is different from LAF. Recent work suggests that a TAF-like molecule in the rat is associated with carboxypeptidase-like activity (B. Waksman, personal communication). This observation is supported by the finding that an elastase-like enzyme from granulocytes also mimics TAF (Havemann and Schmidt 1973). The target of these enzymatic activities has not yet been defined. It is also possible that these enzymes act on precursor molecules for TAF and thus increase its biological activity.

A third nonspecific growth-promoting molecule released by macrophages is transferrin. Transferrin along with albumin has been found to be essential for the growth of lymphocytes *in vitro* and their response to mitogenic lectins (Tormey et al. 1972). Shortly after activation, macrophages produce and secrete detectable amounts of transferrin *in vitro.* The relationship between TAF and transferrin is not clear, but they may act in concert on the peripheral T-cell.

The finding of several different growth-promoting factors derived from macrophages suggests that the initiation of lymphocyte activation requires several distinct signals. The effects of these factors are not well understood; however, it seems likely that one function of macrophages or their soluble products is to maintain essential lymphocyte molecular structures in a reduced state. This hypothesis is based on the finding that reducing agents will augment lymphocyte activation *in vitro,* and partially replace macrophages in some situations (Chen and Hirsch 1972). A number of reducing agents have this capability, including 2-mercaptoethanol (2-ME), α-thioglycerol, and vitamin E. Opitz et al. (1977) have demonstrated that 2-ME alters a serum protein of approximately 70,000 molecular weight. Hoffeld et al. (1980) have shown that reduced glutathione is essential for lymphocyte survival *in vitro*. Since the major glutathione-carrying serum protein "ligandin" is also 70,000 molecular weight in size, it is probable that reducing agents such as 2-ME and macrophage factors function by an exchange reaction to maintain glutathione in its reduced state. The reduced glutathione in turn removes continually generated deleterious oxygen intermediates. It is noteworthy in this regard that transferrin is also a scavenger of oxygen radicals.

All the soluble molecules discussed so far in this section are not antigen specific. In contrast, the antigen-specific macrophage mechanisms seem to depend on the direct contact between syngeneic macrophages and lymphocytes. However, there is one exception to this. Erb and Feldman (1975) have described a soluble macrophage-derived factor that manifests antigen specificity, which they have termed genetically restricted factor (GRF). Genetically restricted factor is found in the supernatants of antigen-pulsed, cultured macrophages. It appears to be a complex that contains Ia antigen plus antigen, and will activate only Ia-identical immune T-lymphocytes. It seems likely that GRF is the molecule recognized by the immune T-cell during the process of cluster formation and subsequent activation. The mechanics of its shedding and secretion are not yet understood.

Macrophage Effector Function

As outlined in Figure 5, activated lymphocytes produce lymphokines that differentiate macrophages into activated effector cells. Macrophages play an important role in the elimination of tumor cells, foreign cells, and many common intracellular pathogens. Much information has been gathered from *in vitro* studies. Nevertheless, several *in vivo* studies have demonstrated the *in vivo* relevance of macrophages in a number of situations. Thus, elimination of macrophages *in vivo* with carrageenan will prevent allogeneic graft rejection (Cudkowicz and Yung 1977). Similarly, treatment of animals with silica to kill or block their macrophages will increase the infectivity or lethality of pathogens, such as herpes virus or *Salmonella typhimurium* (O'Brien et al. 1979).

Tumoricidal Capacity

The *in vivo* role of macrophages in defense against tumors is suggested by several types of experiments (reviewed in Nelson, 1976). Macrophages are found within tumors and can be separated from the tumor cells. When this is done these macrophages, especially when derived from small growing or regressing tumors, are capable of killing tumor cells *in vitro* (Russell et al. 1977). Treatment of animals with antimacrophage serum or silica results in tumor progression. Finally, cytotoxic T-cells when transfused are unable to cause regression of syngeneic tumors, although they are somewhat more effective against allogeneic tumors.

A number of situations will result in the generation of cytotoxic macrophages. Chronic infection of mice with organisms such as BCG, *Corynebacterium parvum*, toxoplasma, or listeria all result in activated peritoneal macrophages capable of killing tumor cells *in vitro*. Studies with T-lymphocytes from such mice strongly suggest that macrophages are activated by MAF released from these immune T-cells in response to the appropriate antigen (Ruco and Meltzer 1977). Normal resident macrophages or monocytes can be rendered tumoricidal with MAF-like molecules derived from either antigen- or mitogen-stimulated lymphocytes (described in a previous section and in Chapter 10). Activated macrophages will kill tumor cells nonspecifically, although evidence suggests that they are selective since they do not kill normal cells. The mechanism by which they discriminate between normal and foreign cells is not known.

Other agents in addition to MAF have been reported to activate macrophages, including endotoxin and double-stranded RNA. The role of endotoxin in macrophage-mediated tumor killing is controversial. Evidence suggests that endotoxin is essential for murine macrophage tumor killing *in vitro* (Weinberg et al. 1978). It appears that the more activated the macrophages, the less endotoxin is required to induce killing. Thus, proteose peptone-stimulated macrophages require >500 ng/ml of endotoxin, while BCG-activated cells require <1 ng/ml. In no case will macrophages kill in the complete absence of endotoxin; however, endotoxin by itself does not induce resident macrophages to kill. These findings, as well as those of several other laboratories, suggest that tumor killing by macrophages requires the interaction of two signals; MAF to activate the macrophage, and endotoxin to trigger the cells to kill (Ruco and Meltzer 1978).

In addition to nonspecific cytotoxicity, macrophages are able to mediate specific killing of tumor cells or allogeneic cells under two conditions. Macrophages in the presence of antitumor cell antibody will become tumoricidal. This mechanism, referred to as antibody-dependent cellular cytotoxicity (ADCC) is discussed in Chapter 7. There have also been reports of a specific factor released from immune T-cells (Pels and den Otter 1974). This specific macrophage-arming factor (SMAF) is reported to be a 50,000 to 300,000-molecular weight molecule that will induce normal macrophages to kill only those tumor cells to which the T-cells were sensitized. However, this is a controversial subject, and this effect has been recently attributed to the synergistic effect of MAF and cytophilic antitumor antibodies on macrophages tumoricidal function.

Some progress has been made in identifying the mechanisms by which macrophages kill tumor cells. Although macrophage culture supernatants are cytotoxic under certain conditions, for the most part macrophage killing of tumor cells is most effective when effector and target are in direct contact (Figure 6). Several products of activated macrophages have been reported to be responsible for killing, including tumor necrosis factor (TNF), hydrolytic enzymes, interferon, thymidine, or the enzyme arginase. However, recent work by Nathan et al. (1979) strongly suggests that killing of lymphoma cells involves macrophage-secreted hydrogen peroxide. Other oxygen metabolites such as superoxide anion may be involved in killing of other target cells.

Microbicidal Capacity

Macrophages deal with those organisms that are able to survive intracellularly (reviewed by Goren 1977; Mogenson 1979). This capacity to survive intracellularly results in infections that are long lasting and persistent. The ultimate control or eradication of these organisms depends on the macrophage becoming activated. Whether activation to kill microbes is identical to "activation" to kill tumor cells is not yet clear; however, evidence showing identical modes of activation and correlation between killing mechanisms suggests that the two functions are intimately related.

Several types of intracellular organisms such as *Mycobacterium tuberculosis, Histoplasma capsulatum, Leishmania donovani* and *Trypanosoma cruzi* are eradicated by activated

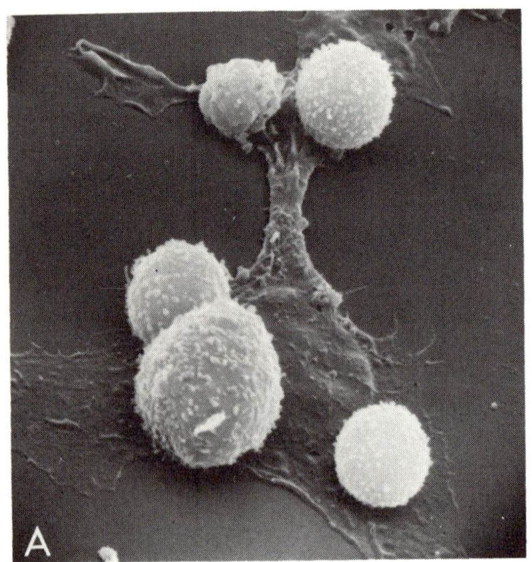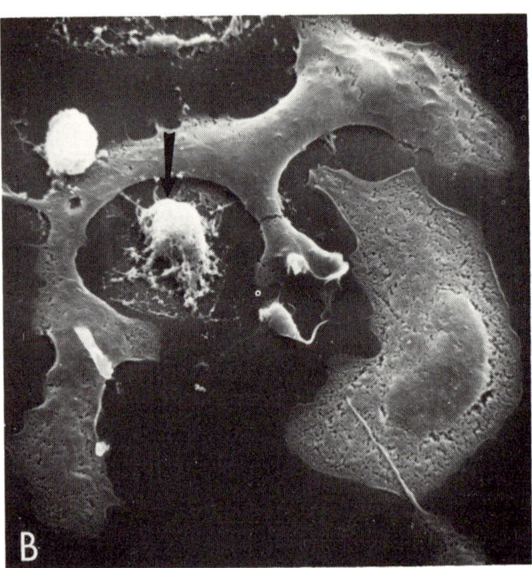

Figure 6. Activated macrophage killing of lymphoma cells *in vitro*. Scanning electron micrograph of BCG-activated murine peritoneal macrophages and lymphoma cells. (**A**) Immediately after explantation, lymphoma cells (round) sit in contact with the macrophage (flattened cell). (**B**) After triggering of macrophage with phorbol myristic acetate (PMA), macrophages kill lymphoma cells; arrow points to a dead lymphoma cell partially enveloped by a macrophage. (Photographs kindly supplied by C. Nathan and taken from Nathan et al. 1979a.)

macrophages. Elimination of these organisms requires the development of specific T-cell immunity and the subsequent production of activated macrophages. Clinically and histopathologically, the combination of chronicity and macrophage activation results in the granulomatous reactions that characterize these infections.

Other infections, such as those caused by rickettsiae and salmonellae, require a combination of antibody and activated macrophages for control. These tend to be of shorter duration and induce less of a chronic granulomatous reaction.

The induction of microbicidal macrophages is similar to the induction of tumoricidal macrophages. Infections such as BCG or Listeria generate macrophages that will kill other organisms nonspecifically (Williams et al. 1976). Killing of intracellular toxoplasma by BCG-activated macrophages is illustrated in Figure 7.

As with the induction of tumoricidal macrophages, the induction of microbicidal macrophages requires the production of a lymphokine by stimulated T-lymphocytes. Such a lymphokine-mediated killing pathway has been demonstrated with *Listeria monocytogenes*, *Toxoplasma gondii*, *L. tropica*, *Trypanosoma cruzi* and *Mycobacterium tuberculosis*. The responsible lymphokine in the listeria and *T. cruzi* system has similar physiochemical properties as MAF based on preliminary purification studies.

As stated earlier, the specific mechanisms used by macrophages to destroy a pathogen vary with the organism in question. Recent work has shown that macrophage killing of *T. gondii* requires the generation of two oxygen intermediates, hydroxyl radical and singlet oxygen (Murray et al. 1979). In contrast, killing of *T. cruzi*, like that of lymphoma cells, requires only hydrogen peroxide (Nathan et al. 1979). All the required oxygen intermediates in question are produced by activated macrophages. It is also possible that some less resistant organisms may be killed by lysosomal hydrolases or proteinases after opsonization and phagocytosis.

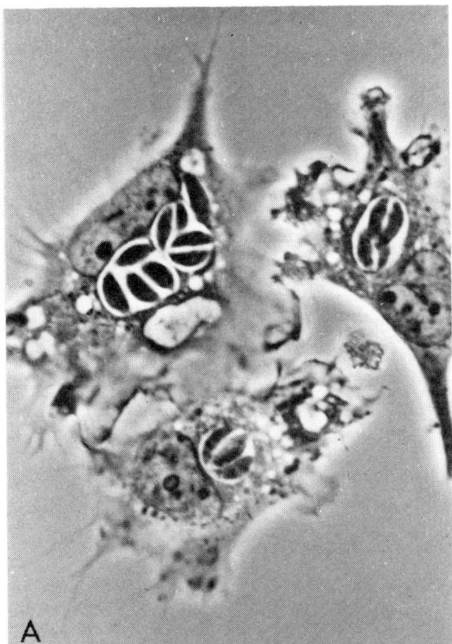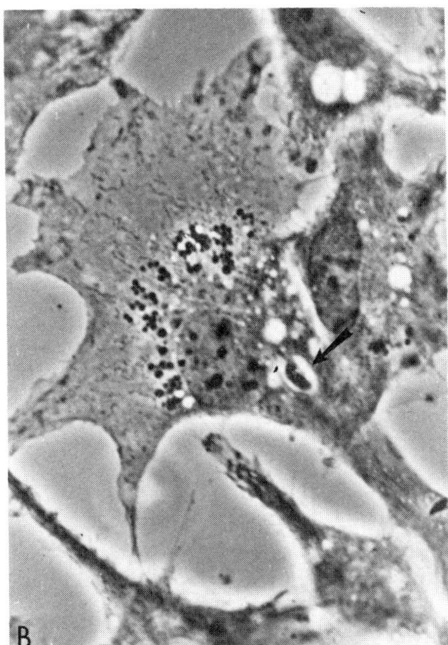

Figure 7. Killing of *toxoplasma gondii* by activated murine macrophages *in vitro*. (A) Normal peritoneal macrophages: Toxoplasma grow and divide unhindered after 18 hours of culture; toxoplasma will continue to divide until the macrophage is killed. (B) Toxoplasma immune macrophages: After 18 hours of culture no live toxoplasma are visible; One dead toxoplasma (arrow) is seen within a vacuole. (Photograph kindly supplied by H. Murray.)

While most work has focused on the role of T-cells in the induction of activated microbicidal macrophages, there are instances where these cells are generated in the absence of T-cells. Thus, both congenitally thymic (nu/nu) and thymectomized mice contain activated bactericidal macrophages. (Cheers and Waller 1975). There must therefore be other mechanisms to induce activated macrophages in addition to the T-cell mediated pathway. Possibilities include release of lymphokines from B-lymphocytes or direct activation of macrophages by bacterial products themselves.

Suppression of Immune Reactions

Conceptually, the immune response can be divided into three distinct stages. In the initial stage, macrophages act on lymphocytes by the mechanisms already described to produce a state of lymphocyte activation. The second stage involves the production of molecules such as MAF by these activated lymphocytes, which induce the differentiation of macrophages to an activated effector state. The third stage involves the suppression of lymphocyte activation by these hyperactivated macrophages. In this way, responses may be prevented from becoming excessively destructive. This area has been reviewed in a succinct and comprehensive fashion by Allison (1978).

The relationship between activated macrophages and immunosuppression has been well documented. *In vivo*, those agents that induce the formation of microbicidal or tumoricidal macrophages (i.e., BCG, *Corynebacterium parvum*, endotoxin, or the presence of tumors) also produce generalized states of immunosuppression. Lymphoid cells from such animals manifest poor immune responses *in vitro* as well, which can be reversed by removal of macrophages. Finally, these macrophages will suppress the *in vitro* immune responsiveness of normal lymphoid cells.

Macrophage suppression seems to involve a number of different mechanisms. Several workers have reported that suppression requires the direct contact of lymphocytes and macrophages. However, others have demonstrated that macrophages produce a number of soluble immunosuppressive factors. Perhaps in this situation, some soluble factors function better over very short distances. One mechanism exists that does seem to demand the direct proximity of cells. Ptak et al. (1977) have demonstrated that macrophage membranes will block immune responses *in vitro*. They postulate that these membranes bind necessary growth factors, thereby preventing lymphocyte activation.

The list of soluble suppressive materials secreted by macrophages is very long and includes many of the substances that are secreted by stimulated macrophages (as described in a previous section). One of the best studied is a low molecular weight dialyzable material. This compound is heat stable, pronase resistant, and produced in increased amounts when macrophages are fed dead cells. Recent work suggests that this substance is the nucleoside thymidine that blocks immune responses by interfering with several enzymes that are essential for DNA replication.

A second group of compounds that are very important inducers of immunosuppression are the prostaglandins. *In vivo*, exogenous prostaglandins will decrease antibody formation and graft rejection. *In vitro*, exogenous prostaglandins will decrease antigen- and mitogen-induced lymphocyte stimulation.

Several other macrophage products are potentially immunosuppressive. Interferon will block a number of immune responses *in vitro* and *in vivo*. Given the requirement for reduced glutathione and the effect of reducing agents such as 2-ME, it is also likely that excess oxidative substances would be immunosuppressive. Thus, the major oxygen metabolites secreted by activated macrophages such as hydrogen peroxide, superoxide anion, and hydroxyl radical may also have immunosuppressive consequences.

The relationship between macrophage activation and immunosuppression is vividly illustrated by work on the soluble immune response suppressor (SIRS) originally described by Pierce and his coworkers. Found in the supernatant of concanavalin A-treated spleen cells, SIRS is a substance that depresses antibody formation *in vitro*. It does not act directly on lymphocytes but instead acts by rendering macrophages suppressive (Tadakuma and Pierce 1976). Preliminary biochemical analysis suggests that SIRS is similar physiochemically to the lymphokine, MIF (Tadakuma et al. 1976).

A role for macrophage-mediated suppression has been suggested in several clinical conditions. A state of immunosuppression or anergy is commonly seen in diseases associated with a high degree of macrophage activation such as miliary tuberculosis or lepromatous leprosy. An antigen-specific (lepromin) adherent suppressor cell has been identified in lymphoid cell cultures derived from patients with lepromatous leprosy, and such a cell could account for the profound state of clinical anergy seen in this disease (Mehra et al. 1979). Similarly, Hodgkins disease is characterized by anergy, and the characteristic abnormal cell of this condition (the Reed–Sternberg giant cell) has been found to possess many macrophage-like properties (Kaplan and Gartner 1977). These patients possess a macrophage-like suppressor cell, and recent work demonstrated that suppression was due to excess secretion of prostaglandins. Thus, *in vitro* suppression was blocked by indomethacin, the inhibitor of the cyclo-oxygenase enzyme that generates prostaglandins.

Conclusion

This chapter has been a relatively condensed discussion about a cell that is extremely complex but very important. Several concepts have been emphasized and are worth repeating. The first is that the macrophage subserves a number of different functions in the body. These range from its probably original primitive function, that of a scavenger, to a more sophisticated function, the enhancement of immunity. The second concept is that the macrophage differentiates and acquires the capacity to mediate certain functions as it differentiates. The last concept is that the macrophage is a cell that interacts with many other components of the immune and host defense system as well as other systems. The important interaction between the macrophage and lymphocytes, complement, and immunoglobulin have been discussed. Interactions with the wound-healing system and the liver have been alluded to and are discussed more fully in other chapters. Undoubtedly, future research will uncover additional interactions and physiological roles for this fascinating cell.

References and Selected Bibliography

Allison, A.C. (1978) Mechanisms by which activated macrophages inhibit lymphocyte responses. Immunol. Rev. 40,3.

Baba, T., Yoshida, T., Yoshida, T., Cohen, S. (1979) Suppression of cell-mediated immune reactions by monosaccharides. J. Immunol. 122, 838.

Beller, D.I., Unanue, E.R. (1977) Thymic maturation *in vitro* by a secretory product from macrophages. J. Immunol. 118,1780–1787.

Ben-Sasson, S.Z., Lipscomb, M.F., Tucker, T.F., Uhr, J.W. (1977) Specific binding of T lymphocytes to macrophages. II. Role of macrophage-associated antigen. J. Immunol. 119,1493–1500.

Bentley, C., Fries, W., Brade, V. (1978) Synthesis of Factors D, B and P of the Alternative Pathway of Complement Activation, as well as of C3, by Guinea-pig peritoneal macrophages *in vitro*. Immunology 35,971.

Bianco, C., Eden, A., Cohn, Z.A. (1976) The induction of macrophage spreading: Role of coagulation factors and the complement system. J. Exp. Med. 144,1531

Bianco, C., Götze, O., Cohn, Z.A. (1979) Regulation of macrophage migration by products of the complement system. Proc. Natl. Acad. Sci. 76,888–891.

Bodel, P. (1978) Spontaneous pyrogen production by mouse histiocytic and myelomonocytic tumor cell lines *in vitro*. J. Exp. Med. 147,1503–1516.

Bodel, P., Miller H. (1977) Differences in pyrogen production by mononuclear phagocytes and by fibroblasts or HeLa cells. J. Exp. Med. 145,607–617.

Bodel, P., Nichols, B.A., Bainton, D.F. (1977) J. Exp. Med. 145,264–274.

Borges, J.S., Johnson, W.D. (1975) Inhibition of multiplication of *Toxoplasma gondii* by human monocytes exposed to T-lymphocyte products. J. Exp. Med. 141,483.

Buchmuller, Y., Mauel, J. (1979) Studies on the mechanisms of macrophage activation. II. Parasite destruction in macrophages activated by supernates from concanavalin A-stimulated lymphocytes. J. Exp. Med. 150,359.

Calderon, J., Unanue, E.R. (1974) The release of antigen molecules from macrophages: Characterization of the phenomena. J. Immunol. 112,1804–1814.

Carswell, E.A., Old, L.I., Kassel, R.L., Green, S., Fiore, N., Williamson, B. (1975) An endotoxin-induced serum factor that causes necrosis of tumors. Proc. Nat. Acad. Sci. 72,3666–3670.

Cheers, C., Waller, R., (1975) Activated macrophages in congenitally athymic "nude" mice and in lethally irradiated mice. J. Immunol. 115, 844–847.

Chen, C., Hirsch, J.C. (1972) The effects of mercaptoethanol and of peritoneal macrophages on the antibody-forming capacity of nonadherent mouse spleen cells *in vivo*. J. Exp. Med. 136,604.

Cline, M.J., Sumner, M.A. (1972) Bone marrow macrophage precursors. I. Some functional characteristics of the early cells of the mouse macrophage series. Blood 40,62–69.

Cohn, Z.A. (1975) Macrophage physiology. Fed. Proc. 34,1725–1975.

Cohn, Z.A. (1978) The maturation and activation of mononuclear phagocytes: Fact, fancy and future. J. Immunol. 121,813.

Cowing, C., Schwartz, B.D., Dickler, H.B. (1978) Macrophage Ia Antigens I. Macrophage populations differ in their expression of Ia antigens. J. Immunol. 120,378–384.

Cudkowicz, G., Yung, Y.P. (1977) Abrogation of resistance to foreign bone marrow grafts by carrageenans. I. Studies with the anti-macrophage agent seakem carrageenan. J. Immunol. 119,483–487.

Edelson, P.J., Cohn, Z.A. (1976) 5′-Nucleotidase activity of mouse peritoneal macrophages. I. synthesis and degradation in resident and inflammatory populations. J. Exp. Med. 144,1581–1595.

Erb, P., Feldmann, M. (1975) The role of macrophages in the generation of T-helper cells. II. The genetic control of the macrophage-T-cell interaction for helper cell induction with soluble antigens. J. Exp. Med. 142,460–472.

Goodwin, J.S., Messner, R.P., Bankhurst, A.D., Peake, G.T., Saiki, J.H., Williams, R.C. (1977) Prostaglandin-producing suppressor cells in Hodgkin's Disease. N. Engl. J. Med. 297,963–968.

Goren, M.B. (1977) Phagocyte lysosomes: Interactions with infectious agents, phagosomes, and experimental perturbations in function. Ann. Rev. Microbiol. 31,507–533.

Götze, O., Bianco, C., Cohn, Z.A. (1979) The induction of macrophage spreading by factor B of the properdin system. J. Exp. Med. 149,372–386.

Griffin, J.A., Griffin, F.M. (1979) Augmentation of macrophage complement receptor function *in vitro*. 1. Characterization of cellular interactions required for the generation of a T-lymphocyte product that enhances macrophage complement receptor function. J. Exp. Med. 150,653–675.

Havemann, K., Schmidt, W. (1973) Potentiating effect of adherent cell supernatants on lymphocyte proliferation. In: *Proc. 8th Leukocyte Culture Confer.*, Lindahl-Kiessling, K. (ed.), New York, Academic Press.

Higgins, T.J., Sabatino, A.P., Remold, H.G., David J.R. (1978) Possible role of macrophage glycolipids as receptors for migration inhibitory factor (MIF). J. Immunol. 121,880–886.

Hoffeld, J.T., Metzger, Z., Oppenheim, J.J. (1980) Oxygen-derived metabolites as suppressors of immune responses *in vitro*. Lymphokine Reports. 2, in press.

Immunological Reviews (1978) Role of Macrophages in the Immune Response. Vol. 40.

Johnston, R.B., Godzik, C.A., Z.A. Cohn. (1978) Increased superoxide anion production by immunologically activated and chemically activated macrophages. J. Exp. Med. 148,115.

Kaplan, H.S., Gartner, S. (1977) Sternberg-Reed Giant cells of Hodgkin's Disease: Cultivation *in vitro*, heterotransplantation, and characterization as neoplastic macrophages. Int. J. Cancer. 19(4),511–525.

Kaplan, A.M., Bear, H.D., Kirk, L., Cummins, C., Mohanakumar, T. (1978) Relationship of expression of a cell-surface antigen on activated murine macrophages to tumor cell cytotoxicity. J. Immunol. 120,2080–2085.

Karnofsky, M.L., Lazdins, J.K. (1978) Biochemical criteria for activated macrophages. J. Immunol. 121:809–813.

Kurland, J.I., Bockman, R. (1978) Prostaglandin E production by human blood monocytes and mouse peritoneal macrophages. J. Exp. Med. 147,952–957.

Kurland, J.I., Pelus, L.M., Ralph, P., Bockman, R.S., Moore, M.A.S. (1979) Induction of prostaglandin E synthesis in normal and neoplastic macrophages: Role for colony-stimulating factor(s) distinct from effects on myeloid progenitor cell proliferation. Proc. Natl. Acad. Sci. U.S.A. 76, 2326–2330.

Leonard, E.J., Skeel, A.H. (1978) Isolation of macrophage stimulating protein (MSP) from human serum. Exp. Cell Res. 114,117–126.

Leonard, E.J., Ruco, L.P., Meltzer, M.S. (1978) Characterization of macrophage activation factor, a lymphokine that causes macrophages to become cytotoxic for tumor cells. Cell Immunol. 41,397.

Lipsky, P.E., Rosenthal, A.S. (1973) Macrophage-lymphocyte interaction. I. Characteristics of the antigen independent binding of guinea pig thymocytes and lymphocytes to syngeneic macrophages. J. Exp. Med. 138,1194.

Lipsky, P.E., Rosenthal, A.S. (1975) Macrophage-lymphocyte interaction. II. Antigen-mediated physical interactions between immune guinea pig lymph node lymphocytes and syngeneic macrophages. J. Exp. Med. 141,138–154.

Littman, B.H., Ruddy, S. (1977) Production of the second component of complement by human monocytes: Stimulation by antigen-activated lymphocytes or lymphokines. J. Exp. Med. 145,1344–1352.

Mackaness, G.B. (1970) The monocyte in cellular immunity. Semin. Hematol. 7,172.

Mehra, V., Mason, L.H., Fields, J.P., Bloom, B.R. (1979) Lepromin-induced suppressor cells in patients with leprosy. J. Immunol. 123(4),1813–1817.

Mizel, S.B., Rosenstreich, D.L. (1979) Regulation of lymphocyte activating factor (LAF) production and secretion in P388D$_1$ cells: Identification of high molecular weight precursors of LAF. J. Immunol. 122,2173–2179.

Mizel, S.B. (1980) Biochemical characteristics of macrophage-derived lymphostimulatory factors. In: *Molecular Mediators of Cellular Immunity: Characterizations and Actions.* Hadden, J.W., Stewart, W.E. New Jersey, Humana Press, in press.

Mogensen, S.C. (1979) Role of macrophages in natural resistance to virus infections. Microbiol. Rev. 43(1),1–26.

Moore, R.M., Goodrum, K.J., Couch, R.E., Berry, L.J. (1978) Elicitation of endotoxemic effects in C3H/HeJ mice with glucocorticoid antagonizing factor and partial characterization of the factor. Infect. Immun. 19,79–86.

Moore, R.M., Vogel, S.N., Wahl, L.M., Mergenhagen, S.E. (1980) Factors Influencing Lipopolysaccharide-induced Interferon Production: Genetic and Cellular Sources and Influences of Prostaglandins and Colony Stimulating Factors. In: *Microbiology,* Schlesinger, D.S. (ed.) Washington, D.C., Am. Soc. of Microbiol.

Mosier, D.E. (1967) A requirement for two cell types for antibody formation *in vitro*. Science 158,1573.

Murray, H.W., Juangbhanich, C.W., Nathan, C.F., Cohn, Z.A. (1979) Macrophage oxygen-dependent anti-microbial activity II. The role of oxygen intermediates. J. Exp. Med. 150,950.

Muschel, R.J., Rosen, N., Bloom, B.R. (1977) Isolation of variants in phagocytosis of a macrophage-like continuous cell line. J. Exp. Med. 145,175–186.

Nathan, C.F., Asofsky, R., Terry, W.D. (1977) Characterization of the nonphagocytic adherent cell from the peritoneal cavity of normal and BCG-treated mice. J. Immunol. 118,1612–1621.

Nathan, C.F., Root, R.K. (1977) Hydrogen peroxide release from mouse peritoneal macrophages. Dependence on sequential activation and triggering. J. Exp. Med. 146,1648–1662.

Nathan, C.F., Nogueira, N., Juangbhanich, C. Ellis, J., Cohn, Z.A. (1979) Activation of macrophages *in vivo* and *in vitro*. Correlation between hydrogen peroxide release and killing of *trypanosoma cruzi*. J. Exp. Med. 149,1056.

Nathan, C.F., Silverstein, S.C., Bruckner, Cohn, Z.A. (1979) Extracellular cytolysis by activated macrophage and granulocytes. I. Pharmacologic triggering of effector cells and the release of hydrogen peroxide. J. Exp. Med. 146,1648.

Nelson, D.S., Ed. (1976) *Immunobiology of the Macrophage.* New York, Academic Press.

Newman, W., Gordon, S., Hammerling, U., Senik, A., Bloom, B.A. (1978) Production of migration inhibition factor (MIF) and an inducer of plasminogen activator (IPA) by subsets of T cells in MLC. J. Immunol. 120,927–931.

Noguiera, N., Cohn, Z.A. (1978) *Trypanosoma cruzi: in vitro* induction of macrophage microbicidal activity. J. Exp. Med. 148,148–288.

North, R.J. (1973) Importance of thymus-derived lymphocytes in cell-mediated immunity to infection. Cell Immunol. 7,166.

Nossal, C.J.V., M.B., Abbot, A., Mitchell, J. (1967) Antigens in immunity. XIV. Electron Microscopic Radioautographic studies of antigen capture in the lymph node medulla. J. Exp. Med. 127,263–275.

O'Brien, A.D., Scher, I., Formal, S.B. (1979) Effect of silica on the innate resistance of inbred mice to *Salmonella typhimurium* infection. Infect. Immun. 25,573.

Opitz, H.G., Opitz, U. Lemke, H., Flad, H.D., Hewlett, G., Schlumberger, H.D. (1977) Humoral primary immune response *in vitro* in a homologous mouse system: Replacement of fetal calf serum by a 2-mercaptoethanol or macrophage-activated fraction of mouse serum. J. Immunol. 119,2089.

Oppenheim, J.J., Leventhal, B.G., Hersch, E.M. (1968) The transformation of column-purified lymphocytes with nonspecific and specific antigenic stimuli. J. Immunol. 101,262.

Pels, E., den Otter, W. (1974) The role of a cytophilic factor from challenged immune peritoneal lymphocytes in specific macrophage cytotoxicity. Cancer Res. 34,3089–3094.

Postlethwaite, A.E., Kang, A.H. (1976) Guinea pig lymphocyte-derived macrophage aggregation factor: Its separation from macrophage migration inhibitory factor. J. Immunol. 117,1651–1655.

Poulter, L.W., Turk, J.L. (1975) Studies on the effect of soluble lymphocyte products (lymphokines) on macrophage physiology I. Early changes in enzyme activity and permeability. Cell Immunol. 20,12–24.

Ptak, W., Naidorf, K.F., Gershon, R.K. (1977) Interference with the transmission of T cell-derived messages by macrophage membranes. J. Immunol. 119,444–449.

Ralph, P., Malcolm, A.S., Moore, Nilsson, K. (1976) Lysozyme synthesis by established human and murine histiocytic lymphoma cell lines. J. Exp. Med. 143,1528–1533.

Reed, W.P., Lucas, Z.J. (1975) Cytotoxic activity of lymphocytes V. Role of soluble toxin in macrophage-inhibited cultures of tumor cells. J. Immunol. 115,395–404.

Remold, H.G. (1973) Requirement for α-L-fucose on the macrophage membrane receptor for MIF. J. Exp. Med. 138,1065.

Rosenstreich, D.L., Oppenheim, J.J. (1976) The role of macrophages in the activation of T and B lymphocytes *in vitro*. In: *Immunobiology of the Macrophage*. Nelson, D.S. (ed.) New York, Academic Press, p. 162.

Rosenstreich, D.L., Mizel, S.B. (1978) The participation of macrophages and macrophage cell lines in the activation of T lymphocytes by mitogens. Imunol. Rev. 40,102–135.

Rosenthal, A.S., Shevach, E.M (1973) The function of macrophages in antigen recognition by guinea pig T-lymphocytes I. Requirement for histocompatible macrophages and lymphocytes. J. Exp. Med. 138,1194.

Rosenthal, A.S., Barcinski, M.A., Blake, J.T. (1977) Determinant selection: a macrophage-dependent immune response gene function. Nature 267,156.

Ruco, L.P., Meltzer, M.S. (1977) Macrophage activation for tumor cytotoxicity: Induction of tumoricidal macrophages by supernatants of PPD-stimulated bacillus calmette-Guerin-immune spleen cell cultures. J. Immunol. 119,889–896.

Ruco, L.P., Meltzer, M.S. (1978) Defective tumoricidal capacity of macrophages from C3H/HeJ mice. J. Immunol. 120L 329–334.

Russell, S.W., Doe, W.F., McIntosh, A.T. (1977) Functional characterization of a stable, noncytolytic stage of macrophage activation in tumors. J. Exp. Med. 146,1511–1520.

Schnyder, J., Baggiolini, M. (1978) Secretion of lysosomal hydrolases by stimulated and nonstimulated macrophages. J. Exp. Med. 148,435–450.

Schultz, R.M., Papamatheailis, J.D., Chirigos, M.A. (1977) Interferon: An inducer of macrophage activation by polyanions. Science 197,674.

Silverstein, S.C., Michl, J., S-S.J. Sung (1978) Phagocytosis. In: *Transport of Macromolecules in Cellular Systems*. Silverstein, S.C. (ed.), Berlin, Dahlem, Konferenzen, p. 261.

Snyderman, R., Pike, M.C., McCarley, D., Lang, L. (1975) Quantification of mouse macrophage chemotaxis *in vitro*: Role of C5 for the production of chemotactic activity. Infect. Immun. 11,488–492.

Stadecker, M.J., Calderon, J., Karnovsky, M.L. Unanue, E.R. (1977) Synthesis and release of thymidine by macrophages. J. Immunol. 119,1738.

Stanley, E.R., Cifone, M., Heard, P.M., Defendi, V. (1976) Factors regulating macrophage production and growth. Identity of colony stimulating factor and macrophage growth factor. J. Exp. Med. 143,63.

Stanley, E.R., Heard, P.M. (1977) Factor regulating macrophage production and growth: Purification and some properties of the colony stimulating factor from medium conditioned by mouse L cells. J. Biol. Chem. 252(12), 4305.

Steinman, R.M., Cohn, Z.A. (1973) Identification of a novel cell type in peripheral lymphoid organs of mice. I. Morphology, quantitation, tissue distribution. J. Exp. Med. 137,1142–1162.

Steinman, R.M, Witmer, M.D. (1978) Lymphoid dendritic cells are potent stimulators of the primary mixed leukocyte reaction in mice. Proc. Natl. Acad. Sci. U.S.A. 75,5152–5136.

Stinnett, J.D., Kaplan, A.M., Morahan, P.S. (1976) Identification of a macrophage specific cell surface antigen. J. Immunol. 116,273–278.

Sutton, J.S. Weiss, L. (1966) Transformation of monocytes in tissue culture into macrophages, epithelioid cells, and multinucleated giant cells. J. Cell Biol. 28,303–332.

Tadakuma, T., Pierce. C.W. (1976) Site of action of a soluble immune response suppressor (SIRS) produced by concanavalin A-activated spleen cells. J. Immunol. 117(3),967–972.

Tadakuma, T., Kuhner, A.L., Rich, R.R., David, J.R., C.W. Pierce. (1976) Biological expressions of lymphocyte activation. V. Characterization of a soluble immune response suppressor (SIRS) produced by concanavalin-A-activated spleen cells. J. Immunol. 117,323.

Thomas, D.W., Yamashita, U., Shevach, E.M. (1977) Nature of the antigenic complex recognized by T lymphocytes. IV. Inhibition of antigen-specific T cell proliferation by antibodies to stimulator macrophage Ia antigens. J. Immunol. 119,223–226.

Tormey, D.C., Imrie, R.C., G.C. Mueller. (1972) Identification of transferrin as a lymphocyte growth promoter in human serum. Exp. Cell Res. 74,163.

Treves, A.J., Schechter, B., Cohen, I.R., Feldman, M. (1976) Sensitization of T lymphocytes *in vitro* by syngeneic macrophages fed with tumor antigens. J. Immunol. 116,1059–1064.

Unanue, E.R., Cerottini, J.C. (1970) The immunogenicity of antigen bound to the plasma membrane of macrophages. J. Exp. Med. 131,711.

Unkeless, J.C., Gordon, S., Reich, E. (1974) Secretion of plasminogen activator by stimulated macrophages. J. Exp. Med. 139,834–850.

Unkeless, J.C. (1977) The presence of two Fc receptors on mouse macrophages: Evidence from a variant cell line and differential trypsin sensitivity. J. Exp. Med. 145,931.

van Furth, R., Ed. (1975) *Mononuclear Phagocytes in Immunity, Infection and Pathology.* London, Blackwell Scientific Publications.

van Furth, R., Ed. (1980) *Mononuclear Phagocytes: Functional Aspects.* The Hague, Mortinus Nijhoff.

van Waarde, D., Hulsing-Hesselink, E., van Furth, R. (1977) Properties of a factor increasing monocytopoiesis (FIM) occurring in serum during the early phase of an inflammatory reaction. Blood 50,727–742.

Vassalli, J.D., Reich, E. (1977) Macrophage plasminogen activator: induction by products of activated lymphoid cells. J. Exp. Med. 145,429–437.

Virolainen, M. (1968) Hematopoietic origin of macrophages as studied by chromosome markers in mice. J. Exp. Med. 127,943–951.

Vogel, S.N., Rosenstreich, D.L. (1979) Defective Fc receptor mediated phagocytosis in C3H/HeJ macrophages. I. Correction by lymphokine induced stimulation. J. Immunol. 123,2842–2850.

Wachsmuth, E.D., Staber, F.G. (1977) Changes in membrane-bound aminopeptidase on bone marrow-derived macrophages during their maturation *in vitro.* Exp. Cell Res. 109,269–276.

Wahl, L.M., Olsen, C.E., Sandberg, A.L., Mergenhagen, S.E. (1977) Prostaglandin regulation of macrophage collagenase production. Proc. Nat. Acad. Sci. U.S.A. 74,4955–4958.

Wahl, L.M., Wahl, S.M., Mergenhagen, S.E., Martin, G. (1974) Collagenase production by endotoxin-activated macrophages. Proc. Natl. Acad. Sci. U.S.A. 71,3598–3601.

Wahl, S.M., Altman, L.C., Oppenheim, J.J., Mergenhagen, S.E. (1974) *In vitro* studies of a chemotactic lymphokine in the guinea pig. Int. Arch. Allergy. 46,768.

Weinberg, J.B., Chapman, H.A., Hibbs, Jr., J.B. (1978) Characterization of the effects of endotoxin on macrophage tumor cell killing. J. Immunol. 121,72–80.

Weiss, S.J., King, G.W., Lobuglio, A.F. (1977) Evidence for hydroxyl radical generation by human monocytes. J. Clin. Invest. 60,370.

Werb, Z., Gordon, S. (1975) Elastase secretion by stimulated macrophages. J. Exp. Med. 142,361.

Williams, D.M., Sawyer, S., Remington, J.S. (1976) Role of activated macrophages in resistance of mice to infection with *Trypanosoma cruzi.* J. Infect. Dis. 134,610–614.

Wood, D.D., Cameron, P.M. (1975) Studies on the mechanism of stimulation of the humoral response of murine spleen cultures by culture fluids from human monocytes. J. Immunol. 114,1094–1100.

Yamashita, U., Shevach, E.M. (1977) The expression of Ia antigens on immunocompetent cells in the guinea pig. II. Ia antigens on macrophages. J. Immunol. 119,1584. .

T-Lymphocyte Development and Function

Ethan M. Shevach, M.D., and Joost J. Oppenheim, M.D.

The definition of the two major subpopulations of lymphocytes was derived from studies in chickens that demonstrated the bursa of Fabricius and the thymus are central lymphoid organs essential for the ontogenetic development of immunity in that species (Cooper et al. 1966). Surgical removal of one or both these organs in the newly hatched chicken permitted the recognition of two functionally distinct cell systems in the peripheral lymphoid tissues. Thymectomized chickens were deficient in their ability to develop delayed reactions, to exhibit graft-versus-host (GVH) reactions, and to reject skin allografts. Bursectomized chickens, on the other hand, rejected skin grafts and showed normal GVH reactions but were completely unable to form circulating antibodies even following strong antigenic stimulation. Analysis of these two major subpopulations of lymphocytes was facilitated by the development of methods for the detection of surface antigenic markers that readily allowed one to distinguish T-lymphocytes from B-lymphocytes (Raff 1971).

The T-lymphocyte plays a critical role in the regulation of the immune response and appears to be the central regulatory cell responsible for focusing inflammatory responses on specific targets in a variety of ways. These include direct cytotoxic effects on target cells (see Chapter 7), collaborative interactions with B-lymphocytes (see Chapter 8), and interactions with many other cell types (see Chapter 10). This chapter will discuss the development, mechanisms of activation, and functions of T-lymphocytes and will focus its attention on the different subclasses of T-lymphocytes identified and the mechanisms that control T–T-lymphocyte interactions. Unless otherwise indicated this chapter will be based on studies of murine T-lymphocytes.

Ontogeny of T-Lymphocytes

The thymus in mammals provides the site for the development of the large and functionally heterogeneous class of T-lymphocytes. T-lymphocyte precursors found in the bone marrow and spleen of the adult and the yolk sac and liver of the fetus must

From the Laboratory of Immunology, National Institute of Allergy and Infectious Diseases, and Laboratory of Microbiology and Immunology, National Institute of Dental Research, National Institutes of Health, Bethesda, Maryland.

migrate to the thymus and differentiate there before acquiring the phenotypic characteristics and immunological reactivity of mature T-lymphocytes (Stutman 1978). The T-lymphocyte stem cells in the hematopoietic tissues already constitute a population of cells distinct from B-lymphocyte precursors. These prothymocytes in bone marrow express the enzyme terminal deoxynucleotidyl transferase (TdT), which is capable of adding mononucleotides to a DNA segment in the absence of a template. The reticuloepithelial cells of the thymus seem to provide a microenvironment needed for further T-lymphocyte maturation. Immature cells in the outer cortex undergo intense proliferation at too high a rate to be accounted for by the export of cells from the thymus.

Antigenic and Functional Markers of Murine T-Lymphocytes

To trace the development of the prothymocyte through the various stages of intrathymic development, it is appropriate to review at this time a series of antigens, markers, and functional properties that allow one to distinguish different populations of T-cells not only within the thymus but also in peripheral lymphoid organs (Table 1) (see Hunter et al. 1978 for review). The Thy-1 antigen is expressed on all murine T-lymphocytes, but it can also be found on brain and some epithelial tissues. The products of the H-2D- and H-2K genes, the so-called classic histocompatibility antigens, are displayed on all cells both lymphoid and nonlymphoid. The products of the TL genes are found only in mice of certain strains and are found exclusively on thymocytes. The Lyt-123 antigens are expressed on thymocytes and peripheral T-lymphocytes (see Section on T-lymphocyte function). Most thymocytes are Lyt-$1^+2^+3^+$, but a small number of Lyt-1^+23^- medullary thymocytes have been identified. Antibodies to the Qa-1 antigen were initially found to be contaminants in some anti-TL sera (Stanton and Boyse 1976). The Qa-1 antigen is found on some thymocytes, but it has proved to be most useful in distinguishing subpopulations of peripheral T-cells (see section on T-lymphocyte function). The gene for the Qa-1 antigens maps between the genes controlling the H-2D and the TL antigens on the murine 17th chromosome. The products of the I-J region of the major histocompatibility complex (MHC) are found on some of the Lyt-1^-23^+ subset of peripheral T-cells (Murphy et al. 1976). The ability to raise alloantisera that react with these differentiation antigens stems from their polymorphic nature; in the case of the Ly systems, the entire inbred mouse population has only two allelic forms; the same holds true for the Thy-1 antigen.

In addition to the antigens listed above, which are primarily detected by cytotoxicity or immunofluorescence, a number of functional characteristics can be used to distinguish thymocyte subpopulations. Although it is not clear how lectin-induced mitogenesis relates to physiological immune function, the lymphocyte proliferative response to various lectins is operationally useful to distinguish immature thymocytes from mature T-cells. The cortical thymocytes fail to respond to phytohemagglutinin (PHA) and respond to a moderate degree to concanavalin A. In contrast, medullary thymocytes react to both PHA and concanavalin A. The lectin, peanut agglutinin (PNA), binds terminal galactose residues with a strong preference for Gal-GalNAc (London et al. 1978). It binds to cortical thymocytes and bone marrow cells but largely fails to bind to medullary thymocytes and mature peripheral T-cells. Treatment of mice with corticosteroids produces a rapid involution of the thymus, with almost complete elimination of the cells in the cortical areas said to be cortisone sensitive. Thymocytes in the medulla suffer some depletion, but in general survive better; peripheral T-cells are also relatively steroid resistant.

Table 1. Antigenic Markers on Murine Lymphocytes

Marker	Coded for by chromosome	Molecular weight	Distribution
Thy-1	9	25,000	All T cells; brain, some epithelial tissues
H-2	17	$40,000 + B_2 - m$	All cells
TL	17	$40,000 + \beta_2 - m$	Thymocytes of some strains
Lyt-1	19	65,000	Thymocytes; some peripheral T-lymphocytes
Lyt-2/3	6	50-60,000 dimer	Thymocytes; some peripheral T-lymphocytes
Qa-1	17	$40,000 + \beta_2 - m$	Some peripheral T-lymphocytes
I-J	17	Unknown	T-suppressor cells

Intrathymic T-Cell Differentiation

In addition to lymphoid cells, the thymus contains cells that make up the thymic stroma including epithelial cells, macrophages, fibroblasts, and myogenic cells. The cell type most important for interaction with lymphocytes is not known and is now the subject of intense investigation (Cantor and Weissman 1976). Three lymphoid cell types (Table 2) can be distinguished in the thymus:

(1) Cortical thymocytes, about 80% of the lymphoid thymocytes, are small dense cells in the thymic cortex, which are in the G_0 or G_1 phases of the cell cycle. They are bound by PNA and express high levels of Thy-1, TL, and the Lyt-123 antigens; low levels of H-2; and do not express Qa-1. They are highly sensitive to steroids and survive poorly in culture. These cells respond only modestly to the mitogenic effects of concanavalin A, not at all to PHA, and have no activity in other assays of mature T-cell function.

(2) Medullary thymocytes, 10 to 15% of the lymphoid cells in the thymus, are in the medulla. These cells are larger and less dense than cortical thymocytes. They display high levels of H-2, low levels of Thy-1, and no TL; they are relatively steroid resistant and survive better in culture than cortical thymocytes. These cells are not bound by PNA; many are $Lyt\text{-}1^+2^+3^+$, but the $Lyt\text{-}1^+23^-$ subpopulation is present in this group. All the immunocompetent thymic cells can be found in this heterogeneous population of medullary cells. These cells are often in the S-phase of the cell cycle and also incorporate ^3H-thymidine and may represent a population that is at least partly self-renewing.

(3) Subcapsular thymocytes, a very small fraction of thymocytes are found around

Table 2. Subpopulations of Lymphocytes in the Thymus

	Percent	H-2	TL	Thy-1	Lyt	Peanut agglutinin	Steroid sensitivity
Cortical thymocytes	80	Low	High	High	High; all Lyt-$1^+2^+3^+$	All positive	High
Medullary thymocytes	10–15	High	Absent	Low	Lyt-$1^+2^+3^+$; a few Lyt-$1^+2^-3^-$	Negative	Low
Subcapsular thymocytes	<5	?	?	High	?	Some positive	High

the outer rim of the cortex. This population is composed of large blast cells with a very high mitotic rate; they are cortisone sensitive and bear large amounts of Thy-1. When these cells are labeled selectively by topical application of ^{3}H-thymidine to the thymic capsule, radioactive label is subsequently found in both small, high Thy-1 cells and in larger, low Thy-1 cells. Thus, the class of subcapsular thymocytes appears to include precursors of both medullary and small cortical thymocytes.

It is believed that the majority of cells generated in the thymus die there. It is not yet known what signal is responsible for the migration of any thymocyte into the periphery. The phenotype of the medullary thymocyte resembles that of the mature peripheral T-cell, and it is the cell that seeds the periphery. It is likely that the noncycling fragile cortical thymocytes might be in the process of differentiating or else marked for intrathymic death. However, it is possible that some immature small cortical cells are exported directly to the spleen where they become a subclass of immunologically incompetent "post-thymic precursor" cells which depend on the presence of a thymus for maintenance and further maturation (Stutman 1978). They disappear within a few weeks in thymectomized animals but can develop into functional T-cells if humoral factors from the thymus are provided.

It should also be considered that different classes of effector T-cells descend from different classes of thymocytes. Some evidence suggests that Lyt-1^-23^+ cells are derived from Lyt-$1^+2^+3^+$ cells. It is likely that Lyt-1^+23^- cells are derived from Lyt-1^+23^+ precursors as preliminary studies indicate that macrophage mediators can induce Lyt-$1^+2^+3^+$, PNA$^+$ thymocytes to change their phenotype to Lyt-1^+23^-, PNA$^-$, which is characteristic of medullary thymocytes.

The mechanism by which prothymocytes become immunocompetent T-cells in the thymus is being actively studied. In Chapter 7, detailed discussion covers how the thymus controls or regulates the development of "self" recognition. In this chapter, we will focus our discussion on a number of *in vitro* systems for studying thymocyte differentiation. One of the methods used to examine thymocyte differentiation *in vitro* is the use of fetal thymus organ cultures (Owen et al. 1977). If the thymus is put into culture at day 13 of gestation, the supply of new stem cells is cut off, but the large lymphoblasts already present in the thymus proliferate extensively. During 2 weeks in culture the proliferating cells undergo changes similar to those seen *in vivo*. The cultures become TL$^+$ then change to a predominantly Thy-1^+, TL$^-$ phenotype. Because the culture is heterogeneous, it is impossible to determine if an individual TL$^+$ cell converts to the TL$^-$ phenotype. Both lymphoid and epithelial cells in these cultures appear to retain some physiological function. Alloreactive cells can be generated after 1 week in culture, and if the cultured gland is grafted into an animal, they appear to organize a vascularized thymus which is histologically normal. Although this model has allowed *in vitro* confirmation of some of the processes taking place *in vivo*, its major disadvantage is that defined populations cannot be added or removed and the sequential changes that the cells undergo remain to be clarified.

Thymic Hormones

Another *in vitro* approach has been to study the effects of thymic extracts on T-cell development. A large body of evidence suggests that T-cell differentiation within the thymus may be regulated by hormones or factors synthesized by the thymic reticular or epithelial cells. A number of these peptides have been isolated and characterized biochemically. Activity in these extracts from thymus is found in fractions varying in molecular weight from < 1000 to 12,600. One of the best characterized of these thymic "hormonal" factors is thymopoietin, a polypeptide with a molecular weight of 5562 with a known sequence of 49 amino acids (Goldstein et al. 1977). It is isolated from thymic epithelial cells and promotes prothymocyte to thymocyte differentiation selectively. A synthetic tridecapeptide corresponding to residues 29 to 41 of thymopoietin has been shown to have all the differentiation-inducing properties of the parent molecule. It has no effect on B-cell differentiation.

One of the major *in vitro* assays for thymic hormonal activity involves the induction of the typical surface phenotype of the differentiated cortical thymocyte (TL^+, Thy-1^+, Lyt-$1^+2^+3^+$) on prothymocytes that do not bear these markers. Prothymocytes are isolated from mouse spleen and bone marrow by using discontinuous albumin gradients in which the less dense fractions (density <26% albumin) constituting less than 20% of the total population demonstrate approximately 15 to 25% Thy-1^- TL^- prothymocytes. Induction takes place rapidly over a period of 2.5 hours in the absence of associated cell proliferation. Brief exposure to thymopoietin suffices to initiate the differentiation processes. If the inducing agent is washed from the cells after 10 minutes, T-cell surface antigens still appear at 2 hours. Induction appears to be mediated by cyclic AMP. Thus, dibutyryl cyclic AMP or cyclic AMP itself is an effective inducing agent, and the induction process can be enhanced by theophylline, which elevates intracellular cyclic AMP and is inhibited by imidazole, which depresses intracellular cyclic AMP. Studies with drugs inhibiting macromolecular synthesis have shown that *in vitro* T-cell differentiation requires transcription of mRNA with new synthesis of protein and can be blocked by actinomycin-D and puromycin. Induction of T-cell differentiation does not require replication of DNA and is not inhibited by cytosine arabinoside or hydroxyurea. The prothymocyte may already be programmed for this differentiation step during embryonic development, and the thymic environment merely provides sufficient inducing agents.

When prothymocytes are treated with thymopoietin they do develop the capacity to respond to the mitogen concanavalin A. However, this response is quite modest when compared to that of fully mature T-cells; therefore, the *in vitro* induction with thymopoietin is capable of generating T-cells with the phenotypic markers of mature cells, but these cells remain largely devoid of functional immunologic competence.

In addition to thymopoietin, another polypeptide, called ubiquitin (molecular weight, 8451), whose amino acid sequence is known, is capable of inducing T-cell differentiation. However, ubiquitin is not a thymic hormone since it is also found in cells of many organisms, including yeasts, bacteria, and higher plants. Ubiquitin is nonselective in its inducing capacity as it can induce both T- and B-cell differentiation. The latter is assessed by the development of B-cell markers, such as the complement receptor, surface immunoglobulin, and surface Ia antigens. The C-terminal hexadecapeptide of ubiquitin has been synthesized and shown to have the inductive activity of the parent molecule.

Recently, an additional model system for T-cell differentiation has been described (Shaw et al. 1978). In these studies cortical thymocytes are purified by density gradient centrifugation or by affinity chromatography on PNA and then cultured with a mitogen such as PHA and a "costimulator" secreted by immunologically stimulated peripheral lymphocytes. The probable identity of costimulator to T-cell growth factor is discussed in Chapter 10. Cortical thymocytes do not proliferate in response to PHA unless costimulator is added and costimulator itself fails to promote proliferation if the mitogen is omitted. The cortical thymocytes also undergo differentiation involving increased surface H-2 expression, increased responsiveness to both PHA and concanavalin A, greater MLR activity, more helper activity for B-cells, and the generation of alloreactive cytotoxic effector cells. Furthermore, recent studies have demonstrated that T-cell growth factor treated athymic nu/nu spleen cells can yield functional alloreactive cytotoxic effector cells (Gillis et al. 1979). Thus, it may be possible to duplicate *in vitro* in the absence of a thymus the full spectrum of maturation of some prethymic lymphocytes.

Nonspecific T-Lymphocyte Activation by Mitogens

Before continuing the discussion of the development and immunologically specific functions of mature T-lymphocytes, we first discuss nonspecific polyclonal T-lymphocyte activation.

Two decades ago Nowell first observed that small "resting" lymphocytes were transformed into dividing lymphoblasts when cultured for several days with extracts of red kidney beans (*Phaseolus vulgaris*). The 180,000 molecular weight lectin, PHA, was the responsible mitogenic agent in the extract. A majority of lymphocytes from different sources are stimulated by PHA to undergo blastogenesis and several cell divisions; therefore it can also be called a polyclonal activator (Ling and Kay 1975). Since then many other plant lectins have been discovered to be mitogenic, including concanavalin A, soybean agglutinin (SBA), PNA, robinia pseudoacacia, and pokeweed mitogen (PWM). There are also many interesting nonlectins with potent mitogenic effects, such as sodium periodate ($NaIO_4$), a calcium ionophore (A23187), some antilymphocyte antisera (ALS), phorbol myristic acetate (PMA), and treatment of lymphocytes with neuraminidase followed by galactose oxidase (Novogrodsky 1976). These agents, in contrast to specific antigenic stimulants, activate sizeable proportions of lymphocytes from most vertebrate species in the absence of prior sensitization and are therefore classified as nonspecific mitogens.

Identity of Responding Lymphocytes

Distinct subpopulations of lymphocytes respond to different mitogens. Studies in which lymphocytes are fractioned into T- and B-cells by various means reveal that mitogens such as PHA, concanavalin A, ALS, $NaIO_4$, A23187, and PMA can activate purified T- but not B-cells. Conversely, mitogens, such as endotoxins and antiimmunoglobulins, stimulate only B- but not T-lymphocytes (see Chapter 8). However, it must be emphasized that when mixtures of T- and B-lymphocytes are stimulated by T-cell mitogens, both T- and B-cells can participate in the proliferative response. This B-cell proliferation is therefore said to be T-dependent. Although generally considered to be a potent B-cell mitogen, PWM cannot stimulate nude mouse splenic B-cells. Pokeweed mitogen functions only in the presence of at least 5 to 10% T-cells and is therefore also actually a T-cell-dependent mitogen. This mitogen-induced T–B collaboration appears to be mediated by lymphokines that can activate purified B-lymphocytes in the absence of T-cells (see Chapter 10).

Further evidence indicating that mitogens stimulate only certain classes of lymphocytes comes from studies of so-called experiments of nature. The lymphocytes from congenitally immunodeficient people and animals that lack B-lymphocytes are unable to produce antibodies (agammaglobulinemia) and can be activated only by T-cell mitogens. Conversely the lymphocytes from T-cell-deficient sources such as nude (nu/nu) mice respond only to B-cell mitogens. These types of observations led to the proposal that the capacity of lymphocytes to react to T-cell mitogens *in vitro* can be used as an indicator of T-cell function. Normal T-cell functions are requisite for the capacity to manifest *in vivo* cell-mediated immune responses such as delayed and contact hypersensitivity, allograft rejection, GVH disease, and chronic granulomatous inflammation. Therefore, the ability of mitogens to induce normal lymphocyte transformation is an *in vitro* correlate of cell mediated immunity (Oppenheim and Rosenstreich 1976).

Characteristics of Lymphocyte Reaction

There is considerable evidence that the degree of lymphocyte activation is a direct function of the number of mitogenic signals. The requirement for viable macrophages and their monokine products (e.g., lymphocyte activating factor) for mitogen induced lymphocyte transformation is discussed elsewhere (see Chapter 5). A minimal exposure time to optimal concentrations of exogenous mitogen is required to induce mitosis. However, once the cells reach S-phase, they become independent of the mitogenic signals. Thus, brief exposure or suboptimal doses of a mitogen will activate lymphocytes to go only from the resting (G_0) into the G_1 phase of the cell cycle without proceeding into the S-phase. The lymphocytes begin to synthesize RNA, proteins, lipids, and lymphokines during the G_1 phase (Resch 1976). More prolonged exposure to appropriate concentrations of the mitogens stimulates lymphocytes to go all the way from G_0 through G_1 into S and G_2 phases of the cell cycle. This is associated with a multiplicity of morphologic changes of small lymphocytes into medium and large lymphoblastoid cells that have increased cytoplasm as well as nuclear size, culminating in mitosis. Mitogens also stimulate the various lymphocyte subpopulations to differentiate and to perform their various functions of lymphokine production, suppression, T-cell-dependent antibody production, and cytotoxicity. The *in vitro* activation of lymphocytes can therefore be assessed by many different means, including microscopy, flow cytometry, the incorporation of various radiolabeled precursors into proteins, lipids, RNA, or DNA by the production of lymphokines or antibodies, or the generation of helper, suppressor, or cytotoxic cells. The most commonly used technique to detect lymphocyte activation assays the incorporation of radiolabeled thymidine into DNA as an indicator of lymphocyte proliferation. Although this parameter has the disadvantage of measuring a relatively late event that peaks after 2 to 4 days of incubation, it has the advantages of being semiautomated, quantitative, and economical, as well as being very sensitive, since the background DNA synthesis by unstimulated lymphocytes is very low (Oppenheim and Schecter 1976).

Mechanism of Lymphocyte Activation

The mechanism of lymphocyte activation represents a challenging problem that has generated many investigations but is still unresolved. We will summarize briefly the known information by discussing the physicochemical events following mitogenic activation that sequentially occur at the outer cell membrane, in the lipid bilayer, and intracellularly. Since mitogens, unlike antigens, stimulate a considerable proportion of lymphocytes, their physicochemical effects are measurable. Mitogens first interact with membranes. In fact, observations that mitogens bound to sepharose beads are still active suggest that the activating signal occurs at the cell membrane. Most mitogens interact with cell surface glycoproteins. However, it also has been shown that anti-DNP can activate thymocytes which had absorbed 2,4-dinitrophenyl (DNP) bound to gangliosides rather than glycoproteins (Spiegel et al. 1979). Both the mitogenicity and binding of many mitogens to cell membranes can be inhibited by simple sugars. For example, methyl-α-D-mannoside inhibits concanavalin A binding and mitogenicity, whereas D-galactose inhibits the activation by PNA, SBA, $NaIO_4$, as well as neuraminidase plus galactose oxidase (Sharon 1976). However, mitogenicity and binding can be dissociated, since lectins such as concanavalin A bind B- as well as T-lymphocytes but activate only T-cells. Furthermore, there are lectins such as wheat germ agglutinin (WGA) that bind equally as well as mitogenic lectins, even though WGA in not mitogenic. Therefore, lectin binding is necessary but not sufficient for mitogenesis. In the case of lectins such as PHA or concanavalin A, binding of only a minority (16 to 25%) of the 10^7 lymphocyte receptors leads to optimal mitogenic responses. Although addition of higher concentrations of mitogens increases the binding to lower affinity lymphocyte receptors fivefold, the resultant mitogenic response is suboptimal (Crumpton et al. 1978).

Studies using fluorochrome-labeled mitogens reveal that they modulate lymphocyte glycoprotein receptors. This interaction results in patching and capping and internalization or occasional shedding of lectin receptor complexes. However, these events can be dissociated from mitogenesis, since nonmitogenic lectins such as WGA can also induce patching and formation of polar caps. Furthermore, doses of lectins insufficient to induce patching and capping can be mitogenic. Conversely, inhibition of capping and patching does not necessarily block mitogenicity. Furthermore, although succinylated, concanavalin A, which is a dimer, will not cap, it is just as mitogenic as the normally tetrameric concanavalin A that does modulate the receptors.

The absolute minimum requirement for mitogenicity appears to be the capacity of a molecule to cross-link cell-surface receptors. It has been shown that anti–T-cell antisera (ATS) and their $F(ab')_2$ divalent fragments are both capable of activating T-lymphocytes. However, even though monovalent F(ab') fragments of ATS bind as well as the fragments and the intact ATS, they are not mitogenic unless re-cross-linked by an antibody to the ATS. Monovalent ligands are incapable of activating either B- or T-lymphocytes, perhaps because they fail to micropatch or immobilize receptors. It has been suggested that such immobilization influences submembranous microtubules and filaments since both colchicine and cytochalasin-B can block lymphocyte activation. These cytoskeletal structures may modulate receptor mobility and be involved in signal transmission (Cunningham et al. 1976). However, the inhibitory effects of cytochalasin-B on mitogenesis have also been attributed to its effects on glucose transport. Moreover it is not clearly established whether divalent cross-linking requires intact microtubules.

The effects of mitogenic activation on the lipid bilayer have been investigated using fluorescence polarization. It has been reported that T-cell mitogens rapidly decrease the fluorescence polarization, presumably by increasing the fluidity of lymphocyte membranes. However, these findings have not been confirmed, and increased membrane fluidity can be detected only in association with blastogenic changes after 24 hours (Hesketh 1978).

Following the interaction of mitogens with the cell membrane, a number of early effects on ion transport have been reported. Calcium is probably necessary for mitogenesis, since its depletion or the addition of EGTA inhibits lymphocyte proliferation.

The cross-linking of membrane receptors is thought to open Ca^{++} channels in the membrane. However, even the mitogenic calcium ionophore A23187 does not cause an immediate measurably increased calcium influx, and can be mitogenic even in calcium-free medium. Perhaps it changes the intracellular distribution of calcium. On the other hand, no other calcium ionophore has been found to be mitogenic. Although calcium influx probably occurs within a few minutes, it is probably smaller than can be measured by currently available techniques (Hesketh 1978).

Mitogens were thought to effect Na^+ and K^+ transport, e.g., to stimulate the ouabain sensitive K^+ pump influx and associated Na^+ efflux within 5 minutes. However, preloaded cells show an efflux of both K^+ and Mg^{++}. Other investigators have been unable to detect any stimulation of $(Na^+-M_2^+)$ ATPase. Thus, although there are detectable changes in cell Na^+ and K^+ content associated with blastogenesis, their role in the initiation of mitogenesis remains unclear.

There is a noncontroversial observation that mitogenic stimulation increases amino acid uptake (e.g., of the nonmetabolizable 2-aminisobutyric acid [AIB]) and glucose transport within 1 hour. The mitogens stimulate transport rapidly across the cell membrane of these small molecules (Hesketh 1978).

Within 3 minutes of concanavalin A stimulation of murine spleen cells, one can detect phospholipid methylation, which peaks at 10 minutes and is followed by degradation of methylated phospholipids by 30 minutes (Hirata et al. 1980). The addition of an inhibitor of this phospholipid methylation (3-deaza-SIBA) also inhibited mitogenesis. Arachidonic acid, a precursor of prostaglandins, is released from the stimulated lymphocytes during the breakdown of the methylated phospholipids. This arachidonic acid generation is presumably a consequence of activation of phospholipase A_2, which also results in an increase of lysophosphatidylcholine release. This product in turn may transiently increase guanylate cyclase and cyclic GMP. The degradation of methylated phospholipids is calcium dependent. This pathway for lymphocyte activation thus appears to account for some of the other biochemical events.

There have been many conflicting investigations of the role of cyclic nucleotides in mitogenesis. Optimal doses of mitogens do not stimulate rapid changes in lymphocyte cyclic AMP levels. The claim that superoptimal doses of mitogens rapidly raise the cyclic AMP level of lymphocytes is also not accepted generally. However, it is agreed that agents that raise cyclic AMP levels also inhibit lymphocyte blastogenesis and proliferation. Conversely, the claims that mitogens elevate lymphocyte cyclic GMP content also have been difficult to confirm. Cyclic GMP agonists have had no effect on blastogenesis either and cyclic GMP itself stimulated only ^{3}H-thymidine transport but not blastogenesis (Parker 1979).

Activated T- and B-lymphocytes show enhanced phosphatidylinositol (PI) turnover (e.g., incorporation of ^{32}PI) within 5 to 10 minutes. Furthermore, colchicine inhibits both mitogenesis and PI turnover, suggesting that PI turnover is vital to mitogenesis. In contrast, although both phospholipid fatty acid turnover (^{14}C oleate or ^{14}C acetate uptake) and RNA synthesis are rapidly stimulated by concanavalin A in 1 to 4 hours neither of these processes is prevented by colchicine. These contradictory data therefore challenge both the relevance of PI turnover and the proposed receptor microtubule association as a mechanism for initial activation.

In conclusion, the salient events in lymphocyte activation remain to be elucidated. Ligand-induced cross-linking of lymphocyte membrane glycoproteins or glycolipids is required. This appears to be followed by facilitated diffusion of ions, amino acids, and glucose. After 3 to 10 minutes, methylation of phospholipids and incorporation of fatty acids into PI occurs. These are followed by synthesis of other lipids, proteins, RNA, and eventually DNA.

T-Lymphocyte Function

The T-lymphocyte population controls or regulates almost all immune functions. In general, the functions of T-lymphocytes can be divided into two broad categories, regulatory and effector. The regulatory functions include T-helper functions, which help B-lymphocytes respond to thymic-dependent antigens and produce antibody; T-amplifier functions, which are capable of amplifying certain T-cell mediated effector functions; and suppressive functions, which suppress antibody synthesis and T-cell effector functions. The effector T-cell functions include delayed-type hypersensitivity reactions, hypersensitivity granulomas, contact sensitivity, cell-mediated cytotoxicity (see Chapter 7), and those involved in transplantation immunity, including allograft rejection and GVH disease. Suppressor, amplifier, and effector functions can be mediated by the production of potent biological substances or lymphokines (see Chapter 10). These lymphokines in turn activate inflammatory cells including macrophages, which play key roles in delay hypersensitivity, contact sensitivity, and resistance to infection with certain facultative intracellular bacteria and viruses.

It is the goal of this section to establish how this large number of T-cell functions is carried out by a number of distinct subsets of T-cells. Thymic-cell subclasses have been separated by a variety of different techniques including velocity gradient sedimentation, limiting dilutions of anti-Thy-1 and complement, sensitivity to antilymphocyte serum, responses to various mitogens, and differential sensitivity to the effects of adult thymectomy. However, none of these approaches yielded a definitive answer, primarily because none of these techniques allows the direct identification and complete separation of relevant T-cell subclasses. A major advance was made by the observation of Cantor (1975 a,b) and Cantor et al. (1976) that a panel of antisera to a group of murine lymphocyte surface differentiation antigens, termed Lyt antigens (Table 1), allowed one to distinguish three major T-lymphocyte subclasses, each of which displays a unique set of biological properties and immune functions (Table 3).

Table 3. Properties of Lyt Subsets

Lyt-1 inducer cells
   30% of the peripheral T-cell pool
   Induce B-cells to secrete antibody
   Induce macrophages to participate in delayed hypersensitivity responses
   Induce a set of nonimmune T-cells to generate feedback inhibiting activity

Lyt-23 cytotoxic and suppressor cells
   5 to 10% of the peripheral T-cell pool
   Develop alloreactive cytotoxic activity
   Suppress both humoral and cell mediated immune responses following immunization

Lyt-123 regulator cells
   50% of the peripheral T-cell pool
   Precursor pool for other subsets
      Present in spleens of mice during first week of life, before Lyt-1 and Lyt-23
      Yield Lyt-23 cytotoxic effector cells after stimulation with chemical cells
      Give rise to Lyt-1 cells after concanavalin A stimulation
   Under certain conditions may be cytotoxic effectors with allogeneic targets
   May exert immunoregulatory effects without differentiation to Lyt-1 or Lyt-23 cells

Inducer and Suppressor T-Cells

The Ly-1$^+$23$^-$ (Lyt-1) cells make up approximately one-third the peripheral T-cell pool. They have been termed the inducer population, because they are especially programmed to help or amplify the functional activity of other cells when activated by antigen associated with I-region determinants. The amplifying effects of this T-cell subclass have been demonstrated for the production of antibody by B-cells, for the differentiation of precursors of killer cells to killer effector cells, and for the activation of macrophages to participate in delayed hypersensitivity responses. In addition, this subclass of cells is capable of inducing the development of a subset of nonimmune precursors of suppressor T-cells to generate feedback inhibitory activity (see below).

It is likely that Lyt-1 cells are already programmed either genetically or by their exposure to environmental antigens, or both, to generate helper function before overt immunization, since polyclonal activation of Lyt-1 cells by the mitogen concanavalin A results in the generation of nonspecific T-helper activity for B-lymphocytes to produce antibody to a variety of antigens, including sheep red blood cells (SRBC) (Jandinski et al. 1976). Helper cell memory activity also resides in the Lyt-1 subset. Thus, Lyt-1 cells are necessary for the capacity to manifest secondary responses. Although Lyt-1 cells account for the major proportion of the proliferative response in the MLR, they do not themselves directly contribute to the cytotoxic effector T-cell pool. The Lyt-1 population is primarily concerned with responding to I-region differences of allogeneic cells as judged by ^{3}H-thymidine incorporation. Lyt-1$^+$ responder cells have been shown to bind allogeneic Ia-molecules specifically from the stimulator cell population.

The Lyt-1$^-$23$^+$ (Lyt-23) population comprises 5 to 10% of the peripheral T-cell pool and includes both cytotoxic and suppressor cells. In contrast to the Lyt-1 class, the Lyt-23 cells do not proliferate in response to I-region differences in the MLR. All the prekiller and killer activity seen after stimulation of lymph node cells with allogeneic lymphocytes and one-half to three-fourths of the splenic prekiller activity are present in this subset. Killer effector activity generated from either lymph node or spleen cells is mediated almost solely by Lyt-23 cells, which also play a major suppressive role after stimulation by polyclonal activators such as concanavalin A. Indeed, the opposing effects of concanavalin A on the Lyt-1 and Lyt-23 sets have eliminated the possibility that suppressor T-cell activity is just due to a population of hyperactivated T-helper cells.

One mechanism of suppression by the Lyt-23 population is the production of nonspecific soluble immune response suppressive factor (SIRS) (see Chapter 10), which inhibits the generation of plaque-forming cells to SRBC. Some cells in the Lyt-23 subset express and antigen coded for in the I-J subregion of the MHC (Murphy et al. 1976). This gene product is only found on Lyt-23 suppressor cells but has not been found on Lyt-23 killer cells or their precursors.

It appears that the Lyt-1 and Lyt-23 subclasses represent the products of separate sublines of thymus-dependent maturation, since separated populations Lyt-1 or Lyt-23 cells that are adoptively transferred to syngeneic B-recipient mice remain phenotypically and functionally stable for as long as 6 months. It has also been shown that cells of the Lyt-1 subclass that mediate helper activity are necessary and sufficient for the production and elicitation of delayed hypersensitivity reactions, while cells of the Lyt-23 subclass do not make any contribution to the DTH reaction and in some cases can inhibit the response (Huber et al. 1976). Thus, the reconstitution of a thymus-deprived host with purified Lyt-1 cells restores their ability to generate a DTH response, while the provision of Lyt-23 cells has no such effect. However, if sufficient numbers of Lyt-23 cells are given in conjunction with Lyt-1 cells, suppression of both the generation and expression of DTH is observed. Thus, the broad division made in the beginning of this section between regulatory and effector T-cell functions does not correspond to the division of labor among T-cell subclasses. While cytotoxic activity, one form of T-effector function, is a property of the Lyt-23 class, DTH (a second T-effector function) is a property of the Lyt-1 population, which also stimulates B-cells to produce antibody.

Regulator T-Cells

Fifty percent of all peripheral T-cells belong to the Lyt-$1^+2^+3^+$ (Lyt-123) regulatory cell population. Three major functional roles for Lyt-123 cells have been described. First, as they are detectable in the spleens of mice within the first week of life, it is likely that Lyt-123 cells function as precursors for both the Lyt-1 and Lyt-23 populations. This view is supported by studies demonstrating that stimulation of purified Lyt-123 cells with concanavalin A gives rise to Lyt-1 cells, while stimulation of Lyt-123 cells with chemically modified syngeneic spleen cells leads to the production of Lyt-23 cytotoxic effectors. Second, under certain circumstances Lyt-123 cells themselves function as cytotoxic effector cells. Third, Lyt-123 cells may exert immuno-regulatory effects directly without giving rise to mature Lyt-1 and Lyt-23 progeny.

Interactions Between T-Cell Subsets

It thus appears that programming for discrete T-cell functions is linked to the expression of either the Lyt-1 or Lyt-23 profile and that the differentiation of precursor lymphocytes into distinct T-cell subpopulations takes place before specific or nonspecific cellular activation, probably in the thymus. Although the studies described above clearly show that the Lyt-1 and Lyt-23 subsets are stable and do not interconvert, recent studies have demonstrated that signals or mediators may be passed among the different types of immunologically competent cells involved in the regulation of the immune response. Thus, Eardley et al. (1979) have shown that antigen-stimulated Lyt-1 cells, in addition to inducing B-cells to secrete antibody, can induce resting or nonimmune Lyt-123 cells to develop profound feedback inhibitory activity. This observation arose out of studies designed to test whether suppressor activity could ever be generated from purified Lyt-1 cells, which were stimulated in culture with very high doses of antigen, a condition that readily induces unselected T-cells to express optimal levels of antigen-specific suppressive activity. When these Lyt-1 cells were tested by coculture with nonimmune purified B-cells, help rather than suppression was observed (Table 4, line 1), thus reinforcing the view that the genetic program for a single differentiated set of cells combines information for both surface phenotype and functions. However, if the effects of these hyperimmune Lyt-1 cells were tested on a mixture of B-cells and unselected T-cells, rather than on purified B-cells, suppression of the response to SRBC was observed (Table 4, line 2).

The surface phenotype of the cells in the nonimmune T-cell population mediating the suppression was Lyt-123, Qa-1^+. The presence of extremely small numbers of these cells was associated with potent inhibitory effects in the face of relatively large numbers of SRBC immune Lyt-1 cells. Fewer than 10^5 nonimmune Lyt-123 cells

Table 4. Feedback Inhibition of the Primary *in vitro* SRBC Response

SRBC-stimulated T-cells	Purified B-cells	Unselected nonimmune T-cells	Anti-SRBC plaque-forming cells
Lyt-1(10^5)	4×10^6	None	2,500
Lyt-1 (10^5)	4×10^6	10^6	400

(Adapted from Eardley et al. 1979.)

were capable of inhibiting 80 to 90% of the PFC response induced by 10^6 Lyt-1 cells. This inhibition was not overcome by increasing the numbers of SRBC immune Lyt-1 cells; in fact, the more Lyt-1 help, the greater the degree of Lyt-123-mediated suppression. It is likely that most of the suppressive effects of the activated Lyt-123 cells reflects Lyt-123 to Lyt-23 differentiation, but it is not clear if all the suppression is mediated in this fashion. The net results of this feedback suppression circuit involve reduction in Lyt-1 T-inducer activity, leading to both a decreased production of antibody by B-cells and to a decreased induction of Lyt-123 cells to become suppressor cells.

It was important to determine whether those Lyt-1 cells that induce suppressive activity are distinct from the Lyt-1 cells that induce B-cells to produce antibody. The presence of the Qa-1 antigen (present on 60% of Lyt-1 cells) allowed Cantor et al. (1978a) to distinguish between these two distinct functions. Thus, Lyt-1, Qa-1^+ cells were responsible for the induction of Lyt-123 cells to exert feedback inhibition, whereas isolated Lyt-1, Qa-1^- cells do not induce detectable suppression. However, signals from both the Lyt-1, Qa-1^+, and Lyt-1, Qa-1^- cells were required for optimal formation of antibody by B-cells. Thus, the ability of an antigenic determinant to induce a detectable antibody response may depend largely on the ratio of Lyt-1, Qa-1^+ and Lyt-1, Qa-1^- clones that bear receptors for that antigen.

Since the Lyt-123 cell, which accepts the feedback signal from the Lyt-1, Qa-1^+, was also shown to be Qa-1^+, it is appropriate to ask how these two populations of cells communicate. It has been shown that both these populations must be specific for the test antigen; therefore, it is possible that both the inducer cell and the feedback acceptor cell both bear receptors specific for the antigen and are brought into association by an antigen bridge. However, it has recently been shown (Eardley et al. 1979) that inducer and acceptor interact only if both T-cell subsets are obtained from donors identical at the Ig-1 locus and that this Ig-1 restriction is independent of the Ig allotype expressed on the B-cells used to assay the level of T-helper activity. Preliminary genetic studies have shown that it is not the products of Ig-C_H genes but probably V_H-like structures that are responsible for this restriction on cell interaction. Thus, it is entirely possible that the nonimmune Lyt-123 cells bear receptors for idiotypic determinants carried on the Lyt-1 inducer population.

Although the feedback inhibitory circuit is probably only one of many that mediate important amplifying and dampening effects on the intensity and duration of an immune response, a number of important conclusions can be derived from this model. The addition of a subpopulation of cells to a complex mixture of lymphoid cells with resultant suppression does not necessarily imply that the added subpopulation contained suppressor cells; it only indicates that the population in question can induce suppressive activity. After immunization T-helper cell activity is always accompanied by T-suppressor-cell activity. The feedback suppressor circuit should always be taken into account in the design of clinical protocols aimed at either specific augmentation of immunity by adoptive cellular immunotherapy or nonspecific augmentation by adjuvant-like methods (Cantor et al. 1978b).

Nature of the T-Lymphocyte Antigen Receptor

It has been relatively easy to understand and to define B-lymphocyte receptors chemically, because B-lymphocytes and plasma cells secrete specific antibody molecules with antigen combining sites that are identical to their surface receptors (see Chapter 8). However, by conventional techniques no such membrane immunoglobulin receptors have been identified on T-lymphocytes and nonimmunoglobulin antigen-specific T-cell products have been identified only recently. Thus antigen recognition by T-lymphocytes is distinct from antigen recognition by B-lymphocytes.

Functional Characteristics of T-Cell Activation

The structural requirements for antigenic stimulation of T-cells are quite different from the requirements for the binding of antigens to the combining site of humoral antibody (Paul 1970; Paul and Benacerraf 1977). Thus T-cells from DNP–GPA immune animals can only be activated by DNP–GPA, the immunizing antigen, for the elicitation of delayed hypersensitivity, T-helper cell activity, or T-cell proliferation and lymphokine production *in vitro*. None of these responses is elicited by DNP conjugates of unrelated proteins, suggesting that the response is carrier specific. In contrast, when guinea pigs are immunized with the DNP derivative of guinea pig albumen (DNP–GPA), they will produce antibodies that will bind to DNP haptens such as DNP-lysine or to DNP derivatives of a large number of proteins. However, it must be noted that the requirements for binding of antigen to the immunoglobulin receptor on the B-cell surface and for activation of B-lymphocytes also appear to differ. The DNP hapten by itself can bind, but it fails to activate B-lymphocytes. The activation of B-cell proliferation, antibody synthesis, and lymphokine production are also specific for both the sensitizing hapten and carrier components of the antigen.

The second major difference between T- and B-cell recognition of antigen is the requirement for another cell type, probably of monocyte–macrophage lineage, to initially interact with antigen and then to present the processed or selected antigen to the T-lymphocyte (see Chapter 5). Thus, when macrophages are briefly exposed (15 to 60 minutes) to soluble protein antigens, washed, and mixed with immune lymphocytes, they induce proliferation to a degree far in excess of that predicted when concentrations of antigen equivalent to that carried by the pulsed macrophage are added continuously to cultures of macrophages and lymphocytes. Of greater significance is that when purified T-lymphocytes are pulse exposed to antigen in similar fashion, they do not proliferate, even when macrophages are added back to them (Rosenthal et al. 1975). This latter experiment excludes the possibility that macrophages function in this response solely to maintain cell viability. It is also unlikely that antigen bound to the T-lymphocyte receptor has induced a state of tolerance because antigen-pulsed macrophages are capable of activating lymphocytes that were previously pulse exposed to antigen.

An additional major constraint on T-lymphocyte activation by antigen is that the antigen-pulsed macrophage and the immune T-lymphocyte must share a common allelic form of the I-region of the MHC to generate a T-cell proliferative response. The nature of the genetic restriction on immunocompetent cell interaction is discussed in great detail in Chapters 7 and 8. In any case, the studies on macrophage–T-cell interaction suggest that the T-lymphocyte recognizes both an I-region gene product of the macrophage and the specific antigen bound to the surface of the macrophage. The T-lymphocyte may have two separate receptors, one for antigen and the other for the Ia antigen, or alternatively may possess a single complex receptor interacting with a combined determinant on the surface of the macrophage formed by the association of the exogenous antigen and the Ia antigen (Thomas et al. 1977).

Idiotype Expression by T-Cells

A large body of evidence has accumulated over the last 5 years that the fundamental mechanism of antigen recognition by T- and B-lymphocytes has many similarities. Thus, it has been proposed that the T-cell receptor for alloantigens and to some soluble antigens has the same idiotype as that of the B-cell receptor. Idiotypes are unique antigenic determinants expressed on the variable region of antibody molecules. In most instances, different antibodies capable of binding to the very same antigen may have distinct idiotypes, but in several instances the same idiotype is represented on a substantial fraction of all the antibody molecules produced by an inbred strain specific for a given antigen (Nisonoff et al. 1977). Such idiotypes are called cross-reactive idiotypes (IdXs), and their expression is controlled by genes linked closely to the genes controlling immunoglobulin heavy chains.

We review several experimental studies that have attempted to demonstrate that specific T-lymphocytes express on their surfaces antigenic determinants similar to the IdXs expressed by antibodies of the same specificity. We will then attempt to correlate the results of those studies with the functional and biochemical properties of the antigen specific products of T-lymphocytes. All three experimental systems to be described have investigated the question of idiotype expression by T-cells in different ways, and each system will be discussed in detail.

The hypothesis tested by Binz and Wigzell (Binz et al. 1974; Binz and Wigzell 1975 a,b) was rather simple. They proposed that T-lymphocytes from strain A must possess recognition structures for certain strain-B alloantigens that, for genetic reasons, are not expressed by the $(A \times B)F_1$ animal (Figure 1). They demonstrated this by the immunization of the F_1 with purified A-strain parental T-cells that have a recognition structure for B-alloantigens. This yielded antibodies directed against receptors on A-strain T-cells for B-strain alloantigens. Rat F_1 hybrid antisera of this type were shown to selectively inhibit the capacity of parental lymphocytes of strain A to react against strain B, both in the MLR and in GVH reactions. Reactivity to a third party strain was left intact.

Of great importance was the observation that these antisera raised against T-cell receptors would also react exclusively with erythrocytes coated with IgG antibodies produced in strain A against strain-B alloantigens. Again, no reaction could be detected using IgG antibodies produced in strain A against other rat histocompatibility antigens. Thus, although these antiidiotype antisera were induced by supposedly pure T-cells, they could be shown to react with B-cell products in a highly specific manner. IgG antibodies of the relevant binding specificity therefore express the full repertoire

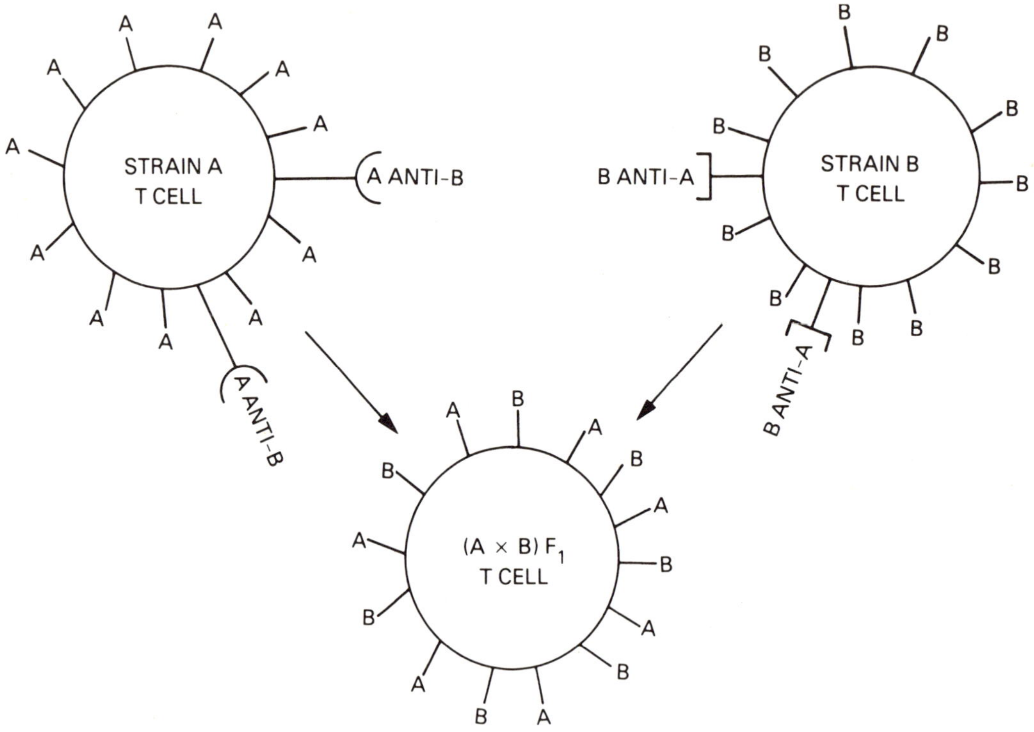

Figure 1. Model used by Binz and Wigzell (1975, a and b) for the design of experiments to prepare antisera to the T-lymphocyte receptor for alloantigens. T-lymphocytes from strain A animals bear a receptor for strain B alloantigens which are not expressed in the (AxB)F₁ animal. Immunization of (AxB)F₁ animals with strain A T-cells yielded an antibody specific for the A anti-B receptors on strain A T-cells.

of idiotypic determinants present on T-lymphocyte receptors reactive against the same antigen. The converse was not true; that is, antibodies to T-cell idiotype-positive receptors could not remove all the idiotypic determinants present on relevant IgG antibodies. The implications of this experiment will be discussed below.

The availability of antibodies to T-cell receptor idiotypes allowed the isolation by affinity chromatography techniques of naturally occurring idiotype positive T-cells in virtually pure form (Binz and Wigzell 1975c). Such pure T-cells were found to express a highly restricted but enhanced immune reactivity when tested for their ability to initiate a GVH reaction. Idiotype-bearing lymphocytes from normal rats could be visualized by immunofluorescent techniques or autoradiography, and the antiidiotype antisera were shown to react with 4 to 6% of peripheral T-lymphocytes and 1% of normal B-lymphocytes from the appropriate strain.

The availability of these antiidiotype antibodies permitted the biochemical analysis of this T-cell receptor (Binz and Wigzell 1975d, 1976). When normal rat serum was filtered through an antiidiotype column and the acid eluate then labeled with ^{125}I, four distinct peaks could be visualized on gel electrophoresis. Peak I was found to have a molecular weight of 180,000, produced by B-lymphocytes and shown by serological techniques to be similar to 8S IgM. In contrast, peak II (molecular weight:

150,000), peak III (molecular weight: 70,000 to 80,000), and peak IV (molecular weight: 35,000) were all shown to be produced by T-lymphocytes. Peak II molecules were shown to dissociate into peak III under reducing conditions, and it is therefore likely that peak II consists of dimers of peak III chains. Peak IV molecules were believed to be proteolytic breakdown products of peak III chains. Nevertheless, both peak III and IV material bore idiotypic markers and retained antigen-binding activity. None of the molecules were shown to react with conventional antiimmunoglobulin reagents specific for the heavy chain constant region or with light chain determinants. Furthermore, no reactivity could be shown with antisera to rat Ia antigens.

In breeding studies, no evidence could be obtained between inheritance of the T-cell idiotypes and the rat MHC. However, T-cell receptors with reactivity to rat strain DA present in Lewis rats could be clearly shown to be coded for by genes linked to the heavy chain constant region genes (Binz et al. 1976). Although these pioneering studies of Binz and Wigzell have greatly influenced immunological thinking about the genetic origin and biochemical characteristics of the T-cell receptor, it should be pointed out that sera such as the one described here have been extremely difficult to produce.

Eichmann and co-workers (Black et al. 1976; Hämmerling et al. 1976) have approached the question of T-cell idiotypes by examining whether antiidiotype serum to an IdX present on serum antibody would have any effect on helper or suppressor T-cell function (Figure 2). Most of these experiments were done using antibodies to group-A streptococcal carbohydrate (A-CHO) generated in strain A/Jax mice by immunization with group-A streptococcal vaccine. A major portion of the antibodies to A-CHO in A/Jax mice represents a homogeneous antibody species defined by its isoelectric focusing spectrum and by idiotype. These investigators predicted that such antiidiotype sera should react with T-cell receptors. The idiotype is detected by guinea pig antiidiotype antibody against clone A5A, and the expression of the A5A idiotype has been shown to be controlled by a single gene linked to the Ig-1^e allotype locus of A/Jax mice. The guinea pig anti-A5A idiotype was fractionated into the IgG$_1$ and IgG$_2$ classes of guinea pig IgG. Mice injected with the IgG$_1$ fraction of the anti-A5A idiotype developed specific T-helper cells, which exhibited specificity for streptococcal antigen, because no helper cell activity was seen with hapten on a different carrier. In contrast to the results with the IgG$_1$ anti-A5A idiotype, the injection of mice with guinea pig IgG$_2$ antiidiotype resulted in the generation of suppressor T-cells, which suppressed the expression of the A5A idiotype in A/Jax mice. The suppressor T-cells could be killed by antiidiotype and complement but not by idiotype and complement. This demonstrates clearly that suppressor T-cells as well as helper T-cells bear an idiotype-bearing antigen-recognizing receptor.

Detailed genetic studies of the ability of antiidiotype to prime for T-helper cell activity could be performed because BALB/c mice bear an idiotype, S117, on antibody molecules that also have specificity for streptococcal A-CHO and is linked to the Ig-1^a allotype of the BALB/c strain. Helper cells can be elicited in both BALB/c and A/Jax mice by priming with strep A, but only A/Jax mice produced helper cells in response to priming with anti-A5A idiotype, and only BALB/c produced helper cells in response to priming with anti-S117 idiotype (Table 5). When a large panel of strains was surveyed, there was a perfect correlation between helper cell responsiveness to antiidiotype stimulation and the presence of the corresponding V_H markers at the immunoglobulin level (Table 5). No correlation was found between responsiveness

A. SERUM ANTIBODY TO A-CHO

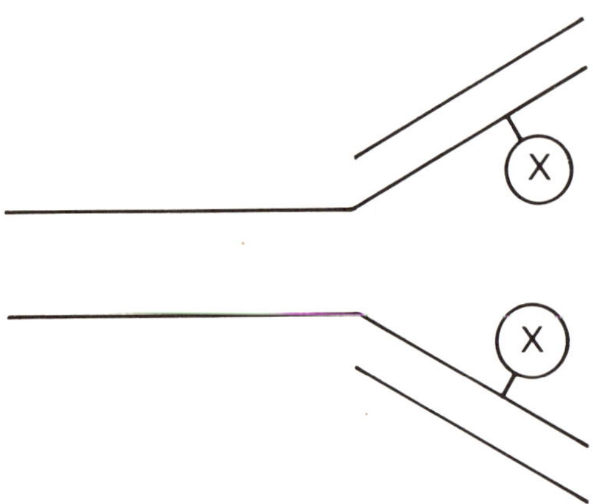

B. T CELL RECEPTOR FOR A-CHO

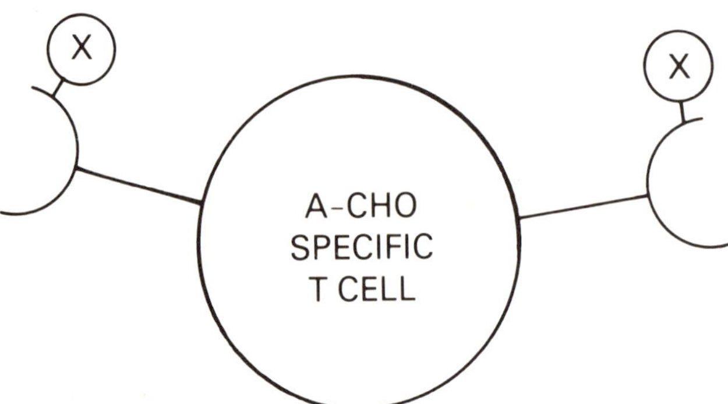

Figure 2.   The T-cell receptor for A-CHO bears the same idiotypic determinant (x) which is expressed by serum antibodies of the same specificity.

and any of the H-2 haplotypes. Thus, helper-cell responsiveness to antiidiotype requires the presence of the gene that encodes the variable region of the heavy chain of the antibody molecule to which the antiidiotype was prepared, but responsiveness was independent of a particular H-2 haplotype.

The third approach, which has been used by Rajewsky and co-workers (Krawinkel et al. 1977a,b) to define the T-cell receptor for antigen, is the direct chemical isolation of the receptor molecules from the lymphocyte surface. This group has shown that T- and B-lymphocytes can be absorbed specifically to hapten-coated nylon mesh at low temperatures; the cells can be released subsequently from the nylon mesh by increasing the temperature to 25°C. Antigen-binding material is apparently left behind by the cells and can be eluted from the nylon disks with acidic buffer or buffer containing free antigen. The antigen-binding specificity of the material is assayed in a very sensitive haptenated phage inactivation assay. Receptors isolated from splenic lym-

Table 5. Evidence for Relationship between Helper-Cell Induction by Antiidiotype and the Presence of V_H Markers

Strain	H-2	Ig-1 complex			Helper cell stimulated with		
		Ig-1	A5A	S117	Strep A	Anti-A5A	Anti-S117
A/Jax	a	e	+	−	+	+	−
A.SW	s	e	+	−	+	+	−
BALB/c	d	a	−	+	+	−	+
B10.A	a	b	−	−	+	−	−
B10.D2	d	b	−	−	+	−	−

(Adapted from Hämmerling et al. 1976.)

phocytes by this method fall into two classes: Seventy to eighty percent of the phage-inactivating material binds to insolubilized antiimmunoglobulin reagents and represents a B-cell product. The second class of receptor material (20 to 30%) does not carry any class-specific antigenic determinants of immunoglobulin heavy chains or κ or λ light chains. It is likely that this material represents a T-cell product as the fraction of the material in the preparation is directly proportional to the fraction of T-cells in the input population. Anti-Ia sera fail to react with these molecules.

Strong evidence has been presented that the T-cell receptor material carries the variable portion of immunoglobulin heavy chains (Krawinkel et al. 1978). When C57BL/6 mice are immunized with the hapten, 4-hydroxy-3-nitro-phenylacetyl (NP), conjugated to a carrier, they produce a rather homogeneous antibody in the primary response. This antibody carries an idiotypic marker, NP[b], linked to a gene in the heavy-chain-linkage group. Thymic-cell receptor material isolated from NP-sensitized C57BL/6 cells also bears the NP[b] idiotype. One difficulty with this model is that the receptor material left behind on the nylon fibers by the T-cells may not actually have been synthesized by the T-cells, but rather could have been passively absorbed to the T-cell surface from serum. However, recent studies of the response of C57BL/6 mice to NP tend to rule out this possibility. Thus, in the secondary immune response the proportion of antibodies carrying the NP[b] idiotype declines with time after immunization from 70 to 90% to 10 to 20%. Thus, receptors on memory B-cells in NP-sensitized animals should be largely devoid of the NP idiotype and in fact a very low frequency of idiotype-positive material can be isolated from memory B-cells. In contrast to the immunoglobulin-positive cell fraction, the fraction of receptor material obtained from immunoglobulin-negative T-cells persists after immunization and continues to bear the NP[b] idiotype. It thus appears that V_H gene expression in T-cell receptors is largely restricted to the major idiotype of the primary anti-NP response and the T-cell may be restricted to the expression of only a few genes which appear at the antibody level as major idiotypes.

The experimental approaches described thus far have demonstrated that both T- and B-lymphocytes reactive against a given antigen share similar idiotypes. However, the studies of Binz and Wigzell (1975a–d) as well as Krawinkel (1978) and co-workers suggest that the T-cell is more restricted than the B-cell in the choice of idiotypes. Therefore T-cell idiotypes may be encoded in the germ line, while B-cells which may express the same genes in their receptors go through a diversification process during ontogeny.

Antigen-Specific T-Helper Factors

Since T-cells have no conventional immunoglobulin determinants, and since none of the material isolated from T-cells in the studies described previously bears markers of immunoglobulin constant region genes, one might conclude that the T-cell receptor is a polypeptide chain coded for by two genes: one gene codes for the same variable region that is present in conventional immunoglobulin, while the other gene codes for an as yet undefined constant region. In fact, several recent studies of antigen-specific products or factors that are produced by T-cells demonstrate that the factors bear both idiotypic determinants and gene products of the MHC. It is assumed that these T-cell-derived factors represent shed T-cell receptors.

One of the first factors that could replace specific T-helper-cell function was described by Taussig, Mozes, and Munro (Taussig and Munro 1974; Munro et al. 1974; Taussig et al. 1975; Mozes et al. 1975). In their experiments, activated T-cells were prepared by injecting irradiated recipients with syngeneic thymocytes and subsequent immunization with the branched polymer (T,G)-A–L in adjuvant (Table 6). T-cells from these animals were placed in culture for 6 hours in the presence of (T,G)-A–L; the supernatants from these cultures were then removed and assayed for their T-cell-replacing capacity in the development of anti-(T,G)-A–L antibody responses *in vivo* when injected with bone marrow. Recipients of bone marrow cells alone developed only a small antibody response; however, recipients of bone marrow cells and the supernatant of the 6-hour culture containing the T-cell factor developed markedly enhanced primary IgM anti-(T,G)-A–L antibody responses comparable to that seen when activated T-cells were transferred with the bone marrow cells (Table 6, lines 1 to 3).

It was shown that T-cells were responsible for the production of the antigen-specific helper factor, since treatment of the activated T-cell population with anti-Thy-1 and complement eliminated production of the factor. The factor had a molecular weight of 50,000 and could be bound specifically to and eluted from a (T,G)-A–L-coated column (Table 6, line 4). Although the factor failed to react with conventional antiimmunoglobulin reagents, it was removed by passage over an anti-H-2 immunoabsorbent column. It was later shown to bear determinants of the I-A subregion of the H-2 complex (Table 6, line 5). It is very likely that the factor was capable of acting directly on B lymphocytes, because B-cells were capable of absorbing the functional activity from the supernatant, while T-cells were incapable of doing so. No genetic restriction in the action of the factor was noted and it could cooperate with allogeneic bone marrow cells. Recent studies show that the factor could be removed by absorption on antiidiotype-affinity columns (Mozes and Haimovich 1979) suggesting that the factor may also bear determinants cross-reactive with idiotypic

Table 6. T-Lymphocytes Produce Specific Helper Factors That Bear Ia Antigens

T-cells or factor	Treatment	Bone marrow cells	Plaque-forming cell response
0	—	+	−
Activated T-cells	—	+	+
T-cell factor	—	+	+
T-cell factor	(T,G)-A–L-Sepharose	+	−
T-cell factor	Anti-Ia-Sepharose	+	−

(Adapted from Taussig et al. 1975.)

determinants of anti-(T,G)-A–L antibodies. Further biochemical studies of this helper factor have proceeded very slowly and similar factors have not been isolated in other laboratories.

Antigen-Specific Suppressor Factors

A number of groups have isolated factors capable of suppressing antibody responses. The first antigen specific T-cell suppressor factor (TSF) was isolated by Takemori and Tada (1975); this factor was capable of suppressing both the primary and secondary IgG antibody responses of mice both *in vivo* and *in vitro* (Takemori and Tada 1975; Taniguchi et al. 1976a). In general, the factor is obtained by physical extraction of thymocytes or spleen cells from mice that have been immunized with a relatively high dose of antigen. The factor exerts a suppressive effect on the antibody response against a hapten coupled to the carrier used for immunization. The factor binds the antigen used for immunization but does not bind unrelated antigens. It does not bear immunoglobulin determinants, and its activity is dependent on the presence of antigen; its molecular weight is between 35,000 and 60,000 (Table 7).

The factor was shown to bear H-2 determinants, since it could be removed by passage over an anti-H-2 immunoabsorbent. Detailed genetic-mapping studies demonstrated that the factor bore determinants coded for by the I–J region of the murine MHC. The I–J region has been shown to code for a class of molecules present on subsets of T-cells with suppressor cell activity; these I-region products are different from serologically detectable B-cell Ia antigens.

Another difference between TSF and helper factor is that the factor does not suppress the antibody response by acting directly on B-cells. TSF acts on T-cells and the activity of TSF can be removed by absorption with normal T-cells. Surprisingly, the T-cells that absorb the factor adhere to nylon wool. The T-cell fraction that fails to adhere to nylon wool cannot absorb the TSF. Strict identity at the I–J subregion of the MHC is required between the donor of the TSF and the acceptor nylon adherent T-cell for effective suppression to be observed (Taniguchi et al. 1976b). The acceptor site for TSF on the nylon adherent T-cell can be blocked by anti-I–J sera.

The suppressor circuit involved in TSF action has recently been analyzed in greater detail (Taniguchi and Tokuhisa 1980). Thus, when nylon-adherent T-cells are cultured with antigen and TSF for 48 hours, a new suppressor cell is generated that suppresses the antibody response in the absence of TSF. The acceptor cell of TSF must be immune to the antigen being tested and has been shown to be of the Lyt-123 subclass. The

Table 7. Characteristics of Antigen-Specific T-Cell Suppressor Factors

Extracted from primed T-cells
Specificity and affinity for antigen
Activity dependent on the presence of antigen
Suppress IgG antibody response
Do not bear immunoglobulin constant region determinants
Protein in nature
Molecular weight: 30,000 to 60,000
Act on primed T-cells (Lyt-123) to generate an Lyt-23 suppressor cell
Absorbed by T-cells, not B-cells
May be I-J restricted, not H-2 restricted
Bear I-J determinants
Bear immunoglobulin V_H determinants

induced suppressor cells obtained after culture with antigen and TSF belong to the Lyt-23 subclass.

Suppressor factors similar to those isolated by Tada have also been isolated by Benacerraf and co-workers (Kapp et al. 1976, 1977; Thèze et al. 1977a,b). These investigators took advantage of the finding that the predominant response of genetic nonresponder mice to immunization with the terpolymer L-glutamic acid, L-alanine, L-tyrosine (GAT) is the production of GAT-specific suppressor cells, which can be detected in both thymus and spleen and which suppress the primary IgG antibody response to GAT coupled to methylated bovine serum albumin (GAT–MBSA); a factor can be isolated from these suppressor cells using methods similar to those described by Tada. Specific for the anti-GAT response, GAT–TSF is effective both *in vivo* and *in vitro*, and can be removed selectively on an antigen sepharose column. It is of interest that GAT–TSF can also be retained by passage over an anti-GAT immunoabsorbent, indicating that the factor in the crude extract is complexed with antigen (Germain et al. 1978); the factor eluted from such a column is not absorbed by an anti-GAT immunoabsorbent and is free of antigen.

Although GAT–TSF is bound by anti-H-2 reagents, precise mapping studies cannot be performed in this system because of a lack of suitable reagents to the I subregions in question. However, a factor raised in certain strains that are nonresponders to the copolymer L-glutamic acid, L-tyrosine (GT) is functionally very similar to GAT–TSF and can be absorbed by anti-I–J reagents (Thèze et al. 1977a). In contrast to the TSF described by Tada, both GAT–TSF (Kapp 1978) and GT–TSF are capable of mediating suppression across an H-2 barrier (Waltenbaugh et al. 1977). In fact, GAT–TSF obtained from nonresponders can generate T-suppressor cells from mice of responder haplotypes (Germain and Benacerraf 1978). However, both GAT–TSF and GT–TSF also appear to function by stimulating the development of suppressor cells, as does the TSF described by Tada. Thus, crude GAT–TSF is capable of inducing T-suppressor cells *in vitro* during a 2-day culture period. Purified GAT–TSF, eluted from an anti-GAT column, fails to induce suppressor cells under the same conditions. The addition of nanogram amounts of GAT (which by itself cannot induce T-suppressor cells) to purified GAT–TSF permits effective T-suppressor cell induction.

An antiidiotype serum prepared against a cross-reactive idiotype on GAT-specific antibodies was capable of binding GAT–TSF (Germain et al. 1979). Thus, this suppressor factor bears idiotypic determinants in common with anti-GAT antibody. In similar studies, a TSF from A/Jax mice specific for azobenzenearsonate (ABA) could be absorbed to and recovered from $F(ab')_2$ fragments of rabbit antibody directed against the IdX of anti-ABA antibodies from A/Jax mice (Bach et al. 1979). Also, ABA–TSF could be specifically absorbed to and eluted from an anti-H-2 column and could then be specifically absorbed to the antiidiotype column. This demonstrates that the same molecular complex that bears I–J determinants also bears V_H gene products.

It is possible that the idiotypic determinants and the I–J determinants are on the same molecule or that TSF represents a multichain molecule coded for by different chromosomes, but with both I–J and V_H gene products required to permit the appropriate function. Recent studies of Tada and co-workers with TSF isolated from monoclonal T-cell lines made by fusing enriched populations of T-suppressor cells with the BW5147 tumor line are strongly supportive of the concept of a multichain molecule (Taniguchi et al. 1979, 1980).

These hybrid lines continuously secrete molecules having both antigen-specific suppressor function and I–J determinants. Thus, the suppressive activity of both the secreted product and an extract prepared by sonication could be removed by absorption on an antigen column or an anti-I–J column. If the antigen binding and the anti-I–J determinants are present on a single molecule, the mixture of the material after absorption with antigen or anti-I–J should not exert suppressive activity. On the other hand, if the activities are on separate molecules, they could combine in the mixture and then act on the target cell. The mixture of the effluents of the secreted TSF gave no suppression. However, the mixture of the effluents of the extracted material produced marked suppression despite the fact that either effluent alone had no activity. Thus, the extract contained molecules that could combine to form the TSF, while the secreted material contained only molecules in the associated form.

The model for the T-cell receptor (at least from T-suppressor cells) that can be drawn from these studies is a multichain molecule with one chain coded for by genes linked to Ig–C_H genes and bearing V_H determinants; this chain has the antigen-binding activity. The second chain associated with the first by covalent or noncovalent interactions is coded for by the I–J region of the MHC. The function of the MHC product in this molecule is unknown, but it might play a role in cell bridging or triggering events during immunocompetent cell interactions.

Conclusion

The resolution of the nature of the T-cell receptor is one of the central problems in modern cellular immunology. Further studies to characterize and define biochemically the model described above are needed. It is still not clear if T-helper cells bear or secrete similar factors with perhaps one chain coded for in the I–A subregion of the MHC, as suggested by the studies of Taussig and Munro. It is also not clear why the contribution of the MHC either at a genetic level or at a biochemical level was not apparent in the elegant studies describing the sharing of idiotypes between specific antibody- and T-lymphocyte receptors. Although the existence of soluble T-cell products may facilitate the biochemical analysis of the T-cell receptor, it is still a complete mystery whether these factors have any function *in vivo*. Furthermore, the genetic restrictions on immunocompetent cell interactions favor the view that direct cell contact may be required. In addition, a multiplicity of nonspecific regulatory lymphokines serve as second signals to augment or suppress lymphocyte reactions independent of genetic restrictions and cell contact.

References

Bach, B.A., Greene, M.I., Benacerraf, B., Nisonoff, A. (1979) Mechanisms of regulation of cell-mediated immunity. IV. Azobenzenearsonate-specific suppressor factor(s) bear cross-reactive idiotypic determinants the expression of which is linked to the heavy-chain allotype linkage group of genes. J. Exp. Med. 149,1084–1098.

Binz, H., Lindenmann, J., Wigzell, H. (1974) Cell-bound receptors for alloantigens on normal lymphocytes. I. Characterization of receptor-carrying cells by the use of antibodies to alloantibodies. J. Exp. Med. 139,877–887.

Binz, H., Wigzell, H., (1975a) Shared idiotypic determinants on B and T lymphocytes reactive against the same antigenic determinants. I. Demonstration of similar or identical idiotypes on IgG molecules and T-cell receptors with specificity for the same alloantigens. J. Exp. Med. 142,197–211.

Binz, H., Wigzell, H.(1975b) Shared idiotypic determinants on B and T lymphocytes reactive against the same antigenic determinants. II. Determination of frequency and characteristics of idiotypic T and B lymphocytes in normal rats using direct visualization. J. Exp. Med. 142,1218–1230.

Binz, H., Wigzell, H. (1975c) Shared idiotypic determinants on B and T lymphocytes reactive against the same antigenic determinants. III. Physical fractionation of specific immunocompetent T lymphocytes by affinity chromatography using anti-idiotypic antibodies. J. Exp. Med. 142,1231–1240.

Binz, H., Wigzell, H. (1975d) Shared idiotypic determinants on B and T lymphocytes reactive against the same antigenic determinants. IV. Isolation of two groups of naturally occurring, idiotypic molecules with specific antigen-binding activity in the serum and urine of normal rats. Scand. J. Immunol. 4,591–600.

Binz, H., Wigzell, H. (1976) Shared idiotypic determinants on B and T lymphocytes reactive against the same antigenic determinants. V. Biochemical and serological characteristics of naturally occurring, soluble antigen-binding T-lymphocyte-derived molecules. Scand. J. Immunol. 5,559–571.

Binz, H., Wigzell, H., Bazin, H. (1976) T-cell idiotypes are linked to immunoglobulin heavy chain genes. Nature 264,639–642.

Black, S.J., Hämmerling, G.J., Berek, C., Rajewsky, K., Eichmann, K. (1976) Idiotypic analysis of lymphocytes in vitro. I. Specificity and heterogeneity of B and T lymphocytes reactive with anti-idiotypic antibody. J. Exp. Med. 143,846–860.

Cantor, H., Boyse, E.A. (1975a) Functional subclasses of T lymphocytes bearing different Ly antigens. I. The generation of functionally distinct T-cell subclasses is a differentiative process independent of antigen. J. Exp. Med. 141,1376–1389.

Cantor, H., Boyse, E.A. (1975b) Functional subclasses of T lymphocytes bearing different Ly antigens. II. Cooperation between subclasses of Ly^+ cells in the generation of killer activity. J. Exp. Med. 141,1390–1399.

Cantor, H., Weissman, I. (1976) Development and function of subpopulations of thymocytes and T lymphocytes. Prog. Allergy 20,1–64.

Cantor, H., Shen, F.W., Boyse, E.A. (1976) Separation of helper T cells from suppressor T cells expressing different Ly components. II. Activation by antigen: after immunization, antigen-specific suppressor and helper activities are mediated by distinct T-cell subclasses. J. Exp. Med. 143,1391–1401.

Cantor, H., Hugenberger, J., McVay-Boudreau, L., Eardley, D.D., Kemp, J., Shen, F.W., Gershon, R.K. (1978a) Immunoregulatory circuits among T-cell sets. Identification of a subpopulation of T-helper cells that induces feedback inhibition. J. Exp. Med. 148,871–877.

Cantor, H., McVay-Boudreau, L., Hugenberger, J., Naidorf, K., Shen, F.W., Gershon, R.K. (1978b) Immunoregulatory circuits among T-cell sets. II. Physiologic role of feedback inhibition in vivo: absence in NZB mice. J. Exp. Med. 147,1116–1125.

Cooper, M.D., Peterson, R.D.A., South, M.A., Good, R.A. (1966) The functions of the thymus system and the bursa system in the chicken. J. Exp. Med. 123,75–102.

Crumpton, M.J., Perles, B., Auger, J. (1978) Mitogen receptors and mitogen responsiveness. In: *International Cell Biology*. Brinkley, B. R., Porter, K. R., (eds.) New York, Rockefeller University Press.

Cunningham, B.A., Sela, B.A., Yahara, I., Edelman, G.M. (1976) Structure and activities of lymphocyte mitogens in mitogens. In: *Mitogens in Immunobiology*. Oppenheim, J.J., Rosenstreich, D.L., (eds.) New York, Academic Press.

Eardley, D.D., Shen, F.W., Cantor, H., Gershon, R.K. (1979) Genetic control of immunoregulatory circuits. Genes linked to the Ig locus govern communication between regulatory T-cell sets. J. Exp. Med. 150,44–50.

Germain, R.N., Benacerraf, B. (1978) Antigen-specific T cell-mediated suppression. III. Induction of antigen-specific suppressor T cells (Ts_2) in L-glutamic acid[60]-L-alanine[30]-L-tyrosine[10] (GAT) responder mice by nonresponder-derived GAT-suppressor factor (GAT-TsF). J. Immunol. 121,608–612.

Germain, R.N., Thèze, J., Kapp, J.A., Benacerraf, B. (1978) Antigen-specific T-cell-mediated suppression. I. Induction of L-glutamic acid[60]-L-alanine[30]-L-tyrosine[10] specific suppressor T cells in vitro requires both antigen-specific T-cell suppressor factor and antigen. J. Exp. Med. 147,123–136.

Germain, R.N., Ju, S.-T., Kipps, T.J., Benacerraf, B., Dorf, M.E. (1979) Shared idiotypic determinants on antibodies and T-cell-derived suppressor factor specific for the random terpolymer L-glutamic acid[60]-L-alanine[30]-L-tyrosine[10]. J. Exp. Med. 149,613–622.

Gillis, S., Union, N.A., Baker, P.E., Smith, K.A. (1979) The in vitro generation and sustained culture of nude mouse cytolytic T-lymphocytes. J. Exp. Med. 149,1460–1476.

Goldstein, G., Scheid, M., Boyse, E.A., Brand, A., Gilmour, D.G. (1977) Thymopoietin and bursopoietin: induction signals regulating early lymphocyte differentiation. Cold Spring Harbor Symp. Quant. Biol. 41,5–8.

Hämmerling, G.J., Black, S.J., Berek, C., Eichmann, K., Rajewsky, K. (1976) Idiotypic analysis of lymphocytes in vitro. II. Genetic control of T-helper cell responsiveness to anti-idiotypic antibody. J. Exp. Med. 143,861–869.

Hesketh, R.L. (1978) Early biochemical event in lymphocyte stimulation. In: *International Review of Biochemistry of Cell Walls and Membranes II*. Metcalf, J. C. (ed.) Baltimore, University Park Press.

Hirata, F., Axelrod, J., Toyoshima, S., Waxdal, M.J. (1980) Phospholipid methylation. A biochemical signal modulating lymphocyte mitogenesis. Proc. Natl. Acad. Sci. U.S.A. 77,862–865.

Huber, B., Devinsky, O., Gershon, R.K., Cantor, H. (1976) Cell-mediated immunity: delayed-type hypersensitivity and cytotoxic responses are mediated by different T-cell subclasses. J. Exp. Med. 143,1534–1539.

Hunter, F., Mole, J.E., Bennett, J.C. (1978) The molecular structure and interrelationships of immune T lymphoid cell surface antigens. Contemp. Top. Molec. Immunol. 1,283–317.

Jandinski, J., Cantor, H., Tadakuma, T., Peavy, D.L., Pierce, C.W. (1976) Separation of helper T cells from suppressor T cells expressing different Ly components. I. Polyclonal activation. Suppressor and helper activities are inherent properties of distinct T-cell subclasses. J. Exp. Med. 143,1382–1390.

Kapp, J.A. (1978) Immunosuppressive factors from lymphoid cells of nonresponder mice primed with L-glutamic acid^{60}-L-alanine^{30}-L-tyrosine^{10}. IV. Lack of strain restriction among allogeneic, nonresponder donors and recipients. J. Exp. Med. 147,997–1006.

Kapp, J.A., Pierce, C.W., de la Croix, F., Benacerraf, B. (1976) Immunosuppressive factor(s) extracted from lymphoid cells of nonresponder mice primed with L-glutamic acid^{60}-L-alanine^{30}-L-tyrosine^{10} (GAT). I. Activity and antigenic specificity. J. Immunol. 116,305–309.

Kapp, J.A., Pierce, C.W., Benacerraf, B. (1977) Immunosuppressive factor(s) extracted from lymphoid cells of nonresponder mice primed with L-glutamic acid^{60}-L-alanine^{30}-L-tyrosine^{10} (GAT). II. Cellular source and effect on responder and nonresponder mice. J. Exp. Med. 145,828–838.

Krawinkel, U., Cramer, M., Imanishi-Kari, T., Jack, R.S., Rajewsky, K., Makela, O. (1977a) Isolated hapten-binding receptors of sensitized lymphocytes. I. Receptors from nylon wool-enriched mouse T lymphocytes lack serological markers of immunoglobulin constant domains but express heavy chain variable portions. Eur. J. Immunol. 1,566–573.

Krawinkel, U., Cramer, M., Mage, R.G., Kelus, A.S., Rajewsky, K. (1977b) Isolated hapten-binding receptors of sensitized lymphocytes. II. Receptors from nylon wool-enriched rabbit T-lymphocytes lack serological determinants of immunoglobulin constant domains but carry the A locus allotypic markers. J. Exp. Med. 146,792–801.

Krawinkel, U., Cramer, M., Melchers, I., Imanishi-Kari, T., Rajewsky, K. (1978) Isolated hapten-binding receptors of sensitized lymphocytes. III. Evidence for idiotypic restriction of T-cell receptors. J. Exp. Med. 147,1341–1347.

Ling, N.R., Kay, J.E. (1975) *Lymphocyte Stimulation*. New York; Elsevier North Holland.

London, J., Berrih, S., Bach, J.-F. (1978) Peanut agglutinin. I. A new tool for studying T lymphocyte subpopulations. J. Immunol. 121,438–443.

Mozes, E., Haimovich, J. (1979) Antigen specific T-cell helper factor cross reacts idiotypically with antibodies of the same specificity. Nature 278,56–57.

Mozes, E., Isac, R., Taussig, M.J. (1975) Antigen-specific T-cell factors in the genetic control of the immune response to poly(Tyr, Glu)-poly-DL-Ala–polyLys. Evidence for T- and B-cell defects in SJL mice. J. Exp. Med. 141,703–707.

Munro, A.J., Taussig, M.J., Campbell, R., Williams, H., Lawson, Y. (1974) Antigen-specific T-cell factor in cell cooperation: physical properties and mapping in the left-hand (K) half of *H-2*. J. Exp. Med. 140,1579–1587.

Murphy, D.B., Herzenberg, L.A., Okumaura, K., Herzenberg, L.A., McDevitt, H.O. (1976) A new *I* subregion (*I-J*) marked by a locus (*Ia-4*) controlling surface determinants on suppressor T lymphocytes. J. Exp. Med. 144,699–712.

Nisonoff, A., Ju, S.-T., Owen, F.L. (1977) Studies of structure and immunosuppression of a cross-reactive idiotype in strain A mice. Immunol. Rev. 34,89–118.

Novogrodsky, A. (1976) A chemical approach for the study of lymphocyte activation. In: *Mitogens in Immunobiology*. Oppenheim, J.J., Rosenstreich, D.L., (eds.) New York; Academic Press.

Oppenheim, J.J., Rosenstreich, D.L. (1976) Signals regulating the *in vitro* activation of lymphocytes. Prog. Allergy 20,65–194.

Oppenheim, J.J., Schecter, B. (1976) Lymphocyte transformation. In: *Manual of Clinical Immunology*. Rose, N.R., Friedman, H., (eds.) Washington D.C., American Society for Microbiology.

Owen, J.J.T., Jordan, R.K., Robinson, J.H., Singh, U., Wilcox, H.N.A. (1977) In vitro studies on the generation of lymphocyte diversity. *Cold Spring Harbor Symp. Quant. Biol.* 41,129–137.

Parker, C.W. (1979) The role of intracellular mediators in the immune response. In: *Biology of the Lymphokines.* Cohen, S., Pick, E., Oppenheim, J.J., (eds.) New York, Academic Press.

Paul, W.E. (1970) Functional specificity for antigen-binding receptors of lymphocytes. Transplant Rev 5,130–166.

Paul, W.E., Benacerraf, B. (1977) Functional specificity of thymus dependent lymphocytes. Science 195,1293–1300.

Raff, M.C. (1971) Surface antigenic markers for distinguishing T and B lymphocytes in mice. Transplant Rev. 6,52–80.

Resch, K.L. (1976) Membrane associated events in lymphocyte activation. In: *Receptors and Recognition,* Series A, vol. 1. Cuatrecasas, P., Greaves, M., (eds.) London, Chapman and Hall Ltd.

Rosenthal, A.S., Lipsky, P.E., Shevach, E.M. (1975) Macrophage-lymphocyte interaction and antigen recognition. Fed. Proc. 34,1743–1748.

Sharon, N. (1976) Lectins as mitogens. In: *Mitogens in Immunobiology.* Oppenheim, J. J., Rosenstreich, D.L., (eds.) New York, Academic Press.

Shaw, J., Monticone, V., Mills, G., Paetkau, V. (1978) Effects of costimulator on immune responses *in vitro.* J. Immunol. 120,1974–1980.

Spiegel, S., Ravid, A., Wilchek, M. (1979) Involvement of gangliosides in lymphocyte stimulation. Proc. Natl. Acad. Sci. U.S.A. 76,5277–5281.

Stanton, T.H., Boyse, E.A. (1976) A new serologically defined locus, *Qa-1,* in the *Tla*-region of the mouse. Immunogenetics 3,525–531.

Stutman, O. (1978) Intrathymic and extrathymic T cell maturation. Immunol. Rev. 42,138–184.

Takemori, T., Tada, T. (1975) Properties of antigen-specific suppressive T-cell factor in the regulation of antibody response of the mouse. I. In vivo activity and immunochemical characterizations. J. Exp. Med. 142,1241–1253.

Taniguchi, M., Tokuhisa, T. (1980) Cellular consequences in the suppression of antibody response by the antigen-specific T-cell factor. J. Exp. Med. 151,517–527.

Taniguchi, M., Hayakawa, K., Tada, T. (1976a) Properties of antigen-specific suppressive T cell factor in the regulation of antibody response of the mouse. II. In vitro activity and evidence for the I region gene product. J. Immunol. 116,542–548.

Taniguchi, M., Tada, T., Tokuhisa, T. (1976b) Properties of antigen-specific suppressive T-cell factor in the regulation of antibody response of the mouse. III. Dual gene control of T-cell-mediated suppression of the antibody response. J. Exp. Med. 144,20–31.

Taniguchi, M., Saito, T., Tada, T. (1979) Antigen-specific suppressive factor produced by a transplant-able I-J bearing T-cell hybridoma. Nature 278,555–558.

Taniguchi, M., Takei, I., Tada, T. (1980) Functional and molecular organisation of an antigen-specific suppressor factor from a T-cell hybridoma. Nature 283,227–228.

Taussig, M.J., Munro, A.J. (1974) Removal of specific cooperative T cell factor by anti-H-2 but not by anti-Ig sera. Nature 251,63–65.

Taussig, M.J., Munro, A.J., Campbell, R., David, C.S., Staines, N.A. (1975) Antigen-specific T-cell factor in cell cooperation. Mapping within the *I* region of the *H-2* complex and ability to cooperate across allogeneic barriers. J. Exp. Med. 142,694–700.

Thèze, J., Waltenbaugh, C., Dorf, M.E., Benacerraf, B. (1977a) Immunosuppressive factor(s) specific for L-glutamic acid[50]-L-tyrosine[50] (GT). II. Presence of *I-J* determinants in the GT-suppressive factor. J. Exp. Med. 146,287–292.

Thèze, J., Kapp, J.A., Benacerraf, B. (1977b) Immunosuppressive factor(s) extracted from lymphoid cells of nonresponder mice primed with L-glutamic acid[60]-L-alaniune[30]-L-tyrosine[10] (GAT). III. Immunochemical properties of the GAT-specific suppressive factor. J. Exp. Med. 145,839–856.

Thomas, D.W., Yamashita, U., Shevach, E.M. (1977) The role of Ia antigens in T cell activation. Immunol. Rev. 35,97–120.

Waltenbaugh, C., Debré, P., Thèze, J., Benacerraf, B. (1977) Immunosuppressive factor(s) specific for L-glutamic acid[50]-L-tyrosine[50] (GT). I. Production, characterization, and lack of H-2 restriction for activity in recipient strain. J. Immunol. 118,2073–2077.

Cell-Mediated Cytotoxicity

David L. Nelson, M.D.

Cellular lysis occurring during the course of a host inflammatory response probably results from a complex interplay of immunologic effector mechanisms. Those cytolytic events that are dependent on the immediate presence of cellular effectors are defined as resulting from cell-mediated cytotoxicity, in contrast to lytic events dependent on humoral effectors such as antibody and complement, which may operate at some distance from the cells that produce them. This classification of cytolytic mechanisms as either cellular or humoral represents an oversimplification of the *in vivo* events, but such a separation provides a relatively convenient operational means of contrasting these types of cytotoxicity.

In vivo, cell-mediated cytotoxic responses probably play a major role in processes such as allograft and tumor rejection, immune surveillance against neoplastic transformations, the elimination of foreign microorganisms infecting host tissues, and deleterious autoimmune reactions (Cerottini and Brunner 1974). Because of the difficulties of studying a complex biologic phenomenon such as cell-mediated cytotoxicity *in vivo*, a variety of methods have been developed to assess the generation and functional capabilities of cytotoxic effector cells *in vitro*. These assays have demonstrated that a variety of cell types can function as cytotoxic effectors, including T-cells, monocytes/macrophages, polymorphonuclear leukocytes, K-cells, and NK-cells. In this chapter, several assays of cell-mediated cytotoxicity are considered that are believed to be models of events occurring *in vivo* (Table 1). These assays are discussed in terms of the requirements for the generation of cytotoxic effector cells, the identities of such effector cells, the potential *in vivo* relevance of these reactivities, and the mechanisms whereby these cells mediate cytolytic effects.

Methods of Assessing Cytotoxicity

Cell-mediated cytotoxic responses are usually assessed *in vitro* by the incubation of putative effector cell populations with a substrate consisting of target cells. Target cell damage, and thus the definition of effector cell competence, may be assessed by

From the Metabolism Branch, National Cancer Institute, National Institutes of Health, Bethesda, Maryland.

Table 1. Major Forms of Cell-Mediated Cytotoxicity

Assay	Effector cells	Targets	*In vivo* relevance
Alloimmune Cell-mediated Cytotoxicity (CMC)	T-Cells	Major Histocompatibility Complex (MHC) Antigen-Bearing Cells	Allograft Rejection, Graft versus Host Disease, Pregnancy
Self-MHC Restricted CMC	T-Cells	Antigen X and MHC Antigen-Bearing Cells	Elimination of Infectious Agents, Immune Surveillance, Tumor Rejection, Autoimmune Reactions
Lectin-dependent CMC	T-Cells	Cell Lines, Tumors, and Erythrocytes	Hypothetical: Could participate in a variety of cytotoxic reactions *in vivo.*
	Macrophages, Monocytes	Erythrocytes	Hypothetical: Could participate in a variety of cytotoxic reactions *in vivo.*
	Polymorphonuclear Leukocytes	Erythrocytes	Hypothetical: Could participate in a variety of cytotoxic reactions *in vivo.*
Antibody-dependent CMC	K-Cells	Cell Lines, Tumors, and Erythrocytes	Allograft Rejection, Elimination of Infectious Agents, Immune Surveillance, Tumor Rejection, Autoimmune Reactions
	Macrophages, Monocytes	Erythrocytes	Allograft Rejection, Elimination of Infectious Agents, Immune Surveillance, Tumor Rejection, Autoimmune Reactions
	Polymorphonuclear Leukocytes	Erythrocytes	Allograft Rejection, Elimination of Infectious Agents, Immune Surveillance, Tumor Rejection, Autoimmune Reactions
Spontaneous (Natural) CMC	NK-Cell	Tumor and/or cell Lines	Immune surveillance, Tumor Rejection
Macrophage-mediated CMC	Macrophages	MHC Antigen-Bearing Cells, Tumors, and Cell Lines	Allograft and Tumor Rejection, Immune Surveillance, Autoimmune Reactions

a variety of techniques. An understanding of the methods whereby such cytolytic effects are measured is critical to any discussion of cell-mediated cytotoxicity. (These methods are reviewed in detail in Bloom and David 1976.) The two most frequently employed methods for determining target cell damage have been: (1) visual assessment of target cell lysis using supravital staining techniques; and (2) radioisotope labeling of target cells prior to (prelabeling) or following (postlabeling) effector cell interaction and measuring isotope release or incorporation, respectively. Of these two methods, radioisotope labeling is currently the most popular and target cell prelabeling is generally preferred since the release of label following effector cell interaction predominantly reflects target cytolysis, whereas postlabeling may reflect target cytolysis and/or cytostasis, and incorporation of label into effector cells may also occur. A variety of radioisotopes and labeled compounds have been employed for target cell prelabeling including $Na_2{}^{51}CrO_2$, ^{125}I-iododeoxyuridine, ^3H-proline, ^{14}C-nicotinamide, and ^{86}Rb. None of these materials satisfies all of the ideal characteristics for target cell prelabeling which include: (1) high specific activity labeling so that small numbers of target and effector cells may be employed for numerous replicate assays; (2) even distribution of isotope occurs among individual target cells regardless of metabolic activity or cell cycle stage so that the lysis of each target cell results in the release of the same amount of label; (3) release of label into the supernatant occurs only with irreversible target cell damage; (4) total isotope release occurs into the supernatant with 100% target cell death; and (5) the released radioisotope is not used. However, $Na_2{}^{51}CrO_2$ satisfies several of these criteria under appropriate assay conditions and this compound has become the most popular prelabeling agent. Following addition to target cells, the $^{51}CrO_2^=$ ion diffuses into the cell cytoplasm and binds to various macromolecules. From studies correlating ^{51}Cr release with the visual assessment of target cytolysis, it appears that the release of ^{51}Cr-labeled molecules occurs predominantly following irreversible alterations of the target cell membrane, which precede, and eventually culminate, in actual cytolysis. Therefore, ^{51}Cr release from prelabeled targets is generally considered a satisfactory measure of direct cytolytic events.

Of the several formulae developed for the expression of target cell damage as measured by ^{51}Cr release, the most commonly employed formula yields a value for percent of target cell lysis, which is calculated as follows: percent of target cell lysis = (E-C)/(T-C) × 100, where E represents the supernatant counts per minute (CPM) from cultures of targets and effector cells, C represents the supernatant CPM from cultures of targets in the presence of medium alone without added effector cells, and T represents the total releasable CPM from the same number of target cells following freeze–thawing or detergent treatment. The numerator of this equation reflects the amount of ^{51}Cr release from the targets that is attributable to the added effector cells, while the denominator reflects the amount of ^{51}Cr available for release by the effector population. While target cell ^{51}Cr release cannot quantitate cytotoxic effector cell activity at the level of an individual effector cell (i.e., ^{51}Cr release does not distinguish one effector lysing two targets versus two effectors each lysing one target), it does allow an estimation of the relative cytotoxic activity of given populations of cells containing cytolytic effectors. When percent of target cell lysis is analyzed as a function of the log of the number of putative effector cells added to a fixed number of target cells, a linear relationship is generally observed (Figure 1). Populations containing effector cells may then be compared on the basis of the number of cells required to produce a given amount of target cell damage. An arbitrary value for percent of lysis is chosen (in this example, 30%) that transects the linear portion of both effector cell titrations, which must be parallel for valid comparisons. The number of effectors required to produce this amount of target cell damage is defined as one lytic unit (LU). As can be seen for population A, one LU represents the activity contained in 4×10^5 added cells, while for population B, one LU represents the activity of 20×10^5 added cells. Expressed as LU per 10^7 effector cells, population A contains 25 LU, while population B contains 5 LU. Thus population A contains

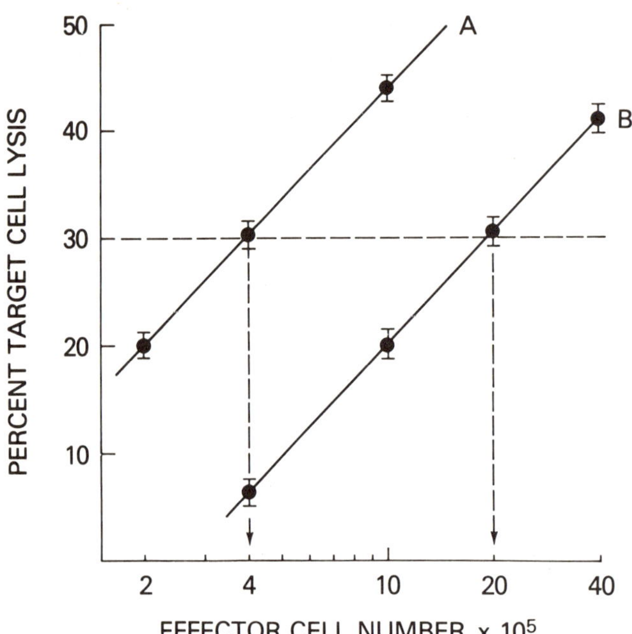

Figure 1. Effector cells from individuals A and B were tested at varying effector-to-target cell ratios in a cytotoxic assay employing 10^4 ^{51}Cr-labeled target cells. In order to compare the relative activities of the effector cells from each individual, the observed values for percent of target cell lysis were plotted against the log of the number of added effector-containing cells producing such lysis. The number of lytic units (LU) in each population was then determined at 30% lysis (see text).

five times more LU per 10^7 effector cells than population B. This activity comparison does not necessarily imply that the actual number of effector cells in A and B are different. Population A may contain the same number of effector cells as population B, but each effector cell in A may be more efficient than those in B in the cytolysis of target cells. The foregoing example relates to activity estimates at the effector cell level. Such techniques may also be employed to estimate the capacity of different cell populations to produce cytotoxic effectors. For example, assume that the same effector populations A and B were each originally derived from 10^7 cells that were induced to mediate cytotoxicity by an identical stimulus. Also assume that following this stimulus, population A yielded a total of 10^7 cells while population B yielded 5×10^7 cells. Then, based on effector cell activity and corrected for fractional cell recovery (i.e., for A, 25 LU $\times$ 1/1 = 25, and for B, 5 LU $\times$ 5/1 = 25), the starting populations of A and B are equivalent in terms of their ability to produce cytotoxic effector cell activity.

One of the hallmarks of the immune response is antigen specificity. In cell-mediated cytotoxic responses, specificity may be assessed at the inductive or lytic phase of the response or both. For example, antigens X and Y may each be employed to produce cytotoxic effectors, which are then tested on target cells bearing either antigen X or antigen Y. If antigens X and Y are distinct, then X-immune cytolytic effectors lyse X-bearing but not Y-bearing targets and vice versa. Specificity at the cytotoxic effector level may also be assessed by "cold target" inhibition. This technique focuses on target cell recognition irrespective of target lysability. In this method, X antigen-immune cytotoxic effectors are tested on radiolabeled X antigen-bearing targets in the presence of nonradiolabeled ("cold") target cells bearing antigen X or antigen Y. Such "cold" targets compete with radiolabeled targets for effector-mediated cytotoxicity and thereby inhibit isotope release. If X and Y antigens are distinct, "cold" X-bearing but not Y-bearing targets inhibit cytotoxicity.

Methods such as these are routinely employed to assess and compare the cytotoxic activity of effector cells, to study the capacity of various cell populations to produce cytotoxic effectors, and to determine the specificity of cytotoxic reactions.

Cytotoxicity Mediated by Thymus-Derived Lymphocytes: Allogeneic Cell-Mediated Cytotoxicity and "Self" Major Histocompatibility Complex Restricted Cell-Mediated Cytotoxicity

Immune cytotoxic responses by thymus-derived lymphocytes (T-cells) involve effector cell recognition of specific antigens expressed on the target cell surface. Immune cytotoxic T-lymphocytes (CTL) invariably recognize target cell surface antigens that are the products of genes coding for histocompatibility antigens. The genetic loci responsible for the expression of histocompatibility antigens are numerous; however, in each species examined, one major locus has been identified of overwhelming importance in graft rejection, and this locus is termed the major histocompatibility complex (MHC). The MHC and its products have been shown to play a major role in several forms of cytotoxicity mediated by thymus-derived lymphocytes (T-cells). Therefore, some understanding of the MHC is critical to the discussion of these forms of cytotoxicity. The MHC is best defined in mice (Klein 1975) and is termed H-2. Within H-2, several major regions have been identified and mapped linearly by genetic recombination events in the order K,I,S,D from the centromere of chromosome number 17. Each region may exist in numerous allelic forms (i.e., is polymorphic), is identified by a testable trait, and the limits of each region are defined by genetic crossover events in intra-H-2 recombinant mice. The set consisting of the alleles at all loci within the complex that were contributed by a single chromosome from either parent is termed a H-2 haplotype. Strains of highly inbred mice that receive the same haplotype from both parents are available and these strains define reference H-2 haplotypes. For example, C57BL/10 mice are H-2^b (KISD = $\frac{bbbb}{bbbb}$), DBA/2 mice are H-2^d ($\frac{dddd}{dddd}$), and strain A mice ($\frac{kkdd}{kkdd}$) are intra-H-2 recombinants of H-2^k and H-2^d mice. A first generation offspring (F$_1$) produced from the mating of strain A and C57BL/10 mice would obtain a different H-2 haplotype from each parent and be $\frac{kkdd}{bbbb}$. The K and D regions of H-2 determine cell surface antigens that were first detected serologically and are present on virtually all murine tissues. The S region codes for serum proteins that may be associated with, or actually constitute, complement components. The I region was identified by immune responsiveness to chemically defined antigens and more recently has been shown to code for serologically detectable cell surface antigens, termed Ia (immune-associated) antigens, which may be of more limited tissue distribution than the antigens coded for by the K and D regions. The human analogs of H-2 K and H-2 D are the A, B, and C regions of HLA, and the human region analogous to H-2 I is probably HLA-D.

Cytotoxic effector cells capable of mediating specific lysis of allogenic target cells are undetectable in nonimmunized animals. Mature CTL, termed alloimmune CTL, may be derived from a pool of precursor cells by immunization with allogeneic cells either *in vivo* or *in vitro*. Alloimmune CTL have been studied in a variety of experimental animals and in human beings. Because of the availability of inbred strains of mice with defined MHC types, much of the experimental work relating to alloimmune CTL has been performed in this species. The two most commonly employed methods for the generation of alloimmune CTL *in vivo* involve host-versus-graft and graft-versus-host reactions. In the former method, an experimental animal is usually grafted by a parenteral injection of a single optimal dose of normal or neoplastic allogeneic cells. The recipient's lymphocytes are then tested for cytolytic activity *in vitro* on target cells sharing MHC-determined antigens with those of the graft donor. For example, following intraperitoneal alloimmunization in mice, cytolytic activity peaks in splenic lymphocytes around day 10 and then gradually falls maintaining detectable levels for several months. In the graft-versus-host model, lethally irradiated mice are injected parenterally with a graft of immunocompetent allogeneic lymphocytes. The spleen cells from such recipients harvested 4 days later consist largely of graft cells that have been sensitized against recipient alloantigens and are tested for lytic activity *in vitro* on targets sharing MHC-determined antigens with the graft recipient. In the *in vitro* method, alloimmune CTL may be generated during unidirectional mixed leukocyte cultures in which responder lymphocytes are cultured with allogeneic lymphocytes whose proliferative capacity has been blocked by irradiation or mitomycin C treatment. The responder cells proliferate and generate alloimmune CTL, which are tested on target cells sharing MHC antigens with the stimulator cell population.

This *in vitro* immunization method has provided a convenient means to examine the requirements for the generation of alloimmune CTL in mice. Provided they have been supplemented by macrophages, T-cells are necessary and sufficient as alloimmune CTL precursors and differentiate into mature cytotoxic effectors that can function in the absence of bone marrow lymphocytes (B-cells) and macrophages. These allogeneic CTL precursors bear receptors for antigens coded for by genes at H-2 K and D. The mature alloimmune CTL also bear easily detectable receptors for antigenic structures coded for by genes at H-2 K and D and bear less readily detectable receptors for surface antigens encoded by H-2 I. These antigenic receptors do not appear to be conventional immunoglobulin. Minimal estimates of the percentage of alloimmune CTL present in *in vitro* immunized murine splenic cells are in the range of 1 to 2% (Lindahl and Wilson 1977). Estimates of the alloimmune CTL precursor frequency for a particular MHC haplotype (i.e., H-2^b in H-2^d mice) are in the range of 1 per 10^3 splenic cells (Lindahl and Wilson 1977b).

A variety of studies have suggested a cooperative interaction among subpopulations of responder T-cells in the generation of alloimmune CTL. The interactions among T-cell subsets occurring during CTL generation have been aided by the discovery of the Lyt series of T-cell differentiation antigens present on the surface of thymocytes and subpopulations of peripheral T-cells (Cantor and Boyse 1977). Three sets of cells have been defined by antisera and complement killing of peripheral T-cells: (1) those that possess large amounts of all three antigens, Lyt 1+, 2+, 3+ cells; (2) those that possess predominantly Lyt 1 and little or no Lyt and 3− Lyt 1+, 2−, 3− cells; and (3) those that possess little Lyt 1 and predominantly Lyt 2 and 3 Lyt 1−, 2+ 3+ cells. With regard to the generation of alloimmune CTL *in vitro,* when the stimulator and responder cell differ at the whole of H-2 (KISD) the following have generally been observed: (1) Ly 1+, Ly 2−, Ly 3− responder cells proliferate in response to H−2 I region determined antigens on the stimulator cells but do not mature into CTL effectors; (2) Ly 1−, Ly 2+, Ly 3+ cells proliferate in response to

H-2 K and D region antigens on the stimulator cell and differentiate into CTL effectors; (3) Ly 1+, Ly 2−, Ly 3− cells, while not required for the differentiation of alloimmune CTL may amplify (help) the response by Ly 1−, Ly 2+, Ly 3+ prekillers; and (4) the mature CTL effector is an Ly 1−, Ly 2+, Ly 3+ cell. In contrast with these results, when alloimmune CTL are induced using stimulator cells possessing only H-2 K or D region differences from the responder cells or both, a somewhat different result was obtained (Bach and Alter 1978). The mature alloimmune CTL effector in this case was also Ly 1−, Ly 2+, Ly 3+. However, Ly 1−, Ly 2+, Ly 3+ cells were no longer sufficient for the generation of alloimmune CTL, and mixtures of Ly 1+, Ly 2−, Ly 3− cells with Ly 1−, Ly 2+, Ly 3+ cells as responders did not reconstitute a response. Thus, with K and D region differences only as opposed to differences at the whole of H-2, a requirement exists for an Ly 1+, Ly 2+, Ly 3+ precursor cell. Some caution must be taken in the interpretation of these results relating to the Lyt phenotype of cells involved in the generation and effector phases of T-cell-mediated cytotoxicity, since recent studies using sensitive fluorescent anti-Lyt sera have failed to demonstrate the existence of a Lyt 1−, 2+, 3+ subset. The availability of monoclonal hybridoma antibodies specific for Lyt antigens may further clarify the phenotype of T-cells involved in the generation and effector phases of cell-mediated cytotoxicity. These results with regard to collaborative interactions among T-cell subsets demonstrate that, depending on the MHC differences between stimulator and responder cells, separate activation mechanisms involving different T-cell subsets may exist for the generation of alloimmune CTL. Soluble factors released by T-cells *in vitro* during alloreactive responses may also amplify (help) CTL generation. One such helper factor was not specific for the MHC type of the stimulator cell type, nor was MHC compatibility required between the T-cell producing the factor and the T-cell responding to the factor (Plate 1977). This helper factor is now termed T-cell growth factor (TCGF) or, more appropriately, interleukin II (IL II).

The expression of alloimmune CTL reactivity may also be under the influence of negative immunoregulatory influences mediated by suppressor cells and/or soluble suppressor factors. Suppressor cells for the generation of alloimmune CTL may be induced *in vitro* by preculture of normal cells with concanavalin A (Con A) prior to the addition of such cells to mixed leukocyte cultures (Peary and Pierce 1974). Suppression is antigen nonspecific (i.e., suppression occurs regardless of the stimulator MHC type) and is T-cell mediated. Suppressor cells may also be induced by alloimmunization either *in vivo* or *in vitro* (Rich and Rich 1974; Hodes et al. 1977). *In vivo* alloimmunization results in suppressor T-cells that produce soluble suppressor factors, and with both cells and factors the observed suppression is antigen nonspecific. *In vitro* alloimmunization in mixed leukocyte cultures results in the generation of suppressor T-cells that do not produce soluble suppressor factors, and the observed suppression may be either antigen specific or nonspecific. Finally, antibodies directed against target cell MHC antigens may also inhibit cytotoxicity.

From these studies, it is apparent that normal individuals possess precursor T-cells committed to polymorphic MHC antigens, which can be efficiently differentiated into cytotoxic effector lymphocytes following alloimmunization. The generation of alloimmune CTL perhaps represents a complex interplay of responder cell types that react with several MHC determined products on the immunizing cell type. Moreover, depending on the precise nature of the antigenic stimulus, several pathways of CTL activation may be operative and the ultimate expression of cytotoxic activity may be dependent on both helper and suppressor T-cells. While such alloimmune cytotoxicity is ultimately cell mediated, both soluble (humoral) amplifier and suppressor factors may influence the expression of this form of cytotoxicity.

Several examples of the *in vivo* relevance of alloreactive CTL are available. When an individual receives an allograft such as a kidney or heart transplant, alloimmune CTL are elicited in the host, and such cells have been demonstrated within the rejecting allograft where they presumably cause tissue cytolysis and inflammation. Another example is graft-versus-host disease in which an immunodeficient individual receives a graft of immunocompetent T-cells that differentiate into alloimmune CTL, which may cause widespread tissue damage and inflammation. Neither of these examples, however, represents a physiological situation, and the true *in vivo* relevance of allo-reactive CTL is unsatisfactorily explained. Curiously, in the one physiologic condition in which allografting occurs, i.e., pregnancy, alloimmune CTL do not appear to be operative in causing abortion. It is possible that certain diseases (viral infectious neoplastic transformation) may alter host expression MHC antigens so as to resemble alloantigens and thereby call for the protective or destructive alloreactive responses. At present, however, the clearly significant *in vivo* relevance of alloreactive CTL is primarily in the field of transplantation biology.

Much of the effort devoted to the study of alloimmune CTL was based on the hypothesis that such reactivity to polymorphic MHC antigens might be of significance in understanding some forms of cell-mediated cytotoxicity directed toward foreign non-MHC antigens or self-antigens, which would be relevant in deleterious autoimmune phenomena, immunologic surveillance against neoplasia, and the elimination of exogenous microorganisms infecting host tissues. Recently, several examples of cell-mediated cytotoxic responses have been described in which antigen-specific cytolytic effects are observed only when the immune effector cells and the antigen-bearing target cells are compatible with respect to the MHC. Two recent excellent reviews of the subject are available (Shearer and Schmitt-Verhulst 1977; Zinkernagel and Doherty 1979).

Such MHC restricted cytotoxic responses have been described for a variety of antigens including viruses (lymphocytic coriomeningitis, ectromelia, influenza, and herpes), chemical antigens (trinitrophenol, dinitrophenol, and fluorescein isothiocyanate), non-MHC-determined minor histocompatibility antigens, including the sex-linked H-Y antigen expressed on male cells, and tumor-associated antigens. These MHC-restricted cytotoxic responses have been observed in mice, rats, chickens, and human beings. These reactions have been generated by immunization *in vivo* or *in vitro,* or by a combination of these methods. The mature cytotoxic lymphocyte is a T-cell whose capacity to produce target cell lysis is both antigen specific and restricted to target cells with which it shares MHC gene products. In mice, the relevant MHC regions required for such effector–target interactions have been predominantly mapped to the K and D regions of H-2. Effector-target homology at H-2 I was generally neither necessary or sufficient for lysis to occur. For example, DBA/2 effector cells $\left(\frac{dddd}{dddd}\right)$ immunized against antigen X would cause the lysis of X antigen-bearing target cells from DBA/2 $\left(\frac{dddd}{dddd}\right)$, strain A $\left(\frac{kkdd}{kkdd}\right)$ and A × C57BL/10 F1 hybrids $\left(\frac{kkdd}{bbbb}\right)$, but not antigen X-bearing targets from C57BL/10 $\left(\frac{bbbb}{bbbb}\right)$ or Z antigen-bearing DBA/2 targets. The cytotoxic effectors are monospecific for both antigen and the restricting K or D structure, i.e., two clones of CTL are present in X-immune DBA/2 cells each clone recognizing X and either K^d or D^d but not both. Studies in mice have shown that the self-MHC restricted mature CTL effector is Ly 1 − , Ly 2 + , Ly 3 + and that the CTL precursor is Ly 1+, Ly 2 +, Ly 3+ (Cantor and Boyse 1977). The precursor frequency for self-MHC restricted CTL for TNP has been estimated to be 1 per 10^4 spleen cells (Lindahl and Wilson 1977b). Antibodies directed against either self-MHC gene products or antigen have been observed to inhibit cytotoxicity.

This dual specificity of T-cells for antigen and self-MHC gene products has led to two major hypotheses regarding the recognition structure(s) on immune CTL (Figure 2). Immune T-cells may possess two distinct receptor specificities, one recognizing self-MHC gene products and the second combining with antigen; alternatively, T-cells may possess a single receptor specificity recognizing a self-MHC product that has been modified into a neoantigenic determinant ("modified-self") by virus, chemicals, or by association with other cell surface constituents such as H-Y or tumor antigens. Currently neither T-cell receptor model is totally consistent with all of the experimental observations in these self-MHC restricted CTL systems.

Given the fact that these restrictions exist at the cytolytic effector level, one might ask how such restrictions are imposed on cytotoxic T-cells. Recent evidence in virus-specific self-MHC restricted cytotoxic systems has suggested that in the process of becoming T-cells, bone marrow stem cells develop receptor specificities for the H-2 K, D, or I region MHC gene products expressed on the radioresistant thymic epithelium before antigenic exposure. These receptor specificities, which then constitute the repertoire of MHC gene products that are considered as "self," are stable in the peripheral T-cell population. From this spectrum of receptor specificities, the phenotypic expression of T-cell specificity for antigen plus "self" is further selected by antigen-presenting radiosensitive cells that must share H-2 I and K or D antigens with the maturing T-cells of the lymphoreticular system. The generation of mature CTL with receptor specificities for H-2 K or D plus antigen may also require the presence of T-helper cells with thymic instructed receptor specificities for H-2 I, which must be identical with those I-region antigens on the host lymphoreticular system and the self-MHC restricted CTL precursor cell.

Figure 2. Receptor models invoked to explain the dual specificities of self-MHC restricted cytolytic T-cells for antigen and a self-MHC gene product: (**A**) the single-receptor model; and (**B**) the two-receptor model. The term "modified self" refers to a neoantigenic determinant distinct from antigen and self-MHC product.

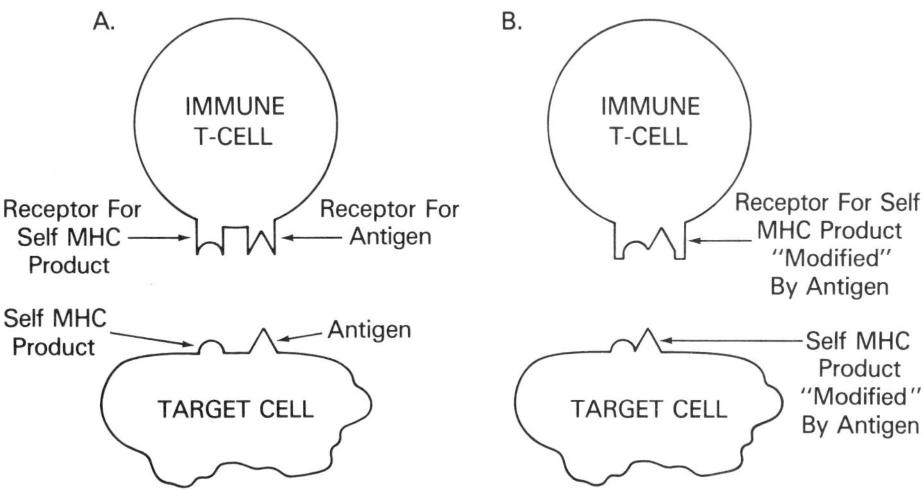

The phenotypic expression of self-MHC restricted cytotoxicity reactions appear to also be modified by the action of immune response genes (i.e., genes that are present with the MHC and control the responsiveness to a given antigen). For example, DBA/2 (H-2 K^d D^d) mice might recognize antigen X in the context of K^d but not D^d. Genes controlling the response to distinct antigens associated with H-2 K or H-2 D region gene products or both have been described in the CTL response to viruses, chemical antigens, and the H-Y antigen, and these genes generally have mapped to the K and D regions of MHC. These immune response gene effects could be expressed at several levels in the generation of self-MHC restricted CTL including: (1) stem cells, (2) the thymic microenvironment for T-cell receptor instruction, (3) T-cell precursors prior to antigenic exposure, (4) T-helper cells and/or those cells that selectively augment self-MHC restricted cytotoxic responses, or (5) T-suppressor cells that could regulate the expression of cytotoxic effector function, or (6) at the level of the cytotoxic effector itself. Many studies are currently in progress to explore these possibilities.

The general concept that emerges from these studies is that cytotoxic T-cells, like other T-cell subsets including T-helper cells and nonlytic T-suppressor cells, are restricted in cell-to-cell interactions by products of the MHC. In the case of cytotoxic T-cells, the K and D regions of the MHC determine both the restriction specificity and the immune response phenotype of the involved cells. Thus, the significance of these gene MHC products can be viewed as facilitating/restricting interactions between cells or serving antigen-presenting recognitive function or both rather than simply providing barriers to allogeneic transplantation. The extreme polymorphism of the MHC could then be considered as a survival mechanism retained by a species to ensure against the high frequency of an MHC allele that is not recognized in association with a particular catastrophic event (i.e., viral infection). Since both alloimmune and self-MHC restricted CTL are T-cells, one might ask what relationship(s) exists between these cells. Obviously, separate T-cell subsets might mediate these cytotoxic activities, or both reactivities might be the function of a single subset. For example, alloimmune cytotoxicity could be the expression of a CTL with self-MHC plus X antigen receptors that produces lysis of allogeneic targets through: (1) recognition of alloantigen as X or a neoantigen (single receptor model); or (2) recognition of alloantigen through anti-X or antiself-MHC receptor alone (dual receptor recognition model). The answer to this question of the relationship of alloimmune CTL and self-MHC restricted CTL is currently under extensive investigation and speculation.

In contrast with alloimmune CTL, the physiologic role of self-MHC restricted CTL is more apparent. Such CTL could survey and lyse host cells bearing exogenous antigens (chemicals or viruses) or host cells possessing minor modifications of non-MHC antigens (tumor antigens). Self-MHC restricted CTL may play a major role in the eradication of infectious virus from host tissues. Following acute viral infections in experimental animals, virus-specific CTL activity becomes maximal at the time when virus titers are rapidly declining, while humoral antibody responses only appear after virus has been eliminated. By lysing host tissues expressing early viral antigens, CTL may prevent further intercellular viral replication leading to viral elimination. Whether or not such lysis of host tissues is beneficial or deleterious must reflect some balance between the pathogenicity of the virus itself, the location of the virus infection, and the magnitude of the host response. Two elegant *in vivo* studies have demonstrated both the deleterious and beneficial potentials of self-MHC restricted CTL. Intracerebral inoculation of lymphocytic choriomeningitis virus (LCMV) into normal but not T-cell-deficient mice results in death of the animal. Transfer of LCMV immune CTL parenterally into T-cell deficient mice inoculated intracerebrally with LCMV results in fatal neurologic disease only if the immune donor and the recipient are compatible at H-2 K or D or both. In this case, inflammation within the brain caused by immune CTL results in cerebral edema and death. In contrast, beneficial effects have been observed when influenza immune CTL were transferred into nonimmune mice challenged with influenza virus. The transfer of immune CTL significantly reduced lung influenza titers, and this protective effect was seen only if the immune donor and the recipient shared H-2 K or D or both. Thus, in this case, inflammation and cytolysis caused by immune CTL within the lung resulted in a protective immune response.

Mitogen-Induced Cellular Cytotoxicity

From the previous sections, it is apparent that the generation of antigen-specific cytotoxic T-lymphocytes represents a series of complex cellular interactions, which culminate in the production of effector cells that exert lytic effects on target cells following interaction of specific effector cell receptors with target cell antigens. Models of cell-mediated cytotoxicity have been sought and developed in which many of these complex events may be bypassed or have occurred before the assessment of cytotoxic effector activity or both. Since lymphocytes activated with plant lectins and/or mitogens possess many characteristics of antigen-activated lymphocytes, such as helper-suppressor cell effects and the production of lymphokines, these mitogen-activated lymphocytes have been extensively studied as models of the functional capacities possessed by antigen-activated lymphocytes. Lymphocytes activated by plant mitogens or lectins or both also exert cytotoxic effector function, and this form of cytolysis has been termed mitogen (lectin)-induced cellular cytotoxicity or mitogen (lectin)-dependent cellular cytotoxicity (Holm and Perlmann P. 1966). Most of the lectins employed in MICC assays are mitogenic such as phytohemagglutinin, concanavalin A, and pokeweed mitogen; however, nonmitogenic lectins such as wheat germ agglutinin may also induce cell-mediated cytotoxicity.

Mitogen-induced cellular cytotoxicity (MICC) contrasts with antigen-specific cellular cytotoxicity in that effectors capable of mediating MICC are present in normal individuals who have not been intentionally immunized. Cytolytic activity is assessed by the incubation of putative effector cells with target cells for relatively brief periods of time (4 to 24 hours) in the presence and absence of mitogen. The increment in target cell damage occurring in cultures containing mitogen over that occurring in cultures of effectors and targets in the absence of mitogen reflects the MICC effector activity of the population being assessed. Lymphocytes preactivated with specific foreign antigens or alloantigens will also mediate cytotoxicity in the presence of plant lectins, and in this case cytotoxicity may be nonspecific. The lectin in MICC probably plays a dual role in: (1) producing cytolytic effector activation in a poorly understood manner; and (2) binding to sites on both the activated effector cell and the target cell, thereby bringing effector and target into close proximity. Cytolysis is not totally nonspecific in that certain target cells lacking binding structures for a particular lectin are not lysed in the presence of effectors and that lectin (Kirchner and Blaese 1973). The binding sites for such lectins on the surfaces of cells consist of relatively simple sugars, and therefore the antigen recognition structures that may be present on these effector cells may be bypassed. Target cytolysis is not self-MHC restricted and may occur with target cells syngeneic, allogeneic, or xenogeneic to the effector cell population. Studies that have employed different types of target cells in MICC have demonstrated that the capacity to effect cell-mediated cytolysis is not confined to T-lymphocytes and that the target cell type employed in MICC plays a critical role in determining which effector(s) may be operative (Nelson et al. 1976). T-cells, macrophages, and polymorphonuclear leukocytes (PMNL) could all mediate cytolysis in the presence of lectin when erythrocytes are employed as target cells. In contrast, with *in vitro* or *in vivo* passaged cell lines as targets, cytolytic effector activity is restricted to T-lymphocytes. The Ly antigen phenotype of the MICC effector is Lyt 1 + , 2 + , 3 + , and these cytolytic T-cells presumably represent the total pool of differentiated antigen-specific cytotoxic effectors whose receptor specificities are bypassed by these lectins. This model of cell-mediated cytotoxicity has demonstrated that cytolytic effector capability is a property of a wide variety of leukocyte types in addition to T-cells and has provided a means of assessing cytotoxic effector activity in a relatively simple fashion.

The *in vivo* relevance of cytotoxicity such as MICC is hypothetical. Given the provision of a suitable activator of leukocyte cytolytic mechanisms (i.e., polyclonal activators or specific antigens or both) such effectors might be bound to cells at inflammatory foci by naturally occurring lectin-like substances and produce cytolysis.

Antibody-Dependent Cell-Mediated Cytotoxicity

The best example of the cooperative interaction of humoral and cellular immune effectors in the expression of cell-mediated cytotoxicity is a phenomenon termed antibody-dependent cellular cytotoxicity (ADCC). In this cytotoxicity model, cellular effectors with receptors for the Fc portion of immunoglobulin molecules produce target cell lysis by attachment to the Fc portion of antibodies bound to target cells via their antigen combining sites. Lymphocytes, monocytes, and polymorphonuclear leukocytes may all function as cytotoxic effectors. Effector cells capable of mediating this form of cytotoxicity are readily demonstrable in various tissues of individuals who have not been intentionally immunized with target cell antigens. Effector cell lysis of syngeneic, allogeneic, and xenogeneic target cells has been observed, although under physiologic *in vivo* conditions cytolysis would only occur with syngeneic cells.

The ADCC assay consists of three components: (1) target cells, (2) antibodies with specificity for target cell antigens, and (3) effector cell populations. A wide variety of target cell types have been employed in ADCC assays, and antigens occurring either as normal cell membrane determinants or resulting from intentional modifications of the cell surface (chemicals and viruses) may serve as antigenic determinants for the sensitizing antibodies. It has become clear that the type of target cells employed in ADCC plays a role in determining which populations of effector cells are capable of mediating lysis. When erythrocytes are employed as targets in the presence of heterologous or homologous antierythrocyte sera, lymphocytes, monocytes, and PMNLs (neutrophiles and eosinophiles) may all mediate lysis (Nelson et al. 1976). In contrast, when most *in vivo* or *in vitro* passaged cell lines are employed as targets with heterologous or homologous antitarget cell sera, lysis is usually restricted to lymphocyte effectors.

The antitarget cell immunoglobulins that are capable of sensitizing targets for ADCC have mainly been limited to the IgG class. Antibodies of other classes have usually inhibited cytotoxicity. Recent evidence, however, has suggested that IgM antibodies may also sensitize targets for ADCC and that the effector lymphocytes operative with IgM are different than those operative with IgG. Complement components are not required for ADCC; however, their presence under certain conditions may augment cytolysis especially for erythrocyte targets. It has been estimated that as few as 20 IgG molecules are sufficient for the sensitization of tumor target cells for lysis in certain ADCC assays. The specificity of ADCC is determined by the antigen combining site of the antitarget cell antibody, while the Fc portion of the immunoglobulin molecule focuses the effector through its Fc receptor onto the target cell, and in poorly understood fashion activates the effector cell to produce lysis. Effector cell activity may be inhibited by antigen–antibody complexes and aggregated immunoglobulins that occupy effector cell Fc receptors, precluding contact with antibody-coated target cells. The relevant structural domains of the anti-target cell IgG molecules for binding to effector cell Fc receptors and ADCC activation appear to be located in C_H3 for both neutrophiles and monocytes and in the C_H2-C_H3 domains for lymphocytes. For comparison, the relevant structural domain of IgG for Clq fixation is in the C_H2 domain.

As mentioned previously, several effector cell types including lymphocytes are capable of mediating ADCC. The lymphocyte mediating ADCC has been termed the K-cell (killer cell). The precise identity of the lymphocyte effector, however, remains the subject of considerable controversy. It is clear that K-cells are nonphagocytic, nonadherent, mononuclear cells lacking the histochemical, ultrastructural, and antigen-presenting properties of monocytes. K-cells also lack high density surface immunoglobulin and large quantities of serologically detectable Ia antigens and thus cannot be considered conventional B-lymphocytes. It is still not known whether some K-lymphocytes bear receptors for complement components and low density surface antigens of thymus-derived lymphocytes. Organ distribution studies of these effector lymphocytes based on their functional capacity have revealed high effector activity in spleen and peripheral blood, moderate effector activity in lymph nodes, and only minimal or absent activity in bone marrow, thymus, and thoracic duct lymph (TDL). Spleen cells and peripheral blood lymphocytes of congenitally athymic nude mice contain more effector cell activity than the same tissues from normal mice. Estimates of effector cell numbers in human peripheral blood lymphocyte populations based on a visual cytolytic plaque assay performed on erythrocyte monolayers revealed that approximately 5% of lymphocytes possessed ADCC effector activity (Wahlin et al. 1976). This proportion of effector cells is in the same range as the proportion of CTL present in *in vivo* or *in vitro* antigen-immunized spleen cell populations from experimental animals.

Thus ADCC represents a potentially powerful cytolytic mechanism in which humoral and cellular immune effectors cooperate. The cell types (lymphocytes, monocytes, and PMNLs) capable of mediating ADCC exist in nonimmunized donors, and these cells may become cytolytic effectors in the presence of small quantities of immunoglobulin molecules bound to target cell antigens. Under physiologic conditions, ADCC could play a role in allograft rejection, the elimination of infectious agents including viruses and parasites, immune surveillance against neoplasia, and autoimmune phenomena such as hemolytic diseases caused by antierythrocyte antibodies.

Natural Cytotoxicity or Spontaneous Cell-Mediated Cellular Cytotoxicity

During investigations of the immune responses to tumors, it was observed that the lymphoid cells of normal individuals who did not have a tumor or who had not been immunized with a tumor or both were capable of producing cytolysis of tumor cells *in vitro*. This phenomenon of target cytolysis *in vitro* by lymphoid cells from normal individuals has been termed "natural" cytotoxicity (NC) or "spontaneous" cell-mediated cytotoxicity. A recent review series has been devoted to this form of cytotoxicity (Moller 1979). The effector cells mediating NC have been termed NK (natural killer) cells. Effector cells are not restricted to lyse MHC-identical targets, and the lysis of syngeneic, allogeneic, and xenogeneic targets has been reported. Effector cell activity has been documented in several species including mice, rats, guinea pigs, and human beings; it is not certain that this phenomena represents the same process in all of these species.

A variety of factors appear to influence the assessment of NC. The choice of target cells employed to assess NC may be of great importance in that: (1) *in vitro* passaged cell lines are routinely more susceptible to natural cytotoxicity than their *in vivo* passaged counterparts; (2) certain target cells but not others may be infected with viruses and other microorganisms and express additional extrinsic antigens to which effectors may react via other cytolytic mechanisms; and (3) considerable variation exists between cell lines and even among the same cell line passaged in different laboratories with regard to susceptibility to NC. These differences among cell line targets are poorly understood but clearly can effect the measurement of NK effector function.

A variety of host factors also appear to play significant roles in the expression of natural cytotoxicity. In some species, age appears to play a critical role in NC. In mice, NK effector function is virtually absent at birth, is first expressed at about 3 to 4 weeks of age, peak activity being attained at 5 to 10 weeks, then activity declines gradually to low but detectable levels thereafter. In humans, such age-related differences in NC have not been observed. It also appears that NC may be under genetic influences. In mice, certain strains exhibit either high or low NK effector activity, and high reactivity has generally been dominant in F_1 offspring produced by mating animals with high and low NK effector activity. These differences in NC reactivity appear to be controlled by multiple genetic loci outside the H-2 complex. In humans, low reactivity in males but not in females with the HLA type A3, B7 has been reported. Moreover, males in general exhibit greater NC effector activity than females. This sex difference has not been observed in experimental animals. Thus, a variety of poorly understood host factors also appear to influence the expression of natural cytotoxicity. It is not known if NK-cells possess specific receptors for target cell antigens that might be widely expressed among cell lines, or whether NC is truly nonspecific. Several investigations have demonstrated certain selective patterns of target cell susceptibility to NC. It is not certain if these selective cytotoxic reactivities represent an antigen-specific component to NC or some other phenomena.

The identity of the effector cells mediating natural cytotoxicity has been the subject of some controversy. Organ distribution studies in experimental animals and human beings have demonstrated comparable and high levels of cytotoxic effector activity in blood and spleen, low levels of activity in lymph node, and virtually undetectable activity in bone marrow, TDL, and thymus. Spleen cells and peripheral blood lymphocytes from congenitally athymic nude mice possess extremely high NC activity. In all species examined thus far, NK-cells are nonphagocytic, nonadherent, mononuclear cells lacking the histochemical staining properties of macrophages and the high density surface immunoglobulin characteristic of mature B-cells. Cell surface Fc receptors for IgG have been demonstrated on NK-cells of all species examined. The presence of complement receptors on the cell surface of NK-cells remains controversial, such receptors being absent from NK-cells of rats and mice, and perhaps being expressed only on a subpopulation of human NK-cells. Some question continues to exist as to whether some NK-cells possess small quantities of T-cell surface products or cell surface characteristics of monocytes/macrophages or both. Thus, the identity of the NK-cell remains elusive and NK-cells may in fact be a heterogenous cell population. However, from the foregoing, it is clear that the NK-cell mediating NC and the K-cell-mediating ADCC possess identical organ distributions and cell surface characteristics. The question thus arises as to whether NC is actually ADCC with the NK-cell being "armed" *in vivo* via its Fc receptors with cytophilic antibody possessing specificity for a broad range of antigens. A definitive answer to this question is not at hand. However, the mechanism(s) of NC and ADCC is known to be different in that NC, but not ADCC, is sensitive to treatment of effector cells and trypsin.

What then might be the relevance of such natural cytotoxicity toward host resistance? The observation that athymic nude mice lacking T-cell-mediated cytotoxic mechanisms, but possessing high NC reactivity, have a low incidence of spontaneous neoplasms and appear to have normal resistance to chemically induced carcinomas has led to speculation that NK-cells may be capable of participating in immune surveillance against certain forms of neoplastic transformation (Stutman 1973). Moreover, the capacity of certain strains of experimental animals to resist tumor grafts has been correlated with NK reactivity to the same tumors, also suggesting a role for NC in the elimination of established tumors. In addition, NK reactivity may play a part in the naturally occurring rejection of nonmalignant hematopoietic grafts such as that described for the rejection of parental bone marrow grafts by F_1 hybrid animals (Hh resistance) (Kiessling et al. 1977). The biological significance and mechanism of NK reactivity is an area of intense investigation at the present time.

Macrophage-Mediated Cytotoxicity

In addition to mediating cytotoxicity in the presence of lectins (MICC) and antitarget cell sera (ADCC), it is clear that macrophages may also be cytolytic in the absence of such agents. The expression of macrophage cytotoxicity may be either nonspecific or specific with respect to the target cells lysed, and cytotoxic effector activity may be induced either *in vitro* or *in vivo*. Cytolytic macrophages are induced *in vivo* during the course of chronic infections with intracellular pathogens such as *Toxoplasma gondii* and listeria monocytogenes. Such macrophages are not antigen specific but selectively lyse tumor cells and not normal cells. Macrophages expressing similar selective cytolytic properties may also be induced *in vitro* by incubation of normal macrophages with lymphokine-containing supernatants produced by T-cells from animals immunized against irrelevent nontumor associated antigens (i.e., ovalbumin) (Piessens et al. 1975). Macrophages capable of exerting antigen-specific cytolytic effects have been reported following *in vivo* immunization with alloantigens and syngeneic tumor cells (Evans and Alexander 1972). Such macrophages may have been "armed" *in vivo* by antitarget cell antibodies (ADCC). T-cells from immune animals are capable of producing a factor(s) that will activate normal macrophages for cytotoxicity that may be specific in the presence of ADCC. This factor does not appear to be conventional immunoglobulin. Certain genetic factors appear to influence the ability of macrophages to be activated by a variety of stimuli, and the genes controlling this responsiveness do not appear to be located in the MHC (Ruco et al. 1978).

Thus, macrophages may mediate several forms of cytolysis and this cytolysis may be either specific or nonspecific. Also, the cytolytic activity of macrophages may in part be modulated by humoral factors secreted by both T-cells and B-cells. Within the milieu of an inflammatory reaction, ample opportunity exists for macrophages to participate in cytolytic events, which may be relatively nonspecific or be initiated by antigen-specific cells of either T- or B-cell lineage that confer antigen specificity on cytotoxic macrophages.

The Mechanism of Target Cell Lysis

Much has been said in this chapter about assays of different forms of cell-mediated cytotoxicity, the identity of the effector cells involved in mediating lysis, and the mechanisms of generating such effectors. The final common pathway of these effector cells is the production of target cytolysis. We shall now consider what is known about the actual mechanism of target cell damage. Most studies of the mechanism of target cytolysis have involved T-cell-mediated lysis of tumor cells by alloimmune CTL. This discussion therefore focuses on this mechanism as the prototype of target cell lysis. This subject has recently been reviewed (Henney 1977; Goldstein and Smith 1977; Martz 1977). Since T-cells lyse targets extracellularly, these studies do not relate to intracellular lytic events that may occur following target cell phagocytosis by macrophages and PMNLs. Where available, most studies of the other models of lymphocyte-mediated cytotoxicity (self-MHC restricted cytotoxicity, MICC, ADCC, NC) have generally yielded similar results.

Certain general observations about the mechanism of lysis have been made. Viable effector cells and direct contact between effectors and targets are required. The CTL themselves are not lysed or inactivated during cytolysis and effector recycling may occur, i.e., one effector may contact and lyse more than one target cell. Lysis of "innocent bystander" cells does not occur; that is to say, target cells possessing antigens for which the effector lacks receptors are not lysed when added to a culture in which target cells are being lysed. Effector-induced lysis is rapid and begins within minutes of effector cell contact. A single collision between effector and target cell is sufficient to produce lysis, but not all such collisions result in cytolysis. Effector cell division is not required and there is no requirement for protein syntheses.

Target cell lysis can generally be divided into three stages (Figure 3): (1) effector–target cell contact or adhesion: (2) induction of the lytic lesion or delivery of the "lethal hit"; and (3) target cell disintegration. Effector–target cell contact is required during the first two stages but is not necessary during the third stage. The adhesion stage is prevented by reduced temperatures ($<15°C$) and by a variety of metabolic inhibitors including sodium azide and sodium cyanide (electron transport), dinitrophenol (aerobic ATP generation), and iodoacetate (anaerobic ATP production). These compounds, however, do not dissociate performed adhesions. Cytocalasin B inhibits adhesions presumably through its effects on microfilament function. Pharmacologic agents influencing intracellular levels of cyclic AMP and GMP do not effect adhesions. Either Ca^{++} or Mg^{++} ions are required for effector–target cell contact. Thus, adhesion is an energy dependent process requiring microfilament function. The next stage of target cell lysis involves the effector cell delivery of the lytic event. Those pharmacologic agents, increasing intracellular cyclic AMP either through increases in adenyl cyclase (isoproterenol, prostaglandins E_1 and E_2) or through inhibition of phosphodiesterase (theophylline), inhibit this step. Histamine also increases intracellular cyclic AMP thereby inhibiting lysis, and this effect is mediated by Type II (metiamide sensitive) histamine receptors. Colchicine and vinblastine also inhibit

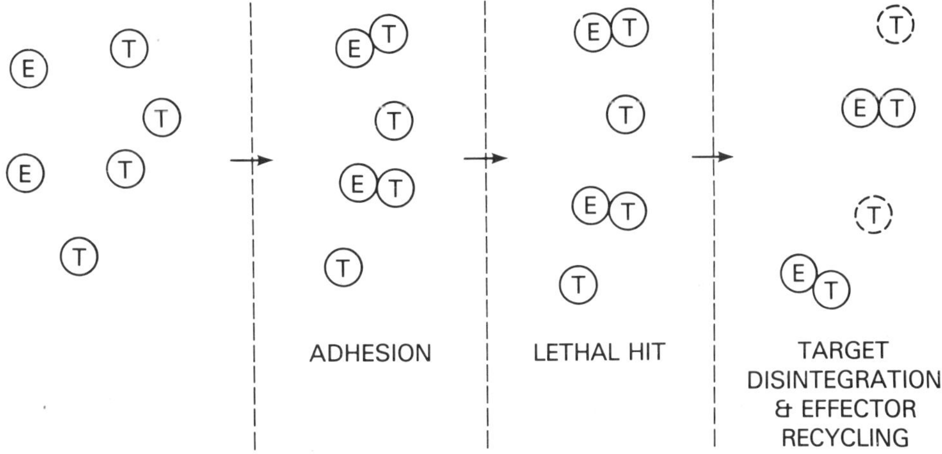

Figure 3. The stages of target cell lysis during cell-mediated cytotoxicity.

during this phase. An absolute requirement exists for Ca^{++} at this stage. This second stage is generally considered to end with the release of intracellular ions (Rb^{86}) and low molecular weight materials such as ^{14}C nicotinamide. The release of these low molecular weight materials is the earliest indication of target cell damage. The precise biochemical nature of this "lethal hit" is not known but those agents inhibiting this phase have suggested some form of secretory process. The role of a family of humoral factors that collectively may be termed lymphotoxins in this process remains controversial. Several factors inhibiting cytolysis by alloimmune CTL do not inhibit lymphotoxin production, and antilymphotoxin antibodies have been observed to inhibit lymphotoxin-dependent cytolysis but not alloimmune CTL-mediated target cell lysis. These studies of lymphotoxins and other humoral factors do not preclude the release of humoral substances locally in the region of effector–target adhesion in the "lethal hit" stage. Thus, even cytolytic events dependent on intimate effector–target contact may ultimately involve humoral factors released from such cells. The final stage of cytolysis is target cell disintegration, which results in the release of cytosol including ^{51}Cr-labeled macromolecules and disruption of target cell integrity. This step is Ca^{++} independent and has been suggested to result from colloid osmotic lysis, which in turn results from defects in small ion permeability caused by the second stage. The size of the lytic lesion induced by alloimune T-cells, as opposed to antibody and complement, is controversial and reports have suggested that this "hole" is either the same or smaller in size than the 90 Å lesion produced by antibody and complement.

Thus, the mechanism of target cell lysis by effector cells represents a sequence of events in which effectors first adhere to, and then lyse targets. The biochemical basis of this "lethal hit" is not well characterized at present.

Conclusion

Cell-mediated cytotoxicity refers to cytolytic reactions that require the immediate participation of cellular effectors. While such reactions may occur independently at inflammatory foci, cytolysis *in vivo* probably results from complex interactions of both cells and the soluble products of immunologic responsiveness. Cell-mediated cytotoxic responses have been studied predominantly *in vitro*. A variety of cell types possess the capacity to produce cellular lysis. T-cells generally produce cytotoxicity that is antigen specific (alloimmune CTL, self-MHC restricted CTL). The genes in the MHC play a major role in cytotoxicity mediated by T-cells. Products of the MHC may themselves serve as antigenic specificities or as specificities recognized in association with foreign antigens. Genes located within the MHC may also control the cytotoxic response to certain antigens (immune-response genes). T-cell subsets and soluble products of T-cells may amplify or suppress cytotoxic reactivity. T-cells may also effect antigen nonspecific cytotoxicity (MICC), given the provision of a stimulus to activate cytolytic mechanisms and lectin-like agents to bind lytic effectors to target cells. Some specificity to these reactions may occur as the result of the binding characteristics of the lectin. Polymorphonuclear leukocytes and macrophages may also mediate MICC. Cells lacking specific antigen receptors but possessing immunoglobulin F_c receptors may mediate specific cytotoxicity (ADCC) through interactions with antibodies bound to target cells via their antigen combining sites. Lymphocytes, macrophages, eosinophiles, and PMNLs may all serve as effectors for ADCC.

Certain lymphoid cells that also possess cell surface F_c receptors appear to be endowed with "natural" cytolytic potential (NC). Macrophages mediate cytotoxicity, which may be either specific or nonspecific, and in most instances specificity is conferred on macrophages by products of T-cells or B-cells or both. Little is known about the actual mechanism whereby effector cells produce cytolysis. This event is energy dependent, possesses many characteristics of a secretory process, and may ultimately lead to osmotic lysis of target cells.

Cell-mediated cytotoxic responses *in vivo* probably play major roles in a variety of immunologic responses. In these reactions, cytotoxic effectors represent only one of a variety of effector mechanisms that may be operative at a given time. These cytotoxic responses are most likely to constitute a predominant effector mechanism in reactivities such as allograft rejections, surveillance against neoplasia and the elimination of tumors, resistance toward infectious agents, and autoimmunity.

References

Bach, F.H., Alter, B.J. (1978) Alternative pathways of T-lymphocyte activation, J. Exp. Med. 148, 829–834.

Bloom, B.R., David, J.R., Eds. (1976) *In Vitro* Methods of Cell-Mediated and Tumor Immunity. New York, Academic Press.

Cantor, H., Boyse, E.A. (1977) Regulation of the immune response by T-cell subclasses. In: *Contemporary Topics in Immunobiology*. O. Stutman (ed.) New York, Plenum Press, (Vol. 7) pp. 47–68.

Cerottini, J-C., Theodore, B.K. (1974) Cell-mediated cytotoxicity, allograft refection, and tumor immunity. In: *Advances in Immunology*. Dixon, J.F., Kunkel, H.G. (eds.) New York, Academic Press, (Vol 18) pp. 67–123.

Evans, R., Alexander P. (1972) Mechanism of immunologically specific killing of tumour cells by macrophages. Nature 236,168–170.

Goldstein P., Smith, E.T. (1977) Mechanism of T-cell mediated cytolysis: The lethal hit stage. In: *Contemporary Topics in Immunobiology*. Stutman, O. (ed.) New York, Plenum Press, (Vol. 7) pp. 273–300.

Henney, C.S. (1977) T-cell mediated cytolysis: An overview of some current issues. In: *Contemporary Topics in Immunobiology*. Stutman, O. (ed.) New York, Plenum Press, (Vol. 7) pp. 245–272.

Hodes, R.J., Nadler, L.M., Hathcock, K.S. (1977) Regulatory mechanisms in cell-mediated immune responses. III. Antigen-specific and non-specific suppressor activities generated during MLC. J. Immunol. 119,961–967.

Holm, G., Perlmann, P. (1967) Cytotoxic potential of stimulated human lymphocytes. J. Exp. Med. 125,721–736.

Kiessling, R., Hochman, P.S., Haller, O., Shearer, G.M., Wigzell, H., Cudkowicz, G. (1977) Evidence for a similar or common mechanism for natural killer cell activity and resistance to hematopoietic grafts. Eur. J. Immmunol. 7,655–663.

Kirchner, H., Blaese, R.M. (1973) Pokeweed mitogen-concanavalin A-, and phytohemagglutinin-induced development of cytotoxic effector lymphocytes: An evaluation of the mechanisms of T cell-mediated immunity. J. Exp. Med. 138,812–824.

Klein, J. (1975) *Biology of the Mouse Histocompatibility-2 Complex. Principles of Immunogenetics Applies to a Single System.* New York, Springer-Verlag.

Lindahl, K.F., Wilson D.B. (1977a) Histocompatibility activated cytotoxic T lymphocytes. I. Estimates of the absolute frequency of killer cells generated *in vitro.* J. Exp. Med. 145,500–507.

Lindahl, K.F., Wilson, D.B. (1977b) Histocompatibility activated cytotoxic T lymphocytes. II. Estimates of the frequency and specificity of precursors. J. Exp. Med. 145,508–522.

Martz, E. (1977) Mechanism of specific tumor-cell lysis by alloimmune T lymphocytes: Resolution and characterization of discrete steps in the cellular interaction. In: *Contemporary Topics in Immunobiology,* Vol. 7. Stutman, O. (ed.) New York, Plenum Press pp. 301–361.

Moller, G., Ed. (1979) *Immunological Reviews* (Entire volume devoted to NK cells). Copenhagen, Munksgaard, (Vol. 44) pp. 1–208.

Nelson, D.L., Bundy, B.M., Pitchon, H.E., Blaese, R.M., Strober, W. (1976) The effector cells in human peripheral blood mediating mitogen-induced cellular cytotoxicity and antibody-dependent cellular cytotoxicity. J. Immunol. 117,1472–1481.

Peavy, D.L., Pierce, C.W. (1974) Cell-mediated immune responses *in vitro.* I. Suppression of the generation of cytotoxic lymphocytes by concanavalin A and concanavalin A-activated spleen cells. J. Exp. Med. 140,356–369.

Piessens, W.F., Churchill, W.H., David, J.R. (1975) Macrophage activation *in vitro* with lymphocyte mediators kill neoplastic but not normal cells. J. Immunol. 114,293–299.

Plate, J.M.D. (1977) Two signals required for the initiation of killer T-cell differentiation. Cell Immunol. 32,183–192.

Rich, S.S., Rich, R.R. (1974) Regulatory mechanisms in cell-mediated immune responses. I. Regulation of mixed lymphocyte reactions by alloantigen-activated thymus-derived lymphocytes. J. Exp. Med. 140,1588–1603.

Ruco, L.P., Meltzer, M., Rosenstreich, D. (1978) Macrophage activation for tumor cytotoxicity: Control of macrophage tumoricidal capacity by the LPS gene. J. Immunol. 121,543–548.

Shearer, G.M., Schmitt-Verhulst, A.M. (1977) Major histocompatibility complex restricted cell-mediated immunity. In: *Advances in Immunology.* Dixon, J.F., Kunkel, H.G. (eds.) New York, Academic Press, (Vol 25) pp. 55–91.

Stutman, O. (1973) Tumor development in immunologically deficient nude mice after exposure to chemical carcinogens. In: *Proceedings of the First International Workshop on Nude Mice.* Rygaard, J., Povlsen, C.O., (eds.) Stuttgart, Fischer-Verlag, pp. 257–264.

Wahlin, B., Perlmann, H., Perlmann, P. (1976) Analysis by a placque assay of IgG- or IgM-dependent cytolytic lymphocytes in human blood. J. Exp. Med. 144,1375–1380.

Yap, K.L., Ada, G.L., McKenzie, I.F.C. (1978) Transfer of specific cytotoxic T lymphocytes protects mice inoculated with influenza virus. Nature 273,238–239.

Zinkernagel, R.M., Doherty, P.C. (1979) MHC-restricted cytotoxic T-cells: studies on the biological role of polymorphic major transplantation antigens determining T-cell restriction-specificity, function, and responsiveness. In: *Advances in Immunology.* Dixon, J.F., Kunkel, H.G., (eds.) New York, Academic Press, (vol. 27) pp. 52–142.

Development and Functions of B-Lymphocytes

Irwin Scher, M.D.

The fundamental cellular unit of the immune system is the lymphocyte. These cells have a characteristic morphology that gives no clues to the fact that they can be subdivided into two distinct groups—the bone marrow-derived, or B-lymphocytes, and the thymus-derived, or T-lymphocytes. The identification of these two classes of lymphocytes has occurred over the past 20 years and is based on differences in their functional and surface membrane characteristics. This distinction is the foundation upon which modern immunology rests, as it has provided a cellular basis for the separation of delayed, or cellular, immunity and humoral, or antibody-mediated, immunity (Landsteiner 1962). Cellular immunity was originally defined by the delayed hypersensitivity reaction, which causes an indurated, erythematous skin reaction that occurs 48 hours after the intradermal inoculation of an antigen into a previously sensitized animal. This reaction can be transferred into animals that have never seen the antigen with lymphocytes (later shown to be T-lymphocytes) from sensitized animals. The ability of T-lymphocytes to induce this response, along with their ability to act as killer cells and as helpers or suppressors of antibody formation, illustrates their functional diversity. By contrast, B-lymphocytes, the effector cells of humoral immunity, serve principally as precursors for antibody-forming cells (AFC), of which the plasma cell represents the terminally differentiated stage. Thus, humoral immunity can be transferred either by B-lymphocytes or their antibody products. It should be noted, however, that B-lymphocytes are able to make certain lymphokines, such as migration inhibitory factor (MIF), the formation of which was originally ascribed exclusively to T-lymphocytes. Furthermore, although direct acting suppressor B-lymphocytes have not been described, antiidiotype antibodies formed during an immune response may act to specifically regulate that response.

Although most immune responses are dependent upon the close interaction and collaboration of T- and B-lymphocytes, these cells differentiate totally independently from one another. Thus, the theta marker of murine T-lymphocytes is not found on B-lymphocytes, whereas surface immunoglobulin (sIg), which is the "trademark" of B-lymphocytes, is absent from T-lymphocytes. Interestingly, the development of B-

From the Naval Medical Research Institute and Uniformed Services University of the Health Sciences, Bethesda, Maryland.

and T-lymphocytes is also independent. Thus, if the bursa of Fabricus is removed early in the development of a bird, B-lymphocytes fail to appear, but T-lymphocyte development is normal (Cooper et al. 1966). Furthermore, B-lymphocyte development proceeds normally in mice that have had their thymic glands removed at birth (Miller 1961).

The independent nature of T- and B-lymphocytes is also reflected in the phylogenetic development of the functional properties attributed to these cells. Thus, whereas T-lymphocyte immunity appears in primitive coelenterates and coelomates with their ability to specifically reject engrafted foreign tissue, these animals show no evidence of a specific antibody response after immunization. Indeed, the most primitive IgM-like antibody response first appears in the much more advanced jawless fish, and more complex antibody responses consisting of IgM, IgG, and/or IgA antibodies are limited to birds and mammals.

Surface Immunoglobulin of B-Lymphocytes

Methods for Identifying B-Lymphocytes

When viable populations of mammalian, avian, or rodent splenic lymphoid cells are incubated with antibodies directed against immunoglobulin (Ig) molecules that have been conjugated with a fluorescence dye, such as fluorescence isothiocyanate (FITC), a certain proportion of the lymphocytes will fluoresce brightly (Möller 1961; Warner 1974). Since antiimmunoglobulin (anti-Ig) antibodies will not penetrate into the cytoplasm of a viable cell, the labeling of these cells by the FITC anti-Ig antibodies indicates that they bear Ig on their surface membranes. The fluorescence staining of such sIg-bearing cells is evenly distributed over the entire plasma membrane of these cells, reflecting the uniform distribution of their sIg (Figure 1a, 1b). Analysis of the amount of sIg present on individual B-lymphocytes with instruments that are able to quantitate the total amount of fluorescence associated with individual cells after labeling with FITC anti-Ig antibodies indicates that considerable heterogeneity exists in the amount of sIg per cell (Figure 1c).

In studies utilizing anti-Ig antibodies to detect the sIg of B-lymphocytes, care must be taken to avoid the interaction of the Fc portion of the anti-Ig with Fc receptors (FcR) (Figure 2). These FcR are present on B- and T-lymphocytes, macrophages, and polymorphonuclear leukocytes (PMNL). The interaction of the Fc portion of the anti-Ig probe with FcR can be avoided by either digestion of the probe with pepsin to isolate its $F(ab')_2$ free of Fc fragments, or by ultracentrifugation to remove aggregated Ig, which has a much greater affinity for the FcR than deaggregated Ig. Surface Ig labeling of nonB-lymphocytes can also occur by the absorption of serum Ig to the FcR of these cells. This FcR-associated Ig can be removed by incubation of the population of cells being studied at 37°C and washing the disassociated Ig free of the cells. Passively acquired Ig has two additional characteristics that allow its distinction from endogenously synthesized B-lymphocyte sIg. First, it is heterogeneous with regard to its light chain, allotype and/or idiotypes on individual cells and, second, it is predominantly of the IgG class.

Under conditions where it is known that anti-Ig antibodies will only react with endogenously synthesized Ig, the labeling of a cell by anti-Ig identifies such cells as B-lymphocytes. In adult mice it is convenient to use anti-κ antibodies to identify B-lymphocytes, since approximately 95% of these cells bear this light-chain class. In

ANALYSIS OF THE SURFACE Ig OF LYMPHOCYTE POPULATIONS

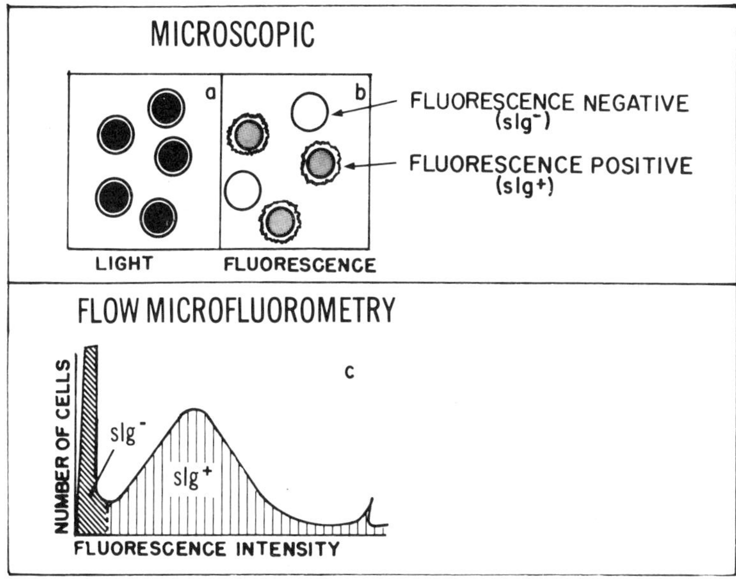

Figure 1. The surface immunoglobulin of B-lymphocytes is usually detected with an antiimmunoglobulin probe, which has had a fluorescence dye coupled to it. The probe is incubated with populations of lymphocytes, which are then washed free of unassociated probe. Fluorescence-labeled cells can then be detected using fluorescent microscopic techniques. Cells are first examined using conventional light microscopy to identify the lymphocytes (**a**), which are then examined with an ultraviolet light so that only those cells that have bound the probe will fluoresce (**b**). In addition to this technique, investigators now are able to examine the fluorescence of individual cells quantitatively using the technique of flow microfluorometry. Instruments utilizing this technique express data in the form of a fluorescence histogram (**c**).

man or monkeys, where one-third of the B-lymphocytes bear Ig of the λ chain type, anti-F(ab')$_2$ antibodies are used. When lymphoid cells derived from different organs are studied with these techniques, the frequency of sIg-bearing cells varies from less than 1% in thymus to greater than 60% in Peyer's patches (Table 1). Although the frequency of B-lymphocytes in spleen is generally less than Peyer's patch, the absolute number of splenic B-lymphocytes is considerably larger than in other lymphoid organs because of its size.

Heterogeneity of Surface Immunoglobulin on B-Lymphocytes

In order to study the classes of sIg on B-lymphocytes, antibodies with Ig class specificity must be utilized. These antibodies are prepared using the Fc portion of IgM, D, A, or G myeloma proteins as immunogens. The resultant antisera are made specific for μ, δ, γ or α heavy chains by absorption with preparations of Ig containing all of the classes, except for the one used to immunize. Analysis of the sIg on B-lymphocytes using such antisera reveals a distribution of Ig classes that is quite distinct from that observed for serum Ig, where IgM and IgG predominate and IgD

ANTI-Ig CAN REACT WITH LYMPHOID CELLS
VIA THEIR sIg OR Fc RECEPTORS

FLUORESCENCE CONJUGATED ANTI-Ig

A) CAN REACT WITH THE sIg OF B LYMPHOCYTES VIA ITS Fab

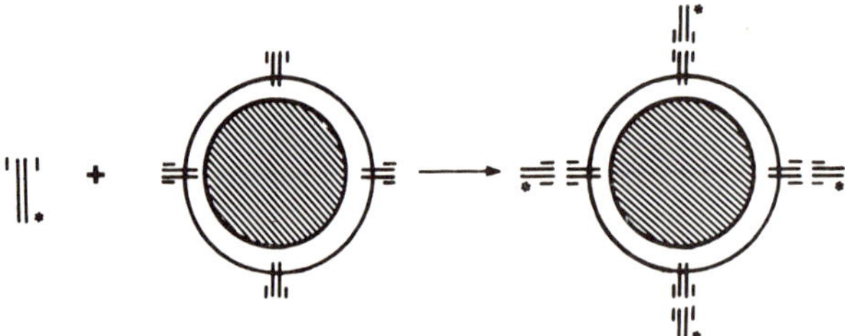

B) OR IT MAY REACT WITH THE Fc RECEPTOR OF B LYMPHOCYTES
   OR OTHER CELLS

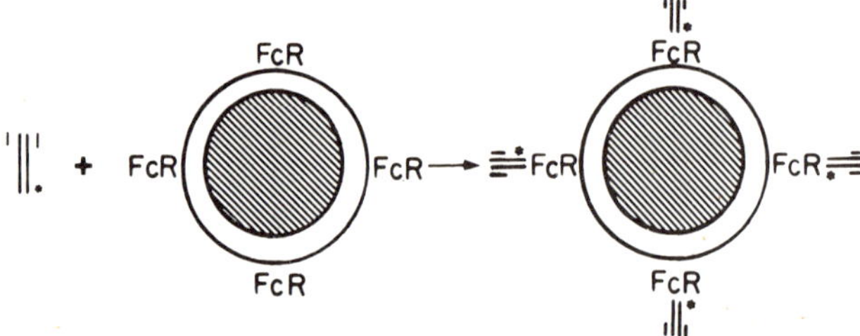

Figure 2. Antiimmunoglobulin probes may react with the surface immunoglobulins of B-lymphocytes via the specific interaction of the antigen-combining region of the probe (**A**), or via the interaction of the Fc portion of the probe with the Fc receptor of B-lymphocytes and/or other cells (**B**).

Table 1. Frequency of sIg[a]-Positive Cells in Different Lymphoid Cells

	Percent sIg-positive cells		
Organ	Mouse	Monkey	Man
Thymus	< 1	< 1	
Bone marrow	19	–	
Lymph node	18	25	
Spleen	55	54	
Peyer's patches	40	–	
Blood	–	30	

[a] sIg = surface immunoglobulin

VIRTUALLY ALL sIgM BEARING B LYMPHOCYTES IN THE SPLEENS OF ADULTS ALSO BEAR sIgD

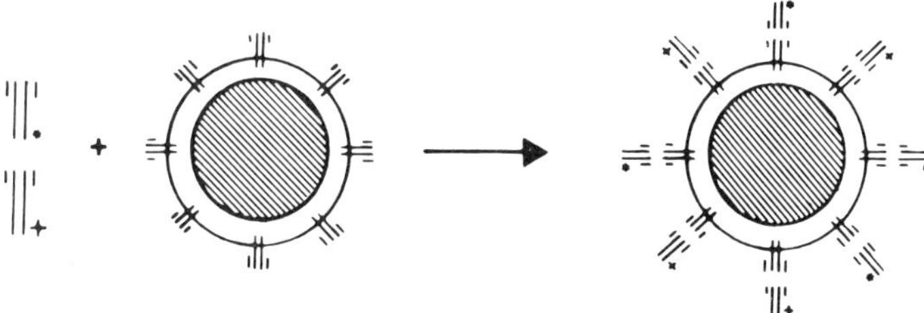

Figure 3. When B-lymphocytes of adults are allowed to react with anti-IgM and anti-IgD antibodies, it can be shown that the majority react with both types of antibodies.

exists as a very minor class. Thus, the frequency of sIgM- and sIgD-bearing cells in the spleens of adult mice, monkeys, or man is similar to each other, and this frequency is generally only slightly less than cells labeled by an anti-Ig reagent. This suggests that the majority of B-lymphocytes in the spleens of these adults bear both sIgM and sIgD. Indeed, studies using anti-μ and anti-δ antibodies, which have been conjugated with fluorescence dyes of different colors, have demonstrated that the majority of sIgM-bearing cells in the spleens of adult animals also bear sIgD (Figure 3). In contrast to the large number of μ- and δ-bearing cells, α- or γ-bearing B-lymphocytes represent a small fraction of the total sIg-bearing cells in the peripheral lymphoid organs of mice, monkeys, or human beings.

Role of Surface Immunoglobulin in B-Lymphocyte Function

Theories of Antibody Formation

The selective nature of the immune response is perhaps its most distinctive feature. This is evident when one considers that while the potential heterogeneity of an individual's antibody response is enormous, after immunization only antibodies directed against the immunizing antigen are formed. As early as 1900, Ehrlich proposed that this specificity was due to the presence of receptors for antigen on immunocompetent cells, which were identical to the antibody molecules he was able to isolate from serum. This hypothesis was rejected for one postulating that the interaction of antigen with immunocompetent cells resulted in the transfer of information to the cell, allowing it to make complementary antibody (Mudd 1932). However, this informational theory of antibody formation had to be abandoned when it became evident that specific antigen-antibody interaction was a function of the antibody's primary amino acid sequence, which is dictated by the genetic code of the cell producing it. As a result, the Clonal Selection Theory was reinforced (Jerne 1955; Burnet 1959; Mitchison 1967). This theory postulated that specific antibody was the product of certain B-lymphocyte clones that were activated to divide, multiply, and produce antibody after the interaction of the antigen with their complementary sIg receptors. This theory stimulated a number of studies designed to establish (a) specific B-lymphocyte antigen binding and (b) if such binding could be inhibited by anti-Ig antibodies.

Evidence That Surface Immunoglobulin is the Antigen-Specific Receptor of B-Lymphocytes

The frequency of B-lymphocytes that bound a particular antigen, antigen-binding B-lymphocytes (ABBL), is very small (Figure 4), a finding that is compatible with the hypothesis that B-lymphocyte clones have unique specificities. Furthermore, it was also shown that the interaction of antigen with its ABBL could be completely inhibited

Figure 4. The clonal selection theory postulated that antigen interacted specifically with B-lymphocyte clones by the selective action of the surface immunoglobulin of the B-lymphocytes of the clone. This hypothesis was supported by the findings that the frequency of antigen-binding B-lymphocytes was small (**A**), that antiimmunoglobulin blocked the interaction of antigen with antigen-binding B-lymphocytes (**B**), and that antibodies directed against the surface immunoglobulin antigen-combining site of B-lymphocytes specifically blocked the interaction of antigen with antigen-binding B-lymphocytes (**C**).

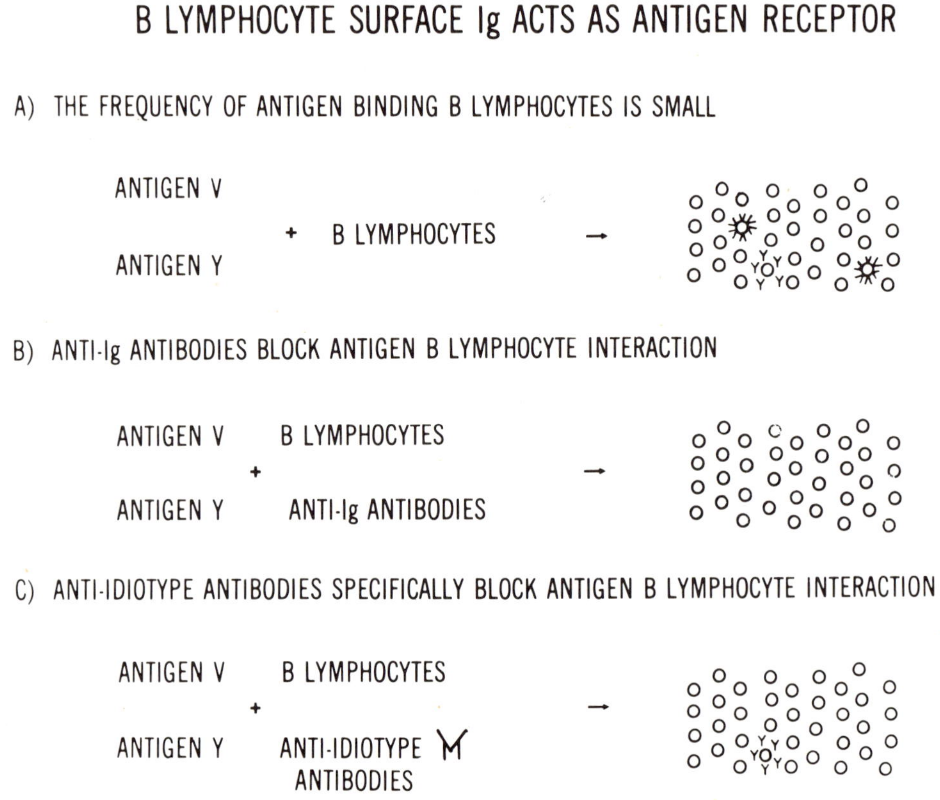

B LYMPHOCYTE SURFACE Ig ACTS AS ANTIGEN RECEPTOR

A) THE FREQUENCY OF ANTIGEN BINDING B LYMPHOCYTES IS SMALL

ANTIGEN V
+ B LYMPHOCYTES →
ANTIGEN Y

B) ANTI-Ig ANTIBODIES BLOCK ANTIGEN B LYMPHOCYTE INTERACTION

ANTIGEN V   B LYMPHOCYTES
+ →
ANTIGEN Y   ANTI-Ig ANTIBODIES

C) ANTI-IDIOTYPE ANTIBODIES SPECIFICALLY BLOCK ANTIGEN B LYMPHOCYTE INTERACTION

ANTIGEN V   B LYMPHOCYTES
+ →
ANTIGEN Y   ANTI-IDIOTYPE M
ANTIBODIES

by anti-Ig antibodies (Warner 1974) (Figure 4), indicating that the molecule serving as the B-lymphocyte antigen receptor shares antigenic determinants with Ig. The nature of the antigen-binding receptor of B-lymphocytes was also studied by taking advantage of the "capping" phenomenon. When B-lymphocytes are incubated with anti-Ig antibodies, the sIg is modulated from the surface in a stepwise fashion, with the formation of patches of sIg, the generation of a cap that contains all of the sIg, and the final shedding and/or ingestion of the Ig from the surface of the cell (Schreiner and Unanue 1976) (Figure 5). If ABBL whose sIg has been modulated into a cap by anti-Ig antibody are incubated with a radioactive antigen, all of the radioactive antigen is confined to the cap (Roelants 1972). Thus, sIg and the antigen receptor of ABBL cocap, suggesting that they are either identical or physically linked on the surface of B-lymphocytes.

Analysis of the inhibitory capacity of antiidiotype antibodies on antigen-ABBL interactions has given additional insights into the nature of the ABBL receptor. Antiidiotype antibodies are prepared by immunizing animals with a myeloma protein

Figure 5. When antiimmunoglobulin antibodies are allowed to interact with B-lymphocytes at 37°C, the crosslinking of the surface immunoglobulin (not shown in Figure 5) results in the modulation of the surface immunoglobulin into patches and then into a cap. Eventually, the modulated surface immunoglobulin is eliminated from the cell surface.

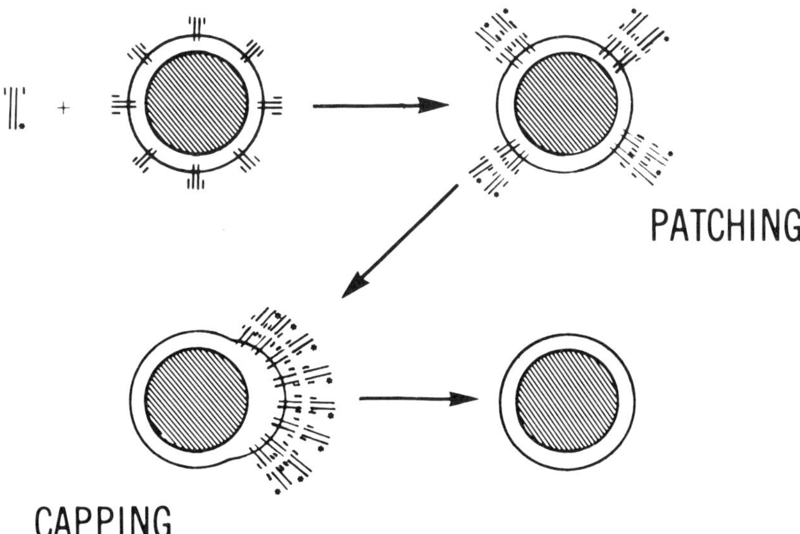

or homogeneous antibody that reacts specifically with a particular antigen. The resultant antibodies are then absorbed with immunoglobulin so that only antibodies directed against the antigen-combining region of the original immunizing myeloma Ig remain. After absorption, these antiidiotype antibodies react only with Ig molecules that share a common idiotype with the original myeloma used for immunization and have no reactivity with Ig molecules as a whole, either as a result of interactions with determinants in their Fc or F(ab')$_2$ regions (Capra and Kehoe 1975). These antiidiotype antibodies therefore provide an important tool for analysis of the properties of the antigen-specific receptor of B-lymphocytes, since their reactivity is limited and highly specific for the unique determinants present in the antigen-binding site of the Ig used as the immunogen. Indeed, antiidiotype antibodies were as effective in inhibiting antigen-ABBL interaction as the nonspecifically inhibitory anti-Ig antibodies (Figure 4). This indicates that the combining sites of serum Ig are similar or identical to those of B-cell sIg.

The observation that sIg acts as the antigen receptor for ABBL is somewhat complicated by the finding that most B-lymphocytes bear two classes of sIg—IgM and IgD. The presence of more than one class of sIg on an individual ABBL raises two important questions: Is the antigen reactivity of these classes of sIg identical on individual B-lymphocytes, and do the different classes of sIg have functional attributes in addition to that of the antigen receptor? The first question has been studied by analyzing the sIg of ABBL after such cells were incubated with antigen. When an antigen containing multiple constant antigenic determinants, such as polymerized flagellin (POL), is incubated with B-lymphocytes, the excess POL removed, and the cells incubated with fluorescence-conjugated anti-POL antibodies, the POL–ABBL can be directly visualized using fluorescence microscopy. Under these conditions and at 37°C, the POL-ABBL will form caps in which the anti-POL-sIg, POL, and fluorescence anti-POL will be located. If these capped cells are subsequently exposed to an anti-Ig that has been conjugated with a fluorescence dye of a different color, it can be shown that all of the sIg of the POL-ABBL is in the cap. Thus, all of the classes of sIg on these POL-ABBL must have had anti-POL specificity.

Other studies addressing the issue of the antigen-binding specificity of the sIgM and sIgD on ABBL have utilized the peripheral lymphocytes of patients with chronic lymphocytic leukemia (CLL). These cells represent the malignant transformation of a single clone of B-lymphocytes and therefore provide a source of large numbers of ABBL with identical antigen-binding characteristics. Certain patients with a type of CLL have circulating monoclonal antibodies of the IgM class that can be used to prepare an antiidiotype antibody. Such antiidiotype antibody was able to react with both the IgM and the IgD on the surface of these CLL cells. Similar analysis using purified sheep red blood cells (SRBC)–ABBL has shown that the antigen-binding specificity of all classes of sIg on these lymphocytes was directed against SRBC. Thus, the antigen-binding specificities of the different classes of sIg on B-lymphocytes are identical.

The Relationship of Antigen-Surface Immunoglobulin Binding to Induction of Antibody Production

Although these studies indicate that the sIg of an ABBL functions as the antigen receptor of that cell, the fundamental question remains: Do ABBL give rise to AFC? This issue has been studied using a number of different experimental techniques.

Early studies utilized a nonimmunogenic hapten molecule, NIP, to determine if the interaction of the hapten with its ABBL would alter subsequent responses to the hapten and crossreactive haptens (Mitchison 1967). Thus, when the hapten NIP was given to animals prior to their challenge with the immunogenic crossreactive hapten DIP coupled to a protein carrier, the resultant antibody was relatively poor in NIP reactivity. Similarly, specific antibody formation was inhibited when antiidiotype antibodies were allowed to interact with populations of ABBL prior to their exposure to antigen (Cosenza and Köhler 1972). Thus, when ABBL are prevented from interacting with antigen by either nonimmunogenic haptens or antiidiotype antibodies, specific antibody formation is prevented. A second approach compared the antigen-binding properties of the antibody formed during an immune response to that of the antigen binding to ABBL during the same period. Such analysis demonstrated that the avidity with which antigen bound to ABBL or to antibody increased in parallel after immunization (Davie and Paul 1972). This is presumably the result of the selective antigen-induced expansion of those ABBL that bound the antigen most avidly. Finally, it has also been shown that depletion of a particular population of antigen-reactive ABBL specifically precluded subsequent antibody formation to that antigen (Julius and Herzenberg 1974). These findings indicate that ABBL act as the progenitors of AFC.

The role of sIg in B-cell activation has been studied in rabbits (Sell and Gell 1965), guinea pigs (Wahl et al. 1974) and, more recently, in mice (Weiner et al. 1976; Parker 1975; Sieckmann et al. 1978). The interaction and crosslinking of whole or F(ab′)$_2$ fragments of anti-Ig antibodies with the sIg of B-lymphocytes in these species induce proliferation, but fail to trigger antibody synthesis. By contrast, Fab fragments, which bind as well as F(ab′)$_2$ fragments, are not stimulatory. In the mouse, whole or F(ab′)$_2$ fragments of anti-μ, anti-δ, or anti-κ antibodies will induce proliferation, which occurs in B-lymphocytes of adult but not neonatal (< 8 weeks) mice. Thus, the interaction of a multivalent ligand (either anti-Ig or an antigen) with the sIgM, sIgD or Fab portion of the sIg of B-lymphocytes leads to proliferation, but fails to induce antibody formation. However, the addition of additional signals provided for by activated T-lymphocytes does enable the anti-Ig antibodies to induce antibody synthesis (Parker et al. 1979). These studies suggest that the binding of a ligand to B-lymphocyte sIg is necessary but not sufficient for B-lymphocyte activation, and that crosslinking of surface receptors may be important for cell triggering.

The role of sIg on B-lymphocyte activation also has been studied by analysis of the effects of anti-μ or δ antibodies on the *in vitro* induction of antibody responses. Interestingly, the addition of anti-μ to such cultures inhibits the response to all classes of antigens, including those which are dependent upon the cooperation of B- and T-lymphocytes for their response (thymic dependent) and those which appear to be less dependent or, in fact, independent of this cooperation (thymic independent). By contrast, whereas anti-δ antibodies suppress the responses to certain thymic-independent and all thymic-dependent antigens, it has little effect on the response to other thymic-independent antigens (Mosier et al. 1977). These findings suggest that those thymic-dependent or thymic-independent antigens that are inhibited by either anti-μ or δ antibodies must interact with both the sIgM and sIgD of B-lymphocytes to trigger antibody synthesis. Alternatively, they could also be interpreted as meaning that the interaction of these antigens with either sIgM alone or sIgD alone, but not with both, leads to a turn-off signal to the cell. Of interest is the fact that certain other thymic-independent antigens appear capable of triggering B-lymphocytes by interaction with only sIgM. Indeed, such antigens are able to induce antibody synthesis *in vitro* in the presence of anti-δ antibodies or in neonatal mice that lack sIgD$^+$ B-lymphocytes.

B-Lymphocyte Ontogeny

Fetal Pluripotential Stem Cells and PreB-Cells

B-lymphocytes can be detected in the liver of fetal mice or man as typical small sIg-bearing cells at approximately 15 to 17 days or 9 to 12 weeks gestation, respectively. These B-lymphocytes are derived from a pluripotential stem cell that also serves as the precursor for T-lymphocytes, megakaryocytes, and erythrocytes. Pluripotential stem cells originate in fetal yolk sac, but migrate to fetal liver where they differentiate into more mature and restricted stem cells. Although the exact developmental sequence from the most primitive pluripotential stem cell, through a presumed restricted stem cell, to a B-lymphocyte is unknown, the immediate precursors of B-lymphocytes have recently been characterized. These so-called preB-cells are large or small lymphoid cells that bear small amounts of IgM in their cytoplasm and lack sIg (Raff 1977). PreB-cells are first detected in mammalian fetal liver some 3 to 4 days before the appearance of B-lymphocytes. However, as fetal development proceeds, the number of hepatic preB-cells diminishes and their numbers increase in bone marrow and spleen. After birth and in the adult, preB-cells are found only in bone marrow. This sequence exactly parallels the appearance of sIg-bearing B-lymphocytes in these organs during ontogeny and suggests that mammalian fetal liver, bone marrow, and spleen can each function as equivalents of the avian bursa of Fabricus. Indeed, it has been demonstrated that isolated mouse fetal liver organ explants are able to generate sIg-bearing B-lymphocytes in *in vitro* cultures, a property that explanted yolk sac does not have (Raff 1977).

Although the sequential appearance of preB- and B-lymphocytes during ontogeny gives strong evidence that these cells are in the same lineage, additional evidence for this has been obtained by studies of the cytoplasm immunoglobulin (cIg) and sIg antigen-binding characteristics of these cells. Thus, antiidiotype antibodies directed against serum monoclonal Ig from certain patients stained the cIg of preB-cells and the sIg of B-lymphocytes from these patients. These data demonstrate that preB-cells do indeed function as the immediate precursors of B-lymphocytes.

The presence of cIg in preB-cells also indicates that the commitment of a B-lymphocyte clone to its unique antigenic specificity must occur at or before the preB-cell stage. This commitment is the product of genetic processes that allow an individual to develop an enormous B-lymphocyte clonal diversity. Thus, when challenged with any of the environmental antigens, an individual will have within their repertoire the appropriate B-lymphocyte clone(s) to make a specific response. Immunologists and molecular biologists have been troubled for some time by the large requirement of genetic information necessary to explain the diversity present in the individual immune repertoire.

The processes involved in the development of an individual immune repertoire have been analyzed by studies of the appearance of this repertoire during ontogeny. Thus, in the chick embryo (Lydyard et al. 1976), hemocyanin or Poly-L (Tyr, Glu)poly-D,L-Ala-poly-L-Lys (TGAL)-binding cells could be detected in the bursa at 16 days gestation, whereas SRBC-binding cell did not appear until day 18. Similar data were obtained in mice (Yung et al. 1973) in studies of flagellin or SRBC-binding cells that appeared 18 to 20 days or 20 to 23 days, respectively, after the transfer of hemopoietic stem cells into irradiated recipients. Although these studies suggest that ABBL appear in a sequential manner, other studies using different techniques (D'Eustachio and Edelman 1975) suggest that this is not the case and indicate that mice have the full range of antibody diversity present by the time of birth or before.

The appearance of antibody repertoire during development has also been studied by analysis of the ability of an individual's potential to make specific antibody responses after immunization. Although these studies have demonstrated the sequential appearance of antibody to a number of antigens in sheep (Silverstein et al. 1963) and other species, these data were criticized because of the influence that immature T-lymphocytes and/or macrophages could have on the ability of a neonatal animal to respond to an antigen. This problem was presumably overcome by isolating neonatal B-lymphocytes and placing them in an adult microenvironment (splenic fragments in microculture trays), where they could receive optimum macrophage and T-lymphocyte cooperation (Sigal and Klinman 1978). This procedure demonstrated that the antibody responses of B-lymphocytes of neonatal mice were in general markedly restricted when compared to the responses of adults. Thus, whereas B-lymphocytes of adults gave a very heterogenous response to the hapten trinitrophenyl (TNP), coupled to a protein carrier, B-lymphocytes of neonates gave a very restricted response. The antibody formed by neonates of the same strain appeared to be derived from B-lymphocyte clones with identical sIg characteristics, since the anti-TNP antibodies produced by these mice were identical with regard to certain of their molecular characteristics. Furthermore, during development, the appearance of new clones occurred in a genetically controlled, highly ordered sequential manner. Thus, individuals of an inbred strain at a particular stage of development all had similar capacities to respond to a battery of antigens. Although the development of the capacity of a B-lymphocyte to respond to an antigen under circumstances where cell cooperation is maximized presumably reflects its repertoire, it must be recognized that this approach may in fact underestimate the repertoire of an individual. However, these findings fit very nicely with observations made by molecular biologists into the nature of the generation of the B-lymphocyte repertoire at the level of the genetic code. Thus, recent studies into the genetic mechanisms responsible for the generation of the immune repertoire have indicated that multiple Ig germ-line genes are present in embryonic deoxyribonucleic acid (DNA) and that these genes undergo both somatic mutation and rearrangement during development (Hood et al. 1979). These processes presumably expand the potential Ig repertoire during ontogeny. Indeed, the restricted responses of neonatal mice may represent their potential to make a response at that time due to the presence of germ-line genes, whereas later more heterogeneous responses may be the result of the somatic mutation and rearrangement of these germ-line genes.

Development of Classes of B-Lymphocytes

The earliest-appearing fetal B-lymphocytes have been characterized as bearing IgM, Ia antigens, FcR, and the mouse B-lymphocyte antigen (MBLA) on their surfaces. Such cells (sIgM$^+$, MBLA$^+$, sIa$^+$, FcR$^+$) are the predominant class of B-lymphocytes in immature mice and reside in bone marrow and spleen (Table 2). These B-lymphocytes are pivotal cells in the development of the humoral immune system, since they act as the precursors of all classes of later-appearing B-lymphocytes. Thus, if these

Table 2. Murine B-Cell Ontogeny and Heterogeneity

Age of first appearance	Organ	Special functional properties	Membrane or cytoplasmic markers[a]
13 days of gestation (- 6)	Fetal liver, adult bone marrow	PreB-cell	cIgM
15 days of gestation (- 4)	Fetal liver, bone marrow and spleen, adult bone marrow	First sIg$^+$ B-cell, easily tolerized, susceptible to anti-μ suppression	sIgM, Fc, MBLA, sIa
1–2 weeks	Adult bone marrow and spleen	Resistant to tolerance, responds to TNP-LPS and TNP-BA	sIgM, Fc, MBLA, sIa, sIgD
3–5 weeks	Adult spleen	Responds to TNP-Ficoll and stimulation by anti-μ antibodies	sIgM, Fc, MBLA, sIa, sIgD, CR, Mls, Lyb 3,5

[a]See text for abbreviations.

cells are suppressed by giving anti-μ antibodies to newborn mice at birth and thereafter on a chronic basis, the appearance of all sIg-bearing cells is suppressed and humoral immunity is absent (Lawton and Cooper 1974). However, such treatment does not affect preB-cells, since such cells are present in anti-μ suppressed mice and such preB-cells are still able to generate B-lymphocytes when transferred into nonsuppressed animals (Raff 1977).

B-lymphocytes bearing sIgD (sIgM$^+$, MBLA$^+$, sIa$^+$, FcR, sIgD$^+$) can be found in small numbers in the spleens of 3-day-old mice, but represent a relatively small population when compared to sIgD$^-$ cells at this time (Vitetta et al. 1977). This subpopulation can be distinguished from later-appearing sIgD-bearing cells by their failure to express both the complement receptor (CR) (Gelfand et al. 1974) and antigens encoded for by the minor lymphocyte-stimulating (Mls) determinants (Table 2). These two classes of sIgD-bearing B-lymphocytes; i.e., the earlier-appearing (sIgM$^+$, MBLA$^+$, sIa$^+$, FcR$^+$, sIgD$^+$) and later-appearing (sIgM$^+$, MBLA, sIa$^+$, FcR, sIgD$^+$, CR$^+$, Mls$^+$) cells, occur with approximately the same frequency in the spleens of adult mice and account for virtually all of the sIg-bearing cells in this organ in adults. Although Cr- and Mls-bearing cells appear later in ontogeny than the other classes of B-lymphocytes, it is not certain if such cells are derived directly from sIgD$^+$, CR$^-$ cells or if they originate from an entirely different lineage.

Development of Other B-Lymphocyte Markers

The non-Ig B-lymphocyte surface markers discussed in this section can be divided into two groups—those that are present on all B-lymphocytes and those that have a limited distribution. The markers found on the earliest-appearing B-lymphocytes (MBLA, Ia, and FcR) fall into the group that is found on all classes of B-lymphocytes. Indeed, Ia antigens and the FcR can also be detected on macrophages and T-lymphocytes. Analysis of the interactions of these cells has indicated that the Ia antigens play an important role in macrophage–T-B-cell interactions during an immune response. Similarly, although early studies suggested that certain B-lymphocyte functions were restricted to CR$^+$ cells, more recent studies have failed to support this view. Interestingly, purified C3b, a product of complement activation, was able to induce B-lymphocyte proliferation *in vitro,* presumably by acting via the CR of CR$^+$ B-lymphocytes.

Development of Lyb Antigens

The availability of large numbers of inbred mouse strains has provided a rich substrate in which to study B-lymphocyte surface alloantigens. Indeed, sIg can itself be considered an alloantigen, and considerable insights into fundamental questions on immunogenetics have been obtained by studies of this system. More recently, investigators have defined the Lyb alloantigens using antisera developed by immunizing mice with lymphocytes derived from mice of a different strain. Seven distinct Lyb alloantigens have been described; four of these, 1, 2, 4, and 6, are present on virtually all the B-lymphocytes of adult mice (McKenzie and Potter 1979). Although the functional role of these four Lyb antigens is unknown, recent studies have demonstrated that the genes controlling Lyb 2, 4, and 6 are closely linked on chromosome 4, suggesting that there may be an important region on this chromosome that influences B-lymphocyte function (Kessler et al. 1979). In contrast to these four Lyb antigens, Lyb 3, 5, and 7 appear to be present only on late-maturing CR$^+$, Mls$^+$ B-lymphocytes. The preparation of antisera that distinguish these latter three Lyb antigens was facilitated by the existence of the CBA/N mouse strain. These mice lack the mature or late B-lymphocyte population that bears Mls determinants (Ahmed and Scher 1979). Analysis of the B-lymphocytes of CBA/N and normal mice has shown that the density of sIgM also acts as a marker that distinguishes B-lymphocyte populations (Scher et al. 1976). Thus, the B-lymphocytes of CBA/N mice bear high densities of sIgM, lack the ability to stimulate an Mls-determined mixed lymphocyte reaction, and do not express Lyb 5 or 7 (Scher et al. 1977; Ahmed and Scher 1979). These properties are identical to that of the majority of splenic B-lymphocytes of immature normal mice at 2 to 3 weeks of age. The defect of CBA/N B-lymphocytes and the properties of the anti-Lyb 3, 5, and 7 reagents have provided important information regarding the functional attributes of adult B-lymphocytes, as will be discussed in a later section.

B-Lymphocyte Tolerance

One of the most striking functional differences between B-lymphocyte subpopulations is their differential susceptibility to the induction of tolerance. Tolerance refers to the inability of an individual or a population of lymphocytes to respond to a particular antigen. This unresponsiveness must be specific and not global, as might be induced by lethal irradiation, in order for it to be considered tolerance. The clonal selection hypothesis predicts that tolerance is associated with the inactivation (deletion) of a specific B-lymphocyte clone (Burnet 1959; Nossal et al. 1979). Such inactivation is critical for the development of immunocompetence, for if it did not occur, individuals would certainly be damaged by the interaction of the immune system with self-antigens. Indeed, a large number of studies have demonstrated the unresponsiveness of an individual to his self-antigens; as early as 1959, it was suggested by Burnet that such tolerance was established during the ontogenetic development of the individual (Burnet 1959).

Enhanced Neonatal B-Lymphocyte Tolerization

Recent studies have indicated that B-lymphocytes play a central role in the development of this tolerance (Nossal et al. 1979; Metcalf et al. 1979). This concept is based on the finding that B-lymphocytes of newborn mice can be made specifically unresponsive to an antigen after a relatively short exposure. This characteristic of immature B-lymphocytes appears to be intrinsic to these cells, since adult B-cells are much more difficult to tolerize than neonatal B-lymphocytes. However, it is also likely that the relative functional deficiency of helper T-lymphocytes in the neonate plays a role in the increased susceptibility of these mice to tolerance induction. Indeed, in the presence of helper T-cells, neonatal B-cells become much less susceptible to tolerance induction. Although tolerance can be induced in adult animals by a number of techniques, these appear to act via the activation of suppressor T-lymphocytes (infectious tolerance) or through a B-lymphocyte sIg receptor blockade. Thus, the tolerance observed in the adult does not appear to be the result of the deletion or specific inactivation of a B-cell clone via a mechanism intrinsic to that clone.

Since the susceptibility of B-lymphocytes to tolerance induction decreases rapidly during the first weeks of life in the mouse, it has been suggested that IgM, MBLA, Ia, FcR, but not IgM, MBLA, Ia, FcR, IgD, B-lymphocytes are susceptible to tolerance (Table 2). This hypothesis has been supported by the finding that, whereas adult splenic B-lymphocytes (which are virtually all sIgD$^+$) are resistant to tolerance induction, approximately 50% of adult bone marrow B-lymphocytes (many of which lack sIgD) are susceptible (Metcalf et al. 1979).

Relationship of Surface Immunoglobulin Capping to Tolerization

As previously noted, when B-lymphocytes are incubated with anti-Ig antibodies, their sIg is modulated with the formation of patches of sIg, which are redistributed to form a cap, leading to the eventual elimination of the sIg from the cell. When the sIg of B-lymphocytes of adult mice is removed under these circumstances and the cells are then allowed to rest in the absence of anti-Ig antibodies, their sIgM is reexpressed within 24 hours. However, under similar conditions, neonatal B-lymphocytes fail to regenerate their sIg. The relative inability of neonatal B-lymphocyte clones to reexpress their sIgM after its modulation may explain the increased susceptibility of neonatal B-lymphocytes to tolerance induction and the phenomenon of tolerance to self-antigens (Raff 1977). It has also been suggested that the ability of anti-μ antibodies to suppress B-lymphocyte development in the neonate, but not in more mature mice, may be due to this characteristic of immature B-lymphocytes.

B-Lymphocyte Activation

Nonspecific Polyclonal Mitogens

A host of agents are capable of inducing polyclonal B-lymphocyte proliferation and/or differentiation in the absence of prior sensitization. Among these so-called B-lymphocyte mitogens are a series of high molecular weight bacterial products that are able to stimulate B-lymphocytes irrespective of their antigen-binding characteristics (Table 3). The most widely studied of these is the endotoxic lipopolysaccharide (LPS), which is found in the cell walls of gram-negative bacteria. When LPS is given to experimental animals, their B-lymphocytes respond in a polyclonal fashion that leads

Table 3. Agents Mitogenic for B-Lymphocytes

Lipopolysaccharide (LPS)	Purified protein derivative
Brucella abortus	Human Fc fragments
Polyinosinic polycytidylic acid	LPS-associated lipoprotein
Dextran sulfate	

to a generalized increase in their serum antibody levels. The stimulatory capacity of LPS for B-lymphocytes is also reflected in its ability to induce proliferation in *in vitro* cultures of these cells. Thus, LPS appears capable of activating one of every three B-lymphocytes in *in vitro* cultures (Melchers 1977). This stimulation does not appear to selectively activate any particular B-lymphocyte subpopulation, since B-lymphocytes of mice of all ages are capable of responding to LPS. Indeed, even B-lymphocytes derived from mouse fetal liver are capable of responding to this agent.

The availability of lymphocyte mitogens that activate either B- or T-lymphocytes has provided investigators with a powerful tool to study these different classes of lymphocytes. Thus, the absence of an LPS response from a population of lymphocytes provides strong evidence that few if any B-lymphocytes are present in that population. B-lymphocyte mitogens, and particularly LPS, have also stimulated a great deal of interest and discussion when used as agents to study B-lymphocyte activation, as will be discussed in the following section.

Thymic-Dependent Antigens and T-B Collaboration

The most important B-lymphocyte triggering agents are, of course, the antigens to which individuals are exposed. Most antigens are thymic dependent and fail to induce antibody responses when cultured *in vitro* with B-lymphocytes in the absence of T-lymphocytes or when used to immunize athymic nu/nu mice. It was recognized early in immunology that small molecules, such as TNP, could induce specific antibody formation by B-lymphocytes only when coupled to a larger "carrier" molecule, such as serum albumin. Analysis of the cellular basis of this phenomenon indicated that TNP was required to be on a carrier (i.e., albumin) to induce the T-lymphocyte help necessary to activate TNP-ABBL. This requirement is one that is shared by many small molecules, which are generally referred to as haptens. Thus, in order for TNP–protein conjugates to induce anti-TNP antibodies, TNP–ABBL must sequentially interact with macrophages and carrier-specific T-lymphocyte helper cells. The T-lymphocyte requirement for these hapten-carrier complexes and for a large group of natural and synthetic antigens has led to their classification as thymic-dependent antigens.

Claman et al. in 1966 first observed that antibody production requires the cooperation of both "helper" T-lymphocytes and the actual antibody-producing B-lymphocytes. Lethally X-irradiated mice, which were reconstituted with either T-lymphocytes or B-lymphocytes, by themselves were unable to produce antibodies in response to SRBC. However, mice reconstituted with both lymphocyte populations produced significant antibody titers. Analysis of the cell responsible for antibody production, using either H-2 antigens or chromosome markers to distinguish the origins of the AFC, demonstrated that although thymus-derived, or theta-positive, cells were required for responses to thymic-dependent antigens, the bone marrow cells were the progenitors of the AFC (Davie and Paul 1972; Miller et al. 1971; Nossal et al. 1979). One issue that has yet to be experimentally resolved is whether T–B collaboration requires direct physical contact between the T-cell and B-cell or whether this interaction is mediated by soluble factors. A large body of experimental evidence has been obtained that supports both viewpoints. For example, Mitchison's studies (1967) showed that T–B collaboration occurs even with inflexible chains of amino acids that prevent intimate contact between T- and B-lymphocytes. Feldmann and Basten (1972) demonstrated that T- and B-cells separated by a nucleopore membrane could still cooperate. A number of other investigators have also described T-cell-derived, soluble antigen-specific helper factors, presumably shed T-cell membrane receptors, which can mediate T–B cooperation without direct physical contact between T- and B-lymphocytes.

On the other hand, a number of groups have described major histocompatibility complex (MHC)-controlled restrictions in T–B cooperation, suggesting cell contact is involved. Katz et al. (1973) observed that when carrier-primed T-cells and hapten-primed B-cells were adoptively transferred to sublethally irradiated recipients, successful T–B interaction was only observed when the interacting populations shared the I–A subregion of the mouse H-2 complex. Sprent (1978) has also shown that F_1 T-cells primed to heterologous erythrocytes in one parental strain develop excellent helper activity for B-cells of this parental strain, but give minimal help for B-cells of the opposite strain. The appropriate control studies demonstrated that the restrictions on macrophage T-cell interaction were distinct from those on T–B cell interaction. Similar studies performed in the guinea pig have led Yamashita and Shevach (1978) to postulate that the primed T-cell is activated by carrier determinants of the nominal antigen in association with Ia antigens on macrophages, and the helper T-cell in turn activates only B-cells that bear the same Ia antigens and determinants of the nominal antigen bound to Ig receptors on their surface. Indeed, the most efficient pathway of T–B collaboration may be the direct activation of B-cells in the immediate vicinity of the helper T-cell.

In contrast to the specific T-helper factors and the MHC-restricted interactions between primed T- and B-cells, a number of nonspecific lymphokines have been described that can mediate T-B collaboration. One of the most thoroughly studied nonspecific factors, T-cell replacing factor (TRF), according to Schimpl and Wecker (1979), functions as a late-acting differentiation signal to already primed B-lymphocytes. However, since these factors are genetically unrestricted and can be obtained from any mouse strain, and since some are not even species specific, these *in vitro* helper effects of soluble factors are difficult to reconcile with the MHC restriction of T–B cooperation. These apparently contradictory pathways of T–B collaboration can be reconciled by observations of Andersson and co-workers (1980). Their experiments suggest that the activation of small, undifferentiated B-lymphocytes requires histocompatible (Ia-identical) T-cells and antigen. However, once a B-cell has been activated by antigens or polyclonal stimulants to become a B-cell blast, it becomes responsive to soluble T-helper factors irrespective of the stimulants or source of the T-cells that are used to generate them. Data suggest that B-cell blasts may develop the appropriate receptor for absorbing such a factor (Gorczynski et al. 1973). Major histocompatibility complex-compatible (Ia-identical) macrophages are necessary only at the earlier phase to activate the T-cells to produce this helper factor (Chapter 6), but macrophages are not needed to activate either the resting B-cells or B-cell blasts. This model encompasses both the observed cell contact-dependent genetically restricted phase of T–B collaboration, as well as the effect of T-helper factors in augmenting and potentiating ongoing immune reactions.

Thymic-Independent Antigens

A second group of antigens, known as thymic-independent antigens, is able to stimulate B-lymphocytes to form specific antibody in the absence of T-lymphocytes. Interestingly, most thymic-independent antigens also act as B-lymphocyte mitogens and are generally large molecules with repeating side chains or epitopes acting as haptenic antigenic determinants. Thus, low concentrations of LPS induce only specific anti-LPS, but not polyclonal, responses in athymic nu/nu mice, whereas high concentrations of LPS induce polyclonal as well as specific antibody synthesis.

These findings suggested to some investigators that the specific stimulation induced by low concentrations of LPS was due to the focusing of the LPS onto LPS–ABBL via their anti-LPS sIg (Coutinho and Möller 1974). Once focused to the LPS–ABBL, the LPS could then interact with putative non-Ig LPS receptors present on these cells and induce activation by this interaction. According to this "one signal" hypothesis, LPS activates B-lymphocytes nonspecifically by virtue of its high concentration, which allows it to interact with LPS receptors of virtually all B-lymphocytes. This theory suggests that B-lymphocyte sIg acts as a passive focusing agent rather than an active participant in B-lymphocyte activation and that activation is the result of a property of the thymic-independent antigen, which is independent from its antigenic properties. While this concept may explain the mechanism involved in the activation of some B-lymphoycte populations to some antigens, it does not deal with the finding that certain thymic-independent antigens, notably TNP–Ficoll, fail to act as B-lymphocyte mitogens and require macrophages in order to induce a specific response. Furthermore, as previously noted, it is now clear that antibodies against sIg induce lymphocyte proliferation (Sieckmann et al. 1978) and can induce antibody formation as well in the presence of additional signals provided by activated T-lymphocytes in the form of lymphokines (Parker et al. 1979). Certainly, the differential effects of antigenic or anti-μ modulation on immature as compared to mature B-lymphocytes indicate that sIg is capable of delivering either a turn-on or turn-off signal to such cells. Thus, as previously noted, immature sIgD$^-$ B-lymphocytes undergo irreversible modulation of their sIg after incubation with anti-μ antibodies, are very susceptible to tolerance induction, and fail to proliferate in response to either anti-μ or anti-κ antibodies. These functional properties, which are quite distinct from more mature B-lymphocyte populations, must be mediated via signals provided to the B-lymphocyte via the interaction of a ligand with its sIg. Indeed, some observers have suggested that the thymic-independent nature of certain antigens is the result of their large size and repeating indentical antigenic determinants, which may align the sIg of an appropriate B-lymphocyte into a critical configuration, leading to activation.

B-Lymphocyte Heterogeneity and Differential Mechanisms of B-Lymphocyte Activation

Analysis of B-lymphocyte activation is complicated by the marked heterogeneity of antigens and by the apparent differences in the requirements for activation of B-lymphocyte subpopulations. Thus, for thymic-dependent responses to antigens, such as TNP-conjugated keyhole limpet hemocyanin (TNP-KLH), accessory cells act to present antigen to helper T-lymphocytes, which then act in turn to activate B-lymphocytes. Indeed, as previously noted, it is likely that T-lymphocytes recognize antigen in the context of the MHC-encoded antigens on accessory cells (Chapter 6).

Accessory cells also appear to be important in the presentation of certain thymic-independent antigens, such as the TNP-Ficoll, to B-lymphocytes, but appear to play a less important role in the presentation of antigens, such as TNP–LPS or TNP–*Brucella abortus* (TNP–BA), to B-cells. Thus, whereas macrophage-depleted populations of spleen cells retain their LPS or TNP–BA responsiveness, they lose their ability to respond to TNP–Ficoll (Boswell et al. 1980). These findings suggest that different antigens induce B-cell activation as the result of distinct cellular interactions; i.e., B-lymphocytes with accessory cells and T-helper cells, B-lymphocytes with accessory cells, or B-lymphocytes with antigen directly.

Additional complexities are apparent when studies are made of the populations capable of responding to different antigens. Thus, thymic-independent antigens have recently been divided into two distinct groups, based on their ability to induce specific immune responses in neonatal normal or adult immune-defective CBA/N mice (Mosier et al. 1977). Thus, the thymic-independent antigens, LPS or BA, gave excellent responses when cultured with B-lymphocytes of neonatal normal or adult CBA/N mice. By contrast, TNP–Ficoll, which induces excellent thymic-independent responses from B-lymphocytes derived from adult mice, gave no response with normal neonatal or adult CBA/N cells. Since, as previously noted, immune-defective CBA/N mice lack the late-developing B-lymphocyte subpopulation characterized as bearing Mls, Lyb 5, and Lyb 3, it was suggested that thymic-independent antigens, such as LPS or BA, which are capable of inducing responses in mice that lack this population (i.e., immature normal and/or immature and adult CBA/N be considered Type I, whereas Ficoll, which only induces responses in adult normal populations), be considered Type II thymic-independent antigens.

Since thymic-dependent antigens require the cooperation of T- and B-lymphocytes, as well as macrophages, whereas thymic-independent Type II antigens require B-lymphocytes and macrophages and little if any T-lymphocyte cooperation, while thymic-independent Type I antigens appear to activate B-lymphocytes with little or no cooperation from either macrophages or T-lymphocytes, the possibility existed that these requirements were the result of the differential activation requirements of different B-lymphocyte subpopulations. This problem was studied by isolating different subpopulations of B-lymphocytes from normal mice using the cytotoxic anti-Lyb 5 antiserum, or by analysis of the B-lymphocytes of CBA/N mice that are deficient in Lyb 5^+ cells. Lyb 5^- B-lymphocytes derived from either normals via anti-Lyb 5 treatment or from CBA/N mice were unresponsive to TNP-Ficoll, but were able to respond to TNP-BA. Furthermore, CBA/N and Lyb 5^- cells were unresponsive to the thymic-dependent antigen TNP-KLH under primary *in vitro* culture conditions (Boswell et al. 1980). However, under conditions where T-helper lymphocytes were maximized; i.e., either *in vivo* or in the presence of primed T-lymphocytes *in vitro,* responses to thymic-dependent antigens by CBA/N B lymphocytes were considerably improved. These findings suggested that Lyb 5^- cells: (1) were incapable of responding to thymic-independent 2 antigens, (2) were responsive to thymic-independent 1 antigens, and (3) could be stimulated by thymic-dependent antigens, but only under circumstances where maximum T-lymphocyte help was available (Scher et al. 1979). The absence of Lyb 3 from CBA/N B-lymphocytes might explain their inability to respond to low concentrations of thymic-dependent antigens, since this marker may be the receptor for a particular type of T-lymphocyte help (Huber et al. 1977). However, since increasing the amount of T-lymphocyte help allows CBA/N B-lymphocytes to respond, such cells must be able to accept T-cell help by some other mechanism. It is not clear if the putative Lyb 3 T-lymphocyte helper factor is distinct from other T-helper factors or if increasing amounts of the Lyb 3-specific helper factors can interact with Lyb 3$^-$ cells via other receptors.

The inability of Lyb$^-$5 B-lymphocytes to respond to TNP-Ficoll and the finding that this thymic-independent antigen requires macrophage presentation in order to induce an antibody response suggested that Lyb 5$^-$ B-lymphocytes might not be able to accept macrophage help. This was recently studied by presenting TNP-BA, an antigen known to stimulate Lyb 5$^-$ cells, via accessory cells (Boswell et al. 1980). Under these circumstances—that is, by pulsing accessory cells with TNP-BA, washing and then incubating with populations bearing Lyb 5$^-$ cells, or Lyb 5$^-$ and Lyb 5$^+$ cells—populations bearing Lyb 5$^+$ cells made excellent responses, whereas those with only Lyb 5$^-$ cells were unresponsive. Thus, Lyb 5$^-$ cells appear to be incapable of responding to macrophage-presented antigen, at least in the presence of submaximal T-lymphocyte help. It would therefore appear that B-lymphocyte activation depends not only on the presence or absence of macrophages and T-lymphocytes and the characteristics of the antigen, but also on the class of B-lymphocytes present in the responding population.

Conclusion

B-lymphocytes are quite independent, both in their phylogenetic and ontogenetic development, from T-lymphocytes. The sIg of B-lymphocytes acts as the antigen-specific receptor for these cells and this specificity defines individual clones of these cells. Although different classes of sIg exist on the surface of B-lymphocytes, the antigenic reactivity of these Ig classes is identical. The antigenic specificity of B-lymphocyte clones is the product of genetic events, which occur in cells that act as precursors of B-lymphocytes. These genetic events produce an antibody repertoire that allows an individual to respond to the host of antigens that he may contact.

B-lymphocytes exist as heterogenous subpopulations, which are defined by analysis of their surface-marker characteristics and functional properties during ontogeny. The acquisition of sIgD by immature sIgM-bearing B-lymphocytes appears to mark a critical change in the functional attributes of these cells. Whereas B-lymphocytes of immature mice are susceptible to tolerance induction, those derived from adult mice are not. The susceptibility of B-lymphocytes of immature mice is associated with the inability of such cells to regenerate their sIgM after it has been removed by capping. The identification of nonimmunoglobulin B-lymphocyte surface markers has allowed further subclassification of B-lymphocyte populations.

B-lymphocytes can be stimulated by mitogens or antigens. Although the mechanism by which mitogens induce B-lymphocyte activation is not entirely understood, these agents have been particularly useful in studies of B-lymphocyte function. Recent studies have shown that antibodies to sIg will also induce a B-lymphocyte proliferation. These findings and the apparent requirement of certain antigens to interact with sIgM and/or sIgD to induce antibody synthesis indicate that sIg must play an active role in the triggering of B-lymphocytes. Antigens have been classified according to their ability to activate B-lymphocytes independently of T-lymphocyte help. These thymic-independent and thymic-dependent antigens appear to address different B-lymphocyte populations. Thymic-independent antigens and mitogens are generally large molecules that have repeating antigenic determinants. Thymic-dependent antigens require the intervention of T-lymphocyte help in order to activate B-lymphocytes.

This work was supported in part by the Naval Medical Research and Development Command Research Task No. M0095-PN.001.1030; in part by the Uniformed Services University of the Health Sciences Research Protocol Nos. C08310, R08307, R08308; and in part by the National Naval Medical Center Clinical Investigation No. 3-06-132. The opinions and assertions contained herein are the private ones of the author and are not to be construed as official or reflecting the views of the Navy Department or the naval service at large. The experiments reported herein were conducted according to the principles set forth in the current edition of the *Guide for the Care and Use of Laboratory Animals,* Institute of Laboratory Animal Resources, National Research Council.

References

Ahmed, A., Scher, I. (1979) Murine B-cell heterogeneity defined by anti-Lyb 5, an alloantiserum specific for a late appearing B-lymphocyte subpopulation. In: *B Lymphocytes in the Immune Response.* Cooper, M., Mosier, D., Scher, I., Vitetta, E., (eds.) New York, Elsevier North-Holland, pp. 117–124,

Andersson, J., Schreier, M.H., Melchers, F. (1980) T-cell dependent B-cell stimulation is H-2 restricted and antigen dependent only at the resting B-cell level. Proc. Natl. Acad. Sci. U.S.A., 77,1612–1616.

Boswell, H. Scott, Ahmed, A., Scher, I., Singer, A. (1980) Role of accessory cells in B cell activation. II. The interaction of B cells with accessory cells results in the exclusive activation of Lyb 5$^+$ B cell subpopulation. J. Immunol., 125,1340–1348.

Burnet, F.M. (1959) The Clonal Selection Theory of Acquired Immunity. Nashville, Vanderbilt University Press.

Capra, J. Donald, Kehoe, J.M. (1975) Hypervariable regions, idiotype, and the antibody-combining site. Adv. Immunol. 20,1–40.

Claman, H.N., Chaperon, E.A., Triplett, R.F. (1966) Thymus-marrow cell combinations. Synergism in antibody production. Proc. Soc. Exp. Biol. Med. 122,1167.

Cosenza, H., Köhler, H. (1972) Specific inhibition of plaque formation to phosphorylcholine by antibody against antibody. Science 176,1027–1029.

Cooper, M.D., Raymond, D.A., Peterson, R.D., South, M.A., Good, R.A. (1966) The functions of the thymus system and the bursa system in the chicken. J. Exp. Med. 123,75–102.

Coutinho, A., Möller, G. (1974) Immune activation of cells: Evidence for "one nonspecific triggering signal" not delivered by the Ig receptors. Scand. J. Immunol. 3,133–146.

Davie, J.M., Paul, W.E. (1972) Receptors on immunocompetent cells. V. Cellular correlates of the "maturation" of the immune response. J. Exp. Med. 135,660–674.

D'Eustachio, P., Edelman, G.M. (1975) Frequency and avidity of specific antigen-binding cells in developing mice. J. Exp. Med. 142,1078.

Feldmann, M., Basten, A. (1972) Specific collaboration between T and B lymphocytes across a cell impermeable membrane *in vitro.* Nature 237,13–15.

Gelfand, M.C., Elfenbein, G.J., Frank, M.M., Paul, W.E. (1974) Ontogeny of B lymphocytes. II. Relative rates of appearance of lymphocytes bearing surface immunoglobulin and complement receptors. J. Exp. Med. 139,1125–1153.

Gorczynski, R.M., Miller, R.G., Phillips, R.A. (1973) Reconstitution of T cell-depleted spleen cell populations by factors derived from T cells. III. Mechanism of action of T cell-derived factors. J. Immunol. 111,900–913.

Hood, L., Schilling, J., Hansburg, D., Davie, J.M. (1979) Antibody diversity. In: *B Lymphocytes in the Immune Response.* Cooper, M., Mosier, D., Scher, I., Vitetta, E., (eds.) New York, Elsevier North-Holland, pp. 173–180.

Huber, B., Gershon, R.K., Cantor, H. (1977) Identification of a B-cell surface structure involved in antigen-dependent triggering: Absence of this structure on B cells from CBA/N mutant mice. J. Exp. Med. 145,10–20.

Jerne, N.K. (1955) The natural-selection theory of antibody formation. Proc. Natl. Acad. Sci. U.S.A. 41,849–857.

Julius, M.H., Herzenberg, L.A. (1974) Isolation of antigen-binding cells from unprimed mice. Demonstration of antibody-forming cell precursor activity and correlation between precursor and secreted antibody avidities. J. Exp. Med. 140,904–920.

Katz, D.H., Hamaoka, T., Dorf, M.E., Benacerraf, B. (1973) Cell interaction between histoincompatible T and B lymphocytes. III. Demonstration that H-2 gene complex determines successful physiologic lymphocyte interaction. Proc. Natl. Acad. Sci. U.S.A. 70,2624.

Kessler, S., Ahmed, A., Scher, I. (1979) Identification and partial molecular and genetic characterization of an alloantigen expressed on murine B cells. In: *B Lymphocytes in the Immune Response.* Cooper, M., Mosier, D., Scher, I., Vitetta, E., (eds.) New York, Elsevier North-Holland, pp. 47–54.

Landsteiner, K. (1962) *The Specificity of Serological Reactions* (rev. ed.) New York, Dover.

Lawton, A.R., Cooper, M.D. (1974) Modification of B lymphocyte differentiation by anti-immunoglobulins. In: *Contemporary Topics in Immunobiology.* Cooper, M., Warner, N.L., (eds.) New York, Plenum, Vol. 3, p. 193.

Lydyard, P.M., Grossi, C.E., Cooper, M.D. (1976) Ontogeny of B cells in the chicken. I. Sequential development of clonal diversity in the bursa. J. Exp. Med. 144,79.

McKenzie, I.F.C., Potter, T. (1979) Murine lymphocyte surface antigens. Adv. Immunol. 27,179–338.

Melchers, F. (1977) Immunoglobulin synthesis and mitogen reactivity: Markers for B lymphocyte differentiation. In: *Development of Host Defenses.* Cooper, M., Dayton, D.H., (eds.) New York, Raven Press, pp. 11–29.

Metcalf, E.S., Schrater, A.F., Klinman, N.R. (1979) Murine models of tolerance induction in developing and mature B cells. Immunol. Rev. 43,142–183.

Miller, J.F.A.P. (1961) Immunological function of the thymus. Lancet 2,748–749.

Miller, J.F.A.P., Basten, A., Sprent, J., Cheers, C. (1971) Interaction between lymphocytes in immune response. Cell Immunol. 2,469.

Mitchison, N.A. (1967) Antigen recognition responsible for the induction in vitro of the secondary response. Cold Spring Harbor Symp. Quant. Biol. 32,431–439.

Möller, G. (1961) Demonstration of mouse isoantigens at the cellular level by the fluorescent antibody technique. J. Exp. Med. 114,415–434.

Mosier, D.E., Zitron, I.M., Mond. J.J., Ahmed, A., Scher, I., Paul, W.E. (1977) Surface IgD as a functional receptor for a subclass of B lymphocytes. Immunol. Rev. 37,89–104.

Mudd, S. (1932) Agglutinin response to typhoid-paratyphoid vaccination. J. Immunol. 23,81–90.

Nossal, G.J., Pike, B.L., Teale, J.M., Layton, J.E., Kay, T.W., Battye, F.L. (1979) Cell fractionation methods and the target cells for clonal abortion of B lymphocytes. Immunol. Rev. 43,184–216.

Parker, D.C. (1975) Stimulation of mouse lymphocytes by insoluble anti-mouse immunoglobulins. Nature 258,361.

Parker, D.C., Fothergill, J.J., Wadsworth, D.C. (1979) B lymphocyte activation by insoluble anti-immunoglobulin: induction of immunoglobulin secretion by a T cell-dependent soluble factor. J. Immunol. 123,931–941.

Raff, M.C. (1977) Development and modulation of B lymphocytes: Studies on newly formed B cells and their putative precursors in the hemopoietic tissues of mice. Cold Spring Harbor Symp. Quant. Biol. 41,159–162.

Roelants, G.E. (1972) Antigen recognition by B and T lymphocytes. Curr. Top. Microbiol. Immunol. 59,135–165.

Scher, I., Sharrow, S.O., Paul, W.E. (1976) X-linked B-lymphocyte defect in CBA/N mice. III. Abnormal development of B-lymphocyte populations defined by their density of surface immunoglobulin. J. Exp. Med. 144,507–518.

Scher, I., Ahmed, A., Sharrow, S.O. (1977) Murine B lymphocyte heterogeneity: Distribution of complement receptor-bearing and minor lymphocyte-stimulating B lymphocytes among cells with different densities of total surface Ig and IgM. J. Immunol. 119,1938–1942.

Scher, I., Berning, A.K., Asofsky, R. (1979) X-linked B lymphocyte defect in CBA/N mice. IV. Cellular and environmental influences on the thymus dependent IgG anti-sheep red blood cell response. J. Immunol. 123,477–486.

Schimpl, A., Wecker, E. (1979) Lymphokines in nonspecific T cell-B cell cooperation. In: *Biology of the Lymphokine.* Cohen, S., Pick, E., Oppenheim, J.J., (eds.) New York, Academic Press, pp. 369–373.

Schreiner, G.F., Unanue, E.R. (1976) Membrane and cytoplasmic changes in B lymphocytes induced by ligand-surface immunoglobulin interaction. Adv. Immunol. 24,37–165.

Sell, S., Gell, P.G.H. (1965) Studies on rabbit lymphocytes *in vitro*. I. Stimulation of blast transformation with an antiallotype serum. J. Exp. Med. 122,423.

Sieckmann, D.G., Scher, I., Asofsky, R., Mosier, D.E., Paul, W.E. (1978) Activation of mouse lymphocytes by anti-immunoglobulin. II. A thymus-independent response by a mature subset of B lymphocytes. J. Exp. Med. 148,1628–1643.

Sigal, N.H., Klinman, N.R. (1978) The B-cell clonotype repertoire. Adv. Immunol. 26,255–328.

Silverstein, A.M., Uhr, J.W., Kraner, K.L., Lukes, R.J. (1963) Fetal response to antigenic stimulus. II. Antibody production by the fetal lamb. J. Exp. Med. 117,799.

Sprent, J. (1978) Restricted helper function of F_1 hybrid T cells positively selected to heterologous erythrocytes in irradiated parental strain mice. I. Failure to collaborate with B cells of the opposite parental strain not associated with active suppression. J. Exp. Med. 147,1142–1158.

Vitetta, E.S., Cambier, J., Forman, J., Kettman, J.R., Yuan, D., Uhr, J.W. (1977) Immunoglobulin receptors on murine B lymphocytes. Cold Spring Harbor Symp. Quant. Biol. 41,185–191.

Wahl, S.M., Ivevson, G.M., Oppenheim, J.J. (1974) Induction of guinea pig B-cell lymphokine synthesis by mitogenic and non-mitogenic signals to Fc, Ig and C3 receptors. J. Exp. Med. 140, 1631–1645.

Warner, N.L. (1974) Membrane immunoglobulins and antigen receptors on B and T lymphocytes. Adv. Immunol. 19,67–216.

Weiner, H.L., Moorhead, J.W., Yamaga, K., Kubo, R.T. (1976) Anti-immunoglobulin stimulation of murine lymphocytes. II. Identification of cell surface target molecules and requirements for cross-linkage. J. Immunol. 117,1527.

Yamashita, U., Shevach, E.M. (1978) The histocompatibility restrictions on macrophage T-helper cell interactions determine the histocompatibility restrictions on T-helper cell B-cell interactions. J. Exp. Med. 148,1171–1185.

Yung, L., Wyn-Evans, T.C., Diener, E. (1973) Ontogeny of the recognition and immune responsiveness. Eur. J. Immunol. 3,224–228.

Immunoglobulins M, G, and A

Michael Potter, M.D.

It is well known that a single organism can produce a vast array of immunoglobulin molecules (estimates vary from 10^4 to 10^7). However, an individual plasma cell is limited to secreting only one molecular form of immunoglobulin. This restricting process is determined by several remarkable clonal differentiation mechanisms. Clonal differentiation begins with the earliest B-lymphocytes and continues into more advanced developmental stages. Clonal differentiation limits the cell and its progeny to producing immunoglobulin with only one kind of antigen binding site. This effect is determined by the selection of V-region genes. Remarkably, though, a V-differentiated clone can further differentiate into subclones that have different heavy chain constant regions. In effect the cells can switch to producing immunoglobulins with the same antigen binding sites hooked to a different heavy chain. This process called heavy chain class switching is controlled by a separate differentiation mechanism.

This chapter describes how the immunoglobulins of the IgG, IgM, and IgA heavy classes work in defending the host, i.e., the effector functions of these immunoglobulins. Effector functions can be grouped into two general categories: those relating to antigen binding specificity and those relating to how the immunoglobulin molecule interacts with tissue and protein components in the organism, i.e., V-region and C-region associated functions, respectively. The binding of immunoglobulin to antigen *in vivo* in several but by no means a majority of situations constitutes an effective action. Situations in which the immunoglobulin binds a toxin or a virus particle the complex can result in neutralization of the toxin or prevent the virus particle from further cell invasions. In many infections, though, the antigenic target is a relatively large organism, e.g., a bacterium, and the binding of immunoglobulin to the organism is insufficient to eliminate it. Ultimately killing or neutralizing the microbe depends upon the participation of another host component, such as complement with its membrane attack complex; a phagocytic cell such as a neutrophil, macrophage, or eosinophil; or a cytotoxic cell. All of these functions, then, depend upon the ability of the immunoglobulin–antigen complex to interact with a host component. Such interactions ultimately depend upon the ability of the immunoglobulin molecule to

From the National Cancer Institute, National Institutes of Health, Bethesda, Maryland.

bind a component of complement or a cell surface receptor specific for a part of the immunoglobulin other than the antigen binding regions (V-regions), i.e., C-regions. These receptors are called Fc-receptors (FcR).

In birds and mammals the immunoglobulin classes can be divided into two functionally distinct groups. One group comprises IgM, IgG, and IgE, for which the primary effector sites of action are in the internal environment of the host. Immunoglobulin-A molecules perform their effector functions at external sites in the lumen of the gut and respiratory tracts and other sites outside the internal environment of the organism. The latter is called the secretory immune system: the IgA molecules are transported across epithelial cells to the effector site. Little is known about the effector action of IgD, but available evidence indicates it functions chiefly as a membrane-bound immunoglobulin, and is involved in B-lymphocyte differentiation.

The discussion of how immunoglobulins work is preceded by details of the structure of immunoglobulins, of the genes that control the structural elements of the molecule, on how diversity is established, and how it is possible to superimpose on this diversity the expression of different heavy chain classes through the IgCH switching process. It is this latter process that in effect further augments the versatility of the immunoglobulin molecule in different physiological situations.

Differentiation of Immunoglobulin Structural Genes and Immunoglobulin-Producing Clones

Immunoglobulin Structure

All functional immunoglobulin molecules are variations of a basic 4-polypeptide chain monomeric unit consisting of two light (L) chains with molecular weights of about 24,000 and 2 heavy (H) chains with molecular weights ranging from 55,000 to 70,000 (see review by Beale and Feinstein 1976).

The immunoglobulin chains fold into three-dimensional subregions of globular structure called domains. The immunoglobulin structure is bifunctional: the parts determined by the "variable" V-region domain form the antigen-binding sites, and the other parts determined by the "constant" C-region domains form the "chassis" or cell-binding parts of the molecule (Figure 1). Light chains have two domains (one V_L and one C_L) and H-chains have four to five domains (one V_H and three to four C_H). The number of domains varies with H-chain class. Mammalian IgG and IgA H-chains each have one V_H-domain and three C_H-domains; IgM has four C_H domains. Domains are connected by short polypeptide segments (Figure 1). Folded domains have strong binding affinities for other domains: V_L for V_H; C_H1 for C_L; C_H3 for C_H3; C_H4 for C_H4. C_H2 domains do not come in close contact with each other. These binding affinities control the ordered assembly of the immunoglobulin molecule from nascent chains.

The immunoglobulin C-domains comprise the bulk of the functional molecule (66 to 75%). Structural features of C_H-domains determine important functions of immunoglobulin molecules: (1) polymer formation—IgM, IgA; (2) interactions with plasma membrane receptors such as Fc receptors (FcR) and complement (C3R) receptors; (3) interactions with other polypeptides such as J-chains, and the secretory component; and (4) binding of complement components, i.e., Clq. These specific

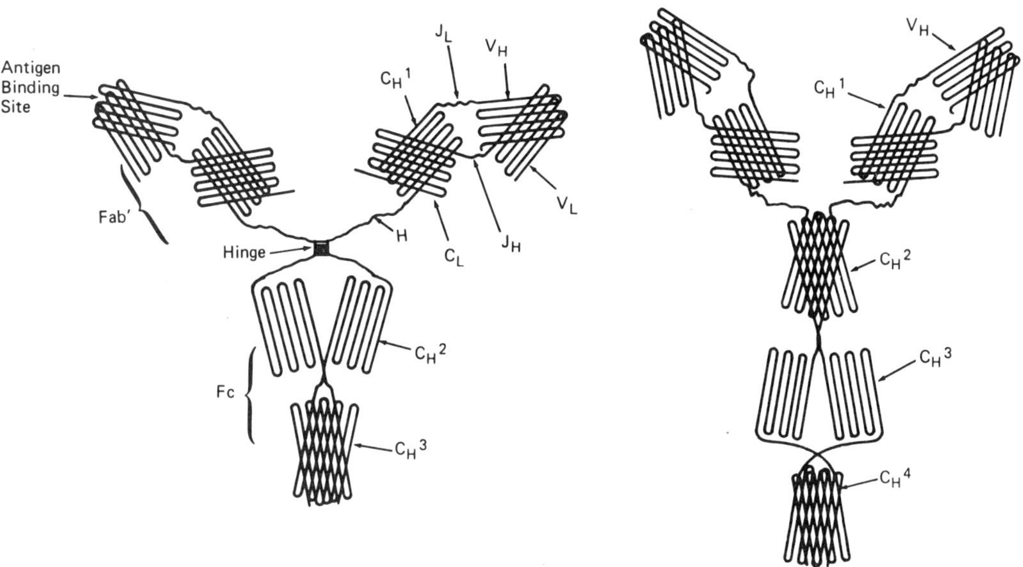

Figure 1. Schematic models of 12 and 14 domain immunoglobulin molecules.

activities govern the structure, movement and localization of immunoglobulin molecules in the organism, the maternal–fetal exchanges of immunoglobulin molecules, and the transmission of signals from immunoglobulin molecules to host inflammatory cells.

A domain contains roughly 120 to 130 amino acids, which in the folded state are arranged in seven strands of β-structure that form two layers, one with four and the other with three antiparallel segments (see Davies et al. 1975). This sandwich-like structure is characteristic of all immunoglobulin domains and has been called the "immunoglobulin fold." The two layers are connected by a disulfide bridge.

The substructure of V-domains is particularly relevant to a discussion of diversity. The V-domain contains two types of subregions: the framework (FR) and the complementarity determining regions (CDR). The framework portions of the polypeptide determine the folding of the domain. Despite extensive sequence differences of regions within a species or between a species, all fold in a similar way, so that if the backbones of their domains could be superimposed, they would not differ in basic architecture. The CDR are the structurally variable parts of the V-domain. Two V-domains controlled by different V-genes within a species may have very different CDR, which vary not only in amino acid sequence but in secondary structure. The CDR subsegments of the V_L and V_H domains form the antigen contacting regions. The parts of the V-region sequence that control the CDR are discontinuous, but in the folded domain they converge to form a continuous surface approximately 20Å × 30Å at the solvent interface (i.e., the interface) that is distal to the junction with the adjacent C domain. The function of the V-region domains then is to determine in immunoglobulin molecules different binding specificities.

Genetic Control of Immunoglobulin Structure

Recent developments have enabled molecular geneticists to map and sequence the DNA for some of the immunoglobulin genes. In well-characterized mammalian species such as house mice and human beings, immunoglobulin structural genes have been located on three unlinked autosomal (complex) loci: the kappa L-chain complex locus, the lambda L-chain complex locus, and the H-chain complex locus. All these loci are considered complex because they contain multiple genes and gene elements (exons or coding genes, and introns or noncoding DNA). Essentially, an immunoglobulin chain is controlled by three separate exons: designated V for variable region; J for joining region; and C for constant region. Kappa and lambda L-chain complex loci appear to contain different numbers of V-genes, depending on the species. The kappa L-chain complex locus of mice, for example, contains a single C_L (C_K) coding gene and at least 38 different V_K genes. The H-chain complex locus contains 8 different C-genes and a large but as yet undetermined number of V-genes. Conservative estimates of the total V-gene number in V_K in the mouse and V_H are 100 or more for each complex locus.

In addition to V-, J-, and C-regions, there are other genetic elements essential for immunoglobulin gene differentiation and biosynthesis that do not code for amino acid sequence in the completed chains, these are: (1) flanking noncoding DNA that contains signals for the differentiation process; (2) introns (i) or noncoding segments; and (3) a short segment that codes for a leader sequence, which is translated into a hydrophobic peptide that anchors the nascent chain in a membrane. Introns are found between domains or other coding segments. The location of introns is shown in Figures 2 to 4. Some intron information can be transcribed into RNA, and this is deleted subsequently by the process of RNA splicing (Figure 2). The DNA controlling the J-region in the mouse kappa immunoglobulin locus has been sequenced completely and contains five different J-coding genes interspersed by introns (Figure 2, See Sakano et al. 1979).

V–J Joining and Clonal Development

During the differentiation process rearrangements of the genes in immunoglobulin complex loci take place on the chromosome. There are two types of rearrangement processes: V–J joining and C_H switching (Figures 2 and 4). In the V–J joining process one V-gene on only one autosome is selected by a mechanism not yet defined and joined at the DNA level to a *cis* J-region gene (Figure 2). The J-region gene is closely linked to the C-gene, although separated from it by introns and or J exons. The V–J translocation process that occurs in chromosomal DNA effectively links two distinct gene elements together to permit transcription of a large message containing V–J–i–C. This large mRNA is subsequently spliced to make V–J–C, which is suitable for translation.

The important consequence of V–J joining is that it restricts a cell to the synthesis of only one type of V_L and one type of V_H. Since there are a large number of different V-genes, one might ask what happens to the chromosomal DNA between the V-gene (being differentiated) and the J-gene? The preliminary evidence supports the hypothesis that the DNA is actually eliminated. Thus, the DNA structure of the complex locus in immunoglobulin-producing cells differs from that in the germ-line state.

From an evolutionary point of view, it is significant that the mechanism of V–J joining permits the formation of a genome with a very large number of retrievable

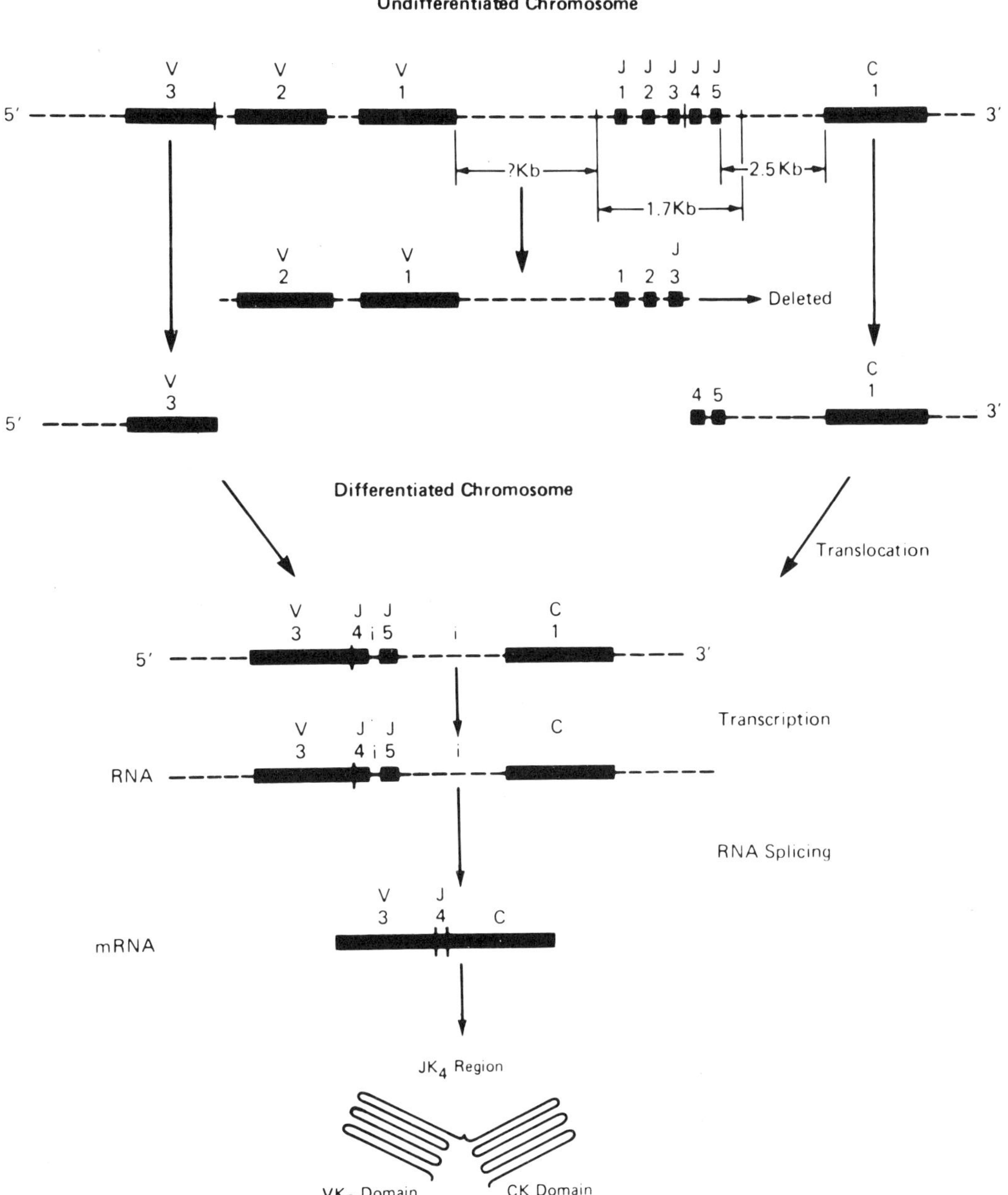

Figure 2. Steps in the differentiation and biosynthesis of a light chain (V-J joining process).

LIGHT CHAIN GENE ELEMENTS

REGION THAT CODES FOR A C_H REGION

Figure 3. Genetic elements that control the light and heavy chains.

different V-genes. Thus, duplication of V-genes followed by subsequent mutations enlarged the capability of the species to generate more diverse kinds of immunoglobulins.

Immunoglobulin C_H Gene Switching

IgC_H switching is another rearrangement process in which one C_H gene (Figure 3) is replaced by another (Figure 4). These switches make it possible for a single V_H gene to be joined to different C_Hs in sequence. Since there are eight or more C_H genes, seven switches could occur. Some C_H gene switching occurs during the postantigenic activation stages of B-lymphocyte development. Thus during division, a V_H-differentiated cell can give rise to subclones of cells that produce different H-chains. For example, in a response to an antigen, one V-differentiated clone that avidly binds that antigen can generate subclones that produce IgM, all IgG subclasses, IgE, and IgA antibody molecules. These have different effector functions and distribution in the body. The C_H gene switch makes it possible to produce highly versatile forms of antibody molecules.

There is considerable evidence that suggests the switching process occurs unidirectionally along a determined sequence. Honjo and Kataoka (1978) have provided evidence of the sequence of immunoglobulin C_H genes in the mouse from hybridization of specific C_H DNA with chromosomal DNA of various class-specific plasmacytomas (tumors of plasma cells). Essentially, they proposed that in immunoglobulin C_H gene differentiation (which involves only one autosome) the unused intervening DNA is deleted. They propose that the order of immunoglobulin C_H genes in the mouse is s'-μ-δ-γ3-γ1-γ2b-γ2a-ε-α-3', Shimizu et al. 1981; Liu et al. 1980. For example, in a cell differentiated to produce γ3, the μ is deleted; in a cell differentiated to produce γ1, μ and γ3 are deleted (Figure 4). This proposal has been confirmed by other workers, and further with the recent availability of a δ-probe, the δ-gene has been located between μ and γ3 (Figure 4). The C_H gene switching process permits one V-differentiated clone (clonotype) to give rise to specialized subclones that perform different functions in the organism.

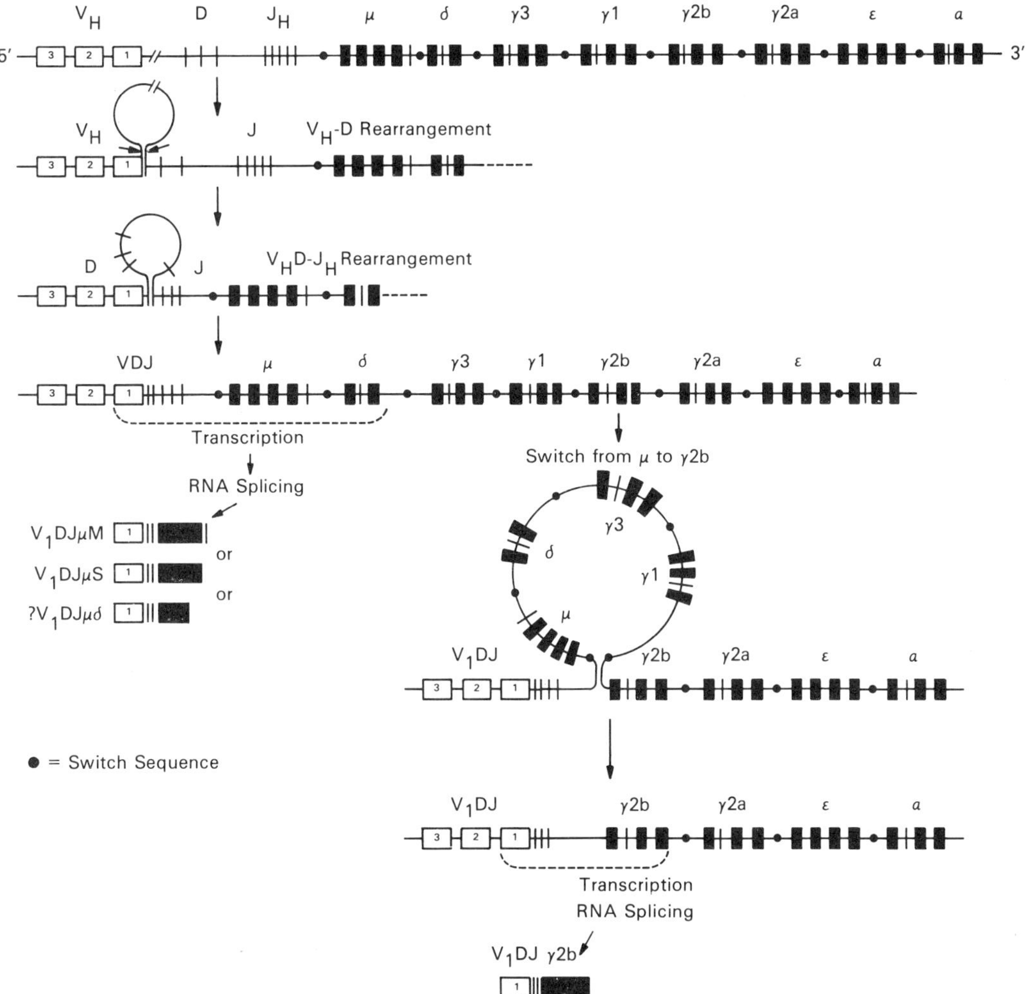

Figure 4. Scheme showing the structure of the mouse heavy chain complex locus. The sequence of the IgC$_H$ genes has been determined by T. Honjo et al. 1978. The location of the ε gene has been determined very recently and communicated to the author courtesy of T. Honjo. The location of the δ gene has been provided by Liu et al. 1980. The joining mechanism in most heavy chains involves an additional small segment called "D", described by Early et al. 1980a. The first transcripts following V-D-J joining are of the μ and possibly δ gene. Different forms of immunoglobulin μM (membrane IgM) MS (secreted IgM) and δ (membrane IgD) are thought to be generated by RNA splicing events, courtesy of P. Early et al. 1980b. Switch sequences located between the C$_H$ genes permit formation of new inverted stem structures and deletion of IgC$_H$ genes. Thus as shown in the figure, the original V$_1$DJ complex can be joined to the C γ 2b gene - or any other C$_H$ gene downstream from μ.

Immunoglobulin Diversity

From the foregoing discussion, it can be seen that the genetic capability of an organism to make a variety of different antigen-binding specificities depends largely upon the repertoires of V$_L$ and V$_H$ genes that are encoded in the genomes. However the V–J joining process amplifies the total number of V_L and V_H structures, as the joining process effects amino acid sequence changes in the third CDR regions of both chain types. For example, in the mouse V_k system there are five J-genes, which when combined to a single V-gene could give rise to hypothetically five different forms. Further, the joining (rearrangement) site can take place at different nucleotides in the $3'$ end triplet of the V-gene, thus producing different amino acids at the junction site. In some unusual cases the triplet can be inserted or deleted at the junction site thus producing a structural change. The joining process in heavy chains is not yet completely understood because there appears to be an additional small segment called the "D" region or diversity region (Early et al. 1980) that is not coded by either V_H or J_H exons.

Immunoglobulin diversity can be defined as the number of different $V_L J$–$V_H DJ$ combinations that can be generated in a single individual. Hypothetically all V_L–V_H pair combinations are possible, thus, if a genome had a library of 10^x V_{L-J} and 10^y V_{H-J}, 10^{xy} different molecular forms could be generated. It has been argued many times that the mammalian genome (defined in terms of total DNA length) could easily accommodate this number of genes and more. It is possible that random V_L–V_H pairing is not possible and that there are favored pairings; if so, this would increase the size of the library needed to generate 10^{xy} different molecules.

For many years, numerous authors have proposed means for increasing the diversity of immunoglobulin molecules from a limited number of germ-line genes by some form of somatic diversification. Mutations in V-region genes that take place in B-lymphocytes and specifically involve codons for amino acids in CDR regions have been proposed as one way to increase the diversity of V-genes. Evidence in the mouse lambda light-chain system supports this notion (Weigert et al. 1977). Available evidence indicates the mouse has only one $V\lambda_1$ gene in its genome, but when 18 different $V\lambda_1$ proteins were sequenced, 6 differed from the common sequence by one to three amino acids. Thus, it could be argued that random mutations involving triplets controlling CDR amino acids modified the genomic $V\lambda_1$ in B-lymphocytes, giving rise to six new forms. It remains to be shown that mutations of this type generate new binding capabilities that are not coded within the genome.

Both somatic mutations and recombinational changes can contribute to immunoglobulin structural diversity; whether these same changes in fact enhance or enlarge the antigen-binding capability remains to be demonstrated. The principal determining factor in the generation of immunoglobulin diversity is the library of inherited V-genes. The great variety of V-genes in the different species suggests the plasticity of the immunoglobulin V-genome in the evolutionary process. New V-genes arise from old by gene duplications and subsequent mutations. As long as these new V-genes remain associated with a C-gene element, many genes can apparently be added onto the complex locus. The suggested evolutionary mechanism for new immunoglobulin V-genes supports the general concept that functional diversity is inherited.

Clonal Selection Theory of Antibody Formation

The clonal selection theory of antibody formation was formulated by Burnet from a modification of Jerne's natural selection theory. This theory has had universal acceptance as the underlying mechanism of antibody formation. Essentially, its present form postulates that immunoglobulin-producing cells are differentiated, so that they can produce a single type of immunoglobulin molecule—as defined by a unique V_L and

V_H region. These first cells, i.e., circulating B-lymphocyte cells, carry the immunoglobulin on the plasma membrane. When the cell encounters antigen and other essential requirements are met, it is triggered to develop into an immunoglobulin-secreting cell. Antigen then selects from a library of predetermined immunoglobulin-producing cells and through mitotic stimulation, specific V_L–V_H differentiated clones are expanded.

Immunoglobulins M and G

Location and Development of IgM- and IgG-Producing Cells

IgM- and IgG-producing cells are discussed together because these cells produce immunoglobulin that functions primarily in the internal environment of the body in the intravascular and tissue spaces. IgM- and IgG-producing plasma cells develop chiefly in the lymph nodes and spleen. These organs act as filters for antigens and particles that enter the lymph and blood. They are also the principal sites where B-lymphocytes are activated to become immunoglobulin-secreting cells.

Antigenic materials enter the blood and lymph from a variety of routes. In addition to the transmucosal route that is discussed below in the section on IgA, antigens enter the body across the dermis, during injuries and infections, or even by direct entry into the bloodstream, for example, following mild trauma to the gums. Antigenic materials that enter the blood and lymph are filtered in the spleen and lymph nodes. The cortex of lymph nodes contains an extensive network of reticular fibers and associated macrophages. The penicilliary arterioles in the white pulp of the spleen are surrounded by a mantle of lymphocytes, macrophages, and reticular fibers. Free antigenic particulates or antigens that have combined with antibody (antigen–antibody complexes) are deposited and bound to the reticular processes of dendritic reticular cells and macrophages. Antigen in this form is potentially immunogenic for T-lymphocytes. At these sites most B-lymphocyte activations take place.

Immunoglobulin-secreting plasma cells, however, are usually seen somewhat "downstream" in the medullary cords of lymph nodes or splenic red pulp. Thus, the activated B-lymphocytes probably move away from activation sites to other locations. Immunoglobulin produced in the medullary cords of lymph nodes or red pulp of the spleen enters the circulation and is distributed throughout the body. Thus, centrally produced IgG and IgM antibody is available locally but also distributed peripherally where effector actions take place. There is another way in which a central (i.e., lymph node or splenic) response to an antigen can be distributed throughout the body; this is by the vehicle of activated B-lymphocytes that leave lymph nodes and spleen and enter the circulation. Activated B-lymphocytes may selectively migrate into inflammatory sites where they undergo further maturation to immunoglobulin-secreting cells.

Comparison of IgM and IgG Production

The immunoglobulin molecules found in the most primitive living vertebrates, the agnatha and fishes are predominantly polymeric high molecular weight or IgM-like molecules. These large polymeric forms of immunoglobulin appear to perform the essential humoral immune functions in these species. Some low molecular weight immunoglobulin is produced in these species but this is the monomeric form of the polymer and its physiologic function is not distinct from high molecular weight immunoglobulin. With the anuran amphibians, there appears for the first time a separate class of low molecular weight immunoglobulin controlled by a different H-chain gene. All vertebrates higher than the anuran amphibians so far studied have one or more classes of low molecular weight immunoglobulin. In some mammalian species, there are four subclasses of immunoglobulins indicating there has been a progressive evolution of low molecular weight IgG subclasses. In human beings, for example, there are the G1, G2, G3, and G4 subclasses and in the mouse G1, G2a, G2b, and G3 subclasses (Table 1).

In mammals, the daily production of IgG exceeds IgM almost 15-fold. In human beings, where immunoglobulin metabolism has been best studied (see Waldmann and Strober 1969), it is estimated that 32 mg of IgG per kg of body weight (the sum of all four subclasses) is produced each day in contrast to 2.2 mg of IgM per kg of body weight. The serum concentration of IgM is approximately 1 mg/ml, almost one-tenth that of IgG.

In human beings, about one-half of the total body pool of IgG is in the intravascular space. It has been estimated that 26 to 39% of the intravascular IgG is transferred to the extravascular pool per day. The half-time survival of most of these molecules in the body has been estimated to be about 21 days (for IgG1, IgG2, and IgG4) in contrast to 5.1 days for IgM.

Since IgM is the first secretory product of a clone, the serum may contain a greater variety of IgM clonotypes at any given time compared with the IgG clonotypes. It is probably safe to make the following correlations: V_L–V_H-differentiated B-lymphocytes are probably being produced continuously from B-cell generative sites and when these surface immunoglobulin (sIg)-bearing lymphocytes are activated by antigen they secrete IgM. Possibly many clones are activated by antigens that only weakly bind the sIg. Thus, this population of B-cells produces relatively low affinity cross-reacting antibodies. Such antibodies may form a part of the so-called natural antibody, which probably is highly polyclonal.

The basis for greater total daily production of IgG would appear to be caused by two factors: (1) the availability of more IgG C_H genes, i.e., four instead of one IgM C_H (in man and mouse); and (2) an expansion of the cell population at the time of IgG differentiation during C_H switching. Thus, a single V-differentiated clone can produce more IgG than IgM by the availability of new switches in C_H and the larger number of cellular progeny in subclones. Further, some of the clonal expansion may take place in the memory stage, increasing the potential availability of cells capable of responding to antigen. This may be particularly relevant to clones that produce antibody that cross-react with many antigens.

Molecular Properties of IgM and IgG Antibodies

The circulating IgM molecule in mammals is a pentamer of five 4-chain units joined together by disulfide linkages and J-chains. The IgM molecule has a molecular weight of about 1×10^6 and has 10 antigen-binding sites. Because of their large size, IgM molecules are retained in blood vessels, and only a small portion normally enters the tissue spaces.

Immunoglobulin-G molecules have physical properties that make them physiologically effective in certain types of *in vivo* antigen–antibody interactions. Size is no doubt the most important of these, since it allows IgG molecules to enter the

Table 1. Functional Properties of IgG Subclasses in Various Mammals

| Species | IgG subclasses identified | Complement fixation C1q binding[a] | Phagocytosis of coated particles by fresh cells | FcR binding function | Transmission to | |
				Homologous passive cutaneous anaphylaxis	Fetus	Neonate
Human beings	G1,G2,G3,G4	G1,G3			G1,G3	
Mouse	G1,G2a,G2b,G3	G2a,G2b	(G1,G2b,G2a)	G1	G2a	G2a
Rat	G1,G2a,G2b,G2c					
Guinea pig	G1,G2	G2				
Rabbit	G	G			G	
Cow	G1,G2	G1 (bovine C)	G2	G1,G2		G1
Goat	G1,G2	G1	G2	G2		
Sheep	G1,G2		G2	G2		

[a] Complement (C) source in parentheses

extravascular tissue spaces, which include sites of acute and chronic inflammation. Unlike IgM, IgG can cross the blood–brain barrier.

The flexibility in the hinge region of the IgG molecule is another important class-specific physical property. The two binding sites on an IgG molecule are able to lie almost 180° apart thus providing maximal cross-linking ability. In IgA and IgM monomers the Fv domains are held closer together, and because of this, these monomers are less efficient in cross-linking functions. The shape of the IgG molecule may permit a greater number of IgG molecules to bind to the surface of a particle or cell, and this property may be particularly important in the process of virus neutralization.

Interactions of Immunoglobulin with Cell Surface Receptors FcR and C3R

The proteolytic digestion of IgG by papain cleaves the molecule into large fragments called Fab (fragment containing antibody activity) and Fc (crystallizable fragment). The Fc contains the C_H2 and C_H3 domains. Structures in both C_H2 and C_H3 domains interact with cell surface receptors (FcR). Another binding site on the C_H2 domain binds the C1q component of complement. Antigen–antibody complex formation involving IgG and IgM classes triggers C1q binding. The complement cascade of reactions via the classical pathway (C1423b) activates C3b binding. Fixed C3b on complexes can now bind another cell surface receptor, C3R. Both FcR and C3R are widely distributed on the major effector cells that appear in inflammations: neutrophils, monocytes, macrophages, eosinophils, mast cells, platelets, and natural killer (NK) cells (Table 2). Further, the FcR on yolk sac endodermal cells, placental syncytiotrophoblasts, and enterocytes of neonates are involved in IgG transport. These membrane FcR bind IgG monomers and form vesicles laden with IgG which are transported across the epithelial cell. Placental syncytiotrophoblasts transfer maternal IgG to the fetal circulation, while neonatal enterocytes ingest colostral or milk IgG.

Several FcR have been described that are specific for different IgC_H classes. An IgE FcR is found on mast cells and basophils, an FcR specific for IgM is formed on T-lymphocytes. The FcRs specific for IgG are studied most widely. It has been suggested there may even be several types of FcR for IgG (Unkeless et al. 1979). FcRs are relatively nonspecies specific; rabbit IgG, for example, binds well to mouse FcR. Unless otherwise specified the term FcR will be used here to define receptors that bind IgG Fc. The FcRs show different affinities for IgG subclasses, which will be discussed below. IgG monomers as they appear in serum are relatively inefficient binders of FcR or complement but when IgG is complexed to antigen, binding efficiency greatly increases for FcR on inflammatory cells.

Immobilized immune complexes can modulate FcR on the surface of macrophages. When macrophages are plated on glass coverslips coated with DNP-rabbit anti-DNP complexes, they stick to the plates but the exposed parts of the cell lose the capacity to ingest or phagocytose erythrocytes coated with IgG. The explanation is that FcRs had moved to occupy the interface between the solid substratum and the binding surface of the macrophages. These adhered macrophages, however, retain their ability to bind IgG antigen–complement-containing complexes. This is one method of demonstrating the independence of FcR and C3R on cells. In many situations, FcR and C3R binding occurs independently, but under certain conditions complexes that contain both IgG and C3b fragments will cross-link these two receptors with potentiating and synergistic consequences. Monomeric IgG binds to FcR with a low affinity

Table 2. Distribution of FcR on C3R on Different Mammalian Cell Types

Cell type	FcR for				C3R	Effector function
	IgG	IgM	IgA	IgE		
Lymphocytes						
T-lymphocyte	−	+	+	−	−	Cytotoxic, cooperation in IgA antibody formation
B-lymphocyte	+				+	Antibody formation, memory cell generation
Null cell	+					Cytotoxic cell
NK-(killer) cell	+					Cytotoxic cell
Mononuclear Phagocyte System						
Monocytes	+				+	Phagocytosis
Peritoneal macrophages	+				+	Phagocytosis
Kupffer cells	+				+	Phagocytosis and removal of particulates from blood
Alveolar macrophages	+				−	Phagocytosis and bacterial clearance in the lung
Neutrophils	+				+	Phagocytosis, lysosomal release in inflammatory sites
Eosinophils	+				+	Parasite destruction
Basophils				+		Histamine, serotonin release
Mast cells				+		Histamine, serotonin release
Platelets	+				+	Release of vasoactive amines
Neonatal enterocyte	+					Transfer maternal IgG to neonate
Yolk sac endodermal cell	+					
Placental syncytiotrophoblast	+					Transfer maternal IgG to fetus
Splenic, lymph node macrophages	+				+	Antibody formation
Dendritic reticular cells	+				+	Germinal center (memory cell formation) antigen retention

$K_A = 5 \times 10^{-5} \ M^{-1}$, and the concentration of IgG in serum is high enough to occupy all the available FcR sites. Oligomeric antigen–antibody complexes of increasing size bind with increasing affinity. In serum, however, the increased affinity of antigen–antibody complexes is still not thought to be high enough to permit effective binding in the presence of a large excess of monomeric IgG. It has been postulated, though, that in inflammatory sites, where the concentration of complexes is higher than free monomeric IgG, effective binding can take place. In particular, in situations where phagocytic cell membranes come into close contact with complexes such as bacteria or virus-infected cells free IgG molecules may be excluded (Segal et al. 1979).

Regulatory Functions of IgG

FcR and C3R are also found on B-lymphocytes. The binding of IgG antigen complexes to B-lymphocytes can exert important regulatory actions on B-lymphocyte mitogenesis and subsequent maturation. Some Ag–Ab complexes can block B-lymphocyte maturation to antibody-secreting cell stages. This cannot be overcome by B-lymphocyte mitogens such as lipopolysaccharides or T-lymphocytes. The blocking effect of Ag–Ab complexes appears to result from the cooperative interaction of antigen and Fc receptors on B-lymphocytes (Sinclair and Chan 1971; Oberbarnscheidt and Kolsch 1978). This blockade, however, apparently does not affect cell proliferation.

Ag-Ab complexes also regulate the memory cell development associated with germinal center formation (White et al. 1975). The latent period in germinal center formation following the intravenous injection of soluble antigen depends upon the formation of antibody and antigen–antibody complexes. The complexes fix complement and activate C3. The Ag–Ab–C3 complex binds to dendritic macrophages via the action of C3 and C3R on macrophages. In mice, Klaus and Humphrey (1977) have abrogated memory cell formation and germinal center formation by depleting immunized mice of C3 with cobra venom factor. Cobra venon factor is a powerful activator of C3, and prolonged treatment depletes mice (and other species) of C3. Ag–Ab complexes can apparently become fixed to dendritic cells via the action of Fc receptors in the absence of C3, but localization in germinal centers requires C3. Paradoxically, C3 is not readily demonstrable in germinal centers. It has been postulated by Embling et al. (1977) that C3 is released from the Ag–Ab–C3 complexes bound to the migrating dendritic macrophages and that the Ag–Ab complex is fixed to the cell by another mechanism (perhaps FcR). It is known that Ag–Ab complex persists apparently in stable form in germinal centers for relatively long periods of time (i.e., several weeks). Antigens can also be shown to persist for extensive periods of time in lymph nodes. Thus, processes that promote persistence of antigen promote immunological memory.

IgG Subclasses

In mammalian species, such as mice and human beings, in which homogeneous immunoglobulins are available in the form of myeloma proteins, four subclasses of IgG have been described. These are called IgG1, G2, G3, and G4 in human beings; IgG1, IgG2a, IgG2b, and IgG3 in mice (Table 2).

In other species, such as rabbits, only one major class of IgG has thus far been described although minor subclasses are suspected. The predominant rabbit IgG can both activate complement and bind to FcR. In guinea pigs, there are two major subclasses of IgG, one of which is an efficient activator of complement, while the other is not.

The exact function of IgG subclasses in mouse and man has not been fully determined but may be related to the binding activities for C1q, C3R and FcR (Michl et al. 1979). For example, in mice the G1 and G3 subclasses fix complement inefficiently. However, G1 can bind Fc receptors (Segal and Titus 1978) and is the most cytophilic of the mouse IgG subclasses. Michl et al. 1979 have shown that insoluble and immobilized immune complexes of rabbit IgG can modulate the distribution of FcR on mouse macrophages and, further, that complexes to which complement is fixed are unable to bind to FcR. They postulate that the noncomplement binding classes of IgG will be able to bind to FcR even in the presence of complement and because of this may be able to "mediate a specific subset of effector functions" (Michl et al. 1979).

Maternal Transmission of IgG

In mammals, subclasses of IgG are transferred somewhat selectively to fetus and neonate. There are considerable species variations in maternal transmission of immunoglobulins (Table 3), as first described by Brambell (1966). In all of these

Table 3. Maternal Transmission of IgG

Species (IgG subclass)[a]	Recipient		Absorbing cell
	Fetus	Neonate (duration)	
Cow $(IgG_1)^{+b}$	0	+ + + (36 hrs)	
Goat	0	+ + + (36 hrs)	
Sheep	0	+ + + (36 hrs)	Enterocyte fetal mucosal cell
Pig	0	+ + + (36 hrs)	
Horse	0	+ + + (36 hrs)	
Dog	+	+ + (10 days)	Enterocyte
Mouse (IgG_{2a})	+	+ + (16 days)	Yolk sac endoderm
Rat	+	+ + (20 days)	Placental syncytiotrophoblast
Guinea pig	+ + +	0	
Rabbit	+ + +	0	Placental syncytiotrophoblast
Human beings (IgG_1, IgG_3)	+ + +	0	

(Table adapted from Brambell 1966.)

[a]IgG subclass that is preferentially transported.

[b]In the last 5 weeks of gestation there is a decline in the serum IgG_1 level and a progressive increase in IgG_1 content in the colostrum (Brandon 1977).

processes, IgG is taken up by pinocytosis and transported across maternal or fetal cells and finally released in the fetal or neonatal circulation. Brambell in 1966 proposed this was a selective process, since pinocytotic membranes could take up many types of proteins, most of which were destroyed in phagolysosomes. He postulated that the binding of the IgG molecule to the membrane (via FcR), protected these immunoglobulin molecules from proteolytic degradation during transport across the absorbing cell. As may be seen from Table 3 in species such as guinea pigs, rabbits, and human beings, all the maternal transmission of IgG is to the fetus, while in the other extreme (horses, sheep, cows, goats, and pigs) very little of any IgG is transferred to the fetus but is transferred to the neonate. In rats, mice, and dogs, IgG is transmitted to both fetus and neonate, and this process continues for as long as 3 weeks in rats.

Effector Functions of IgM and IgG

Individuals who are incapable of generating immunoglobulins, e.g., those with congenital Bruton-type agammaglobulinemia, develop repeated infections of the kind most normal persons experience infrequently, such as sinusitis, pharyngitis, bronchitis, otitis media, pneumonia, skin infections. These are caused by pyogenic microorganisms such as species of *Streptococcus, Neisseria, Hemophilus,* etc. The inability to limit invasion by these organisms not only initiates local infection but leads to more serious complications, such as septicemia, meningitis, bronchiectasis. Since individuals with congenital agammaglobulinemia possess complement and a phagocytic system of inflammatory cells, the lack of ability of these effector mechanisms is attributable to the absence of antibody molecules. The passive administration of human gammaglobulin provides protection to these individuals.

The ability of immunoglobulin molecules or antibody to cooperate with effector cells and complement is the major function of IgM and IgG. Immunoglobulin or antibody alone, i.e., without complement or a phagocytic system, is not able to control infections because the binding of antibody to a microorganism does not affect the viability of that organism.

Opsonization. One of the major effector functions mediated by IgM and IgG is the facilitation of the phagocytic process. Phagocytosis has several steps: (1) particle binding; (2) formation of vesicles or phagosomes around the particles; (3) ingestion; and finally (4) enzymatic digestion and other forms of destruction of phagolysosome contents. IgM and IgG antibodies facilitate different steps in the phagocytic process. First, they aid in the binding process (this is the opsonic function). Second, they can increase the intensity of phagocytic activity. Third, they facilitate digestion in the phagolysosomes.

The mammalian body contains an extensive system of mononuclear phagocytes. In addition to the macrophages of the lymphoid tissues, there are macrophages throughout the connective tissues. There are strategic concentrations of macrophages in the sinusoids of the liver (Kupffer cells) that actively remove particles from the blood. Pulmonary alveolar macrophages are continuously migrating across the alveoli and ascending the respiratory tract. Their presence in the lower respiratory tract and lung provides policing of these tissues to remove inhaled bacteria and other particles. One kind of special effector function of IgG and IgM is to cooperate with these specialized cells.

In the organism, IgM avidly activates complement by the classical pathway. One IgM molecule on the surface of an erythrocyte is sufficient to trigger all complement components including the final assembly of C5-9, the membrane attack complex. Complement activation by the classical pathway evoked by IgM is a highly efficient mechanism for promoting opsonization and ultimate bacteriolysis or bacteriocidal effects mediated by the membrane attack complex. In this form of opsonization, a few IgM molecules on the surface of a bacterium or particle can trigger a much more extensive deposition of C3 molecules and the membrane attack pathway proteins of the complement system C5, C6, C7, C8, and C9. Macrophages and PMNL attach to C3-coated particles via C3 receptors. Rowley and Turner (1966) estimated that IgM was 275 times more efficient as an opsonin than IgG. Thus at low concentrations, IgM can potentially be highly effective since only a few molecules are sufficient to activate complement. The location of IgM molecules in intravascular sites suggests its chief function may be to aid in the clearance of particles from the blood. This is facilitated by the system of fixed macrophages, e.g., Kupffer cells in the liver sinusoids. In addition, IgM plays an important role in localizing antigen on macrophage surfaces where it can initiate further humoral immune responses by interacting with T- and B-lymphocytes.

IgM may also interact with T-lymphocyte receptors for IgM Fc regions. Andersson et al. (1979) have found that IgM antibodies can bind to both cytotoxic and helper T-lymphocytes. This type of interaction has not been as intensively studied as the IgG Fc receptor but may nonetheless provide an important effector pathway for IgM molecules.

IgG exerts its major effector functions via binding to FcR and C3R of inflammatory cells and Clq, which leads to complement activation. Two major forms of effector function are opsonization and cell-mediated cytotoxicity. There is also evidence that IgG antigen complexes stimulate the release of some lysosomal enzymes (Cardella et al. 1974).

Opsonin is a term derived from the Greek meaning "to flavor." In many bacterial and microbial infections the surfaces of the organism such as slimes, capsules, and other surfaces are coated with large amounts of polysaccharides. Many of these polysaccharides cannot be bound directly by macrophages, thus such organisms (many of which are pathogenic bacteria) escape the major effector actions of inflammatory cells. However, antibody attached to the surfaces of bacteria can provide binding sites for phagocytes to attack the offending organisms. These complexes contain both polyvalent Fc or C3. In overwhelming infections, such as pneumococcal pneumonia, cryptococcosis, etc., large amounts of free polysaccharide may be produced, which further can prevent phagocytosis by (1) consuming available antibody or (2) blocking the development of specific antibody-forming cells (via complexes). Rowley and Turner (1966) have determined the relative efficiency of IgG and IgM antibodies to kill living *Salmonella adelaide* organisms. This was studied by injecting 10^6 viable organisms mixed with antibodies into the peritoneal cavities of mice and determining the viability over a 90-minute period. They purified IgG and IgM antibodies in rabbits; 0.1 ml of a 1/500,000 dilution of the whole antiserum gave 99% kill of the organisms in 90 minutes. They estimated that eight molecules of IgM per bacterium and 2200 molecules of IgG per bacterium were essential for intraperitoneal killing. According to this evidence IgM is clearly more efficient than IgG in opsonization.

Using monolayers of human PMNL, Menzel et al. (1978) studied the ability of IgG, IgM, and complement on opsonization and intracellular killing of *Escherichia coli*. IgG but not IgM opsonized *E. coli* in the absence of complement. IgM alone lacked opsonic activity. When complement was present, IgM became an opsonin and was more efficient than IgG. Intracellular killing was promoted by both IgG and IgM.

The relative inability of IgM to gain access to tissue inflammatory sites and the higher concentration of IgG in serum and the extracellular compartment probably render IgG the more important immunoglobulin effector in many *in vivo* situations. In addition, effective opsonization can be mediated in the absence of complement with IgG-sensitized organisms.

Cytotoxic Processes. Attention is currently being focused on the potential role of cytotoxic mechanisms in bacterial infections. Lowell et al. 1979 have demonstrated antibody-dependent cellular cytotoxicity (ADCC) killing of meningococci *in vitro*. In their system, human peripheral blood lymphocytes (K-cells) and specific antibody were found to be cytotoxic to meningococci. Antibody dilutions of 1/160 were optimal but some sera could be diluted to 1/1280 and still induce killing. Prozone effects were observed. The effector/target ratio used was 400:1, which suggests some inefficiency of the system. Further, these authors point out that *in vivo* such a mechanism may operate in tissue sites where complement is in short supply, such as in the cerebrospinal fluid and the lower respiratory tract.

IgG antibody can play a role in parasitic infections, in particular in situations in which continuous reinfection occurs. Antibody is not able to completely eject parasites from their niches but can prevent reinfections. One case in point is experimental schistosomiasis in rats and mice. Infected animals develop antibodies to an immature larval stage of the organism, the schistosomula. These antibodies cooperate with eosinophils and complement to bring about schistosomula destruction. Eosinophils as effector cells are required for this type of effector action since they contain special cytotoxic properties for parasites. Antibodies attached to the schistosomula surface form an effective complex for FcR of eosinophils and complement. McLaren and

Ramalho-Pinto (1979) have shown a synergistic effect of IgG class antischistosomula antibody and C3 in promoting eosinophil-mediated cytotoxicity. Low concentrations of antibody promoted high eosinophil-mediated killing of schistosomula, suggesting that their *in vitro* system reflects a potential *in vivo* mechanism.

Virus Neutralization. Pathogenic viruses of different species vary greatly in size and morphology. The particle sizes of the smaller viruses range from 22 nm (parvoviruses) to 28 nm for pico RNA viruses, while the largest pox viruses have particles of 250 to 300 nm. In many viruses, the nucleic acids are contained in cores covered only by nucleocapsid shells, while in others the cores and their shells are enveloped with membranous outer coats. Viruses spread from cell to cell by infectious particles released from one cell, attaching and penetrating a new cell. Viruses can also be spread from one cell to another by cell fusion as, for example, in the paramyxoviruses. In these, an outer coat viral glycoprotein can cause cells to fuse, thereby allowing virus to pass from one cell to another. Viral infections evoke cellular inflammatory responses and immune responses. The limitation of a viral infection by the host involves blocking viral spread from cell to cell and cytotoxic killing of cells producing virus. The neutralization of viruses by antibodies can involve different mechanisms. First, antibody acts as an opsonin by combining with virus to form a complex, which is taken up and destroyed in phagolysosomes of macrophages. Second, antibodies by combining with viral particles can limit viral proteins from attaching and penetrating uninfected cells. In such complexes, the virus is not necessarily inactivated but rather rendered less efficient or incapable of spreading from cell to cell or throughout the body. Paramyxovirus spread by fusion can be blocked effectively by antibody. In the internal environment, IgG antibodies carry out these functions. In the external environment, IgA antibodies act by blocking viral attachment to epithelial cells.

The mechanics of effective neutralization is not completely understood, but requires the attachment of immunoglobulin molecules to the viral particle. The actual density of immunoglobulin molecules may be important. In fact, virus neutralization can be mediated by a few specific antiviral antibodies and anti-IgG cross-linking molecules, suggesting patches on the particle surface are modified. A list of viral neutralization mechanisms is shown in Table 4.

Summary. IgG appears to be the predominant class of immunoglobulins that enter the tissue spaces, and that cross various vascular barriers (e.g., as in maternal–fetal transmission, or the blood–brain barrier). Some effector functions are the direct consequence of the binding of IgG to antigen (antitoxic properties, some forms of viral neutralization). Many effector functions are mediated by IgG Ag complexes:

Table 4. Mechanisms of Antibody-mediated Viral Neutralization

Promotes phagocytosis of virus-Ab complexes via FcR and C3R by macrophages.

Promotes intracellular destruction of virus-Ab complexes in phagolysosomes.

Mediates lysis of membrane enveloped virus particles by complement.

Virus-Ab complex blocks:
    absorption to uninfected cell
    uncoating of virus particle
    fusion of virus infected cells.

Causes deposition of virus-Ab complexes on cell membranes limiting spread across protected sites,
    e.g., blood-brain barrier, fetus.

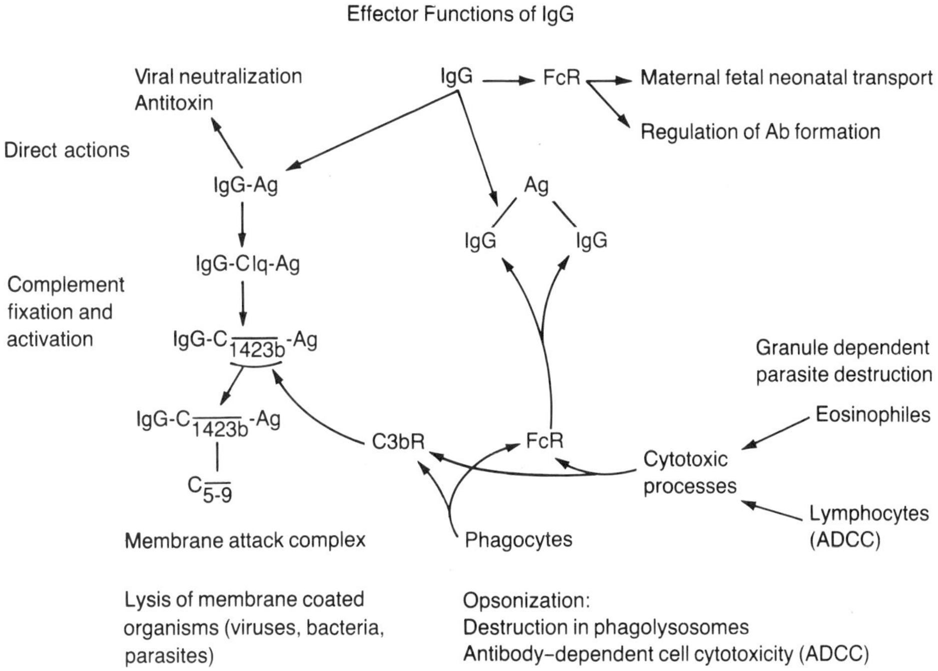

Effector Functions of IgG

Direct actions

Viral neutralization
Antitoxin

IgG $\longrightarrow$ FcR $\longrightarrow$ Maternal fetal neonatal transport

Regulation of Ab formation

IgG-Ag

Ag

IgG   IgG

Complement
fixation and
activation

IgG-CIq-Ag

IgG-C$\overline{1423b}$-Ag

IgG-C$\overline{1423b}$-Ag

C$\overline{5\text{-}9}$

Membrane attack complex

C3bR   FcR

Phagocytes

Cytotoxic
processes

Granule dependent
parasite destruction

Eosinophiles

Lymphocytes
(ADCC)

Lysis of membrane coated
organisms (viruses, bacteria,
parasites)

Opsonization:
Destruction in phagolysosomes
Antibody–dependent cell cytotoxicity (ADCC)

Figure 5. Effector functions of IgG.

opsonization, complement binding, cytotoxic processes (see Figure 5). Finally, IgG molecules perform some regulative functions relating antibody formation and memory.

Immunoglobulin A

Location of IgA-Producing Cells

IgA-secreting plasma cells are found in subepithelial sites along the gut, upper respiratory tract, salivary glands, lacrimal glands, lactating mammary glands, and tonsils. Thus, unlike IgG and IgM production, which occurs in lymph nodes and spleen and is delivered to the circulation, IgA molecules are secreted close to epithelial cells, where they can be more readily transported across these cells into the external environment. Not all of the IgA molecules are secreted immediately. About one-half have been estimated to enter the circulation. Many of these molecules are taken up by liver cells and secreted in the bile (Birbeck et al. 1979). IgA molecules are primarily destined to operate in extracorporeal sites and defend the mucosal and internal linings of the body. In addition, IgA is secreted in relatively large quantity in milk, thereby passively transmitting maternal immunity to the suckling neonate.

The life history of the IgA-producing cell is a complex process, and not all phases of it are completely understood.

Antigen Entry

An extensive system of lymphoid tissues lines the gut and respiratory tracts, e.g., appendix, bronchial-associated lymphoid aggregates (Waksman and Ozer 1976). These lymphoepithelial organs appear to be associated with antigen intake—that is, antigenic materials such as proteins, polysaccharides, particulates, bacteria—other organisms and materials found in the lumen of these tracts are transported across the epithelial lining into lymphoid tissue where they come into immediate contact with lymphocytes (Figure 6). Specialized cells in the epithelium over lymphoid aggregates function to facilitate the transport of luminal material (Bockman and Cooper 1973). In much of the gut-associated lymphoid tissue, there are specialized columnar cells called microfold or M-cells (Owen and Jones 1974). These lack the usual ordered microvillus configuration of the plasma membrane as seen in columnar epithelium and have plasma membranes that extend in folds or ruffles. These membranes form vesicles around luminal materials and by pinocytosis transport the vesicles in and across the M-cells where they are discharged into the intracellular spaces around lymphocytes (Owen 1977) (Figure 6). Microfold-cell cytoplasm also forms tight junctions between surrounding microvillus columnar cells. Over regions where there are underlying lymphocytes, the M-cell cytoplasm forms a very thin bridge between the columnar cells, thus reducing the cytoplasmic distance for tubulovesicular transport in the lymph nodes.

Live bacteria have been cultured from Peyer's patches, mesenteric lymph nodes, and even in the spleen after colonization of the gastrointestinal tract by new types of microorganisms. For example, when germ-free mice are associated with the flora of pathogen-free mice, bacteria (e.g., indigenous *Escherichia coli*, lactobacilli) have been found to traverse the epithelium and enter the internal environment (Berg and Garlington 1979). This process usually involves only Peyer's patches but can occur across the epithelial mucosa as well. Penetration of bacteria across intestinal epithelium is thought to occur more frequently with pathogens (Takeuchi 1971). Regardless of the route, the penetration of the epithelium by bacteria stimulates immune responses. It is well known that natural antibodies to blood group substances are generated by responses to cross-reacting antigens on bacteria (Springer et al. 1961), and natural antibodies in general may actually arise due to such repeated bacterial antigen exposures.

Development of IgA-Producing Cells

While it is generally thought that the entry of antigens across epithelial surfaces promotes the development of IgA differentiation, all the steps in this process are not defined. It is clear though that the introduction of antigens into the lumen of the gastrointestinal tract induces the formation of specific IgA-producing cells locally and in distant sites.

The IgA-producing cell population along the subepithelium (lamina propria) of the gut is very large and, further, is dynamically replaced within relatively short periods of time (days). The lamina propria plasma cell population in the gut comes from two sources: (1) the influx of new IgA lymphoblasts from the circulation; and (2) by local proliferation of cells in the lamina propria *in situ*. It has been estimated in the rat that about 1.2×10^7 IgA lymphoblasts are produced daily. This represents the potential pool of cells that can migrate into the lamina propria, and a large percentage of these enter the lamina propria every day.

The evidence for local proliferation of IgA-committed cells or even possibly IgA-secreting cells is less direct. For example, in nu/nu mice, which have relatively deficient numbers of IgA-producing cells in lymph nodes, the bronchial tract, parotid gland, and lactating mammary gland, appear to have near normal levels of IgA-

Development of IgA Producing Cells

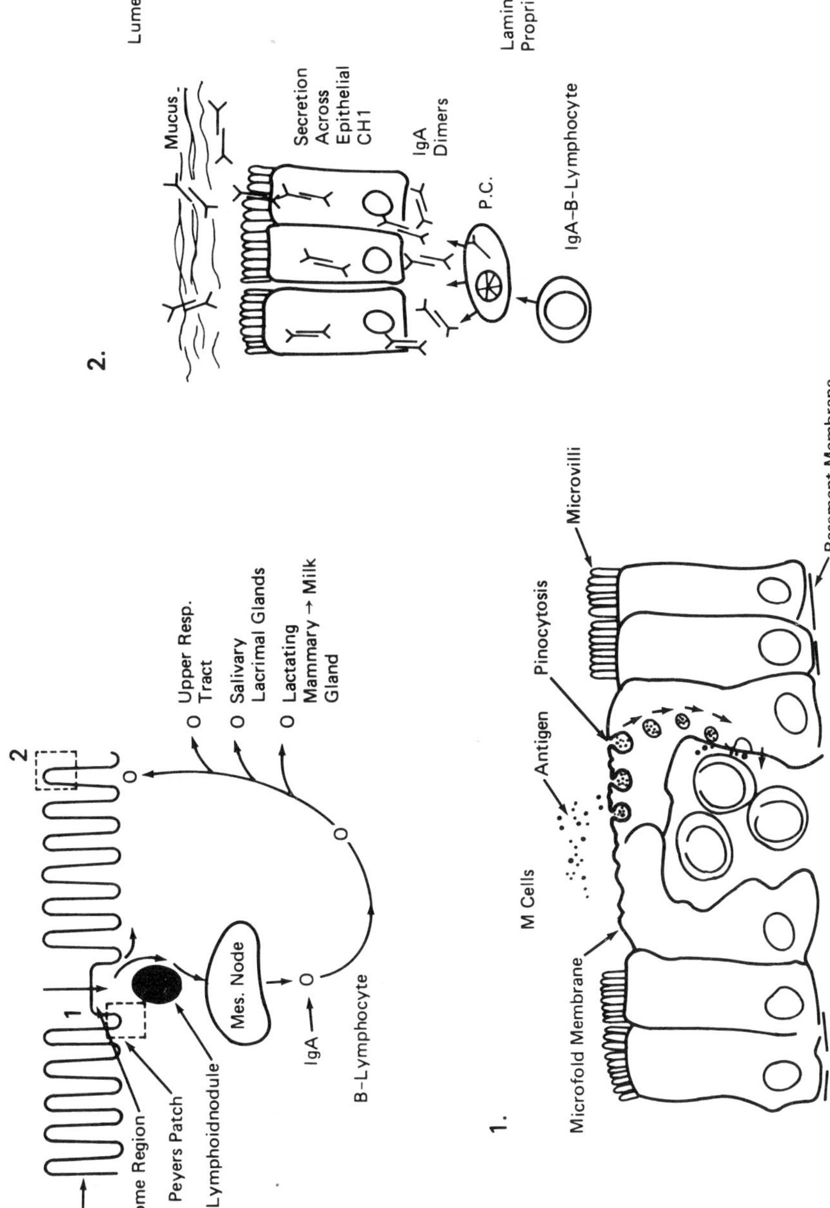

Figure 6. Schematic diagram of the development of IgA-producing cells, the secretion of IgA, and the relation to the entry of antigens across the specialized epithelium over Peyer's patches.

secreting cells in the gastrointestinal tract suggesting that local stimulation of IgA-producing cells by mitogens coming from the gut is taking place locally. This and other evidence supports the idea that a limited number of divisions of IgA-committed cells occur in the lamina propria and further that these are stimulated by the presence of high concentrations of antigens and mitogens in the lumen of the gastrointestinal tract.

While these facts reflect the magnitude of the IgA populations in mammals, they do not explain how the IgA-producing cells are specifically produced. The first clue that a special mechanism was involved came from the findings of Craig and Cebra (1971) who showed that the Peyer's patches of rabbits contained a large number of precursors for IgA-producing lamina propria cells. This suggests that Peyer's patches are sites where cells programmed to produce IgA specifically proliferate perhaps as memory B-lymphocytes.

Despite the presence of all of the necessary components, IgA antibody formation does not occur locally in the lymphoid aggregate regions such as Peyer's patches but rather in the regional draining mesenteric node. Precisely what occurs in the dome region (Figure 6) is not clear. Particularly perplexing is the proliferation of B-lymphocytes in close proximity to regions of antigen intake. Possibly, these areas represent collecting channels for both antigen and antigen reactive cells. Both T- and B-lymphocytes thereafter flow rapidly into the afferent lymphatics of the mesenteric lymph node. The presence of B-lymphocytes could increase the likelihood that specific responses to antigens will take place.

One cannot conclude from these observations that all IgA-secreting lamina propria cells have been stimulated by antigens that have traversed the gut or respiratory epithelium. Beh et al. (1979) have shown that immunization of sheep with intraperitoneal injections of ovalbumin in complete Freunds adjuvants induces appearance of specific IgA B-lymphocytes in the thoracic duct lymph followed by homing and formation of IgA-secreting cells in the intestinal lamina propria.

Redistribution of IgA-Producing B-Cells

The selective migration of activated IgA-producing lymphocytes into lamina propria microenvironments permits cells sensitized by antigens entering at one site to migrate to a variety of sites. Thus, an infection occurring in one part of the gastrointestinal tract can result in the migration of IgA-producing B-lymphocytes throughout the length of the gut or even to other tissue sites such as the respiratory tract (Table 5). A high concentration of intraluminal antigen can influence the kinetics of entry of cells in the lamina propria and the production of specific antibody. In experiments in which two Thiry–Vella loops were prepared in the same rat and antigen was introduced into only one, the migration of IgA-producing cells initially developed in the lamina propria regions of both loops but persisted in considerable intensity only in the injected loop. These results indicate local antigen plays a role in maintaining the local population. The large numbers of specific IgA-producing cells at sites where intraluminal antigen was available was attributed to cell division of the IgA-producing cells at the site, since 10% of these antibody-producing cells incorporated ^{3}H-thymidine, suggesting they were still capable of dividing. It was concluded that local antigen profoundly influences the "localization, magnitude, and persistence of the response" to a local challenge. Cell divisions of IgA-producing lamina propria cells might also have been facilitated by specific local T-helper cells.

Table 5. Number of Cells Involved in Different Levels of the Human Gastrointestinal Tract

	IgA	IgM	IgG
Duodenum–jejunum	352,000	52,000	16,000
Colon	346,000	17,000	15,000
Rectum	156,000	17,000	10,000

(Data taken from Crabbé and Heremans 1966.)

The migration of specific IgA-committed lymphocytes can also involve mammary epithelium. Thus, in mammals, IgA plays an important role in maternal–neonatal immunization. IgA B-lymphocytes of mesenteric node origin migrate to mammary tissues in late pregnancy and during lactation, thus providing the colostrum and milk with antibodies to gut flora antigens of the mother. The IgA of colostrum and milk is not absorbed but exerts its effects intraluminally.

IgA Effector Functions

The C_α domains have free disulfides that facilitate the formation of IgA dimers and higher polymers. The C_α domains also have affinities for two other polypeptides, J-(joining) chain and secretory component (SC). The J-chain (not to be confused with J-region minigene or the J-segment of the immunoglobulin molecule) is a 15,000 molecular weight polypeptide that also aids in the formation of IgA polymers. The SC is a 35,000-molecular weight polypeptide that interacts with $C_\alpha 2$ and $C_\alpha 3$ domains. The dimeric IgA SC complex has two functions. First, it is involved in the actual transport of IgA through epithelium. Second, SC protects IgA molecules in the gut lumen from proteases.

IgA molecules can gain access to the lumen of the gastrointestinal tract by three routes: (1) the transepithelial transport process, (2) transhepatic secretion into the bile, and (3) by ingestion of colostrum and milk (in neonates). Heremans (1974) estimated that in the average man, the gut lamina propria population produced approximately 2 g of IgA daily, one-half of which was secreted transepithelially. A second source of intraluminal IgA in adults is through the clearance of serum IgA through the liver and the excretion in the bile. Orlans et al. (1978) studied the kinetics of IgA clearance in rats injected with [125]IgA myeloma proteins. These IgA molecules were rapidly cleared, only 10% remained after 3 hours. When the bile ducts were cannulated, 25% of the total injected dose was recovered in the bile in 3 hours. IgA molecules are secreted actively across mammary epithelium into the colostrum and milk. Human colostrum contains 3 mg/ml of IgA and milk 1 mg/ml.

IgA antibodies are not opsonins nor do they bind to IgG FcR or activate complement by the classical pathway. IgA antibodies exert their major effects by binding with antigens outside the internal environment of the organism. These actions take place in the lumens of the gut, upper respiratory tract, oral and pharyngeal cavities, and in the fluids bathing the eye. Although IgA is not secreted across the epithelium of the lower urinary tract, IgA antibodies do appear in urethral inflammation. IgA antibodies carry out their major effects then along the internal epithelial linings of the body and probably in direct relation to epithelial cells themselves. Since there are many different histological types of epithelia represented in these tissues (i.e., ciliated, columnar,

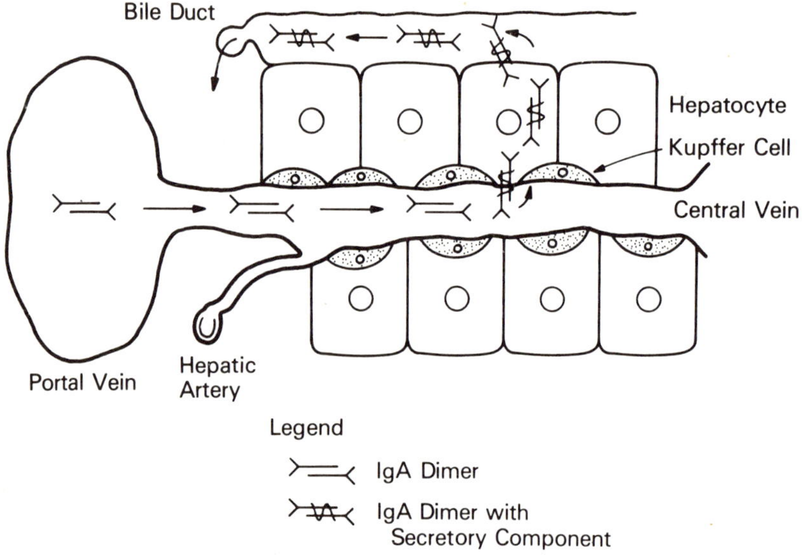

Figure 7. Schematic diagram of transhepatic migration of IgA across the liver sinusoid.

respiratory, microvillous, squamous) there are no doubt different ways in which IgA operates. Although IgA molecules are transported across columnar epithelial cells proper, IgA is present in the intercellular spaces above the level of the basement membrane. IgA antibody in these locations may be important in preventing the spread of viruses from one epithelial cell to another.

It has been hypothesized that secreted IgA has an affinity for mucus via the SC, although the details of this interaction are not known. Mucus, as is well known, provides a protective coating and can act as a barrier that limits access of luminal materials to intestinal epithelium. It may also act as a brush that cleans the epithelial surface. Florey (1933) noted many years ago that the movement of the intestine causes the villi to lengthen and shorten. This motion causes particles on the epithelium to brush against the less mobile mucous strands. The presence of IgA could plausibly cause an increase in the specific adhesive properties of mucus for bacteria and proteins. IgA antibodies lodged in mucus can have several effects: (1) they can interfere with the adhesion of organisms to epithelial surfaces, (2) they can retain potential immunogenic molecules longer in the mucous matrix thereby permitting longer time of action of digestive enzymes, and (3) by binding the antigenic molecule or organism to the mucus they cause it to be swept along by the passage of materials down the gastrointestinal tract. Bacteria can penetrate the epithelium covering lymphoid aggregates as mentioned above. In normal conventional animals, viable bacteria apparently do not spread beyond these portal sites. The situation is different in the immunologically immature germ-free mice that are infected with normal indigenous flora. Berg and Garlington (1979) inoculated germ-free mice with the cecal microflora from a pathogen-free strain and recovered viable organisms from the mesenteric node.

Thus, ordinary indigenous microbial species are able to invade and infect immuno-deficient hosts. Presumably, IgA antibodies apparently prevent these invasions in the normal host (Table 6).

Intraluminal IgA antibodies then may play an active role in limiting the entry of microbes into the internal environment at the Peyer's patch and other related sites in the gut- and bronchial-associated lymphoid tissues. Passively transferred IgA antibodies have been shown to protect mice from oral infection of the metacestode *Taenia taeniaformis* (Lloyd and Soulsby 1978). The mechanism of this hypothetical process is not known; since IgA is not opsonic, it is unlikely to involve macrophages. Evidence indicates bacterial surfaces become coated with IgA antibody. Further, Strober et al. (1978) have shown there are T-lymphoid cells in the spleen that have IgA receptors. Peyer's patches and the lamina propria of the gut are known to contain large numbers of T-lymphocytes. Though not yet proven it is possible that T-lymphocytes could directly interact with IgA-coated microbial cells and thereby play a role in controlling the viability of bacteria that do traverse the epithelium.

IgA antibodies play a role in preventing the absorption of inert antigenic macro-molecules across the epithelial lining. These include, of course, potential toxins and allergens.

Selective IgA deficiency of humans, a common trait that occurs in 1 of every 700 persons, is frequently associated with the formation of IgG and IgM antibodies to dietary antigens including proteins and allergens. In some of these persons, it is postulated antigenic molecules traverse the gut epithelium in sufficient concentration to induce the formation of IgG and IgM antibodies.

Novel actions of IgA antibodies involve cooperative effects with other proteins present in the luminal spaces. IgA antibodies are known to act in concert with iron-binding proteins (transferrin, lactoferrin). The growth of many bacteria depends upon the trapping of available iron ions from the environment by an active chelating protein on the bacterial cell surface. Iron ions are in critical supply in the gut lumen, for example, and thus the competition for available ions can be influenced by the concentration of iron-binding proteins plus IgA antibody. Lactoferrin is secreted by epithelial cells of the gut. In human milk, the concentration of lactoferrin is 2 to 6 mg/ml,

Table 6. Effector Functions of IgA

General		Specific examples	Host cooperating component
Impedes absorption of macromolecules	Dietary macromolecules	Slows absorption through epithelium	
	Toxins	Neutralizes toxins	
		Enterotoxin absorption	
	Allergens	Retards absorption	
Impedes entry of microbe	Virus	Prevents viral invasion and spread	
	Bacteria	Prevents adhesion	Mucus, peristalsis
		Bacteriostasis	Transferrin
			Lactoferrin
	Parasite	Retards adhesion of organism to gut epithelium	Intraluminal proteolytic enzymes
		Prevents reinfestations	

transferrin is 10 to 15 μg/ml, and IgA is 1 mg/ml. Neither iron-binding proteins nor IgA antibodies are bacteriostatic *in vitro,* but together they are so. The exact immunochemical mechanism is not known but may involve interaction of the antibody with the iron bacterial binding and trapping system. Human milk has been shown *in vivo* to be effective in controlling epidemics of intestinal infections in newborns, e.g., diarrheal disease caused by *E. coli.* Since human milk contains bacteriostatic IgA antibodies to *E. coli,* this evidence, though indirect, strongly supports the *in vitro* studies.

The binding of IgA antibody to a living target may retard entry and permit damaging secondary action of a cooperating protein such as a digestive enzyme. Evidence for cooperative interactions is provided by the study of infections in chickens by coccidial organisms such as *Eimeria tenella.* These coccidial parasites proliferate in the gut lumen, and the sporozoites must rapidly attach to epithelial cells and invade the cells where they undergo further stages of development. In immune chickens sporozoites have been observed to invade epithelial cells but lack the ability to further differentiate. Davies and Porter (1979) speculate IgA antibody slows the attachment absorption process, thereby allowing nonspecific intraluminal proteolytic enzymes to damage the sporozoites so they can no longer differentiate. Thus, by binding to macromolecules or microorganisms, IgA impedes the entry of molecule or organism into the epithelial cell. As this occurs, other effects, such as trapping in mucus, proteolysis, or iron deprivation, damage or digest the offender.

Summary. IgA antibodies operate outside the internal environment. Most effector functions probably depend upon direct antigen binding and are probably sufficient enough to prevent invasion or antigen entry. Local host factors interacting with IgA-Ag° complexes (e.g., digestive enzymes, mucus, etc.) further aid in impeding antigen entry.

Conclusion

The capability of generating immunoglobulin molecules that can bind virtually every type of polar grouping on the macromolecules and organelles of eukaryotic and prokaryotic origin is controlled in large part by multiple immunoglobulin V-region genes. Recent studies of the molecular genetics of immunoglobulins have revealed how genes can create such a diverse system of immunoglobulins through the process of V-region–J–C-region joining of chromosomal DNA, which permits the retrieval of a large library of V-genes.

The V–J–C-joining mechanism occurs in genetic loci controlling heavy and light chains. Since this process appears to involve deletion of DNA and therefore to be irreversible, cells in which one V_H and one V_L region are differentiated are committed to making immunoglobulins with only one structural binding site, hence the immunoglobulin clone. This clone can further differentiate into subclones that are specialized to produce different H-chain classes by the immunoglobulin C_H switching process.

The submolecular structures (domains) controlled by V-regions determine antigen binding sites, while the C_H genes determine constant region domains that interact with other components such as IgG FcR, complement, (Clq), and SC. Each immunoglobulin class, as defined by the C_H chain class, is associated with a specific set of effector functions.

In the internal environment, IgM and IgG molecules that bind to antigens permit the inflammatory system to focus on selected targets by binding to C-region domains. Effector actions of the inflammatory system, such as phagocytosis, cell lysis, and cytotoxic processes, are carried out by effector cells and complement but in most cases not by the antibody molecule per se.

The secretory immune system (represented by IgA molecules), which operates on the internal epithelial linings of the body, cannot utilize phagocytic- and complement-mediated effectors. Instead, IgA antibodies bind to targets and limit their entry into the internal environment.

References

Ablas, A.B., van Furth, R. (1979) Origin kinetics and characteristics of pulmonary macrophages in the normal steady state. J. Exp. Med. 149,1504.

Andersson, B., Skoglund, A-C, Rosen, A. (1979) Functional characterization of mouse T-lymphocytes with IgM-Fc receptors. I. Studies on ADCC and helper cell function. J. Immunol. 123,1936–1939.

Beale, D., Feinstein, A. (1976) Structure and function of the constant regions of immunoglobulins. Quart. Rev. Biophys. 9,135–180.

Beh, K.J., Husband, A.J., Lascelles, A.K. (1979) Intestinal response of sheep to intraperitoneal immunization. Immunology 37,389.

Berg, R.D., Garlington, A.W. (1979) Translocation of certain indigenous bacteria from the gastrointestinal tract to the mesenteric lymph nodes and other organs in a gnotobiotic mouse model. Infect. Immun. 23,403–411.

Berg, R.D., Savage, D.C. (1975) Immune responses of specific pathogen free and gnotobiotic mice to antigens of indigenous and non-indigenous microorganisms. Infect. Immun. 11,320–329.

Birbeck, M.S.C., Cartwright, P., Hall, J.G., Orlans, E., and Peppard, J. (1979) The transport by hepatocytes of immunoglobulin A from blood to bile visualized by autoradiography and electron microscopy. Immunology 37,477.

Bockman, D.E., Cooper, M.D. (1973) Pinocytosis by epithelium associated with lymphoid follicles in the bursa of fabricius; appendix and Peyer's patches. An electromicroscopic study. Am. J. Anat. 136,455–477.

Bradish, C.J., Fitzgeorge, R., Titmuss, R.F.D. Baskerville, A. (1980) The responses of nude athymic mice to nominally avirulent togavirus infections. J. of Gen. Virology 42,555–566.

Brambell, F.W. (1966) *The transmission of passive immunity from mother to young.* Neuberger, A., Tatum, E.L. (eds.) Amsterdam, North Holland Publishing Co.

Brandon, M.R. (1976) Selective transfer and catabolism of IgG in the ruminant. In: *Maternofetal Transmission of Immunoglobulins,* Hemmings, W.A. (ed.), Cambridge, Cambridge University Press, pp. 437–448.

Bullen, J.J., Rogers, H.J., Griffiths, E. (1977) Role of iron in bacterial infection. Curr. Top. Microbiol. Immunol. Vol.

Cardella, C.J., Davies, P., Allison, A.C. (1979) Immune complexes induce selective release of lysosomal hydrolases from macrophages. Nature (Lond.) 247,40.

Crabbé, P.A., Heremans, J.F. (1966) The distribution of immunoglobulin containing cells along the human gastrointestinal tract. Gastroenterology 51,305.

Craig, S.W., Cebra, J.J. (1971) Peyer's patches: an enriched source of precursors for IgA producing immunocytes in the rabbit. J. Exp. Med. 134,188.

Davies, D.R., Padlan, E.A., Segal, D.M. (1975) Immunoglobulin structures at high resolution. In: *Contemporary Topics in Molecular Immunology* (Vol. 4) Inman, F.P., Mandy, W.J. (eds.) New York, Plenum Press, pp. 127–155.

Davies, P.J., Porter, P. (1979) A mechanism for secretory IgA-mediated inhibition of the cell penetration and intracellular development of eimeria tenella. Immunology 36,471.

Daniels, C.A. (1975) Mechanisms of viral neutralization. In: *Viral Immunology and Immunopathology,* Notkins, A.L. (ed.) New York, Academic Press, pp. 79–97.

Della-Porta, A.J., Westaway, E.G. (1977) A multi-hit model for the neutralization of animal viruses (a review). J. Gen. Virol. 38,1–19.

Dorrington, K.J. (1976) Properties of the Fc receptor on macrophages and monocytes. Immunol. Comm. 5,263–280.

Early, P., Rogers, J., Davis, M., Calame, K., Bond, M., Wall, R., Hood, L. (1980a) Two mRNA's can be produced from a single immunoglobulin µ gene by alternative RNA processing pathways. Cell 20,313.

Early, P., Huang, H., Davis, M., Calame, K., Hood, L. (1980b) An immunoglobulin heavy chain variable region gene is generated from three segments of DNA: V_H, D and J_H. Cell 19,981–992.

Edwards, P.A.W. (1978) Ig mucous, a selective barrier to macromolecules. Br. Med. Bull. 34,55–56.

Embling, P.H., Evans, H., Guttierez, C., Holobrow, E.J., Johns, P., Johnson, P.M., Papamichial, M., Stanworth, D.R. (1977) The structural requirements for immunoglobulin aggregates to localize in germinal centers. Immunology, 781–786.

Florey, H.W. (1933) Observations on the functions of mucous and the early stages of bacterial invasion of the intestinal mucosa. J. Pathol. Bact. 37,283–289.

Goldstein, E., Lippert, W., Warshauer, D. (1974) Pulmonary alveolar macrophage. Defender against bacterial infection of the lung. J. Clin. Invest. 54,519.

Goldstein, I.M., Kaplan, H.B., Radin, A., Frosch, M. (1976) Independent effects of IgG and complement upon human polymorphonuclear leukocyte function. J. Immunol. 117,1292.

Griffen, J.A., Griffen, F.M. (1979) Augmentation of macrophage complement receptor in vitro. I. Characterization of the cellular interactions required for the generation of a T-lymphocyte product that enhances macrophage complement receptor function. J. Exp. Med. 150,653.

Guyer, R., Koshland, M.E., Knopf, P.M. (1976) Immunoglobulin binding by mouse intestinal epithelial receptors. J. Immunol. 117,587.

Hanson, L.A., Ahlstedt, S., Fatsth, A., Jodal, U., Kaijser, B., Larsson, P., Lundberg, U., Olling, S., Sohl-Akerlund, A., Svanborg-Eden, C. (1977) Antigens of escherichia coli human immune response and the pathogenesis of urinary tract infections. J. Infect. Dis. 136, (Suppl), S144.

Hemmings, W.A., Williams, E.W. (1976) The attachment of IgG to cell components of transporting membranes. In: *Maternofetal Transmission of Immunoglobulins*, Hemmings, W.A. (ed.), Cambridge, Cambridge University Press, 91–108.

Heremans, J.F. (1974) Immunoglobulin A. In: *The Antigens* (Vol. II), Sela, M. (ed.) New York, Academic Press.

Honjo, T., Kataoka, T. (1978) Organization of immunoglobulin heavy chain genes and allelic deletion model. Proc. Natl. Acad. Sci. U.S.A. 75,2140.

Hohmann, A., Schmidt, G., Rowley, D. (1979) Intestinal and serum antibody responses in mice after oral immunization with salmonella, escherichia coli, and salmonella-escherichia coli hybrid strains. Infect. Immun. 25,27–33.

Hildemann, W.H. (1979) Immunocompetence and allogeneic polymorphism among invertebrates. Transplantation 27,1.

Jones, J.F., Plotz, P.H., Segal, D.M. (1979) Complement and cell mediated lysis of haptenated erythrocytes sensitized with oligomers of rabbit IgG antibody. J. Mol. Immunol. in press.

Kaltreider, H.B. (1976) Expression of immune mechanisms in the lung. Am. Rev. Resp. Dis. 113,347.

Klaus, G.G.B., Humphrey, J.H. (1977) The generation of memory cells. I. The role of C3 in the generation of B memory cells. Immunology 33,31–40.

Lamm, M.E. (1973) Cellular aspects of immunoglobulin A. Adv. Immunol. 22,223.

Liu, C.P., Tucker, P.W., Mushinski, J.F., Blattner, F.R. (1980) Mapping of heavy chain genes for mouse immunoglobulins M and D. Science 209,1348.

Lloyd, S., Soulsby, E.J.L. (1978) The role of IgA immunoglobulins in the passive transfer of protection to Taenia taeniaformis in the mouse. Immunology 34,939.

Lowell, G.H., Smith, L.F., Artenstein, M.S., Nash, G.S., MacDermoth, R.P., Jr. (1979) Antibody dependent cell mediated antibacterial activity of human mononuclear cells. I. K-lymphocytes and monocytes are effective against meningococci in cooperation with human immune sera. J. Exp. Med. 150,127–137.

Mantovani, B., Rabinovitch, M., Nussenzweig, V. (1972) Phagocytosis of immune complexes by macrophages. Different roles of the macrophage receptor sites for complement (C3) and for immunoglobulin (IgG). J. Exp. Med. 135,780.

Mattioli, C.A., Tomasi, J.B. Jr. (1978) The life span of IgA plasma cells from the mouse intestine. J. Exp. Med. 138,452–460.

Max, E.E., Seidman, J.G., Leder, P. (1979) Sequences of five potential recombination sites encoded close to an immunoglobulin K constant region gene. Proc. Natl. Acad. Sci. U.S.A. 76,3450.

McConnell, I., Lachmann, P.J. (1977) Complement receptors and cell associated complement components. Immunol. Comm. 6,111.

McGhee, J.R., Mestecky, J., Babb, J.L. eds. (1978) *Secretory Immunity and Infection.* Advances in Experimental Medicine and Biology, (Vols. 107), New York, Plenum Press, pp. 903.

McLaren, D.J., Ramalgo-Pinto, F.J. (1979) Eosinophil-mediated killing of schistosomula of schistosoma mansoni *in vitro*: Synergistic effects of antibody and complement. J. Immunol. 123:1431–1438.

McNabb, T., Koh, T.Y., Dorrington, K.J., Painter, R.H. (1976) Structure and function of immunoglobulin domains. V. Binding of immunoglobulin G and fragments to placental membrane preparations. J. Immunol. 117,882–888.

Menzel, J., Jungfer, H., Gemsa, D. (1978) Contribution of immunoglobulins M and G, complement and propertin to the intracellular killing of escherichia coli by polymorphonucleus leukocytes. Infect. Immun. 19,659–666.

Merz, D.C., Scheid, A., Choppin, P.W. (1980) Importance of antibodies to the fusion glycoprotein of paramyxoviruses in the prevention of spread of infection. J. Exp. Med. 151,275–288.

Michl, J., Pieczonka, M.M., Unkeless, J.C., Silverstein, S.C. (1979) Effects of immobilized immune complexes on Fc and complement-receptor function in resident and thioglycollate-elicited mouse peritoneal macrophages. J. Exp. Med. 150,607–621.

Mitchell, G.F. (1979) Effector cells molecules and mechanisms in host protective immunity to parasites. Immunology 38,209.

Munthe-Kaas, A.C. (1977) Endocytosis studies on cultures rat Kupffer cells. In: *Kupffer cells and other liver sinusoidal cells.* Wisse, E., Knook, D.L. (eds.) Amsterdam, Elsevier/North Holland Biomedical Press, pp. 325–332.

Oberbarnscheidt, J., Kolsch, E. (1978) Direct blockade of antigen reactive B-lymphocytes by immune complexes. An "off " signal for precursors of IgM-producing cells provided by the linkage of antigen and Fc-receptors. Immunology 35,151–157.

Orlans, E., Peppard, J., Reynolds, J., Hall, J. (1978) Rapid active transport of immunoglobulin A from blood to bile. J. Exp. Med. 147,588.

Owen, R.L. (1977) Sequential uptake of horseradish peroxidase by lymphoid follicle epithelium of Peyer's patches in the normal unobstructed mouse intestine. An ultrastructural study. Gastroenterology 72,440–451.

Owen, R.L., Jones, A.L. (1974) Epithelial cell specialization within human Peyer's patches. An ultrastructural study of intestinal lymphoid follicles. Gastroenterology 66,189–203.

Owen, R.L., Nemanic, P. (1978) Antigen processing structures of the mammalian intestinal tract: An SEM study of lymphoepithelial organs. In: *Scanning Electron Microscopy* (VII). O'Hare, Illinois, SEM, Inc., AMF, pp. 367–378.

Potter, M. (1977) Antigen binding myeloma proteins of mice. Adv. Imm. 25,141.

Rowley, D., Turner, K.J. (1966) Number of molecules of antibody required to promote phagocytosis of one bacterium. Nature 210,496–498.

Reynolds, H.Y., Thompson, R.E. (1973) Pulmonary host defenses. II. Interaction of respiratory antibodies with pseudomonas aeruginosa and alveolar macrophages. J. Immunol. 111,369.

Sainte-Marie, G. (1964) Study of plasmacytopoiesis. I. Description of plasmocytes and of their mitoses in the mediastinal lymph nodes of ten week old rats. Am. J. Anat. 114,207–233.

Sakano, H., Rogers, J.H., Huppi, K., Brack, C., Traunecker, A., Maki, R., Wall, R., Tonegawa, S. (1979) Domains and the hinge region of an immunoglobulin heavy chain are encoded in separate DNA segments. Nature 277,627.

Sakano, H., Huppi, K., Heinrich, G., Tonegawa, S. (1979) Sequences at somatic recombination sites of immunoglobulin light chain genes: Their implications for the mode of recombination antibody diversity and evolution. Nature 280,288.

Schlamowitz, M. (1979) Fc specificity and properties of IgG receptors of foetal rabbit yolk sac membranes. In: *Protein Transmission through Living Membranes,* Hemmings, W.A. (ed.) Amsterdam, Elsevier-North Holland Biomedical Press.

Segal, D.M., Titus, J.A. (1978) The subclass specificity for the binding of murine myeloma proteins to macrophage and lymphocyte cell lines and to normal spleen cells. J. Immunol. 120,1395–1403.

Segal, D.M., Titus, J.A., Jones, J.F. (1979) Cell surface receptors for immunoglobulin G. In: *Physical Chemical Aspects of Cell Surface Events in Cellular Regulation.* DeLisi, C., Blumenthal, R. (eds.) New York, Elsevier-North Holland, pp. 307–323.

Shimizu, A., Takahashi, N., Yamawaki-Kataoka, Y., Nishida, Y., Kataoka, T., and Honjo, T. (1981) Ordering of mouse immunoglobulin heavy chain genes by molecular cloning. Nature (Lond.) 289, 149–153.

Silverstein, S. (1970) Macrophages and viral immunity. Sem. in Hematol. 7,185–214.

Sinclair, N.R.S.C., Chan, P.L. (1971) Regulation of the immune response. IV. Role of the Fc-fragment in feedback inhibition by antibody. And. Exp. Med. Biol. 12,609.

Springer, G.F., Williamson, P., Brandos, W.C. (1961) Blood group activity of gram-negative bacteria. J. Exp. Med. 113,1077–1093.

Strober, W., Hague, N.E., Lum, L.G., Henkart, P.A. (1978) IgA-Fc receptors on mouse lymphoid cells. J. Immunol. 121,2440.

Takeuchi, A. (1971) Penetration of intestinal epithelium by various microorganisms. Curr. Top. Pathol. 54,1–27.

Unkeless, J.C., Kaplan, G., Plutner, H., Cohn, Z.A. (1979) Fc receptor variants of a mouse macrophage line. Proc. Natl. Acad. Sci. U.S.A. 76,1400.

Van Saene, H.K.F., Van Der Waaij, D. (1979) A novel technique for detecting IgA coated potentially pathogenic microorganisms in the human intestine. J. Immunol. Meth. 30,87.

Waksman, B.H., Ozer, H. (1976) Specialized amplification elements in the immune system. Progr. Aller. 21,1–113.

Waldmann, T.A., Strober, W. (1969) Metabolism of immunoglobulins. Progr. Aller. 13,1–110.

Wild, A.E. (1976) Mechanism of protein transport across the rabbit yolk sac endoderm. In: *Maternofetal Transmission of Immunoglobulins.* Hemmings, W.A. (ed.) Cambridge, Cambridge University Press, pp. 155–165.

Weigert, M., Riblet, R. (1977) Genetic control of antibody variable regions. Cold Spring Harbor Symp. Quant. Biol. 41,837–846.

Weller, P.F., Smith, A.L., Smith, D.H., Anderson, P. (1978) Role of immunity in the clearance of bacteria due to haemophilus influenzae. J. Infect. Dis. 138,427.

White, R.G., Henderson, D.C., Eslami, M.B., Nielsenk, H. (1975) Localization of a protein antigen in the chicken spleen. Effect of various manipulative procedures on the morphogenesis of the germinal center. Immunology 28,1–21.

Lymphokines

Joost J. Oppenheim, M.D.

Lymphokines can be defined as nonimmunoglobulin secretory products of activated lymphocytes with a wide range of potent physiological effects in inflammation and immunity. In general lymphokines are glycoproteins of varying molecular weights, which are usually named descriptively for their effects. The physiological activities of lymphokines appear to fall into two broad categories of true "kines" that affect motility, or of regulatory substances that modulate the proliferation and differentiation of target cells. Lymphokines have biochemical and functional similarities to other growth factors including nerve growth and epidermal growth factors, as well as to some hormones. Lymphokines resemble hormones in that they have a multiplicity of intercellular effects and are biologically active at extremely low concentrations.

Historically, the discovery of lymphokines followed the key observation by George and Vaughan in 1962 that antigens inhibited the migration of mononuclear cells from immunized animals. It was subsequently shown that the cells that were inhibited from migrating were macrophages and that this inhibitory effect required the presence of both sensitized lymphocytes and the sensitizing antigen. The pivotal experiment, subsequently performed independently by David (1966) and by Bloom and Bennett (1966), demonstrated that antigen-stimulated lymphocytes from immune animals elaborated a factor called migration inhibitory factor (MIF), which inhibited the migration of macrophages from unsensitized animals.

The term lymphokine was first coined by Dumonde et al. (1969) to encompass nonimmunoglobulin factors produced by activated lymphocytes with biological effects on cells involved in immunological reactions. Since the discovery of the first lymphokine, many others have been described (Waksman 1978). It is not clear how many lymphokines there are, since some lymphokines have multiple biological effects. Essentially lymphokines are regulatory signals produced by lymphocytes that modulate the activities of cells participating in immune and inflammatory processes. Although lymphokines at first were thought to be responsible only for delayed hypersensitivity reactions, they are now also known to participate in a wide spectrum of inflammatory and reparative processes and are involved in both afferent and efferent limbs of both

From the National Institute of Dental Research, National Institutes of Health, Bethesda, Maryland.

cell and antibody-mediated immune responses. Lymphokines mediate immune and inflammatory processes by affecting the movement of cells as well as cell growth. They also influence some noninflammatory processes and may therefore function as extracellular signals that regulate normal cell growth and differentiation. During the past decade it has become apparent that fibroblasts, endothelial cells, and osteoclasts, as well as cells more directly involved in inflammatory reactions, can serve as targets of various lymphokines. Furthermore, factors that resemble lymphokines closely are also produced by monocytes or macrophages and a number of fibroblast cell lines. It has therefore been proposed by Cohen (1977) that this class of substances be renamed "cytokines."

There are a number of nonimmunoglobulin factors produced by thymic-derived (T) lymphocytes with enhancing and suppressive immunological effects that exhibit antigenic specificity. These substances, which probably consist of shed membrane components of T-cells that contain specific receptors for antigens, are discussed in Chapter 6. Antibodies, because of their capacity to react specifically with antigens, have on occasion been considered specific lymphokines. However, they differ from lymphokines in that they are produced in greater quantities and are readily detected in body fluids. They also have so many unusual and complex roles in inflammation and immunity that they require coverage in separate chapters. Factors produced by monocytes and macrophages, which are sometimes termed "monokines," have important roles in inflammation and immunity that are considered in Chapter 5.

General Characteristics of Lymphokines

Conditions for Production, Inducers, and Sources

Lymphokines (LK) have a number of characteristics in common (Cohen et al. 1979). Before considering some of the more important LK in detail, we will consider those properties, mechanisms of action, cell sources, stimulants, and rate of production that LK have in common (Table 1). Peripheral immunocompetent lymphocytes from the blood, lymph nodes, spleen, or thoracic duct can all be stimulated successfully to produce LK, but thymic lymphocytes produce little or no LK activity. Lymphokines are soluble substances produced by cultured lymphoid cells, which are usually detectable in the culture supernatants within 6 to 8 hours following activation and generally reach maximum levels by 24 to 72 hours of incubation. In general, more LK activity is produced in serum-containing than by serum-free lymphocyte cultures. However, LK produced in the absence of serum are easier to purify and characterize. The same culture conditions that result in optimal lymphocyte proliferation will usually also yield LK production. However, maximum LK production is usually obtained at lymphocyte concentrations of 5 to 10 $\times$ 10^6/ml, which are two- to five-fold higher than those yielding optimal lymphocyte proliferation. Some activities such as chemotactic factors are generated by doses of stimulants that are too low to stimulate lymphocyte proliferation and therefore provide more sensitive indicators of lymphocyte activation. In contrast, induction of other LK such as lymphotoxin requires higher doses of stimulants than needed to induce proliferation and are therefore less sensitive indicators.

Lymphokines are not stored but are actively synthesized and secreted by lymphocytes that have been stimulated by agents interacting with lymphocyte membrane receptors. Extraction of resting lymphocytes fails to yield significant LK activity, and treatment

Table 1. Characteristics of Lymphokines

Spontaneously produced by lymphoid cell lines
Produced by T- and B-lymphocytes in response to
    Nonspecific mitogenic stimulants
    Specific antigenic stimulants (if previously sensitized)
Actively synthesized and secreted
Usually glycoproteins with molecular weights varying from 10,000–200,000
High specific activity: great biological potency at low concentrations
Immunological effects of factors:
    Nonspecific: not genetically restricted and independent of sensitizing antigens
    Specific: genetically restricted and limited to the inducing antigen
React directly with wide variety of target cells
    Inflammatory cells: leukocytes and lymphocytes
    Noninflammatory cells: fibroblasts, osteoclasts, and endothelial cells
Inhibitable by antisera
Are active *in vivo* as well as *in vitro*
Biochemically and functionally similar factors are also produced by monocytes and fibroblast
    cell lines

of activated lymphocytes by actinomycin D or cycloheximide inhibits the appearance of extracellular LK activity.

Agents that can stimulate LK production include antigens that specifically stimulate only lymphocytes from sensitized donors and mitogens that nonspecifically stimulate lymphocytes independently of any prior sensitization. Under the appropriate conditions, both T- and B-lymphocytes can be activated by appropriate agents to produce LK. T-lymphocyte can be activated by lectins, such as phytohemagglutinin or concanavalin A or by mitogenic agents, such as sodium periodate or antilymphocyte serum, to produce LK. T-lymphocytes from immunized donors also produce LK when stimulated by sensitizing antigens. The activation of T-lymphocytes to produce LK requires the presence of a few viable macrophages to present the antigen. In contrast, B-lymphocytes are activated to produce lymphokines by B-cell mitogens such as endotoxin lipopolysaccharide, antiimmunoglobulin antisera, or polyvalent long chain carbohydrate molecules such as polymerized flagellin. This process of B-cell activation is thought to occur independently of macrophages. Nevertheless, the LK produced by B- and T-lymphocytes are apparently identical. However B- unlike T-lymphocytes are not capable of producing mitogenic amplification factors. Lymphokines are also spontaneously produced in continuous cultures by some B- and T-lymphoid lines as well as some nonlymphoid cell lines.

Biochemical and Immunological Properties

All LK are proteins (as shown by their protease sensitivity) that possess molecular weights varying from 10,000 up to 200,000. They are not immunoglobulins and do not exhibit any homology with immunoglobulin molecules. At least one LK exhibits enzymatic activity (see following section on leukocyte inhibitory factor). To date, all attempts to purify and sequence LK have been frustrated by the extremely low yield of factors obtained from even large volumes of supernatants. These LK are extremely active molecules and are estimated to be effective at picogram levels.

Most LK, even if antigen induced, act in an immunologically nonspecific manner. They neither bind nor exhibit specificity for the inducing antigens. The nonspecific factors are genetically unrestricted and do not react with inducing antigens, antihistocompatibility antigens (including the immune response gene product, Ia), antiimmunoglobulins, or antiidiotypic antibodies.

Effects of Lymphokines on Target Cells

Lymphokines affect a wide variety of cells including lymphocytes, macrophages, neutrophils, basophils, eosinophils, thymocytes, bone marrow cells, platelets, fibroblasts, osteoclasts, and endothelial cells. The nature of the interaction of LK with target cells is unclear. Most of the LK are not readily absorbed by the target cell. They may bind only weakly or, alternatively, interact with a cytoplasmic or nuclear rather than membrane site.

In Vivo Activity

Several lines of evidence favor an *in vivo* role of LK. First, various LK have been isolated from tissues or body fluids from sites with an on-going immunologically based inflammatory reaction. Lymphokines, particularly MIF, interferon, colony stimulating factor (CSF), and chemotactic factors have been demonstrated in serum, in lymph from efferent lymphatics, in fluid from peritoneal, middle ear, joint, and pleural cavities during delayed hypersensitivity reactions, in tissues undergoing graft-versus-host reactions, and in schistosome granulomas. Second, injection of LK containing culture supernatants or chromatography fractions will induce inflammatory skin reactions. This biological activity has been called skin reactive factor (SRF). Another factor causes the disappearance of macrophages from the peritoneal cavity in the so-called macrophage disappearance reaction (MDR). It is believed that this LK increases the adherence of macrophages to the mesentery and may be identical to MIF. Injection of other factors has resulted in demyelinization, germinal center proliferation, augmented antibody production, and promotion of cell-mediated immunity (CMI). Third, anti-LK antibodies have been obtained that can block the *in vitro* biological effects of lymphokines, but also can inhibit the appropriate *in vivo* biological manifestations of the lymphokines. For example, an antiserum with inhibitory effects on *in vitro* assays for MIF and monocyte chemotaxis, when injected locally, will inhibit antigen-induced cutaneous delayed hypersensitivity reactions.

The participation of LK has been demonstrated or inferred from *in vitro* studies in virtually all components of immune and inflammatory responses. These multitudinous biological activities are most readily considered by discussion of a number of the more well-characterized factors. The chapter on wound healing (see Chapter 19) considers the fibroblast growth factors, and interferon will be covered in depth elsewhere (see Chapter 11).

Classification of Lymphokines

In some instances, one factor may act on several different target cells and may therefore have widely differing biological effects involving both afferent and efferent immune functions. However, many of the activities of LK have not been characterized adequately, so that their precise biochemical identity remains unclear. The factors that affect nonlymphocytic leukocytes and other target cells have been the most thoroughly studied. Generally they function to recruit and activate target cells and thus augment inflammatory host reactions to invasive organisms, foreign bodies, allografted tissues, immunostimulating neoplasias, and to organ-specific antigens in autoimmune reactions. Therefore, for purposes of presentation, the factors will be grouped on the basis of their biological effects. They predominantly affect cells engaged in either a regulatory role in the afferent or efferent arm of the immune response or they act on target cells not directly involved in immunity or inflammation. The immunoregulatory factors can have either augmenting or suppressive effects on immune responses.

Nonspecific Effector Lymphokines

Macrophages Migration Inhibitory Factor and Macrophage Activating Factor. Although the biological activities of MIF and macrophage activating factor (MAF) are distinct, they have not been separated by physicochemical means[1]. Both MIF and MAF are produced by Lyt-1 cells, B-lymphocytes, L_2C leukemia cells, and growing fibroblast cell lines. The effect of MIF on macrophages is detected earlier than the effects of MAF. Migration inhibitory factor activity is usually detected by its ability to inhibit the migration of macrophages or monocytes out of the ends of a capillary tube. The area of cell migration out of the tubes is measured with a planimeter and compared with the area of migration of cells incubated without MIF. The inhibition of migration, which is due in part to increased macrophage adherence, is usually assayed after 24 hours of exposure of macrophages to the LK. In contrast, the assays of macrophage activation such as increased phagocytosis, oxidation of ^{14}C glucose-1-C, increased ^{14}C glucosamine incorporation, bacteriostasis, and tumoricidal effects are usually measured after 72 hours of incubation of macrophages with MAF. The capacity of MAF to induce macrophages to become tumoricidal is attributed to its unique capacity to induce macrophages to produce peroxides. Tumor killing is assayed by the usual radioisotope release methods.

---

[1] See Table 2 for detailed list of nonspecific effector lymphokines.

Table 2. Nature of Nonspecific Efferent Lymphokines

Migration inhibitory factor (MIF) or macrophage activation factor (MAF)	15,000–70,000 molecular weight, inhibitable by α-L fucose	Inhibits macrophage migration Promotes polymerization of tubulin Activates macrophages metabolically to become bactericidal and tumoricidal
Leukocyte inhibitory factor (LIF)	68,000 molecular weight: has esterase activity	Inhibits neutrophil mobility
Chemotactic factors (CTX)	12,000 molecular weight	Attract leukocytes
Lymphotoxin (LT)	10,000–200,000 molecular weight	Cytostatic and cytolytic for nonleukocyte target cells
Vascular permeability factor (VPF)	12,000 molecular weight	Increases extravascular edema formation
Osteoclast activating factor (OAF)	17,000 molecular weight	Increases bone-resorbing activity of osteoclasts

Even though MIF has been characterized more thoroughly than any of the other LK there are considerable species differences and multiple active forms of MIF. MIF and MAF are heterogeneous (molecular weight: 12,000 to 70,000) heat stable glycopeptides that increase the adhesiveness of macrophages, inhibit their migration, and subsequently activate macrophages metabolically as well as immunologically (see Chapter 5). Remold and Mednis (1979) reported guinea pig MIF to be heat stable at 56°C for 30 minutes but labile at 80°C for 30 minutes, sensitive to chymotrypsin and neuraminidase, to elute in the molecular weight range of 25,000 to 65,000 from Sephadex G100 columns, and to migrate anodal to albumin on disk electrophoresis. Guinea pig MIF activity exhibited isoelectric point (pI) values of 5.0 and 3.0. The pI 5.0 material possessed a molecular weight of 20,000 and 65,000 whereas the pI 3.0 species had a molecular weight of 38,000 and was rich in carbohydrate.

Sorg and Bloom (1973) successfully labeled newly synthesized guinea pig LK in cultures of mitogen-activated lymphocyte cultures using a double-labeling technique. These labeled LK-containing secreted proteins were reacted with antibodies to MIF and subjected to sodium dodecyl sulfate polyacrylamide gel electrophoresis (SDS-PAGE). Three peaks of radioactivity were obtained which are perhaps oligomers of a common subunit with a molecular weight of 15,000 (Sorge and Geczy 1976). Furthermore, preliminary evidence that human but not guinea pig MIF may actually consist of a small dialyzable fragment (molecular weight: 5,000 to 10,000), which is inactive until mixed with a larger nondialyzable component. It has been proposed that one subunit interacts with the receptor of the target cell, whereas the other is the biologically active moiety (Cohen et al. 1980).

Other physicochemical studies indicate that MIFs have similar isoelectric points of 5.2, but different molecular weights of 30,000, 50,000 and 70,000 on SDS gel electrophoresis. Although human and guinea pig MIF are quite similar, they differ from the chromatographic properties of murine MIF, which has four molecular weights in the range of 14,000 to 55,000 but possess the same pI 5.0. Therefore, MIF shows considerable species differences and is only partially active across species lines.

The mechanisms of action of MIF and MAF have been investigated intensively. Both the effects of MIF on migration and MAF on macrophage activity are ATP dependent but independent of oxidative phosphorylation. However, there are some differences in the metabolic requirements for MIF and MAF activities. In contrast to MAF, the effect of MIF is not blocked by cycloheximide or actinomycin-D and thus appears to act independent of protein or RNA synthesis.

Several reports indicate that MIF binds to macrophage membrane receptors and can be absorbed selectively by macrophages. Trypsinization of macrophages temporarily (12 to 14 hours) blocks them from responding to MIF. The observation that α-L-fucose and rhamnose interfere competitively with the effect of MIF but not with leukocyte inhibitory factor (LIF) on their respective target cells suggests that α-L-fucose may be an essential component of the guinea pig macrophage receptor for MIF. Moreover, treating of macrophages with the Fucose-binding lectin *Lotus tetragonolobus* or with L-fucosidase blocks their responses to MIF. Guinea pig macrophage glycolipids that contain L-fucose increase the response of macrophages to subthreshold concentrations of MIF, whereas other glycolipids do not. This effect is blocked by treating the glycolipids with L-fucosidase, which suggests that fucose-containing gangliosides may be components of the macrophage receptor for MIF. Addition of L-fucose or L-rhamnose to LK-containing supernatants blocks the induction of the intraperitoneal macrophage disappearance reaction (MDR) and cutaneous inflammatory responses to the MIF-containing supernatants (Baba et al. 1979).

The MIF-induced inhibition of migration of macrophages does not occur simply because of an increase in their membrane adhesiveness. In fact, an LK that promotes the aggregation of macrophages with a molecular weight of 150,000 to 200,000 and is separable from MIF has been described (Godfrey and Geczy 1978). The intracellular effects of MIF on macrophages include increased polymerization of tubulin, which increases cellular rigidity and decreases mobility. Deuterium oxide (D_2O), by stabilizing microtubules, enhances MIF activity. Cytochalasin B also emulates the effect of MIF by reversibly depolymerizing the microfilaments, which in turn inhibits cell mobility and increases the surface adherence of macrophages. Conversely, colchicine and vinblastine by disrupting microtubules enhance the random mobility of macrophages and counteract the effect of MIF. MIF also induces an influx of calcium into macrophages promoting the polarization of the membrane and decreases membrane resistance.

The role of cyclic nucleotides in macrophage mobility is complex. It has been shown that cyclic AMP, or agents that increase the intracellular level of cyclic AMP, inhibits macrophage migration. Paradoxically, pretreatment of macrophages with isoproterenol and theophylline, which elevate intracellular cyclic AMP, blocks the effect of MIF on macrophages. Thus, elevation of intracellular cyclic AMP interferes with the action of MIF. Conversely, pretreatment of macrophages with MIF lowers their cyclic AMP content and makes them refractory to the cyclic AMP elevating effects of agents that stimulate adenyl cyclase such as PGE_2 and isoproterenol. This refractory state can be blocked by colchicine but is enhanced by D_2O. Thus, this refractory state is associated with decreased levels of cyclic AMP and may be secondary to the hyperpolymerized state of the microtubules of macrophages exposed to MIF. The polymerized state would limit receptor mobility in the plane of the membrane and inhibit the activation of adenyl cyclase. Whether cyclic GMP contributes to the action of MIF remains to be established.

Macrophage activation factor stimulates macrophage bactericidal and tumoricidal functions and thus promotes host defense mechanisms directly (Figure 1). L-fucose and L-fucosidase, which reduce the tumoricidal capacity of macrophages, can also block MAF-induced guinea pig macrophage cytotoxicity, suggesting that the receptors for MIF and MAF are similar. Macrophages treated with inhibitors of surface esterases acquire enhanced reactivity for both MIF and MAF, suggesting that surface enzymes normally degrade these LK. Despite the evident similarity between MIF and MAF, there is a preliminary report suggesting they may be separable. Supernatants of concanavalin-A-stimulated mouse spleen cells, which induce macrophages to become cytotoxic, when sequentially chromatographed on hydrophobic phenylalanine-Sepharose, phenyl Sepharose and Ultragel yield MAF activity at a molecular weight of 20,000, which is resistant to neuraminidase and has isoelectric points of 7.6 and 8.6 on isoelectric focusing (IEF). This factor therefore may be distinct from neuraminidase-sensitive MIF. However, this activity may not be MAF but related to other mediators that activate macrophages such as CSF, SIRS, or the recently described heat-labile LK that activates macrophages to produce components of complement, monocyte complement stimulator (Littman and Ruddy 1979). Thus, there are a number of cytokines that activate macrophages, only one of which also has MIF activity.

Leukocyte Inhibitory Factor (LIF). Leukocyte inhibitory factor (LIF) activity was initially described and differentiated from MIF in 1967. Leukocyte inhibitory factor is heat stable (56° for 30′) with a molecular weight of 68,000, which inhibits

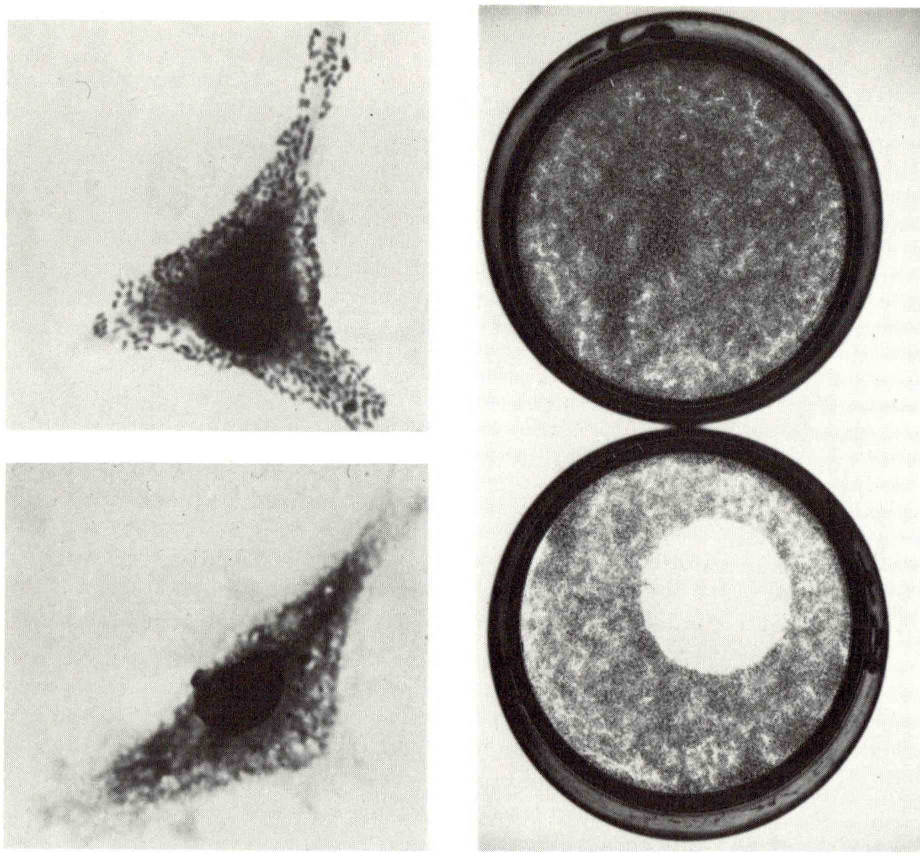

Figure 1. Demonstration of enhanced bacteriocidal and tumoricidal activities of murine macrophages incubated with MAF produced by lectin- or antigen-activated mouse lymphocytes. The left upper panel shows that exposure of macrophages to rickettsia results in phagocytic uptake of numerous organisms. The left lower panel shows that macrophages exposed to MAF destroy the intracellular rickettsia. The right upper panel shows the effect of control macrophages on a tumor cell monolayer whereas activated macrophages in the bottom dish have a lytic effect. (Courtesy of Monte Meltzer, NCI, NIH.)

selectively the random migration of polymorphonuclear leukocytes (PMN). Leukocyte inhibitory factor also enhances PMNL adherence, phagocytosis, metabolism, and electrophoretic mobility. Assay of LIF activity can be accomplished by measuring its effect on migration of peripheral blood leukocytes or polymorphonuclear neutrophils (PMN) out of capillary tubes. Usually a more sensitive agarose assay is used, and the inhibitory effect of LIF on the migration of PMN from a well into the surrounding agarose is determined. As with MIF, both T- and B-lymphocytes produce LIF, and antigen-induced LIF production is a correlate of delayed hypersensitivity. Leukocyte inhibitory factor is unique among the LK because of its documented esterolytic activity (Rocklin 1978). Unlike MIF, LIF activity is blocked by the serine esterase inhibitors diisopropylphosphofluoridate (DFP), eserine, and phenylmethylsulfonyl fluoride. The inactivation of LIF activity by DFP follows standard enzyme-substrate kinetics. Leukocyte inhibitory factor is activity chromatographic with and is not separable from an activity that hydrolyses a synthetic amino acid derivative, ^{3}H benzoyl arginine ethyl ester (^{3}HBAEE). Conversely ^{3}HBAEE protects LIF from inactivation by DFP. Also, ^{3}HBAEE blocks the biological activity of LIF on PMNL. Thus, ^{3}HBAEE can serve as a synthetic substrate, and its hydrolysis may provide an easy assay for LIF.

Chemotactic Factors. Macrophages and PMN are particularly important as phagocytic wandering cells that eliminate invasive organisms and promote wound healing by eliminating tissue debris, fibrin deposits, and degrading antigens. These processes are crucial in maintaining host integrity and depend upon the rapid accumulation of leukocytes at local inflammatory sites. There are innumerable substances in nature that will attract the various types of leukocytes to migrate directionally along a concentration gradient. Leukocytes are attracted by chemotactic factors produced by foreign organisms, tissue breakdown, or immunological reactions. The localization of leukocytes at sites of immune reactions follows the binding of antigen by antibody, which generates the chemotactic cleavage product of the fifth component of complement, C_5a. Alternatively, activation of either T- or B-lymphocytes by mitogen or an antigen can result in the production of LK with chemoattractant properties. Lymphocyte-derived chemotactic factors (LDCF) in turn can attract all the other nonlymphocytic types of leukocytes. Lymphocyte-derived chemotactic factor was described by Ward and coworkers shortly after MIF in 1969. Activity of LDCF is best assayed using a Boyden chamber that has two chambers separated by a filter. A new multichamber device has been developed as well (Figure 2). The target cells in the upper chamber migrate through the filter pores in the direction of the chemoattractant LK present in the lower chamber. The number of cells penetrating the filter relative to the control number during a defined period is determined (Figure 3). Lymphocytes can be stimulated to produce chemotactic factors by doses of stimulants that are too low to induce lymphocyte proliferation. This assay therefore provides a more sensitive indicator of lymphocyte activation. There are specific chemotactic LK for each type of granulocyte, e.g., neutrophils, basophils, eosinophiles, and monocytes. The chemotactic factor that is selective for human and guinea pig monocytes has a molecular weight of 12,500 and is a heat-stable protein that is destroyed by proteolytic enzymes but is resistant to neuraminidase. In contrast, the chemotactic factor for mouse macrophages has a molecular weight of 38,000 and a lower isoelectric point on diethyl-amino-ethyl (DEAE) cellulose than does the human factor.

The mechanism of action of chemotactic factors on PMN is discussed in detail in Chapter 3. The chemotactic factor for monocytes, even though separable from MIF, is also inhibited competitively by α-L-fucose and α-L-rhamnose. Chemoattractants therefore presumably interact with cell membranes and may enhance the influx or redistribution of divalent cations in the cytoplasm. This, in turn, may control the contractility of the microfilaments, which influences cellular locomotion.

Factors have been shown recently to be elaborated by tumor cells that interfere with chemotaxis. Tumors elaborate several nondialyzable as well as dialyzable peptides capable of suppressing directional macrophage mobility. Tumors may thereby evade host defense mechanisms by interfering with macrophage accumulation and tumoricidal functions.

Lymphotoxin (LT). Lymphotoxin (LT) is a product of highly activated mammalian T- and B-lymphocytes that lyses susceptible target cells such as 3T3 and L929 murine fibroblasts. The necrotic ulceration seen with the more active cell-mediated inflammatory reactions has been attributed to LT. However, even though LT has been

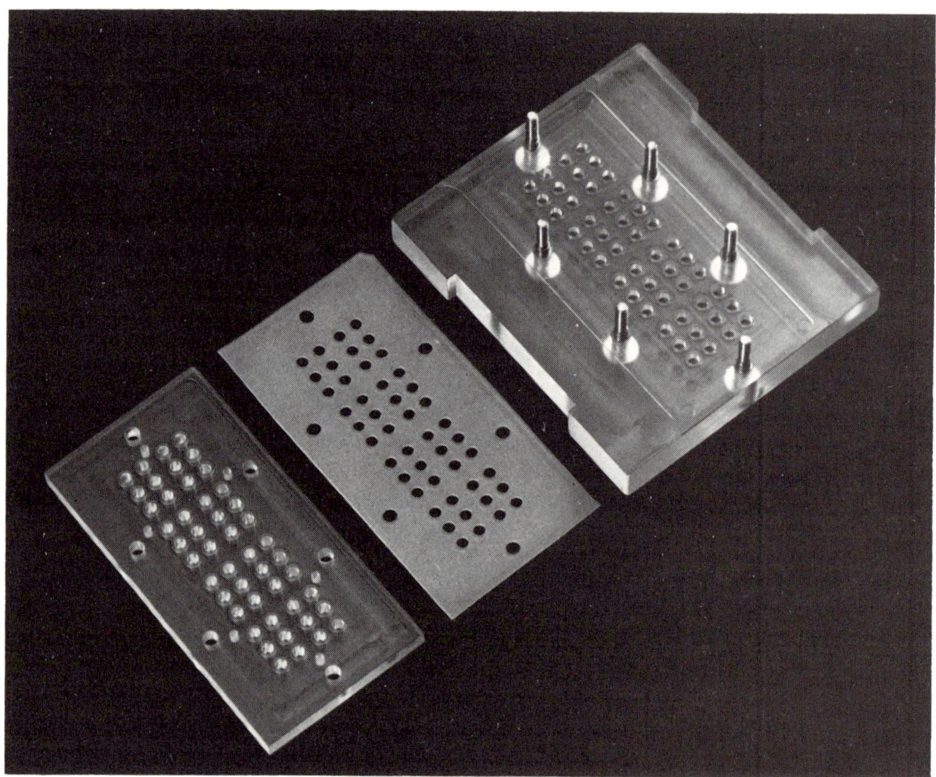

Figure 2. New multichamber device for chemotactic assays. View of the multiwell chemotaxis chamber. Twenty-five microliters of a solution containing a chemoattractant are pipetted into the wells in the base plate. A single rectangular filter membrane (Nucleopore) is then placed over the set of wells. This is followed by the silicone gasket and the top plate, which is then screwed down. Into the upper wells are pipetted 30 μl of responding cell suspensions. After incubation at 37°C, the chamber is disassembled, the filter is removed, cells are wiped off the upper side of the filter and the filter is air dried and stained with Wright's stain (or DifQuick). (Courtesy of W. Falk and E. J. Leonard, NCI, NIH.)

intensively studied since its discovery in 1968, it has not yet been recovered from any *in vivo* inflammatory sites. Several investigators have postulated that the nonspecific LT activity is the effector molecule for specific cell-mediated cytotoxic reactions, but this assertion remains very controversial.

The presence of LT activity can be assayed by inhibition of isotope uptake, release from target cells, or simply by direct counting of target cells (Figure 4). There is wide variation in the degree of sensitivity of various cells lines to the cytotoxic effects of LT. Only some of the L929 lines are sensitive, whereas most lymphoblast cell lines and primary fibroblast lines are quite resistant. Treatment of target cells with mitomycin C or actinomycin D can be used to increase their susceptibility to LT. The steps in LT–target cell interaction involve rapid binding to a trypsin-sensitive cell surface receptor. This binding appears to be irreversible and consumes LT. Lymphotoxin-mediated cytolysis can occur in 3 to 5 minutes or be associated with slower cell swelling for 1 to 2 hours. These events are associated with increased Ca^{2+} flux into the target cells.

The activities of LT are associated with a family of protein molecules that are trypsin

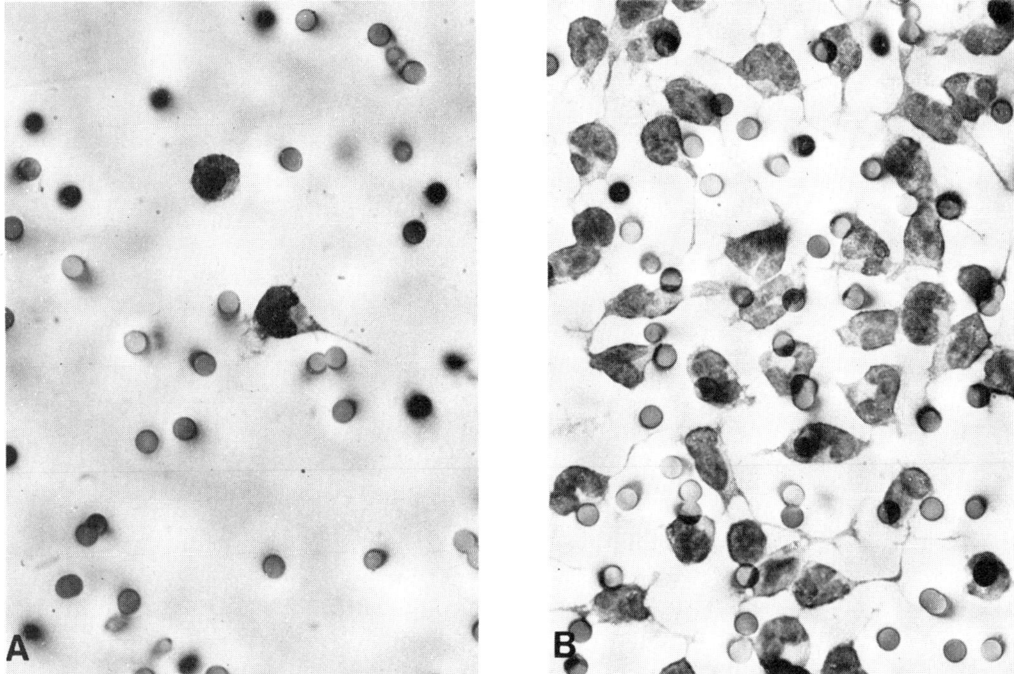

Figure 3. Monocyte chemotactic response to A supernatant of unstimulated lymphocyte culture B. Supernatant of antigen stimulated culture of sensitized lymphocytes. (Courtesy of Sharon Wahl, NIDR,NIH.)

sensitive but inactivated by heating to 85°C for 15 minutes. Human LT exists as four physicochemically related but heterogenous forms. These include: α_H (molecular weight: 110,000 to 140,000); α_L (60,000 to 90,000); β (40,000 to 50,000); and γ LT (10,000 to 20,000). Antisera against the unfractionated supernatant or the α_H form of LT react with the three other types of murine LT. In fact, it is easier to obtain

Figure 4. Effect of lymphotoxin on the growth and survival of 7493 murine fibrosarcoma cells. 15,000 fibrosarcoma cells were cultured for 72 hours in **(a)** medium or **(d)** medium with 1:50 dilution of PHA-stimulated spleen cell supernatant. (Magnification 100 ×, courtesy of Robert Aksamit, NCI, NIH.)

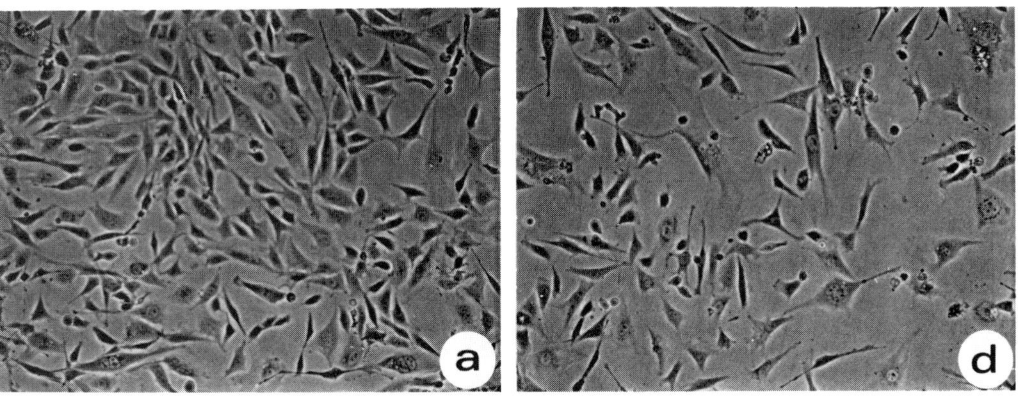

antibodies to LT than to any of the other lymphokines. In addition to these forms, a high molecular weight ($>$200,000) unstable form with very high specific activity has been observed and has been reported to be involved as an effector molecule of cell-mediated cytotoxicity. This form of LT may contain a complex consisting of T-cell-derived antigen-binding receptor, gene products of the MHC locus, and the more stable α_H (molecular weight: 110,000 to 140,000) subunits of LT. The specific target cells can selectively absorb and remove this lytic activity from supernatants. Antibodies made against the unstable LT-containing complex will not only block LT activity but can also inhibit cell-mediated cytotoxic reactions. However, the anti-α_H, anti-β or anti-γ LT did not block cell-mediated cytotoxic reactions. Therefore, antibodies to the antigen-binding component rather than to the LT in the large complex form of LT may be responsible for inhibiting cell-mediated cytotoxic reactions. Only lower molecular weight moieties with LT activity have been described in guinea pigs and mice.

Several cytostatic or growth inhibitory LK have been reported and termed cloning inhibitory factor (CIF) and proliferation inhibitory factor (PIF). However, higher concentrations of these cytostatic LK are cytotoxic and lower concentrations of LT are cytostatic. Furthermore, PIF, CIF, and LT sequentially cochromatograph on DEAE cellulose, Sephadex G100, and hydroxylapatite, suggesting they are identical.

Skin Reactive and Vascular Permeability Factors. The intradermal injection of supernatants of activated lymphocytes results in erythema and induration as well as perivascular mononuclear infiltration, fibrin deposition, and increased capillary–venule permeability, beginning after 1 hour and peaking by 24 hours. In contrast with other LK, mediators that affect vascular permeability are identified using *in vivo* bioassays. The active supernatants are said to contain skin reactive factor (SRF). The SRF activity has not been successfully isolated and is probably due to a mixture of LK in the supernatants, especially chemotactic factors for macrophages, MIF, and VPF. The intradermal injection of chemotactic factor also induces a delayed hypersensitivity type of skin reaction. Activity of SRF can be blocked by specific inhibitors of kinin generation and is enhanced by certain inhibitors of kininase as well as serum or plasma. These findings suggest the presence of a kinin generator in SRF. This may be attributable to a vascular permeability factor (VPF) with a molecular weight of 12,000 isolated from SRF-containing supernatants, which can also be inhibited by agents that block the generation of kinin.

In addition, lymphocytes can be stimulated to produce a platelet-aggregating factor (PAF), which is antagonized by heparin. This activity has the potential to decrease clotting time, accelerate clot retraction, and coagulate fibrinogen and thus can account for the fibrin deposits seen in chronic inflammatory reactions. Finally, there is a lymphocyte-derived activity that initiates angiogenesis *in vivo*. It may be responsible for the endothelial proliferation observed in delayed hypersensitivity reactions.

Factors Regulating Connective Tissues: Osteoclast Activating Factor and Others. In 1972, Horton et al. discovered that T- and B-lymphocytes produce an LK with a molecular weight of 17,000 that stimulates osteoclasts to resorb bone. This osteoclast activating factor (OAF) has also been detected in the supernatants of long term B-lymphocyte lines. In addition, OAF has been extracted from biopsy specimens of plasmacytomas, implicating its involvement in bone resorption by neoplasms that cause pathological fractures.

Bone resorption is quantitated using organ cultures of fetal rat bone. The fetal rat bone can be radiolabeled with ^{45}Ca. Incubation of bone cultures in OAF-containing supernatants for 3 to 6 days results in an increase in the number and size of multinucleated osteoclasts in the bone. Osteoclasts presumably contain degradative enzymes that resorb bone because their appearance is associated with a concomitant loss of bone matrix and greater release of ^{45}Ca by the fetal bone. Since osteoclasts and macrophages may have a common precursor cell and some investigators find that macrophages also resorb bone, OAF may be related to other LK that affect macrophages. However, OAF has not been shown to have MIF activity. The role of OAF in normal bone development remains to be established, but there is preliminary evidence that osteopetrosis may develop in an animal model that has a genetic defect resulting in dysfunctional macrophages and osteoclasts.

Human OAF production by phytohemagglutinin-stimulated lymphocytes is monocyte dependent. Indomethacin, which blocks prostaglandin synthesis, also blocks OAF production by leukocytes. Addition of prostaglandin E_1 or E_2 or monocyte-conditioned media can reverse the inhibition of OAF production by indomethacin. This suggests that monocytes can mediate OAF production by synthesizing prostaglandins (Yoneda and Mundy 1979). Osteoclast-activating factor has a molecular weight of 13,000 and is a polypeptide that has been purified using DEAE cellulose and polyacrylamide gel electrophoresis. Hybridomas have been produced that make monoclonal antibodies against human OAF (Luben et al. 1979). These antibodies do not react with other bone resorbing agents such as parathormone and vitamin D. Using these antibodies, a radioimmunoassay has been developed that detects nanogram levels of OAF in the sera of some patients with multiple myeloma.

There are several other LK that regulate connective tissue metabolism that have not as yet been well characterized. Lymphokines are involved in regulating fibroblast functions. Antigen-stimulated guinea pig lymphocytes produce a fibroblast activating factor (FAF) that induces resting fibroblasts to proliferate, synthesize proteins including collagen, secrete prostaglandins, and elevate their intracellular cyclic AMP (Wahl et al. 1978). This fibroblast activation may contribute to fibrotic reactions seen in chronic inflammatory lesions as well as in normal wound-healing processes.

Another activity that has been observed in culture fluids of mitogen-activated mononuclear cells appears to degrade cartilage matrix. It is assessed by its capacity to release proteoglycan from heterologous cartilage substrates *in vitro*. A similar activity has been detected in synovial fluid of |inflamed joints, suggesting this factor may be involved in the pathogenesis of inflammatory articular disease.

Thus, a wide variety of mediators exist that participate at various stages of immunologically induced inflammatory responses involving a wide variety of distinct target tissues.

Nonspecific Stimulating Lymphokines

Colony Stimulating Factors. Colony stimulating factor (CSF) signals precursor stem cells to proliferate and leads to the production of more differentiated and functional inflammatory cells.[2] In fact, CSF can determine the route of differentiation of multipotential precursor cells. Therefore, CSF can be classified as the first of the afferent

---

[2]See Table 3 for detailed list of nonspecific stimulating lymphokines.

signals that increase the number of host effector cells capable of engaging in inflammatory reactions. Different CSF activities promote the growth of colonies on soft agar consisting of mixtures of granulocyte and macrophages (GM-CSF), of more differentiated macrophages only (CSF-1), eosinophils (EO-CSF), erythroid cells (E-CSF), and megakaryocyte percursor cells (MEG-CSF), which produce more red blood cells and platelets, respectively. The best characterized CSF–LK is GM–CSF, which induces bone marrow cells to develop mixed colonies *in vitro* in 7 days containing 5 to 10^3 granulocytes and or monocytes. Lower concentrations of GM–CSF favor macrophage, whereas higher concentrations promote granulocyte production.

Although GM–CSF is produced by stimulated T- and B-lymphocytes and macrophages, it is also produced by cells that presumably do not initiate inflammatory reactions, e.g., placental and fetal tissues, endothelial cells, fibroblast, hybridoma, and melanoma cell lines. Also, GM–CSF is present in serum and urine and can be extracted from all bodily organs. The production of CSF is modulated by exposure of macrophages to endotoxin, purified protein derivative (PPD), zymosan, glucan, or polyinosine:cytidine (polyI.C.). This suggests that granulocytopoiesis and monocytopoiesis may be positively regulated by the products of the responding cells themselves (Cline and Golde 1979). Negative feedback control for CSF production may be mediated by chalones, which are tissue but not species-specific products of differentiated cells that inhibit precursor cells of the same lineage. Alternatively, factors from mature cells may inhibit CSF production rather than inhibit its effects on precursor cells. Finally, prostaglandins of the E series produced by activated monocytes can inhibit colony growth and can in turn be overcome by indomethacin.

T-lymphocytes when stimulated by concanavalin A or phytohemagglutinin produce GM–CSF. However, pokeweed mitogen-stimulated lymphocytes in addition to GM-CSF also produce EO–CSF, MEG–CSF, and E–CSF. Still, GM–CSF is the predominant activity in pokeweed-mitogen-stimulated lymphocyte cultures, since a 10–fold lower concentration of this supernatant loses its EO–CSF, MEG–CSF, and E–CSF activities but retains its GM–CSF activity. All four CSF activities chromatograph at a molecular weight of about 23,000. They all bind to concanavalin A-Sepharose and are eluted with 0.1 M alpha-methyl-glycopyranoside suggesting that they are all glycoproteins. The activities, however, exhibit charge heterogeneity and can be separated by IEF. Antiserum against mouse lung GM–CSF exhibits greater inhibitory activity for spleen cell-derived GM–CSF than for EO–CSF, which further supports the view that these are distinct factors.

The CSF-1 stimulates more mature macrophages. The more mature the macrophages, the smaller the resultant colonies. The CSF-1 (molecular weight: 70,000) can be extracted from human urine or supernatants of mouse L-cells. Iodine-125-labeled CSF-1 binds to receptors on macrophages, whereas GM–CSF binds best to bone marrow precursor cells. They can be distinguished by radioreceptor assays or by radioimmunoassays since anti-CSF-1 and anti-GM–CSF are only minimally cross-reactive. After binding, the CSF-1 and GM–CSF are internalized and degraded.

The primary control of erythropoiesis is mediated by erythropoietin produced by the kidneys and liver in conjunction with T-lymphocyte-derived E–CSF. The more primitive erythroid stem cell responds to E–CSF, but not erythropoietin. It gives rise to the precursor for normoblasts, which becomes sensitive to erythropoietin. Thus, conditioned media from a T-cell line can potentiate the effects of erythropoietin.

An *in vivo* role for GM–CSF is suggested by the observations that (1) injections of endotoxin or bacterial antigens induce both generalized lymphopoiesis and a 50- to

Table 3. Nature of Nonspecific Regulatory Lymphokines

Factors	Biochemical characteristics	Biological activities
Stimulating		
Colony-stimulating factors (CSF)	20,000–70,000 molecular weight, exhibits charge heterogeneity	Stimulates precursor stem cells for leukocytes, platelets, and erythrocyte; stimulates macrophage growth, replication, and function
Murine amplification factors	30,000 molecular weight	Mitogenic for murine thymocytes
(IL-2 with LMF, TCGF, KHF, and T-helper activity)	15,000–60,000 molecular weight (human) Inhibitable by IBF	Promotes antibody production Generates cytotoxic T-cells
Allogeneic effect factor (AEF)	45,000 molecular weight Ia positive and contains B_2 microglobulin	Promotes antibody production Associated with KHF activity
TRF	30,000–40,000 molecular weight protein	Promotes B-cell antibody production
Suppressive		
Interferon (IF)	25,000–75,000 molecular weight	Antiviral and antiproliferative factor. Low doses stimulate, high doses suppress humoral and cell mediated immunity. Promotes differentiation of cell markers. Increases killer cell activity.
Soluble immune suppressive substance (SIRS)	48,000–67,000 molecular weight	Activates macrophages to inhibit B-cell proliferation and antibody production
Antibody inhibitory material (AIM)	10,000–50,000 molecular weight RNase sensitive	Inhibits *in vitro* 2° antibody responses
Inhibitor of DNA synthesis (IDS)	35,000–50,000 molecular weight; raises target cell cAMP	Inhibits T-cell proliferation and antibody production; arrests cells late in G_1 by binding DNA polymerase
Chalones	Complexed with RNA; <10,000 $\bar{s}$ RNase	Inhibits lymphocyte DNA synthesis Tissue but not species specific
Immunoglobulin binding factor (IBF)	Shed T-cell Fc receptor with 38,000 and 18,000 molecular weight chains	Inhibits *in vitro* antibody production and may bind TRF

500-fold increased serum GM-CSF level; (2) graft versus host disease is also associated with both elevation of serum GM–CSF and lymphoid hyperplasia; (3) elevated serum GM–CSF has also been observed in some patients and mice with lymphoid leukemia and lymphoid tumors, respectively; (4) the injection of both theta-positive lymphocytes and bone marrow precursors are required to correct the macrocytic anemia of W/W V mice. Thus, T-cells may supply a factor, presumed to be CSF, that corrects the anemia.

Although considerations of the role of CSF have focused on its capacity to increase the production of more mature functional inflammatory cells, another major effect of CSF may be to increase the functional activity of mature inflammatory cells. The evidence supporting this latter view is that GM–CSF (1) increases prostaglandin and

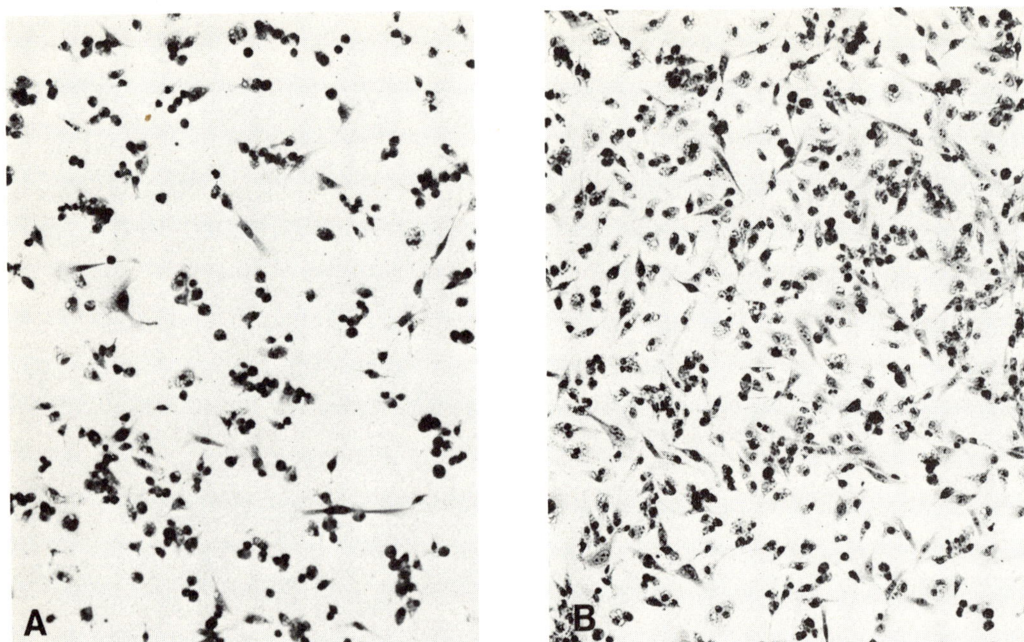

Figure 5. Macrophage growth inducing activity of CSF. Thioglycolate induced adherent murine peritoneal exudate cells were cultured for 72 hours in A (McCoys 5A medium with 15% FCS B) medium supplemented with 5000 units/ml of L-cell-derived CSF. (Magnified 130 ×, courtesy of Robert Moore, NIDR, NIH.)

lymphocyte activating factor (LAF) production by macrophages (Moore et al. 1980); (2) increases the proliferative activity of activated peritoneal macrophages, i.e., macrophage growth factor, (MGF) (Stanley et al. 1976) (see Figure 5); (3) activates macrophages metabolically and increases their phagocytic capacity and adherence to substrates. Thus, the various CSF activities appear to have multiple biological consequences through their stimulating effects on different cell types at a number of stages in their development. However, other lymphokines such as interferon and MIF are also found in cell-line supernatants along with CSF and may be responsible for some of these biological effects rather than CSF. This problem can only be resolved by further purification and characterization of these cytokines.

Amplification Factors. Some LK induce proliferation as well as differentiation of lymphocytes. They amplify *in vitro* immunological reactions through their lymphoproliferative effects as well as by promoting diverse lymphocyte effector functions, such as the induction of killer T-cells and antibody-producing B-cells. The best characterized of the growth-stimulating LK is the one produced by murine T-lymphocytes. It was renamed interleukin-2 (IL-2) at the Second International Lymphokine Workshop and has been shown to be active in four distinct biological assays. Interleukin-2 is mitogenic for thymocytes, promotes antibody production by B-cells in response to T-dependent antigens (T-helper activity), promotes the development of functional cytotoxic-T-effector lymphocytes or the killer helper factor (KHF), and can support the continued growth of T-cell lines (TCGF activity). The mitogenic effect of IL-2 is assessed by its capacity to stimulate tritiated thymidine incorporation by microcultures of murine thymocytes or peripheral lymphocytes. The helper activity of IL-2 is determined from increases in the number of antibody-forming lymphocytes in cultures of murine spleen cells. The KHF activity of IL-2 is established by its induction of killer cells in 5-day cultures containing mixtures of parental murine thymocytes and stimulatory spleen cells. Interleukin-2 induces cytotoxic lymphocytes to develop in a mixed lymphocyte reaction that specifically lyse only ^{51}Cr-labeled tumor cells that are syngeneic with the spleen cells. The repeated addition of IL-2 to lectin or antigen-activated lymphocytes results in the proliferation and long-term survival of normal cytotoxic T-lymphocyte lines (CTLL). Interleukin-2 is produced by murine spleen cells in response to concanavalin A and elutes at a molecular weight of 30,000 to 38,000 on Sephadex G100. These four activities cannot be chromatographically separated by successive two-cycle gel filtration, hydroxyl apatite and phenyl sepharose columns or by isoelectric focusing. Interleukin-2 can be separated by hydrophobic chromatography from MG–CSF and by hydroxyl apatite chromatography from interferon, which also has the capacity to induce specific T-killer cells but has antiproliferative effects (Simon et al. 1979).

In the case of the killer cell induction, Lyt-1^{+} T-cells have been shown to be the source of IL-2 and Lyt-2^{+}, 3^{+} T-cells the targets (Wagner and Rollinghoff 1978). Although IL-2 resembles monocyte-derived thymocyte mitogenic factor (LAF or IL-1) in that they both have T-helper activity, they differ because LAF, unlike IL-2, does not directly induce T-killer cells and does not support the growth of CTLL (see Chapter 5).

The T-helper activity of IL-2 is produced by T- but not B-lymphocytes and requires the presence of a small number of macrophages. The T-helper activity of IL-2 appears to require the presence of T-cells and it may therefore not act directly on B-lymphocytes. Interleukin-2 may be involved in T–T collaboration and acts on some intermediary T-cells with multiple biological consequences. This T-helper activity appears to be active at concentrations of less than 10^{-9}M.

The murine helper activity of IL-2 is very similar to some other T-cell helper activities that have been called costimulator, nonspecific mediator (NSM) and nonspecific factor (NSF) by other investigators. Each of these factors can restore both IgM as well as secondary IgG types of antibody responses. Murine IL-2, NSM as well as NSF in the presence of antigens and T-cells can stimulate B-lymphocyte proliferation and promote B-cell antibody production.

The molecular properties of IL-2 show some species differences. Human and rat IL-2 are 15,000 in molecular weight, whereas mouse IL-2 is 30,000. The TCGF activity of IL-2 can immortalize human or mouse cytolytic T-lymphocytes (CTLL). It can therefore be used to develop specific cytotoxic cells in quantity. However, TCGF activity can not initiate the CTLL growth. An alloantigenic, antigenic, or mitogenic stimulant is required to initiate the proliferation by the CTLL. Thereafter, the TCGF activity of IL-2 is required as a second signal to amplify and maintain the cell growth. As is the case for the T-helper activity of IL-2, the TCGF activity therefore depends on an initial exogenous signal. The nonspecific IL-2 can therefore serve only to augment specific immune responses and selectively amplify ongoing cell-mediated and humoral immune reactions (Gillis et al. 1978). Although both the TCGF and KHF activities of IL-2 are not antigen specific they can increase the size of the clone of cells that reacts specifically to the sensitizing or initiating antigen.

In addition, CTLL cells and activated nonresting lymphocytes are capable of absorbing the factor from the medium. Activated lymphocytes presumably develop a receptor for the factor which enables them to absorb the factor.

There is another nonspecific T-helper activity called T-replacing factor (TRF), first described by Schimpl and coworkers (1977). In contrast, to IL-2, TRF is not mitogenic for T- or B-lymphocytes and is a late acting B-cell differentiation signal. A T-replacing factor is a nonantigen-specific T-helper activity which is not genetically restricted and participates in T–B collaboration. T-replacing factor functions late in cultures as a differentiation signal which promotes the maturation of B-lymphocytes into antibody secreting cells in the primary response. Lyt-1^+2^- (Fc receptor negative) T-lymphocytes produce TRF and it is thought to act directly on B-cells. A T-replacing factor is not a polyclonal stimulant, but serves as a second signal dependent on a prior antigenic signal for its augmenting influence on B-cell stimulation.

T-replacing factor may interact with Fc receptors on T-suppressor cells and or B-cells since TRF is inhibited in the presence of purified preparations of the Fc part of IgG. Lower doses of Fc were needed on day 2 to inhibit the primary antibody response, whereas higher doses of Fc were needed from the time of initiation to inhibit secondary antibody responses. Thus, free Fc pieces or alternatively Fc receptors on T-cell suppressors may both prevent the induction of B-cell maturation by TRF and may thus modulate complex immunoregulatory processes.

Another helper factor, that deserves to be mentioned, is produced by *in vivo* sensitization of thymocytes for 6 to 7 days to alloantigens. These educated thymocytes when cultured for 24 hours with irradiated allogeneic spleen cells produce an allogeneic effect factor (AEF) (Katz 1977). The principal biological activity of AEF resembles TRF in that it can replace the requirement for helper T-cells in *in vitro* antibody responses. Thus, AEF can reconstitute the *in vitro* antibody-forming B-cell response of T-cell-depleted primed spleen cells to particulate erythrocyte or soluble haptein-protein conjugated antigens. Allogeneic effect factor consists of two polypeptide chains of molecular weight of 12,000 and 35,000. Supernatants with AEF activity also contain KHF and T-helper activities and are directly mitogenic for splenic T-lymphocytes. The generation of AEF from the primed lymphocytes is macrophage independent. In contrast, with the other helper factors AEF reacts with antisera to Ia and contains β_2 microglobulins. Nevertheless AEF behaves in a genetically unrestricted fashion and may therefore consist of a complex or mixture of IL-2 and histocompatibility antigens (Altman et al. 1980).

Transfer Factor. A discussion of factors that transfer a specific response to an antigen should include transfer factor. In 1953, Lawrence discovered that delayed hypersensitivity reactions could be transferred to humans by injections of the dialyzates of extracts of sonicated lymphocytes from immune donors. Progress in the characterization of this dialyzable extract has been slow because it has been difficult to adapt the phenomenon to either animal models or to tissue cultures. The factor is thought to have a molecular weight of about 3500, is heat unstable (56°C for 30″), resistant to digestion with DNase and RNase and contains amino acids and polynucleotides. Transfer factor does not cross-react with antiimmunoglobulin, and is not immunogenic. Despite this, transfer factor is reported to transfer specific immune responses. Extracts from hyperimmune donors but not from nonimmune donors will convert the delayed cutaneous reaction of an antigen unreactive recipient from negative to positive. The small size and nonimmunogenic nature of the factor make this an inexplicable and provocative finding. Some investigators believe that transfer factor functions merely as an adjuvant that augments the recipient's "natural" immune responses to ubiquitous antigens nonspecifically. Despite these controversies about the mechanism of action of transfer factor, it is being tested clinically in various immunodeficiency states. Some of these studies suggest that transfer factor promotes the maturation of T-lymphocytes, and may thus promote T-dependent antibody production, suppressor cell activities, cytotoxicity, and delayed hypersensitivity reactions. However, the therapeutic benefits of transfer factor to date have been modest at best.

Nonspecific Suppressive Factors

It has been established that just as some subpopulations of thymus-derived lymphocytes augment, others suppress immunological reactions. Some of the suppressive effects have been shown to be cell mediated, while others involve the participation of a variety of factors. The importance of suppressor cells and their products in regulating *in vivo* inflammatory responses are well established. Soluble suppressive factors are of interest for their potential utility in investigations of the mechanisms of immune regulation and for their therapeutic potential in autoimmune diseases and transplant rejection.

Soluble Immune Responses Suppressor (SIRS). Concanavalin A activated mouse spleen cell cultures elaborate a soluble immune response suppressor (SIRS) which nonspecifically inhibits *in vitro* B-cell proliferation and antibody responses to both thymic-dependent as well as to thymic independent antigens (Pierce and Tadakuma 1977). Soluble immune response suppressor is not cytotoxic and is effective even after the residual concanavalin A is all absorbed with Sephadex. Soluble immune response suppressor does not inhibit the MLR, the induction of cytotoxic T-cells, or other T-cell functions. Soluble immune response suppressor appears within 6 hours after concanavalin A activation and is maximal in 12 to 24 hours. The $LY\text{-}1^-2^+$ subpopulations of T-lymphocytes produces SIRS and is maximally suppressive when added within the first 6 hours; it no longer inhibits antibody production if added 48 hours after initiation to the 5 to 7 day cultures. In addition to acting early, the effect of SIRS is not seen until after 4 days of incubation and it depends on the presence of splenic macrophages to exert its suppressive effect. Soluble immune response suppressor does not act directly on B-cells but induces macrophages to become suppressive. The delayed effect of SIRS is attributable to the 48 to 72 hours required for the SIRS to activate macrophages to become suppressive. Macrophages preincubated with SIRS can suppress even if added late to the cultures.

The mechanism by which the macrophages mediate this suppressive effect on B-cells is unclear. Macrophages are known to produce a number of nondialyzable suppressive factors such as interferon and tumor necrosing factor (TNF). Dialyzable factors produced by macrophages with suppressive activity include prostaglandins, thymidine analog, and oxygen intermediates. In addition, macrophage-suppression may be mediated by cell contact dependent processes (see Chapter 5).

Physicochemical studies indicate that SIRS is a glycoprotein with a molecular weight of between 48,000 and 67,000 which is stable to 56°C for one hour but not at 70°C, and unstable at pH 2.0. Antigens, immunoglobulins, and antiimmunoglobulins do not react with SIRS, which is not genetically restricted and does not contain histocompatibility antigens. Mouse spleen can absorb SIRS activity but kidney cells cannot. *In vivo* administration of SIRS has been reported to delay the onset of autoimmune reactions in NZB/NZW mice and produce hypogammaglobulinemia in normal mice.

The relationship of SIRS to other lymphokines that act on macrophages is not resolved. Supernatants that contain SIRS also usually contain MIF activity. To date SIRS has not been separated from MIF. However, SIRS can be distinguished from MIF in that it is not inhibited by L-fucose. The fact that SIRS does not inhibit T-cell proliferative reactions distinguishes it from the cytostatic activity of lymphotoxin and the antiproliferative effects of interferon. Soluble immune response suppressor is also distinguishable from another nonspecific suppressive factor that can be recovered from the supernatants of murine MLRs involving primed responding T-lymphocytes. This MLR-suppressive factor, unlike SIRS, bears antigenic determinants encoded for by the I–C subregion of the H-2 histocompatibility complex. This factor suppresses only MLR of T-cells syngeneic with the I-C antigen present in the suppressor cells, can be absorbed only by concanavalin A activated lymphoblasts and therefore differs from SIRS.

Immunoglobulin Binding Factor. A soluble factor is secreted by alloantigen activated mouse T-cells, which binds to the Fc fragment of IgG and inhibits complement activation by IgG. This immunoglobulin binding factor (IBF), which consists of shed Fc receptors of suppressor T-cells, inhibits *in vitro* antibody synthesis by mouse splenic B-cells to both T-dependent and T-independent antigens (Neauport-Sautes et al. 1979). Murine IBF has a molecular weight of 80,000 and is a glycoprotein which is thermolabile, acid pH resistant, trypsin-sensitive, and neuraminidase resistant. It is composed of chains with molecular weights of 38,000 and 18,000 bound to a 25,000 molecular weight component that contains Ia determinants. However, IBF is not genetically restricted in its suppressive capacity. Immunoglobulin binding factor does not have any suppressive effect during the initial 60 hours of incubation, but a 2-hour exposure at 72 hours is inhibitory. It has been proposed that this late-acting suppressive effect of IBF may be exerted through its inhibitory effect on late-acting TRF. The immunoglobulin binding factor or the Fc receptor on T-suppressor cells may interact with TRF and prevent it from inducing the differentiation of B-cells. The immunoglobulin binding factor may actually require cell contact to be effective. It can be blocked by serum IgG and its *in vivo* role remains to be established.

Miscellaneous Suppressive Factors. A soluble factor from stimulated T-cell cultures has been detected, which suppresses MIF production by B-lymphocytes. This activity was therefore called MIF inhibitory factor (MIFIF) (Cohen and Yoshida 1977). This inhibitor has not been well characterized and its relationship to other inhibitors is unclear, but as with many other LK it is heat labile and nondialyzable. Cell-mediated immunity involving lymphokine participation may be limited to a T-dependent process. It may account for the observation that even though B-lymphocytes can make lymphokines, LK have been detected only in chronic T-cell mediated reactions. Thus, MIFIF may serve as a negative feedback regulator of B-cell lymphokine production.

Another suppressive factor is the inhibitor of DNA synthesis (IDS). Inhibitor of DNA synthesis is produced by mitogen or antigen activated lymphocytes. It is also produced by spleen or thymus cells from rodents given large "tolerogenic" doses of antigens. A subpopulation of "adherent" T-lymphocytes, that is, mitomycin-C and hydrocortisone sensitive, produces IDS and it chromatographs in the molecular weight range of 35,000 to 50,000. Inhibitor of DNA synthesis inhibits DNA synthesis and mitosis of lymphocytes in responses to T- or B-cell mitogens and also suppresses the primary and secondary antibody response. IDS inhibits the proliferation of a fibroblast cell line (L-cells). Inhibitor of DNA synthesis arrest cells in G_1 presumably by activating adenyl cyclase and raising cyclic AMP (Waksman and Wagshall 1978). Inhibitor of DNA synthesis has been reported to bind to αDNA polymerase, which provides a mechanism by which it may selectively inhibit cells at the onset of S phase. Inhibitor of DNA synthesis does not inhibit the ability of lymphocytes to produce LK such as lymphotoxin, but it does block further clonal expansion of the antigen-reactive population. Inhibitor of DNA synthesis does not inhibit target cells until 12 hours after they have been activated, and therefore it acts as a second signal only on previously activated cells.

Antigenic competition results in the inhibition of an immune response to any antigen administered during an ongoing response to a previously administered unrelated antigen. Since the competing antigens can be injected into different sites, this may be mediated by nonspecific suppressive factors. Conversely, lymphocytes removed from the suppressive environment of the host can develop normal *in vitro* antibody responses to subsequently administered antigens. Several soluble suppressors have been implicated in antigenic competition, including an α_2 globulin fraction of human, or rabbit serum. This serum α_2 globulin is related to the immunoregulatory activity (IRA) in serum (see Chapter 16). In addition, an antibody inhibitory material (AIM) appears in organ cultures of rabbit lymph nodes undergoing secondary responses. It is made to a greater degree by lymphocytes from donors injected with competing antigens. The antibody inhibitory material is probably too short-lived to be detected in serum and is therefore distinct from IRA, but AIM may mediate cell-contact-dependent suppression of competing antigens.

Chalones are defined as suppressive factors that are tissue but not species specific (Houck 1977). Most chalones arrest target cells in G_1 by raising cyclic AMP. Lymphocytic chalones are cationic glycopeptides that complex with RNA. After treatment with RNAse they become dialyzable (molecular weight $< 10,000$) but remain sensitive to tryptic digestion. A chalone that suppresses lymphocyte-mediated reactions, the lymphocyte inhibiting factor-thymus (LIFT), has been extracted from thymic tissue. Lymphocyte inhibiting factor-thymus and IRA are both similar to IDS in that they elevate cyclic AMP and selectively inhibit lymphocyte DNA synthesis. The relationship of these nontoxic reversible inhibitors of lymphocyte proliferation to AIM is unknown.

Conclusion

Lymphokines are the intercellular messengers that regulate interactions between monocytes, T- and B-lymphocytes as well as nonlymphoid cells. Lymphokines are involved in the homeostatic regulation and expression of immunity and inflammation. Despite our abysmal ignorance regarding the complex interactions between factors and cells a model integrating some of the relationships, is presented in Figure 6. This model illustrates how sequential stimulating effects may lead to feedback inhibitory signals. A wide variety of cells such as lymphocytes, macrophages, and fibroblasts produce colony stimulating factors (CSF) which promote the growth and functions of macrophages. Macrophages are scavenger cells that process and present antigens in a

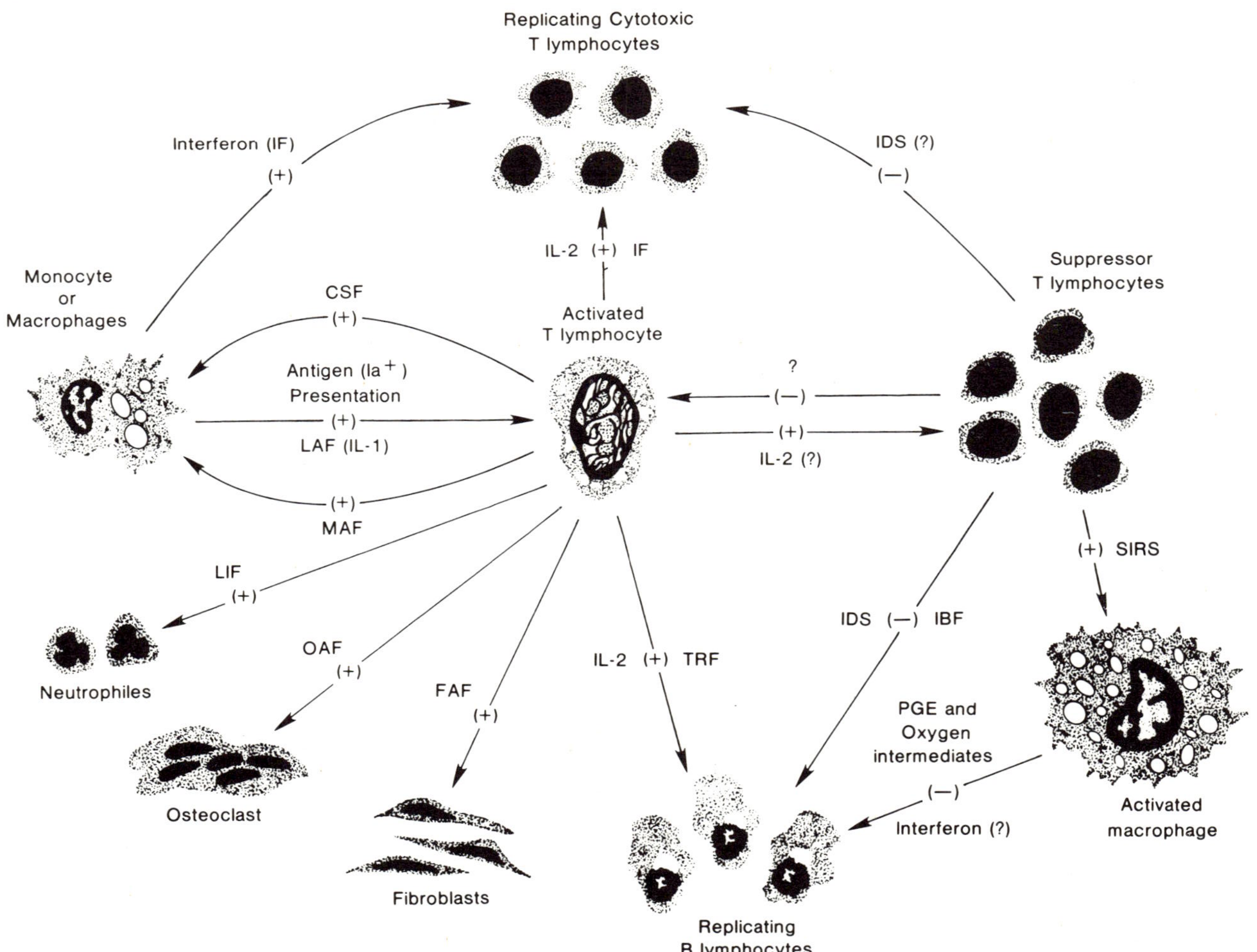

Figure 6. Interactions between cells and nonspecific cytokines. See Conclusion for definitions of the abbreviations and description of the relationships.

manner that is capable of activating T-lymphocytes. They also produce lymphocyte-activating factor (LAF or IL-1), which augment T-helper functions of activated T-lymphocytes. These activated T-cells in turn produce many LK, one of which, MAF, activates macrophages to degrade antigens more effectively and to develop bactericidal and tumoricidal capacity. Thus, MAF and LAF serve as bidirectional amplifying signals in the immune response. Another T-lymphocyte product, IL-2 has three distinct effects: it has T-helper activity and thus promotes B-cell proliferation and antibody production, it also has mitogenic effects on T-cells and furthers the growth and differentiation of cytotoxic T-cells. T-cells also produce effector lymphokines such as leukocyte inhibitory factor (LIF) that acts on neutrophiles, osteoclast activation factor (OAF) that promotes bone resorption and fibroblast activation factor (FAF) involved in wound repair. Interferon is produced by many cells and exerts feedback inhibitory effects on proliferation by T- and B-lymphocytes and concomitantly promotes the differentiation of cytotoxic T-cell functions. The T-suppressor subpopulation produces inhibitory signals. They probably produce soluble immune response suppressor (SIRS) that activates macrophage to inhibit B-cell proliferation and antibody production. T-suppressor cells probably also produce the inhibitor of DNA synthesis (IDS) that blocks T-cell replication, an inhibitor (MIFIF) which blocks MIF production by B-cells and shed their Fc receptors which then form an immunoglobulin binding factor (IBF) that inhibit antibody production by B-cells.

We are grateful for the constructive criticisms of this manuscript by Igal Gery, William Farrar, Robert Moore, Steven Mizel and John Farrar and the superb secretarial assistance of Sybil Ceja.

References

Altman, A., Katz, D.H. (1980) The spontaneous induction of primary cytotoxic T-lymphocyte responses in vitro by allogeneic effect factor (AEF). In: *Biochemistry of Lymphokines*. de Weck A., Kristensen F., Landy M. (eds.) New York, Academic Press, pp. 519–526.

Ascher, M.S., Gottlieb, A.A., Kirkpatrick, C.M. (1976) Transfer Factor, New York, Pub. Acad. Press.

Baba, T., Yoshida, T., Yoshida, T. Cohen, S. (1979) Suppression of cell mediated immune reactions by monosaccharides. J. Immunol. 122,838–847.

Bloom, B.R., Bennett, B. (1966) Mechanism of a reaction *in vitro* associated with delayed-type hypersensitivity. Science 153,180–182.

Cline, M.J., Golde, D.W. (1979) Cellular interactions in haematopoiesis. Nature 277,177–181.

Cohen, M.C., Possanza, G., Yoshida, T., Cohen, S. (1980) Evidence for the existence of a cofactor necessary for human MIF activity In : *The Biochemistry of Lymphokines*. de Weck, A., Kristensen, F., Landy, M. (eds.) Academic Press, New York, pp. 7–11.

Cohen, S. (1977) The role of cell-mediated immunity in the induction of inflammatory responses. Am. J. Pathol. 88,501–528.

Cohen, S., Pick, E., Oppenheim, J.J. (1979) The biology of the lymphokines. New York, Academic Press, pp. 7–14.

Cohen, S., Yoshida, T. (1977) Suppression of B-cell MIF production by T-cells and soluble T-cell-derived factors. J. Immunol. 119,719–721.

David, J.R. (1966) Delayed hypersensitivity *in vitro*: Its mediation by cell-free substances formed by lymphoid-cell-antigen interactions. Proc. Natl. Acad. Sci. U.S.A. 56,73–77.

Dumonde, D.C., Wolstencroft,R.A., Panayi, G.S., Mathew, M., Morley, J., Howson, W.T. (1969) Lymphokines: Nonantibody mediators of cellular immunity generated by lymphocyte activation. Nature 224,38–42.

George, M., Vaughan, J.H. (1962) *In vitro* cell migration as a model for delayed hypersensitivity. Proc. Soc. Exp. Biol. Med. 111,514–521.

Gillis, S., Ferm, M.M., Ou, W., Smith, K.A. (1978) T-cell growth factor: Parameters of production and a quantitative microassay for activity. J. Immunol. 120,2027–2032.

Godfrey, H.P., Geczy, C.L. (1978) Guinea pig macrophage agglutination factor is antigenically distinct from migration inhibitory factor and immunoglobulin. J. Immunol. 121,1428–1431.

Horton, J.E., Raisz, L.G., Simmons, H.A., Oppenheim, J.J., Mergenhagen, S.E. (1972) Bone resorbing activity in supernatant fluid from cultured human peripheral blood leukocytes. Science 177, 793–795.

Houck, J.C. (ed) (1976) *Chalones,* North Holland Amsterdam/Oxford.

Houck, J.C. et al. (1977) Lymphocyte and fibroblast chalones, some chemical properties. Science 196,896–897.

Katz, D.H. (1977) *Lymphocyte Differentiation, Recognition and Regulation.* New York, Pub. Acad. Press.

Littman, B.H., Ruddy, S. (1979) Monocyte complement stimulator: Lymphocyte product which stimulates synthesis of the second component (C_2) of complement. Cell Immunol. 43,388–397.

Luben, R.A., Mohler, M.A., Nedwin, N.E. (1979) Production of hybridomas secreting monoclonal antibodies against the lymphokine osteoclast activating factor. J. Clin. Invest. 64,337.

Moore, R.N., Oppenheim, J.J., Farrar, J.J., Carter, C.S., Waheed, A., Shadduck, R.K. (In press) LAF production by macrophages stimulated with colony stimulating factors. J. Immunol.

Neauport-Sautes, C., Rabourdin-Combe, C., Fridman, W.H. (1979) T-cell hybrids bear Fc gamma receptors and secrete suppressor immunoglobulin binding factor (IBF). Nature 277,656–659.

Pierce, C.W., Tadakuma, C.W. (1977) The role of macrophages and their soluble products in immune regulation. Prog. Immunol. 3,413–428.

Remold, H.G., Mednis, A.D. (1979) Two migration inhibitory factors differ in density and susceptibility to neuraminidase and proteinase. J. Immunol. 122,1920–1925.

Rocklin, R. (1978) Human leucocyte inhibitory factor (LIF) a lymphocyte mediator with esteratic properties. Fed. Proc. 37,2743–2747.

Schimpl, A., Wecker, E., Hübner, L., Hunig, Th., Müller, G. (1977) Properties and mode of action of T-Cell replacing factor (TRF). Prog. Immunol. 3,397–404.

Simon, P.L., Farrar, J.J., Kind, P.D. (1979) Biochemical relationship between murine immune interferon and a killer cell helper factor 122,127.

Sorg, C., Bloom, B.R. (1973) Products of activated lymphocytes. The used of radiolabeling techniques in the characterization and partial purification of the migration inhibitory factor of the guinea pig. J. Exp. Med. 137,148–170.

Sorg, C., Geczy, C.L. (1976) Antibodies to guinea pig lymphokines. III. Reactions with radiolabeled lymphocyte activation products. Eur. J. Immunol. 6,688–693.

Stanley, E.R., Cifone, M., Heard, P., Defendi, V. (1976) Factors regulating macrophage production and growth: Identity of Colony stimulating factor and macrophage growth factor. J. Exp. Med. 143,631–647.

Wagner, H., Rollinghoff, M. (1978) T-T cell interactions during *in vitro* cytotoxic allograft responses. I. Soluble products from activated Ly 1[+] T-cells trigger autonomously antigen primed Ly 23[+] T-cells to cell proliferation and cytolytic activity. J. Exp. Med. 148,1523–1538.

Wahl, S.M., Wahl, L.M., McCarthy, J.B. (1978) Lymphocyte mediated activation of fibroblast proliferation and collagen production. J. Immunol. 121,942–946.

Waksman, B.H. (1978) Modulation of immunity by soluble mediators. Pharmacol. Therapeut. 2, 623–672.

Waksman, B.H., Wagshal, A.B. (1978) Lymphocytic functions acted on by immunoregulatory cytokines: Significance of the cell cycle. Cell. Immunol. 36,180–196.

Ward, P.A., Remold, H.G., David, J.R. (1969) Leukotactic factor produced by sensitized lymphocytes. Science 163,1079–1081.

Yoneda, T., Mundy, G.R. (1979) Prostaglandins are necessary for osteoclast-activating factor production by activated peripheral blood leucocytes. J. Exp. Med. 149,279–283.

Interferons

Robert M. Friedman, M.D.

Interferons are proteins whose production can be induced in animal cells by a variety of stimulating substances; interferons inhibit a wide range of viruses by inducing an intracellular antiviral state, yet many interferons are species specific in their range of antiviral activity (Isaacs and Lindenmann, 1957). Although interferons were first described in 1957, there is as yet no entirely satisfactory explanation of their induction, biological role, or biological activities. One reason for this is the impressive potency of interferons: the specific antiviral activity of human interferon, for instance, is more than 2×10^9 IU/mg of protein (Rubinstein et al. 1978). With a molecular weight of about 20,000 this means that biological activity resides in 0.4 pgr or 10^7 molecules (2×10^{-15} M). This suggests that one or very few molecules of interferon are necessary to induce an antiviral state in a cell. Since some of the most potent bacterial toxins are marginally active at several orders of magnitude higher concentration, on a molar basis interferon seems to be one of the most active biological substances known. As a consequence, interferon preparations containing very large amounts of antiviral activity contain extremely small amounts of interferon, and until recently it has been difficult to purify interferons.

Interferon assays are also a problem, because they have been exclusively biological and are based on the ability of a preparation to inhibit the production of a virus or a viral product in infected cells (Buckler 1977). They are time consuming and, although sensitive, relatively imprecise. Because of the inherent inaccuracies of biological assays, a two- or three-fold inhibition of a viral function is considered barely significant to define a unit; therefore, a level of uncertainty is present that is usually intolerable in a biochemical or biophysical system. It is to be hoped that recent progress in the purification of interferons will soon lead to the design of accurate immunological or biochemical assays for it.

One other problem in interferon research has been that multiple substances bear the name interferon in man and probably in other species (Burke 1977; Youngner 1977). There are several types of human interferon (Table 1). Human leukocytes in cultures produce, when stimulated with virus, predominantly a species called alpha

From the Laboratory of Experimental Pathology, National Institutes of Arthritis, Metabolism, and Digestive Diseases, National Institutes of Health, Bethesda, Maryland interferon. There are at least eight distinct genes for human alpha interferon (Nagata et al. 1980). Most human fibroblast cultures when stimulated with a double-stranded inducer of interferon, such as the polymer polyriboinosinic acid: polyribocytidylic acid (poly I:C) produce an interferon immunologically distinct from alpha interferons and designated as beta (fibroblast) interferon. It is important to note, however, that leukocytes can produce beta interferon under some conditions and that fibroblasts can be stimulated to produce alpha interferon. Beta interferon is more hydrophobic than is alpha interferon, and, therefore, adheres to hydrophobic ligands such as hydrocarbons, which do not interact with alpha interferons. Beta, but not alpha, interferons also bind to lectins such as concanavalin A. One other striking difference between alpha and beta interferons is in the species specificity of antiviral activity. While beta interferon is quite species specific and induces antiviral activity only in human cells, alpha interferons induce activity in human as well as in bovine and porcine cell cultures.

The third type of interferon, gamma, immune, Type II, or T interferon, differs from alpha or beta interferons in several fundamental respects (Youngner 1977). It is antigenically distinct from them and is more labile to heat and acid. Alpha and beta interferons are stable at pH 2.0, while the antiviral activity of gamma interferon is destroyed by this treatment. Gamma interferon is produced by T (thymus-derived) lymphocytes in response to mitogens or exposure to an antigen to which the cell had already been sensitized. Gamma interferon is, therefore, a lymphokine that appears to play a role in the regulation of immune response, and possibly in the antitumor effects of interferon treatment. Macrophages must be present for the production of gamma interferon by T lymphocytes. Also, a T-to-T-cell interaction mediated by interleukin 2 seems to be required (Farrar and Farrar, in press). B lymphocytes produce alpha, rather than gamma, interferons, whereas lymphoblastoid cell lines produce mixtures of alpha and beta interferons.

Interferon Production

The production of alpha and beta interferons is considered first. Alpha and beta interferons can be induced in animals or in cell cultures by living or killed RNA or DNA viruses (Burke 1977); in addition, a number of natural and artificial nonviral substances are effective inducers of interferon production. The best studied of these are double-stranded RNA forms from fungi or synthetically produced molecules such as poly I:C (Vilcek and Kohase 1977). It is possible that double-stranded RNA forms are such excellent interferon inducers because they are similar in structure to a natural product that is the actual cell signal for turning on interferon production. Many viruses contain double-stranded RNA as a structural element or produce it during the course of their replication processes. A diverse group of substance also induces interferon production in cell cultures or in animals (Merigan 1973a; Grossberg 1977). These include bacteria that grow intracellularly, such as *Brucella abortus, Listeria monocytogenes* or *Haemophilis influenzae,* rickettsiae, mycoplasmae, protozoae, clamydiae, microbial products such as lipopolysaccharides, pyran copolymers, polyvinvyl sulfate, and a variety of low molecular weight substances, such as cycloheximide, kanamycin, tilorone, and toluidine blue. Some of these induce interferon production only in animals; others, *in vivo* and in tissue cultures. This group is so divergent in structure, however, that the inducers may act by causing the production and release of a common intracellular substance such as a cellular double-stranded RNA. There

Table 1. Induction of Human Interferons in Various Cell Types

Human cell type	Inducer	Percent of type of interferon induced		
		Alpha	Beta	Gamma
Leukocytes (buffy coat)	Viruses	99	1	0
Namalva cells (Lymphoblastoid)	Viruses	85	15	0
Skin fibroblasts (FS4)	Poly I:C	<0.1	>99.9	0
	Viruses	2	98	0
Skin fibroblasts (GM258)	Poly I:C	<0.1	>99.9	0
	Viruses	20	80	0
T-cells	*C. parvum*	>95	<5	<5
	Tuberculin	10	<1	90
	Mitogens (e.g., PHA)	?	?	>80

are naturally occurring double-stranded RNA forms in animals cells, and these forms are themselves good interferon inducers (Stern and Friedman 1970). Alternatively, the diverse substances which induce interferons and double-stranded RNA forms may interact with similar receptors on the cell surface.

Many types of cells make interferons (Table 1). Both T and B lymphocytes produce alpha interferons; macrophages and most fibroblast cultures can also be induced to make both alpha and beta interferons. The inability of some cell lines to produce interferons has not been explained, but one interesting system involves mouse teratocarcinoma cell cultures which, when undifferentiated, do not produce interferons; however, after differentiation, production of interferons can be induced (Burke et al. 1978). The genetic control of interferon production is not well understood, but it is likely that it involves a repressor mechanism (Figure 1). The genetic locus for production of human beta interferon is probably on chromosome 9. Since the cellular

Figure 1. Beta (F) interferon production and its control.

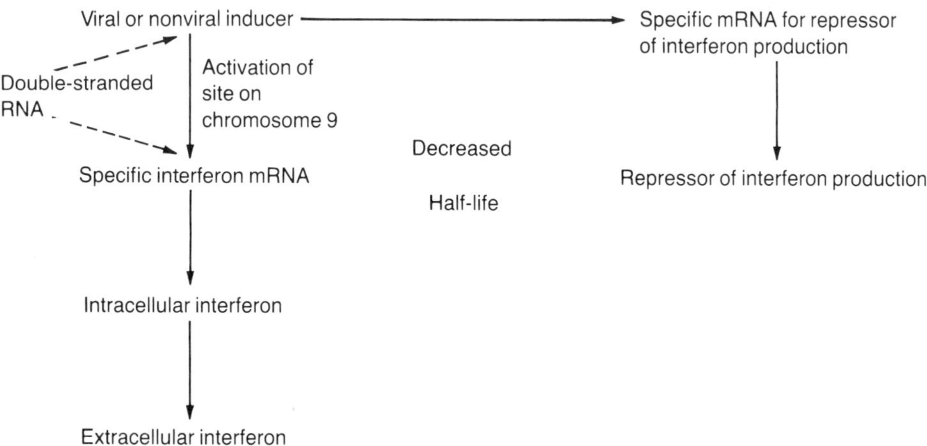

content of interferon mRNA can be assayed by several methods, the mRNA forms for alpha and beta interferons have been shown to be poly A rich, distinct 8-12S molecular forms (Cavalieri and Pestka 1977). The best evidence for a repressor mechanism for interferon production is the phenomenon of superinduction in human fibroblast cultures (Vilcek and Kohase 1977). Superinduction is the greater than normal production of beta interferon in cells induced to form interferon in the presence of inhibitors of protein synthesis (cycloheximide or puromycin), of RNA synthesis (actinomycin D), or of processing (dichloro-1-β-D-ribofuranosylbenzimidazole, DRB). Increased production of interferon is observed when the inhibitors of protein synthesis are removed. This is, at least in part, due to a prolongation of the half-life of interferon mRNA, which may in turn be related to an inhibition of the production of a repressor of interferon mRNA function; the repressor may normally be made so that interferon synthesis is not usually constitutive. There does appear, however, to be a cell line in which interferon synthesis is semiconstitutive; in this case the proposed repressor of interferon production will probably turn out to be deficient (Jarvis and Colby 1978).

Several antimetabolites or antibiotics inhibit interferon production. Interferon production involves new cellular RNA and protein synthesis, so it is not too surprising that pretreatment with actinomycin D and inhibitors of protein synthesis decreased interferon production markedly (Burke 1977). Since interferon is a glycoprotein, inhibitors of glycosylation such as tunicamycin, 2-deoxy-D-glucose, or D-glucosamine alter interferon production; beta interferon made in the presence of either of the later two compounds, for instance, was more heat labile, less efficiently neutralized by antibody, and had less charge heterogeneity than did beta interferon produced in their absence (Havell et al. 1977).

Gamma interferon production on the other hand is induced by substances that stimulate a mitogenic response in lymphocytes; these include antigens to which the donor of the lymphocytes had been exposed. The antigens can be viral or nonviral in origin and may be bacterial products such as tetanus and diphtheria toxoids or PPD. Plant-derived mitogens, e.g., phytohemagglutinin A (PHA), pokeweed mitogen (PWM), or concanavalin A, as well as other mitogenic stimulants, such as *Staphylococcus* enterotoxin A (SEA) or antilymphocyte serum, are also effective in stimulating *in vitro* gamma interferon production. In practice, a good way to stimulate production of large quantities of gamma interferon in serum has proved to be intravenous OT or PPD inoculation of mice previously sensitized by infection with BCG. Unfortunately, the preparation of gamma interferon from serum has resulted in difficulties in its purification that to date has not been achieved (Youngner 1977).

As noted, T lymphocytes produce immune interferon; so far, its production has not been reported in cultures of cells other than T lymphocytes. T cells produce gamma interferon within 3 days of exposure to mitogens; interferon production in this system depends on the presence of macrophages in the culture, so pure cultures of lymphocytes cannot be stimulated to produce gamma interferon; however, lymphocytes depleted of macrophages can still be stimulated by viruses to produce alpha interferons. It is not known what factor the macrophages supply, but the species of origin is determined by the lymphocytes that produce the interferon and not by macrophages in the culture (Youngner 1977).

Substances that increase the intracellular level of cyclic AMP such as cholera toxin, or cyclic AMP itself and its chemical derivatives inhibit gamma interferon production.

Mechanisms of Action

The Antiviral State

The mechanism of interferon action, for the sake of discussion, may be divided into two phases (Friedman 1977). The first relates to how interferon treatment induces an antiviral state or other activities in cells; the second relates to how these induced states are expressed on various cellular activities, such as virus growth, cell replication, or the immune response.

To bring about its effects on cells, interferon must first interact with the plasma membrane (Grollman et al. 1978). The significant reaction is a binding of interferon, which is not an energy requiring process. The bound interferon can be released from the cell surface without inhibiting the later development of an antiviral state. Interferon seems to have a specific cell surface receptor; evidence for the specificity of the binding site is the finding that interferon action can be blocked in human cells by an antibody to a product of chromosome 21 (Revel et al. 1976) or competitively by thyroid-stimulating hormone (TSH), chorionic gonadotropin, or cholera toxin. Since the last three appear to have a similar or identical binding site, it follows that interferon must bind to the same or a very similar site as substances in this general group. There is some evidence that the species specificity of interferon action is not determined by its binding to a species-specific site, since mouse interferon binds to cells in which it does not induce antiviral activity (Grollman et al. 1978). Binding is, therefore, necessary but not sufficient for activity.

There is some information on the location and chemical nature of the putative interferon binding site (Besancon and Ankel 1977; Chang et al. 1977). It appears to be on the outer surface of the plasma membrane. This was determined by stimulating human fibroblasts to produce interferon in the presence of antibody to beta interferon. Antiviral activity failed to develop in cells that were producing interferon; this meant that interferon had to be externalized for it to induce an intracellular antiviral state (Vengris et al. 1975). Other studies indicated that the chemical nature of the interferon receptor is complex and consists of both ganglioside and glycoprotein components. How these interact to bind interferon on the cell surface and transmit information to an intracellular site is not clear, since in *in vitro* experiments interferon binds to either gangliosides or glycoprotein components. It is possible that the glycoprotein component represents an activation or amplification site for the induction of intracellular antiviral activity; similar mechanisms have been proposed for polypeptide hormone action (Grollman et al. 1978).

Once interferon has interacted with its receptors on the cell surface it is not clear what steps follow immediately. For instance, it is uncertain whether interferon is taken up by the cell. Some evidence indicates that it is not, because interferons bound to Sepharose beads are active in inducing an antiviral state. It is not clear in such studies, however, how tightly the interferons are bound to the carrier. For the most part, the activity of interferons can be accounted for by reactions initiated at the cell surface, but studies with purified, radioactive interferons will be necessary to answer definitively the question of whether interferon uptake is required for its biologic activities. The great specific activity of interferon suggests that no more than one molecule per cell can induce an antiviral state.

The Inhibition of Virus Replication

Cellular protein and RNA synthesis are required for the development of antiviral activity; also, in human cells chromosome 21 is necessary (Tan et al. 1977). The following steps seem necessary to the development of the antiviral state: after human interferons are bound to a specific receptor, a number of biochemical reactions occur that cause, among other things, activation of a site on the distal portion of the long arm of chromosome 21. This results in the production of specific mRNA forms that in turn are translated to give rise to at least two proteins related to the antiviral state (Figure 2).

The recent development of some understanding of the activities of these proteins is one of the most important aspects of studies on interferon. It was recognized several years ago that in cultures of interferon-treated, virus-infected cells, viral messenger RNA was not efficiently translated. In some cases in which virion-associated transcriptases were present, interferon treatment did not inhibit viral mRNA synthesis. In cell-free protein synthesizing systems derived from mouse cells treated with low,

Figure 2. Actions of interferon.

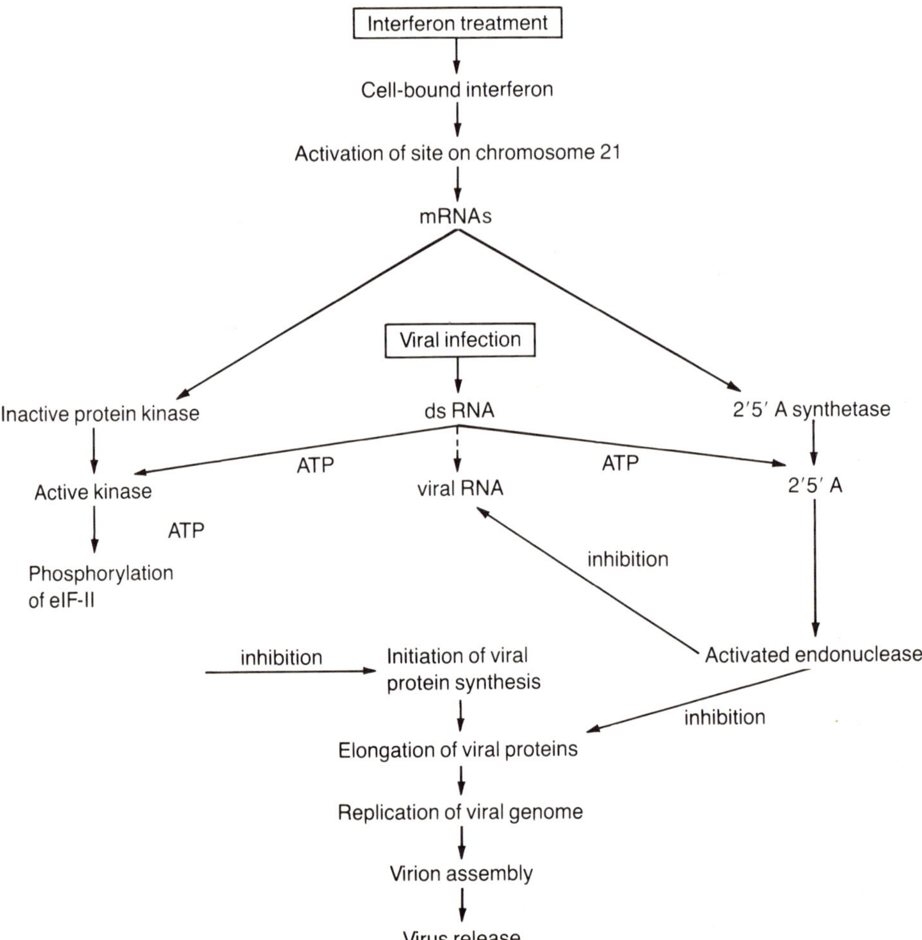

but still highly effective, concentrations of interferon, there was no inhibition of viral mRNA translation unless the cells had also been infected with a virus. This suggested that interferon treatment induced a potential antiviral state that was not fully developed until the cells were virus infected (Friedman 1977).

In considering what factors might be involved in the activation of the antiviral state in interferon-treated cells, it was found that addition of minute quantities of double-stranded RNA resulted in the inhibition of viral protein synthesis (Kerr et al. 1974). This might be related to the requirement for viral infection of interferon-treated cells to demonstrate an inhibition of translation of viral mRNA, because double-stranded RNA species are produced in many viral infections.

Treatment of cells with interferon causes the induction of two enzymatic activities that are activated by double-stranded RNA. These are the $2'5'$-oligoadenylate synthetase and a protein kinase. The synthetase catalyzes the formation of adenylate oligomers linked by a $2',5'$-phosphodiester linkage and terminated at the $5'$-end with a triphosphate; the predominant product of the synthetase of extracts of interferon-treated cells is the trimer, $pppA2'p5'A2'p5'A_{OH}$ (called $2'5'A$). The protein kinase activated by double-stranded RNA in interferon-treated cells phosphorylates with adenosine triphosphate (ATP) the small subunit of the protein synthesis initiation factor elF-2. This is consistent with several observations strongly suggesting that initiation of viral protein synthesis is inhibited following interferon treatment.

The $2'5'A$ and an endoribonuclease, on the other hand, are closely related; $2'5'$ adenylate synthetase, which is induced following interferon treatment, is activated by double-stranded RNA. It uses ATP as a substrate to form $2'5'A$, which in turn inhibits virus protein synthesis by activating the endoribonuclease. Thus, there are at least two ways in which interferon treatment inhibits viral protein synthesis, and it is uncertain which of these is the more significant in a given virus infection of interferon-treated cells. It is possible that both contribute to an antiviral state. It is interesting to note that rather than being unique antiviral mechanisms, the processes employed in the interferon system may well be adaptations of the normal systems that control cell growth and differentiation. It was not entirely unexpected, therefore, to find that interferons also have effects on the immune system and on the growth of uninfected cells.

There are, however, several reports of interferon inhibition of steps in the virus replication cycle other than at the level of virus protein synthesis. In several systems, for instance, interferon treatment seems to inhibit virus-directed transcription (Oxman 1977). The best-documented case of this is the treatment of AGMK cells with interferon before their infection with SV40. In this study early SV40 mRNA synthesis was inhibited, although under other conditions the site of interferon action appeared to be on the translation of SV40 mRNA. The apparent inhibition of virus-directed transcription by interferon could, of course, be related to an interferon-induced endoribonuclease activity.

The effect of interferon treatment on RNA tumor virus replication was, however, unanticipated (Friedman 1977). It has been known for many years that interferon treatment inhibits yields of murine leukemia viruses (MLV), but recent studies indicate that the synthesis at least some (and possibly all) the virus-specific proteins of MLV is not inhibited in interferon-treated cells. In one study, early steps in virus maturation were inhibited in interferon-treated cells (Figure 3a). The final stages in virus maturation were also the site of inhibition in some systems; therefore, virus was not released efficiently from the plasma membrane of the cell. In scanning electron

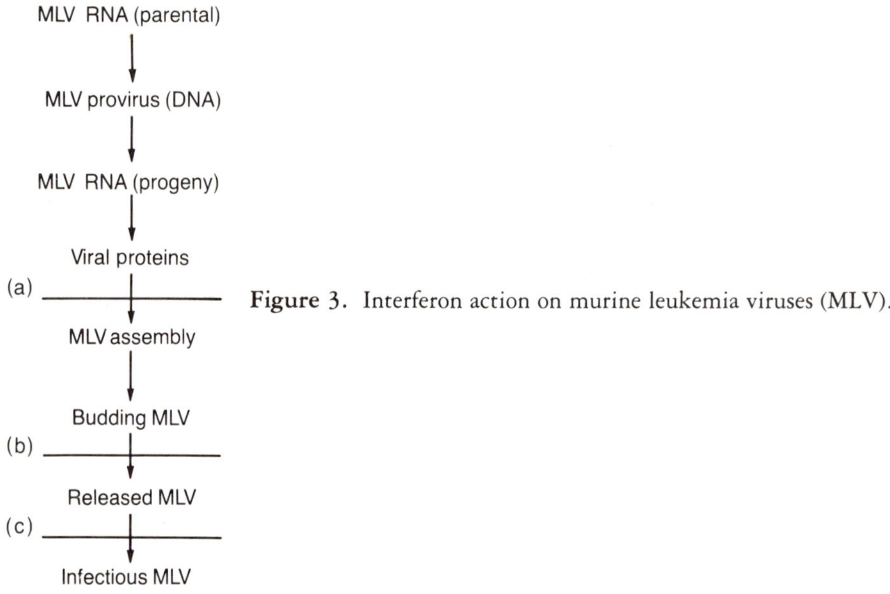

Figure 3. Interferon action on murine leukemia viruses (MLV).

micrographs of the surface of interferon-treated MLV-producing cells, apparently morphologically normal virus particles were attached to the cell surface in much higher concentrations than were normally present (Figure 3b).

In other systems MLV morphogenesis and release were normal, but the MLV released from interferon-treated cells was markedly deficient in infectivity (Figure 3c). The MLV produced by interferon-treated cells is deficient in the viral glycoprotein gp69/71; this may well account for the deficiency in infectivity of such viral particles. Similar decreases in the glycoprotein content of a virus from another group, the rhabdovirus vesicular stomatitis virus (VSV) have been reported. Vesicular stomatitis virus from interferon-treated cells is deficient in glycoprotein G. This may be responsible for the decrease infectivity of VSV from interferon-treated cells.

These findings may be related to alterations in the mechanism of cell protein synthesis; for instance, translational errors could be induced by interferon treatment. They could also be primarily associated with changes in the cell plasma membrane following interferon treatment (see below).

Restriction of Interferon Effect in Virus-Directed Translation Where the Viral Genome Is Integrated into an Interferon-Resistant Virus or Host Genome

As noted above, in cells chronically producing murine leukemia virus, interferon treatment does not inhibit protein synthesis directed by the virus. In view of what is discussed in this section, it is important to note that the mRNA of MLV in chronic infections is generated from a provirus integrated into the genome of the host.

Another special case in which interferon is inactive when normal inhibitory activity might be expected requires additional consideration (Friedman 1977). Interferon treatment inhibited lytic infection with SV40 virus. Not only was infectious virion formation inhibited, but production of T antigen, an early gene product, was decreased also as was viral-induced cell transformation. The stimulation of cellular DNA synthesis in BHK-21 hamster cells after infection with polyoma virus (which is closely related to SV40) was also inhibited by treatment with hamster interferon. All these findings indicated clearly that SV40 and polyoma virus functions are sensitive to interferon treatment.

It was therefore surprising that in SV40 virus-transformed mouse cells the production of SV40 T antigen was insensitive to interferon treatment, in spite of the fact that such treatment made them resistant to VSV infection. That is, the same interferon system that recognized and inhibited new lytic infections with SV40 and other viruses apparently failed to inhibit viral function when it was exercised by an integrated tumor virus genome. This interpretation was strengthened by findings in cells infected with an adeno-SV40 hybrid virus, which is infectious and presents an interesting genetic combination of an interferon-sensitive (SV40 virus) with an interferon-insensitive (adenovirus) genome. In simultaneous infection of cells with both complete viruses, the sensitivity of adenovirus or SV40 T-antigen production was characteristic of infection with either virus separately; however, in infection with SV40-adenovirus hybrid, production of both T antigens was as resistant as adenovirus T-antigen production in infection with adenovirus alone (Oxman 1977).

Evidence suggests that the SV40 genome is linked covalently to the adenovirus genome in the hybrid or to cellular DNA in SV40-transformed cells. The mRNA produced by the integrated SV40 genome contains host sequences, and the mRNA of the hybrid contains both adenovirus and SV40 sequences. The resistance of SV40 T-antigen production to interferon treatment in the case of integrated genomes may indicate that the primary sequence of nucleotides in the genome does not determine sensitivity to interferon. Other sites on the genome, such as those concerned with initiation or control of genetic expression, would then have to be the loci of interferon action, and the presence of host or interferon-resistant virus RNA sequences would render the virus mRNA resistant to interferon action. This interpretation is quite consistent with a mechanism of interferon action involving phosphorylation of protein synthesis initiation factor eIF-2.

Plasma Membrane Alterations

Several alterations in the plasma membrane of cells (Table 2) have been reported to follow treatment with interferon (Friedman 1977). In mouse L-cells there is a change in the electrical charge on the cell surface, so that the treated cells migrate toward the anode more rapidly than do control cells. These findings are consistent with interferon-induced alterations reported in the density of the plasma membrane of mouse AKR cells. The later changes seem to be related to an increase in the concentration of plasma membrane glycoproteins as determined by chemical, physical, and morphological alterations.

Table 2. Cell Surface Alterations after Interferon Treatment

Increase in migration in electrical field toward anode
Increase in high density fraction of cell membranes
Increase in intramembraneous granules
Increase in glycoprotein concentration
Increase in cytotoxicity of lymphocytes
Increase in histocompatibility antigens

Changes in specific glycoproteins have also been reported. In mouse cells an increase in H-2D and H-2K histocompatibility antigens follows interferon treatment (see below). Also, the cytotoxicity of mouse lymphocytes is increased after interferon treatment. Taken together, these results suggest that interferon treatment brings about significant alterations in the cell surface. These changes may be important in establishing the antiviral state. They may also be related to the inhibitory effect of interferon on cell multiplication and to alterations in the immune system.

Inhibition of Cell Growth

Interferon treatment inhibited the growth of both animal cells and intracellular parasites. There are examples of the latter among several biologic groups (Merigan 1973b). The agents that cause psittacosis or trachoma (chlamydiae) do not replicate well in interferon-treated cells. Similar results have been reported in protozoae with *Toxoplasma gondii*; in addition, exogenous interferon inhibited transmission of *Plasmodium berghei* in mice.

Interferon treatment also inhibited growth of a number of animal cells in culture (Gresser 1977). This action was due to both a decrease in the rate of cell division and a low saturation density in the cultures of the treated cells. The time of mitosis was increased and the G_1 and G_2 phases of the cycle prolonged, so that the entry of the cells into all phases of the growth cycle was also prolonged. These effects appear due to interferon since they are seen in mouse cells treated with purified mouse interferon and do not occur in cells resistant to the antiviral action of interferon; however, they are readily induced in interferon-sensitive cells.

The growth inhibitory properties of interferons have been noted in developing mouse embryos, in newborn mice, and probably in neonatal humans. Interferon treatment also inhibited the growth of virus, radiation, or chemically induced tumors *in vitro* or *in vivo*. Gamma interferon was reported to be a more potent inhibitor of sarcoma MC-36 growth in mice than L cell interferon (which is analogous to both alpha and beta human interferons), but this comparison was based on antiviral activity and not on molecular concentration of the interferons. It is possible that the specific activities of the various interferons will be found to differ.

Effects of Interferons on the Immune System

Interferon has inhibitory as well as stimulatory effects on many different immunological reactions. Interferon treatment appears to affect all the cells of the immune system and thus to influence many humoral and cellular aspects of the immune response (Johnson 1977). For instance, interferon treatment has varied affects on antibody-forming B cells. The primary or secondary antibody response can be initiated or enhanced; the particular effect depends on the concentration of interferon employed and whether the exposure of the B cells is before or after sensitization. Interferon treatment was effective in both *in vivo* and *in vitro* systems. Interferon treatment of B cells before exposure to antigens resulted in very significant inhibition of antibody production. The primary and secondary responses to thymus-dependent antigens (sheep red blood cells [SRBC]) or to a thymus-independent antigen (lypopolysaccharide) were inhibited. This may have been due to an inhibition in the number of proliferating clones; thus, it may be related to the growth-inhibiting properties of interferon.

Treatment with interferon before sensitization with antigen also inhibited the function of memory cells in the immune system. When mice were exposed to interferon before immunization with SRBC, there was a marked suppressive effect on secondary responses to SRBC even very long after primary interferon treatment.

Administration of interferon after sensitization with antigen resulted in an altogether different effect, however. With interferon treatment of animals or of lymphocyte cultures 3 to 4 days after antigenic stimulation there was significant enhancement of the immune response. This appeared to be due to an effect of interferon on immune-cell proliferation or function; in this case the effect was on suppressor T cells, which modulate the immune response. Although late addition of interferon did not seriously affect the functions of B cells, T cells, which appear to moderate the amplitude of the immune response, were apparently inhibited. This resulted in production of larger quantities of antibody than were produced in the absence of interferon.

The response of the antibody-producing system was also dose dependent. Very low concentrations of interferon caused enhancement of antibody production, while high concentrations resulted in inhibition of the antibody response. It is interesting to note that the same observations were made several years ago on the effect of interferon treatment on interferon production (Fleischmann 1977). Low or moderate concentrations of interferon often enhanced interferon production, while high concentrations resulted in decreased production. The similarity to the effect on antibody production is obvious: both effects could be secondary to inteferon inhibition of cell growth.

The type of interferon used to influence antibody production was important. On the basis of antiviral activity, gamma interferon was reported to be 250 times more effective on the immune mechanism than alpha or beta interferons. It should again be noted, however, that such differences between gamma interferon, on one hand, and alpha or beta interferons, on the other, may not mean gamma interferon is a specific modulator for the immune system, since the specific activity of gamma interferon is unknown and could be significantly lower than that of the other human interferons. Therefore, on a molar basis the activity of the three interferons may not differ in the immune system, and the antiviral activity of gamma interferon could be 250 times less than alpha or beta interferons.

Interferon also affected the IgE system. Treatment of mouse spleen cells with interferon resulted in a decreased ablity to transfer cutaneous anaphylaxis. Interferon treatment also increased the release of histamine from basophils after exposure to ragweed antigen or to anti-IgE antibody; therefore, production of interferon may play a role in the development of asthma during viral respiratory infections (Sonnenfeld 1980).

Interferon treatment also altered many T-cell-mediated immune mechanisms (De Maeyer and De Maeyer-Guignard 1977). *In vitro* studies indicated that pretreatment with interferon inhibited cell proliferation in the mixed lymphocyte response. As noted previously, interferon also increased the expression of H-2D and H-2K antigens on mouse spleen cells, splenic T lymphocytes, and thymic lymphocytes. This observation may be related to the finding that interferon enhanced the expression of tumor antigens on the surface of tumor cells. Since interferon treatment also enhanced the activity of natural killer T cells, it may be mediating this effect by promoting antigen binding by T cells; in addition, interferon treatment enhanced both H-2 expression and antigen binding of suppressor or killer T cells. Thus interferon concomitantly had negative effects on both cell differentiation and cell proliferation.

In *in vivo* studies in the mouse, interferon treatment also affected many aspects of cell mediated immunity. In the graft-versus-host reaction, both the cell proliferation and murine leukemia virus activation reactions were inhibited by interferon treatment. As in *in vitro* studies, H-2 antigen expression was enhanced by interferon treatment. Other cell-mediated immune responses were altered by interferon treatment. Allograft rejection was delayed in mice receiving exogenous interferon. In the delayed hypersensitivity response, interferon treatment inhibited both the sensitization to antigen and the tissue response to a test antigen.

In contrast to most of the reports of inhibitory effects of interferon treatment on antibody production and cell-mediated immunity, interferon treatment apparently stimulated all of the macrophage functions that have been studied (Huang 1977; Chirigos 1977). The results of these studies suggested that interferon is a monocyte or macrophage activating factor. *In vivo* and *in vitro* phagocytosis by macrophages were increased by interferon treatment; this appeared to be due to both an increase in the number of particles taken up per cell and to an increase in the percentage of cells engaged in phagocytosis. Interferon treatment inhibited the maturation of monocytes into macrophages; however, interferon treatment enhanced spreading of macrophages and the tumor cell growth-inhibiting properties of resting mouse peritoneal macrophages. Interferons are, however, difficult to separate from macrophage-activating factors, so it is not clear that these effects are due necessarily to interferon and not to a contaminant in the interferon preparation (Schultz 1980).

Interferons have also been reported to stimulate cytotoxic cells. This activity may well contribute to the antitumor effects of interferons. Specific T-cell cytotoxicity by lymphocytes was enhanced against tumor cells in mice by interferon. Addition of human beta interferon to mixed lymphocyte cultures increased the cytotoxic response of the T cells to allogeneic cells.

In addition, antibody-dependent cell cytotoxicity (ADCC) was enhanced by inter-

Table 3. Effects of Interferons on Various Aspects of the Immune System.

Effector Cell	Function and effect	
	In vitro	*In vivo*
B cell	Antibody formation Late: increased[a] Early: Decreased	Antibody formation Late: increased Early: Decreased
T cell (cell-mediated immunity)	Mixed lymphocyte reaction: Decreased H-2 antigen expression: Decreased Mitogenic response: increased	Graft-versus-host reaction: Decreased H_2 expression: Increased DTH (sensitization and response): Decreased Allograft rejection: delayed
Macrophage (nonspecific resistance)	Phagocytes: Increased Antitumor cell activity: Increased	Phagocytes: Increased Antitumor cell activity: Increased
Cytotoxic cells	Natural killer (NK) cell activity: increased Specific T-cell cytotoxicity: increased Antibody-dependent cell Cytotoxicity: increased	NK cell activity: increased

[a]Capitalization indicates large increase or decrease; lower case means relatively small effects.

feron treatment. Cytotoxic cells (K cells) from interferon-treated mice had increased cytotoxicity. Finally, natural killer cell (NK) activity was increased by *in vivo* and *in vitro* by interferon treatment in several systems (Sonnenfeld 1980).

Interferon appears, therefore, to be an important regulatory factor in many aspects of the immune response (Table 3). Some of these activities may be related to the general growth inhibitory properties of interferon, but others seem to be especially adapted to regulation for the immune system. It is possible that T interferon is a specific regulatory substance of the immune system; additional research on T interferon should aid in establishing its significance in the immune response.

Interferons and the Pathogenesis of Lymphocytic Choriomeningitis Virus Disease

Lymphocytic choriomeningitis virus (LCMV), an arenavirus, has long been known to cause two sorts of diseases in mice. The first is a rapidly fatal form of meningitis in adult mice; the second is related to the establishment of a chronic state of infection in which high titers of virus are present in the mouse blood and tissues. The mice with chronic infections are inoculated at birth with virus; after about 6 months, such mice develop a chronic renal disease, glomerulonephritis. Immune complexes consisting of complement, LCMV, and antibody to LCMV have been reported in the glomerular basement membrane of these mice; therefore, it was assumed that such complexes were related to the pathogenesis of the renal disease associated with chronic LCMV infection.

Some doubt was shed on the significance of these findings by studies involving the administration of antibody to mouse interferon at the time of infection of newborn mice with LCMV. The mice that had received the antibody to interferon did better than did controls, in spite of the fact that there were high titers of LCMV in the mice receiving the antiinterferon antibody. One possible interpretation of this result was that interferon production induced by LCMV infection had something to do with the pathogenesis of the late renal disease. To determine whether this was a reasonable possibility, newborn mice infected with LCMV were checked for interferon production; indeed, these mice were found to have high titers of circulating interferon. The next step was to inject newborn mice with interferon in order to check whether interferon itself could cause the renal disease. The mice receiving interferon did develop glomerulonephitis at about the same time as did the mice injected with LCMV at birth. This study therefore suggests that late renal disease may result from production of high titers of interferon in newborn mice (Riviere et al. 1980). Subsequent work has indicated that the pathogenesis of attenuated LCMV strains is related to their ability to induce the production of interferon; those strains of virus that are better interferon inducers are more virulent. It has also been noted that treatment with an interferon inducer of monkeys infected with arenaviruses results in exacerbation of disease rather than its mitigation (Jacobson et al. unpublished data).

Role of Interferon as a Defense Mechanism Against Viral Infections

The immune response plays a critical role in preventing many viral infections, thus the importance of many vaccines directed against the spread of viral diseases. It also seems important in the recovery of animals from some primary virus infections; however, the ability of patients with severe immunodeficiences and of animals with

Table 4. Evidence That Interferons Contribute to Recovery from Primary Virus Infections.

1. Interferons are produced at the site of virus infections quite early in the infection
2. Interferons are produced in sufficient quantity to account for recovery from primary infections
3. Inhibition of interferon production increases the severity of viral infections
4. Administration of exogenous interferon inhibits the development of viral infections

blocked cellular and humoral immune responses to recover from many virus infections indicates that factors other than the immune response are involved in recovery from such infections (Baron 1973). Administration of exogenous interferons can inhibit the development of virus diseases. Interferons are produced early and in large quantities in viral diseases; furthermore, studies employing antibody-to-mouse interferon suggested that interferons do indeed have an important function in recovery from primary infections with picornaviruses and herpesviruses. When animals infected with agents from these groups were treated with antibody-to-mouse interferon they developed diseases with uncharacteristically short latent periods, unusual severity, and high mortality rates (Gresser, 1976). The severity of all viral infections was not increased by treatment with antibody to interferon. For instance, antiinterferon antibody did not activate chronic herpes simplex infections in mice, a result that suggested immune responses are critical in the control of some chronic viral infections; however, the modulation of the immune response by interferon indicates it may play an important role in infections in which the immune response is involved.

Therefore, in one way or another, interferons seem important in recovery of animals from some virus infections (Table 4).

Interactions Between Interferons and Other Mechanisms Related to Immunity and Inflammation

While the antiviral activity of interferons is indeed important in recovery from some primary viral infections, there are other mechanisms which seem to play roles of varying importance respectively in recovery from different viral diseases. Cell-mediated immunity is thought to act as an antiviral mechanisms in that viruses, which alter the plasma membrane by insertion of viral proteins, bring about changes that make these cells targets for natural killer (NK) or killer (K) cell lysis. This mechanism is therefore likely to be important in infections with herpesviruses and other viruses that alter the cell surfaces so they can bud from it. This would include such agents as influenza virus, paramyxoviruses, most arthropod-borne viruses, and RNA tumor viruses (Sonnenfeld 1980).

The febrile response also seems to play a role in recovery from some virus infections. Many wild-type viruses do not grow as well at 40°C as they do at body temperature (37°C); indeed, some attenuated virus strains act as temperature-sensitive mutants. Elevation of body temperature or local increases in temperature, such as is seen in an area of inflammation may, therefore, serve as a potent antiviral mechanism. Other aspects of the inflammatory response may also act as significant antiviral factors. For instance, during the inflammatory reaction, acid products are generated and the pH is often lowered locally. Also, white blood cells and tissue monocytes are brought into the area of infection during the inflammatory reaction. The granulocytes and monocytes provide phagocytic activity that may be quite important in engulfing viruses

Table 5. Interaction of Various Antiviral Mechanisms with Interferon

Mechanism	Action of Interferon
Phagocytosis by macrophages	Increased
Antibody production	Stimulated by low concentration and late administration
Inflammatory response	Increased
Fever	Induced
Cell-mediated immunity: NK, K, and ADCC; cell-mediated lysis of infected cells	Stimulated

and thus removing them from the site of infection, if the phagocytic cell is capable of killing the virus. The inflammatory response and the febrile reaction may also act to hasten the antibody response and to increase the level of antibody produced. This may prove of importance in infections in which antibody is thought to play a role in recovery (Baron 1973).

It is interesting to note that all of these mechanisms (e.g., fever, phagocytosis, inflammation, and cell-mediated immunity) were originally considered to be quite independent of one another and of the interferon mechanism in acting to aid recovery from virus infection. As some of the nonantiviral activities of interferons have come to light, it has become apparent that interferons may stimulate many of the mechanisms which were previously thought to be independent (Table 5). For instance, interferons stimulate the phagocytic activity of macrophages, an activity that may eventuate in the destruction of infecting viruses. Antibody production itself can under some conditions be stimulated by interferon treatment, when low concentrations of interferon are employed, or when interferon appears after viral antigen, certainly the sequence that occurs during infections with viruses. In addition, investigators have reported that interferons directly or indirectly stimulate the inflammatory and the febrile responses. Finally, some aspects of the cell-mediated immune response, which are stimulated by interferon, may play an important role in recovery from viral infections. These include the stimulation of cell lysis by K and NK cells, activities that can be directed against infected cells in which the plasma membrane has been altered by viral infection. It is possible that the action of interferon to prevent murine RNA tumor viruses from budding off the cell surface has the same effect, in that a cell surface, which has so much viral antigen associated with it, is likely to interest surveillance cells with which the immune system monitors for the presence in the body of foreign antigens. Therefore, without attempting to overplay the importance that interferons may have in recovery from primary virus infections, the interactions with the other systems discussed above provide evidence for a significant role for interferons.

Clinical Studies with Human Interferon

Therapeutic trials in humans have employed alpha interferons. This was because of the availability of quantities of these interferons, and, in very limited trials, the failure so far of beta interferon to effect significant amelioration of viral diseases.

Exogenous alpha interferon decreases virus growth and severity of disease in a rhinovirus respiratory infection, in vaccinia virus skin infection, and in ocular herpes simplex (Tyrrell 1977; Guerra et al. 1977). Prompt treatment with 5 × 10^5 U/kg/ day after onset of herpes zoster in patients with lymphomas decreased viral spread; it also alleviated the most distressing symptom of this disease: pain. Interferon-treated patients recovered more quickly and had less postherpetic neuralgia than did untreated controls (Jordan 1974).

Interferon treatment may also be effective in selected patients with certain forms of chronic hepatitis B virus infection. Clinical studies employing interferon treatment in patients with long-standing static disease have demonstrated a decrease in circulating hepatitis B virus surface antigen, Dane particles, and infectious virus. The liver biopsies of such patients showed less chronic inflammation and some healing after treatment (Greenberg 1976).

Because of the effects of interferon on tumor virus replication and cell multiplication, attempts have been made to treat human cancers with alpha interferon preparations. In one such study Strander has, for several years, treated osteogenic sarcoma patients with interferon. The results so far suggest that the recurrence and mortality rates after surgical removal of primary tumors is decreased in the patients receiving long-term interferon therapy (Strander 1977). In addition, preliminary studies in patients with poorly differentiated lymphomas suggested that interferon treatment caused radiologically detectable decreases in the size of tumors. This effect appeared to be specific, because interferon treatment was ineffective in similarly selected patients with a diagnosis of histiocytic lymphoma (Merigan 1978).

It is interesting that long-term treatment with interferon in human beings has not produced any pronounced effect on the immune system or on cellular metabolism. The most common effect has been a moderate decrease in circulating leukocytes, but in no case has this been so profound that therapy had to be abandoned.

Conclusion

Interferons have been shown to be highly active in a number of biological systems. The most important of these relate to virus growth, cell replication, and the immune system. It is not unlikely that interferons will be found to play a significant role in normal regulation of several aspects of cell metabolism, and their discovery as antiviral substances may prove to be fortuitous. So active a biological substance should prove to be of great interest to many medical scientists and to physicians.

Several types of human interferon have been described. It is not clear why there are in human beings at least 10 distinct genes related to interferon production (8 to production of alpha interferons and at least 1 each to beta and gamma interferons). It is possible that the multiplicity of interferon genes may relate to the multiple functions ascribed to interferons and such heterogeneity may be due to different basic functions for each specific type of interferon.

Interferon production can be induced by a wide variety of chemical substances that have little in common structurally. This suggests that interferon induction is a common biological response to a wide variety of stimuli. The actual production of interferons, on the other hand, is a tightly controlled cell process. The ability to superinduce human beta interferon production suggests the existence of natural repressor molecules for interferon induction.

The mechanism of action of interferons is complex. Interferons must be bound to the cell surface to initiate their biological functions. Metabolic activity is necessary for the development of interferon action. The actions of interferon appear related to the induction of two specific enzyme systems, a protein kinase that phosphorylates the initiation factor eIF-2 and an oligoadenylate synthetase that produces $2'5'A$ from ATP. The $2'5'A$ polymers in turn activate a specific endoribonuclease. These enzyme systems probably function to inhibit the initiation and elongation phases of viral protein synthesis, respectively. It is uncertain how these systems relate to the effects of interferons on cell growth and the immune system.

Interferons appear to be growth control factors that inhibit cell growth. Interferons also exert varying immunoregulatory effects on B cells, T cells, macrophages, and cytotoxic cells. They appear to be powerful lymphokines that play an important role in immunobiology.

References

Baron, S. (1973) The defensive and biological roles of the interferon system. In: *Interferon and Interferon Inducers*. Finter, N.B. (ed.) New York, Elsevier North Holland, pp. 267–294.

Besancon, F., Ankel, H. (1977) Membrane receptors for interferon. Texas Rep. Biol. Med. 35,282–289.

Buckler, C.E. (1977) Interferon assays: general considerations. Texas Rep. Biol. Med. 35,150–153.

Burke, D.C. (1977) Biological properties of virus-induced interferons. Texas Rep. Biol. Med. 35,11–16.

Burke, D.C., Graham, C.F., Lehman, J.M. (1978) Appearance of interferon inducibility and sensitivity during differentiation of murine teratocarcinoma cell *in vitro*. Cell 13,243–248.

Cavalieri, R.L., Pestka, S. (1977) Synthesis of interferon in heterologous cells, cell-free extracts, and *Xenopus laevis* oocytes. Texas Rep. Biol. Med. 35,117–125.

Chang, C., Pauloin, A., Chany-Foutnier, F. (1977) Role of the membrane-bound receptor system in biological activity of interferons. Texas Biol. Med. 35,330–335.

Chirigos, M. (1977) Antitumor action of interferon: animal systems. Texas Rep. Biol. Med. 35,399–405.

De Maeyer, E., De Maeyer-Guignard, J. (1977) Effects of interferon on cell-mediated immunity. Texas Rep. Biol. Med. 35,370–374.

Farrell, P.J., Sen, G.C., Dubois, M.F., Ratner, L., Slattery, E., Lengyel, P. (1978) Interferon action: Two distinct pathways for inhibition of protein synthesis by double-stranded RNA. Proc. Natl. Acad. Sci. U.S.A. 75,5893–5897.

Fleischmann, Jr., W.R. (1977) Priming of interferon action. Texas Rep. Biol. Med. 35,316–325.

Friedman, R.M. (1977) Antiviral activity of interferons. Bacteriol. Rev. 41,543–567.

Greenberg, H.B., Polland, R.B., Lutwick, L.I., Gregory, P.B., Robinson, W.S., Merigan, T.C. (1976) Effect of human leukocyte interferon on hepatitis B virus infections in patients with chronic hepatitis. N. Engl. J. Med. 295,517–522.

Gresser, I., Tovey, M.G., Maury, C., Bandu, M.T. (1976) Role of interferon in the pathogenesis of virus diseases in mice as demonstrated by the use of anti-interferon serum II. Studies with *Herpes simplex* Maloney sarcoma, vesicular stomatitis, Newcastle disease, and influenza viruses. J. Exp. Med. 144,1316–1323.

Gresser, I. (1977) On the mechanism of the antitumor effects of interferon. Texas Rep. Biol. Med. 35,394–399.

Grollman, E.F., Lee, G., Ramos, S., Lazo, P.S., Kaback, R., Friedman, R.M., Kohn, L.D. (1978) Relationships of the structure and function of the interferon receptor to hormone receptors and establishment of the antiviral state. Cancer Res. 38,4172–4185.

Grossberg, S.E. (1977) Nonviral interferon inducers: natural and synthetic products. Texas Rep. Biol. Med. 35,111–116.

Guerra, R., Galin, M., Frezzotti, R. (1977) Use of interferon in eye infections in man. Texas Rep. Biol. Med. 35,497–500.

Havell, E.A., Yamazaki, S., Vilcek, J. (1977) Altered molecular species of human interferon produced in the presence of inhibition of glycosylation. J. Biol. Chem. 252,4425–4427.

Huang, K. (1977) Effect of interferon on phagocytes. Texas Rep. Biol. Med. 35,350–356.

Isaacs, A., Lindenmann, J. (1957) Virus interference. I. The interferon. Proc. Roy. Soc. Ser. B 147,258–267.

Jarvis, A.P., Colby, C. (1978) Murine interferon system regulation: isolation and characterization of a mutant 3T6 cell engaged in the semi-constitutive synthesis of interferon. Cell 14,355–363.

Johnson, H.M. (1977) Effect of interferon on antibody formation. Texas Rep. Biol. Med. 35,357–369.

Jordan, G.W., Fried, R.P., Merigan, T.C. (1974) Administration of human leukocyte interferon in *Herpes zoster*. I. Safety, circulating antiviral activity, and host responses to infection. J. Inf. Dis. 130,56–62.

Kerr, I.M., Brown, R.E., Ball, L.A. (1974) Increased sensitivity to double-stranded RNA of protein synthesis in cell-free systems from interferon-treated cells. Nature 250,57–59.

Merigan, T.C. (1973a) Non-viral substance which induce interferons. In: *Intereferons and Interferon-Inducers*. Finter, N.B. (ed.) Amsterdam, Elsevier North-Holland, pp. 45–72.

Merigan, T.C. (1973b) The actions of interferon and inducers in infections caused by agents other than viruses in some other conditions. In: *Interferons and Interferon-Inducers*. Finter, N.B. (ed.) Amsterdam, Elsevier North Holland, pp. 251–262.

Merigan, T.C., Sikora, K., Breeden, J.H., Levy, R., Rosenberg, S.A. (1978) Preliminary observations on the effect of human leukocyte interferon in non-Hodgkin's lymphoma. N. Engl. J. Med. 299, 1449–1453.

Nagata, S., Mantei, N., Weissman, C. (1980) The structure of one of the eight or more distinct chromosomal genes for human interferon-alpha. Nature 287,401–408.

Oxman, M.N. (1977) Molecular mechanisms of the antiviral action of interferon: Effects of interferon on the transcription of viral messenger RNA. Tex. Rep. Biol. Med. 35,230–238.

Revel, M., Bash, D., Ruddle, F.H. (1976) Antibodies to cell-surface components coded by human chromosome 21 inhibit action of interferon. Nature 260,139–141.

Riviere, Y., Gresser, I., Guillon, J-C., Bandu, M-T., Ronco, P., Morel-Maroger, L., Verroust, P. (1980) Severity of lymphocytic choriomeningitis virus disease in different strains of suckling mice correlates with increasing amounts of endogenous interferon. J. Exp. Med. 152,633–640.

Rubinstein, M., Rubinstein, S., Familletti, P.C., Gross, M.S., Miller, R.S., Waldman, A.A., Pestka, S. (1978) Human leukocyte interferon purified to homogeneity. Science 202,1289–1290.

Schultz, R.M. (1980) Macrophage activation by interferon. Lymphokine Rep. 1,63–97.

Sonnenfeld, G. (1980) Modulation of immunity by interferon. Lymphokine Rep. 1,113–131.

Stern, R., Friedman, R.M. (1970) Apparent double-stranded RNA synthesized in animal cells in the presence of actinomycin D. Nature 226,612–616.

Strander, H. (1977) Interferons: Anti-neoplastic drugs? Blut. 35,277–288.

Tan, Y.H., Tan, C., Berthold, W. (1977) Genetic control of the interferon system. Texas Rep. Biol. Med. 35,63–68.

Tyrell, D. (1977) Trials of interferon in respiratory infections of man. Texas Rep. Biol. Med. 35,486–490.

Vengris, V.E., Stollar, B.D., Pitha, P.M. (1975) Interferon externalization by producing cells before induction of antiviral state. Virology 65,410–417.

Vileck, J., Kohase, M. (1977) Regulation of interferon production: cell culture studies. Texas Rep. Biol. Med. 35,57–62.

Youngner, J.S. (1977) Properties of interferon induced by specific antigens. Texas Rep. Biol. Med. 35,17–22.

Mast Cells and Basophils

Dean D. Metcalfe, M.D., and Michael Kaliner, M.D.

The mast cell and the basophil participate in a wide variety of inflammatory processes as do all inflammatory cell types. The unique aspect of both mast cell and basophil function is "exocytosis," the release of granular contents to the outside of the cell on specific external stimulation. The pathophysiologic consequences of this degranulation depend on the action of mast cell and basophil-dependent chemical mediators on specific surrounding tissues. To understand the mast cell and basophil, this chapter first reviews mast cell and basophil morphology, their origin and distribution, and their participation in specific disease states. This general information is followed by a more detailed examination of the mediators responsible for mast cell and basophil-dependent inflammatory processes.

Morphology and Distribution

The mammalian mast cell is generally ovoid or irregularly elongate in tissue and varies in size from 10 to 30 μm. In suspension, the rat peritoneal mast cell averages 12.6 μm in diameter. The characteristic feature of the mast cell is the presence of dense cytoplasmic granules, which occupy the cytoplasm to such a degree as to obscure the nucleus (Figure 1) and stain intensely with a variety of metachromatic dyes such as toluidine blue. In most mammals, each granule averages 0.2 to 0.4 μm in diameter. There is species specificity for granule morphology. In rats, the granule is amorphous in character. In human beings, the granule contains lattice or scroll-like structures that become amorphous during degranulation (Caulfield et al. 1980). The nucleus of the mast cell in hematoxylin- and eosin-stained sections is reminiscent of the plasma cell. It is only after demonstration of the metachromatic nature of the cytoplasmic granules that the identity of the mast cell becomes evident. The usual array of subcellular organelles is present in the mast cell and includes mitochondria, rough endoplasmic reticulum, a sparce Golgi complex, and abundant submembranous filaments.

From the Laboratory of Clinical Investigation, National Institute of Allergy and Infectious Diseases, National Institutes of Health, Bethesda, Maryland.

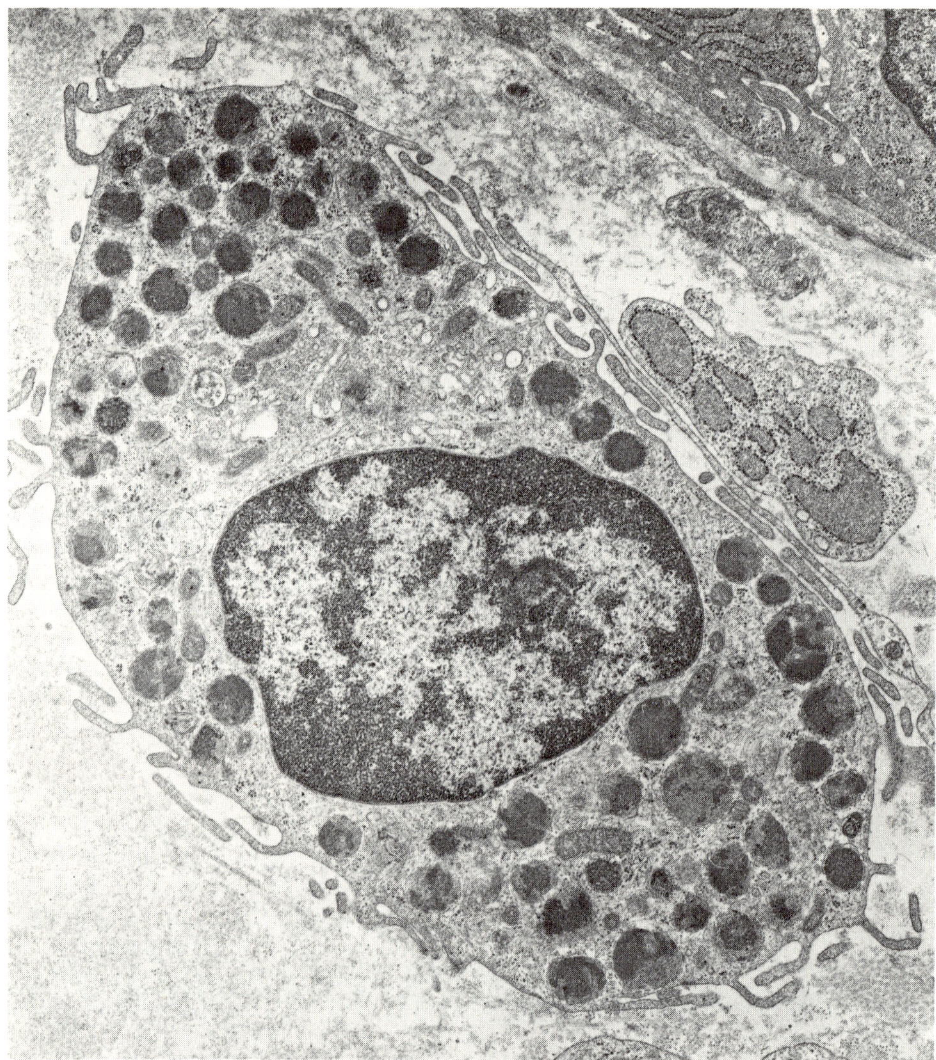

Figure 1. Human mast cell. Rectal biopsy ($\times$ 51,000). Courtesy of D. Lagunoff.

In warm-blooded vertebrates, mast cells appear chiefly in the loose connective tissue surrounding blood vessels, nerves, and glandular ducts and under epithelial, serous, and synovial membranes. In human tissues, the mast cells are relatively abundant in the skin, lymphoid tissue, uterus, urinary bladder, tongue, synovia, mesentery, subserosal and submucosal layers of the digestive tract, and around large and small blood vessels (Selye 1965). In the lungs, mast cells are found both in bronchial airway connective tissues and in peripheral intraalveolar spaces. In the skin, mast cells appear in greatest number near blood vessels, hair follicles, sebaceous glands, and sweat glands. Human skin contains approximately 10,000 mast cells/mm^3. As a generalization, mast cells are relatively infrequent in muscle, bone, and cartilage, yet are found nearby in the connective tissues surrounding cartilage and the synovium of joints. The kidney parenchyma and its internal stroma are poor in mast cells, whereas

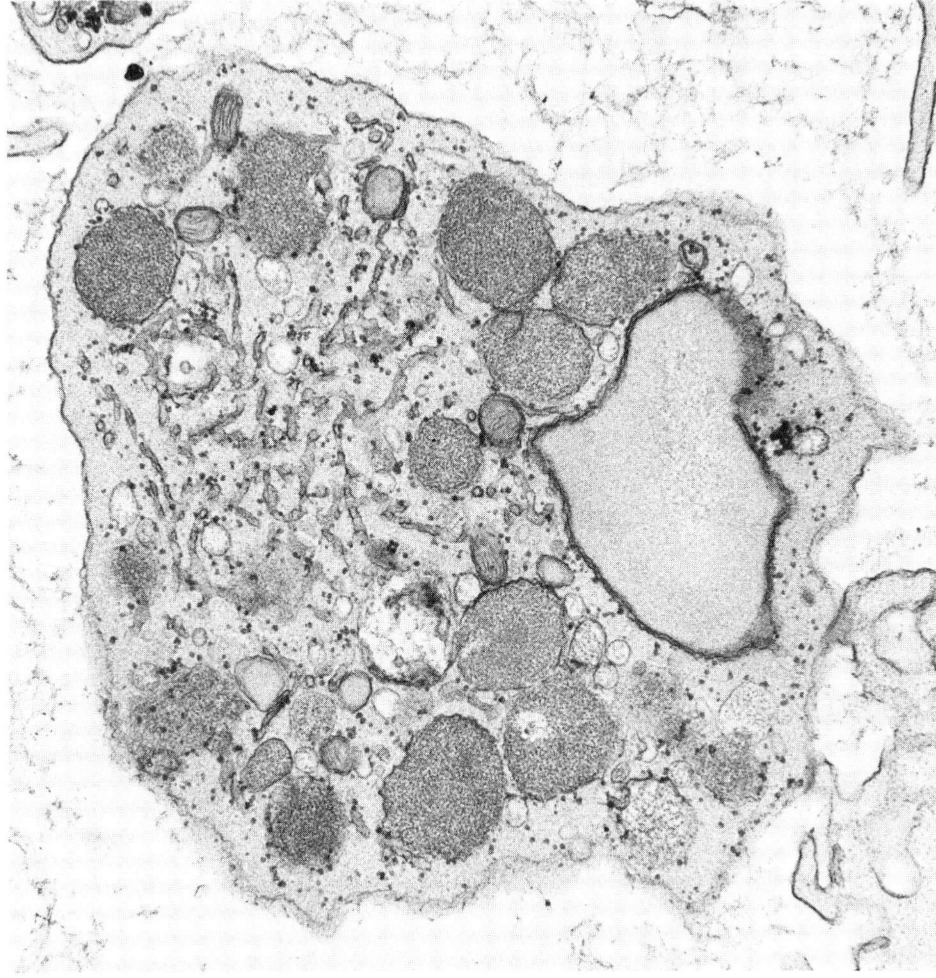

Figure 2. Human basophil (× 29,000). Courtesy of A. M. Dvorak (Dvorak et al. 1980).

the capsule, connective tissue, and fat around the pelvis of the kidney contain many mast cells.

Basophils are circulating blood cells that rarely occur in extravascular sites other than bone marrow (Dvorak 1978). In human beings, their absolute numbers in peripheral blood vary between 20 and 45 cells/mm^3. The half-life of normal human blood basophils is unknown, although studies in a patient with a basophil leukemia have suggested a half-life of about 2.5 days (Lichtenstein et al. 1978). Basophil leukocytes are polymorphonuclear cells with a diameter of 10 to 14 μm. Human basophils have fewer granules than are contained in mast cells (Figure 2) (Dvorak et al. 1980). The granules are somewhat larger than mast cell granules, ranging up to 1.2 μm. Basophil granules stain with metachromatic dyes, but to a lesser degree than mast cell granules. The granule itself is composed of dense particles appearing in a matrix of less density (Dvorak 1978).

Origin

Although the mast cell is a unique cell, distributed widely, it has never been clear from which precursor the mast cell is derived. It has been suggested that the mast cell precursor may be a histiocyte, fibroblast, undifferentiated mesenchymal cell, endothelial cell, eosinophil, clasmatocyte, lymphocyte, plasma cell, thymocyte, or reticular cell. Until recently, the mast cell was believed to be derived from undifferentiated mesenchymal cells. This concept was based on the perivascular location of the mast cell and the observation that restoration of mast cells appeared to occur through differentiation of small mesenchymal cells following extensive mast cell destruction (Fawcett 1955). Bone-marrow–derived cells have also been suggested as mast cell precursors. Beige mice, as well as human beings with Chediak–Higashi syndrome, have granulocytes and mast cells with abnormally large granules. Transplants of bone marrow from beige mice have been used to replenish mast cell pools in irradiated recipients (Kitamura et al. 1977). By identifying donor mast cells by their characteristic giant cytoplasmic granules, donor mast cells can be detected in the recipient's cecum and stomach after 60 days. However, no mast cells of the beige mouse type appear in the mesentery or skin. This experiment suggests mast cells in certain organs may be derived from bone marrow precursors. Mice of the W/W^v genotype are congenitally anemic and extremely deficient in mast cells. When normal bone marrow is transplanted into W/W^v mice, it results in the appearance of mast cells in the skin, stomach, cecum, and mesentery after 3 to 4 months (Kitamura and Hatanaka 1978). Thus, mast cells in mice may be derived from bone marrow precursors. Thymus cells in rats appear to differentiate into mast cells under a variety of conditions. Tissue culture of rat or mouse thymus glands leads to the appearance of mast cells in the culture, supporting the concept that the thymus may contain mast cell precursors (Ginsburg and Sachs 1962). Lymphocyte-derived growth factors promote mast cell proliferation in culture. This suggests that irregardless of the nature of the mast cell precursor, the number of mast cells may be regulated by lymphocyte-derived factors (Ginsburg et al. 1978).

Human basophils appear to originate from azurophil granulated promyelocytes in the bone marrow as do other blood leukocytes (Parwaresch 1976). As basophils develop, there is a gradual accumulation of characteristic metachromatic granules and segmented nuclei. Basophils are first detected in the third month of embryonic life. The clear distinction between basophils and mast cells observed in man is not present in all species. In some lower vertebrates, as well as in chickens, it appears that a single type of metachromatic-staining cell both circulates in blood and populates the connective tissues (Michels 1938).

Immunopathology

The mast cell and basophil may be recruited into inflammatory reactions primarily through IgE-mediated and complement-dependent pathways or by lymphocyte-derived mediators. The subsequent degranulation of the mast cell or basophil both releases a multiplicity of preformed mediators and generates new substances that may promote inflammation. The concerted effect of these mast cell and basophil-dependent molecules can be observed in the participation of the mast cell and basophil in immediate hypersensitivity reactions, including hay fever and asthma, the delayed-type hypersensitivity reaction, and in host response to parasites.

Historical Perspective

Both the mast cell and basophil were described by Ehrlich between 1877 and 1879. The deeply staining mast cell was considered to be "overfed," thus the word "mast" is derived from the German "mastung," which means to masticate or chew. The search for a physiologic function for the mast cell was stimulated by an early observation (Webb 1931) that the intraperitoneal injection of various irritants, including India ink and egg white into rats, induced mast cell degranulation and generalized urticaria. This reaction, which included hyperemia and edema, especially of the lips, nose, tongue, ears, paws, and genitalia, resembled anaphylaxis. At about the same time it was discovered that heparin, an anticoagulant, was the constituent that imparted the metachromatic-staining property to the mast cell (Jorpes et al. 1937). The relationship between heparin and the mast cell produced a new name for the mast cell, the "heparinocyte." The association of heparin with the mast cell supported the earlier observations that dogs undergoing anaphylactoid reactions become anticoagulated. Further evidence of mast cell function was derived from the observation that compound 48/80, a polyamine that induces mast cell degranulation, produced urticaria when injected locally into the skin of human beings, hypotension when injected intravenously in cats, and anaphylactic shock when injected intravenously in dogs. Histamine was known to be capable of inducing contraction of the guinea pig ileum *in vitro*, suggesting that histamine might participate in anaphylaxis. Histamine was associated with the mast cell when it was observed that lethal doses of fluorescent histamine-releasing agents, such as stilbamidine, localized to the mast cells noted to undergo degranulation. Subsequently, the close correlation between the mast cell number and histamine content of several tissues in several species, including human beings, was reported (Riley and West 1953). These observations were followed shortly by a number of reports detailing other specific bioactive constituents of the mast cell.

The appreciation of the role of the mast cell in immunologic reactions evolved through a series of observations that led to the discovery of IgE. These studies were carried out concurrently with the biologic and biochemical studies of the mast cell. In 1902 Portier and Richet (1902) first described the phenomenon of anaphylaxis. They observed that not all dogs injected with sea anemone toxin died. Animals that survived exhibited a dramatic reaction if subsequently injected with much smaller quantities of toxin. The manifestations included shortness of breath, diarrhea, bloody vomitus, and death—all occurring within several minutes. Using this model of anaphylaxis, Portier and Richet defined two basic principles: First, initial inoculation of proteins leads to the development of sensitivity. Second, an incubation period between the initial inoculation and the development of sensitivity is required. It was shown subsequently that serum obtained from animals after an initial immunization contained a circulating factor capable of transferring this hypersensitivity to normal animals. The biologic consequences of anaphylaxis were determined by the site of antigen administration; thus, intravenous administration of antigen resulted in systemic anaphylaxis, whereas intracutaneous injection produced a local acute inflammatory response leading to local necrosis. It was observed subsequently that the factors produced by sensitization adhered to tissue and induced anaphylactic responses on combining with antigen.

The initial report, which indicated that anaphylactic sensitivity could be passively transferred in human beings, involved a physician who was transfused with the serum from a subject allergic to horses. The recipient subsequently developed asthma while grooming his own horse (Ramirez 1919). In 1921 Prausnitz and Küstner demonstrated that serum transferred into the skin of a normal recipient induced an allergic reaction on contacting antigen to which the original donor was sensitive. The Prausnitz–Küstner test demonstrates two principles fundamental to allergy. First, serum contains a sensitizing factor that can be transferred passively to normal recipients; and, second, sensitization persists for a long period after transfer. The search for this factor, known as reagin, continued for an additional 45 years before final identification of the immunoglobulin-E (IgE) molecule. In the late 1960s, reagin was isolated from normal serum and from a subject with multiple myeloma and found to be a unique class of immunoglobulin designated IgE that bound preferentially to mast cells and basophils (Ishizaka et al. 1966; Johansson and Bennich 1967). Investigations have also established non-IgE pathways of basophil and mast cell degranulation including that mediated by complement.

Immediate Hypersensitivity

Immediate hypersensitivity reactions are those in which an antigen interacts with corresponding antigen-specific IgE on the surface of both tissue mast cells and circulating blood basophils with resultant granule discharge and release of specific mediators, in particular histamine (see Chapter 13). The reaction occurs within seconds of exposure to antigen. The allergic diseases, including "hay fever," asthma, hives (urticaria), anaphylaxis, food allergy, and drug sensitivity, are clinical examples of immediate hypersensitivity reactions. The antigens involved include such substances as pollens, animal danders, foods, and medications. The exact clinical expression of antigen–IgE-mediated mast cell or basophil degranulation depends on the site of this degranulation. Thus, degranulation of mast cells in the conjunctiva of the eye and membranes of the nose and mouth leads to pruritis of the eyes, nose, and mouth; lacrimation; and rhinorrhea. Degranulation of the skin mast cells leads to urticaria and angioedema. Involvement of pulmonary mast cells leads to bronchospasm and excessive mucus production, whereas involvement of gastrointestinal mast cells leads to nausea, abdominal pain, and diarrhea. Massive degranulation of blood basophils and tissue mast cells, particularly after systemic administration of medications, leads to hypotension often associated with bronchospasm and generalized urticaria (i.e., anaphylaxis).

In addition to immediate reactions following tissue mast cell degranulation, a delayed component may also be present (Kline et al. 1932). This biphasic reaction may best be illustrated by events in skin following mast cell degranulation. The first or "immediate" phase consists of an acute vascular response characterized by a wheal and flare occurring within minutes, which subsides within 30 to 60 minutes. This is followed by "late-phase" responses, including erythema, edema, and induration beginning within 2 hours and lasting for 12 hours; a subset of patients have reactions for 12 to 24 hours. Histologically, the late-phase reaction is characterized by a prominent neutrophil, eosinophil, basophil, and mononuclear cell infiltrate.

Treatment of allergic disorders is directed toward blocking the action of the mediator histamine after its release from mast cells and basophils, preventing the release of mediators from these cells by preventing degranulation, or stopping the interaction between antigens and mast cell surface IgE. The action of histamine on tissues can be blocked by antihistamines, a treatment particularly effective in hay fever and urticaria. In other allergic diseases such as asthma, however, antihistamines are much less effective, probably because of the importance of prostaglandins and slow-reacting substances (SRS) (leukotrienes). Thus, in asthma in particular, treatment is aimed at preventing degranulation through the use of such agents as aminophylline, epinephrine, and cromolyn sodium. In specific cases, such as ragweed-pollen-dependent hay fever, injection of small amounts of pollen antigen can lead to the production of IgG antibodies. This IgG or "blocking antibody" binds to ragweed pollen before the pollen can interact with mast cell bound IgE and, thus, prevents mast cell degranulation. This is the basis for immunotherapy or "allergy shots."

Delayed Hypersensitivity

There is evidence that basophils and mast cells participate in a variety of delayed reactions, in addition to the late-phase reaction following mast cell degranulation in skin. In many forms of chronic inflammation and neoplasia mast cell hyperplasia is associated with lymphocytic infiltration. Partial mast cell degranulation has been observed in delayed contact reactions. During the rejection of renal allografts in man, both basophilic infiltration and mast cell hyperplasia have been reported. The function of the basophil and mast cell in these inflammatory reactions is unknown, although it has been suggested that their granule-derived enzymes might contribute to tissue remodeling and regulate vascular tone and permeability.

The role of the basophil in delayed reactions has been studied extensively using a guinea pig model of delayed inflammation termed cutaneous basophil hypersensitivity (CBH). In this reaction, guinea pigs immunized with antigen by procedures that avoid the use of mycobacterial adjuvants develop a form of delayed-onset reactivity at the site of antigen injection, characterized by extensive infiltrations of basophils in addition to mononuclear cells. In typical guinea pig CBH, basophils comprise 20 to 60% of the infiltrating cells. In contrast, lesions of contact dermatitis in man generally have less than 5% infiltrating basophils. Guinea pig CBH may be transferred passively in guinea pigs by sensitized T- (or possibly B-) cells or, in some cases, by IgG_1 antibodies. Cutaneous basophil hypersensitivity is usually transient and, after several weeks, this cell-mediated reaction is replaced by an antibody-mediated (IgG) immune response. It has been proposed that histamine release by basophils may, by interacting with H_2 receptors on T-suppressor cells, be inhibiting this cell-mediated immunity.

Parasitic Disease

The host defense mechanisms against parasites are concentrated at body surfaces, because once parasites gain access to internal organs, they are difficult for the infected host to control. Dermal and mucosal mast cells, in particular, appear to participate in this early phase of defense by acting as "sentinel cells" to initiate an inflammatory response at the site of parasite penetration. In nonimmunized animals, entry of the parasite at dermal or mucosal surfaces activates the alternative complement pathway with generation of the C3a and C5a anaphylatoxins (see Chapter 13), which causes local mast cells to degranulate. In immunized animals, mast cell degranulation additionally follows interaction of parasite antigen with specific IgE, IgG, or IgM antibodies. Interaction of parasite antigen with IgE bound to mast cells leads directly to mast cell activation, whereas parasite antigen–IgM or antigen–IgG complexes activate the classical complement pathway and generate anaphylatoxins. With repeated exposure to parasites, the number of mast cells in skin and mucous membranes may actually increase and thus amplify subsequent parasite induced inflammatory processes. Basophils, in some instances, may be drawn into tissues by chemotactic factors and further amplify the inflammatory responses.

Activation of dermal or mucosal mast cells and recruited basophils results in release or generation of a number of mediators with inflammation-promoting potential. Increased local permeability caused by histamine leads to accumulation of tissue fluids containing antibodies, complement components, fibrinogen, and kinin precursors. Chemotactic factors attract effector cells that act in concert with antibody and complement to damage parasites. Prostaglandins and leukotrienes can stimulate gastrointestinal smooth-muscle contraction, leading to peristalsis and expulsion of parasites. Nematode parasites in experimental animals are in part expelled from the gastrointestinal tract through increased production of mucus by goblet cells. This enhanced mucus secretion appears to provide a slippery wall to which parasites cannot attach, making them vulnerable to peristaltic elimination. Mucosal mast cells can be involved in goblet cell discharge of mucus through IgE-dependent mechanisms. Mast cells may directly contribute to the death of parasites through synergistic action with effector eosinophils. Mast cells adhere to antibody-coated parasites, but by themselves cause little cytotoxicity *in vitro*. However, addition of eosinophils to this reaction leads to antibody-dependent eosinophil-mediated cytotoxicity greater than that seen using eosinophils alone (Askenase 1980).

Basophils appear particularly important in the expulsion of arthropod ectoparasites such as ticks. In immunized guinea pigs, a local basophil rich infiltrate occurs within 6 hours of tick feeding. With time, an intraepidermal abscess forms and basophil degranulation is seen, with basophil granules adhering to tick mouthparts. A role for basophil-derived inflammatory mediators is suggested by the observation that ticks survive longer on antihistamine-treated hosts (Askenase 1980). In summary, cumulative evidence suggests strongly that mast cells and basophils are involved in the rejection of parasites during attempted entry. This rejection involves immunoglobulin and complement-mediated mast cell or basophil degranulation and release of materials that lead to local inflammation and, in the case of the gastrointestinal tract, goblet cell secretion and increased peristalsis.

Mediators

The effects of mast cells and basophils in inflammation depends on the secretion of cytoplasmic granules, which both contain and generate biologic molecules termed mediators. These mediators lead to specific biologic effects, which vary with the tissue in which the reaction takes place. For example, the release of the mediator histamine in the vascular system may produce hypotension and in the skin and nose may produce vasopermeability, whereas in the lung it may induce bronchoconstriction.

The primary source of information on mediators has been the rat peritoneal mast cell, although considerable information is known about the mediators from human lung, skin, and nasal mast cells; the guinea pig basophil and lung mast cell; and the human basophil. Using the rat mast cell as a model of degranulation, three distinct sources of mediators can be identified: (1) preformed mediators that are rapidly released from the granule; (2) secondarily formed mediators generated by the interaction of primary mediators and nearby cells and tissues; and (3) granule matrix derived mediators that are preformed but remain associated with the granule after discharge

Table 1. Mast-Cell-Dependent Mediators

Preformed, eluted rapidly under physiologic conditions
  Histamine
  Eosinophil chemotactic factors
  Neutrophil chemotactic factors
  O_2^-
  Arylsulfatase A
  Exoglycosidases
  Serotonin[a]

Preformed, associated firmly with the granule under physiologic conditions
  Heparin
  Chymotrypsin/trypsin
  Peroxidase[a]
  SOD[a]
  Arylsulfatase B

Mediators generated as a consequence of mast cell activation/secretion
  SRS (leukotrienes)
  Prostaglandins
  Thromboxanes
  PAF[a]

[a] Demonstrated in the mast cells of species other than human.

(Table 1). The physicochemical nature of the interaction of mediators with the mast cell granule matrix allows a temporal sequence of mediator dissociation from the granule. Preformed mediators such as histamine are released into the surrounding fluid immediately. Several of the secondary mediators are formed at once (prostaglandins) or within minutes (SRS). The granule matrix and its associated mediators remain complexed for hours until dissolution occurs or they are phagocytosed.

Mediator discharge clearly initiates the events associated with immediate hypersensitivity, which are apparent within minutes and explosive in nature. Late-phase reactions occur in tissues in response to granule derived mediators as well. Thus, 2 to 8 hours after degranulation, there is an influx of polymorphonuclear leukocytes followed by mononuclear cells over the subsequent 24 hours. In summary, the events following mast cell or basophil degranulation occur in response to mediators, are spread over a lengthy period of time due both to granule dissolution with delayed mediator release and the time course of action of mediators, involve a complicated interaction between several cell types involved in propagating the inflammation or inhibiting the degranulation process, and produce the signs and symptoms associated with allergic inflammation. In this section, the mast cell and basophil mediators (Table 1) are discussed individually from both a biochemical and a physiologic viewpoint. The order of presentation begins with proteoglycans, which are the major structural components of the granule matrix, followed by individual reviews on each of the preformed molecules bound to the proteoglycan. The preformed and proteoglycan-bound molecules discussed include histamine, serotonin, and a number of lysosomal enzymes. Chemotactic factors, including both preformed and newly generated molecules, are discussed next, followed by a discussion of the newly formed mediators including prostaglandins, SRS, and platelet-activating factor (PAF).

Proteoglycans

Proteoglycans are a special class of glycoproteins that consist of a protein core covalently linked to sugar side chains (glycosaminoglycans) of similar size and composition. The class of proteoglycan (chondroitin sulfates, dermatan sulfate, keratan sulfate, heparan sulfate, heparin) is defined by the alternating sugar groups that make up the glycosaminoglycan side chain. With the single exception of keratan sulfate, a uronic acid always alternates with a hexosamine. Proteoglycans are polyanions by virtue of the presence of negatively charged side groups on the carbohydrate portion of the molecule. This polyanionic character of proteoglycans in large part determines the nature of the interaction between proteoglycans and other components of the cell granule. Heparin contains more negative charge units than chondroitin sulfates do, principally because of a higher degree of sulfation. Mast cells contain primarily a proteoglycan heparin, whereas basophils contain more chondroitin sulfates.

Heparin. In the rat mast cell, heparin makes up approximately one-third of the granule by dry weight. Rat mast cells contain approximately 25 μg of heparin/10^6 mast cells, whereas human lung mast cells contain 2.4 to 7.8 μg of heparin per 10^6 mast cells (Metcalfe et al. 1979). Rat mast cell proteoglycan heparin has an average molecular weight of 750,000 and contains a unique protease resistant protein core made up of a polymer consisting of serine and glycine to which are attached 40,000 to 60,000 molecular weight heparin side chains (glycosaminoglycans) (Yurt et al. 1977a). Degranulation of rat peritoneal mast cells releases heparin (Yurt et al. 1977b). Human lung and skin contain heparin as identified by the use of selective polysaccharidases (Metcalfe et al. 1979; Metcalfe et al. 1980a). Human lung heparin has an anticoagulant activity of 137 U/mg, similar to commercial preparations of bovine and porcine lung used clinically to prevent clot formation. Human lung mast cells, after partial purification, could be shown to contain a small proteoglycan heparin with an approximate average molecular weight of 60,000.

There is convincing evidence that mast cell proteoglycan heparin, by virtue of its high-charge density, functions as the storage matrix for histamine, as well as enzymes and chemotactic factors. Exposure of membrane free rat peritoneal mast cell granules to increasing amounts of sodium results in dissociation of histamine from the heparin granule matrix by cation exchange (Uvnäs 1978). Zinc, known to be present in rat mast cell granules (Angyal and Archer 1968), may facilitate binding of histamine to heparin. In physiologic buffer (0.15 M NaCl), not only is histamine dissociated from the heparin–granule matrix, but the binding between proteins with lower affinities for heparin such as beta-glucuronidase is also interrupted. As successively higher salt concentrations (to 1 M NaCl) are employed, further granule proteins are dissociated, including chymase, and finally the granule heparin itself is solubilized. Thus, in the rat mast cell, exposure of the granule to physiologic buffer results in a selective elution of certain constituents, whereas others remain granule bound, offering a unique mechanism to regulate local bioavailability of inflammatory mediators.

Heparin, following release from the mast cell, influences both fluid-phase and cell-mediated events as determined in studies using commercial porcine and bovine heparins. In the fluid phase, heparin binds to a number of cationic molecules by virtue of its negative charge. Heparin binds to eosinophil major basic protein with the loss of heparin's anticoagulant activity and the abolition of the capacity of major basic protein to kill parasites (Butterworth et al. 1979). Platelet factor 4, a protein released from the platelet alpha-granule, binds to heparin and neutralizes heparin's anticoagulant activity (Broekman et al. 1975). Heparin, by charge interactions, may bind to various enzymes and influence their function. This is illustrated by the demonstration of binding between heparin and granulocyte elastase, which results in stimulation of the elastase activity (Lonky et al. 1978). Heparin may alter cell adherence. For example, heparin potentiates fibronectin binding to native collagen. Thus, fibroblasts, normally coated with fibronectin, would be expected to bind to collagen fibrils more easily in the presence of heparin (Jilek and Hormann 1979).

Heparin has been shown to interact with cell membranes with subsequent displacement of cell surface associated components. Thus, heparin induces the release of plasminogen activator from rat vascular endothelium, both a triglyceride lipase and a phospholipase A from plasma membranes of rat liver, and both a triglyercide lipase and a phospholipase into serum in man after intravenous injection (Markwardt and Klocking 1977; Waite and Sisson 1973; Ehnholm et al. 1975).

A prominent effect of glycosaminoglycan pig, bovine, and human heparins is that of anticoagulation. Heparin binds to antithrombin, increasing the inhibitory capability of antithrombin toward thrombin. Because thrombin is critical to clot formation, the final effect of heparin is that of an anticoagulant. The fraction of human heparin with antithrombin binding capacity comprises approximately one-third of the total glycosaminoglycan pool (Metcalfe et al. 1979).

The glycosaminoglycan heparin–antithrombin complex inhibits thrombin by blocking the active serine site. In a similar fashion, heparin inhibits other serine proteases, including factor XII low molecular weight fragments; factors XIa, Xa, and IXa; kallikrein; and plasmin. Antithrombin alone increases the potency of macrophage inhibitory factor by neutralizing the activity of a serine protease on the macrophage surface (Remold and Rosenberg 1975). Conversely, heparin added to the system binds to antithrombin, and the inactivation of the serine protease is prevented. The net effect is to decrease macrophage inhibitory factor activity by decreasing the encapsulation of macrophages in a fibrin clot.

Heparin has long been known to possess anticomplementary activity. In the classical complement pathway, heparin interferes with C1q binding to immune complexes and the interaction of $\overline{C1s}$ with C4 and C2 (Loos et al. 1976). The effect of $\overline{C1}$ inhibitor on $\overline{C1}$ has been reported to be potentiated by heparin (Rent et al. 1976). The alternative complement pathway is influenced by heparin. Heparin inhibits cobra venom factor dependent C3 inactivation in serum (Brai and Osler 1972). Heparin both inhibits generation of cell bound or fluid phase amplification convertase C3b,Bb and interferes with β1H-mediated decay dissociation of C3b,Bb (Weiler et al. 1978). Isolated rat mast cell granules also inhibit C3b,Bb, suggesting that mast cell granules may regulate amplification convertase. The coupling of heparin to zymosan changes the zymosan surface so that it no longer functions as an activator of the alternative complement pathway (Kazatchkine et al. 1979). Finally, commercial heparin inhibits formation or binding of the trimolecular complex C5b67 as measured by the reactive lysis phenomenon (Baker et al. 1975).

Chondroitin Sulfates. Mast cells contain primarily proteoglycan heparin, whereas basophils appear to contain proteoglycan chondroitin sulfates. Thus, purified guinea pig basophils contain approximately 38% chondroitin 6-sulfate, 12% chondroitin 4-sulfate, 35% dermatan sulfate, and 15% heparin sulfate (Orenstein et al. 1978).

Similarly, rat leukemic basophils contain approximately 55% chondroitin 4-sulfate, 8% chondroitin 6-sulfate, 25% dermatan sulfate, and 12% heparin (Metcalfe et al. 1980b). The proteoglycans of normal human basophils have not been identified unequivocally; the proteoglycans, synthesized *in vitro* by a human leukemic cell preparation consisting of 50% mature basophils, are 75% chondroitin sulfates, 9% dermatan sulfate, and 11% heparan sulfate (Galli and Dvorak 1979).

In contrast to heparin, little is known concerning the extracellular function of basophil proteoglycans. They have been established as granule associated by autoradiography and presumably can complex to cationic granule proteins. Further, morphologic studies have demonstrated that basophil granules can be discharged. It may be assumed that basophil granule mediators dissociate from chondroitin sulfate when exposed to extracellular ions.

Histamine

The main sources of histamine are the mast cell and basophil. In rat peritoneal mast cell granules, histamine is known to constitute about 10% of the dry weight (Lagunoff et al. 1964). Using purified rat peritoneal mast cell suspensions, the amount of histamine per cell has been shown to be 7 to 10 pg. This compares to a histamine content in human lung mast cells of 1.0 to 5.5 pg/cell and to a human basophil histamine content of 1.0 to 1.2 pg/cell. The mast cells in a few nonhuman species contain additional amines. Serotonin (5-hydroxytryptamine) has been identified in rat and mouse peritoneal mast cells and transplantable mastocytomas, and dopamine has been reported in ungulates. Histamine is formed by decarboxylation of L-histidine and degraded by deamination with diamine oxidase (histaminase) or by methylation through histamine-N-methyl transferase. Both of these enzymes are present in tissues, and diamine oxidase is present in both eosinophils and neutrophils (Zeiger and Colten 1977). The primary excretory products of histamine catabolism are the ribosides of imidazole and methylimidazole acetic acids.

The biologic effects of histamine following cell degranulation are mediated by at least two distinct histamine receptors now conventionally termed H_1 and H_2 receptors. Drugs that inhibit the pharmacological effects of histamine are thus termed either H_1 receptor or H_2 receptor antagonists. H_1 receptor antagonists make up the "classical" antihistamines, including ethanolamines, ethylenediamines, alkylamines, piperazines, phenothiazines, and cyproheptadine. The H_2 receptor inhibitors include cimetidine, metiamide, and burimamide. There are also compounds that are selective agonists for H_1 and H_2 receptors. For example, 2-methylhistamine and 2-pyridylethylamine are selective H_1 agonists, and impromidine and dimaprit are H_2 agonists. Following the release of histamine from the mast cell or basophil, a number of local and systemic effects may be seen, as summarized in Table 2.

The effect of histamine on hypersensitivity reactions may be divided into proinflammatory and antiinflammatory actions (Plaut 1979). Histamine release leads to increases in vascular permeability through venular dilation and creation of interendothelial pores (Majno et al. 1967). This proinflammatory phenomenon is primarily an H_1 effect and facilitates ingress of plasma proteins and leukocytes. The effects of histamine on both eosinophil and neutrophil motility are discussed elsewhere in this chapter. A number of antiinflammatory effects of histamine mediated through the H_2 receptor have been reported, including inhibition of mouse T-lymphocyte-mediated cytotoxicity, inhibition of human neutrophil lysosomal enzyme release, inhibition of

Table 2. H_1- and H_2-Dependent Actions of Histamine

H_1 receptor mediated	H_2 receptor mediated
Smooth-muscle constriction	Gastric-acid secretion
Increased vascular permeability	Mucous secretion
Pruritis	Increases in cyclic AMP
Irritant-receptor stimulation	Inhibition of basophil histamine release
Prostaglandin generation	Inhibition of lymphokine release
	Inhibition of neutrophil enzyme release
Increases in cyclic GMP	Inhibition of eosinophil migration
	Inhibition of T-lymphocyte mediated cytotoxicity
Antagonized by "classical" antihistamines	Antagonized by Cimetidine

H_1 and H_2

Vasodilation (hypotension)
Flush
Headache

antigen-induced IgE mediator release of histamine from human peripheral leukocytes but not human mast cells, and suppression of antigen induced migration inhibitory factor production by T-cells that possess histamine receptors (Rocklin et al. 1979).

The diverse physiologic responses of various target tissues and cell populations to histamine may be mediated secondarily through cellular changes in cyclic nucleotide levels or through subsequent prostaglandin generation. Histamine has been reported to increase cyclic AMP in guinea pig and human lung, human peripheral basophilic leukocytes, and human neutrophils. The rise in cyclic AMP is due to H_2 receptor stimulation (Platshon and Kaliner 1978). In a similar fashion, histamine increases cyclic GMP in guinea pig and human lung through an H_1 receptor. Further, histamine stimulation of the guinea pig or human lung generates prostaglandin production in part through an H_1 receptor.

Serotonin

Serotonin in mammals is synthesized from dietary tryptophan, which is first hydroxylated to 5-hydroxytryptophan and then decarboxylated to 5-hydroxytryptamine. In human beings, serotonin is detoxified primarily by oxidative deamination by monoamine oxidase to form 5-hydroxyindoleacetaldehyde, which is further degraded by oxidation to 5-hydroxyindoleacetic acid or reduction to 5-hydroxytryptophol. The principal metabolite, 5-hydroxyindoleacetic acid, is excreted in the urine along with much smaller amounts of 5-hydroxytryptophol, mainly as the glucuronide or sulfate. In mice and rats, serotonin is contained in mast cells with a concentration of 0.8 to 1.3 pg/cell. Although serotonin is not present to any extent in mast cells from human mastocytosis skin, it has been reported in a human mastocytoma. Serotonin is found in human platelets at a concentration of 0.4 to 0.8 pg/platelet. The presence of serotonin in the platelet would allow serotonin to be recruited into the allergy reaction through agents that lead to platelet degranulation. In human beings, serotonin is additionally located in the enteroendocrine (enterochromaffin) cells of the mucosal layer of the gastrointestinal tract and in nervous tissues. There are a number of compounds that antagonize one or more of the effects of serotonin. These include ergot alkaloids; indole antihistamines of the ethylenediamine type, as well as cyproheptadine and phenothiazines; and sympathetic blocking drugs.

The pharmacological actions of serotonin include the stimulation of smooth muscle and nerves. In human beings, serotonin can cause bronchoconstriction; affect blood vessel tone; induce vasoconstriction of renal, meningeal, and pulmonary vasculature; cause vasodilation in skeletal muscles and superficial vessels; increase gastrointestinal motility; stimulate sensory nerves and autonomic ganglia; and produce discharge of catecholamines from the adrenal medulla. The sum total of these effects may be seen in individuals with serotonin-secreting carcinoid tumors. In this disease, discharge of serotonin from the tumor is sporadic and results in cutaneous flushes, bronchial constriction, fall in blood pressure, and diarrhea. Serotonin stimulation of human monocytes increases cyclic GMP with no change in cyclic AMP and induces the production of a monocyte-specific chemotactic factor from human peripheral blood leukocytes (Foon et al. 1976).

Lysosomal Enzymes

The mast cell in particular has been shown to contain a number of enzymes associated specifically with the granule, including arylsulfatase, β-hexosaminidase, β-glucuronidase, β-D-galactosidase, chymotrypsin, trypsin, acid phosphatase, superoxide dismutase (SOD), and peroxidase. In most instances, the enzymes were first shown to be associated with the mast cell granule by histochemical techniques. Subsequently, using the rat peritoneal mast cell model, a number of these enzymes have been isolated and characterized using synthetic substrates. Arylsulfatases consist of a family of enzymes with the ability to hydrolyze a variety of aromatic sulfate esters. Arylsulfatases have been reported to inactivate SRS of anaphylaxis. β-Hexosaminidase, β-glucuronidase, and β-D-galactosidase are exoglycosidases, enzymes that cleave terminal sugars from parent molecules (Schwartz et al. 1979). Mast cells and basophils, depending on the cell source, contain trypsin or chymotrypsin-like activity or both. A number of possible substrates for these proteases have been reported, including collagen, bradykinin, and cartilage proteoglycan. Chymase remains complexed to heparin after degranulation of the rat mast cell *in vivo*. This macromolecular heparin-chymase complex is able to hydrolyze smaller molecules, whereas hydrolysis of large molecules is inhibited (Yurt and Austen 1977). These enzymes are not limited to the mast cell. However, the release of the granule enzymes from the mast cell during degranulation makes them available uniquely to the extracellular environment.

Superoxide Dismutase, Peroxidase, and Superoxide

Rat peritoneal mast cells, human lung mast cells, and human basophils have been shown to produce superoxide (O_2^-) during granule expression (Henderson and Kaliner 1978). Rat mast cells also contain a granule-associated peroxidase. In phagocytic cells, the O_2^-/peroxidase system appears to play a role in the microbiocidal activity. The O_2^- production in rat mast cells parallels histamine release after immunologic activation of the cells, and isolated secretory granules release O_2^- on exposure to cations. The enzyme superoxide dismutase facilitates the dismutation of O_2^- to hydrogen peroxide and may protect tissues from this radical's toxicity. Superoxide dismutase

(SOD) has been shown to be present in rat granules in these same studies. Superoxide dismutase remains bound to the granule–heparin matrix in physiologic buffer, although its activity can be measured, and is displaced from heparin in 1 M NaCl.

A mast cell peroxidase has been identified that is tightly bound to the granule matrix and dissociates from the granule in 3 M NaCl (Henderson and Kaliner 1979). In neutrophils and macrophages, a similar peroxidase (myeloperoxidase) in the presence of hydrogen peroxide and halide ions contributes to the microbiocidal activity of these cells. The function of O_2^- production and the presence of a peroxidase in the mast cell may, as in the neutrophil and macrophage, relate to the intracellular killing of ingested bacteria; rat mast cells phagocytose *Salmonella*. The consequence of the production of O_2^- during mast cell degranulation would depend on the relative effect of SOD. The extracellular function of mast cell granule peroxidase is unclear, although it does inactivate SRS of anaphylaxis in physiologic buffer (Henderson and Kaliner 1979). Further, myeloperoxidase is capable of destroying the chemotactic activity associated with C5a.

Chemotactic Factors

A number of chemotactic factors generated or released from mast cells during degranulation have been identified, which may account for subsequent cellular infiltrates. Preformed factors reported to have chemotactic activity include histamine, neutrophil chemotactic factor (NCF), and eosinophil chemotactic factors of anaphylaxis (ECF-A) of the tetrapeptide and oligopeptide types. Additional chemotactic factors are generated by antigen challenge and are lipid in nature.

Histamine. Histamine has been reported to be chemotactic *in vivo* and *in vitro* (Clark et al. 1975). Histamine is selectively chemotactic for eosinophils *in vitro* at micromolar concentrations, whereas higher concentrations of histamine inhibit eosinophil migration. The attractant activity of histamine is not inhibited by H_1 or H_2 receptor antagonists; however, the inhibition of migration observed at higher histamine concentrations is reversed by metiamide, an H_2 receptor antagonist. Histamine, at 1-5 $\times$ 10^{-5} M, will stimulate the random migration (chemokinesis) of neutrophils, an effect also eliminated by pretreatment with metiamide. Imidazole acetic acid, a major catabolite of histamine, has also been reported to be chemotactic *in vitro*.

Neutrophil Chemotactic Factors. NCF has been described in the venous effluents of cold-challenged arms of patients with idiopathic cold urticaria and in the serum of ragweed sensitive subjects following bronchial provocation challenge using ragweed and mecholyl. This factor has a molecular weight of 750,000 and is chemotactic for neutrophils and, to a lesser extent, eosinophils. Neutrophil chemotactic factor will deactivate both neutrophils and eosinophils at high concentrations (Atkins et al. 1977; Wasserman et al. 1977). In this example, deactivation refers to the failure of leukocytes to migrate along a concentration gradient of a chemotactic stimulus as a result of a previous interaction with an active chemotactic factor.

Inflammatory Factor of Anaphylaxis. A mast cell granule derived chemotactic factor that will attract neutrophils and mononuclear cells in a pattern reminiscent of the late-phase allergic reaction when injected into rat skin has been described (Tannenbaum et al. 1980). This factor, designated the inflammatory factor of anaphylaxis, appears to be a protein with a molecular weight of 1350 and consists of 13 amino acids.

Eosinophil Chemotactic Factor of Anaphylaxis. Two tetrapeptides, Ala-Gly-Ser-Glu and Val-Gly-Ser-Glu, differing only in their N-terminal amino acid, have been identified as the active substances in the ECF-A (Goetzl and Austen 1975). Eosinophil chemotactic factor of anaphylaxis is selectively chemotactic for eosinophils and has been released by IgE dependent mechanisms from human leukemic basophils, human lung fragments, dispersed concentrated human pulmonary mast cells, and purified rat peritoneal mast cells. In the latter study, ECF-A was shown to be largely associated with the mast cell granule, and ECF-A will deactivate eosinophils to subsequent attraction by either ECF-A or C5a. According to reports, ECF-A tetrapeptide (Val-Gly-Ser-Glu) recruits eosinophils locally in abraded human skin. An "intermediate" molecular weight (oligopeptide) eosinophil chemotactic factor of a 1500 to 2500 molecular weight, associated with rat mast cell granules and released by immunologic challenge of rat peritoneal mast cells has been described (Boswell et al. 1978). This material is capable of rendering eosinophils unresponsive to homologous or heterologous chemotactic stimuli and is weakly chemotactic for neutrophils.

Lipid Chemotactic Factors. Monohydroxy-eicosatetraenoic acids (HETE) are metabolites of arachidonic acid through the lipoxygenase pathway. These compounds may be generated by many tissues and cells, including rat mast cells and human leukocytes. Depending on the concentration, HETE can elicit either chemotactic or chemokinetic responses in human neutrophils and eosinophils (Goetzl et al. 1980).

Newly Formed Factors

Prostaglandins and Thromboxanes. Prostaglandins and thromboxanes may be formed by almost every mammalian tissue, in general appearing in greater quantity following stimulation than detectable by simple extraction (see Chapter 14). The mast cell is capable both of generating prostaglandins itself and recruiting prostaglandin generation by action of mast cell granule constituents on surrounding tissues. In rat mast cells, degranulation initiated by either calcium ionophore A23187 or rabbit antirat $F(ab')_2$ results in the production primarily of PGD_2 and PGI_2 (Roberts et al. 1979). Other products derived from arachidonic acid in lesser quantity in ionophore-stimulated rat mast cells are PGE_2, $PGF_{2\alpha}$, thromboxane B_2 (TxB_2), and 12-HETE. The evidence for mast-cell-mediated recruitment of prostaglandin generation is taken from the observations that the mast-cell-dependent reaction in whole tissue generates a number of prostaglandins; that certain mast cell mediators themselves, when added to fresh tissue, result in prostaglandin formation; and that the addition of histamine antagonists prevents a portion of prostaglandin generation accompanying anaphylaxis.

The possible physiological and pathological effects of arachidonic acid metabolites, whether generated during mast cell degranulation by the mast cell itself or resulting from action of mast-cell-derived mediators on other tissues, are extensive (Table 3). The specific effects of mast cell degranulation would depend on the number of mast cells activated, the quantity of mediators released, the nature of the surrounding

Table 3. Selected Physiological Effects of Specific Arachidonic Acid Metabolites

Cyclooxygenase pathway
   Cyclic endoperoxides (PGG_2 and PGH_2)
      Bronchospasm
   HHT
      Chemotaxis
   TxA_2
      Irreversible platelet aggregation, bronchospasm
   Prostaglandin I_2 (PGI_2)
      Inhibition of platelet aggregation
      Neutrophil chemotaxis, bronchodilation
      Increased cyclic AMP, pain potentiation
   Prostaglandin $F_{2\alpha}$ (PGF_2)
      Bronchoconstriction, mucous secretion, increased cyclic GMP, augmented mediator release
   Prostaglandin D_2 (PGD_2)
      Bronchospasm, increased cyclic AMP
   Prostaglandin E_2 (PGE_2)
      Increased vascular permeability, vasodilation, potentiation of edema and pain produced by other
         mediators, increased cyclic AMP, inhibition of mediator release by mast cells and basophils
      Inhibition of lymphocyte transformation and lymphocyte-mediated cytotoxicity
      Inhibition of T-cell secretion of lymphokines

Lipoxygenase pathway
   HETE
      Chemotaxis
   HPETE
      Enhancement of mediator release
   Leukotrienes
      Smooth-muscle constriction, increased vascular permeability, chemotaxis, hypotension

tissues and their prostaglandin generating capacity, and the efficiency of reaction limiting compensation mechanisms (see Chapter 14).

Slow-Reacting Substances (Leukotrienes). The term SRS, meaning slow-reacting substance, was introduced originally for a slow onset, prolonged smooth-muscle-contracting activity appearing in the effluent of perfused guinea pig and cat lung following treatment with cobra venom (Feldberg and Kellaway 1938). Slow-reacting substance was shown to be functionally distinct from histamine by demonstrating that the contraction it produced on the isolated guinea pig ileum was not inhibited by histamine receptor antagonists. SRS activity has been generated subsequently by the immunologic challenge of a number of tissues and cell populations including the rat peritoneal cavity; bovine, monkey, and human lung; human nasal polyps; peripheral blood leukocytes; and partially purified human lung mast cells. The source of SRS in these tissues is unknown, but SRS production may require the participation of leukocytes (Orange et al. 1968). During the purification of dispersed human lung mast cells, the release of SRS, but not of histamine, diminishes with the increasing relative purity of the mast cell isolation. Further, immunologic challenge of purified rat mast cells fails to yield SRS in easily detectable amounts, and IgG_a-induced SRS production by antigen *in vivo* by rats requires the presence of polymorphonuclear leukocytes. Thus, although immunologically mediated SRS production can follow activation of the mast cell, it appears that the mast cell may not be the primary source of SRS generation. In line with these observations, an SRS-like activity has been generated by the calcium ionophore A23187 from rat peritoneal mononuclear cells and mast cells, rat and human leukemia basophils, and human peripheral blood leukocytes.

Recently, SRS has been reported to be related to a group of compounds called leukotrienes derived from arachidonic acid through the lipoxygenase pathway (see Chapter 14). Leukotrienes elicit a number of biologic events that follow mast cell degranulation, including bronchospasm, hypotension, and increased vascular permeability. Histamine, in particular, has similar biologic effects, and the presence of both histamine and SRS leads to synergistic responses, especially in bronchospasm.

Platelet-Activating Factor. Platelet-activating factor (PAF) is documented as a basophil-derived mediator of anaphylaxis in rabbits and can be released from mixed leukocytes in human beings. Platelet-activating factor has been reported released from rabbit skin and lungs, rat peritoneal cavities, and human lungs and kidneys where the cellular source is uncertain. Although both the mast cell and basophil are implicated in the generation of PAF activity by the IgE dependence of the reactions, the principal cellular sources of PAF appear to be monocytes or polymorphonuclear cells in both rodents and man. Rat mast cells do not appear to be able to generate PAF.

Platelet-activating factor generated during IgE-mediated basophil-dependent reactions has been studied extensively in rabbits. Rabbit PAF appears to be a lipid mediator with the chemical structure of acetyl-glyceryl-ether-phosphorylcholine (Demopoulos et al. 1979). This PAF acts on platelets by inducing a calcium-independent shape change and calcium-dependent aggregation and secretion of granular constituents, including platelet factor 4. The release of PAF during IgE-induced systemic anaphylaxis in rabbits is possibly responsible for the intravascular platelet aggregation and the subsequent platelet sequestration in various organs, particularly the pulmonary vasculature (Pinckard et al. 1977).

Systems in addition to PAF that may be capable of recruiting platelets in allergic reactions include complement and metabolites of arachidonic acid, particularly thromboxane A_2 (TxA_2). The recruitment of the platelet at the allergic reaction site thus adds an additional potential source of inflammatory mediators, including serotonin, permeability factors, proteases, and acid hydrolases.

Fate of Mast Cell Granules

Mast cells are located generally in loose connective tissue, and the cellular constituents ordinarily found in this tissue are the first cells to be exposed to discharged mast cell granules. The mast cell granule varies from species to species in its solubility and, thus, would vary in its rate of dissolution once released. Rodent mast cell granules remain visible in connective tissue for up to 24 hours after discharge, and this is the only model in which the interaction of granules with connective tissue has been examined (Fawcett 1955; Tannenbaum et al. 1980). Rat peritoneal and cutaneous mast cells treated with compound 48/80 or subjected to lysis release their granules, which are phagocytosed by fibroblasts, macrophages, and, to a lesser extent, neutrophils and eosinophils over an 8 to 24 hour period.

Mast cell granules released into rat skin, after either reversed anaphylaxis or compound 48/80, induce the accumulation of neutrophils over a 2 to 8 hour period followed by a mononuclear infiltration. A purified component of the granule, termed inflammatory factor of anaphylaxis, will induce a similar cellular infiltrate. These late-phase responses following mast cell degranulation are seen in rodent and human skin and human lung. Phagocytosis of particles by neutrophils has been shown to be accompanied by lysosomal enzyme release and inflammation. Thus, it is possible that phagocytosis of mast cell granules by neutrophils might be partly responsible for causing the inflammation. However, the capacity of a purified mast-cell-derived mediator to induce the inflammation indicates that the phagocytosis of granules occurs as a result of the infiltration rather than causing it. In summary, mast cell degranulation leads to an influx of inflammatory cells, which in turn phagocytose residual granules and then degrade the granule matrix.

Conclusion

The biologic functions of mast cells and basophils are not known with certainty, although these cells are associated with both immediate and delayed hypersensitivity immunologic reactions. Many investigators believe that additional biologic functions exist for mast cells and basophils. Due to the perivascular location of the mast cell and the vasoactive and clot-inhibiting effects of several mediators, it is tempting to speculate that this cell may play a role in the regulation of vascular permeability and tone as well as blood coagulation. There are other suggested functions of mast cells and basophils, including regulation of gastric acid secretion and synthesis of endocrine hormones.

It is in the area of immunologic defense against parasites that one function of mast cells and basophils is apparent. Most parasites that have a tissue phase in their life cycle elicit a large IgE antibody response. The IgE that forms sensitizes both tissue mast cells and circulating basophils. On encountering a parasite migrating through skin, lung, or gastrointestinal tract, sensitized mast cells degranulate and invoke an inflammatory response that includes the recruitment of polymorphonuclear leukocytes and the leakage of vascular proteins that contribute to the host defense against these large foreign invaders. One may speculate that a population of mast cells and basophils evolved as "sentinel" cells, widely dispersed and concentrated in areas exposed to the external environment. These "sentinel" cells, ideally located for initiating acute and chronic inflammatory lesions, developed a complex secretory granule containing numerous mediators capable of initiating inflammation. Thus, certain mast cells may be fixed connective tissue outposts of the immune system, whereas basophils may serve a similar role in the vascular compartment.

References

Angyal, A.M., Archer, G.T. (1968) The zinc content of rat mast cells. Aust. J. Exp. Biol. Med. Sci. 46,119–121.

Askenase, P.W. (1980) Immunopathology of parasitic diseases: Involvement of basophils and mast cells. Springer Semin. Immunopathol. 2,417–442.

Atkins, P.G., Norman, M., Weiner, H., Zweiman, B. (1977) Release of neutrophil chemotactic activity during immediate hypersensitivity reactions in humans. Ann. Intern. Med. 86,415–418.

Baker, P.J., Lint, T.F., McLeod, B.C., Behrends, C.L., Gewurz, H. (1975) Studies on the inhibition of $\overline{C56}$-induced lysis (reactive lysis). VI. Modulation of $\overline{C56}$-induced lysis by polyanions and polycations. J. Immunol. 114,554–558.

Boswell, R.N., Austen, K.F., Goetzl, E.J. (1978) Intermediate molecular weight eosinophil chemotactic factors in rat peritoneal mast cells: Immunologic release, granule association and demonstration of structural heterogeneity. J. Immunol. 120,16–20.

Brai, M., Osler, A.G. (1972) Studies of the C3 shunt activation in cobra venom induced lysis of unsensitized erythrocytes. Proc. Soc. Exp. Biol. Med. 140,1116–1121.

Broekman, M.J., Handin, R.I., Cohen, P. (1975) Distribution of fibrinogen, and platelet factors 4 and XIII in subcellular fractions of human platelets. Br. J. Haematol. 31,51–55.

Butterworth, H.E., Wassom, D.L., Gleich, G.J., Loegering, D.A., David, J.R. (1979) Damage to schistosomula of Schistosoma mansoni induced directly by eosinophil, major basic protein. J. Immunol. 122,221–229.

Caulfield, J.P., Lewis, R.A., Hein, A., Austen, K.F. (1980) Secretion in dissociated human pulmonary mast cells. Evidence for solubilization before discharge. J. Cell. Biol. 85,299–311.

Clark, R.A., Gallin, J.I., Kaplan, A.P. (1975) The selective eosinophil chemotactic activity of histamine. J. Exp. Med. 142,1462–1476.

Demopoulos, C.H., Pinckard, R.N., Hanahan, D.J. (1979) Platelet-activating factor (PAF): Evidence for 1-0-alkyl-2-acetyl-sn-glyceryl-3-phosphorylcholine as the active component (a new class of lipid chemical mediators). J. Biol. Chem. 254,9355–9358.

Dvorak, A.M. (1978) Biology and morphology of basophilic leukocytes. In: *Immediate Hypersensitivity: Modern Concepts and Developments.* Bach, M.K. (ed.) New York, Dekker, pp. 369–405.

Dvorak, A.M., Newball, H.H., Dvorak, H.F., Lichtenstein, L.M. (1980) Antigen-induced IgE-mediated degranulation of human basophils. Lab. Invest. 43,126–139.

Ehnholm, C., Shaw, W., Greten, H., Brown, W.V. (1975) Purification from human plasma of a heparin-released lipase with activity against triglyceride and phospholipids. J. Biol. Chem. 250,6756–6764.

Fawcett, D.W. (1955) An experimental study of mast cell degranulation and regeneration. Anat. Rec. 121,29–50.

Feldberg, W., Kellaway, C.H. (1938) Liberation of histamine and formation of a lecithin-like substance by cobra venom. J. Physiol. 94,186–226.

Foon, K.A., Wahl, S.M., Oppenheim, J.J., Rosenstreich, D.L. (1976) Serotonin-induced production of a monocyte chemotactic factor by human peripheral blood leukocytes. J. Immunol. 117,1545–1552.

Galli, S.J., Dvorak, H.F. (1979) Basophils and mast cells: Structure, function, and role in hypersensitivity. In: *Comprehensive Immunology 6: Cellular, Molecular, and Clinical Aspects of Allergic Disorders.* Gupta, S., Good, R.A. (eds.) New York, Plenum, pp. 1–53.

Ginsburg, H., Sachs, L. (1962) Formation of pure suspensions of mast cells in tissue culture by differentiation of lymphoid cells from the mouse thymus. J. Natl. Cancer. Inst. 31,1–39.

Ginsburg, H., Hammel, N.I., Eren, R., Weissman, B.A., Yehudith. (1978) Differentiation and activity of mast cells following immunization in cultures of lymph-node cells. Immunology 35,485–502.

Goetzl, E.J., Austen, K.F. (1975) Purification and synthesis of eosinophilotactic tetrapeptides of human lung tissue: identification as eosinophil chemotactic factor of anaphylaxis. Proc. Natl. Acad. Sci. U.S.A. 72,4123–4127.

Goetzl, E.J., Brash, A.R., Tauber, A.I., Oates, J.A., Hubbard, W.C. (1980) Modulation of human neutrophil function by monohydroxy-eicosatetraenoic acids. Immunology 39,491–501.

Henderson, W.R., Kaliner, M. (1978) Immunologic and nonimmunologic generation of superoxide from mast cells and basophils. J. Clin. Invest. 61,187–196.

Henderson, W.R., Kaliner, M. (1979) Mast cell granule peroxidase: location, secretion, and SRS-A inactivation. J. Immunol. 122,1322–1328.

Ishizaka, K., Ishizaka, T., Hornbrook, M.M. (1966) Physicochemical properties of reagenic antibody: V correlation of reaginic activity with $\gamma\epsilon$ globulin antibody. J. Immunol. 97,840–853.

Jilek, F., Hormann, H. (1979) Fibronectin (cold-insoluble globulin). VI. Influence of heparin and hyaluronic acid on the binding of native collagen. Z. Physiol. Chem. 360,597–603.

Johansson, S.G.O., Bennich, H. (1967) Immunological studies of an atypical myeloma immunoglobulin. Immunology 13,381–394.

Jorpes, J.E., Holmgren, H., Wilander, O. (1937) Uber das vorkommen von heparin in den Gefäbwanden und in den Augen. Z. Mikrosk. Anat. Forsch. 42,279–285.

Kazatchkine, M.D., Fearon, D.T., Silbert, J.E., Austen, K.F. (1979) Surface-associated heparin inhibits zymosan-induced activation of the human alternative complement pathway by augmenting the regulatory action of the control proteins on particle-bound C3b. J. Exp. Med. 150,1202–1215.

Kitamura, Y., Hatanaka, K. (1978) Decrease in mast cells in W/W^v mice and their increase by bone marrow transplantation. Blood 52,446–452.

Kitamura, Y., Shimada, M., Hatanaka, K., Miyano, Y. (1977) Development of mast cells from grafted bone marrow cells in irradiated mice. Nature 268,442–443.

Kline, B., Cohen, M., Rudolph, A. (1932) Histologic changes in allergic and non-allergic wheals. J. Allergy 3,531–541.

Lagunoff, D., Phillips, M.T., Iseri, O.A., Benditt, E.P. (1964) Isolation and preliminary characterization of rat mast cell granules. Lab. Invest. 13,1331–1344.

Lichtenstein, L.M., Sobotka, A.K., Malveaux, F.J., Gillespie, E. (1978) IgE-induced changes in human basophil cyclic AMP levels. Int. Arch. Allergy Appl. Immunol. 56,473–478.

Lonky, S.A., Marsh, J., Wohl, H. (1978) Stimulation of human granulocyte elastase by platelet factor 4 and heparin. Biochem. Biophys. Res. Commun. 85,1113–1118.

Loos, M., Volanakis, J.E., Stroud, R.M. (1976) Mode of interaction of different polyanions with the first (C1, C̄1), the second (C2) and the fourth (C4) component of complement. III. Inhibition of C4 and C2 binding site(s) on C1s̄ by polyanions. Immunochemistry 13,789–791.

Majno, G., Gilmore, V., Leventhal, M. (1967) On the mechanism of vascular leakage caused by histamine-type mediators. Circ. Res. 21,833–847.

Markwardt, F., Klocking, H.P. (1977) Heparin-induced release of plasminogen activator. Haemostasis 6,370–374.

Metcalfe, D.D., Lewis, R.A., Silbert, J.E., Rosenberg, R.D., Wasserman, S.I., Austen, K.F. (1979) Isolation and characterization of heparin from human lung. J. Clin. Invest. 64,1537–1543.

Metcalfe, D.D., Soter, N.A., Wasserman, S.I., Austen, K.F. (1980a) Identification of sulfated mucopolysaccharides including heparin in the lesional skin of a patient with mastocytosis. J. Invest. Dermatol. 74,210–215.

Metcalfe, D.D., Wasserman, S.I., Austen, K.F. (1980b) Isolation and characterization of sulphated mucopolysaccharides from rat leukaemic (RBL-1) basophils. Biochem. J. 185,367–372.

Michels, N.A. (1938) The mast cells. Ann. N.Y. Acad. Sci. (Appendix) 103,1–372. (1963).

Orange, R.P., Valentine, M.D., Austen, K.F. (1968) Antigen-induced release of slow reacting substance of anaphylaxis (SRS-A^{Rat}) in rats prepared with homologous antibody. J. Exp. Med. 127,767–776.

Orenstein, N.S., Galli, S.J., Dvorak, A.M., Silbert, J.E., Dvorak, H.F. (1978) Sulfated glycosaminoglycans of guinea pig basophilic leukocytes. J. Immunol. 121,586–592.

Parwaresch, M.R. (1976) *The Human Basophil.* New York, Springer-Verlag, p. 1.

Pinckard, R.N., Halonen, M., Palmer, J.D., Butler, C., Shaw, J.O., Henson, P.M. (1977) Intravascular aggregation and pulmonary sequestration of platelets during IgE-induced systemic anaphylaxis in the rabbit. Abrogation of lethal anaphylactic shock by platelet depletion. J. Immunol. 119,2185–2193.

Platshon, L., Kaliner, M. (1978) The effect of the immunologic release of histamine upon human lung cyclic nucleotide levels and prostaglandin generation. J. Clin. Invest. 62,1113–1121.

Plaut, M. (1979) Histamine, H1 and H2 antihistimines and immediate hypersensitivity reactions. J. Allergy Clin. Immunol. 63,371–375.

Portier, P., Richet, C. (1902) De l'action anaphylactique de certaine venins. C. R. Soc. Biol. 54,170–180.

Prausnitz, C., Küstner, H. (1921) Studien über die VeberempFindlichkeit. Z. Bakt. 86,160–168.

Ramirez, M.A. (1919) Horse asthma following blood transfusion. J.A.M.A. 73,984–985.

Remold, H.G., Rosenberg, R.D. (1975) Enhancement of migration inhibitory factor (MIF) by plasma esterase inhibitor. J. Biol. Chem. 250,6608–6613.

Rent, R., Myhrman, R., Fiedel, B.A., Gewurz, H. (1976) Potentiation of C1-esterase inhibitor activity by heparin. Clin. Exp. Immunol. 23,264–271.

Riley, J.F., West, G.B. (1953) The presence of histamine in tissue mast cells. J. Physiol. 120,528–537.

Roberts, L.J., Lewis, R.A., Oates, J.A., Austen, K.F. (1979) Prostaglandin, thromboxane, and 12-hydroxy-5,8,10,14-eicosatetraenoic acid production by ionophore-stimulated rat serosal mast cells. Biochim. Biophys. Acta 575,185–192.

Rocklin, R.E., Greineder, D.K., Melmon, K.L. (1979) Histamine-induced suppressor factor (HSF):

further studies on the nature of the stimulus and the cell which produces it. Cell Immunol. 44,405–415.

Schwartz, L.B., Austen, K.F., Wasserman, S.I. (1979) Immunologic release of β-glucuronidase from purified rat serosal mast cells. J. Immunol. 123,1445–1450.

Selye, H. (1965) *The Mast Cell.* Washington D.C.: Butterworth, p. 1.

Tannenbaum, S., Oertel, H., Henderson, W., Kaliner, M. (1980) The biologic activity of mast cell granules. I. Elicitation of inflammatory responses in rat skin. J. Immunol. 125,325–335.

Uvnäs, B. (1978) Chemistry and storage function of mast cell granules. J. Invest. Dermatol. 71,76–80.

Waite, M., Sisson, P. (1973) Solubilization by heparin of the phospholipase A_1 from the plasma membranes of rat liver. J. Biol. Chem. 248,7201–7206.

Wasserman, S.I., Soter, N.A., Center, D.M., Austen, K.F. (1977) Cold urticaria. Recognition and characterization of a neutrophil chemotactic factor which appears in serum during experimental cold challenge. J. Clin. Invest. 60,189–196.

Webb, R.L. (1931) Peritoneal reactions in the white rat, with special reference to the mast cells. Am. J. Anat. 49,283–334.

Weiler, J.M., Yurt, R.W., Fearon, D.T., Austen, K.F. (1978) Modulation of the formation of the amplification convertase of complement, C3b, Bb, by native and commercial heparin. J. Exp. Med. 147,409–421.

Yurt, R.W., Austen, K.F. (1977) Preparative purification of the rat mast cell chymase. Characterization and interaction with granule components. J. Exp. Med. 146,1405–1419.

Yurt, R.W., Leid, R.W., Austen, K.F., Silbert, J.E. (1977a) Native heparin from rat peritoneal mast cells. J. Biol. Chem. 252,518–521.

Yurt, R.W., Leid, R.W., Spragg, J., Austen, K.F. (1977b) Immunologic release of heparin from purified rat peritoneal mast cells. J. Immunol. 118,1201–1207.

Zeiger, R.S., Colten, H.R. (1977) Histaminase release from human eosinophils. J. Immunol. 118, 540–543.

Immediate Hypersensitivity Reactions

Reuben P. Siraganian, M.D., Ph.D.

Immediate hypersensitivity reactions occur when previous exposure to an antigen has elicited antibody that binds to basophils and mast cells. The reexposure to the specific antigen results in interaction of antigen with the cell-bound antibody, causing the rapid release of short-lived pharmacologically active mediators. The release of these mediators from mast cells and their pharmacological effect results in immediate hypersensitivity reactions. Similarly, the activation of the basophil or mast cell by other mechanisms results in the release of mediators and identical symptoms. Therefore, combination of antigen with antibody alone does not cause any damage, unless the immune complex triggers cells to release various mediators.

The types of mediators released and their pathological manifestations differ in various species. For example, systemic anaphylaxis in man is due to release of histamine and slow-reacting substance of anaphylaxis (SRS-A), with the pharmacologic effects caused by circulatory collapse and laryngeal edema. In contrast, although histamine is released in rats during anaphylaxis, it plays a minor role in the symptoms that develop in these animals. This is related to the ineffectiveness of histamine in causing pharmacological change in this species, i.e., to the susceptibility of different animals to known agents (Table 1).

Assays for Immediate Hypersensitivity

Injecting antigen into the skin of a sensitized person results within minutes in itching at the site followed by a wheal surrounded by a zone of redness (erythema). This wheal-and-erythema reaction peaks in about 15 minutes and gradually subsides during the next hour. This is an example of a localized anaphylactic reaction (Table 2). If the antigen enters the circulation, mediators are released in large amounts, resulting in a generalized anaphylactic reaction.

Another assay for immediate hypersensitivity is the passive transfer of serum from allergic donors into the skin of a normal individual. Small amounts of serum (or

From the Clinical Immunology Section, Laboratory of Microbiology and Immunology, National Institute of Dental Research, National Institutes of Health, Bethesda, Maryland.

Table 1. Anaphylaxis in Different Species

Species	Important pharmacological mediator	Site of reaction	Manifestations
Humans	Histamine SRS-A Kinin (?)	Lungs Larynx	Dyspnea Hypotension Urticaria Laryngeal edema
Guinea pig	Histamine SRS-A Kinin	Lungs	Respiratory distress
Rat	Serotonin Kinin	Intestine	Circulatory collapse Hemorrhage in intestine and lungs
Dog	Histamine Kinin Serotonin (?)	Hepatic veins	Hepatic engorgement Hemorrhage in abdominal viscera

dilutions of serum) are injected into the skin and the site is injected 24 to 48 hours later with the corresponding antigen. Such reactions are called the Prausnitz–Kustner (P–K) reaction in man or passive cutaneous anaphylaxis in animals.

In vitro studies of immediate hypersensitivity use smooth muscle, basophils, or mast cells from either actively sensitized animals or cells from nonsensitized animals that are passively sensitized with antibody *in vitro*. The challenge of these cells or tissues with antigen results in the contraction of the smooth muscle or the release of mediators from the mast cells or basophils.

The Antibody Involved

The presence of antibodies in the serum of allergic patients mediating the immediate hypersensitivity reaction was first reported in 1921. However, it was only after progress in the characterization and classification of the immunoglobulins that a unique class of immunoglobulin was recognized as responsible for causing allergic reactions and was named IgE (Ishizaka and Ishizaka 1975). The original purification from serum was accomplished by the painstaking work of the Ishizakas and was

Table 2. Immediate Hypersensitivity Reactions

Actively immunized	Passively immunized

In vivo
  Generalized anaphylactic reaction
  Localized—cutaneous wheal-and-erythema
	In vivo
	  Generalized anaphylactic reaction
	  Localized—cutaneous

In vitro
  Mast cells ⎫
  Basophils ⎭ Mediator release
  Smooth muscle contraction (Schultz–Dale)
	Prausnitz–Kustner reaction
	    Passive cutaneous anaphylaxis
	In vitro
	  Same as actively immunized

Table 3. Characteristics of IgE

Structural	Physiological
Molecular weight: 190,000	Concentration in serum: 10 to 250 ng/ml
Carbohydrate: 10 to 12%	Half-life in serum: 2 to 4 days
2 L chains (κ or λ)	Concentration for P–K reaction (in skin) 0.2 ng antibody
2 H chains (ϵ): no subclasses known	Activates complement through the alternative pathway at high concentrations
$\kappa_2\epsilon_2$ or $\lambda_2\epsilon_2$	
Molecular weight of ϵ chain: 71,000	Species specificity for binding to mast cells (homologous and closely related species only)
An extra heavy chain domain	

confirmed by the recognition of a myeloma belonging to this unique immunoglobulin class by Johannson and Bennich (1971). Our present understanding of this immunoglobulin class has been due to the availability of a number of IgE myelomas in man and more recently by the discovery by Bazan of a strain of rats which spontaneously develop myelomas at a high frequency, a large number of which are of the IgE class. More recently the development of the hybridoma technology has made it possible to produce IgE hybridoma antibodies with defined specificity.

The most important characteristic of IgE is its capacity to bind reversibly with high affinity to specific receptors on basophils and mast cells. Some of the characteristics of IgE are summarized in Table 3. The basic structure of IgE is similar to all immunoglobulins having two light chains (κ or λ) and two heavy chains. The heavy chains are unique in distinguishing its class specificity. The IgE molecule has a molecular weight of 190,000 with an 8.2 sedimentation coefficient; 10 to 12% of the molecule is carbohydrate. The molecular weight of the ϵ chain is the same as the μ chain, but greater than γ or α chains by a molecular weight of about 11,000. This is because the IgE molecule, like IgM, has a 5-domain polypeptide chain; this extra domain is the C-2 region of the molecule.

The Fab and Fc regions of IgE are compact and linked by a flexible hinge region. There is some evidence to suggest that the IgE molecule is less flexible than IgG. The unique ability of IgE to bind with high affinity to specific receptors on mast cells and basophils is mediated through the Fc portion of the molecule. This cell-binding ability of IgE is lost on heating at 56°C for 1 to 2 hours; thermal denaturation involves changes in both the $C_\epsilon 3$ and $C_\epsilon 4$ domains of the molecule. Biological activity of IgE is also lost on reduction of the molecule with reducing agents (e.g., 2-mercaptoethanol). The loss of the cell-binding activity accompanying reduction of IgE correlates with the cleavage of one interchain disulfide bond between the paired $C_\epsilon 2$ domains. A characterisitic of IgE binding to cells is that it is of high affinity only to mast cells and basophils of the species from which the IgE was obtained (Dorrington and Bennich 1978).

IgE is present in serum at very low concentrations (less than 250 ng/ml in most normal individuals). Although it has a short half-life in the circulation (about 3 days) it has a much longer half-life when bound to mast cells. Thus, it is estimated that IgE can persist at skin sites for several weeks.

Whereas human allergic reactions appear to be predominantly, if not totally, due to the IgE class of immunoglobulins, in animals, antibody of the IgG class also induces immediate hypersensitivity reactions (Stechuschulte 1978). The characteris-

Table 4. Comparison of IgE- and IgG-mediated allergic reactions.

	IgE-mediated	IgG-mediated
Antibody class	IgE	IgG
Molecular weight of antibody	190,000	155,000
Antibody stability to 56°C	Activity lost	Stable
Antibody stability to 2-ME (reduction)	Activity lost	Stable
Species specificity	Homologous	Homologous and heterologous
Amount required for a positive skin reaction	$\sim$1 ng	$\sim$100 ng
Affinity for cell receptor	Very high $(K\sim10^{-11}\,M)$	Low $\sim10^{-6}\,M$

itics of this immunoglobulin are different, the two reactions are compared in Table 4. The presence of such IgG immunoglobulins mediating allergic reactions has been observed in some human diseases (Parish 1978).

Abnormal Serum IgE Levels

Serum IgE levels are very low throughout fetal life, and most newborns have less than 1 ng/ml in their cord serum. Maternal IgE does not cross the placental barrier. Serum IgE levels continue to be low during the neonatal period; there is a progressive increase with adult levels being reached by the teens.

The serum IgE level is influenced by a number of factors including age, genetic, immunological factors, and environmental exposures. Significantly elevated levels (over 250 ng/ml) have been found in some patients with atopic disease; however, the serum levels of IgE are not clearly different and overlap with the IgE levels of normal individuals. The highest IgE levels have been noted in patients with atopic eczema, parasitic infestations, and the IgE myelomas. Elevated IgE levels have also been found in patients with some immunodeficiency disorders. Patients with Wiskott–Aldrich syndrome are characterized by a lack of a humoral response to polysaccharide antigens, susceptibility to infections, depressed cell-mediated immunity, thrombocytopenic purpura, and pruritic eczema; such patients have elevated IgE levels. Similarly elevated IgE levels have been noted in some patients with thymic hypoplasia or dysplasia syndromes (DiGeorge's and Nezelof's syndrome). Abnormalities in T-cell function may be the cause of these changes in IgE levels. Although some patients with selective IgA deficiency have elevated IgE levels, most IgA-deficient patients have normal IgE levels.

A group of pediatric patients have been described who have very high levels of IgE. They have recurrent severe infections of both the skin and respiratory tract. The infections are commonly due to *Staphylococcus aureus;* the patients also have atypical pruritic rashes. There is usually an accompanying eosinophilia and *in vitro* cell-mediated immunity to common antigens are depressed although responses to mitogens are normal (Buckley and Becker 1978).

Immunoglobulin E levels are depressed in a number of immunodeficiency states e.g., in the serum of patients with agammaglobulinemia or hypogammaglobulinemia and ataxia telangectasia. A number of healthy individuals also have been observed who have very low or undetectable serum IgE levels.

The Radioallergoabsorbent Test

The radioallergoabsorbent solid phase radioimmunoassay (RAST) is used to measure IgE antibody to antigens. Allergen is covalently linked to a solid matrix, the allergen-coated particles are incubated with the patient's serum, washed, and then exposed to radiolabeled anti-IgE. Although the allergen-coated particles absorb antibodies of all classes, only the IgE class is measured by this assay. After washing, the number of counts remaining becomes a measure of the specific IgE bound to the particles (Bennich and Johannson 1971).

Antigens Capable of Inducing Immediate Hypersensitivity Reactions

The term allergen refers to those antigens that cause allergic symptoms (King 1976). The known allergens are usually proteins and range in molecular weight from 2,800 to 65,000. However, most of the important allergens tend to be globular proteins and have a molecular weight between 20,000 and 40,000 (Marsh 1975; King 1976). In animal experiments, IgE-antibody responses have been induced to a variety of proteins and haptens with the possible exception of non-T-cell-dependent antigens (e.g., polysaccharides). Dextrans and polysaccharides in a small number of individuals can cause IgE-mediated allergic reactions. The complete amino acid sequence of several allergens is known; inspection of these sequences does not reveal any unique characteristics which could explain their allergenicity. Parasitic infestations result in high IgE levels, and a number of parasitic antigens are potent allergens. Allergic reactions also occur to haptens (e.g., drugs) which couple to proteins *in vivo;* one example of this type of reactivity is allergic reactions to penicillin (Levine 1966).

Most allergens are either inhaled, ingested, or injected antigens. Among the inhaled allergens, wind-borne pollens predominate; the most common pollen is ragweed, whose major allergen is antigen E. It has been estimated that the annual dose of antigen E absorbed by individuals is in the microgram range. The most common ingested allergens are cows' milk (which contains a number of proteins including casein, β-lactoglobulin, lactalbumin, serum albumin), fish allergens, and eggs. Injected allergens include medications, e.g., polypeptide hormones or drugs that act as haptenic determinants after *in vivo* conjugation to proteins. Insect stings inject their allergen; among the major allergens in bee venom are phospholipase A_2 and hyaluronidase.

Genetics of the IgE Responses

Familial predisposition to develop allergies has been long recognized. A genetic basis for this was postulated more than 50 years ago, although it appeared to be multifactorial.

Genetic studies in mice have suggested at least two mechanisms for the control of IgE production (Levine 1971). First was the recognition of the role of the histocompatibility-linked Ir-immune response genes. Complex antigens (e.g., ovalbumin, bovine gamma globulin) injected in limiting quantities resulted in immune response gene control of the antibody response. For example, $H-2^a$ mice are good responders to DNP-bovine gamma globulin whereas the $H-2^b$ mice are poor responders. Not only the IgM and IgG responses, but also the IgE response was controlled by the Ir-genes.

The second level of control was at the level of the total IgE response: the SJL mouse strain was incapable of an IgE response to all antigens tested except after very specialized immunization procedures. This defect is a recessive trait, probably a single gene, not sex linked nor linked to H-2 or to the allotype markers. The studies of Ovary and coworkers have demonstrated that this is due to the presence of a non–antigen-specific IgE suppressor T-cell with the Lyt-1 surface marker (Ovary et al. 1978).

Studies in mono- and dizygotic twins suggest that total serum IgE levels in man is determined genetically. A study by Marsh et al. (1976) found that low serum IgE levels are inherited as a single dominant gene. A number of studies have found an association between HLA type and allergic reactivity to some of the ragweed allergens. There was better association with HLA among the individuals with the low serum IgE phenotype (Marsh et al. 1979). The data suggest that the development of allergies is controlled by multiple interacting genes and cannot be explained on the basis of a single gene.

The Cells Involved

Mast cells and basophils play a critical role in immediate hypersensitivity reactions and have several features in common: they both have IgE and IgG receptors, they have granules containing histamine formed in the cell by a cytoplasmic enzyme, histidine decarboxylase. In general, both basophils and mast cells seem to have similar receptors sensitive to pharmacologic modulation of their secretory activity and their pathways for secretion are similar.

The basophils are polymorphonuclear granulocytes that differentiate in bone marrow and enter the circulation. They comprise 0.5 to 1.0% of the circulating leukocytes, but in human beings basophils contain all the histamine in blood. In some species, e.g., rabbits, there is also histamine in platelets. Within the different animal species there is usually an inverse relationship between the frequency of basophils in the circulation and the number of mast cells in tissues. The basophils are relatively short lived, smaller than mast cells, with granules that are, however, larger than those of the mast cells.

Mast cells are long lived and are usually abundant near epithelial surfaces, and connective tissues and are close to blood vessels and peripheral nerves (Table 5). The mast cells divide even when mature (Padawer 1978). Although human mast cells contain histamine but no serotonin, rat and mouse mast cells contain both histamine and serotonin. The origin of mast cells is in dispute: studies suggest that they can come from mononuclear cells of the thymus, spleen, or lymph nodes. Recent studies suggest that the mast cell precursors are derived from hematopoietic tissues coming predominantly from the bone marrow (Kitamura et al. 1979) (Figure 1).

Basophils have the capacity for chemotactic migration and respond chemotactically to a number of factors including C5a, bacterial culture supernatant, antigen-induced lymphokine, and fMet-peptide (Ward 1978). These factors also attract monocytes and neutrophils and are not specific chemoattractants for basophils. However, the response of the basophil appears to be faster than that of neutrophils. The C5a-induced chemotactic response is augmented by lymphokine-containing fluids (Table 5).

Basophils have been observed in large numbers in the delayed-onset reaction called cutaneous basophil hypersensitivity reaction, especially in guinea pigs immunized with antigens in adjuvants in the absence of mycobacteria (Askenase 1977). This

Table 5. Comparison of Basophils and Mast Cells

	Basophil	Mast cell
Origin	Bone marrow	Bone marrow precursor (?)
Mitosis in mature cell	No	Yes
Distribution	Blood 0.5 to 1% of leukocytes	Connective tissue especially epithelial tissue and perivascular
Life span	Short (8 to 12 days)	Long: months to years
Size	10 to 14 μ	10 to 30 μ
Metachromasia	Yes	Yes
Granule size	0.8 to 1.2 μ	0.1 to 4 μ
Granule structure	Dense particles	Whorls
Pinocytosis	Yes	Yes
Phagocytosis	Yes (?)	Yes
Locomotion	Yes	No
Histidine decarboxylase (cytoplasmic)	Yes	Yes
Histamine content	1 to 2 pg/cell	20 to 30 pg per cell (rat) 3 to 8 pg per cell (guinea pig)
IgE receptors	Yes	Yes
IgG receptors	Yes	Yes
Degranulation (mediator release)	Yes	Yes
Granule proteoglycan	Chrondroitin sulfate (dermatan sulfate)	Heparin
Peroxidase (granular)	Yes	No
N-acetyl-β-D-glucosamidase (granular)	No	Yes
Zinc (granular)	Yes	Yes

reaction can be transferred by sensitized T- or B-cells or by IgG_1 antibodies. Evidence also suggests that mast cells can be triggered by T-cells to release vasoactive amines and these contribute to the pathogenesis of cell-mediated reactions of the delayed type. Thus, mast cell- and basophil-released mediators could play an important role in both immediate and delayed hypersensitivity reactions.

The IgE Receptor (IgE–Cell Interactions)

IgE has the unique characteristic of binding very firmly to surface receptors on basophils and mast cells. Antigens then react with the cell bound IgE, triggering the release of histamine and other mediators. Recently, binding of IgE to lymphocytes (predominantly B- but also T-cells) has been described; however, this binding is of lower affinity than IgE binding to the basophil-mast cell receptors.

The binding of IgE with its receptor is by a simple bimolecular forward reaction with first order dissociation (Metzger 1978). In studies with rat mast cells, the forward rate constant of IgE binding was found to be $\simeq 1 \times 10^5$ M^{-1} sec^{-1} at 37°C. The reverse rate constant (i.e., the dissociation rate) from whole cells is very slow, nonlinear, and therefore difficult to measure accurately. However, this rate could be estimated for the free receptor at 4°C and was found to be $\simeq 3 \times 10^{-7}$ S^{-1}. Therefore, the binding of IgE to the cell surface is extremely tight with a $K_A > /10^{-10}$ M^{-1}. This tight binding suggests multiple noncovalent interactions. Binding of IgE is through

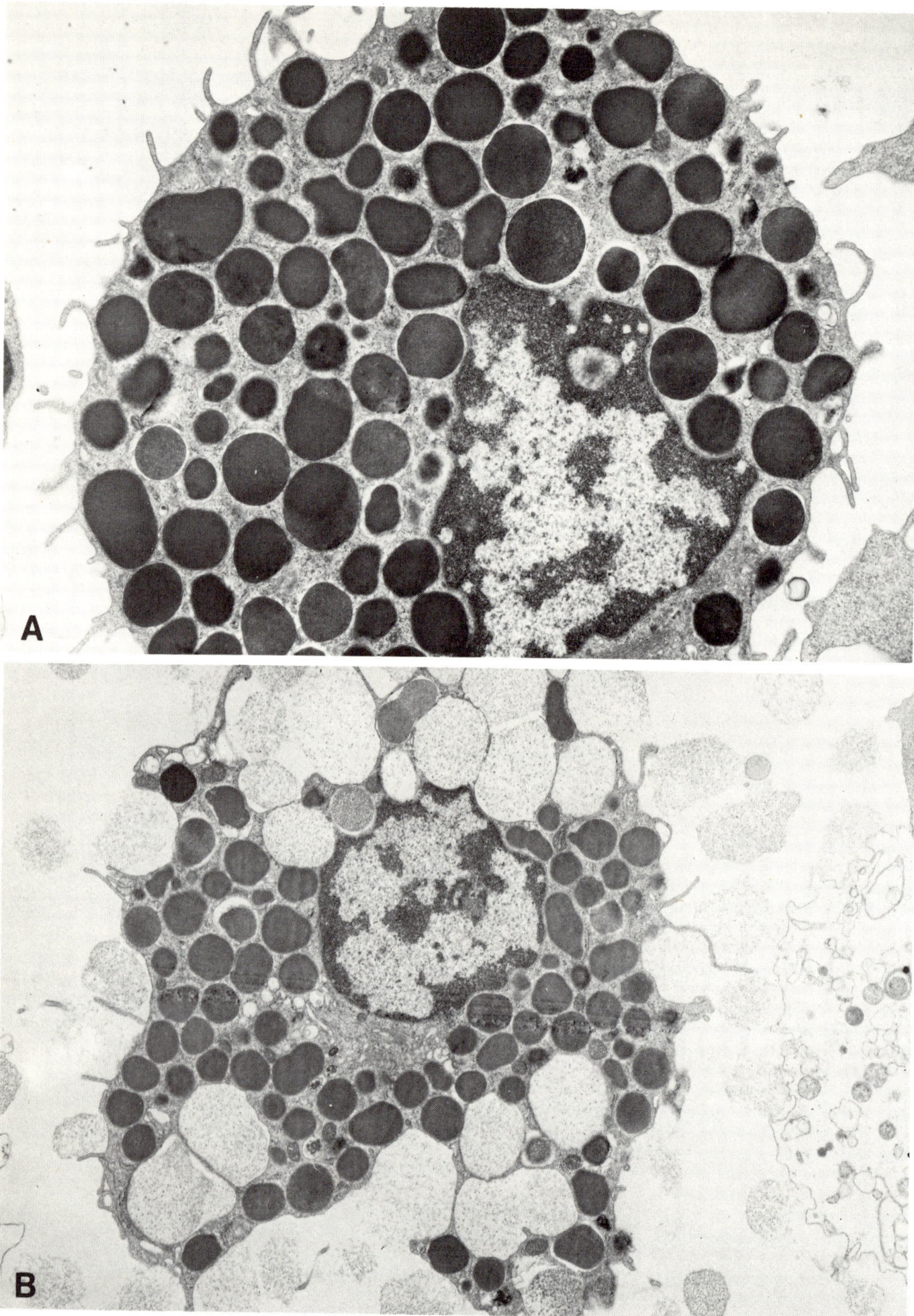

Figure 1. (A) Rat mast cell with many granules and surface protrusions ($\times$ 40,000). (B) Rat mast cell stimulated with concanavalin A for 15 minutes at 37°C in the presence of phosphatidylserine. A large number of granules have released their contents and several appear to be interconnected ($\times$ 34,000). (Courtesy D. Lagunoff and Emil Y. Chi.)

the Fc portion of the molecule. The isolated Fc fragment of the IgE binds to cells, whereas subfragments of the Fc lack this activity. $F(ab)_2$ fragments that contain the $C_\epsilon 2$ domain but lack the Fc fragment fail to bind to the IgE receptor. It has been recognized for a long time that heating of IgE results in a loss of its skin sensitizing (Fc_ϵ receptor) binding characteristics. This is correlated with structural changes within the Fc portion of the molecule probably involving both the $C_\epsilon 3$ and $C_\epsilon 4$ but not the $C_\epsilon 2$ domains. It has been proposed that the cell-binding site may be formed jointly by more than one domain. The reduction of IgE by reducing agents such as mercaptoethanol also results in loss of its cell-binding site. This appears to be due to a cleavage of one interchain disulfide bond between the paired $C_\epsilon 2$ domains, causing a loss in the quarternary structure of Fc. The IgE–Fc receptor interaction also demonstrates considerable species specificity, e.g., rat but not human IgE can bind to the rat mast cell receptor. Therefore, the integrity of the whole molecule and the close fit are essential for binding (Dorrington and Bennich 1978).

The IgE receptors are distributed diffusely over the surface membrane of basophils or mast cells and by autoradiography can only be visualized on the plasma membrane. All the active receptors are available on the cell surface and no additional receptors are found when the cell is solubilized. The IgE bound to the cell surface can be eluted by acid pH. The receptor is probably deeply embedded in the membrane with little of it exposed to the exterior and therefore not sensitive to proteolytic enzymes or to surface iodination in the presence of cell bound IgE. There is no evidence that the IgE receptor is exposed on the cytoplasmic side of the plasma membrane.

The number of IgE receptors on normal rat mast cells has been reported to be about $500,000 \pm 300,000$/cell, and about the same number of receptors are on a rat basophilic leukemia cell line. The number of receptors on the cell line varies with the cell cycle, with the highest number being present on stationary cells. The number of IgE receptors on human basophils is probably similar, but the actual determination is more prone to experimental error because basophil preparations are usually contaminated with other cells. Estimates of the number of IgE receptors on human basophils have suggested the presence of between 40,000 and 700,000 sites per cell; surprisingly, the number correlates directly with the concentration of serum IgE. Subjects with low serum IgE levels have basophils with a low number of IgE receptors, whereas atopic individuals have high serum IgE and high number of IgE receptors (Conroy et al. 1977; Malveaux et al. 1978). In several hormone receptor systems there is a decrease in the number of receptors when the hormone level is increased. This phenomenon has been called "down regulation" and frequently is due to a decreased synthesis of the receptors. Such changes have not been observed in the IgE receptor of cultured rat basophilic leukemia cells when the cells were exposed to high levels of IgE.

The IgE-receptor complexes are predominantly freely mobile in the membrane and uniformly distributed on the basophil or mast cell surface. Photobleaching studies demonstrate that a small fraction (about 20%) are immobile and that the mobility of the IgE-receptor is about 40 times less than that of lipids in the membrane. When cells are activated with antigen to release histamine, there is no gross redistribution of the IgE receptors into patches. Furthermore anti-IgE results in histamine release before or in the absence of redistribution. By photobleaching techniques no measurable changes in IgE-receptor mobility were observed at concentrations of anti-IgE sufficient to activate histamine secretion. However, higher levels of anti-IgE result in patch and cap formation and decreased receptor mobility with inhibition of histamine release. Therefore, the two events of cell triggering and patch–cap formation can be dissociated. Two IgEs with different fluorescent markers are capped independently of one another on rat mast cells, suggesting that the IgE receptors are univalent and independently mobile. However, some cocapping experiments with human basophils suggest that the IgE receptors are either multivalent or interconnected.

Mast cell or basophil membranes can be surface iodinated, reacted with IgE, the cell membranes solubilized, reacted with anti-IgE, and the specifically precipitated proteins analyzed by gel electrophoresis. Such studies have indicated that the IgE receptor has an apparent molecular weight of about 55,000. The receptor is a glycoprotein, and recent data suggest that another approximately 30,000 molecular weight non-surface-labeled protein copurifies with the receptor. There are a number of laboratories attempting to purify material to sequence the IgE receptor.

Regulation of IgE Synthesis

The collaboration of bone-marrow-derived precursors of antibody-forming cells (B-cells) and of thymus-derived lymphocytes (T-cells) is essential for the induction of antibody production to most antigens. This is true for the production of not only IgM and IgG but also the IgE class of antibodies. Athymic nude mice do not produce an IgE response, and T-cell-independent antigens do not induce an IgE response in experimental animals.

This section emphasizes the unique features in the control mechanisms of the IgE antibody response. There are two main features: (1) IgE synthesis seems to be more susceptible to regulatory control than the other immunoglobulin classes and (2) the T-helper and pre–B-cells involved in the IgE response are distinct from those involved in the response to other immunoglobulin classes.

The production of an IgE response in experimental animals depends on a number of factors (Tada 1975; Ishizaka 1976). Some antigens produce better IgE responses than others, e.g., protein antigen from the helminthic parasites. For example, *Ascaris* antigens are excellent for inducing an IgE response in mice and rats. Low doses of antigen favor IgE responses, whereas high doses result in IgG antibody formation. The use of alum as adjuvant favors IgE production, whereas complete Freund's adjuvant results predominantly in an IgG response. Helminthic infections lead to high levels of serum IgE and in the rat augment the IgE response to other previously injected antigens. Genetic factors are also important, e.g., SJL mice are incapable of an IgE response under most immunizing conditions. Immunoglobulin E antibody synthesis is regulated in part by non-IgE humoral antibody against the same antigen. Passively administered IgG antibody before immunization will suppress an IgE antibody response; whereas IgM antibody will enhance the IgE response. (Table 6).

Table 6. Factors that Favor IgE Production

Low dose of antigen
Alum as adjuvant
Genetic predisposition, e.g., low versus high IgE-producing mouse strains
Certain antigens, e.g., helminthic
Helminthic infection
Route of immunization

Developmental studies suggest that the earliest B-cells have only μ chain on their surface. The early stages of the differentiation of the IgE producing cells are T-cell independent. During maturation these cells acquire δ on their surface and therefore have both μ and δ on their membrane. Some of these cells then acquire ϵ (IgE) on their surface. In the adult rat spleen about 6% of the cells have IgE on their surface and more than 95% of these cells also carry surface IgM. Surface IgE-bearing lymphocytes comprise 2.5% of the lymph node cells and 10 to 20% of the B-lymphocytes. After parasitization there is a marked increase in the IgE-bearing lymphocytes to as high as 50% of the B-lymphocytes of the mesenteric lymph nodes. A 10,000 to 20,000 molecular weight factor is released from the T-cells, which induces this B-cell maturation. This factor has affinity for IgE and binds selectively to IgE-bearing B-cells. T-cells are not required for the initial development of IgE surface-positive cells; however, they are essential for the conversion of these IgE surface-positive cells to IgE-secreting cells. Experiments indicate that B-memory cells committed for IgE are made of either only IgE-bearing cells or also IgE-IgM-double-bearing cells. The IgE-producing cells are both radioresistent and long lived (Ishizaka and Ishizaka 1978).

Immunoglobulin-E-forming lymphoid cells are in high density in the human respiratory and gastrointestinal mucosa and in tonsil and adenoid tissue. In contrast, IgG-forming cells predominate in peripheral lymph nodes and in the spleen. However, in rabbits and mice, IgE-forming cells are common in the spleen. The preferential presence of IgE-forming cells in the respiratory–gastrointestinal tract of humans might be due to the more frequent exposure of these surfaces to environmental allergens. Some evidence indicates that IgE is synthesized and acts locally.

T-Cell Regulation of IgE Production

T-Cell Help. T-helper cells (Lyt-1) are essential for mounting an IgE antibody response. The data from *in vitro* secondary responses of rabbit or mouse lymph node cells suggest that there is a specific T-helper subpopulation for IgE production, which is distinct from the T-helper subpopulation for IgG formation. Further evidence for this is the isolation of a factor from a T-cell hybridoma that can specifically enhance the IgE but not the IgG response. *In vitro* studies have allowed the isolation of a factor that enhanced the IgE but not the IgG response; this factor has a molecular weight of 20,000 to 40,000, did not have immunoglobulin determinants, and was not absorbed by antigen-coated immunoabsorbent. Unlike this factor, which is non–antigen-specific, an antigen-specific factor can be extracted from rat thymocytes capable of restoring the ability of thymectomized rats to form IgE. The factor has a molecular weight of about 150,000 and has immunoglobulin determinants and carrier specificity. The possibility that this might be monomeric IgM has not been ruled out.

Non–Antigen-Specific T-Cell Suppression. Several lines of evidence support the existence of nonspecific T-suppressor cells: IgE responses can be suppressed by the transfer of normal thymocytes or spleen cells from hyperimmunized mice into animals producing an IgE response. Conversely, experimental procedures known to interfere with suppressor cell function, such as low dose irradiation, cycloheximide, or anti-lymphocyte treatment, will enhance the IgE response in both low responder, e.g., SJL, AKR mouse strains, and normal responder mice. The SJL mouse strain fails to produce an IgE response following normal immunization because of a hyperactive T-cell mediated suppression by the Lyt-1 subset of T-cells (Ovary et al. 1978), whereas in other systems suppressor cells belong to the Lyt-2 subclass of T-cells. It is therefore possible that the Lyt-1 cells in SJL mice act through an amplification loop to recruit other T-cells. Allogeneic splenic T-cells in low responder mice will cause IgE-suppression.

The non–antigen-specific T-cell suppression of IgE responses is enhanced by some adjuvants. In mice injected with complete Freund's adjuvant, a factor appears in the serum that causes IgE suppression. The factor is strain specific, heat stable, non-antigen specific, with a molecular weight of 150,000 but is not an immunoglobulin. It is, however, removed by antibodies to β-2 microglobulin. Whereas the serum of low–IgE-responder mice contain this suppressor factor, the serum of high IgE responders develop the factor only after challenge with complete Freund's adjuvant (Katz 1978).

A non–antigen-specific but IgE class-specific suppressor factor has also been generated *in vitro* from spleen cells of mice immunized with DNP coupled to mycobacteria. The factor does not have antigen or immunoglobulin determinants. The factor is absorbed by anti-H-2, its effect is strain specific and acts on IgE B-cells. A T-cell hybridoma has been established that secretes IgE class-specific and antigen-nonspecific suppressor factors. The suppressive effect is mediated through the inactivation of normal or antigen-primed B-cells responsible for IgE production.

Antigen-Specific T-Suppressor Cells. Antigen-specific T-suppressor cells can be generated by repeated immunization with large doses of antigen or by the injection of denatured or chemically modified antigen. *In vitro* antigen-specific T-suppressor cells can be generated by the presentation of antigen to the lymphocyte in the absence of macrophages. The coupling of the protein antigen to mouse gamma globulin or to polyethylene glycol induces T-suppressor cells to the protein antigen. Hapten coupled to polyvinyl alcohols or to N-acetyl muramyl peptides results in the activation of hapten-specific T-suppressor cells.

Tada and co-workers (1975) isolated a factor in extracts of mechanically disrupted thymocytes or spleen cells that suppressed IgE responses. The factor has rat thymocyte determinants, was removed by antirat immunoglobulin, and has a molecular weight of 35,000 to 60,000, three findings suggesting that the molecule might contain the antigen receptor for T-cells.

There are multiple sites at which a T-suppressor cell could influence an IgE response: (1) it could inactivate IgE producing cells directly, (2) it could block the interaction of T-helper with the B-cells, or (3) it could prevent the differentiation of the T-helper cells responsible for IgE formation. The suppressive effects occur at the T-dependent part of the maturation of B_ϵ cells (Table 7).

Tolerance of the IgE-B-cells can be induced by a number of experimental procedures. Hapten coupled to nonimmunogenic carriers like isologous gamma globulin, ficoll, or the copolymers of D-glutamic acid or D-lysine result in suppression of IgE antibodies to the haptens. This is due either to the elimination or direct inactivation of hapten-specific B-cells. Table 8 summarizes the different experimental approaches that have resulted in the suppression of IgE responses (Lee and Sehon 1978).

Recently a large percentage of B-lymphocytes have been shown to carry Fc_ϵ receptor, which will bind IgE molecules onto their surface (Spiegelberg and Melewicz 1980). The binding of IgE to these receptors is of much lower affinity than the binding to receptors on basophils or mast cells. Fc_ϵ receptors on T-cells have also been found: these Fc_ϵ-positive T-cells play a role in the control of IgE differentiation and release

Table 7. Suppression of IgE Response

By humoral IgE antibodies: negative feedback; only for primary responses
By suppressor T-cells
Antigen nonspecific
Complete Freund's adjuvant immunization
Genetically determined hyperactive T-suppressors
Antigen specific
Hapten specific
Carrier specific
Denatured antigen
Multiple large doses of antigen
In vitro antigen presentation without macrophages
By tolerance of IgE B-cells:
Hapten-DGAL
Hapten-mouse gammaglobulin
Hapten-Ficoll
Hapten-mycobacteria

a factor that binds selectively to IgE-bearing B-cells and enhances the differentiation of the precursors of the IgE-forming cell.

Therefore, although IgE is a major immunoglobulin on the B-cell surface, the differentiation of these cells to IgE production is under a number of controls. The level of IgE produced is due to the balance between the opposing forces of the T-helper cells and the T-antigen-specific and nonspecific suppressor cells, as well as the negative feedback control of IgG antibody.

Mediators of Immediate Hypersensitivity

Immediate hypersensitivity reactions are caused by secretory products released from mast cells and basophils. These reactions can be subdivided into those present pre-

Table 8. Modification of IgE Response

	Response abrogated			Mechanism	
Compound used	Primary	Secondary	IgE class specific	T-suppressor	B-cell tolerance
Responses to Hapten DNP–DGL	+	+	No	No	+
DNP–gamma globulin	+	+	No	No	+
DNP–polyvinyl alcohol	+	+	No	Yes	(+)
DNP–mycoloyl muramyl peptide	+	+	Yes	Yes	?
DNP–mycobacteria	+	+	Yes	Yes	?
Carrier responses					
Denatured antigen	+	+	+	Yes	No
Antigen–polyethylene glycol	+	+	+	Yes	No
Antigen–myramyl peptide	+	±	+	Yes	?
Oral antigen administration	+	No	No	Yes	Yes
Nebulized antigen	+	No	Yes	?	?

Table 9. Mediators of Immediate Hypersensitivity: Preformed Mediators

Mediator	Physicochemical characteristics	Biological effect	Inhibitor or antagonist	Mechanism of inactivation
Histamine	MW[a]: 111	Contracts smooth muscle; increases vascular permeability; inhibits histamine release	H-1: mepyramine H-2: cimetidine	Histaminase Histamine N-methyltransferase
Serotonin	MW: 176	Increases vascular permeability in the rat	LSD Methysergide Cyproheptidine	Monoamine oxidase
ECF-A	Val-Gly-Ser-Glu Ala-Gly-Ser-Glu MW: ~380	Chemotactic for eosinophils and neutrophils	N- and C-terminal tripeptides	Pronase, subtilisin aminopeptidase-M carboxypeptidase A
Intermediate MW ECF-A	Peptide MW., 1200–2500	Same as ECF-A	?	Carboxypeptidase
Neutrophil chemotactic factor	Protein MW: ~750,000	Chemotactic for neutrophils and eosinophils	?	?
Basophil kallikrein of anaphylaxis	Protein MW: 1,200,000	Generates bradykinin	Trypsin inhibitors	
Heparin	Proteoglycan MW: 750,000	Anticoagulant inhibits complement	Protamine	Heparinase
Chymase	Protein MW: 29,000	Chymotrypsin	Serotonin Heparin Chymotrypsin inhibitors	?
N-acetyl-β-d glucosaminidase	Protein MW: 150,000	Cleavage of glucosamine residue	Glurosamine	?
Arylsulfatase	Protein MW· 116,000	Inactivates SRS-A	Phosphate, sulfate	?

[a]MW, molecular weight

formed within the cell and those mediators generated after cell activation. In this section we will consider the different mediators, their physiological action, their inhibitors, and the mechanisms of their inactivation *in vivo* (Table 9).

Preformed Mediators

Histamine. Histamine is formed from the amino acid L-histidine by the action of the enzyme L-histidine decarboxylase. This enzyme is located in the cytoplasm of basophils and mast cells, whereas the cellular histamine is stored within the granules. In the granule, histamine is bound to a small molecular weight protein by a cation exchange mechanism.

Histamine enhances vascular permeability by acting on the endothelial cells of the postcapillary venules. The effect is to allow transudation into tissue spaces of plasma proteins. The bronchial constrictive effects of histamine are mediated by both irritant receptors and by reflex vagal action. Histamine is also a chemoattractant for eosinophils (Beaven 1976). The ability of histamine to act through H-2 receptors to modulate a number of lymphocyte functions including T-cell cytotoxicity, MIF production, T-cell proliferation, and lymphokine production has been described (Plaut 1979; Bourne et al. 1974).

There are two classes of histamine receptors on cells to which histamine binds, the H-1 and H-2. The H-1 receptors are blocked by the classical antihistamines such as mepyramine and are involved in the histamine-induced contraction of intestinal and bronchial smooth muscle. The H-2 receptors are involved prominantly in gastric acid secretion and are present on suppressor lymphocytes; these receptors are blocked by the H-2 blockers, e.g., cimetidine. The activation of the H-2 receptor by histamine on neutrophils, basophils, and lymphocytes results in the suppression of lysozomal enzyme or histamine release and allogeneic cell killing. Thus histamine has a negative feedback effect on its own secretion. This inhibition is probably through the elevation of the intracellular cyclic AMP level, since histamine has been shown to elevate cyclic AMP levels (Plaut 1979).

Histamine is rapidly inactivated *in vivo* by two pathways: one pathway involves methylation to form methylhistamine and then deamination by monoamine oxidase. The second route involves the oxidation of histamine by histaminase (a diamine oxidase) to imidazole-4-acetic acid, which is then conjugated and excreted in the urine as 1-ribosyl-imidazole-4-acetic acid. Whereas monocytes have the methylation pathway, granulocytes use the histaminase to inactivate histamine (Beaven 1976).

Serotonin. Serotonin is present in the mast cells of some species especially the rat and mouse: it is absent from human mast cells, but present in platelet granules. Serotonin is formed by decarboxylation of 5-hydroxytryptophan. It is stored in rat mast cell granules bound to the enzymatic site of the chymotryptic enzyme-chymase.

There is considerable species variation in the physiological responses to serotonin: in rats and mice it causes edema formation and plays an important role in anaphylactic reactions. Human bronchial tissue is not constricted by serotonin. Serotonin antagonists include LSD, methysergide, and cyproheptadine.

Serotonin is catabolized by monoamine oxidase present in a number of tissues, including monocytes and pulmonary endothelial cells.

Eosinophil Chemotactic Factors. A number of different factors chemotactic for eosinophils are released by activated mast cells. A low molecular weight factor called eosinophil chemotactic factor of anaphylaxis (ECF-A) has been isolated, characterized, and synthesized by Goetzl and Austen (1975). These two peptides have the following structure: Ala-Gly-Ser-Glu and Val-Gly-Ser-Glu. Both of these peptides act as preferential chemoattractants for human eosinophils and at higher concentrations cause directed migration of human neutrophils. Evidence suggests that the eosinophil receptor for these peptides binds to the N-terminal region and the cell is activated or inactivated by the C-terminal region. Intermediate molecular weight ECF-peptides have a molecular weight of 1200 to 2500 and are present in mast cells of both rats and humans.

The peptides attract eosinophils and neutrophils to the site of an immediate hypersensitivity reaction. The peptides are degraded by the action of proteinases.

Neutrophil Chemotactic Factor. Neutrophil chemotactic factor (NCF) has a molecular weight of 750,000 and is present in the venous return of cold-challenged arms of patients with cold urticaria. It is selectively chemoattractive for neutrophils.

Other Granular Material. A number of different molecules present in the granules are released together with histamine. (1) Heparin: rat mast cell heparin has a molecular weight of 750,000, whereas the heparin isolated from human lung (mast cells [?]) is much smaller. This heparin has only 10 to 20% of the anticoagulant activity of commercial heparin, which has a much lower molecular weight. It might function as a regulatory antiinflammatory agent. (2) Chymase: an α-chymotrypsin-like enzyme with a molecular weight of 29,000. (3) N-acetyl-b-D-glucosaminidase: a 150,000 molecular weight enzyme that degrades polysaccharides. (4) Basophil kallikrein of anaphylaxis: a 1,200,000 molecular weight enzyme released from human basophils is an arginine esterase that reacts with plasma kininogen to generate kinin (Newball et al. 1979a, 1979b). (5) Arylsulfatase A: an enzyme with the ability to inactivate SRS-A.

Newly Generated Mediators

Newly generated mediators are generated in the cell only after an appropriate stimulus (Table 1Q). Unlike the mediators just discussed they are not present preformed in the cell and therefore cannot be released by breaking up the cells by physical means (Lewis and Austen 1977).

Slow Reacting Substance of Anaphylaxis (SRS-A). A low-molecular weight sulfur-containing lipid, SRS-A is generated during mast cell activation. Recent structural studies of rat mast-cell-derived SRS-A have shown that arachidonic acid is a precursor of this molecule (Jakschik et al. 1977) and that SRS-A is formed by the newly described leukotriene pathway of prostaglandin synthesis as shown in Figure 2 (Murphy et al. 1979; Goetzl 1980). The molecules with SRS-A activity have been called leukotriene C (LTC$_4$), leukotriene D (LTD$_4$), and leukotriene E (LTE$_4$). Leukotriene C (LTC$_4$) is a glutathione-containing derivative of 5-hydroxy7,9,11,14 icosatetraenoic acid (Hammerstrom et al. 1979); the loss of the glutamic acid group results in the formation of leukotriene D (LTD$_4$), and the loss of the glycine residue results in the formation of leukotriene E (LTE$_4$). The ratio of the different leukotrienes released varies, depending on the tissue stimulated and the activating agent.

Slow protracted contraction of guinea-pig ileum or uterus is caused by SRS-A. This

Table 10. Mediators of Immediate Hypersensitivity: Newly Generated Mediators

Mediator	Physicochemical characteristics	Biological effect	Inhibitor or antagonist	Mechanism of inactivation
SRS-A	Leukotriene C,D,E Molecular weight: ~400	Contracts smooth muscles Increases vascular permeability	FPL 55712	Arylsulfatase A and B
PAF	Phospholipid(?) 1-0-alkyl-2 acetyl-sn- glyceryl-3 phosphoryl-choline	Aggregates rabbit platelets; causes vasoamine release	?	Phospholipase A, C, and D
Prostaglandin derivatives	Arachidonic acid PgE_2, TXA_2, PGD_2, PGI_2, HHT, HETE	Bronchial muscle contraction and relaxation, platelet aggregation, vasodilation, chemotaxis	?	Dehydrogenases
Superoxide	O_2^-	Cell damage		Superoxide dismutase

activity is not blocked by H-1 or H-2 anti-histamines, is not blocked by serotonin antagonists, but is inhibited by the specific anti-SRS-A antagonist FPL 55712. On a molar basis LTC_4, LTD_4, and LTE_4 are 50- to 500-fold as active as histamine in contracting the guinea pig ileum; SRS-A is more active in contracting the small rather than the larger airways. Human lung smooth muscle is very sensitive to SRS-A, which has been detected in the plasma of asthmatic children following bronchial provocation. Increased vascular permeability is also caused by intracutaneously injected SRS-A; synthetic LTD_4 has this activity but LTC_4 does not (Orange and Austen 1969).

Basophils, mast cells, neutrophils, and macrophages generate SRS-A following activation with an appropriate stimulus. The stimuli for the generation of SRS-A have included IgE- and IgG-mediated reactions with allergen and the calcium ionophore A23187. Unlike histamine, SRS-A is generated only after an appropriate stimulus and cannot be extracted from nonstimulated cells; the SRS-A is first detectable intracellularly and then is released into the medium.

Arylsulfatase B present in eosinophils and other arylsulfatase A and B enzymes present in mast cells and lung tissues inactivate SRS-A.

Platelet-Activating Factor. Platelet-activating factor (PAF) is a small molecular weight phospholipid released primarily from rabbit basophils following immunological stimulation by allergens (Osler and Siraganian 1972). It results both in the aggregation of rabbit platelets and the release of the histamine and serotonin from the platelets. Recent studies by Demapoulus et al. (1979) suggest that the rabbit PAF is 1-0-alkyl-2-acetyl-sn-glyceryl-3 phosphorylcholine (Figure 3). There is, however, suggestive evidence that the PAFs obtained from different sources are not identical on the basis of different susceptibilities to enzymatic degradation.

Rabbit platelets are aggregated by PAF and activated to release vasoactive amines. These two activities are independent of each other and function as an amplifying step

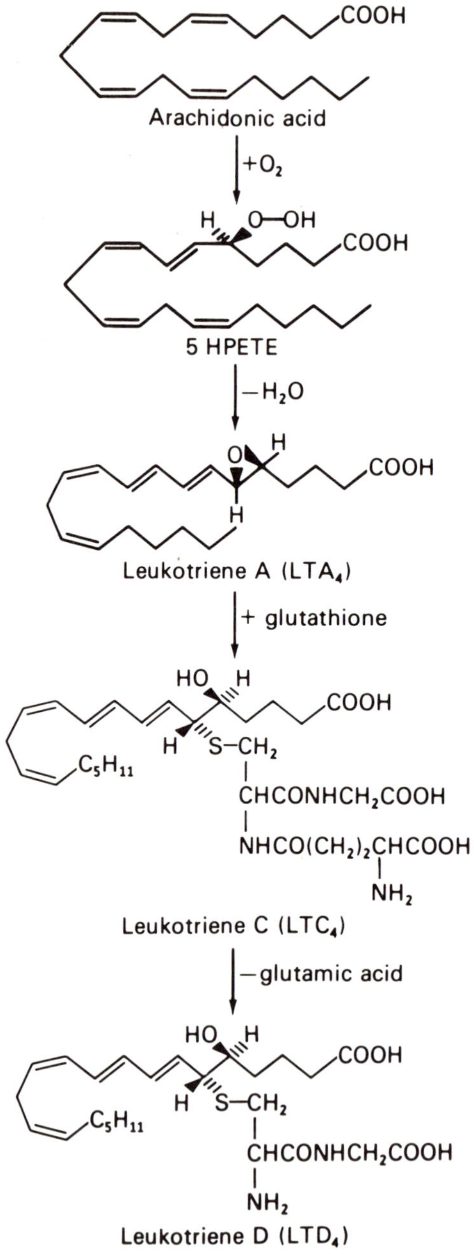

Figure 2. Proposed scheme for the biosynthesis of the leukotrienes. Subscript 4 indicates the total number of double bonds in the molecules.

for immediate hypersensitivity reactions. During acute anaphylactic reactions PAF is present in the circulation of rabbits, resulting in platelet aggregation and platelet segregation in the pulmonary vasculature. The release of platelet factor 4(PF4) is also induced by PAF and could play an important role in the deposition of immune complexes (Cochrane and Koffler 1973).

An IgE-dependent mechanism releases PAF from rabbit basophils; it has also been released from human basophilic leukemia cells and other cells by the calcium ionophore A23187.

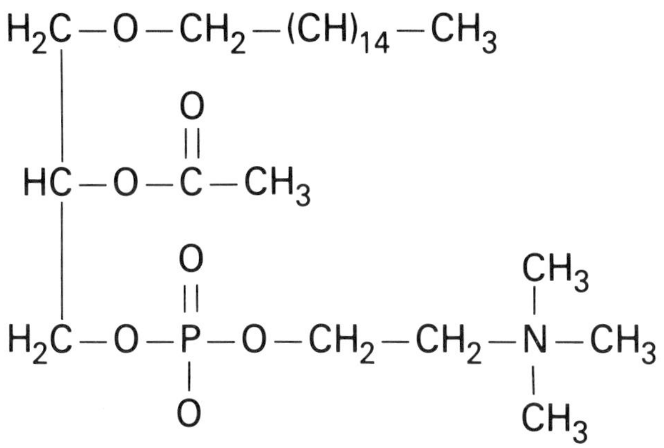

$$H_2C-O-CH_2-(CH)_{14}-CH_3$$

$$HC-O-\overset{\overset{\textstyle O}{\textstyle \|}}{C}-CH_3$$

$$H_2C-O-\overset{O}{\underset{O}{\overset{\|}{P}}}-O-CH_2-CH_2-\overset{\overset{\textstyle CH_3}{|}}{\underset{\underset{\textstyle CH_3}{|}}{N}}-CH_3$$

Figure 3. Structure of platelet-activating factor: 1—0-alkyl-2 acetyl-sn-glyceryl—3 phosphorylcholine.

Phospholipases inactivate PAF: some studies suggest that phospholipase D is active, whereas others found phospholipase A and C to be capable of this action. Human eosinophils contain polyholipases D, and therefore eosinophils might be important in breaking down this molecule.

Prostaglandins. Membrane perturbation causes activation of phospholipase A_2, which results in the release of arachidonic acid. The arachidonic acid may then be metabolized either by the cyclo-oxygenase pathway resulting finally in PGE_2, PGD_2, PGF_2, TXA_2, and PGI_2 or by the lipoxygenase pathway generating H-PETE, HETE and SRS-A.

A number of these metabolites have been shown to be released during cell activation including arachidonic acid, prostaglandins E_1, E_2, F_2, D_2, and TXA_2. A wide variety of biological effects for these compounds have been described including contraction or relaxation of smooth muscles and chemotactic activity (see also Chapter 14).

Superoxide. Mast cells generate superoxide in response to immunological or non-immunological secretory stimuli. The superoxide radical is formed by the univalent reduction of oxygen. The effect of the superoxide would be to potentiate the inflammatory reactions accompanying immediate hypersensitivity lesions. Superoxide radicals are removed by superoxide dismutase, which is also present in the mast cells (Henderson and Kaliner 1978).

Stimuli for Mast Cell or Basophil Degranulation

Basophils and mast cells can be triggered by a variety of different stimuli to degranulate. There are differences in the response of basophils and mast cells both from the same and different species.

Immunologic Stimuli

IgE-Mediated Stimulus. Both mast cells and basophils have high affinity receptors for the Fc part of the IgE molecules bound to the cell surface (see The IgE Receptor). The interaction of antigen with the cell-bound IgE triggers the cell to release its endogenous mediators. Similarly the cells can be activated by the use of heterologous antisera to IgE (Ishizaka and Ishizaka 1975). Mitogens, e.g., concanavlin A can also activate the mast cell to release histamine by cross-linking the carbohydrates of the IgE molecules (Siraganian 1976). Protein A from *Staphylococcus aureus* reacts with the Fc portion of IgE and releases histamine.

IgG-Receptor Stimulus. A mast cell and basophil receptor for IgG with lower affinity than the receptor for IgE has been described (Ishizaka et al. 1979; Segal et al. 1981). The binding of IgG is of low affinity and IgG is removed easily by washing the cells. However, the cells can be activated through this receptor for histamine release, but this appears to be a less efficient system than IgE-mediated cell triggering.

Antialloantisera Stimulation. Mouse mast cells can be activated by alloantisera for histamine release by a mechanism that requires the presence of the Fc fragment of the antialloantiserum. This activation depends on the antibody binding to the mast cell by its antibody binding sites and triggering the cell through the Fc portion of the molecule.

Complement Stimuli

C5a and C3a. Both anaphylatoxins release histamine from mast cells and basophils. With human basophils, C5a is much more active than C3a on a molar basis. The C5a-induced release involves a different receptor than the IgE-mediated release. The basophil and mast cell appear to have both C5a and C3a receptors (Hook et al. 1975; Siraganian and Hook 1976; Hugli and Müller-Eberhard 1978).

C3b Receptors. Although C3b receptors have been demonstrated on rat mast cells, they do not appear to mediate histamine release. A similar receptor on human basophils does modulate IgE-mediated histamine release (Thomas and Lichtenstein 1979).

Lymphocyte Mediators

Lymphokines may both directly induce basophil histamine release and modulate the level of release induced by other activators, e.g., IgE–antigen interaction. Viral and immune interferon modulate the level of histamine release by inducing metabolic changes in the basophil (Ida et al. 1978). This may be of clinical relevance in the exacerbations of asthma observed during viral infections.

Ionophores (A23187)

Ionophores are lipophilic compounds that insert themselves into membranes and transport Ca^{++} and Mg^{++} across the membrane. The ionophore A23187 triggers histamine release from basophils and mast cells when exogenous calcium is present in the medium (Foreman et al. 1973).

fMet-tri-Peptides

By stimulating basophils but not mast cells fMet-tri-peptides release histamine by binding to specific cell-surface receptors (Siraganian and Hook 1977). These compounds are also chemotactic for basophils, polymorphonuclear cells, and monocytes.

Other Polycationic Amines

Compound 48/80, which is a polymerization product of equimolar concentrations of methoxy-N-methylphenethylamine and formaldehyde, is very active in triggering rat mast cells. Other active compounds include polymyxin B, another cationic peptide antibiotic; morphine; tubocurarine; protamine; and ATP. Other compounds that activate mast cells include dextrans of 4×10^5 to 2×10^6 molecular weight, chymotrypsin, mellitin from bee venom, mast-cell–degranulating peptide isolated from cobra venom, and a cationic lysozomal granule protein.

Mechanism of Mediator Release

The release of histamine and other mediators from basophils and mast cells is a rapid process that is over in less than 20 minutes; the release is a secretory event that is noncytotoxic and does not require serum or other cofactors. There is, however, an absolute requirement for the presence of divalent cations. Allergic reactions depend on the presence of IgE on the basophil surface; therefore the concentration of specific antibody on the cell surface and its antigen-binding affinity are important parameters in the triggering of the cells. Release depends on the concentration of antibody, antigen, and their binding affinity. *In vitro* studies have shown that the system is very sensitive; with 1×10^6 washed human leukocytes, of which 1% are basophils, triggering occurs by adding as little as 1 ng/ml of ragweed antigen E (2.5×10^{-11} M). Although basophils have provided an important *in vitro* model for studies of immediate hypersensitivity, they have a serious limitation since basophils cannot be obtained as a pure cell population. Therefore biochemical studies with unpurified cell populations are difficult to interpret (Osler et al. 1968; Becker and Henson 1973; Ishizaka and Ishizaka 1975; Kaliner and Austen 1975; Kazimierczak and Diamant 1978).

Bridging Hypothesis

The studies by Levine (1966) demonstrated that bridging of two IgE molecules initiates the signal for histamine release. Studies with defined-length bivalent haptens have shown that the bridging occurs between two adjacent IgE molecules (Siraganian et al. 1975). Preformed IgE-dimers are active in triggering the cell indicating that this is the minimal size aggregate required (Segal et al. 1977). Mathematical analyses of the dose response curves indicate that the number of IgEs that have to be cross-linked for cell activation is only a small fraction of the total available specific IgE on the cell surface (DeLisi and Siraganian 1979a, 1979b). Recently it has been shown that antibodies to the IgE receptor trigger the cells; therefore, IgE only functions in conferring specificity to the mechanism of cell activation (Ishizaka and Ishizaka 1978; Isersky et al. 1978).

Activation of cells depends strongly on the concentration of the stimulus; excess (supraoptimal concentration) of the antigen results in significant decrease of secretion. However, unlike the antigen excess curves in the precipitin reactions, this is not due to lack of bridging of the antibody molecules. The mechanism for this phenomenon is not clear.

The fluid mosaic model of cell membranes postulates that proteins are fully diffusible in the plane of the lipid membrane and therefore randomly distributed across the cell surface. The IgE receptors appear to be freely mobile. The addition of divalent antibody to IgE results in some patch formation on the cell surface. At higher concentrations the patches coalesce to form caps; this cap formation probably requires the action of microtubules and microfilaments. The mobility of the IgE receptors as well as histamine release is inhibited by colchicine.

The mediators present in cells that can be activated for release are characteristically contained in granules. The activation of the cells brings about a series of events that are not completely characterized but probably involves the following steps:

1. Activation of the cell by crosslinking of receptors
2. Transmethylation of lipids
3. Ca^{++} fluxes across the membrane into the cytoplasm
4. A contractile process which moves the granules by an energy-dependent step to the membrane
5. Membrane fusion (fusion of the granule membrane to the plasma membrane)
6. Membrane fission with the granule opening to the extracellular space
7. Release of the mediators by an ion-exchange mechanism (Uvnas 1978).

Morphologic Aspects of Degranulation

Electron-microscopic studies have detected the changes occurring in mast cells and basophils during secretion. Scanning electron microscopy reveals a rat mast cell surface that is normally smooth; on stimulation of the IgE-mediated release there is rapid increase in cell size. The final event is the fusion of the granule and plasma membrane that is then followed by "fission" of this membrane and the opening of the granules to the outside (Lagunoff 1972; Anderson et al. 1973). As fusion progresses, the intramembranous particles are cleared and the membrane-associated protein molecules (including IgE-receptors) are displaced laterally (Chi et al. 1976; Lawson et al. 1977). As the granule approaches the cell surface, bulges appear on the surface membrane. The granule matrix is released as a whole to the outside, but the granule membrane is left behind. The histamine is released from the granule as an ion exchange between granule histamine and the cation Na^+ in the extracellular fluid. Granules deep in the mast cell open to the outside through interconnecting channels between several granules.

Biochemical Modulation of Mediator Release

Role of Divalent Cations

A requirement for divalent cations has long been recognized as essential for mediator release. The evidence for the essential role of Ca^{++} in the release process includes: (1) the ability of Ca^{++} chelators to completely inhibit these reactions; (2) the ability of Ca^{++} ionophore A23187 to transport Ca^{++} into the cell and release histamine; (3) the ability of intracellularly introduced Ca^{++} either by microinjection or by fusion with Ca^{++}-containing vesicles to activate the cells for secretion. The data therefore support the hypothesis proposed by Douglas that calcium influx into cells results in "stimulus–secretion coupling" (Theoharides and Douglas 1978). The hypothesis postulates that increased Ca^{++} concentration in the cytoplasm result in cell activation; the sources for this Ca^{++} could be the extracellular medium, the mitochondria, or the cell membrane (Rubin 1974).

Other steps in the secretory process also require Ca^{++}: basophils require Ca^{++}

during the late stages of release. In studies with rat mast cells, both early and late requirements for Ca^{++} have been described (Kaliner and Austen 1975).

Phospholipid Methylation

Transmethylation reactions play important roles in biological phenomena, membrane function, and cell activation. Two methyltransferase enzymes have been described to be asymmetrically distributed in the rat erythrocyte membrane. One enzyme located on the cytoplasmic surface converts phosphatidylethanolamine to phosphatidyl-N-monomethylethanolamine. The second enzyme is on the external surface and transfers two additional methyl groups from S-adenosyl-L-methionine to form phosphatidyl-choline. The methylation of these phospholipids precedes the Ca^{++} influx into the cells and occurs prior to histamine release, i.e., it is an early step in histamine release (Figure 4). Inhibitors of transmethylation reactions cause inhibition of histamine release from rat mast cells and from human basophils. However, these inhibitors do not effect the release induced by 48/80, ionophore, C5a, or fMet-peptides. The results suggest that methylation of lipids is a crucial step only in the IgE-mediated histamine release reaction (Crews et al. 1980; Morita et al. 1980).

Requirement for Metabolic Energy

Energy is required for the release of mediators from basophils or mast cells. A number of different inhibitors have been used to block energy use, including anoxia in the absence of glucose, cyanide, 2,4-dinitrophenol, 2-deoxyglucose, and antimycin A. Energy use during mediator release has also been demonstrated in a number of isolated cell systems. Whereas some cell systems require glycolysis, oxidative phosphorylation plays the dominant role in other cells.

Serine Esterase

A number of inhibitors (e.g., di-isopropyl fluorophosphate) phosphorylate serine residues in proteins and inactivate enzymes that contain serine in their active site. In some of the experimental systems the serine esterase inhibitors must be present during actual cell activation, suggesting their interaction with a proenzyme activated only after the start of the secretory process. Calcium is required for the proenzyme to become inhibitable by these esterase inhibitors, and the activation of these enzymes occurs early. However, some recent studies with a series of these inhibitors have suggested that the action of these compounds on the cell might not be caused by the interaction with an esterase (Taurog et al. 1979).

Phospholipid Metabolism

A number of changes in mast cell phospholipids occur during mediator release (Kennerly et al. 1979a, 1979b, 1979c). There is increased incorporation of ^{32}P label into phosphatidic acid, phosphatidylcholine, and phosphatidylinositol. The changes in phosphatidic acid occur before the onset of mediator release. Parallel changes in mediator release and phosphate incorporation into these lipids occur with a number of different inhibitors. The accelerated formation of these phospholipids could be due to the increased formation of their precursor, 1,2 diacylglycerol. The enzyme diglyc-

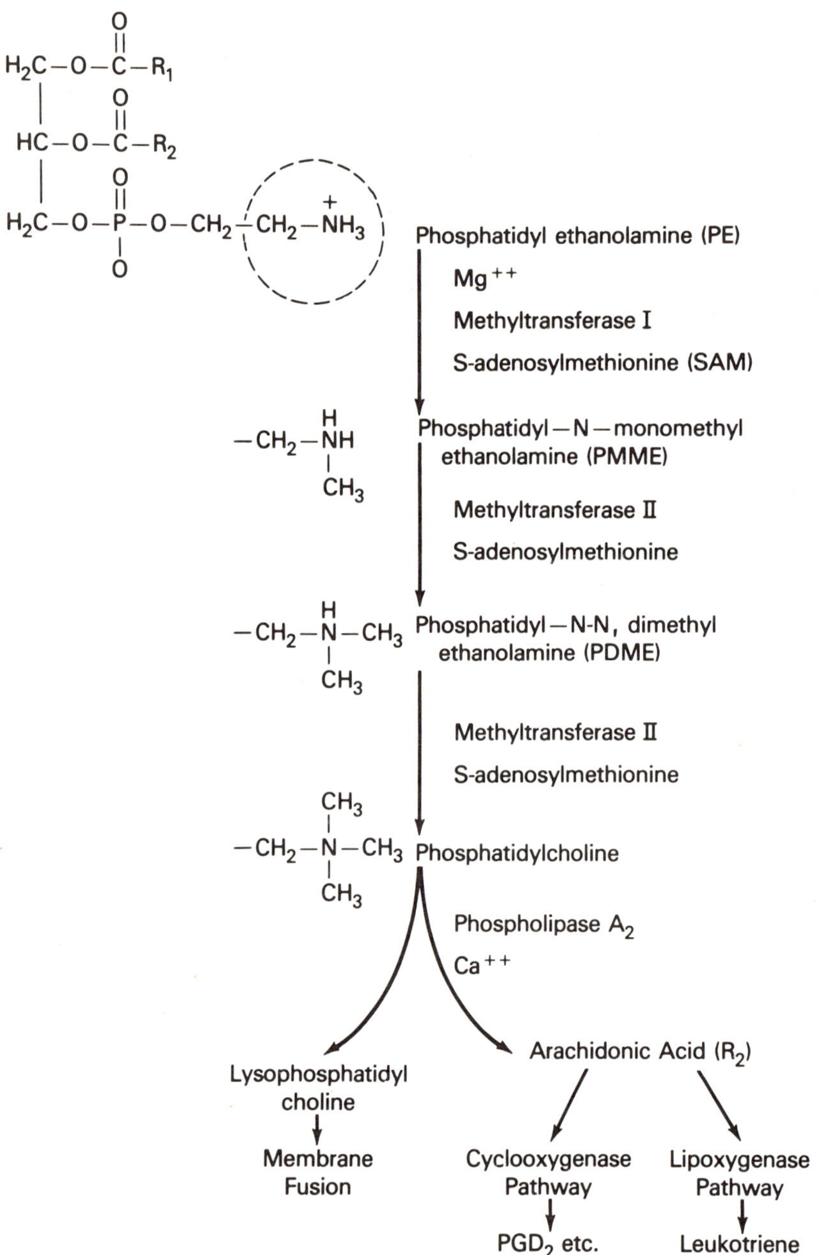

Figure 4. Phospholipid methylation pathways in the cell membrane and pathways of phospholipid breakdown.

eride lipase present in mast cells can convert 1,2 diacylglycerol to 1-monoacylglycerol and release a fatty acid, usually arachidonic acid, from the 2-position of the 1,2 diacylglycerol molecule. Both 1,2 diacylglycerol and monoacylglycerol enhance membrane fusion (Kennerly et al. 1979b).

During secretion, arachidonic acid is released from the cell; the source of this is probably the 2-position of phospholipids, e.g., phosphatidylcholine (Crews et al. 1980). The activation of phospholipase A2 would be crucial for this reaction. Inhibition of this enzyme with p-bromophenacyl bromide or mepacrine inhibits the release of histamine and arachidonic acid from the cell. The activation of phospholipase A2 would also result in the formation of lysophosphatidyl choline, which is known to cause membrane fusion, a crucial step in the secretory process. The arachidonic acid released from either pathway could then be the source for the production of biologically active metabolites, including prostaglandin D_2 and for the formation of SRS-A.

Both aspirin and indomethacin inhibit the cycloxygenase prostaglandin pathway but have no effect on histamine release from basophils or mast cells. However, another inhibitor (ETYA), which blocks both the cycloxygenase and lipoxygenase pathways, inhibits histamine release. This would suggest the importance of lipoxygenase pathway products in the secretory process (Sullivan and Parker 1979; Marone et al. 1979).

Cyclic Nucleotides

Cyclic nucleotides appear to modulate mediator secretion from cells. Such a modulating role for cyclic AMP and cyclic GMP has been postulated in a number of secretory systems. The intracellular levels of cyclic AMP can be elevated by the addition of exogenous cyclic AMP or by dibutyryl cyclic AMP, which penetrates membranes better and is more stable. The intracellular cyclic AMP level can also be raised by the interaction of a number of different agonists with their specific cell surface receptors. This interaction results in the activation of adenylate cyclase and the formation of cyclic AMP. Some prostaglandins (e.g., PGE_1), beta adrenergic agents, histamine acting through H-2 receptors, adenosine, and cholera toxin increase intracellular cyclic AMP levels, leading to inhibition of mediator release. Similarly, the level of cyclic AMP is elevated by the addition of inhibitors of the phosphodiesterase enzyme, e.g., theophylline. This also results in inhibition of histamine release. In contrast, the action of imidazole is to inhibit this phosphodiesterase and decrease the cyclic AMP levels. Alpha-adrenergic stimulants acting through a cell surface receptor decrease cyclic AMP levels in mast cells and result in enhanced mediator release (Bourne et al. 1974; Kaliner and Austen 1975; Sullivan et al. 1975a, 1975b).

So far it has not been possible to demonstrate a drop in the cyclic AMP level following cell activation, a prerequisite to the hypothesis that cyclic AMP is the second messenger in the mast cells. In fact, there appears to be an elevation in cellular cyclic AMP levels in mast cells following activation for IgE-mediated histamine release (Sullivan et al. 1976).

The level of cyclic GMP may also modulate the level of histamine release. The activation of muscarinic receptors by acetylcholine or carbamylcholine increases tissue concentrations of cyclic GMP and enhances IgE-mediated secretion from human lung fragments. However, no effects of these agents have been observed with human basophils.

Role of Microtubules and Microfilaments

Microtubules are intracellular structures found in most eukaryotic cells, including mast cells and basophils. They have a characteristic tubular structure formed from the protein tubulin. Microtubules are labile structures and in the living cells are in a state of constant dynamic equilibrium with a pool of tubulin subunits. The assembly of the tubules is controlled by the level of Ca^{++} and cyclic nucleotides. Colchicine and vinblastine binds to tubulin dimers and prevents their assembly into microtubules. There is some evidence that colchicine can also bind to membrane and other proteins.

Colchicine has been reported to inhibit histamine release from both basophils and mast cells. Conversely, deuterium oxide (heavy water) is known to stabilize microtubules and enhance IgE-mediated histamine release from human basophils, but it has no effect on human lung mast-cell-mediator release. The action of deuterium oxide is reversed by colchicine (Gillespie 1975).

Microfilaments are filamentous structures observed in many cells and are usually distributed close to the cell surface. They are thought to be actin-like polymers. The tubules are passive direction markers or are actively involved in the secretory or motile process. It is postulated that microtubules are inserted into both the cell membrane and vesicles. It has been suggested that receptors on the cell surface may exist in two interconvertible states: free, where they can diffuse in the cell membrane, or bound, where they interact directly or indirectly with microtubular or microfilament structures within the cell. The microtubules would therefore modulate the mobility of the cell receptors and play an essential role in histamine release.

Microfilaments are filamentous structures observed in many cells and are usually distributed close to the cell surface. They are thought to be actin-like polymers. The cytochalasins are fungal products that inhibit cellular movements; although they were originally thought to act directly on microfilaments, more recent studies suggest they have a number of effects, including inhibition of membrane transport systems. The effect of these agents on mediator release are quite variable. Both IgE- and C5a-mediated histamine release from human basophils is usually enhanced by the addition of cytochalasin B. With human lung fragments, histamine release is enhanced, whereas SRS-A formation is inhibited. Cytochalasin B also inhibits lateral diffusion of the IgE molecules. Therefore, the integrity of the microfilaments is correlated inversely with the ability of the cells to release histamine (Table 11).

A Hypothetical Scheme of the Release Reaction

A speculative scheme of the steps in histamine release is shown in Figure 5. The initial event is the binding of antigen by cell-bound IgE; the antigen interacts subsequently with another cell-bound IgE to result in bridging of the IgE molecules. The IgE-receptor aggregation activates the transmethylation enzymes resulting in phospholipid methylation to phosphatidylcholine. This results in localized membrane perturbations (with changes in membrane fluidity) and opens up Ca^{++} channels. The entry of Ca^{++} into the cell then results in the activation of a phospholipase A-2, which cleaves phosphatidylcholine or phosphatidylinositol to release arachidonic acid and form lysolecithin. The lysolecithin could function as a fusogen, promoting membrane fusion. The esterase-inhibitable step could be this phospholipase A-2 enzyme. The released arachidonic acid would be metabolized by either the cyclo-oxygenase or the lipoxygenase pathways, leading to metabolic products like prostaglandins (predominantly PGD-2 for the mast cells) and leukotriene C (SRS-A).

The entry of Ca^{++} into the cell together with the cellular cyclic AMP and cyclic GMP levels would regulate the next step in the release process. During this step the phosphorylation of proteins probably plays an important role (Sieghard et al. 1979). The microtubular aggregation is probably important during this phase of the release process; for these reactions to occur, metabolic energy is required that could have modulating effects at multiple steps.

The biochemical pathway for the generation and release of SRS-A has some unique differences from that involved in histamine release. Serine esterase is present in the activated state, diethylcarbamazine selectively suppresses the release of SRS-A but not

Table 11. Inhibitors and Enhancers of Mediator Secretion from Cells

Drug	Agonist (inhibitor)	Antagonist	Mechanism of action*	Effect on histamine release*	
				Basophils	Mast cell
Adenylate cyclase activators					
α-Adrenergic agents	Norepinephrine	Phenoxylbenzamine	↓ cyclic AMP	—	↑
β-Adrenergic agents	Isoproterenol	Propranolol	↑ cyclic AMP	↓	↓
Prostaglandins	PGE_1		↑ cyclic AMP	↓	↓
Histamine	Histamine 2-Methyl-histamine	H-2 blockers cimetidine	↑ cyclic AMP	↓	↓
Cholera toxin	Cholera toxin	Antitoxin	↑ cyclic AMP	↓	↓
Phosphodiesterase					
Inhibitor	Theophylline	—	↑ cyclic AMP	↓	↓
Activator	Imidazole	—	↓ cyclic AMP	—	↑
Guanylate cyclase activators					
Cholinergic	Carbachol	Atropine	↑ cyclic GMP	—	↑
Microtubular active agents					
Disaggregators	Colchicine	Heavy water	↓ Microtubulin aggregation	↓	↓
Microfilament					
Disaggregation	Cytochalasin B	?	↓ Microfilaments	↑	↑
Transmethylation inhibitors					
	3-Deazaadenosine	?	↓ Phospholipid methylation	↓	↓
Metabolic inhibitors					
	2-Deoxyglucose	Glucose	↓ ATP	↓	↓
Serine esterase					
	DFP	?	↓ Esterase activity	↓	↓
Phospholipase inhibitor					
	Mepacrine	?	↓ Phospholipase A_2	↓	↓
Mechanism unclear					
	Disodium chromoglycate	?	↓ Ca^{++} (?) channels	—	↓

*—No effect. ↑ —Enhances. ↓ —Decreases.

histamine, and cytochalasin B suppresses the release of SRS-A but enhances the release of histamine. In contrast, thiol compounds enhance SRS-A selectively but do not enhance histamine release. Because SRS-A release is more susceptible than histamine release to inhibition or enhancement by agents capable of modifying the levels of intracellular cyclic AMP or cyclic GMP, differences in the pathways for the release of different mediators exist within the cell.

Conclusion

The role of the immediate hypersensitivity system is to protect the organism against parasitic–environmental antigens. The mast cell with a specific IgE for a foreign substance is located in tissues; the IgE functions as the recognition unit for foreign

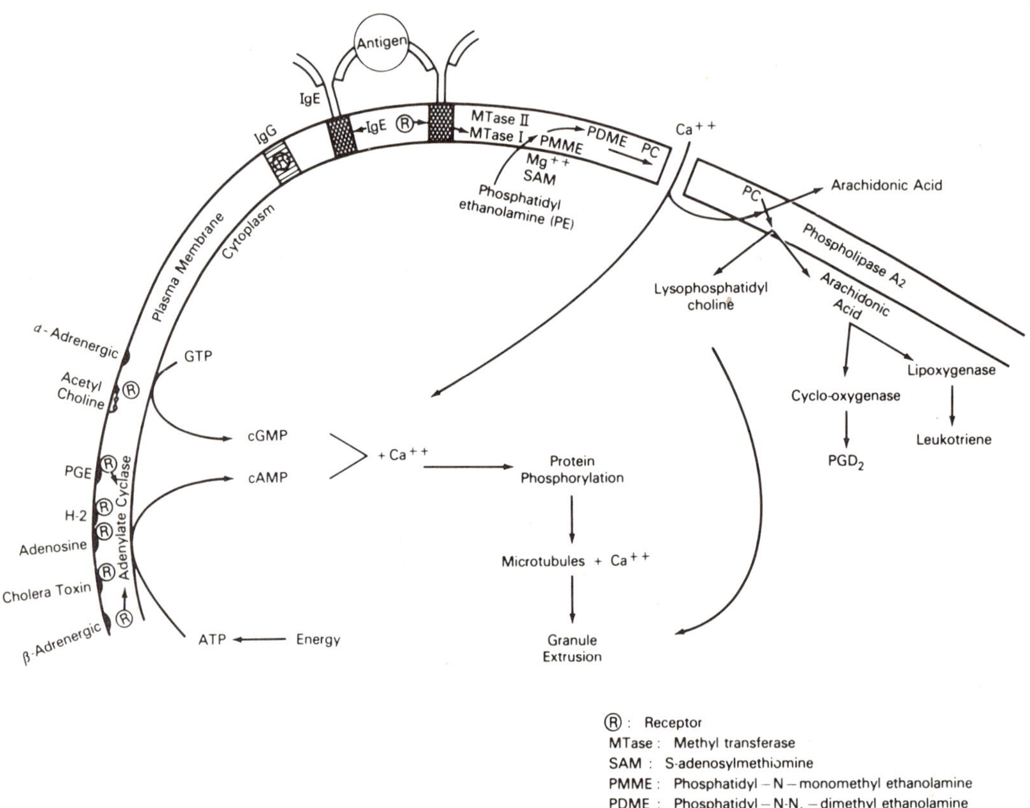

Figure 5. Hypothetical scheme of the steps in the release process in the mast cell.

substances. Once the mast cell is activated, it will release mediators that will recruit proteins and cells from the circulation to the reaction site (e.g., immunoglobulin of other classes, complement, phagocytic cells, and eosinophils). The arrival of all these cells results in both control of the immediate hypersensitivity reaction due to the degradation of the mediators and in the control of the foreign antigens, e.g., by the antibodies or phagocytic cells. In the case of parasites, the IgG antibodies and eosinophils could result in parasitic killing (Mitchell 1979).

Although immediate hypersensitivity reactions are classically short lived, prolonged reactions to the intracutaneous injection of allergens has been observed (Solley et al. 1976). In some patients, a wheal-and-flare progresses to pruritic erythematous-indurated lesions, peaking at 6 to 12 hours and subsiding in 24 hours. These responses are due to IgE-mediated reactions, not due to histamine. Histologically, the site contains mononuclear cells including neutrophils, basophils, and eosinophils with some fibrin deposition (de Shazo et al. 1979). Therefore, an IgE-initiated reaction can summon other cells into a site, and these cells cause a more pronounced inflammatory reaction and also clear the site of the mediators which had been released.

The therapy of immediate hypersensitivity reactions can take a number of different approaches (Table 12). The common end goal is to suppress the release of mediators or to prevent by pharmacological means the effects of these mediators. One therapeutic approach is to treat allergic patients with repeated small and increasing injections of the allergen. The effect of this is to eventually result in T-cell suppression and blocking

Table 12. Pharmacological Control of Mediator Release

Decreased IgE production Immunosuppressive drugs Immunotherapy (hyposensitization) Immunologic desensitization	*Block effect of mediators* Antihistaminics Antiserotonin agents Anti-SRS-A agents
Deplete basophils Corticosteroids	*Counteract effect of mediators* β-adrenergic agents Steroids Theophylline
Decreased mediator synthesis Histamine synthesis: steroids Prostaglandin synthesis: aspirin, indomethacin	
Decreased mediator release See Table 11	

antibodies. These antibodies are of the IgG class, do not bind to mast cells or basophils but compete with the cell-bound IgE for the allergen. Therefore, by binding the allergen before it can activate the cells they prevent the release of mediators. Blocking antibodies might also depress the synthesis of IgE specific for that antigen by a feedback inhibition type of mechanism. This desensitization therapy demonstrates only modest effectiveness. It is hoped that in the next few years, modified allergens that activate T-suppressor mechanisms will become available and useful for the treatment of patients.

References

Anderson, P., Slorach, S.A., Uvnas, B. (1973) Sequential exocytosis of storage granules during antigen-induced histamine release from sensitized mast cells *in vitro*. An electron microscopic study. Acta Physiol. Scand. 88,359–372.

Askenase, P.W. (1977) Role of basophils, mast cells and vasoamines in hypersensitivity reactions with a delayed time course. Prog. Allergy 23,199–320.

Beaven, M.A. (1976) Histamine. New Engl. J. Med. 294:30–36; 320–325.

Becker, E.L., Henson, P.M. (1973) *In vitro* studies of immunologically induced secretion of mediators from cells and related phenomena. Adv. Immunol. 17,93–193.

Bennich, H., Johansson, S.G.O. (1971) Structure and function of human immunoglobulin E. Adv. Immunol. 13,1–55.

Bourne, H.R., Lichtenstein, L.M., Melmon, K.L., Henney, C.S., Weinstein, Y., Shearer, G.M. (1974) Modulation of inflammation and immunity by cyclic AMP. Science 184,19–28.

Buckley, R.H., Becker, W.G. (1978) Abnormalities in the regulation of human IgE synthesis. Immunol. Rev. 41,288–314.

Chi, E., Lagunoff, D., Koehler, J.K. (1976) Freeze-fracture study of mast cell secretion. Proc. Natl. Acad. Sci. U.S.A. 73,2823–2827.

Cochrane, C.G., Koffler, D. (1973) Immune complex disease in experimental animals and man. Adv. Immunol. 16,185–264.

Conroy, M.C., Adkinson, N.F., Jr., Lichtenstein, L.M. (1977) Measurement of IgE on human basophils; relation to serum IgE and anti-IgE-induced histamine release. J. Immunol. 118:1317–1321.

Crews, F.T., Morita, Y., Hirata, F., Axelrod, J., Siraganian, R.P. (1980) Phospholipid methylation affects immunoglobulin E-mediated histamine and arachidonic acid release in rat leukemic basophils. Biochemical Biophys. Res. Comm. 93,42–49.

DeLisi, C., Siraganian, R.P. (1979) Receptor cross-linking and histamine release. I. The quantitative dependence of basophil degranulation on the number of receptor doublets. J. Immunol. 122,2286–2292.

DeLisi, C., Siraganian, R.P. (1979) Receptor cross-linking and histamine release. II. Interpretation and analysis of anomalous dose response patterns. J. Immunol. 122,2293–2299.

Demopoulos, C.A., Pinckard, R.N., Hanahan, D.J. (1979) Platelet-activating factor. Evidence for 1-0-alkyl-2-acetyl-sn-glyceryl-3-phosphorylcholine as the active component (a new class of lipid chemical mediators). J. Biol. Chem. 254,9355–9358.

de Shazo, R.D., Levinson, A.I., Dvorak, H.F., Davis, R.W. (1979) The late phase skin reaction; evidence for activation of the coagulation system in an IgE-dependent reaction in man. J. Immunol. 122,692–698.

Dorrington, K.J., Bennich, H.H. (1978) Structure-function relationships in human immunoglobulin E. Immunol. Rev. 41,3–25.

Foreman, J.C., Mongar, J.L., Gomperts, B.D. (1973) Calcium ionophores and movement of calcium ions following the physiological stimulus to a secretory process. Nature 245,249–251.

Gillespie, E. (1975) Microtubules, cyclic AMP, calcium and secretion. Ann. N.Y. Acad. Sci. 243, 771–779.

Goetzl, E.J. (1980) Mediators of immediate hypersensitivity derived from arachidonic acid. N. Engl. J. Med. 303,822–825.

Goetzl, E.J., Austen, K.F. (1975) Purification and synthesis of eosinophilotactic tetrapeptides of human lung tissue: Identification as eosinophil chemotactic factor of anaphylaxis. Proc. Natl. Acad. Sci. U.S.A. 72,4123–4127.

Hammarström, S., Murphy, R.C., Samuelsson, B., Clark, D.A., Mioskowski, C., Corey, E.J. (1979) Structure of leukotriene C: Identification of the amino acid part. Biochem. Biophys. Res. Comm. 91,1266–1272.

Henderson, W.R., Kaliner, M. (1978) Immunologic and nonimmunologic generation of superoxide from mast cells and basophils. J. Clin. Invest. 61,187–196.

Hook, W.A., Siraganian, R.P., Wahl, S.M. (1975) Complement-induced histamine release from human basophils. I. Generation of activity in human serum. J. Immunol. 114,1185–1190.

Hugli, T.E., Müller-Eberhard, H.J. (1978) Anaphylatoxins: C3a and C5a. Adv. Immunol. 26,1–53.

Ida, S., Hooks, J.J., Siraganian, R.P., Notkins, A.L. (1977) Enhancement of IgE-mediated histamine release from human basophils by viruses: Role of interferon. J. Exp. Med. 145,892–906.

Isersky, C., Taurog, J.D., Poy, G., Metzger, H. (1978) Triggering of cultured neoplastic mast cells by antibodies to the receptor for IgE. J. Immunol. 121,549–558.

Ishizaka, K. (1976) Cellular events in the IgE antibody response. Adv. Immunol. 23,1–75.

Ishizaka, K., Ishizaka, T. (1978) Mechanisms of reagenic hypersensitivity and IgE antibody response. Immunol. Reviews 41,109–148.

Ishizaka, T., Ishizaka, K. (1975) Biology of immunoglobulin E. Molecular basis of reaginic hypersensitivity. Prog. Allergy 19,60–121.

Ishizaka, T., Sterk, A.R., Ishizaka, K. (1979) Demonstration of Fc_γ receptors on human basophil granulocytes. J. Immunol. 123,578–583.

Jakschik, B.A., Falkenhein, S., Parker, C.W. (1977) Precursor role of arachidonic acid in release of slow reacting substance from rat basophilic leukemia cells. Proc. Natl. Acad. Sci. U.S.A. 74,4577–4581.

Kaliner, M., Austen, K.F. (1975) Immunologic release of chemical mediators from human tissue. Ann. Rev. Pharmacol. 15,177–189.

Katz, D.H. (1978) The allergic phenotype: manifestation of 'allergic breakthrough' and imbalance in normal 'damping' of IgE antibody production. Immunol. Reviews 41,77–108.

Kazimierczak, W., Diamant, B. (1978) Mechanisms of histamine release in anaphylactic and anaphylactoid reactions. Prog. Allergy 24,295–365.

Kennerly, D.A., Secosan, C.J., Parker, C.W., Sullivan, T.J. (1979a) Modulation of stimulated phospholipid metabolism in mast cells by pharmacologic agents that increase cyclic 3', 5' adenosine monophosphate levels. J. Immunol. 123,1519–1524.

Kennerly, D.A., Sullivan, T.J., Parker, C.W. (1979b) Activation of phospholipid metabolism during mediator release from stimulated rat mast cells. J. Immunol. 122,152–159.

Kennerly, D.A., Sullivan, T.J., Sylvester, P., Parker, C.W. (1979c) Diacylglycerol metabolism in mast cells: A potential role in membrane fusion and arachidonic acid release. J. Exp. Med. 150,1039–1044.

King, T.P. (1976) Chemical and biological properties of some atopic allergens. Adv. Immunol. 23,77–105.

Kitamura, Y., Shimada, M., Go, S., Matsuda, H., Hatanaka, K., Seki, M. (1979) Distribution of mast-cell precursors in hematopoietic and lymphopoietic tissues of mice. J. Expt. Med. 150,482–490.

Lagunoff, D. (1973) Membrane fusion during mast cell secretion. J. Cell Biol. 57,252–259.

Lawson, D., Raff, M.C., Gomperts, B., Fewtrell, C., Gilula, N.B. (1977) Molecular events during membrane fusion. J. Cell Biol. 72,242–259.

Lee, W.Y., Sehon, A.H. (1978) Suppression of reaginic antibodies. Immunol. Reviews 41,200–224.

Levine, B.B. (1966) Immunologic mechanisms of penicillin allergy. A haptenic model system for the study of allergic diseases of man. N. Engl. J. Med. 275:1115–1125.

Levine, B.B. (1971) Genetic factors in reagin production in mice. In: *Biochemistry of the Acute Allergic Reactions,* Second International Symposium. Austen, K.F., Becker, E.L. (eds.) Oxford, Blackwell Scientific, pp. 1–7.

Lewis, R.A., Austen, K.F. (1977) Non-respiratory functions of pulmonary cells: The mast cell. Fed. Proc. 36,2676–2683.

Lichtenstein, L.M., Sobotka, A.K., Malveáux, F.J., Gillespie, E. (1978) IgE-induced changes in human basophil cyclic AMP levels. Int. Arch. Allergy Appl. Immunol. 56,473–478.

Malveáux, F.J., Conroy, M.C., Adkinson, N.F., Jr., Lichtenstein, L.M. (1978) IgE receptors on human basophils: Relationship to serum IgE concentration. J. Clin. Invest. 63,176–181.

Marone, G., Kagey-Sobotka, A., Lichtenstein, L.M. (1979) Effects of arachidonic acid and its metabolites on antigen-induced histamine release from human basophils *in vitro*. J. Immunol. 123:1669–1677.

Marsh, D.G. (1975) Allergens and the genetics of allergy. In: *The Antigens,* vol. 3. Sela, M. (ed.) New York, Academic Press, pp. 271–359.

Marsh, D.G., Bias, W.B., Ishizaka, K. (1974) Genetic control of basal serum immunoglobulin E level and its effect on specific reaginic sensitivity. Proc. Natl. Acad. Sci. U.S.A. 71,3588–3592.

Marsh, D.G., Chase, G.A., Freidhoff, L.R., Meyers, D.A., Bias, W.B. (1979) Association of HLA antigens and total serum immunoglobulin E level with allergic response and failure to respond to ragweed allergen Ra3. Proc. Natl. Acad. Sci. U.S.A. 76,2903–2907.

Metzger, H. (1978) The IgE-mast cell system as a paradigm for the study of antibody mechanisms. *Immunol. Rev.* 41,186–199.

Mitchell, G.F. (1979) Effector cells, molecules and mechanisms in host-protective immunity to parasites. Immunology 38,209–223.

Morita, Y., Chiang, P.K., Siraganian, R.P. (1981) Effects of inhibitors of transmethylation on histamine release from human basophils. Biochem. Pharmacol. 30. In press.

Murphy, R.C., Hammarström, S., Samuelsson, B. (1979) Leukotriene C: A slow-reacting substance from murine mastocytoma cells. Proc. Natl. Acad. Sci. U.S.A. 76,4275–4279.

Newball, H.H., Berninger, R.W., Talamo, R.C., Lichtenstein, L.M. (1979) Anaphylactic release of a basophil kallikrein-like activity. I. Purification and characterization. J. Clin. Invest. 64,457–465.

Newball, H.H., Talamo, R.C., Lichtenstein, L.M. (1979) Anaphylactic release of a basophil kallikrein-like activity. II. A mediator of immediate hypersensitivity reactions. J. Clin. Invest. 64,466–475.

Orange, R.P., Austen, K.F. (1969) Slow reacting substance of anaphylaxis. Adv. Immunol. 10,106–144.

Osler, A.G., Lichtenstein, L.M., Levy, D.A. (1968) *In vitro* studies of human reaginic allergy. Adv. Immunol. 8,183–231.

Osler, A.G., Siraganian, R.P. (1972) Immunologic mechanisms of platelet damage. Prog. Allergy 16,450–498.

Ovary, Z., Itaya,T., Watanabe, N., Kojima, S. (1978) Regulation of IgE in mice. Immunol. Reviews 41,26–51.

Padawer, J. (1978) The mast cell and immediate hypersensitivity. In: *Immediate Hypersensitivity: Modern Concepts and Developments.* Bach, M.K. (ed.) New York, Marcel Dekker, pp. 301–367.

Parish, W.E. (1978) Evidence for human IgG antibodies anaphylactically sensitizing man. In: *Immediate Hypersensitivity: Modern Concepts and Development.* Bach, M.K. (ed.) New York, Marcel Dekker, pp. 277–299.

Plaut, M. (1979) Histamine, H-1 and H-2 antihistamines, and immediate hypersensitivity reactions. J. Allergy Clin. Immunol. 63,371–375.

Rubin, R.P. (1974) *Calcium and the Secretory Process.* New York, Plenum Press.

Segal, D.M., Sharrow, S.O., Jones, J.F., Siraganian, R.P. (1981) Fc (IgG) receptors on rat basophilic leukemia cells. J. Immunol. 126,138–145.

Segal, D.M., Taurog, J.D., Metzger, H. (1977) Dimeric immunoglobulin E serves as a unit signal for mast cell degranulation. Proc. Natl. Acad. Sci. U.S.A. 74,2933–2997.

Sieghard, W., Theoharides, T.C., Alper, S.L., Douglas, W.W., Greengard, P. (1979) Calcium-dependent protein phosphorylation during secretion by exocytosis in the mast cell. Nature 275, 329–331.

Siraganian, R.P. (1976) Basophil activation by concanavalin A. In: *Mitogens in Immunobiology.* Oppenheim, J.J., Rosenstreich, D.L. (eds.) New York, Academic Press, pp. 69–84.

Siraganian, R.P., Hook, W.A. (1976) Complement-induced histamine release from human basophils. II. Mechanism of the histamine release reaction. J. Immunol. 116:639–646.

Siraganian, R.P., Hook, W.A. (1977) Mechanism of histamine release by formyl methionine-containing peptides. J. Immunol. 119,2078–2083.

Siraganian, R.P., Hook, W.A., Levine, B.B. (1975) Specific *in vitro* histamine release from basophils by bivalent haptens: Evidence for activation by simple bridging of membrane bound antibody. Immunochemistry. 12,149–157.

Solley, G.O., Gleich, G.J., Jordan, R.E., Schroeter, A.L. (1976) The late phase of the immediate wheal and flare skin reaction—Its dependence upon IgE antibodies. J. Clin. Invest. 58,408–420.

Spiegelberg, H.L., Melewicz, F.M. (1980) Fc receptors specific for IgE on subpopulations of human lymphocytes and monocytes. Clin. Immunol. Immunopathol. 15,424–433.

Stechschulte, D.J. (1978) Non-IgE homocytotropic antibody in animals. In: *Immediate Hypersensitivity: Modern Concepts and Developments.* Bach, M.K. (ed.) New York, Dekker, pp. 237–276.

Sullivan, T.J., Parker, C.W. (1979) Possible role of arachidonic acid and its metabolites in mediator release from rat mast cells. J. Immunol. 122,431–436.

Sullivan, T.J., Parker, K.L., Eisen, S.A., Parker, C.W. (1975a) Modulation of cyclic AMP in purified rat mast cells. I. Responses to pharmacologic, metabolic and physical stimuli. J. Immunol. 114, 1473–1479.

Sullivan, T.J., Parker, K.L., Eisen, S.A., Parker, C.W. (1975b) Modulation of cyclic AMP in purified rat mast cells. II. Studies on the relationship between intracellular cyclic AMP concentration and histamine release. J. Immunol. 114,1480–1485.

Sullivan, T.J., Parker, K.L., Kulczycki, A., Jr., Parker, C.W. (1976) Modulation of cyclic AMP in purified rat mast cells. III. Studies on the effects of concanavalin A and anti-IgE on cyclic AMP concentrations during histamine release. J. Immunol. 117,713–716.

Tada, T. (1975) Regulation of reaginic antibody formation in animals. Prog. Allergy 19,122–194.

Taurog, J.D., Fewtrell, C., Becker, E.L. (1979) IgE mediated triggering of rat basophil leukemia cells: Lack of evidence for serine esterase activation. J. Immunol. 122,2150–2153.

Theoharides, T.C., Douglas, W.W. (1978) Secretion in mast cells induced by calcium entrapped within phospholipid vesicles. Science 201,1143–1145.

Thomas, L.L., Lichtenstein, L.M. (1979) Augmentation of antigen-stimulated histamine release from human basophils by serum-treated zymosan particles. I. Characteristics of enhancement. J. Immunol. 123,1462–1467.

Uvnäs, B. (1979) Chemistry and storage function of mast cell granules J. of Invest. Dermatol. 71,76–80.

Ward, P.A. (1978) Chemotactic factors specific for cells involved in anaphylaxis. In: *Immediate Hypersensitivity: Modern Concepts and Developments.* Bach, M.K. (ed.) New York, Marcel Dekker, pp. 649–657.

The Role of Prostaglandins
in Allergic Inflammation

Dean D. Metcalfe, M.D., and Michael Kaliner, M.D.

The cardinal signs of inflammation include edema, erythema, and pain. Interest in prostaglandins (PG) as mediators of inflammation stems from their capacity to reproduce these signs of inflammation, their production during various inflammatory processes, and the ability of nonsteroidal antiinflammatory drugs (NSAID) to reduce inflammation by preventing PG production. The four classical immunologic effector mechanisms (immediate hypersensitivity, cytotoxicity, immune complexes, and lymphocyte mediated reactions) lead to inflammatory states that are also associated with PG production. This review focuses initially on the PGs themselves (structure, synthesis, and metabolism) then discusses their role in immediate hypersensitivity reactions and cellular immune responses. A major recurring theme is the opposing proinflammatory and antiinflammatory properties of various PGs. The data reviewed suggest that although PGs have myriad proinflammatory properties, findings suggest that PGs act primarily as negative modulators of immunologic effector systems.

Concise History of the Prostaglandins

The PGs were initially observed as components of seminal plasma that had the capacity to cause hypotension and smooth-muscle spasm (von Euler 1934). Despite the observations that the prostate is neither the sole nor primary source of seminal PG and that several well-defined biochemical pathways leading to formation of products in addition to PGs are also stimulated during the course of PG production, the name prostaglandin has come to be accepted as the general term for this family of 20 carbon unsaturated fatty acids.

The first isolation, purification, and structural description of natural PG was accomplished in 1962 (Bergström et al. 1962). The structure identified suggested that unsaturated C-20 fatty acids such as arachidonic acid (AA) might serve as natural precursors. In 1965, Samuelsson suggested that an unstable endoperoxide intermediate structure was an essential step in PG synthesis, based on the observation that both

From the Laboratory of Clinical Investigation, National Institute of Allergy and Infectious Diseases, National Institutes of Health, Bethesda, Maryland.

oxygen molecules in the cyclopentane ring structure of PGE_1 derive from the same O_2 molecule. These endoperoxide intermediates (PGG_2 and PGH_2) subsequently have been isolated and characterized biologically (Hamberg and Samuelsson 1973).

More recently, two biologically important and structurally distinct PGs have been discovered: thromboxane A_2 (TxA_2) and the prostacyclin PGI_2. Thromboxane-A_2 was first observed in 1969 as a portion of the rabbit-aorta-constricting substance (RCS) noted to be generated during lung anaphylaxis (Piper and Vane 1969). RCS has been determined subsequently to consist of the endoperoxides PGG_2 and PGH_2, and TxA_2; all three are unstable and break down within minutes in aqueous solutions. Comparison of the half-life of PGG_2 and PGH_2 (5 minutes) with the 2-minute half-life of RCS led to the purification of TxA_2 (Svensson et al. 1975). Finally, the most recently discovered PG is the prostacyclin PGI_2, which is formed by arterial walls and has the capacity to inhibit platelet aggregation (Moncada et al. 1976).

Structure, Synthesis, and Metabolism

Prostaglandins are oxidized derivatives of prostanoic acid, a C-20 unsaturated acid (Figure 1). The alphabetical nomenclature of the PG is based on the potential oxidative patterns of the 5-carbon cyclopentane ring structure of prostanoic acid and are demonstrated in Figure 1. The structure of TxA_2 is distinct in that an oxygen molecule is incorporated into the ring structure (hence, the oxane in thromboxane). PGI_2 is described chemically as Δ^7-6(9) - oxy - $PGF_{1\alpha}$.

The numerical subscript in each PG indicates the number of unsaturated bonds: PGE_2 has unsaturations at the 5-6 and 13-14 positions. Prostaglandin F may exist in either an alpha or beta configuration; however, only the alpha configuration exists in nature. There are three natural substrates for PG formation: dihomo-γ-linolenic acid (from which PGs with one unsaturation are derived), arachidonic acid (AA) (from which the 2 series derives), and 5,8,11,14,17-eicosapentaenoic acid (the 3 unsaturated bond series). Arachidonic acid is the most common substrate in nature and is described chemically as 5,8,11,14-eicosatetraenoic acid. AA generally is derived from membrane phospholipids, primarily phosphatidylcholine (70%), phosphatidylinosital (25%), and phosphatidylserine (5%) (Bills et al. 1977). Prostaglandins are not stored in tissue; instead, virtually all cells have the ability to synthesize various PGs in response to diverse stimuli.

Because the PGs are synthesized within seconds in response to a great variety of stimuli, a remarkably efficient synthetic system must be involved. Apparently, many membrane interactions (hormone-receptor binding, ion flux, perturbations of many sorts) lead to liberation of membrane phospholipids. Released phospholipids, but not phospholipids within intact membranes, are susceptible to the actions of membrane-associated enzyme phospholipase A_2, which hydrolyzes the C-2 fatty acid from the glycerol chain. Among the fatty acids liberated is AA, which is converted to the PG endoperoxides, PGG_2 and PGH_2, by the action of the membrane-associated enzyme fatty acid cyclooxygenase (Figures 2 and 3). This cyclooxygenase requires heme as a cofactor, is inactivated by peroxides (and as such irreversibly catalyzes its own destruction), and is inhibited by NSAID as well as poly-ynoic fatty acids. Prostaglandin-G_2 is converted to PGH_2 by action of a peroxidase-like enzyme. An additional enzyme, prostaglandin endoperoxide isomerase, converts PGG_2 and PGH_2 to PGE_2; PGE_2 may be converted to $PGF_{2\alpha}$ by a 9-keto reductase enzyme. Alternatively, PGG_2 can be converted directly to $PGF_{2\alpha}$ without an intervening PGE stage. 9-Keto reductases

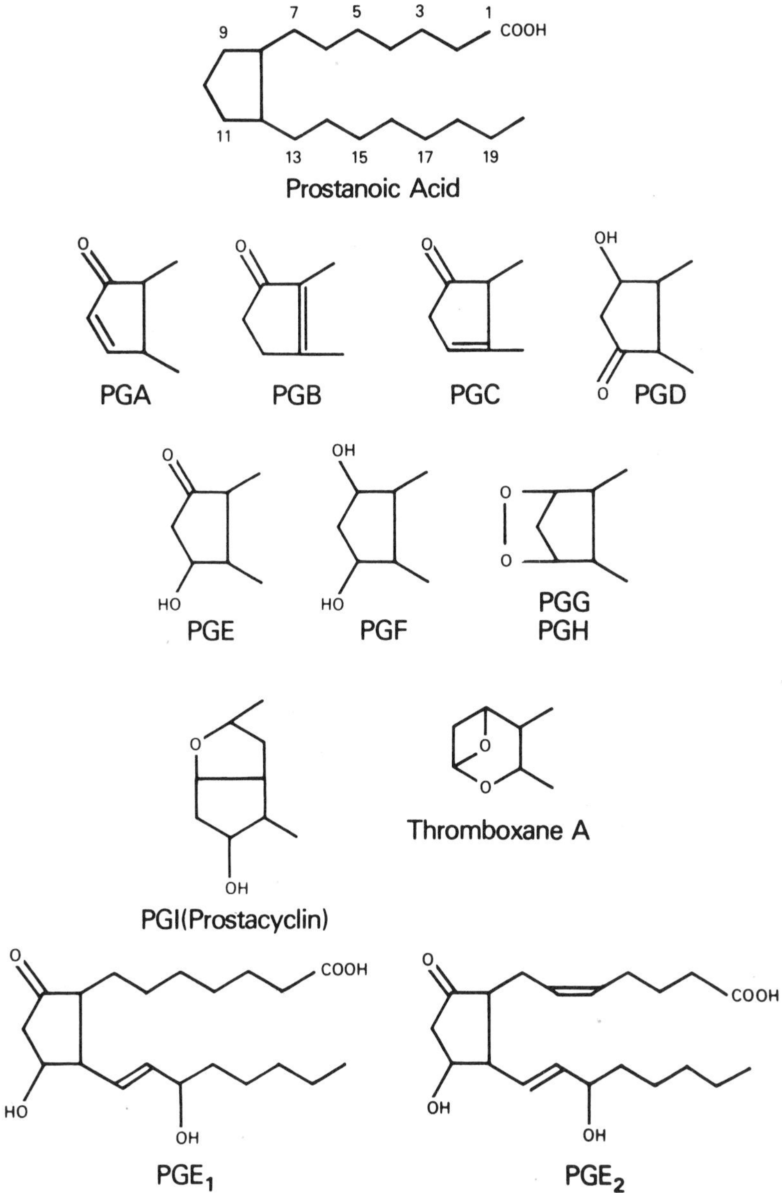

Figure 1. Structure and nomenclature of prostaglandins.

are distributed widely in many tissues and interestingly are activated by cyclic GMP. Prostaglandin-G$_2$ may also be reduced to form PGD$_2$. A unique product of the endoperoxides is hydroxyheptadecatrienoic acid (HHT), which is formed by splitting a malondialdehyde from PGG$_2$; both HHT and malondialdehyde have low biologic activity.

Recently, a major derivative of thrombin-activated platelets has been identified. This compound, TxA$_2$, causes irreversible platelet aggregation, differs structurally from PG by having a tetrahydropyranyl ring, and is formed from PGG$_2$ by the action

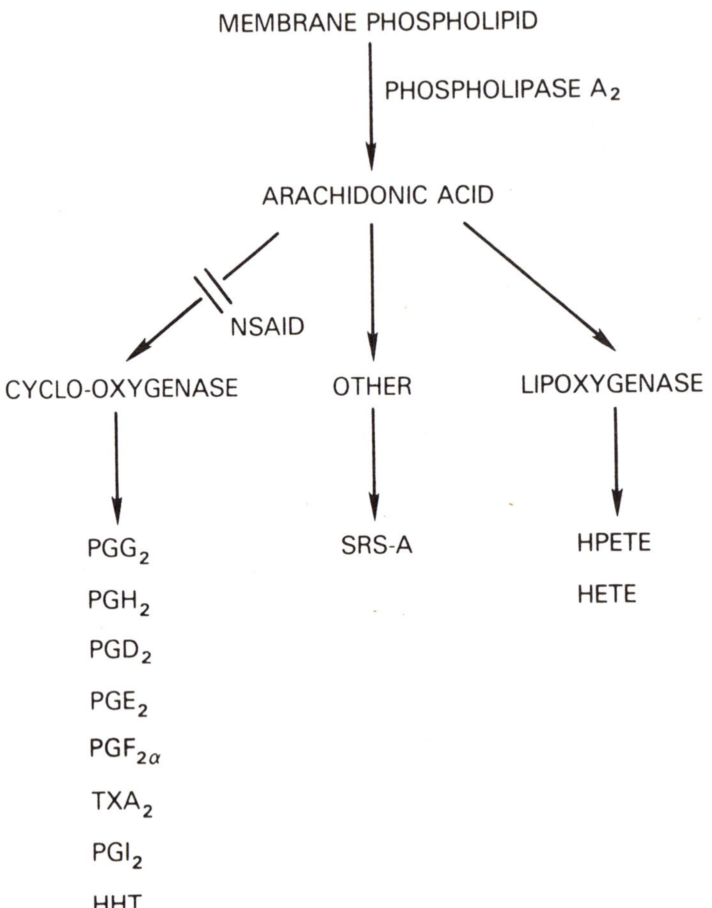

Figure 2. Generation of arachidonic acid derivatives.

of the enzyme thromboxane synthetase. In aqueous solutions, TxA_2 has a half-life of 36 seconds and must be captured in nucleophilic solutions (methanol) to be studied. Prostaglandin I_2 is also formed from the endoperoxides but by the enzyme prostacyclin synthetase.

Arachidonic acid may also be metabolized by a fatty acid lipoxygenase enzyme (Figures 2 and 3). The major products of this system are hydroperoxyeicosatetraenoic acids (HPETE) and hydroxyeicosatetraenoic acids (HETE). Recent studies (Borgeat et al. 1976) suggest that the position of AA to which the hydroxyl groups are added may vary; indeed, dihydro-HETE has now been described (Borgeat and Samuelsson 1979). The search for biologic activities for the various HETE compounds has not been reported extensively as yet; however, chemotactic (Turner et al. 1975) and proinflammatory (Adcock et al. 1978) effects may be demonstrated.

The degradative patterns and products of the classic PGs, PGE_2 and $PGF_{2\alpha}$, have been examined more extensively than the newer derivatives of AA. Neither PGE_2 nor $PGF_{2\alpha}$ is a true circulating hormone, since 80 to 95% of them are removed from the circulation during one passage through the pulmonary system. The primary deactivation steps involve oxidation of the 15-hydroxy position by 15-hydroxy-PG dehydrogenase, reduction of the 13,14-double bond, two sequences of beta-oxidation of the acid side chain, and omega-oxidation of the alkyl side chain. The additional

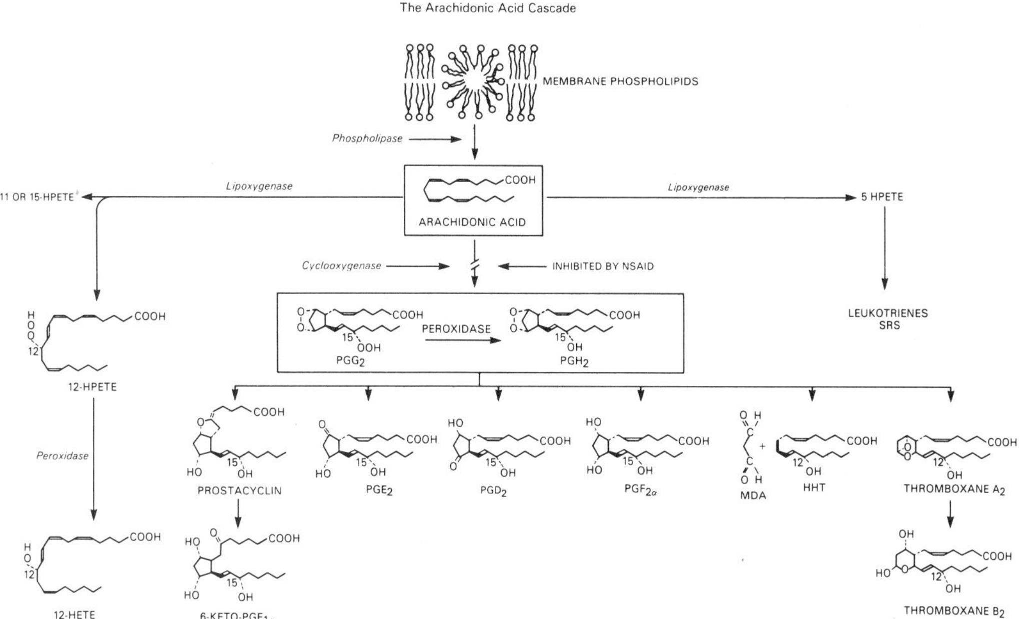

Figure 3. Structure of arachidonic acid derivatives.

enzymes in this sequence are 13-PG reductase and 15-Keto-PG reductase. Thromboxane-A_2 is converted rapidly to thromboxane B_2 (TxB_2) and excreted in the urine after an additional beta-oxidation step. Prostaglandin I_2 is converted to 6-Keto-$PGF_{1\alpha}$.

The Role of Prostaglandins in Immediate Hypersensitivity

Of the various immunologic effector mechanisms, immediate hypersensitivity has received the most attention regarding PG. Thus, a large amount of literature deals with various parameters of PG and allergic manifestations. This section reviews selected aspects of PG and allergy and attempts to draw the firmest conclusions possible based on the data available. The following observations, each of which is discussed separately, underscore the confusion inherent in the potential role PGs play in allergy: (1) Various PGs have biologic activities that reproduce many signs and symptoms found in allergic diseases. Many of these activities appear to be antagonistic to one another. (2) A large spectrum of PGs are generated during anaphylaxis of isolated tissues. These PGs may be formed both as a direct consequence of anaphylaxis or as a consequence of tissue responses to anaphylaxis. In the former instance, PGs may be proinflammatory and in the latter, antiinflammatory. (3) Prostaglandins may act as potent inhibitors of allergic reactions by virtue of their effect on cyclic AMP or augmentors of the reaction by increasing cyclic GMP. (4) Suppression of PG generation *in vivo* by administration of NSAID affects only a small portion of allergic subjects.

Biologic Effects of Prostaglandins Relevant to Allergy

The most common diseases in which allergy plays a major role are allergic rhinitis (involving about 15 to 20% of the United States population or more than 35 million people), asthma (about 7% or 16 million people), and urticaria (estimated as involving somewhat more than 10% of the citizenry). These diseases have common features: mucosal edema, eosinophilic leukocyte infiltration, vasodilation, and excessive mucus secretion. Among the unique features of these syndromes are pruritis in both rhinitis and urticaria, bronchospasm in asthma, and postcapillary sinusoidal spasm in rhinitis. The pharmacologic administration of PG causes the following changes in man: increased vascular permeability (PGE, $PGF_{2\alpha}$), pain (PGE), vasodilation (PGE, $PGF_{2\alpha}$), potentiation of edema and pain produced by other mediators (PGE, PGA, $PGF_{2\alpha}$), bronchospasm ($PGF_{2\alpha}$, TxA_2, PGG_2, PGH_2, PGD_2), bronchodilation ($PGF_{2\alpha}$, PGI_2), mucus secretion ($PGF_{2\alpha}$, PGA, PGB), and eosinophil chemotaxis (HHT, HETE) (Vane 1976). Based on the relevance of this multitude of responses, it is easy to see why the PGs are suggested as potentially important mediators of allergy.

Prostaglandins Generated During Anaphylaxis

Many investigations of anaphylaxis employ an isolated organ, frequently lung, from an actively sensitized animal or after passive sensitization in antibody-rich sera. The model most commonly used to study PG generation has been actively sensitized guinea pig lung, although human lung and rat mast cell preparations also have been studied. In each instance the reaction is triggered by the addition of antigen. The exact mechanisms involve antigen interaction with IgE class immunoglobulins fixed to the cell surface of mast cells and resulting in the discharge of mast cell secretory granules (Figure 4). The granules contain the mediators of anaphylaxis including histamine, eosinophil chemotactic factors of anaphylaxis (ECF-A), neutrophil chemotactic factors, superoxide-generating enzymes, superoxide dismutase, mast cell peroxidase, heparin, and chymotrypsin, as well as factors that generate slow-reacting substance of anaphylaxis (SRS-A) and the PG. Once released or generated, these mediators interact with the surrounding tissues causing the pathologic changes recognized as allergic inflammation. Other stimuli may also cause mast cell degranulation (anaphylatoxins, physical injury, certain drugs, and chemicals), which leads to exactly the same responses (Figure 4).

Presently, recognized PGs generated by allergic reactions of lung tissue include PGF, PGE, TxA, PGI, PGD, PGG, PGH, HHT, HPETE, HETE, PGA, and PGB. Purified mast cells stimulated in the presence of arachidonic acid generate a series of PGs, most appreciably PGD (Jakschik et al. 1978; Roberts et al. 1978). A PG-releasing factor has been described as an anaphylactic product from guinea pig lung, which can cause PG generation independent of other pharmacologic responses (Nijkamp et al. 1976). Histamine can induce PG generation in two fashions, both of which involve stimulation of the H-1 receptor site (Platshon and Kaliner 1978) (Figure 5). The first mechanism involves induction of smooth-muscle contraction and results in the generation of PGE. This action is reproduced by any agent that causes smooth muscles to contract and reflects a response to muscle contraction. The second mechanism of histamine-induced PG generation is by stimulation of lung parenchyma, which results in the generation of both $PGF_{2\alpha}$ and PGE and is not reproduced by muscle stimulators. In total, PGs are generated by three separate mechanisms during anaphylaxis: directly from mast cells, as a consequence of histamine and other factors interacting with target cells, and as a consequence of smooth-muscle spasm. The exact PG formed after anaphylaxis is determined by the tissue stimulated. Airway-free human lung parenchymal preparations generate both $PGF_{2\alpha}$ and PGE, whereas parenchyma-free bronchial airway smooth-muscle preparations generate exclusively PGE (Figure 5).

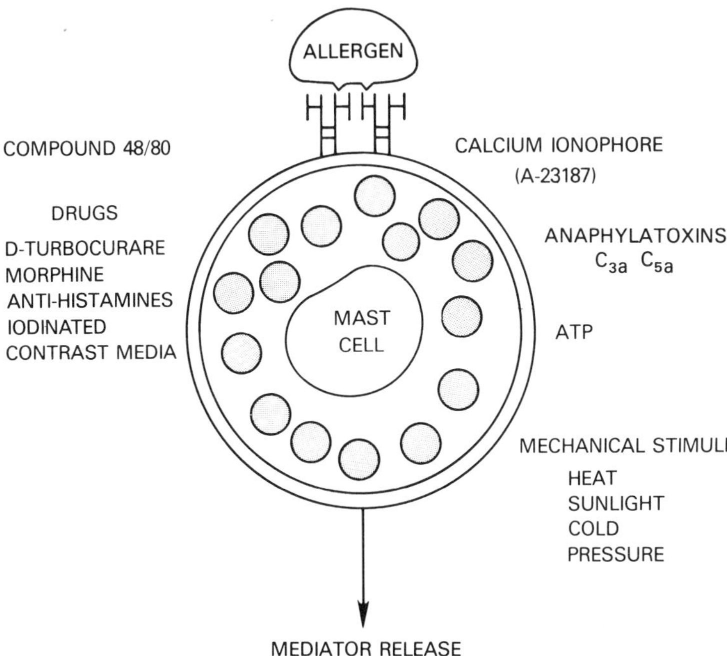

Figure 4. Mast cell degranulation in response to pharmacologic, immunologic, and mechanical stimuli.

Slow-reacting substance of anaphylaxis may be an important mediator of bronchial asthma. This possibility is based upon the following observations (Orange and Austen 1969): human bronchial smooth muscle is exquisitely sensitive to the contracting properties of SRS-A; pretreatment of human IgE sensitized airways with antihistamines fails to prevent antigen induced muscle contraction, and the supernatant from this reaction contains SRS-A; and antihistamines are relatively ineffective in treating bronchial asthma. Slow-reacting substance of anaphylaxis is a small (~500 molecular weight), lipid-like molecule whose exact molecular structure has only recently been uncovered. Like the PGs, SRS-A is synthesized and released after anaphylaxis. The exact mechanisms responsible for SRS-A generation are as yet unidentified. Several studies have demonstrated recently that an SRS-A-like molecule can be synthesized from AA by a pathway that appears to involve neither the cyclooxygenase nor lipoxygenase pathways (Figure 2) (Bach et al. 1977; Jakschik et al. 1977). Indeed, some studies have suggested that NSAID treatment of tissue may result in augmented SRS-A production, possibly because of a shifting of AA from PG generating pathways to SRS-A production (Walker 1973).

The cellular source of SRS-A is unclear. It may be formed by cells found within the rat peritoneal cavity; lung tissue obtained from rodents, ungulates, and humans; human nasal mucosa; peripheral blood leukocytes; and partially purified dispersed human lung mast cells. However, during partial purification of human lung mast cells, the capacity of these cells to synthesize SRS is reduced with the increasing purity of the mast cell preparation (Paterson et al. 1976). Thus, although the mast cell is required to initiate SRS-A formation, it is likely that additional cell types contribute to its synthesis. In line with these observations, an SRS-like activity has been generated from rat peritoneal mononuclear cells (Bach et al. 1979) and human peripheral blood leukocytes (Conroy et al. 1976) by the calcium ionophore A23187.

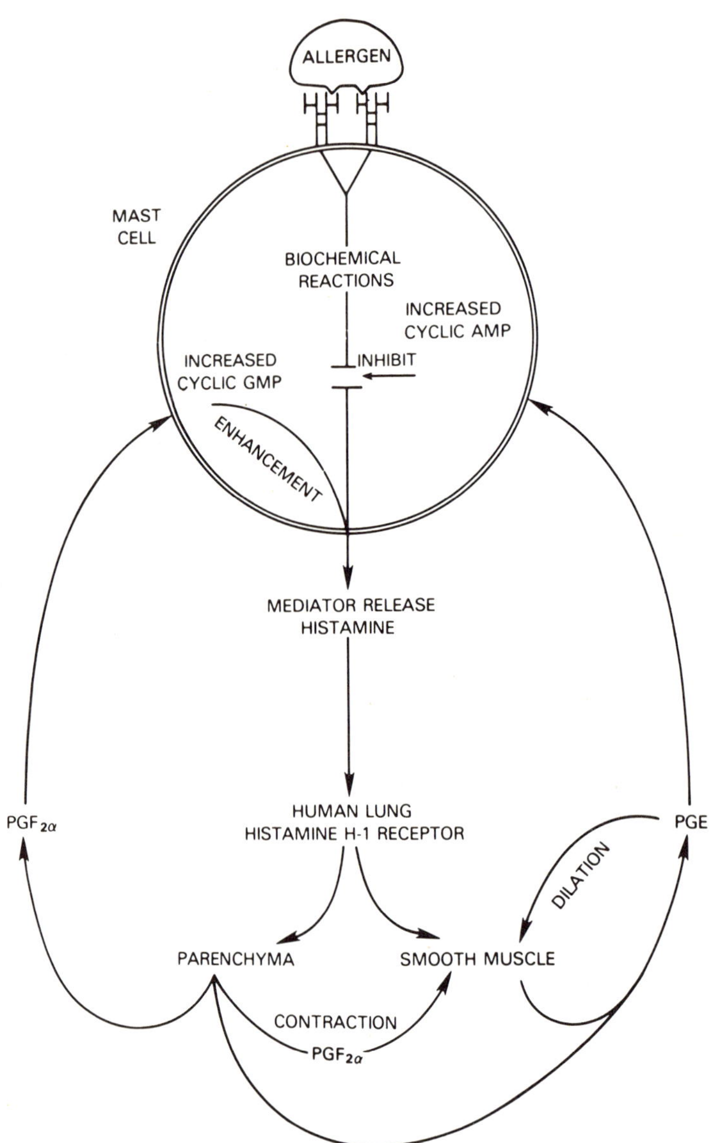

Figure 5. Complex interactions among mast-cell-derived mediator release, prostaglandin generation by human lung, and modulation of mast cell responses.

Recently, SRS has been reported to be related to a group of compounds called leukotrienes (Borgeat and Samuelsson 1979; Murphy et al. 1979) (Figure 6). Before this observation, the absence of a structural description of SRS necessitated that a number of biologic and physicochemical characteristics be used to define SRS. These descriptive characteristics included contractile activity for guinea pig ileum without action on gerbil colon or rat uterus in estrus, inactivation by arylsulfatases, inhibition of smooth-muscle contractile activity for the guinea pig ileum with the antagonist FPL-55712, resistance to inactivation at 37°C in 0.2 M NaOH, and characteristic behavior over Amberlite XAD-2, silicic acid, and DE-52 cellulose with defined solvents. The pharmacologic actions of SRS have been examined previously using SRS

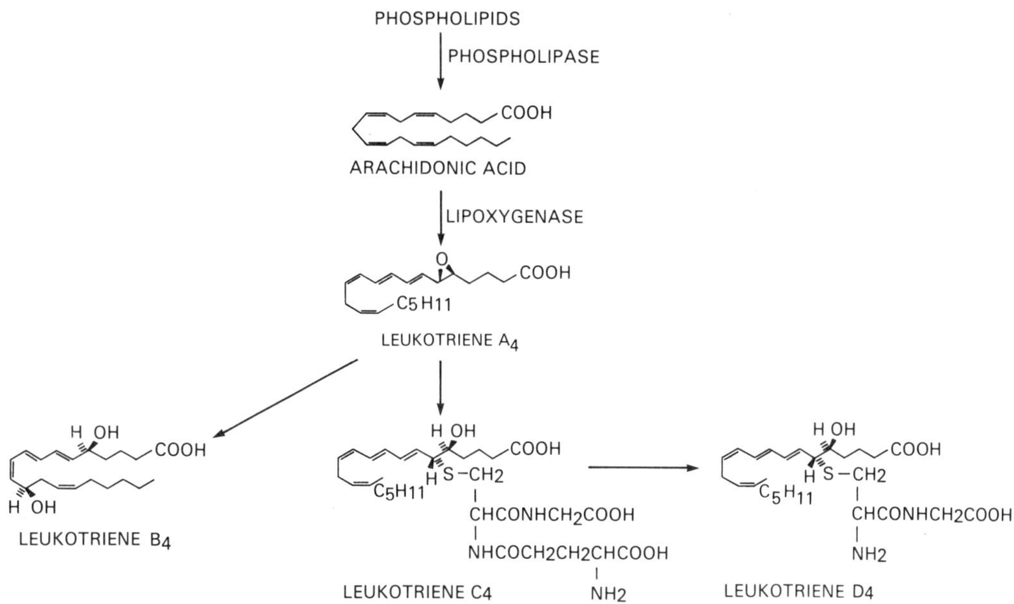

Figure 6. Leukotriene pathway of arachidonic acid metabolism.

of variable purity and thus are subject to reevaluation as structurally defined material becomes available. With this limitation in mind, SRS has been shown to alter cutaneous vascular permeability in the guinea pig and cow and to have a preferential contractile activity on peripheral guinea pig airways as compared with guinea pig tracheal spirals. This *in vitro* observation appears to agree with *in vivo* guinea pig data in which intravenous SRS infusion into unanesthetized guinea pigs produced a marked fall in dynamic compliance but only a small increase in pulmonary resistance, findings compatible with a predominant effect on peripheral airways.

Leukotrienes are a series of closely related compounds derived from AA through a lipoxygenase pathway and appear to include several SRSs (Figure 6). Structurally, leukotrienes are conjugated trienes and were given the designation leukotriene because of their initial generation by leukocytes (Borgeat and Samuelsson 1979; Samuelsson, in press). The leukotriene pathway (Figure 6) consists of leukotriene A_4 (LTA$_4$), an unstable epoxide intermediate; leukotriene B_4 (LTB$_4$), the enzymatically formed 5(S),12(R)-dihydroxy acid; leukotriene C_4 (LTC$_4$), a γ-L-glutamyl-L-cysteinylglycine (glutathione) derivative; and leukotriene D_4 (LTD$_4$), a cysteinylglycine derivative (Samuelsson, in press). The subscript denotes the number of double bonds in the molecule reflecting the structure of the precursor fatty acid. Murine mastocytoma cells treated with calcium ionophore A23187 produce an SRS that appears identical to LTC$_4$, and rat basophil leukemia cells treated with calcium ionophore A23187 produce primarily an SRS that appears to be LTD$_4$. Slow-reacting substance prepared from immunologically challenged lungs of guinea pigs produce LTD$_4$. The two SRSs produced by reversed anaphylaxis in human lung are LTD$_4$ and a lesser amount of LTC$_4$. Using guinea pig lung *in vitro*, LTC$_4$ and LTD$_4$ are selective peripheral airway agonists with LTD$_4$ 100 times as active as LTC$_4$. *In vivo* LTC$_4$ and LTD$_4$ have approximately equal effects on pulmonary mechanics. The antagonist FPL 55712 acts on LTD$_4$ but not on LTC$_4$. Further, both LTD$_4$ and LTC$_4$ have direct hypotensive effects, whereas LTD$_4$ increases vascular permeability in skin (Drazen et al. 1980).

Modulation of Immediate Hypersensitivity Reactions by Prostaglandins

The sequence of biologic events involved in the release of the mediators of asthma involves a series of relatively well-defined intracellular events (Kaliner and Austen 1973). The release process is modulated by the intracellular levels of cyclic nucleotides: increases in cyclic AMP inhibit mediator release, whereas increases in cyclic GMP augment the release reaction. Both of these modulating influences act by altering the functional state of microtubules, which are cytoplasmic microorganelles involved in the secretory process. Increased levels of cyclic GMP cause the microtubules to assume a functional, polymerized state whereas elevated levels of cyclic AMP change the microtubules to a nonfunctional, depolymerized state (Kaliner 1977).

Prostaglandins are capable of increasing intracellular cyclic nucleotide levels by activating adenylate cyclase (PGE) or guanylate cyclase ($PGF_{2\alpha}$). PGE has been found to prevent the IgE-mediated release of histamine from human lung, rat peritoneal mast cells, and human basophils by increasing elevations in cyclic AMP (Tauber et al. 1973). Likewise, $PGF_{2\alpha}$ may induce augmentation of allergic mediator release in association with increases in cyclic GMP. Because both PGE and $PGF_{2\alpha}$ are formed during the allergic response, it is possible that endogenous PG formation provides a controlling influence on mediator discharge (Figure 5). In support of this possibility, a factor derived from eosinophils, attracted to the site of allergic reactions, is capable of suppressing mast cell degranulation and appears to be PGE (Hubscher 1975).

An additional source of PG related modulation of allergic responses is through autonomic nervous system influences. Mediator release is associated with stimulation of autonomic afferent and efferent discharges. Specifically in asthma, histamine (and probably other mediators) stimulates vagal afferent nerve endings known as irritant receptors (Nadel 1973), thereby initiating a reflex arc resulting in an efferent discharge over the vagus nerve. Parasympathetic stimulation of lung tissues may contribute to smooth-muscle contraction and mucus secretion and, through increases in cyclic GMP levels, may augment further histamine release. Beta-adrenergic stimulation, produced by the action of circulating epinephrine, will tend to counteract all these actions. Alpha-adrenergic stimulation, caused by norepinephrine release from sympathetic nerves located in the vascular bed of the lung, may contribute both to smooth-muscle contraction and increased mediator release as well as decreased pulmonary blood flow. Prostaglandins have profound influences on neural transmission across ganglia (Brody and Kadowitz 1974), and as such might contribute to the autonomic abnormalities so clearly described in asthma.

There is also the suggestion that asthmatic subjects are unduly responsive to the inhalation of $PGF_{2\alpha}$ (up to 800 times more sensitive than controls) (Mathé et al. 1973). However, asthmatic airways are hyperresponsive to the irritant properties of many inhaled compounds (histamine, methacholine, SO_2, O_3, etc.), so the relevance of this observation is not clear.

Effects of Nonsteroidal Antiinflammatory Drugs on Allergic Reactions

Suppression of PG generation by NSAID *in vitro* results in either no change in mediator release (Platshon and Kaliner 1978) or augmentation of SRS-A release (Walker 1973). Interestingly, acetylenic acids such as eicosatetraynoic acid (ETYA), an AA-like molecule with triple bonds in place of the double bonds, inhibit both the cyclooxygenase and lipoxygenase enzyme systems. Eicosatetraynoic acid both prevents PG formation during anaphylaxis *in vitro* and inhibits mediator release itself (Sullivan and Parker 1979), suggesting that AA metabolism may be an essential step in the biochemical reactions intrinsic to mast cell degranulation. In support of this possibility, HPETE (a lipoxygenase product identified as an additional product formed during anaphylaxis) has the capacity to cause an enhancement of histamine and SRS-A release (Adcock et al. 1978).

Aspirin and other NSAID may cause a severe asthmatic response in a minority of asthmatic subjects. These patients characteristically have rather refractory asthma, a long history of nasal polyps and sinusitis, and blood eosinophilia (Samter and Beers 1968). Little evidence exists for IgE-related mechanisms in this disease, although the onset and symptoms after NSAID ingestion suggest that underlying mechanisms are "anaphylactoid" in nature. The capacity of chemically diverse NSAID to both cause the disease as well as prevent PG generation suggests that these two effects may be related causally. The exact causal mechanism is still not clear, although there are several possibilities. Nonsteroidal antiinflammatory drugs interfere with cyclooxygenase enzymes, thereby permitting other noncyclooxygenase derivatives to be generated in an unabated, or even augmented, fashion. Indeed, SRS-A generation as well as certain lipoxygenase products (Adcock et al. 1978) may be increased in the presence of NSAID. These lipoxygenase products may act on the mast cell to further potentiate the release process. Thus, interference with the cyclooxygenase system may result in increased SRS-A production.

Prostaglandins may also play a critically important bronchodilatory role in certain severe asthmatics. There are autonomic nervous system abnormalities in all asthmatics: excessive parasympathetic and alpha adrenergic responsiveness as well as diminished beta adrenergic responsitivity. This abnormal pattern of homeostatic balances may contribute toward the hyperresponsive airways universally found in asthma. Contraction of bronchial airways generates PGE (Figure 5), which is a bronchodilator. It is possible that aspirin-sensitive asthmatic subjects have a significant degree of autonomic nervous system dysfunction leading toward excessive dependence on the bronchodilating influence of PGE. In such subjects, the removal of the bronchodilating influence of PGE by aspirin may be sufficient to trigger their disease. In such individuals, it may not be an abnormal response (allergy) to NSAID that triggers their disease but an excessive dependence on the production of PGE that is at fault.

Summary of Prostaglandin Interactions Relevant to Immediate Hypersensitivity

Prostaglandins have been demonstrated to have biologic activities potentially capable of contributing to allergic inflammation. Further, a wide spectrum of PGs are generated during anaphylaxis of tissues *in vitro,* and evidence exists demonstrating their generation *in vivo* (Svanborg et al. 1973). The mechanism responsible for PG generation is complex and involves at least three separate pathways. Furthermore, specific tissues have the propensity to generate certain PGs selectively. An essential mediator of asthma, SRS-A may also be a derivative of AA, synthesized through a newly defined lipoxygenase pathway. Prostaglandins generated during the allergic response have the ability to act as either pro- or antiinflammatory factors. For instance, PGE suppresses mediator release whereas $PGF_{2\alpha}$ augments its release. These complex interactions are summarized in part in Figure 5. Although PGs have the potential for making a major contribution in allergic reactions, the effects of NSAID in asthmatic subjects suggest a more limited contribution. Thus, only a small portion of asthmatics are adversely or beneficially influenced by NSAID.

Prostaglandins in Other Immune Responses

Other immune mechanisms responsible for the mediation of inflammation that involve PGs include immune complex activation of complement and lymphocyte-mediated inflammation.

Complement Activation by Immune Complexes

The complement system consists of 12 proteins involved in the classical pathway, an additional four proteins intrinsic to the alternative pathway, and three inhibitory proteins (see Chapter 15). Each of these proteins circulates in plasma. Appropriate stimulation activates either the classical (antigen–antibody complexes) or alternative (bacterial lipopolysaccharides) complement pathway, and this activation leads to the generation of a series of biologically active components (Frank 1975). Among these generated activities are the anaphylatoxins C3a and C5a, which both cause mast cell degranulation (Figure 4), and, in the case of C5a, chemotaxis. The attracted polymorphonuclear leukocytes (PMN) ingest the immune complexes, simultaneously releasing a portion of their proteolytic enzymes and generating prostaglandins (Goldstein et al. 1977). The PGs formed consist primarily of PGE and TxB_2, although PGI_2 may also be formed. The PGs generated have several activities of potential relevance to this form of inflammation: PGE may contribute to the chemotaxis of PMN, whereas HHT and HETE are reported to attract PMN with a preference for eosinophils, PGE suppresses the secretion of lysosomal enzymes accompanying phagocytosis by PMN and may therefore act as a negative feedback mechanism for this system, and PGE is a potent vasodilator that would favor the increased local deposition of immune complexes, thereby augmenting the entire reaction. Local mast cells degranulated by the anaphylatoxins also contribute to this inflammation through release and generation of inflammatory molecules (see Chapter 12). Therefore, a logical sequence of events in immune-complex-mediated injury involves activation of complement with the generation of anaphylatoxins and chemotactic factors (Figure 7). The resulting mast cell degranulation plus PMN infiltration account for the acute inflammatory response, the sum total of which is an Arthus-type lesion consisting of a vasculitis with hemorrhage and complement deposition. The generated PGs may act to augment or suppress the inflammation. Specifically, PGE may prevent both mast cell degranulation and PMN lysosomal enzyme release, whereas $PGF_{2\alpha}$ (and probably TxA_2) may augment both reactions.

Lymphocyte-Mediated Inflammation

Of particular interest in the context of PG involvement in cellular immune responses is the potential modulation by macrophages of T-cell proliferation and lymphokine generation. Exogenously added PGE has been found to inhibit the generation of migration inhibition factor and presumably other lymphokines from guinea pig lymphocytes as well as leukocyte inhibition factor from human T-cells. Furthermore,

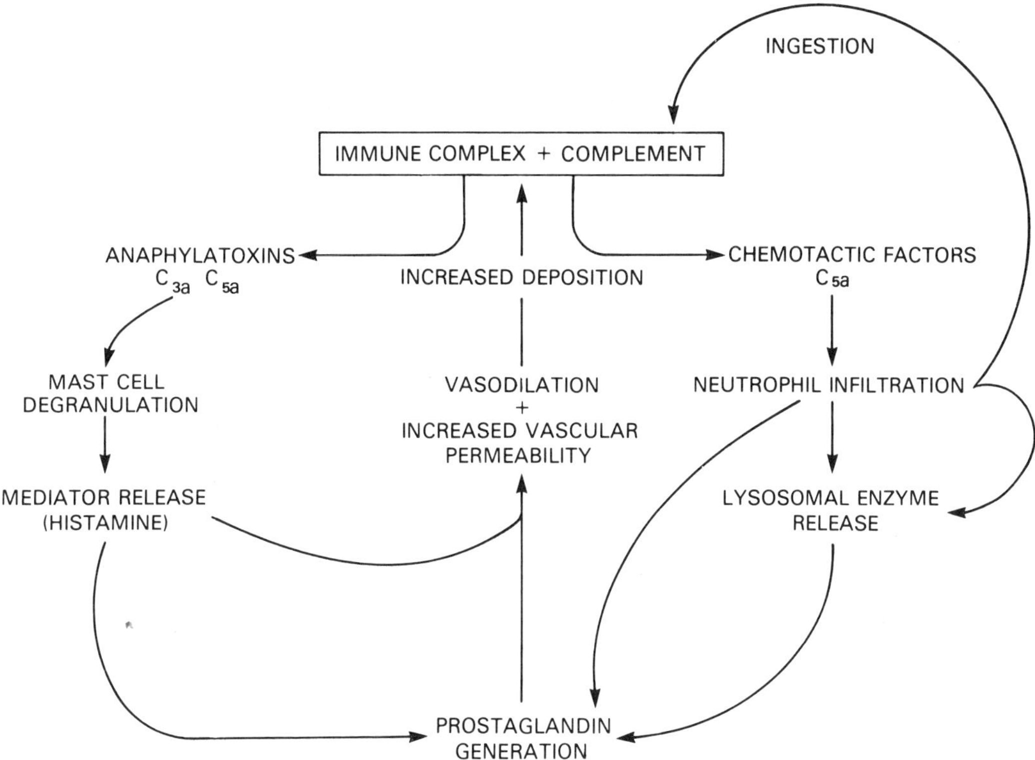

Figure 7. Complex interactions between immune complexes and complement mediated mast cell degranulation and PMN infiltration.

PGE suppresses T-lymphocyte-mediated cytotoxicity in association with increased levels of cyclic AMP, hemolytic plaque formation by murine lymphocytes, and human T-cell thymidine incorporation in response to mitogens. Thus, PGE appears to have the capacity to inhibit T-cell processes (Figure 8).

Both human and animal monocytes and macrophages produce PGE in culture (Kurland and Bockman 1978). Indeed, cultures containing activated macrophages can suppress both antigen and mitogen stimulated lymphocyte proliferation; NSAID prevent this inhibition, suggesting that monocyte/macrophage PGE is responsible (Goodwin et al. 1977b). Recent analysis of the skin test anergy displayed by patients with Hodgkin's disease reveals that monocytes from these subjects produce excessive PGE in culture (up to four-fold control levels). Furthermore, macrophages obtained from these subjects with Hodgkin's disease suppress T-cell responses in culture (Goodwin et al. 1977a). These data suggest that macrophages derived from patients with Hodgkin's disease who are anergic may produce excessive PGE, which negatively influences T-cell responses (Figure 8). However, administration of NSAID to normal subjects *in vivo* has no influence on delayed skin test responses to a battery of antigens (Duncan et al. 1977) nor on subsequent culture of lymphocytes and their response to mitogens. These observations suggest that under normal circumstances macrophage-derived PGE fails to dampen T-cell responses, whereas some lymphoreticular malignancies may be associated with abnormal macrophage–PG-mediated suppression of T-lymphocyte functions.

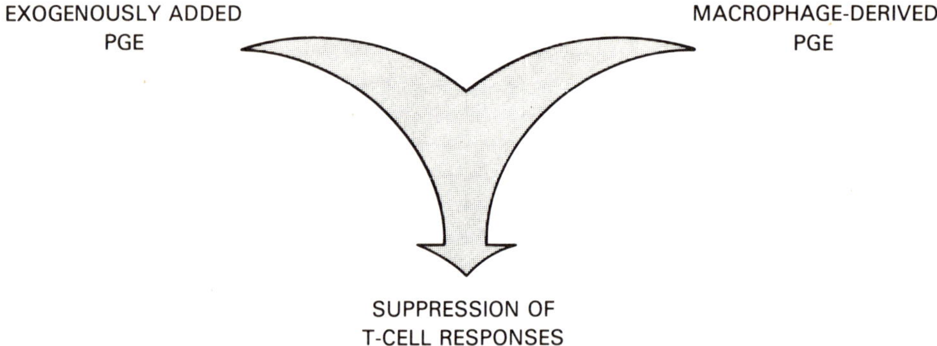

Figure 8. PGE modulation of T-cell functions.

Role of Prostaglandins in Rheumatic Disorders

The coincidental elevation of prostaglandins and the destruction of connective tissue in some chronic rheumatic lesions has prompted the suggestion that PGs may have a regulatory role in the destruction of collagen, the major component of connective tissue. Evidence to support this possibility comes from experiments with macrophages, a prominent cell type in such lesions, in which the production of the collagen degrading enzyme collagenase was demonstrated to be regulated by prostaglandins. Macrophages produce prostaglandins and collagenase in response to agents that activate these cells, including bacterial cell walls and lymphokines (Wahl et al. 1977a,b). The addition of the NSAID indomethacin to macrophage cultures prevents collagenase production, which is reversed by the addition of PGE_1 or PGE_2 but not $PGF_{2\alpha}$. The synovial fluid and tissue obtained at synovectomy from patients with rheumatoid arthritis contain elevated levels of prostaglandins and collagenase. A population of unclassified adherent cells obtained from such synovial specimens produce high levels of PG and collagenase (Dayer et al. 1976). Peripheral blood monocytes and the monocyte-derived mediator LAF are able to stimulate this adherent cell population to even greater production of PG and collagenase. Further, PG added to fetal rat bone cultures induces significant bone resorption, possibly by stimulating osteoclast activity (Klein and Raisz 1970). A mediator released by lymphocytes in culture, osteoclast-activating factor (Horton et al. 1977), requires PG generation for its release (Tonedo and Mundy 1979).

Thus, PGs may be involved in inflammatory processes in the rheumatic diseases by modulating enzyme release, bone resorption, and collagen destruction. The clinical usefulness of NSAID in treating the rheumatic diseases supports the suggestion that PGs are, overall, proinflammatory in rheumatic diseases.

Conclusion

In this review, the focus has been on immunologic inflammation and the role of PGs. Such experimental models as carageenin-induced inflammation or urate-crystal-induced responses in which a role for PGs has been suggested have not been discussed. Likewise, thermal burn injury and ultraviolet light injury, in which PG generation plays an essential role, were not covered. Emphasis has been placed on the potential roles of PGs in allergic disease states. The conclusions drawn suggest that generally more evidence exists for antiiinflammatory effects of PGs than for proinflammatory actions in immunologic processes.

The interest in PGs as mediators of inflammation stems from the observations that certain PGs cause the cardinal signs of injury: erythema (PGE, PGF), edema (PGE), pain (PGE, HETE, HPETE, possibly TxA_2), and fever (PGE), whereas NSAID appear to work by interfering with the cyclooxygenase enzyme system. The potential role of PGs in immediate hypersensitivity follows from the observations that PGE, $PGF_{2\alpha}$, TxA_2, and PGI_2 are produced by anaphylactic reactions and that these agents cause a variety of responses associated characteristically with allergic reactions. However, NSAID have an effect on only a few allergic subjects, and this effect generally exacerbates allergic disease rather than ameliorates it. Thus, greater evidence exists for an antiallergic role of PGs in allergy than for a proinflammatory role. Likewise, despite biologic activities that might augment cellular immune responses, current evidence suggests that PGs act as negative feedback modulators of PMN and T-lymphocyte responses. Thus, PGs appear to regulate immune responses rather than to mediate them.

Innumerable areas remain to be explored in the interrelationships between PG generation and immune responses. Clearly, these fascinating compounds are generated rapidly in response to diverse stimuli and have many potent and relevant activities. It is equally clear, however, that the demonstration of biologic responses to pharmacologic concentrations of PGs is far different than the actual demonstration of endogenous PG-mediated responses. The current interest in PGs is sufficient to allow the prediction that a clearer picture of PGs in allergy will be available in the foreseeable future. It is also evident that this search will uncover new and presently unrecognized functions for these molecules in human disease.

References

Adcock, J.J., Garland, L.G., Moncada, S., Salmon, J.A. (1978) The mechanism of enhancement by fatty acid hydroperoxides of anaphylactic mediator release. Prostaglandins 16,179–187.

Bach, M.K., Brashler, J.R., Gorman, R.R. (1977) On the structure of slow reacting substance of anaphylaxis: Evidence of biosynthesis from arachidonic acid. Prostaglandins 14,21–38.

Bach, M.K., Brashler, J.R., Brooks, C.D., Neerken, A.J. (1979) Slow-reacting substances: Comparison of some properties of human lung SRS-A and two distinct fractions from ionophore-induced rat mononuclear cell SRS. J. Immunol. 122,160–165.

Bergström, S., Dressler, F., Krabisch, L., Ryhage, R., Sjovall, J. (1962) The isolation and structure of a smooth muscle stimulating factor in normal sheep and pig lungs: prostaglandin and related factors 9. Arki. Kemi. 20,63–66.

Bills, T.K., Smith, J.B., Silver, M.J. (1977) Selective release of arachidonic acid from the phospholipids of human platelets in response to thrombin. J. Clin. Invest. 60,1–6.

Borgeat, P., Samuelsson, P. (1979) Transformation of arachidonic acid by rabbit polymorphonuclear leukocytes. J. Biol. Chem. 254, 2643–2646.

Borgeat, P., Hamberg, M., Samuelsson, B. (1976) Transformation of arachidonic acid by rabbit polymorphonuclear leukocytes. Monohydroxy acids from novel lipoxygenases. J. Biol. Chem. 251, 7816–7820.

Brody, M.J., Kadowitz, P.J. (1974) Prostaglandins as modulators of the autonomic nervous system. Fed. Proc. 33,48–54.

Conroy, M.C., Orange, R.P., Lichtenstein, L.M. (1976) Release of slow reacting substance of anaphylaxis (SRS-A) from human leukocytes by the calcium ionophore A23187. J. Immunol. 116, 1677–1681.

Dayer, J.M., Krane, S.M., Russell, R.G.G., Robinson, D.R. (1976) Production of collagenase and prostaglandins by isolated adherent rheumatoid synovial cells. Proc. Natl. Acad. Sci. U.S.A. 73, 945–949.

Drazen, J.M., Austen, K.F., Lewis, R.A., Clark, D.A., Goto, G., Marfat, A., Corey, E.J. (1980) Comparative airway and vascular activities of leukotrienes C-1 and D *in vivo* and *in vitro*. Proc. Natl. Acad. Sci. U.S.A. 77,4354–4358.

Duncan, M.W., Person, D.A., Rich, R.R., Sharp, J.T. (1977) Aspirin and delayed type hypersensitivity. Arthritis Rheum. 20,1174–1178.

Frank, M.M. (1975) Complement. In: *Curr. Concepts,* Upjohn Co., pp. 1–48.

Goldstein, I.M., Malmsten, C.L., Samuelsson, B., Weissmann, G. (1977) Prostaglandins, thromboxanes and polymorphonuclear leukocytes. Inflammation. 2,309–317.

Goodwin, J.S., Messner, R.P., Bankhurst, A.D., Peake, G.T., Saiki, J.H., Williams, R.C., Jr. (1977a) Prostaglandin-producing suppressor cells in Hodgkin's disease. N. Engl. J. Med. 257, 963–968.

Goodwin, J.S., Bankhurst, A.D., Messner, R.P. (1977b) Suppression of human T-cell mitogenesis by prostaglandin. J. Exp. Med. 146,1719–1734.

Hamberg, M., Samuelsson, B. (1973) Detection and isolation of an endoperoxide intermediate in prostaglandin biosynthesis. Proc. Natl. Acad. Sci. U.S.A. 70,899–903.

Horton, J.E., Raisz, L.G., Simmons, H.A., Oppenheim, J.J., Mergenhagen, S.E. (1977) Bone resorbing activity in supernatant fluid from cultured human peripheral blood leukocytes. Science 177,- 793–795.

Hubscher, T. (1975) Role of the eosinophil in the allergic reactions. II. Release of prostaglandins from eosinophilic leukocytes. J. Immunol. 114,1389–1393.

Jakschik, B.A., Falkenheim, S., Parker, C.W. (1977) Precursor role of arachidonic acid in release of slow reacting substance from rat basophilic leukemia cells. Proc. Natl. Acad. Sci. U.S.A. 74,4577–4581.

Jakschik, B.A., Lee, L.H., Shuffer, G., Parker, C.W. (1978) Arachidonic acid metabolism in rat basophilic leukemia (RBL-1) cells. Prostaglandins 16,733–749.

Kaliner, M. (1977) Human lung tissue and anaphylaxis: evidence that cyclic nucleotides modulate the immunologic release of mediators through effects on microtubule assembly. J. Clin. Invest. 60, 951–959.

Kaliner, M., Austen, K.F. (1973) A sequence of biochemical events in the antigen-induced release of chemical mediators from sensitized human lung tissue. J. Exp. Med. 138,1077–1094.

Klein, D.C., Raisz, L.G. (1970) Prostaglandins: Stimulation of bone resorption in tissue culture. Endocrinology 86,1436–1442.

Kurland, J.I., Bockman, R. (1978) Prostaglandin E production by human blood monocytes and mouse peritoneal macrophages. J. Exp. Med. 147,952–957.

Mathé, A.A., Hedqvist, P., Holmgren, A., Svanborg, N. (1973) Prostaglandin $F_{2\alpha}$: Effect on airway conductance in healthy subjects and patients with bronchial asthma. Adv. Biosci. 9,241–247.

Moncada, S., Gryglewski, R., Bunting, S., Vane, J.R. (1976) An enzyme isolated from arteries transforms prostaglandin endoperoxides to an unstable substance that inhibits platelet aggregation. Nature 263,663–665.

Murphy, R.C., Hammarstrom, S., Samuelsson, B. (1979) Leukotriene C: a slow-reacting substance from murine mastocytoma cells. Proc. Natl. Acad. Sci. U.S.A. 76,4275–4279.

Nadel, J.A. (1973) Neurophysiologic aspects of asthma. In: *Asthma—Physiology, Immunopharmacology and Treatment.* Austen, K.F., Lichtenstein, L.M. (eds.) New York: Academic Press, pp. 29–39.

Nijkamp, F.P., Flower, R.J., Moncada, S., Vane, J.R. (1976) Partial identification of rabbit aorta contracting substance releasing factor and inhibition of its activity by anti-inflammatory steroids. Nature 263,479–482.

Orange, R.P., Austen, K.F. (1969) Slow-reacting substance of anaphylaxis. Adv. Immunol. 10, 106–145.

Paterson, N.A., Wasserman, S.I., Said, J.W., Austen, K.F. (1976) Release of mediators from partially purified human lung mast cells. J. Immunol. 117,1356–1362.

Piper, P.J., Vane, J.R. (1969) Release of additional factors in anaphylaxis and its antagonism by anti-inflammatory drugs. Nature 223, 29–35.

Platshon, L., Kaliner, M. (1978) The effects of the immunologic release of histamine upon lung cyclic nucleotide levels and prostaglandin generation. J. Clin. Invest. 62, 1113–1121.

Roberts, L.J., Lewis, R.A., Hansbrough, R., Austen, K.F., Oates, J.A. (1978) Biosynthesis of prostaglandins, thromboxanes and 12-hydroxy-5,8,10,14-eicosatetraenoic acid by rat mast cells. Fed. Proc. (Abstr.). 37,384.

Samter, M., Beers, R.F. (1968) Intolerance to aspirin. Ann. Intern. Med. 68,975–983.

Samuelsson, B. (1965) On the incorporation of oxygen in the conversion of 8,11,14-eicosatrienoic acid to prostaglandin E_1. J. Am. Chem. Soc. 87,3011–3013.

Samuelsson, B. (in press) Oxidative products of arachiodonate: leukotrienes, a new group of compounds including SRS-A. In: *Biochemistry of the Acute Allergic Reaction, Vol. 4.* Austen, K.F., Becker, E.L. (eds.) New York: Alan R. Liss.

Sullivan, T.J., Parker, C.W. (1979) Possible role of arachidonic acid and its metabolites in mediator release from rat mast cells. J. Immunol. 122, 431–436.

Svanborg, N., Hamberg, M., Hedqvist, P. (1973) Aspects of prostaglandin action in asthma. Acta Physiol. Scand. (Suppl. 396) 89: 22.

Svensson, J., Hamberg, M., Samuelsson, B. (1975) Prostaglandin endoperoxides. IX. Characterization of rabbit aorta contracting substance (RCS) from guinea pig lung and human platelets. Acta Physiol. Scand. 94:222–228.

Tauber, A.I., Kaliner, M., Stechschulte, D.J., Austen, K.F. (1973) Immunologic release of histamine and slow reacting substance of anaphylaxis from human lung. V. Effect of prostaglandins on release of histamine. J. Immunol. 111:27–32.

Toneda, T., Mundy, G.R. (1979) Prostaglandins are necessary for osteoclast activating factor production by activated peripheral blood leukocytes. J. Exp. Med. 149, 279–292.

Turner, S.R., Campbell, J.A., Lynn, W.S. (1975) Polymorphonuclear leukocyte chemotaxis toward oxidized lipid components of cell membrane. J. Exp. Med. 141,1437–1441.

Vane, J.R. (1976) The mode of action of aspirin and similar compounds. J. Allergy Clin. Immunol. 58,691–712.

von Euler, U.S. (1934) Zur Kenntnis der pharmakologischen Wirkungen von Nativsekreten und Extrakten männlicher accessorischer Geschlectsdrüsen. Naunyn-Schmiedebergs Arch. Exp. Pathol. Pharmacol. 175,78–84.

Wahl, L.M., Olsen, C.E., Wahl, S.M., Sandberg, A.L., Mergenhagen, S.E. (1977a) Prostaglandin-regulated macrophage collagenase. *Proceedings, Mechanisms of Localized Bone Loss.* Horton, J.E., Tarpley, T.M., and Davis, W.F. (eds.) Supplement to Calcified Tissue Abstracts, pp. 181–190.

Wahl, L.M., Olsen, C.E., Sandberg, A.L., Mergenhagen, S.E. (1977b) Prostaglandin regulation of macrophage collagenase production. Proc. Natl. Acad. Sci. U.S.A. 74, 4955–4958.

Walker, J.L. (1973) The regulatory function of prostaglandins in the release of histamine and SRS-A from passively sensitized human lung tissue. In: *Advances in the Biosciences,* Vol. 9. International Conference on Prostaglandins, Braunschweig. Bergström, S., Bernhard, S. (eds.) Vieweg; Pergamon Press pp. 235–240.

Complement

Ann L. Sandberg, Ph.D.

Numerous inflammatory reactions and host defense mechanisms are now recognized to be attributable to activation of a humoral mediator system, the complement system. Complement was functionally described at the turn of the century, when it was discovered that, in addition to antibody, heat labile serum factors were required to kill bacteria and lyse erythrocytes. This latter biological effect has provided the basis for the highly sensitive assay still in current use, in which complement activity is quantitated by its ability to induce the release of hemoglobin from sheep erythrocytes sensitized with rabbit antisheep erythrocyte antibody (Rapp and Borsos 1970; Mayer 1961). Although it was originally thought that complement function was related primarily to lytic reactions, findings throughout the intervening years have extended the biological consequences of complement activation to include viral neutralization, smooth muscle contraction, release of vasoactive amines and lysosomal enzymes, the directed migration of inflammatory cells, enhanced phagocytosis, leukocyte mobilization, and interaction with the cellular immune system.

Precise definition of the mechanisms involved in these complement-mediated effects was not achieved until advanced techniques in protein chemistry were developed and applied to the purification of the individual complement components in the 1950s and 1960s (Nelson et al. 1966; Müeller-Eberhard 1975). The complement system is composed of at least 15 distinct serum proteins, many of which exist in trace amounts and are easily denatured. These complement components exist in blood as functionally inactive precursors that must interact in a specific sequence for expression of biological function. As the complement sequence is traversed, some of the components are endowed with enzymatic activity for later acting components. The activation of complement is specific in that it requires an initiating agent. Complement may be directed to a localized site by such diverse factors as deposits of immune complexes; antibody bound to cell, bacterial or viral surface antigens; or bacteria and their products. The action of complement must be limited, and several control mechanisms exist to subserve this function.

From the Humoral Immunity Section, Laboratory of Microbiology and Immunology, National Institute of Dental Research, National Institutes of Health, Bethesda, Maryland.

Table 1. Properties of the Complement Components

Classical pathway component	Molecular weight	Approximate serum concentration (μg/ml)
C1q	400,000	180
r	198,000	—
s	80,000	110
C2	117,000	25
C3	180,000	1600
C4	200,000	600
C5	185,000	80
C6	128,000	75
C7	121,000	55
C8	174,000	80
C9	80,000	200
Alternative pathway component		
Properdin	185,000	25
Factor B	95,000	200
Factor D	25,000	trace

Figure 1. Schematic representation of the reaction sequences of the complement system.

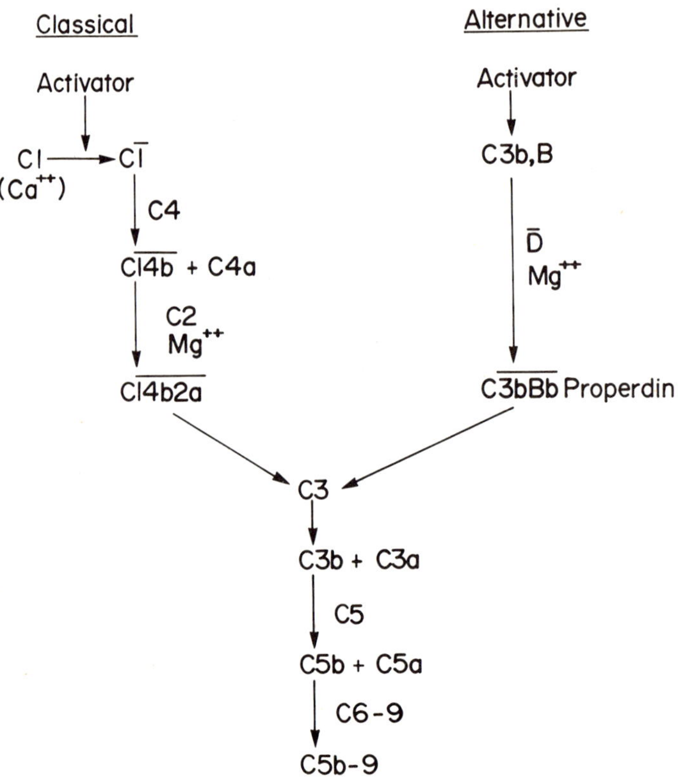

Two pathways of complement activation have been described and have been the subject of several reviews (Müeller-Eberhard 1975; Müeller-Eberhard 1977; Mayer 1978). The most extensively studied, the classical pathway, consists of 11 serum proteins. Each of these components has been assigned a number in the order of its discovery and is designated by this number preceded by the letter C (C1, C4, C2, C3, C5, C6, C7, C8, and C9). The activated components are identified by a bar over the component number (e.g., $\overline{C1}$) and fragments arising by enzymatic cleavage are designated by a letter (e.g., C3a or C3b). The existence of a second pathway, the alternative complement pathway, was proposed in the 1950s to account for host defense against bacterial and viral attacks prior to or in the absence of the evocation of an antibody response. The early components of this system are denoted by symbols or common names (e.g., factor B, properdin). As discussed later, these two pathways converge, using the same terminal complement components. Some properties of the complement components are presented in Table 1 and the reaction sequences are summarized in Figure 1.

Pathways of Complement Activation

Classical Complement Pathway

Activation of the classical complement pathway is initiated by immune complexes containing immunoglobulins of several IgG subclasses or the IgM class. The effects of complement are most easily studied when the antigenic component of the immune complex is present on a cell surface, usually an erythrocyte. A single molecule of IgM antibody attached to a surface antigen will bind the first component of complement, C1. However, two adjacent molecules of IgG are required for this event (Borsos and Rapp 1965). This spatial requirement accounts for the relative inefficiency of IgG antibodies to mediate cell destruction. Although many IgG molecules may combine with surface antigens they may not be in appropriate proximity to bind C1.

A trimolecular complex, C1 consists of the discrete proteins C1q, C1r and C1s, which are held together in the presence of calcium ions. After binding of C1 to the Fc portion of the immunoglobulin through the C1q subunit, C1 is endowed with esteratic activity, a property that resides in the C1s subunit. Enzymatically active C1 has two complement component substrates, C4 and C2. Cleavage of C4 into two fragments, C4a and C4b, results in the exposure of a labile site on the larger fragment, C4b, by which it binds to the cell membrane (Figure 2). C4a is released into the fluid phase and has recently been reported to have the biological properties of the other complement-derived anaphylatoxins (Gorski et al. 1979), which are discussed later. In this reaction, multiple C4 molecules are cleaved and a cluster of C4b fragments is formed around the antibody–C1 site. If the C4b fragments are not rapidly bound to the cell membrane, they lose the ability to do so due to decay of the binding site and are unable to participate in the subsequent steps of complement activation. In the presence of bound C4b and magnesium ions, activated C1 enzymatically cleaves C2 into two fragments exposing a labile binding site on the larger fragment, C2a, by which it complexes with C4b on the membrane (Figure 3). C4b2a is unstable and decays via the release of C2a in an inactive form. The bimolecular complex, C4b2a, also possesses enzymatic activity. Its natural substrate is the next complement component in the sequence, C3, and it has, therefore, been termed C3 convertase, the proteolytic site of which is present in C2a. Its action on C3 is to cleave this molecule

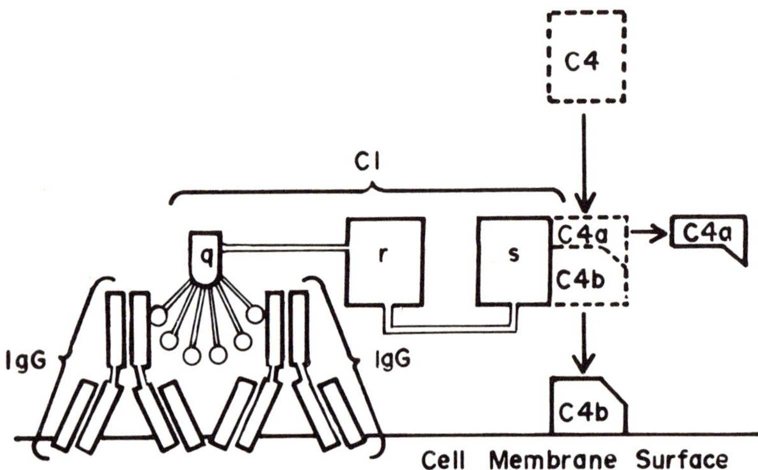

Figure 2. Binding and activation of C1 and activation of C4. The C1q subunit of the C1 molecule attaches to the immunoglobulin bound to the cell membrane. This results in activation of C1r, which in turn activates C1s. C4 is cleaved by C1s and the larger fragment, C4b, binds to the cell surface, while the smaller anaphylatoxic fragment, C4a, is released.

into two fragments (Figure 4). The larger fragment, C3b, associates with the C4b2a complex and, in addition, a labile binding site is exposed by which it attaches to the cell membrane. Many molecules of C3 are cleaved by each C4b2a complex and the numerous C3b fragments generated are deposited around each site where the attached antibody has activated the earlier acting complement components. Due to the lability of the membrane binding site on C3b, many of these fragments exist in an inactive form in the fluid phase. The smaller C3 fragment, C3a, is released into the fluid phase and acts as a mediator of several anaphylatoxic reactions. The attachment of C3b to the cell membrane carrying C4b2a results in the expression of the final enzymatic activity generated during complement activation. It is specific for the next complement

Figure 3. Activation of C2. C2 is cleaved by C1s and the larger fragment, C2a combines with C4b to form an enzyme on the cell membrane. The C2b fragment is released.

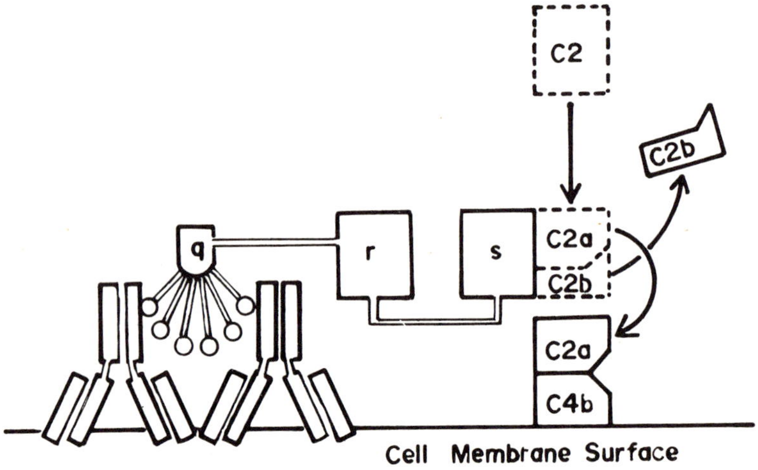

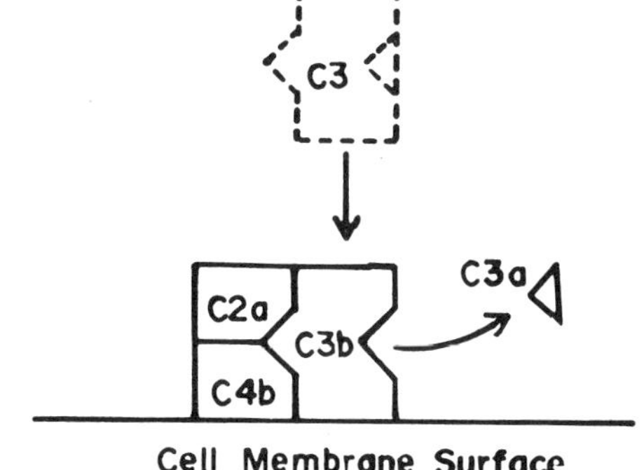

Figure 4. Activation of C3. The enzyme, C4b2a, cleaves C3 and the larger fragment C3b, attaches to the membrane while the smaller anaphylatoxic fragment, C3a, is released.

component in the sequence, C5. C4b2a3b is an unstable enzyme which, like C4b2a, decays by releasing C2a. Two fragments of C5 are produced by the action of C4b2a3b, the larger of which, C5b, can attach to the membrane for a short period of time and the smaller fragment, C5a, is released (Figure 5). The latter fragment possesses both anaphylatoxic and chemotactic properties. Sites on C5b fragments serve as receptors for the next two complement components, C6 and C7 (Figure 6). Then C8 and C9 are bound (Figure 7), a membrane lesion is produced and lysis of the cell ensues. Minimal lysis occurs after the attachment of C8 and the function of C9 is apparently to augment this reaction.

Several factors regulate the activation of the complement system. Amplification occurs by the cleavage and deposition of numerous C4 and C3 molecules. As previously mentioned, highly labile binding sites exposed upon cleavage of C2, C3, C4, and C5 may limit the attachment of C2a, C3b, C4b, and C5b to the cell membrane or to earlier acting components. The decay of the enzymatic complexes, C4b2a and C4b2a3b terminates further progression of the complement sequence. Several serum proteins which specifically inhibit the activated forms of some of the complement components have also been identified. The C1 inhibitor (C1 Inh) binds to activated C1 and destroys its enzymatic site, thus blocking further cleavage of C4 and C2. A second inhibitor, the C3b inactivator, cleaves C3b into two fragments: C3c and C3d. This enzymatic process is accelerated by another serum protein, β1H (Whaley and Ruddy 1976) and in addition requires another enzyme, probably elastase. The C3c fragment, which possesses no known unique biological function, is released while the C3d fragment remains bound to the cell. Cells carrying C4b2a3d on their surface are incapable of cleaving C5, and, therefore, the complement sequence cannot be completed. The C3

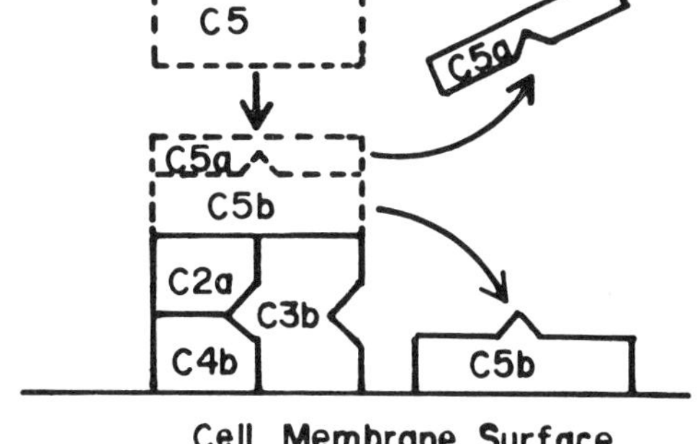

Figure 5. Activation of C5. The enzyme, C4b2a3b, cleaves C5 and the larger fragment, C5b, attaches to the membrane while the smaller anaphylatoxic and chemotactic fragment, C5a is released.

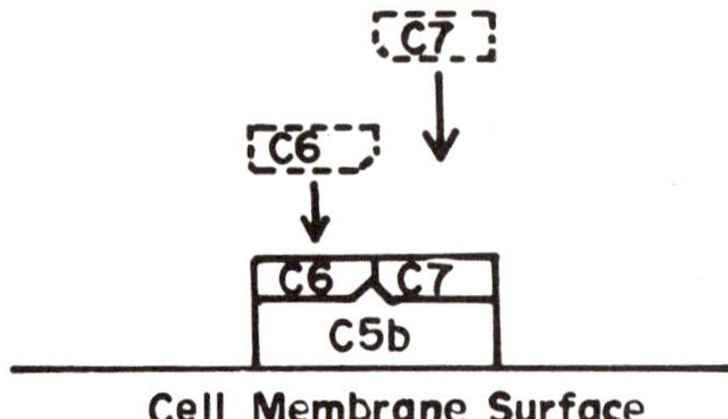

Figure 6. Activation of C6 and C7. In a non-enzymatic process C6 and C7 attach to the membrane bound C5b.

inactivator also cleaves C4b into fragments which are inactive. Another serum protein, the C4-binding protein, functions as a cofactor in this inactivation step (Shiraishi and Stroud 1975; Fujita et al. 1978).

Alternative Complement Pathway

It was postulated in 1954 that the late complement components (C3 to C9) could be activated in the absence of antibody and the early complement components C1, C4, and C2, utilizing instead the serum proteins designated properdin and factors A and B (Pillemer et al. 1954). Controversy arose whether properdin was, in fact, distinct from antibody, and since some physicochemical properties of factors A and B were similar to those of the early complement components, interest in this pathway declined. However, in the late 1960s evidence began to accumulate which definitively established the existence of an alternative complement pathway. Cell-wall preparations from gram-negative bacteria (endotoxins) were shown to activate the late complement components beginning at C3 with minimal consumption of C1, C4, and C2 (Gewurz et al. 1968). A similar profile of complement component consumption was observed when serum was incubated with a protein from cobra venom (Götze and Müeller-Eberhard 1971). In addition, immune complexes containing a subclass of guinea pig immunoglobulins (IgG1), which does not activate the classical pathway, depleted serum of late complement components without utilizing the early components (Sandberg et al. 1970). The F(ab')$_2$ fragments of these antibodies as well as other IgG immunoglobulins, which lack the C1 binding site on the Fc fragments, also consumed only the late components. An important impetus to these studies was the discovery and development of a strain of guinea pigs totally deficient in the early component, C4 (Ellman et al. 1970). These animals provided a model for the direct evaluation of alternative pathway function. In the complete absence of C4, endotoxin and immune complexes activated the late complement components.

In addition to the agents mentioned above, other aggregated immunoglobulins of the IgG and IgA classes and complex polysaccharides have been shown to activate the

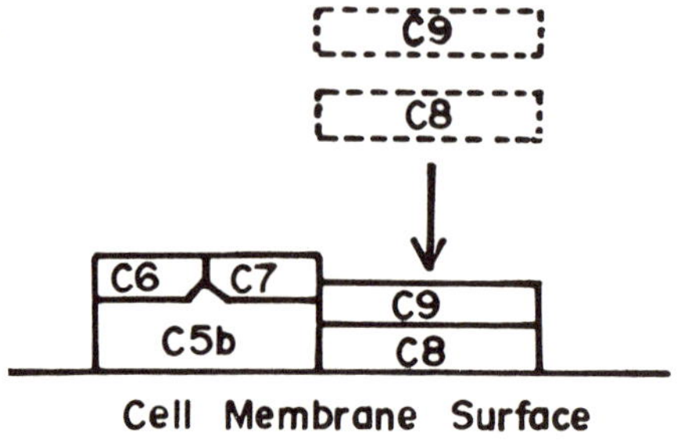

Figure 7. Activation of C8 and C9. C8 and C9 attach to the C5b67 complex and to the cell membrane. Lysis of the cell ensues.

alternative complement pathway. The early components of this pathway have been purified and include factor D, factor B, C3b, and properdin. Factor D is similar to the C1s subcomponent of C1. The alternative pathway component, which shares the hydrazine sensitive properties of C4, is C3b. Factor B has characteristics in common with C2 of the classical pathway in that the heat sensitivity of the two proteins is similar.

The sequence of events which occurs during activation of the alternative pathway is as follows (Figure 1): In the presence of magnesium, factor D cleaves factor B, which is bound to preformed C3b, generating the C3 cleaving enzyme of the alternative pathway, C3bBb (Fearon et al. 1973). Activation of the alternative pathway in serum by cobra venom factor is attributable to its forming a C3 cleaving enzyme when complexed with activated factor B (Bb), since this venom protein is apparently the cobra analog of C3b (Alper and Balavitch 1976). Like the C3 convertase, C4b2a, of the classical pathway, C3bBb is also subject to decay, which occurs by the release of Bb (Weiler et al. 1976). The activity of C3bBb is identical to that of C4b2a. It also cleaves C3 into the two fragments, C3a and C3b. The function of properdin in this pathway is that of stabilizing the C3bBb enzyme by binding to C3b (Fearon and Austen 1975). The point of convergence of the two pathways is C3. Cleavage of C5 and activation of the later components is similar regardless of the pathway traversed.

In addition to the decay of the C3bBb enzyme, the alternative pathway is also subject to regulation by cleavage of the C3b generated into the inactive fragments C3c and C3d by the C3b inactivator, β1H, and an additional serum protease. In addition, β1H dissociates Bb from the C3bBb enzyme complex rendering this C3b susceptible to inactivation by the C3b inactivator.

Biological Consequences of Complement Activation

The biological activities of the classical and alternative complement pathways are summarized in Tables 2 and 3, respectively.

Table 2. Biological Activities of the Classical Complement Pathway

Components	
C14b	Viral neutralization (e.g., herpes simplex)
C14b2a3b	Viral neutralization (e.g., Newcastle disease virus, polyoma virus)
	Adherence to lymphocytes and phagocytic cells
	Enhanced phagocytosis
	Generation of lymphokines
C14b2a3d	Adherence to lymphocytes and phagocytic cells
C1-9	Lysis of cells, bacteria and viruses
Component fragments	
C3a	Anaphylatoxin
C3b	Complexes with factor B to form C3 convertase of the alternative pathway
	Generation of lymphokines
C3e	Leukocyte mobilization
C4a	Anaphylatoxin
C5a	Anaphylatoxin
	Chemotactic factor

Table 3. Biological Activities of the Alternative Complement Pathway

Cytolysis
Viral neutralization
Enhanced phagocytosis
Macrophage activation
Generation of anaphylatoxins (C3a, C5a)
Generation of chemotactic factor (C5a)

Cytolysis

In addition to antibody-coated erythrocytes, numerous other cells such as bacteria, lymphocytes, and certain viruses are subject to complement mediated lysis. However, cells vary greatly in their susceptibility to complement attack. This may reflect the distribution of antigens on the cell surface, or it may be related to the structure of the cell wall or membrane. Factors in addition to complement may also be required as is the case of some bacteria that are lysed only in the presence of both complement and lysozyme.

Cytotoxic damage is dependent on the activation of the entire complement sequence. Although the majority of cytolytic reactions can be attributed to classical pathway activation, the alternative pathway can also function in this regard. For example, human lymphocytes sensitized with some human antihistocompatibility antibodies are destroyed exclusively by this pathway (Ferrone et al. 1973). Of clinical relevance is the lysis of an abnormal subpopulation of erythrocytes from patients with paroxysmal nocturnal hemoglobinuria by activation of the alternative pathway, to be discussed below.

Complement may also contribute to the destruction of cells that have not themselves activated either pathway. By reactive lysis, C5b67 may transfer to the membrane of an unsensitized bystander cell, which can then undergo cytolysis owing to its subsequent interaction with C8 and C9 (Lachmann and Thompson 1970; Goldman et al. 1972).

Recent studies have been directed toward elucidating the mechanism by which complement induces cell lysis. These endeavors have been greatly facilitated by the use of liposomes and planar lipid bilayers, both of which are susceptible to lysis by complement. As measured by conductance in a reactive lysis system, permeability changes are detected after addition of C5b6 to rigid lipid bilayers, which contain oxidized cholesterol. The conductance is greatly augmented by the sequential addition of C7, C8, and C9. These findings as well as several other lines of evidence indicate that portions of the late complement components, C5b-9, intercalate into the lipid bilayer of the cell membrane. Radiolabeled C5, C7, C8, and C9 resist tryptic removal from sensitized erythrocytes incubated with C1-7, C1-7, C1-8, and C1-9, respectively, and C5 and C9 are not dissociated from erythrocyte-complement intermediates by high salt concentrations (Mayer 1978). Analysis of electron spin resonance spectra of lipid bilayers treated with the late components indicates their insertion into the bilayer and further suggests that C5b-7 penetrates the hydrophobic region of the bilayer only slightly, whereas C5b-8 and C5b-9 penetrate more deeply (Esser et al. 1979). The hydrophilic nature of the late complement components would seem to preclude their insertion into the lipid membrane. However, direct evidence has been obtained that hydrophobic sites are exposed during the formation of the C5b-9 complex which can interact with membrane lipids. Thus, C5b-7 binds to phospholipids, and this reaction is greatly enhanced by the sequential addition of C8 and C9 (Podack et al. 1979).

Two hypotheses have been proposed to explain the complement-dependent disruption of the cell membrane that leads to osmotic swelling and resultant destruction of the cell. One of these attributes this effect to the penetration of the lipid bilayer by a hollow doughnut-shaped structure composed of C5b-9. The hydrophilic interior of this cylinder forms a transmembrane channel that permits the passage of ions and water (Mayer 1978). The other proposes that the late complement components disorient the membrane by binding to the phospholipids in the membrane, forming protein-phospholipid micelles and creating hydrophilic lipid channels in the vicinity of the inserted C5b-9 complex (Podack et al. 1979).

As first observed by Humphrey and Dourmashkin (1965), electron microscopic examination of erythrocyte membranes exposed to complement reveals specific ring-like lesions with an inner diameter of approximately 110 Å and an outer diameter of 200 Å. Similar lesions are induced by complement on gram-negative bacterial cell wall preparations (Figure 8). These rings are now interpreted as top views of the membrane-bound C5b-9.

Figure 8. Complement lesions on a cell wall preparation from *Veillonella alcalescens*. Bladen, H., Hageage, G., Harr, R., Pollock, F. (1973) Lysis of certain organisms by the synergistic action of complement and lysozyme. J. Dent. Res. 52,371.

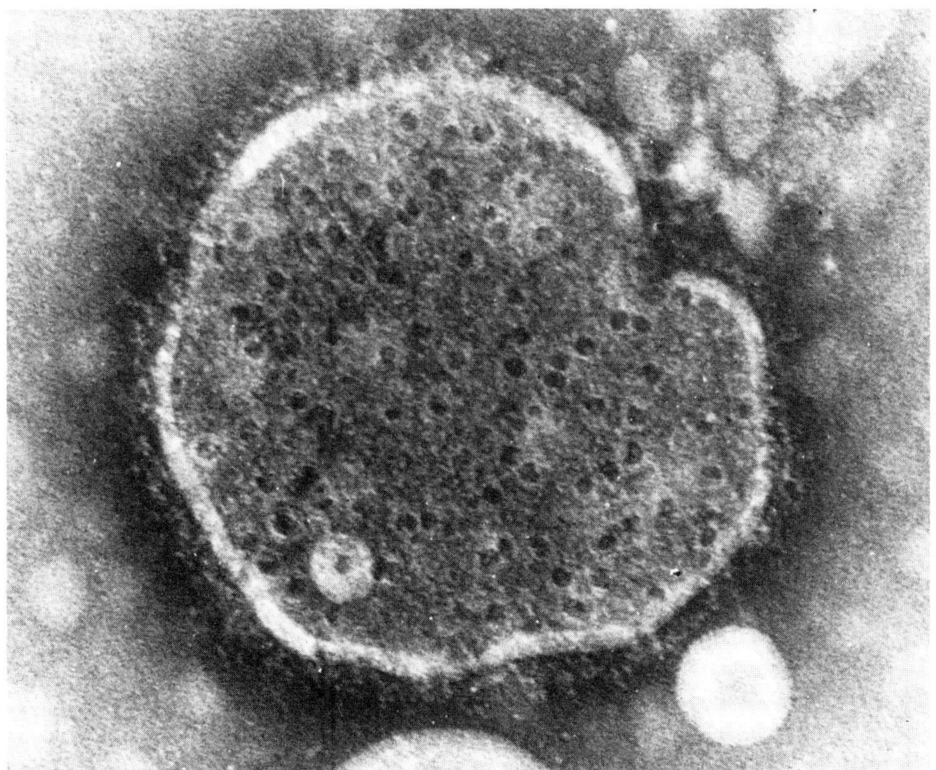

Viral Neutralization and Destruction of Viral Infected Cells

Several mechanisms appear to mediate complement-dependent neutralization of virus. A simple coating of the virus by antibody and complement components results in neutralization of several viruses; e.g., herpes simplex virus, avian infectious bronchitis virus, Newcastle disease virus, and vaccinia virus. The complement component requirements vary for the different viruses. Herpes simplex virus can be neutralized by antibody and optimal concentrations of C1 and C4 (Daniels et al. 1970). However, if C1 and C4 are limiting, C2 and C3 enhance neutralization. The alternative complement pathway is also effective in neutralizing this virus, presumably due to viral coating by the early components of this pathway. Newcastle disease virus is neutralized by C1, C4, C2, and C3 or by components of the alternative pathway (Linscott and Levinson 1969).

A second mechanism operative in viral neutralization occurs with polyoma virus. Interaction of this virus with specific antibody and C1, C4, C2, and C3 results in cross-linking of the viral particles accompanied by loss of infectivity (Oldstone et al. 1974).

Specific antibody complexed with AKR leukemia virus, equine arteritis virus, lymphocytic choriomeningitis virus, Sindbis virus, or Moloney leukemia virus initiates complement dependent lysis (Oldstone 1974).

Complement dependent inactivation of other viruses apparently occurs in the absence of specific antibody. Thus, vesicular stomatitis virus is neutralized by C1, C4, C2, and C3 in a reaction which requires an additional nonimmunoglobulin serum factor (Mills and Cooper 1978). Complement-mediated lysis of RNA tumor viruses is also independent of specific antibody and is apparently a function only of the classical pathway (Welsh et al. 1976).

Following infection, viral antigens are expressed on the surface of mammalian cells. Viral-specific antibodies interact with these antigens and initiate complement-dependent lysis of the infected cells. This function, like other complement-mediated cytolytic reactions, requires completion of the entire complement sequence. The classical pathway, the alternative pathway, and the cooperation of both pathways have been implicated in the destruction of viral infected cells (Hicks et al. 1976; Sissons et al. 1979).

Contraction of Smooth Muscle, Histamine Release, Enhanced Capillary Permeability, and Enzyme Release

As mentioned, the anaphylatoxins, C3a and C5a, are generated during complement activation by the highly selective proteolytic cleavage of C3 and C5. The chemical and biological properties of the anaphylatoxins have been reviewed recently (Hugli and Müeller-Eberhard 1978). C3a and C5a may be produced by the classical pathway, the alternative pathway, or by direct digestion of the parent complement component by cellular derived proteases. These anaphylatoxins cause the contraction of smooth muscle, including that in the intestine, uterus, trachea, arteries, and atrium. A property of both C3a and C5a is their ability to produce tachyphylaxis (desensitization) in these tissues. The tachyphylaxis is highly specific in that C3a desensitizes the tissue to subsequent stimulation by C3a but not by C5a; C5a desensitization is specific for this polypeptide, indicating interaction of C3a and C5a with independent tissue sites.

This contractile process in several tissues is a direct effect on the musculature rather than a secondary event that occurs as a result of histamine release, since the tissues retain their responsiveness to the anaphylatoxins after desensitization to histamine. The peptide derived by cleavage of C4 by C1s, C4a, has also been shown recently to be spasmogenic (Gorski et al. 1979). This third anaphylatoxin of the complement system apparently is related to C3a since it confers tachyphylaxis on smooth muscle treated with C3a.

In addition to their direct action on smooth muscle, the anaphylatoxins exert their biological effects by inducing the release of histamine from mast cells and basophils (Siraganian and Hook 1976; Glovsky et al. 1979). These cells possess receptors which are specific for either C3a or C5a. The cells can be rendered unresponsive to C3a or C5a by their prior exposure to the respective complement component fragment reminiscent of the specific tachyphylaxis that occurs with smooth muscle contraction. The release of histamine occurs in the absence of cytotoxicity and results in contraction of smooth muscle as well as enhanced capillary permeability with the consequent production of edema. *In vivo,* C4a also enhances vascular permeability.

Another inflammatory consequence of C3a and C5a stimulation of cells is the release of lysosomal enzymes from polymorphonuclear leukocytes (Becker et al. 1974).

Since the anaphylatoxins are extremely potent mediators of inflammation, a strict control mechanism must exist to terminate their biological function. This is achieved by the action of a serum carboxypeptidase which rapidly removes the carboxyterminal arginine from C3a and C5a (Bokisch et al. 1969). The fragments lacking this amino acid are incapable of inducing smooth muscle contraction, the release of histamine, and the release of enzymes.

Chemotaxis

An initial event in inflammation is the migration of characteristic cells to a localized site. Polymorphonuclear leukocytes, macrophages, basophils, and eosinophils have been identified in association with certain inflammatory lesions. The complement-derived factor that directs the influx of these cells to a specific area is C5a (Shin et al. 1968). Neither C3a nor C4a possesses this biological activity. In areas of inflammation induced by endotoxins or immune complexes, activation of the alternative or classical complement pathways may lead to the cleavage of C5 with the resultant generation of this chemotactic factor. C5a may also be the stimulus for the migration of inflammatory cells to the site of virus infections since supernatants from certain viral infected cells generate a serum derived chemotactic factor which has been identified as C5a (Brier et al. 1970). The accumulation of polymorphonuclear leukocytes and macrophages in these lesions may lead to the further influx of inflammatory cells since both of these cell types produce proteases which cleave C5 (Ward and Hill 1970; Snyderman et al. 1972).

In contrast to the previously discussed inactivation of the anaphylatoxic properties of C5a by the serum carboxypeptidase, removal of the terminal arginine of C5a does not result in a loss of chemotactic activity. Regulation of this biological response may, instead, be attributed to the fact that cell migration occurs over a narrrow concentration of C5a. Since high concentrations of C5a and desarginine C5a are inhibitory, their continued production at a specific site may limit the duration of inflammatory cell accumulation.

Enhanced Phagocytosis

A major function of the complement system in host defense is that of enhancing phagocytosis (Gigli and Nelson 1968). Phagocytic cells (polymorphonuclear leukocytes, monocytes, and macrophages) possess surface receptors for both the Fc portion of IgG immunoglobulins and a stable binding site on the C3b fragment of C3 (Lay and Nussenzweig 1968). Immune complexes containing IgG are ingested by these cells after attachment via the Fc receptor. However, if the immune complexes have activated the complement system, the C3b bound to them markedly increases both the rate and extent of phagocytosis. The role of C3b in this reaction is related primarily to enhanced adherence of the immune complex to the phagocytic cell whereas the IgG is required for ingestion. In the absence of IgG, bound C3b can mediate both adherence to and ingestion by macrophages if the cells have been previously activated (Bianco et al. 1975). C5b has also been implicated in the enhancement of phagocytosis of some microorganisms (i.e., yeast and pneumococci) (Miller and Nilsson 1974; Shin et al. 1969). Phagocytosis can occur by activation of either the classical or alternative complement pathway, although it takes place more slowly via the latter pathway.

Interaction with the Cellular Immune System

C3b provides a point of juncture between the humoral and cellular immune systems. In addition to polymorphonuclear leukocytes and macrophages, B-lymphocytes have receptors for C3b (Lay and Nussenzweig 1968). Interaction of this complement fragment with B-cells results in the production of at least one lymphokine, a cellular derived factor chemotactic for inflammatory cells (Sandberg et al. 1975). B-lymphocytes also possess a receptor for a stable binding site on the C3d fragment (Ross et al. 1973). In contrast to lymphocyte activation by C3b, as determined by lymphokine production, C3d, although bound, is incapable of cellular stimulation (Koopman et al. 1976). The current concept of cell activation implicates a cross-linking of receptors. Whereas C3b binds to both the C3b and C3d receptors and would, therefore, satisfy this criterion, C3d binds only to its specific receptors. Both the classical and alternative complement pathways can generate C3b, so, by inference, either pathway could be functional. In addition, cellular-derived proteolytic enzymes can cleave C3, resulting in C3b production which, in turn, could activate lymphocytes.

A role for C3 has also been suggested in the antibody response to T-cell-dependent but not T-cell-independent antigens (Pepys 1974). The mechanism by which C3 participates in immunoglobulin production has not been delineated and may be multifactorial; influencing antigen localization, the cooperation between the cell types responsible for antibody synthesis, or the generation of memory cells. Although the complement pathway involved in this response has not been defined, evidence exists that in animals lacking the early complement component, C4, antibody synthesis may be impaired (Ellman et al. 1971).

Leukocyte Mobilization

The release of leukocytes from the bone marrow may serve as a prelude to the accumulation of inflammatory cells in a local site. This biological activity has been ascribed to another C3 fragment, C3e (Rother 1972; Ghebrehiwet and Müeller-Eberhard 1979). Although it is not chemotactic *in vitro,* this fragment mobilizes leukocytes from the marrow of isolated bones, and it produces leukocytosis *in vivo*. This fragment presumably arises as a late cleavage product of C3. Following inactivation of C3b with the resultant production of C3c and C3d, C3c is further degraded, liberating C3e. The enzymatic nature of this degradation is unknown.

Macrophage Activation

The expression of macrophage function (monokine production, enzyme production, bactericidal activity, and cytotoxicity), which results as a consequence of prior activation of these cells, may also be influenced by the complement system. An early *in vitro* morphological correlate of *in vivo* macrophage activation is the ability of these cells to spread rapidly on a surface to which they have become attached. Activated factor B, Bb, of the alternative complement pathway can initiate this reaction (Götze et al. 1979). Thus, in an area of macrophage infiltration the generation of this fragment may be a contributor to the inflammatory response.

Phylogeny and Ontogeny of the Complement System

Phylogeny

The available evidence indicates strongly that the alternative complement pathway evolved before both the classical pathway and the specific humoral antibody response. Components of the alternative pathway have been detected in invertebrate species (Day et al. 1970). In the starfish, a factor, presumably factor B, is present that, when complexed with cobra venom factor, activates the late complement components in frog sera. An intact alternative pathway exists in the horseshoe crab and sipunculid worm, as evidenced by the ability of cobra venom factor to activate the terminal complement components in these invertebrates. Although immunity first appears in the sea lamprey, apparently it lacks the classical complement pathway. A hemolysin is present in this species but it is physicochemically distinct from complement. The nurse shark is the lowest vertebrate shown to possess a complement system analogous to the classical complement system of higher vertebrates (Jensen 1969).

Ontogeny

Human newborns have only one-half to two-thirds the adult complement activity. Human fetal cord blood contains a hemolytically active complement system, and the fetal synthesis of the complement components is initiated at or slightly before the onset of fetal immunoglobulin synthesis. In cultures of human fetal tissues, biologically active C2 and C4 have been detected as early as the eighth week of gestation. The rate of C2 synthesis remains approximately constant throughout fetal life, whereas that of C4 apparently increases during this period. C3 can be found in human liver cultured at 4 weeks after gestation and in fetal serum between 5 and 10 weeks of gestation. Fetal synthesis of C1, C5, C6, C8, and factor B has also been established. The complement components do not cross the placental barrier. Several of these components exhibit genetic polymorphism and discrepancies between the fetal and maternal allotypes have been demonstrated. Further evidence of the lack of transplacental passage is finding of a component, which is genetically absent in maternal serum in the serum of the fetus.

Synthesis of the Complement Components

The biosynthesis of complement has been reviewed by Colten (1976). The small and large intestine in *in vitro* organ culture appear to be the primary tissues involved in the production of macromolecular hemolytically active C1. This synthetic function resides primarily in the columnar epithelial cells. Macrophages and fibroblasts also synthesize C1, but the levels are markedly lower than those produced by epithelial cells (Morris et al. 1978).

De novo synthesis of C2 has been detected in a variety of organs including spleen, lung, bone marrow, and liver. Since mononuclear cells from peritoneal exudates in culture and long-term primary cultures of peripheral blood monocytes synthesize C2, it is probable that this component is produced by the macrophage population within these organs.

Spleen and liver have both been shown to synthesize C4. Macrophages produce C4 and at least some of these cells are capable of synthesizing both C4 and C2. In serum, hemolytically active C4 consists of three polypeptide chains linked by disulfide bridges. In contrast to the immunoglobulins, in which the polypeptide chains are synthesized separately and then assembled, C4 is synthesized as a single polypeptide chain (pro-C4), which is subsequently cleaved (Hall and Colten 1977). Pro-C4 has been detected in trace amounts in serum (Gigli 1978). It is antigenically identical to C4 but possesses no hemolytic activity.

The liver is the principal site of C3 synthesis as evidenced by a complete conversion from recipient to donor C3 allotype in the serum of a patient receiving a liver transplant following hepatectomy. The parenchymal cells of the liver produce C3, but *in vitro* studies indicate macrophages are also capable of its synthesis. Serum C3 comprises two polypeptide chains, which are disulfide linked. C3 is synthesized in a precursor form (pro-C3) as a single polypeptide chain which is then cleaved (Brade et al. 1977). Small amounts of pro-C3 have been found in plasma (Minta 1979).

The spleen, lung, liver, and intestine synthesize C5. The macrophage appears to be the cell that produces it. Like C4 and C3, C5 is synthesized as a single polypeptide chain and is cleaved into two chains that are covalently linked by disulfide bridges.

Although C6 can be synthesized by liver, its synthesis by other organs has not been excluded. The cell that produces this component has not been identified.

No studies have been carried out concerning the site of C7 biosynthesis.

Although C8 has been detected in spleen, liver, lung, intestine, and kidney, the cell type involved has not been investigated.

The liver synthesizes C9, and the available evidence indicates that it is produced by parenchymal cells.

The C1 inhibitor is also synthesized by the liver, and immunofluorescent studies localize its source as the parenchymal cells.

Functionally active factors B and D and properdin of the alternative pathway have been detected in macrophage culture supernatants (Bentley et al. 1978).

Complement Component Deficiencies

Animal Models

As mentioned previously, a strain of guinea pigs has been developed that is completely deficient in functional or antigenically identifiable C4 (Ellman et al. 1970; Ellman et al. 1971; Frank et al. 1971). These animals, therefore, lack the classical complement pathway and must depend on the alternative pathway for complement-dependent biological functions. This deficiency is linked to the major histocompatibility locus of the guinea pig. Although immune clearance, as detected by the loss of labeled erythrocytes coated with antibody from the circulation, and, under some circumstances, antibody production, are impaired in these guinea pigs, they exhibit normal survival.

A genetic C4 deficiency has also been described in a strain of Wistar rats (Arroyave 1977). Like the C4-deficient guinea pigs, these animals do not appear to have an increased incidence of infection. The existence of an intact alternative pathway apparently suffices in complement-dependent host defense in these animals.

About 40% of the inbred strains of mice genetically lack C5 activity as well as the C5 protein (Cinader et al. 1964). These mice are not subject to early death, although some evidence exists of defective phagocytosis in these animals.

Rabbits have been described which are genetically deficient in C6 (Rother et al. 1966). The only obvious defective functions in these animals are enhanced susceptibility to endotoxin shock and a prolonged blood clotting time (Zimmerman et al. 1971). This latter effect is highly species specific and is attributable to the requirement for late complement components for platelet lysis and the release of platelet factor III of the clotting system (Siraganian 1972).

Human Deficiencies

The human complement deficiencies have been reviewed by Alper and Rosen (1976) and Agnello (1978). Congenital deficiencies of a single component of the complement system have been found in all the classical components. The total or near-total absence of C1r, C4, C2, C3, C5, C6, C7, C8, or C9 is inherited as autosomal recessive traits. For the most part, the sera of heterozygotes contain approximately half-normal levels of the single components, and these are sufficient to mediate complement-dependent functions. Various relationships have been made between the homozygous component deficiencies and clinical abnormalities.

Although the genetic deficiency of C1q has not been clearly established, its absence or depression is associated with hypogammaglobulinemia. Bone marrow transplantation, which restores the gamma globulin levels, has also been found to restore C1q levels. It has been postulated that in the absence of normal immunoglobulin levels, C1q leaves the intravascular space and is degraded. The increased susceptibility to infections in patients with this deficiency may be primarily attributable to the defect in antibody production.

Deficiencies of C1r and C1s are associated with glomerulonephritis and autoimmune diseases such as systemic lupus erythematosus. In some patients lacking C1r, low levels of C1s have been detected, possibly indicating linkage in the synthesis of these two subunits or regulation of the anabolism or catabolism of one subunit by the other.

Undetectable serum C4 has also been associated with lupus erythematosus. This deficiency is linked to the major histocompatibility complex located on chromosome 6 (Ochs et al. 1977).

The most common of the complement deficiencies is C2. This defect is also linked to the major histocompatibility complex but is located outside it (Raum et al. 1979). Although C2 deficiency may be associated with lupus erythematosus, chronic glomerulonephritis or polymyositis, it is also present in healthy individuals. The absence of functional and immunochemically detectable C2 in this deficiency state is attributable to the inability of macrophages to produce this specific component, although these cells are normal in other respects in that they possess membrane receptors for IgG and C3 and are phagocytic and bactericidal. Factor B of the alternative pathway is closely linked genetically to C2. Decreased levels of this factor have been found in conjunction with some C2 deficiencies. However, inconsistencies between serum factor B and C2 levels indicate that the low factor B levels are not attributable to the linkage of the genes coding for the two proteins but that they are acquired rather than inherited.

In view of the function of C3 in enhancing phagocytosis, mobilizing leukocytes, and serving as the point of juncture between the two complement pathways, it might be anticipated that a deficiency of this component would have serious clinical effects. In patients with barely detectable levels of C3, extremely severe recurrent infections have been prevalent, emphasizing the vital importance of C3 in host defense against bacteria.

The absence of serum C5 has been associated with lupus erythematosus and bacterial infections including those caused by *Neisseria* (Rosenfeld et al. 1976; Snyderman et al. 1977). No, or markedly depressed, chemotactic activity for phagocytic cells is generated in the sera of these patients. Another C5 abnormality exists in which this serum component is normal by hemolytic and immunochemical analysis but is dysfunctional in terms of phagocytosis of yeast and staphylococci. Although this defect has been detected in healthy individuals, others exhibit severe dermatitis, diarrhea, and enhanced susceptibility to infection.

Recurrent neisserial infections occur in patients with deficiencies in C6 (Leddy et al. 1974), C7 (Lee et al. 1978), and C8 (Peterson et al. 1976). Although sera from these patients possess normal opsonic activity, they lack bactericidal activity for *Neisseria,* indicating that phagocytosis is not sufficient for host defense against these organisms but that late complement-component-mediated killing is required.

The total absence of C9 is apparently unrelated to any pathologic states (Lint et al. 1978). Hemolysis and killing of bacteria occur in C9 deficient sera but at a slower rate than in normal sera. This might be predicted from earlier studies that defined the function of C9 to be an enhancement of the rate of hemolysis.

Loss of the rigid control under which complement activation takes place also has serious clinical consequences. A deficiency of the C1 inhibitor, which is inherited as an autosomal dominant trait, results in hereditary angioneurotic edema. Symptoms of this disease include peripheral edematous swelling, which may be fatal if the upper respiratory tract is involved and intestinal swelling with resultant severe abdominal cramps. In the absence of this regulatory protein, episodic activation of C1 leads to continued C4 and C2 cleavage with termination of complement activation at this stage. It has been postulated that a kinin-like fragment of C2 is generated, which causes the edema. Two forms of hereditary angioneurotic edema exist. In about 85% of the affected individuals the serum levels of the C1 inhibitor are markedly depressed. In the remaining 15% this regulatory protein is present in normal or elevated amounts but is nonfunctional.

An inherited deficiency of the C3b inactivator has been described in a patient with a history of multiple severe infections (Ziegler et al. 1975). The serum of this patient contained low C3 and factor B levels and the majority of the C3 was in the form of the C3 conversion product, C3b, which is normally not present in serum. Since C3b could not be cleaved to the inactive fragments, C3c and C3d, its persistence resulted in continued activation of the alternative pathway. Administration of the purified C3b inactivator led to the disappearance of C3b from the serum, elevation of C3 and factor B levels, and restoration of serum phagocytic and bactericidal functions.

Complement-Mediated Pathology in Human Disease

Although complement is involved intimately in host defense it may also serve a pathobiological function. Innumerable correlations have been made between serum total complement and component levels, the anabolism and catabolism of these components, and disease states. This discussion is limited, however, to examples of those areas in which complement activation apparently contributes directly to cellular and tissue destruction.

Several of these pathological conditions are attributable to antibody or immune complex initiation of complement activation. In systemic lupus erythematosus, DNA–anti-DNA antibody complexes are detected in the circulation. They are also deposited in the skin and in the glomeruli, as demonstrated by immunofluorescence, and specific anti-DNA antibodies have been eluted from involved kidney tissue. Immunofluorescent examination of the renal and skin lesions has revealed, in addition to antibody, the presence of C1q, C3, properdin, and factor B, implicating both the classical and alternative pathways as mediators of the tissue damage that occurs in lupus (Rothfield et al. 1972).

In seropositive rheumatoid arthritis, rheumatoid factor (an antibody usually of the IgM class but possibly of the IgG class) reactive with IgG immunoglobulins is produced. The interaction of the rheumatoid-factor–IgG-immune complexes with the complement system in the synovium probably contributes to the connective tissue damage observed in this disease. *In vitro,* synovial tissues from affected individuals synthesize immunoglobulins, C2, C4, C3, C5, and factor B (Ruddy and Colten 1974). Yet total hemolytic complement, C1, C4, C2, C3, properdin, and factor B activities in rheumatoid joint fluid are depressed, indicating local complement activation (Ruddy and Austen 1970; Ruddy et al. 1975); a hypothesis further strengthened by the finding of a fragment of C5 with chemotactic activity (C5a) (Ward and Zvaifler 1971) and C3 fragments (Zvaifler 1969) in this fluid. The depression of components of the classical and alternative pathways implies they both participate in this pathological event. The entire complement sequence is completed, since C9 levels are also decreased. Complement-bearing immune complexes have been detected in inclusions in inflammatory cells in the involved synovia. The phagocytosis of these complexes probably results in the release of the lysosomal enzymes, which are found locally. In addition to enzyme release, another mechanism by which the complement system damages connective tissue has been defined *in vitro* using bones (rat ulnae and radii) in organ culture (Sandberg et al. 1977). Antibody-dependent complement activation results in the resorption of these bones via production of prostaglandin E_2 (a potent bone-resorbing agent) by some cellular element present in the bone. As in rheumatoid arthritis, both complement pathways are effective, and the late complement components are required, a finding that may provide an explanation for the depressed C9 levels observed in rheumatoid arthritis synovial fluid.

Immune complex activation of complement may also be involved in the arteritis associated with viral hepatitis since Australia antigen, IgG, and C3 have been found on affected vascular endothelium.

In addition to the pathological consequences of complement activation occurring at a site of immune complex deposition, autoantibodies with a specificity for normal cell surface constituents may be produced and destroy these cells by complement-mediated mechanisms. The antigenic determinants may be present on erythrocytes, platelets, or basement membranes. In autoimmune hemolytic anemia antibodies of either the IgG (warm-antibody-mediated disease) or the IgM (cold-agglutinin disease) class are produced. The specificity of the IgG antibodies is usually anti-Rh, while that of the IgM antibodies is anti-I. Whereas the presence of IgG on erythrocytes is readily identified by agglutination by antihuman IgG antisera (Coombs test), IgM may be undetectable, since the antibody attaches to the red cell at the lower temperatures in the peripheral circulation, particularly following exposure to cold external temperatures, and may dissociate on warming. The complement system participates in these diseases, primarily by activation of the complement sequence, with resultant deposition of C3b on the erythrocyte membranes and, to a far lesser extent, by lysis. The C3b-coated cells are phagocytosed extravascularly. The organs of sequestration and destruction of the erythrocytes is dependent on the class of antibody bound to the red cell (Atkinson and Frank 1974). Erythrocytes coated with IgG are attached by their Fc fragments to the Fc receptors on macrophages in the spleen and are phagocytosed, an event greatly augmented by the simultaneous binding of C3b on the erythrocytes to the C3b receptors on the macrophages. In contrast, IgM-coated cells are cleared primarily by macrophages (Kupffer cells) in the liver. Since macrophages do not possess a receptor for IgM, attachment to and ingestion of the erythrocytes is solely a function of the interaction of erythrocyte-bound C3b to the macrophage C3b receptor.

Complement also participates in drug-induced hemolytic anemias. Drugs (e.g., quinidine) may be antigenic and complex with antibodies produced against them. These complexes attach to erythrocyte membranes where they activate complement, resulting in cell destruction primarily by phagocytosis.

The destruction of platelets in idiopathic thrombocytopenic purpura is attributable to the production of IgG autoantibodies that react with the platelets and activate the complement cascade. Although some of the platelets may be lysed, the majority are phagocytosed by macrophages in the spleen because of interaction of platelet-bound IgG and C3b with their respective receptors on the macrophages.

Antibodies directed against antigenic determinants on subepithelial basement membranes may also initiate complement activation at this site. These antibodies have been detected by immunofluorescence in the dermatologic disorder, bullous pemphigoid, as have deposits of C1q, C3, properdin, and factor B, indicating the involvement of both pathways (Provost and Tomasi 1973).

In the absence of antibody, activation of the alternative pathway can also be pathogenic. Although human erythrocytes are resistant to complement-mediated lysis, subpopulations of these cells in patients with paroxysmal nocturnal hemoglobinuria are extremely susceptible to intravascular lysis by complement. No anti-erythrocyte antibodies have been detected in patients with this disorder, and hemolysis may be caused by a slight increase in serum magnesium which activates the alternative pathway (May et al. 1973).

In patients with membranoproliferative glomerulonephritis, immune complexes have not been found, but both properdin and C3 are deposited in the kidney. The presence of these components in the renal lesion (Westberg et al. 1971) as well as the finding of depressed serum levels of properdin, factor B and C3, but normal C1, C4, and C2 levels indicate alternative pathway activation (Gewurz et al. 1968). The sera from some of these patients contains a protein, C3 nephritic factor (C3NeF), which is an autoantibody directed against the alternative pathway C3 convertase, C3bBb, that serves to stabilize C3bBb (Daha et al. 1978). C3b receptors are present on the epithelial surface of glomerular capillary walls which could bind C3bBb (Shin et al. 1977). Stabilization of this convertase by C3NeF would permit efficient activation of the alternative pathway with resultant local tissue damage.

Some Interactions of the Complement System with Other Mediator Systems

It is becoming apparent that complement not only functions as a discrete mediator system but may also regulate or be modulated by other humoral mediators such as the Hageman-factor-dependent coagulation and fibrinolytic and kinin-generating systems (Wooldridge et al. 1974). In the intrinsic clotting pathway, the C1 inhibitor blocks the conversion of precursor plasma thromboplastin antecedent to plasma thromboplastin antecedent, the latter of which is also susceptible to the C1 inhibitor. Clot dissolution, which occurs from the fibrinolytic system, may also be blocked by the C1 inhibitor that inhibits plasmin. The vasodilation, enhanced microvascular permeability, and contraction of smooth muscle attributable to kinin-generation is also influenced by the C1 inhibitor, which prevents the conversion of kininogen to kinin by inhibition of kallikrein.

Components of the kinin and fibrinolytic systems may modulate the complement system in that kallikrein destroys C1 activity by cleavage of the C1s subunit. In contrast, activation of C1 can be initiated by plasmin.

Conclusion

Awareness of the importance of the complement system, both in host defense and in the pathogenesis of human disease, is continually increasing. The purification of the components and delineation of many of the mechanistic aspects of complement activation have greatly contributed and continue to contribute to the clarification of the numerous biological functions of this mediator system and the definition of new activities attributed to it. From the original observations concerning the lytic potential of complement, research has progressed to the sophisticated analysis of the interactions of the complement components with specific cellular receptors and the membrane alterations that occur after attack by the terminal components, areas that have yet to be explored fully. The current emphasis on the correlation of disease with serum component levels, component deficiencies and dysfunctions, deposition of specific components in areas of tissue damage, and altered component anabolism or catabolism are furthering the understanding of the role of complement in immunopathology, thus providing a basis for its future therapeutic manipulation.

References

Alper, C.A., Balavitch, D. (1976) Cobra venom factor: Evidence for its being altered cobra C3 (the third component of complement). Science 191,1275.

Alper, C.A., Rosen, F.S. (1976) Genetics of the complement system. Adv. Human Genetics 7,141.

Agnello, V. (1978) Complement deficiency states. Medicine 57,1.

Arroyave, C.M., Levy, R.M., Johnson, J.S. (1977) Genetic deficiency of the fourth component of complement (C4) in Wistar rats. Immunology 33,453.

Atkinson, J.P., Frank, M.M. (1974) Studies on the *in vivo* effects of antibody. Interaction of IgM antibody and complement in the immune clearance and destruction of erythrocytes in man. J. Clin. Invest. 54,339.

Becker, E.L., Showell, H.J., Henson, P.M., Hsu, L.S. (1974) The ability of chemotactic factors to induce lysosomal enzyme release. I. The characteristics of the release, the importance of surfaces and the relation of enzyme release to chemotactic responsiveness. J. Immunol. 112,2047.

Bentley, C., Fries, W., Brade, V. (1978) Synthesis of factors D, B and P of the alternative pathway of complement activation, as well as C3, by guinea-pig peritoneal macrophages *in vitro*. Immunology 35,971.

Bianco, C., Griffin, F., Silverstein, S.C. (1975) Studies of the macrophage complement receptor. Alteration of receptor function upon macrophage activation. J. Exp. Med. 141,1278.

Bokisch, V.A., Müller-Eberhard, H.J., Cochrane, C.G. (1969) Isolation of a fragment (C3a) of the third component of human complement containing anaphylatoxin and chemotactic activity and description of an anaphylatoxin inactivator of human serum. J. Exp. Med. 129,1109.

Borsos, T., Rapp, H.J. (1965) Complement fixation on cell surfaces by 19S and 7S antibodies. Science 150,505.

Brade, V., Hall, R.E., Colten, H.R. (1977) Biosynthesis of pro-C3, a precursor of the third component of complement. J. Exp. Med. 146,759.

Brier, A.M., Synderman, R., Mergenhagen, S.E., Notkins, A.L. (1970) Inflammation and herpes simplex virus: release of a chemotaxis-generating factor from infected cells. Science 170,1104.

Cinader, B., Duviski, S., Wardlaw, A.C. (1964) Distribution, inheritance and properties of an antigen, MuBl, and its relation of hemolytic complement. J. Exp. Med. 120,897.

Colten, H.R. (1976) Biosynthesis of complement. Adv. Immunol. 22,67.

Daha, M.R., Austen, K.F., Fearon, D.T. (1978) Heterogeneity, polypeptide chain composition and antigenic reactivity of C3 nephritic factor. J. Immunol. 120,1389.

Daniels, C.A., Borsos, T., Rapp, H.J., Synderman, R., Notkins, A.L. (1970) Neutralization of sensitized virus by purified components of complement. Proc. Natl. Acad. Sci. U.S.A. 65,528.

Day, N.K.B., Gewurz, H., Johannsen, R., Finstad, J., Good, R.A. (1970) Complement and complement-like activity in lower vertebrates and invertebrates. J. Exp. Med. 132,941.

Ellman, L., Green, I., Frank, M.M. (1970) Genetically controlled total deficiency of the fourth component of complement in the guinea pig. Science 170,74.

Ellman, L., Green, I., Judge, F., Frank, M.M. (1971) *In vivo* studies in C4-deficient guinea pigs. J. Exp. Med. 134,162.

Esser, A.F., Kolb, W.P., Podack, E.R., Müller-Eberhard, H.J. (1979) Molecular reorganization of lipid bilayers by complement: A possible mechanism for membranolysis. Proc. Natl. Acad. Sci. U.S.A. 76,1410.

Fearon, D.T., Austen, K.F., Ruddy, S. (1973) Formation of a hemolytically active cellular intermediate by the interaction between properdin factors B and D and the activated third component of complement. J. Exp. Med. 138,1305.

Fearon, D.T., Austen, K.F. (1975) Properdin binding to C3b and stabilization of the C3-dependent C3 convertase. J. Exp. Med. 142,856.

Ferrone, S., Cooper, N.R., Pellegrino, M.A., Reisfeld, R.A. (1973) Activation of human complement by human lymphoid cells sensitized with histocompatibility alloantisera. Proc. Natl. Acad. Sci. U.S.A. 70,3665.

Frank, M.M., May, J., Gaither, T., Ellman, L. (1971) *In vitro* studies of complement function in sera of C4-deficient guinea pigs. J. Exp. Med. 134,176.

Fujita, T., Gigli, I., Nussenzweig, V. (1978) Human C4-binding protein. II. Role in proteolysis of C4b by C3b-inactivator. J. Exp. Med. 148,1044.

Gewurz, H., Pickering, R.J., Mergenhagen, S.E., Good, R.A. (1968) The complement profile in acute glomerulonephritis, systemic lupus erythematosus, and hypocomplementemic glomerulonephritis. Int. Arch. Allergy Appl. Immunol. 34,556.

Gewurz, H., Shin, H.S., Mergenhagen, S.E. (1968) Interactions of the complement system with endotoxic lipopolysaccharide. Consumption of each of the six terminal complement components. J. Exp. Med. 128,1049.

Ghebrehiwet, B., Müller-Eberhard, H.J. (1979) C3e: an acidic fragment of human C3 with leukocytosis inducing activity. J. Immunol. 123,616.

Gigli, I. (1978) A single chain precursor of C4 in human serum. Nature 272,836.

Gigli, I., Nelson, R.A. (1968) Complement dependent immune phagocytosis. Exp. Cell Res. 51,45.

Glovsky, M.M., Hugli, T.E., Ishizaka, T., Lichtenstein, L.M., Erickson, B.W. (1979) Anaphylatoxin-induced histamine release with human leukocytes. Studies of C3a leukocyte binding and histamine release. J. Clin. Invest. 64,804.

Goldman, J.N., Ruddy, S., Austen, K.F. (1972) Reaction mechanism of nascent $\overline{C567}$ (reactive lyses) I. Reaction characteristics for production of EC567 and lysis by C8 and C9. J. Immunol. 109,353.

Gorski, J.P., Hugli, T.E., Müller-Eberhard, H.J. (1979) C4a: The third anaphylatoxin of the human complement system. Proc. Natl. Acad. Sci. U.S.A. 76,5299.

Götze, O., Müller-Eberhard, H.J. (1971) The C3-activator system: An alternate pathway of complement activation. J. Exp. Med. Suppl. 134,90.

Götze, O., Bianco, C., Cohn, Z.A. (1979) The induction of macrophage spreading by factor B of the properdin system. J. Exp. Med. 149,372.

Hall, R.E., Colten, H.R. (1977) Cell-free synthesis of the fourth component of guinea pig complement (C4): Identification of a precursor of serum C4 (pro-C4). Proc. Natl. Acad. Sci. U.S.A. 74,1707.

Hicks, J.T., Klutch, M.J., Albrecht, P., Frank, M.M. (1976) Analysis of complement-dependent antibody-mediated lysis of target cells acutely infected with measles. J. Immunol. 117,208.

Hugli, T.E., Müller-Eberhard, H.J. (1978) Anaphylatoxins: C3a and C5a. Adv. Immunol. 26,1.

Humphrey, J.H., Dourmashkin, R.R. (1965) Electron Microscope Studies of Immune Cell Lysis. Ciba Found. Sym., Complement. London, Churchill, p. 175.

Jensen, J.A. (1969) A specific inactivator of mammalian C4 isolated from nurse shark (Ginglymostoma cirratum) serum. J. Exp. Med. 130,217.

Koopman, W.J., Sandberg, A.L., Wahl, S.M., Mergenhagen, S.E. (1976) Interaction of soluble C3 fragments with guinea pig lymphocytes. Comparison of effects of C3a, C3b, C3c and C3d on lymphokine production and lymphocyte proliferation. J. Immunol. 117,333.

Lachman, P.J., Thompson, R.A. (1970) Reactive lysis: the complement-mediated lysis of unsensitized cells. II. The characterization of activated reactor as C56 and the participation of C8 and C9. J. Exp. Med. 131,643.

Lay, W.H., Nussenzweig, V. (1968) Receptors for complement on leukocytes. J. Exp. Med. 128,991.

Leddy, J.P., Frank, M.M., Gaither, T., Baum, J., Klemperer, M.R. (1974) Hereditary deficiency of the sixth component of complement in man. I. Immunochemical, biologic and family studies. J. Clin. Invest. 53,544.

Lee, T.J., Utsinger, P.D., Synderman, R., Yount, W.J., Sparling, P.F. (1978) Familial deficiency of the seventh component of complement associated with recurrent bacteremic infections due to *Neisseria*. J. Infec. Dis. 138,359.

Linscott, W.D., Levinson, W.E. (1969) Complement components required for virus neutralization by early immunoglobulin antibody. Proc. Natl. Acad. Sci. U.S.A. 64,520.

Lint, T.F., Zeitz, H.J., Scott, D., Malkinson, J.R., Gewurz, H. (1978) Hereditary deficiency of the ninth component of complement (C) in man. Clin. Res. 26,714A.

Mayer, M.M. (1961) Complement and complement fixation. In: Kabat, E.A. *Experimental Immunochemistry*, ed. 2. Charles C Thomas, Springfield, Illinois, p. 133.

Mayer, M.M. (1978) *Complement, past and present*. The Harvey Lectures, series 72. New York, Academic Press, p. 139.

May, J.E., Rosse, W., Frank, M.M. (1973) Paroxysmal nocturnal hemoglobinuria: Alternate-complement pathway mediated lysis induced by magnesium. N. Engl. J. Med. 289,705.

Miller, M.E., Nilsson, U.R. (1974) A major role of the fifth component of complement (C5) in the opsonization of yeast particles. Partial dichotomy of function and immunochemical measurement. Clin. Immunol. Immunopathol. 2,246.

Mills, B.J., Cooper, N.R. (1978) Antibody-independent neutralization of vesicular stomatitis virus by human complement. I. Complement requirements. J. Immunol. 121,1549.

Minta, J.O., Ngan, B., Pang, A.S.D. (1979) Purification and characterization of a single chain precursor C3-protein (pro-C3) from normal human plasma. J. Immunol. 123,2415.

Morris, K.M., Colten, H.R., Bing, D.H. (1978) The first component of complement. A quantitative comparison of its biosynthesis in culture by human epithelial and mesenchymal cells. J. Exp. Med. 148,1007.

Müller-Eberhard, H.J. (1975) Complement. Ann. Rev. Biochem. 44,697.

Müller-Eberhard, H.J. (1977) Chemistry and function of the complement system. Hospital Practice (Aug.), p. 33.

Nelson, R.A., Jensen, J., Gigli, I., Tamura, N. (1966) Methods for the separation, purification and measurement of nine components of hemolytic complement in guinea pig serum. Immunochemistry 3,111.

Ochs, H.D., Rosenfeld, S.I., Thomas, E.D., Giblett, E.R., Alper, C.A., Dupont, B., Schaller, J.G., Gilliland, B.C., Hansen, J.A., Wedgwood, R.J. (1977) Linkage between the gene (or genes) controlling synthesis of the fourth component of complement and the major histocompatibility complex. N. Engl. J. Med. 296,470.

Oldstone, M.B.A. (1975) Virus neutralization and virus-induced immune complex diseases. Virus-antibody union resulting in immunoprotection or immunologic injury: two sides of the same coin. Prog. Med. Virol. 19,84.

Oldstone, M.B.A., Cooper, N.R., Larson, D.L. (1974) Formation and biologic role of polyoma virus-antibody complexes. A critical role for complement. J. Exp. Med. 140,549.

Peterson, B.H., Graham, J.A., Brooks, G.F. (1976) Human deficiency of the 8th component of complement. The requirement of C8 for serum *Nesseria gonorrhoeae* bactericidal activity. J. Clin. Invest. 57,283.

Pepys, M.B. (1974) Role of complement in induction of antibody production *in vivo*. J. Exp. Med. 140,126.

Pillemer, L., Blum, L., Lepow, I.H. (1954) The properdin system and immunity. I. Demonstration and isolation of a new serum protein, properdin, and its role in immune phenomena. Science 120,279.

Podack, E.R., Biesecker, G., Müller-Eberhard, H.J. (1979) Membrane attack complex of complement: Generation of high-affinity phospholipid binding sites by fusion of five hydrophilic plasma proteins. Proc. Natl. Acad. Sci. U.S.A. 76,897.

Provost, T.T., Tomasi, T.B. (1973) Evidence for complement activation via the alternate pathway in skin disease: I. Herpes gestationis, systemic lupus erythematosus and bullous pemphigoid. J. Clin. Invest. 52,1779.

Rapp, H.J., Borsos, T. (1970) *Molecular basis of complement action.* New York, Appleton Century Crofts.

Raum, D., Glass, D., Carpenter, C.B., Schur, P.H., Alper, C.A. (1979) Mapping of the structural gene for the second component of complement with respect to the human major histocompatibility complex. Am. J. Hum. Genet. 31,35.

Rosenfeld, S.I., Kelly, M.E., Leddy, J.P. (1976) Hereditary deficiency of the fifth component of complement in man. J. Clin. Invest. 57,1626.

Ross, G.D., Polley, M.J., Rabellino, E.M., Grey, H.M. (1973) Two different complement receptors on human lymphocytes. J. Exp. Med. 138,798.

Rother, K. (1972) Leukocyte mobilizing factor: a new biological activity from the third component of complement. Eur. J. Immunol. 2,550.

Rother, K., Rother, U., Müller-Eberhard, H.J., Nilsson, U.R. (1966) Deficiency of C6 in rabbits with an inherited complement defect. J. Exp. Med. 124,773.

Rothfield, N., Ross, H.A., Minta, J.O., Lepow, I.H. (1972) Glomerular and dermal deposition of properdin in systemic lupus erythematosus. N. Engl. J. Med. 287,681.

Ruddy, S., Austen, K.F. (1970) Activation of the complement system in rheumatoid synovitis. I. An analysis of complement component activities in rheumatoid synovial fluids. Arthritis Rheum. 13,713.

Ruddy, S., Colten, H.R. (1974) Rheumatoid arthritis: Biosynthesis of complement proteins by synovial tissues. N. Engl. J. Med. 290,1284.

Ruddy, S., Fearon, D.T., Austen, K.F. (1975) Depressed synovial fluid levels of properdin and properdin factor B in patients with rheumatoid arthritis. Arthritis Rheum. 18,289.

Sandberg, A.L., Osler, A.G., Shin, H.S., Oliveira, B. (1970) The biologic activity of guinea pig antibodies. II. Modes of complement interaction with γ-1 and γ-2 immunoglobulins. J. Immunol. 104,329.

Sandberg, A.L., Wahl, S.M., Mergenhagen, S.E. (1975) Lymphokine production by C3b-stimulated B cells. J. Immunol. 115,139.

Sandberg, A.L., Raisz, L.G., Goodson, J.M., Simmons, H.A., Mergenhagen, S.E. (1977) Initiation of bone resorption by the classical and alternative complement pathways and its mediation by prostaglandins. J. Immunol. 119,1378.

Shin, H.S., Snyderman, R., Freidman, E., Mellors, A., Mayer, M.M. (1968) Chemotactic and ana-phylatoxic fragment cleaved from the fifth component of guinea pig complement. Science 162,361.

Shin, H.S., Smith, M.R., Wood, W.B., Jr. (1969) Heat labile opsonins to pneumococcus. II. Involvement of C3 and C5. J. Exp. Med. 130,1229.

Shin, M.L., Gelfand, M.C., Nagle, R.B., Carlo, J.R., Green, I., Frank, M.M. (1977) Localization of receptors for activated complement on visceral epithelial cells of the human renal glomerulus. J. Immunol. 118,869.

Shiraishi, S., Stroud, R.M. (1975) Cleavage products of C4b produced by enzymes in human serum. Immunochemistry. 12,935.

Siraganian, R.P. (1972) Platelet requirement in the interaction of the complement and clotting systems. Nature 239,208.

Siraganian, R.P., Hook, W.A. (1976) Complement induced histamine release from human basophils. II. Mechanism of the histamine reaction. J. Immunol. 116,639.

Sissons, J.G.P., Cooper, N.R., Oldstone, M.B.A. (1979) Alternative complement pathway-mediated lysis of measles virus infected cells. Induction by IgG antibody bound to individual viral glycoproteins and comparative efficacy of F(ab')$_2$ and Fab' fragments. J. Immunol. 123,2144.

Snyderman, R., Shin, H.S., Dannenberg, A.M., Jr. (1972) Macrophage proteinase and inflammation: the production of chemotactic activity from the fifth component of complement by macrophage proteinase. J. Immunol. 109,896.

Snyderman, R., Pike, M.C., Szatolowicz, V., Meadows, L.M. (1977) A familial deficiency of the fifth component of complement. Clin. Res. 25,368A.

Ward, P.A., Hill, J.H. (1970) C5 Chemotactic fragments produced by an enzyme in lysosomal granules of neutrophils. J. Immunol. 104,535.

Ward, P.A., Zvaifler, N.J., (1971) Complement-derived leukotactic factors in inflammatory synovial fluids of humans. J. Clin. Invest. 50,606.

Weiler, J.M., Daha, M.R., Austen, K.F., Fearon, D.T. (1976) Control of the amplification convertase of complement by the plasma protein β1H. Proc. Natl. Acad. Sci. U.S.A. 73,3268.

Welsh, R.M., Jr., Jensen, F.C., Cooper, N.R., Oldstone, M.B.A. (1976) Inactivation and lysis of oncornoviruses by human serum. Virology 74,432.

Westberg, N.G., Naff, G.B., Boyer, J.T., Michael, A.F. (1971) Glomerular deposition of properdin in acute and chronic glomerulonephritis with hypocomplementemia. J. Clin. Invest. 50,642.

Whaley, K., Ruddy, S. (1976) Modulation of C3b hemolytic activity by a plasma protein distinct from C3b inactivator. Science 193,1011.

Wooldridge, D., Spragg, J., Austen, K.F. (1974) Interaction of Hageman-factor-dependent systems with the complement sequence. Chemistry and biology of the kallikrein-kinin system in health and disease. In: Pisano, J.J., Austen, K.F. (eds.) *Fogarty International Center Proceedings* No. 27. p. 221.

Ziegler, J.B., Alper, C.A., Rosen, F.S., Lachmann, P.J., Sherington, L. (1975) Restoration by purified C3b inactivator of complement-mediated function *in vivo* in a patient with C3b inactivator deficiency. J. Clin. Invest. 55,668.

Zimmerman, T.S., Arroyave, C.M., Müller-Eberhard, H.J. (1971) A blood coagulation abnormality in rabbits deficient in the sixth component of complement (C6) and its correction by purified C6. J. Exp. Med. 134,1591.

Zvaifler, N.J. (1969) Breakdown products of C'3 in human synovial fluids. J. Clin. Invest. 48,1532.

Coagulation, Kinins, and Inflammation

Allen P. Kaplan, M.D.

Bradykinin is a nine-amino-acid polypeptide of molecular weight 900 that can cause vasodilation, an increase in vascular permeability, and a burning-type pain upon contact with sensory nerve endings. It is active at approximately the same molarity as histamine and therefore has the potential to function as a mediator of inflammation. While histamine release from basophils or mast cells can be a direct consequence of the reaction of IgE antibody and antigen, direct immunologic release or generation of bradykinin has not been definitely demonstrated. Rather, its formation appears to be a consequence of the tissue damage that has resulted from immunologic- or nonimmunologic-mediated tissue injury, and its local formation may then contribute to the inflammatory response that results. For example, local vasodilation and increased vascular permeability will facilitate both the local deposition of immune complexes in Type III immunologic reactions and the influx of cells stimulated by formation of chemotactic factors. In this fashion, bradykinin would tend to perpetrate inflammation within tissues. Bathing of injured tissues with plasma proteins will also activate both the intrinsic and extrinsic coagulation pathways so that local coagulation and fibrinolysis is initiated and deposition of fibrin and fibrin-degradation products results. This is particularly evident in the collagen–vascular or autoimmune diseases such as systemic lupus erythematosus in which fibrinoid deposition is a prominent characteristic of involved tissues. As is seen in the following sections, bradykinin formation can be a byproduct of activation of the intrinsic coagulation pathway, and in this fashion the vasoactive properties of bradykinin help recruit the proteins needed for its own formation.

Bradykinin is generated from plasma proteins known as kininogens, of which there are two major forms: the predominant one is a low molecular weight (LMW) kininogen that accounts for approximately 85% of the total plasma kininogen and the minor component is a high molecular weight (HMW) kininogen that represents the remaining 15%. Kallikreins are tryptic-like enzymes present in many body tissues defined as such by their ability to generate bradykinin when in a kininogen-containing milieu. Tissue kallikreins are all extremely similar, if not identical, and are derived from kidney (urine), pancreas, submaxillary and submandibular glands, colon, and other

From the Division of Allergy, Rheumatology and Clinical Immunology, Department of Medicine, State University of New York at Stony Brook, Health Sciences Center, Stony Brook, New York.

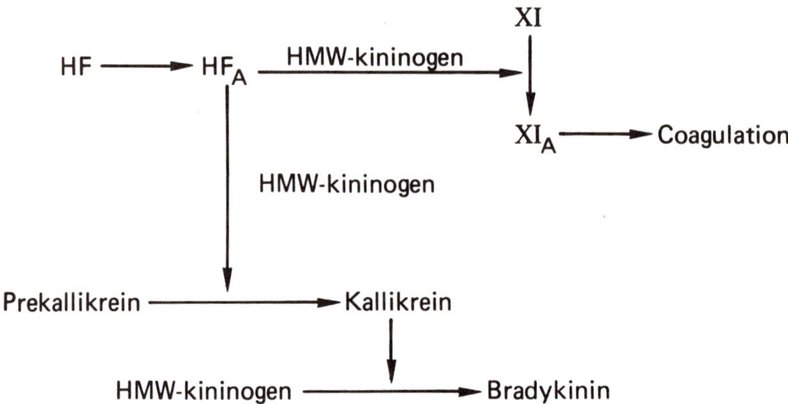

Figure 1. Pathway for bradykinin generation in plasma as initiated by activated Hageman factor (HFa).

organs (Nustad et al. 1974). They appear capable of digesting both forms of kininogen with equal avidity (Pierce and Guimaraes 1974) thus, because of their quantitative difference, most bradykinin generated on exposure of plasma to tissue kallikreins would be expected to be derived from LMW kininogen.

Plasma also contains a kinin-forming pathway. However, active kallikrein does not exist in plasma, but instead circulates in a precursor form called prekallikrein. The activation of prekallikrein is accomplished by activated Hageman factor (factor XIIa or HFa), which is formed as a consequence of activation of the intrinsic coagulation pathway. Thus, factor XIIa converts prekallikrein to kallikrein and plasma kallikrein digests kininogen to generate bradykinin. However, this enzyme preferentially digests HMW kininogen, a property that distinguishes it from tissue kallikreins (Pierce and Guimaraes 1974). This sequence of events is shown in Figure 1.

As shown in Figure 1, HFa also converts coagulation factor XI to factor XIa to continue the intrinsic coagulation pathway. However, in normal plasma, Hageman factor is a proenzyme (HF), and initiation occurs on binding of Hageman factor as well as other critical proteins to negatively charged surfaces, which results in the conversion of HF to HFa. In fact, the three proteins of the plasma-kinin-forming system, i.e., Hageman factor, prekallikrein, and HMW kininogen, have been shown to be the major factors required for contact activation of plasma to initiate both coagulation and fibrinolysis. Thus, bradykinin formation is a byproduct of the first step of the intrinsic coagulation pathway. In this review, I summarize our present concept of the early events of these pathways. A detailed analysis of the molecular interactions occurring in each step of the pathway has been reported recently (Kaplan 1978).

The Requirement for Interaction of Plasma Proteins with Negatively Charged Surfaces to Initiate Coagulation and Generate Bradykinin

Initiation of the intrinsic coagulation pathway depends on binding of factor XII (Hageman factor), HMW kininogen, prekallikrein, and factor XI to negatively charged surfaces. For example, when blood is allowed to clot in a glass test tube, the intrinsic coagulation pathway is initiated. Negative charges present along the surface of the glass tube are required for this reaction to proceed and are supplied by the silicate moiety present in glass. In coagulation laboratories, kaolin, a chalk-like negatively charged powdered silicate, is added to plasma to activate the pathway. In this fashion, a large surface area is supplied by a small quantity of material. After a 2-minute incubation, the plasma is recalcified (the citrated plasma routinely used has its calcium ions chelated to inhitit coagulation) and the time for clot formation is recorded. This test is called the partial thromboplastin time (PTT). The PTT would be expected to be prolonged if a deficiency of any protein of the intrinsic coagulation cascade were present. Although initiating surfaces all appear to be negatively charged, the precise charge distribution required for initiation to proceed is unknown. Thus, there are many surfaces or macromolecular particles that do not initiate coagulation or kinin formation. An example might be DNA, which has a backbone containing a long string of negatively charged phosphate groups; but when DNA is added to plasma, no activation of these pathways occurs.

It is generally assumed that tissue injury exposes macromolecules that can act as initiating surfaces, and there is some evidence to support this contention. Crude preparations of collagen (containing other connective tissue elements) appear to initiate the intrinsic coagulation cascade, but purified collagen is inactive. Thus other connective tissue elements, perhaps certain proteoglycans that are bound to collagen, can provide an initiating surface. However, the precise biochemical identity of this material is unknown. In support of this idea, preparations of rabbit vascular basement membrane have been shown to initiate the Hageman-factor-dependent pathways. In addition, activation of Hageman factor has been reported to occur on contact with certain bacterial lipopolysaccharides (endotoxins) in which the lipid-A fraction containing negatively charged phosphate groups appears to be the functional moiety. Evidence of activation of the Hageman-factor–dependent pathways has been observed during episodes of endotoxic shock, and the products of this reaction may contribute to the intravascular coagulation and hypotension seen in this condition (Mason et al. 1970). Of particular interest is the observation that endotoxic shock in rabbits could not be inhibited by prior complement depletion, although much of the tissue inflammatory reaction was diminished (Ulevitch et al. 1975). It is possible that the generation of vasoactive peptides such as bradykinin is responsible for the vasodilation and pooling of fluid into tissues that is characteristic of endotoxic shock.

Crystals of sodium urate, calcium pyrophosphate, or L-homocysteine (which are produced by patients with gout, pseudogout, or homocysteinuria, respectively) may act as initiating surfaces, while exposure of human plasma to articular cartilage generates kinin-like activity that appears to be Hageman-factor dependent. These examples of *in vitro* activation with biologically relevant materials can potentially occur *in vivo* in patients with pathologic conditions. However, we cannot be certain whether any contribution made by activating these pathways is important for the pathogenesis of the disease. The data with regard to endotoxic shock, however, are quite convincing and suggest a role for bradykinin as a mediator of hypotension.

Interaction of Prekallikrein, Factor XI, and High Molecular Weight Kininogen in Plasma

Early studies have suggested that kallikrein could be purified as a complex with another plasma protein and that prekallikrein also appeared to circulate in plasma as a complex. This observation was reexamined in 1976, soon after the discovery of three unrelated patients who were congenitally deficient in HMW kininogen. Purified prekallikrein was known to have a molecular weight of about 85,000 to 90,000. However, its molecular weight in normal plasma appeared to be 260,000, a discrepancy of over 160,000. However, when the molecular weight of prekallikrein was determined in plasma that was congenitally deficient in HMW kininogen, a value of 90,000 was obtained, just as is found for the purified material. This result suggested that prekallikrein circulates in plasma, bound to HMW kininogen, and experiments demonstrating that this is in fact the case were of two kinds. First, when HMW kininogen–deficient plasma was reconstituted with purified HMW kininogen, the molecular weight of prekallikrein in the plasma shifted from 90,000 to 260,000. Second, this same change in molecular weight was seen when purified prekallikrein and HMW kininogen were mixed. When the molecular weight of each protein was considered, it appeared that one molecule of prekallikrein was complexed to one molecule of HMW kininogen (Mandle Jr. et al. 1976).

The same observations just described were made when the molecular weight of factor XI was determined in the purified state and in various plasmas. Thus factor XI was also shown to form a 1:1 complex with HMW kininogen (Thompson et al. 1977). Since the size of factor XI is twice that of prekallikrein, the complexes of prekallikrein HMW kininogen and factor XI–HMW kininogen were separable in plasma, and there appeared to be no complex in which a molecule of HMW kininogen was bound to both prekallikrein and factor XI. We have shown recently that HMW kininogen probably has only one effective binding site and can complex with either a molecule of prekallikrein or a molecule of factor XI, but not both. Since normal plasma contains more than twice the concentration of HMW kininogen as the sum of the prekallikrein and factor XI concentrations, there is theoretically enough kininogen present to bind all of prekallikrein and factor XI. The amount that is actually bound depends on the binding or association constant and the relative concentration of the reactants in plasma. For factor XI, over 99% circulates bound to HMW kininogen, while for prekallikrein about 90% is bound and 10% remains free. Figure 2 is a schematic diagram showing the critical constituents bound to the surface. Hageman factor (factor XII) binds directly to the surface, where it interacts with two bimolecular complexes. The main binding of prekallikrein and factor XI to the surface occurs by attachment through the HMW kininogen. Thus, the binding of prekallikrein and factor XI to kaolin in plasma that is deficient in HMW kininogen is less than is seen when kaolin is added to normal plasma (Wiggins et al. 1977). One of the functions

Figure 2. Interaction of surface-bound Hageman factor with complexes of HMW kininogen and prekallikrein and complexes of HMW kininogen and factor XI.

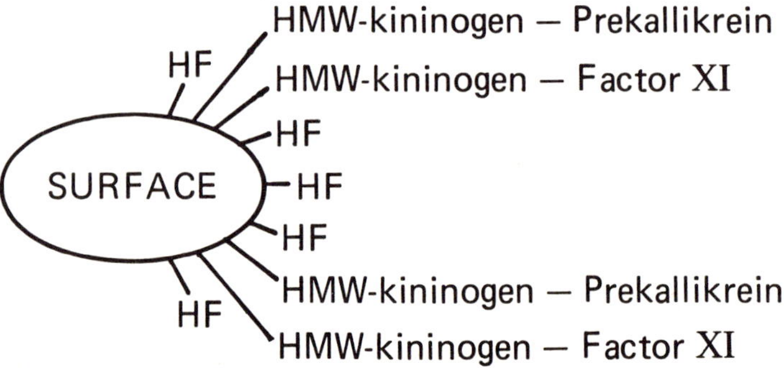

of HMW kininogen in the coagulation-kinin-generating pathway is to bind prekallikrein and factor XI and facilitate their attachment to negatively charged surfaces.

Functional Abnormalities Associated with Congenital Deficiencies of Proteins of the Plasma-Kinin-Forming Pathway

Figure 3 demonstrates the abnormalities in coagulation and fibrinolysis that result from deficiencies of Hageman factor, prekallikrein, and HMW kininogen. Plasma was incubated with kaolin, and the rate of activation of coagulation and generation of the fibrinolytic enzyme plasmin were determined. In each instance, no generation

Figure 3. Time course of activation of the Hageman-factor-dependent coagulation and fibrinolytic pathways in normal plasma and plasmas deficient in Hageman factor, prekallikrein, and HMW kininogen.

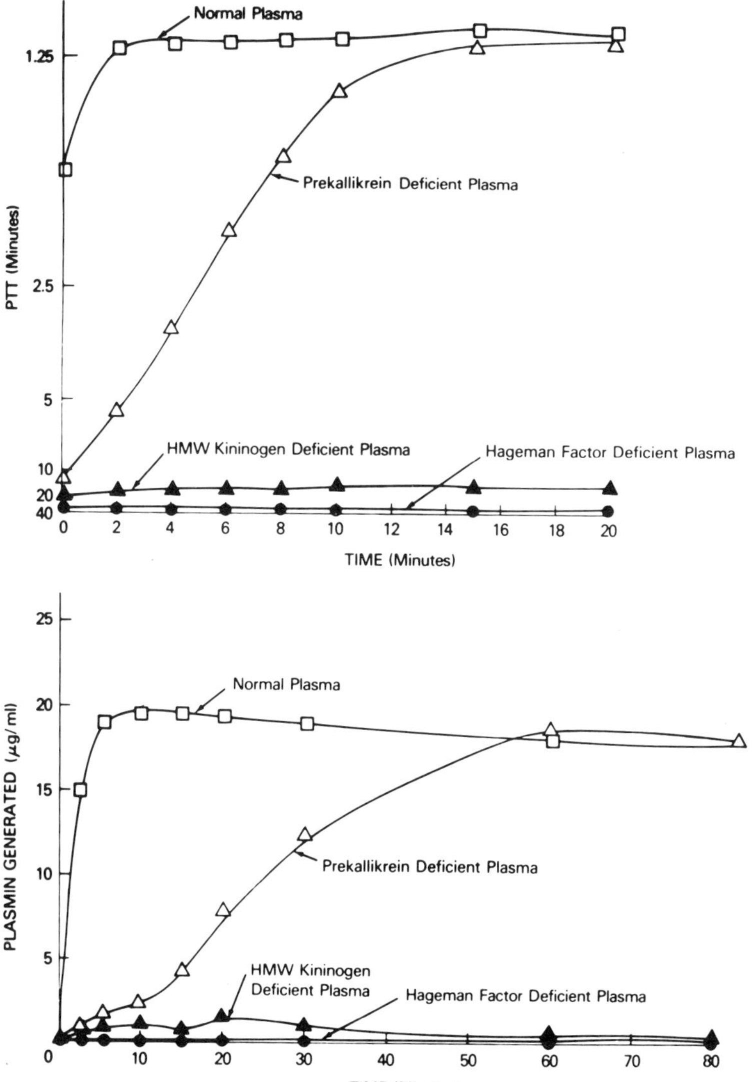

of bradykinin is obtained (Figure 3). When Hageman-factor–deficient plasma was activated, no shortening of the partial thromboplastin time (PTT) was observed and no plasmin was generated. Normal plasma, in contrast, clotted rapidly, resulting in a shortened PTT that was maximal by 2 minutes, and a rapid evolution of plasmin was observed. The HMW-kininogen–deficient plasma was markedly abnormal and was similar to Hageman-factor deficiency. Thus HMW kininogen, although not an enzyme, has an important function as a coagulation cofactor that acts at the first step of the coagulation cascade; it is not just a substrate from which bradykinin is derived. Prekallikrein-deficient plasma was also abnormal; however, as the time of incubation with kaolin was prolonged, its clotting time became progressively shorter and increasing quantities of plasmin were generated. Thus prekallikrein-deficient plasma was activatable, but its rate of activation was markedly impaired. This autocorrection is characteristic of prekallikrein deficiency and demonstrates that prekallikrein also plays a role in the initiation of coagulation; however, in contrast to Hageman factor and HMW kininogen, it is not an absolute requirement.

Activation and Fragmentation of Hageman Factor: The Role of the Surface, High Molecular Weight Kininogen, and Prekallikrein

Although the precise mechanism of initiation of the coagulation cascade is not completely understood, many of the molecular interactions involving these various proteins have been described. A major problem that is not yet resolved, however, is the identification of the site of formation of the first enzymatic site so that the cascade can be initiated. The major forms of activated Hageman factor that have been isolated or described are the products of proteolytic digestion. But all of the enzymes involved in this pathway circulate in plasma as unactivated proenzymes. It is possible, however, that what we call unactivated Hageman factor is actually a protein with measurable intrinsic enzymatic activity augmented markedly when it is subsequently cleaved. Another possibility is that on binding to the surface, a conformational change occurs that exposes (or augments) an otherwise hidden proteolytic site. A third possibility is that traces of active enzyme (i.e., the product of enzymatic activation) normally exist in plasma and are not expressed until bound to a surface where optimal interaction with other critical proteins can occur. There is no evidence to support this last possibility, and present studies are focused on the first two alternatives.

There are two general mechanisms by which Hageman factor can be activated enzymatically. The first involves cleavage within a disulfide bond without obvious fragmentation in which the active enzyme consists of chains linked by disulfide bonds (Revak et al. 1977) and the second, a cleavage that liberates one or more active fragments (Kaplan and Austen 1970). The main enzyme responsible for this digestion is plasma kallikrein; thus, kallikrein not only digests kininogen to yield bradykinin, but it also cleaves and activates Hageman factor. This represents a reciprocal interaction in which kallikrein activates Hageman factor and activated Hageman factor converts prekallikrein to kallikrein. The cleavages in Hageman factor, which result in an activated product, are shown in Figure 4. The Hageman factor is shown to bind to the surface and an initial cleavage by kallikrein is shown. This converts Hageman factor, which is a single chain, to a two-chain molecule. The heavy chain is the part attached to the surface; it is disulfide linked to a light chain. The active enzymatic site of Hageman factor is present in the light chain. In this fashion, although the molecule has been digested, its molecular weight has not changed. A second cleavage

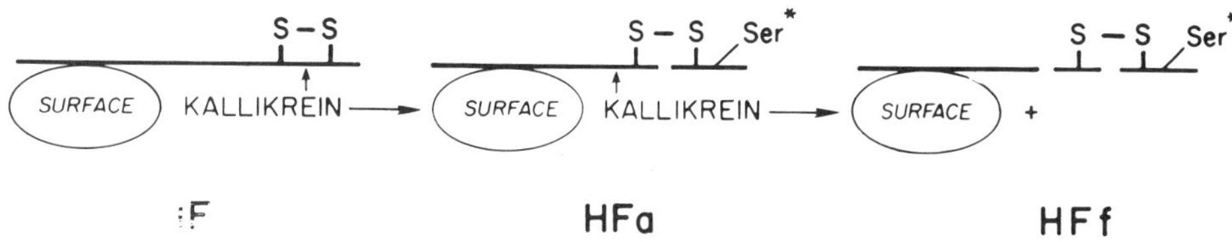

Figure 4. Cleavages seen on activation and fragmentation of human Hageman factor.

then occurs external to the disulfide bond, thereby breaking the molecule in two. The heavier portion remains attached to the surface, but the lighter portion, containing the active site of the enzyme, is then no longer attached to the surface and is found in the surrounding fluid. Cleavages of active enzymes in this fashion often inactivates them; however, this case is an exception. The product, called Hageman factor fragment (HFf), loses 95% of its ability to activate factor XI but retains fully its ability to convert prekallikrein to kallikrein (Kaplan and Austen 1970). This represents an intrinsic mechanism by which surface coagulation is inhibited while kinin formation in the fluid phase is perpetuated. This fragment continues to function until it is inactivated by plasma proteolytic inhibitors. An example of the consequences of elevated plasma HFf levels have been demonstrated when plasma expanders used to treat hypotensive or volume-depleted patients were found to contain significant quantities of HFf. Infusion of this material into patients caused profound hypotension (Alving et al. 1978).

With the discovery of HMW-kininogen deficiency, it became clear that the proposed reciprocal mechanism by which activated Hageman factor converts prekallikrein to kallikrein, and kallikrein in turn activates Hageman factor, was incomplete. This is evident in Figure 3, since the HMW kininogen-deficient plasma is markedly abnormal in both the coagulation and fibrinolytic assays. When the function of HMW kininogen in this reciprocal interaction was examined, it was found to augment the reaction in both directions, i.e., it augmented the conversion of prekallikrein to kallikrein by activated Hageman factor, and it also augmented the activation of Hageman factor by kallikrein (Griffin and Cochrane 1976; Meier et al. 1977). This is shown in Figure 5, in which the HMW kininogen is shown along the arrows to indicate that it acts as a cofactor or catalyst. Its contribution to these reactions has been quantitated. Thus, it augments the activation of prekallikrein by activated Hageman factor three- to four-fold, and it augments the activation of Hageman factor

Figure 5. Reciprocal activation of surface-bound Hageman factor and prekallikrein catalyzed by HMW kininogen.

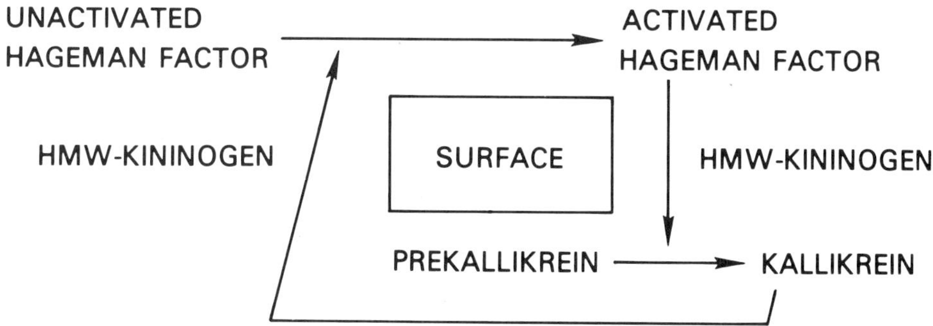

40-fold. Together then, it augments the reciprocal reaction by 120 to 160-fold. If the surface is left out, the rate of activation of Hageman factor by kallikrein is diminished 10 to 20-fold (Griffin 1978). Thus the surface, plus HMW kininogen, augments the rate of this reciprocal interaction somewhere between 1200- and 3200-fold. This augmentation allows the pathways to overcome whatever control mechanisms exist in plasma, and activation ensues.

Although plasma deficient in either Hageman factor or HMW kininogen does not activate (Figure 3), gradual activation of prekallikrein-deficient plasma is seen. Thus it is likely that one or more alternative enzymes exist that can activate Hageman factor. Both plasmin and factor XIa have been shown to activate and cleave Hageman factor but they are about 10% as effective as kallikrein and, in contrast to kallikrein, little or no augmentation by HMW kininogen is seen. Another possibility that could contribute to the gradual activation seen in prekallikrein-deficient plasma is auto-activation of Hageman factor. For example, if native Hageman factor possesses some limited proteolytic capacity, it is possible that one molecule could digest another, once one of them is bound to the surface. This would form the two-chain molecule we call activated Hageman factor. Thus, it is possible that native Hageman factor (or activated Hageman factor), factor XIa, and plasmin could slowly cleave and activate additional Hageman factor molecules so that with prolonged incubation with the surface, autocorrection is seen. These possibilities are presently being investigated. In all cases, the surface augments the reaction rate seen; however, the molecular mechanism by which this occurs is speculative. Perhaps an initiating surface places the reactants in proximity to each other and in optimal configuration so that cleavage and activation are facilitated.

Activation of Prekallikrein and Factor XI

Human plasma prekallikrein is a trace gammaglobulin separable from the bulk of IgG. It is found as a single chain present in two molecular forms having molecular weights of 88,000 and 85,000 and designated prekallikreins I and II, respectively. Activation to kallikrein occurs by limited proteolytic digestion in which the single-chain prekallikrein is cleaved within a disulfide bridge such that a two-chain kallikrein is formed in which the heavy chain is disulfide linked to the light chain (Mandle Jr. and Kaplan 1977). This process is diagrammed in Figure 6. The mechanism is essentially identical to that seen during Hageman-factor activation, and the active site of kallikrein is also found in the light chain. If one carefully examines the cleavage products obtained on digestion of prekallikrein under conditions that disrupt the disulfide bonds, a single heavy chain with a molecular weight of 56,000 as well as two light chains with molecular weights of 36,000 and 33,000 are seen. It appears, therefore, that the heterogeneity present in the initial prekallikrein preparation is present in the light chain. Thus, as shown in Figure 7, prekallikrein I at a molecular weight of 88,000 is converted to a heavy chain disulfide linked to light chain I at a molecular weight of 36,000, while prekallikrein II at a molecular weight of 85,000 is converted to the same heavy chain disulfide linked to light chain II at a molecular weight of 33,000. The evidence that both bands represent prekallikrein includes the following observations: (1) both are cleaved by HFf and yield similar activation patterns; (2) they are not interconvertible on activation by Hageman factor or on incubation of prekallikrein with kallikrein; (3) they contain the same antigenic determinants; (4) an active site is found in light chain I and light chain II; (5) neither

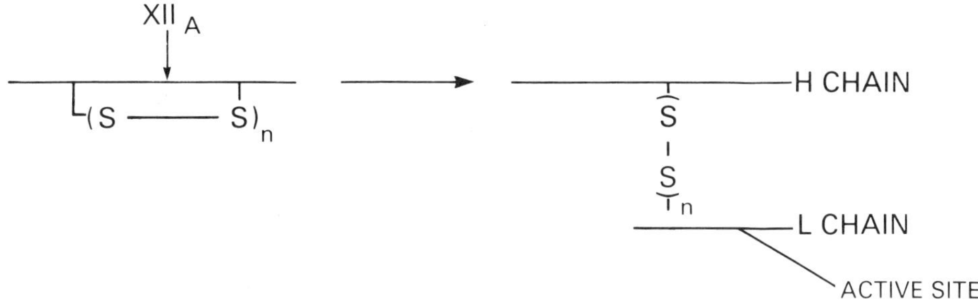

Figure 6. Conversion of prekallikrein to kallikrein. Cleavage within a disulfide bound by HFa results in the single-chain prekallikrein being converted to a two-chain disulfide-linked kallikrein in which the active site is located in the light chain.

band is seen when prekallikrein deficient plasma is fractionated. However, the difference between these two forms of prekallikrein has not yet been elucidated in molecular terms.

Factor XI, the other substrate of activated Hageman factor, is also a gammaglobulin; however, it is larger than prekallikrein and has a molecular weight of approximately 160,000. Although its size is therefore strikingly similar to that of IgG, it is separable from IgG, its structure is different from that of immunoglobulin, and it shares no antigenic determinants with IgG. This molecule is particularly interesting because its molecular weight halves if it is assessed under reducing conditions that break all disulfide bands, i.e., its molecular weight goes from 160,000 unreduced to 80,000 when reduced. Thus, it appears that factor XI is a two-chain molecule and the two constituent chains are virtually the same size. When the molecule is activated by factor XIIa, each 80,000 molecular weight chain is cleaved within a disulfide bond,

Figure 7. Two forms of prekallikrein and kallikrein on sodium dodecyl sulfate (SDS) gel electrophoresis.

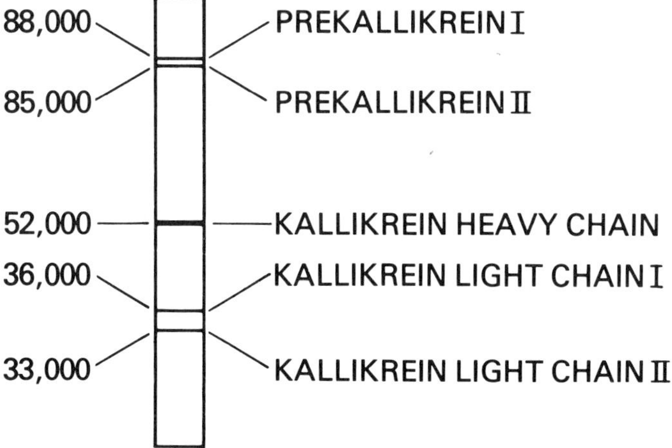

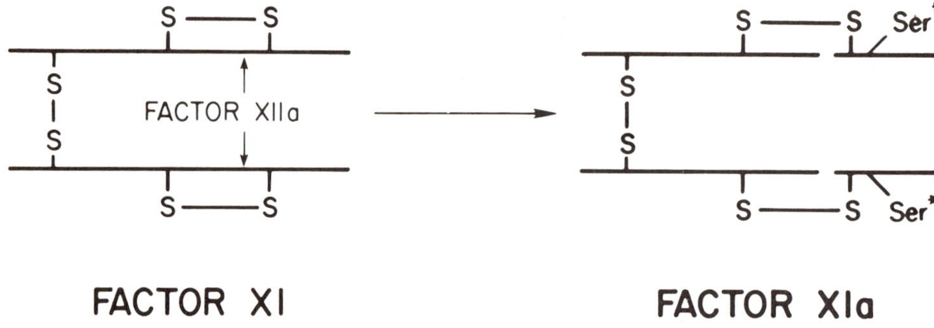

FACTOR XI   **FACTOR XIa**

Figure 8. Mechanism of activation of factor XI.

so that on reduction, a heavy chain is disulfide linked to a light chain. Once again, when the active site of factor XIa is located, it is found on the light chain. However, when the number of active sites is quantitated, there are two sites per mole of starting factor XI. Thus it appears that the two 80,000 molecular weight chains of the molecule are identical, and each one can be activated to create an active site (Kurachi and Davie 1977). Figure 8 shows factor XI to be a two-chain molecule, while factor XIa is a four-chain molecule with each active site associated with a light chain of molecular weight 30,000.

The rate of conversion of factor XI to factor XIa not only depends on activated factor XII, but the conversion of factor XI to factor XIa by factor XIIa is augmented by HMW kininogen in the same fashion seen for the conversion of prekallikrein to kallikrein. Yet the rate of factor XI activation is also prekallikrein dependent, since prekallikrein-deficient plasma has a diminished rate of initiation (Figure 3). Since factor XI and prekallikrein circulate complexed to different molecules of HMW kininogen (Figure 2), a mechanisms must exist in which Hageman factor activated by kallikrein can interact with the factor XI–HMW kininogen complex. This may require (1) movement of either kallikrein or activated Hageman factor along the surface or (2) release and readsorption of activated molecules at the surface–fluid-phase interface. The fact that the association constant of prekallikrein and HMW kininogen is such that at least 10% of prekallikrein (and presumably kallikrein) is found free, the kallikrein may then be able to activate molecules of Hageman factor that are in proximity to the factor XI-HMW kininogen complex.

Digestion of Human High Molecular Weight Kininogen
by Plasma Kallikrein

When HMW kininogen is cleaved by plasma kallikrein, two bonds close to the center of the molecule are cleaved and the 900 molecular weight bradykinin moiety is liberated. The resulting kinin-free kininogen is a two-chain molecule, while native HMW kininogen is a single chain. Thus, once again, cleavage has occurred within a critical disulfide bridge, however, in this case, two cleavage sites are attacked so that a small polypeptide is liberated. Of particular interest, however, is the observation that the coagulant activity of HMW kininogen is completely retained. When the resultant heavy and light chains are separated by reduction and the individual chains are purified, all coagulant activity is found in association with the light chain (Thompson et al. 1978). This light chain has been shown to be the part of the molecule that binds prekallikrein and factor XI and attaches them to initiating surfaces. Also of interest is the structural relationships of HMW kininogen and LMW kininogen. Although the latter is the more plentiful kininogen in plasma, it has no coagulant activity and plasma kallikrein preferentially digests HMW kininogen. When structural studies of these kininogen are performed, their heavy chains are found to be extremely similar if not identical, but their light chains are completely dissimilar. Thus, it is the light chain of kinin-free HMW kininogen that distinguishes it from LMW kininogen and is responsible for the functions of HMW kininogen as a coagulation cofactor.

The Role of the Early Steps of the Intrinsic Coagulation Cascade in Fibrinolysis

Hageman-factor-dependent conversions of plasminogen to plasmin is readily demonstrable in whole plasma; however, the identification of the enzyme or enzymes responsible for plasminogen activation has been particularly difficult. This appears to be because multiple enzymes are capable of activating plasminogen, but none of them is a potent activator such as urokinase or streptokinase. The major plasminogen activator formed by activation of those pathways appears to be kallikrein (Mandle Jr. and Kaplan 1977). However, factors XIa (Mandle Jr. and Kaplan 1979) XIIa and XIIf have been reported to be capable of catalyzing the same reaction, but when assessing their relative contribution in plasma, the effect of kallikrein on fibrinolysis through its ability to activate Hageman factor must be distinguished from its ability to convert plasminogen to plasmin. One estimate of the direct contribution of kallikrein to Hageman-factor-dependent plasminogen activation in plasma was approximately 50%, although this observation has not yet been confirmed. However, the fact that factors XIIa, XIIf, and XIa can also function in a similar manner may explain the autocorrection of the fibrinolytic defect seen in prekallikrein-deficient plasma at the time of incubation when the surface is prolonged.

Figure 9 summarizes the various interactions described herein. When initiation occurs, the components bound to the surface consist of Hageman factor and complexes of HMW kininogen with prekallikrein and factor XI. In a functional sense, the first step of the intrinsic coagulation and fibrinolytic pathways consists of a reciprocal enzymatic reaction between Hageman factor and prekallikrein, in which HMW kininogen functions as a reciprocal cofactor. Thus, it augments the rate of activation of prekallikrein by activated Hageman factor as well as the rate of activation of Hageman factor by kallikrein. Once Hageman factor has been activated, it acts on its other substrate, i.e., factor XI and converts it to factor XIa. High molecular weight kininogen also functions as a cofactor at this point and augments the rate of this reaction. The rate of formation of factor XIa therefore not only depends on Hageman factor and HMW kininogen but also on prekallikrein. However, as described earlier, this requirement is not an absolute one, and prolonged incubation with the surface leads to gradual evolution of factor XIa as is seen upon activation of prekallikrein deficient plasma.

Kallikrein digests the cofactor, HMW kininogen, to liberate the vasoactive peptide bradykinin. Further digestion of activated Hageman factor by kallikrein leads to formation of Hageman factor fragments, which leave the surface (indicated by the dashed line in Figure 9). This enzyme lacks significant coagulant activity on factor XI but readily activates prekallikrein in the fluid phase. Activation of plasminogen is

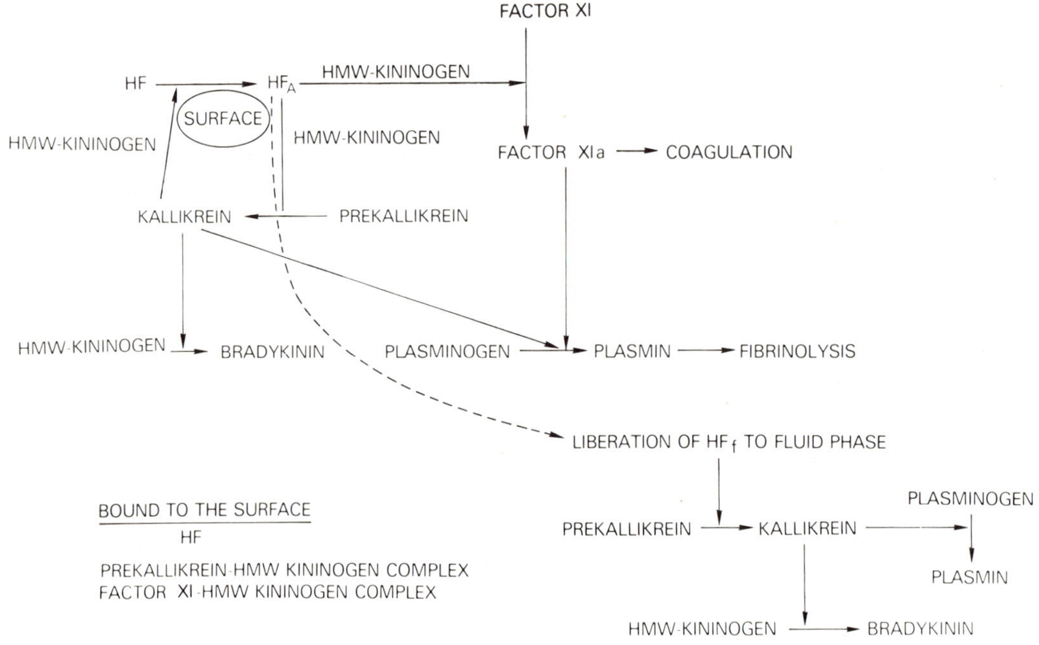

Figure 9. Known interactions leading to Hageman-factor–dependent coagulation, fibrinolysis, and generation of bradykinin.

also coupled to these events; this conversion appears to be mediated by multiple enzymes. The primary enzyme is kallikrein; however, both factor XIa and activated Hageman factor (factors XIIa and XIIf) can also contribute. Since the process is a surface-dependent phenomenon, it has an absolute requirement for Hageman factor and HMW kininogen. Prekallikrein has a dual effect because it is needed to activate the Hageman factor optionally and to interact directly with plasminogen.

Conclusion

The generation of bradykinin in plasma is a byproduct of the first step of the intrinsic coagulation pathway involving interactions among Hageman factor, prekallikrein, and HMW kininogen. Deficiencies of each of these plasma proteins have been described, yet the patients are quite healthy and, more specifically, do not have a bleeding abnormality. Congenital deficiencies of factor XI or subsequent proteins of the coagulation cascade are associated with signficant bleeding disorders. This difference has not been explained; however, it may be that the primary role of these proteins may relate to functions other than hemostasis. We are not, however, able to define the function of this cascade in normal physiologic states, although recent work has suggested that bradykinin can act as a local hormone that regulates blood flow at the tissue level and that kallikreins may act as activators of hormones.

Similarly, the functions of the kallikrein–kinin system in inflammation are somewhat speculative, and its contributions in this regard have not been demonstrated directly. Nevertheless, not only is bradykinin a likely mediator, based on its interactions with small blood vessels (primarily venules), but plasma kallikrein appears to be chemotactic for neutrophils (Kaplan and Austen 1971). In this fashion, local formation of kallikrein would direct its cells to the vicinity of an inflammatory reaction, while bradykinin would facilitate migration from the circulation into tissues. Finally, the possible link between IgE-mediated reactions in the lung, release of a tissue kallikrein, and release of a lung prekallikrein activator (Meier et al. 1979) may provide a link between inflammatory lung disease and kinin formation.

References

Alving, B.M., Hojima, Y., Pisano, J.J., Mason, B.L., Buckingham, R.E., Jr., Mozen, M.M., Finlayson, J.S. (1978) Hypotension associated with prekallikrein activator (Hageman-factor fragments) in plasma protein fraction. The New Eng. J. Med. 299,66.

Griffin, J.H. (1978) Role of surface in surface-dependent activation of Hageman factor (Blood coagulation factor XII). Proc. Natl. Acad. Sci. U.S.A. 75,1998–2002.

Griffin, J.H., Cochrane, C.G. (1976) Mechanisms for the involvement of high molecular weight kininogen in surface-dependent reactions of Hageman factor. Proc. Nat. Acad. Sci. U.S.A., 8, 2554–2558.

Kaplan, A.P. (1978) Initiation of the intrinsic coagulation and fibrinolytic pathways of man: The role of surfaces, Hageman factor, prekallikrein, high molecular weight kininogen and factor XIa In: *Progress in Hemostasis and Thrombosis*. Spacet, T.H. (ed.) New York, Grune and Stratton, p. 127.

Kaplan, A.P., Austen, K.F. (1970) A prealbumin activator of prekallikrein. J. Immunol. 105,802–811.

Kaplan, A.P., Austen, K.F. (1971) A prealbumin activator of prekallikrein II. Derivation of activators of prekallikrein from active Hageman factor by digestion with plasmin. J. Exp. Med. 133,672–712.

Kurachi, K., Davie, E.W. (1977) Activation of human factor XI (plasma thromboplastin antecedent) by factor XIIa (activated Hageman factor). Biochemistry 16,5831.

Mandle, R., Jr., Colman, R.W., Kaplan, A.P. (1976) Identification of prekallikrein and HMW-kininogen as a circulating complex in human plasma. Proc. Natl. Acad. Sci. U.S.A. 73,4179–4183.

Mandle, R., Jr., Kaplan, A.P. (1977) Hageman factor substrates. Human plasma prekallikrein: mechanism of activation by Hageman factor and participation in Hageman factor-dependent fibrinolysis. J. Biol. Chem. 252,6097–6104.

Mandle, R., Jr., Kaplan, A.P. (1979) Hageman factor-dependent fibrinolysis: Generation of fibrinolytic activity by the interaction of human activated factor XI and plasminogen. Blood 54, 850–862.

Mason, J.W., Kleeberg, U., Dolan, P., Colman, R.W. (1970) Plasma kallikrein and Hageman factor in gram-negative bacteremia. Ann. Intern. Med. 73,545.

Meier, H. L., Pierce, J.V., Colman, R.W., Kaplan, A.P. (1977) Activation and function of human Hageman factor. The role of high molecular weight kininogen and prekallikrein. J. Clin. Invest. 60,18–31.

Meier, H.L., Newball, H.H., Lichtenstein, L.M., Kaplan, A.P. (1979) The discovery of a prekallikrein activator released from antigen challenged passively sensitized human lung. Blood 54,292a.

Nustad, K., Gautvik, K.M., Pierce, J.V. (1974) Glandular kallikreins: Purification, characterization, and biosynthesis. In: *Chemistry and Biology of the Kallikrein–Kinin System in Health and Disease*. Pisano, J.J., Austen, K.F. (eds.) Fogerty International Center Proceedings no. 27. Washington, D.C., U.S. Government Printing Office, pp. 77–92.

Pierce, J.V., Guimaraes, J.A. (1974) Further characterization of highly purified human plasma kininogens. In: *Chemistry and Biology of the Kallikrein–Kinin System in Health and Disease*. Pisano, J., Austen, K.F., (eds.) Fogerty International Center Proceedings No. 27, U.S. Government Printing Office, Washington, D.C., pp. 121–127.

Revak, S.D., Cochrane, C.G., Griffin, J.H. (1977) The binding and cleavage characteristics of human Hageman factor during contact activation. A comparison of normal plasma with plasmas deficient in factor XI, prekallikrein, or high molecular weight kininogen. J. Clin. Invest. 59,1167–1175.

Thompson, R.E., Mandle, R., Jr., Kaplan, A.P. (1977) Association of factor XI and high molecular weight kininogen in human plasma. J. Clin. Invest. 60, 1376–1380.

Thompson, R.E., Mandle, R., Jr., Kaplan, A.P. (1978) Characterization of human high molecular weight kininogen. Procoagulant activity associated with the light chain of kinin-free high molecular weight kininogen. J. Exp. Med. 147,488–499.

Ulevitch, R.J., Cochrane, C.G., Henson, P.M., Morrison, D.C., Doe, W.F. (1975) Mediation systems in bacterial lipopolysaccharide-induced hypotension and disseminated intravascular coagulation I. The role of complement. J. Exp. Med. 142, 1570–1590.

Wiggins, R.C., Bouma, B.N., Cochrane, C.G., Griffin, J.H. (1977) Role of high molecular weight kininogen in surface-binding and activation of coagulation factor XI and prekallikrein. Proc. Natl. Acad. Sci. U.S.A., 77,4636–4640.

Serum Factors Associated with Inflammation

Jean D. Sipe, Ph.D. and David L. Rosenstreich, M.D.

The process of inflammation produces rapid and profound changes in a large number of serum components. The existence of these changes, which have been collectively called the acute phase response, has been appreciated for many years. However, only recently have we begun to understand the nature of these changes and their relationship to the cellular components of immunity and inflammation.

The serum changes induced by inflammation can be divided into early and late responses. The early response (1 to 2 hours) involves elevations in substances such as endogenous pyrogen thought to be derived from phagocytic cells. The late response (24 hours and beyond) involves elevations in substances such as C-reactive protein (CRP), fibrinogen and, serum amyloid A protein (SAA) produced by the liver.

In this chapter, we discuss the nature of the substances that compose these early and late changes, their physicochemical composition, cell of origin, and physiological functions. Above all, the close relationship between these substances and the cellular elements of immunity and inflammation should be noted. The early components produced by inflammatory cells can serve as potential mediators of inflammation, and the late components produced in response to the early ones can serve as feedback regulators of immunity and inflammation.

The Acute Phase Response

Characteristics

Cell and tissue injury results in significant changes in the concentration of a number of serum constituents. These fluctuations, which can be detected within a few hours of the inflammatory stimulus and are manifest for 2 or 3 days, are known collectively as the acute phase response. During the acute phase response, the concentrations of most serum proteins change, except for immunoglobulins (Koj 1974; Gordon 1976).

Two inducible proteins, C-reactive protein (CRP) and serum amyloid A (SAA), appear in as much as 1000-fold greater concentrations within 12 to 36 hours of an

From the National Institutes of Health, Bethesda, Maryland.

Table 1. Acute Phase Reactants

Early (macrophage products)

Endogenous pyrogen (EP)
    Related closely to leukocytic endogenous mediator (LEM) and lymphocyte-activating factor (LAF)
SAA inducer
    Probably related closely to CRP inducer
Tumor necrosis factor (TNF)
    May be related closely to glucocorticoid antagonizing factor (GAF)
Glucocorticoid antagonizing factor (GAF)

Late (hepatocyte products)	
Nonglycoproteins	Glycoproteins
C-reactive protein (CRP)	Fibrinogen
Similar temporal response to SAA; sequence homology with the glycoprotein SAP, which is acute phase reactant in mouse but not man	Ceruloplasmin
Serum amyloid A (SAA)	Transferrin
May be precursor of or is closely related to precursor of amyloid fibril protein in inflammation associated amyloidosis	Decreases because of excess catabolism over synthesis; appears to be produced by lymphoid cells as well as parenchymal cells in some species
Prealbumin	Alpha globulins
Concentration appears to be decreased during acute phase	Haptoglobin, Alpha 1-antitrypsin, Alpha 1-glycoprotein, etc.
Albumin	
Concentration appears to be decreased during acute phase	

inflammatory stimulus. There is also a heterogeneous group of glycoproteins for which concentration increases ranging from 50 to 300% are observed. These two groups of proteins, as outlined in Table 1, are known as acute phase reactants. The concentration of cations such as zinc, iron, and copper also changes during the acute phase response, and alterations in serum substances that modulate host defense are as yet chemically undefined (Gordon 1976; Kushner 1980).

The concentration of CRP in serum and the erythrocyte sedimentation rate (ESR) are time-honored indicators of clinical inflammation. Their measurement is useful for initially evaluating and monitoring the course of disease in a given patient. The ESR is an indirect measure of increased concentrations of fibrinogen, primarily, and also of other acute phase proteins. A high concentration of these serum proteins causes increased aggregation and rapid sedimentation of suspended erythrocytes [see Kushner (1980) for a detailed discussion of ESR and the acute phase response].

Koj (1974) has outlined the biosynthesis of acute phase reactants as part of the systemic response of an animal to injury. A complex chain of events is triggered by an inflammatory stimulus. First, there is the local reaction to cell or tissue injury, which involves such diverse phenomena as edema, fibrin formation, platelet aggregation, leukocyte accumulation, and release of lysosomal enzymes and small molecules such as histamine, kinins, superoxide, and hydrogen peroxide. The sequelae to the localized reactions, which involve cell injury and death, are systemic reactions that exert protective and antiinflammatory effects and result in restitution. It has recently become evident that mediators that arise during the initial response to an inflammatory stimulus subsequently bring about systemic changes in the host, such as fever, alterations in plasma composition, and synthesis of acute phase reactants. Such mediators appear within a few hours of an inflammatory stimulus; their presence is relatively short lived, and they are followed by the appearance of acute phase reactants in the blood.

The Inflammatory Stimulus

Various aspects of the acute phase response have been studied following injury and trauma such as surgery, acute myocardial infarction, acute disease, and neoplasia (as reviewed by Gordon 1976; Kushner 1980). Experimentally, numerous agents and methods have been employed to induce the acute phase response, including lipopolysaccharides, casein, Freund's adjuvant, and phagocytic stimulation and antigenic stimulation of sensitized animals with subsequent formation of lymphokines. The various regimens for experimentally inducing inflammation have been reviewed by Koj (1974), Gordon (1976), and Kushner (1980). The pyrogenic steroid etiocholanolone has been employed for experimental studies of fever and the acute phase response in human beings (Dinarello and Wolff 1978; McAdam et al. 1978).

Macrophage-Derived Mediators

The following discussion focuses primarily on the multiplicity of reactions that occur in response to a frequently employed and well-studied inflammatory stimulus, endotoxin. The broad spectrum of the biological effects of endotoxin has recently been reviewed by Berry (1977).

Endotoxin is known to bring about many alterations of the hematopoetic system. Initially there is an influx of granulocytes into the inflammatory exudate and a decrease in the number of circulating lymphocytes and granulocytes. This is followed by leukocytosis, which is due almost exclusively to an increase in the granulocyte formation. This secondary granulocytosis depends on the availability of cells from the bone marrow pool and thus, clinically, endotoxin has been used to assess bone marrow reserves.

The early part of the response to endotoxin, which is observed for approximately 3 to 4 hours following administration, appears to consist of effects caused by the interaction of endotoxin with various phagocytic cells such as monocytes, Kupffer cells, alveolar macrophages, and lymphoreticular cells in spleen and lymph nodes. The polymorphonuclear leukocyte is a special phagocytic cell with minimal ability to synthesize protein. It may have a less significant role in the early inflammatory response than does the monocyte, which is observed to have considerable capacity for protein synthesis.

One of the first responses to LPS and other stimuli is the production of various oxygen intermediates that have such a short half-life that they act only locally and thus are not detected in serum. Polymorphonuclear leukocytes or granulocytes accumulate at sites of cell and tissue injury and have been shown to liberate large amounts of hydrogen peroxide, superoxide, and hydroxyl radicals during the burst of respiration that accompanies phagocytosis (Badwey et al. 1979). This accumulation of superoxide-producing phagocytic cells at the site of injury may result in damage to other granulocytes and surrounding tissue during the expression of microbicidal activity. This kind of localized tissue injury during an inflammatory reaction may occur in response to various inflammatory stimuli such as adjuvants, bacteria, viruses, and burns.

Fridovich (1978) has reviewed present concepts of the role of superoxide in inflammation. The production of superoxide as part of the microbicidal activity of activated phagocytes can be expected to prolong and exacerbate the inflammatory process. Superoxide is known to depolymerize hyaluronate, a lubricating component of synovial fluid, and it is thought that superoxide interacts with hydrogen peroxide or with organic peroxides to form the hydroxyl radical and other reactive species, which are difficult for enzymatic scavengers to remove and can damage DNA and membrane components of cells.

Superoxide dismutases catalyze the conversion of superoxide radicals to hydrogen peroxide plus oxygen. The superoxide dismutases are extraordinarily efficient catalysts, but their concentration in extracellular fluids is normally low. There are several reports, as reviewed by Fridovich (1978), of antiinflammatory effects resulting from passive administration of superoxide dismutase.

Dormandy (1978) has pointed out there must be stringent control of free radicals during life, because of the extreme susceptibility of unsaturated fats to rancidification (autooxidation, peroxidation, and free radical oxidation). Although most protection against free radicals would occur locally, some may occur systemically as well. Human plasma is a powerful antioxidant; two of its more well-studied constituents, ceruloplasmin and transferrin, are involved in the oxidation and reduction of iron. Both proteins are acute phase reactants (Table 1); however, transferrin concentration normally decreases during the acute phase due to an excess of catabolism over synthesis.

Early Acute Phase Reactants

The most well studied of the serum constitutents that appear during the early acute response to endotoxin is endogenous pyrogen. The chemistry and production of endogenous pyrogen (EP) has been studied from the perspective of its biological activity, as reviewed by Dinarello (1979). Leukocyte endogenous mediator (LEM) is a preparation derived from leukocytic cells and is similar if not identical to EP. Leukocyte endogenous mediator has been studied in terms of its capacity to induce the acute phase response as well as in terms of its pyrogenicity (Kampschmidt 1974). Three other less well-defined early reactants are Tumor Necrosis Factor (TNF), Glucocorticoid Antagonizing Factor (GAF), and the SAA Inducer. All these factors appear to be products of macrophages, and inflammatory agents such as LPS can activate macrophages to stimulate production of these mediators.

Endogenous Pyrogen

Rabbits have been used for experimental studies of fever for more than 30 years. Beeson (1948) first observed a rapidly acting pyrogenic material in sterile rabbit peritoneal exudates; he called it leukocytic or granulocytic pyrogen. Atkins and Wood (1955) detected a pyrogenic mediator in the circulation of rabbits injected with endotoxin. They called the mediator endogenous pyrogen, to distinguish it from the exogenous pyrogen, endotoxin. Human blood incubated with endotoxin was observed to bring about rapid-onset fevers when reinjected into human subjects (Atkins and

Bodel 1972). This endogenous human leukocytic pyrogen has been characterized chemically, as reviewed by Dinarello and Wolff (1979), using fever production in rabbits as a bioassay. Circulating endogenous pyrogen has not yet been detected in man, perhaps as a result of its low concentration or lability. The recently developed radioimmunoassay for human endogenous pyrogen may facilitate its detection *in vivo* (Dinarello and Wolff 1979).

The pathogenesis of endotoxin-induced fever in man has been reviewed by Dinarello and Wolff (1978). An increase in total body oxygen consumption can be observed approximately 90 minutes after an intravenous injection of endotoxin into normal volunteers; the length of the latent period is inversely proportional to the amount of endotoxin. The oxygen increase is followed by peripheral vasoconstriction, which causes the subject to feel cold and shiver. The combination of increased metabolic activity and peripheral vasoconstriction results in decreased peripheral dissipation of heat and, therefore, the core body temperature rises. Subsequently, there is vasodilatation and sweating, which allows heat dissipation through the skin, and then body temperature returns to normal. In view of the fact that fever is initiated in the thermoregulatory center of the hypothalamus, it is unlikely that exogenous pyrogens act directly, since their size, complexity, and varied physical forms could preclude their gaining direct access to the brain. Moreover, abundant evidence now exists that exogenous pyrogens (such as LPS) act to induce a small mediator, endogenous pyrogen (EP). This molecule circulates during fever; it is not clear whether it acts directly on the thermosensitive neurons of the hypothalamus or through a small molecule made by endothelial cells lining blood vessels in that region.

Endogenous pyrogen is thought to induce fever by triggering a series of events in the brain, as reviewed by Dinarello and Wolff (1979). It is noteworthy that radioiodine-labeled human EP, when injected into rabbits, is rapidly cleared into the urine before maximum fever occurs. It is known that the preoptic region of the anterior hypothalamus is rich in thermosensitive neurons and that other areas such as the medulla and brainstem have a secondary role. Animal studies have shown that neurons sensitive to EP are remarkably specific and localized. It is known that prostaglandins are involved in stimulation of thermosensitive neurons by EP, because prostaglandin-E stimulates fever when injected into the hypothalamus, and drugs that inhibit prostaglandin synthesis are antipyretic. The cellular source of the prostaglandins is not known, and EP itself may not actually penetrate the blood brain barrier. In addition to monoamines and prostaglandins, cyclic nucleotides are also involved in fever production by EP. It has been observed that cyclic AMP concentrations are doubled in the cerebral spinal fluid during fever. Thus, EP is believed to enhance prostaglandin production, which in turn acts to produce cyclic AMP, which then acts on the thermogenic neurons. It is now recognized that EP can originate from either fixed tissue or circulating phagocytic cells. Cells that can be stimulated to produce EP are bone-marrow-derived phagocytes. Lymphocytes do not produce EP, nor do cells such as fibroblasts, even after particle ingestion. Macrophages produce 20 times more EP than neutrophils, and some investigators believe that macrophage contamination accounts for EP production by neutrophils. Lymphocyte activating factor (LAF), a substance first identified in the culture medium of activated cultured macrophages, appears to be very similar, if not identical, to endogenous pyrogen in man (Rosenwasser et al. 1979), rabbit (Murphy et al. 1980), and mouse (Mannel et al. 1980a). It is noteworthy that LAF can act on cultured nonlymphoid cells as evidenced by the observation of Mizel and coworkers (personal communication) that LAF, newly termed interleukin I, stimulates synovial cells to produce prostaglandin and collagenase. Thus, LAF/EP may be a major mediator of inflammation in the body.

Human endogenous pyrogen has been shown to retain its biological activity after treatment to remove lipid and carbohydrates. It is sensitive to heat denaturation and proteolytic digestion, and, in its unpurified form, is stable over the pH range 3 to 9. Human EP in the monomeric form has a molecular weight of 15,000 and has an isoelectric point of 7. Human EP has been shown to exist in another form, a larger protein of 45,000 molecular weight with an isoelectric point of 5.1. Mouse EP from tumor cell lines has molecular weights of 30,000 and 60,000. Rabbit EP differs from human EP in its pH stability, and antigenic differences between rabbit, guinea pig, and human EPs have been shown with antiserum to human EP.

It is known, as cited by Dinarello and Wolff (1979) that EP is newly synthesized in response to phagocytosis and to endotoxin. Early studies with human blood showed that there was no preformed EP in the peripheral leukocytes, and this was subsequently confirmed in other studies with human and rabbit leukocytes. Moreover, pyrogen release from leukocytes requires synthesis of new messenger RNA and of protein; the production of endogenous pyrogen from human leukocytes can be blocked by actinomycin-D as well as by cycloheximide or puromycin.

Thus, synthesis of EP is consistent with the well-defined concepts of gene expression. Initially, exogenous pyrogen brings about gene activation, which is followed by transcription of messenger RNA and by protein synthesis. Endogenous pyrogen is released quickly and there is no appreciable accumulation in the cell. It is not yet understood how EP production is terminated. Termination may result from accumulation of cellular products or perhaps of EP itself. If the mechanism of EP induction is consistent with the induction of other inducible proteins, perhaps shutoff is specified at the time of EP induction by the exogenous pyrogen.

Leukocytic Endogenous Mediator

Kampschmidt (1974) and colleagues have for a number of years studied the biologic effects of endogenous pyrogen prepared according to a modification of the method of Beeson (1948). Peritoneal granulocytes produced by the intraperitoneal injection of large amounts of glycogen into rats or rabbits release into the culture medium a low molecular weight, pyrogenic protein called leukocytic endogenous mediator (LEM), first identified during studies of iron metabolism in tumor-bearing rats, when it was observed that a leukocyte product was involved in the decreased plasma iron concentration associated with neoplasia and endotoxin shock. In addition to pyrogenicity, numerous biologic effects have been ascribed to LEM. They include decreased concentrations of iron and zinc, increased blood copper concentration, increased concentration of acute phase proteins, increased amino acid flux from muscle to liver, decreased liver catalase activity, and increased nonspecific resistance to infection. Attempts to separate LEM functionally from EP were unsuccessful (Merriman et al. 1977). The chemical and functional differences between EP, LAF, and LEM are now being explored intensively, and the three substances can be expected to be very similar, if not identical, proteins.

Tumor Necrosis Factor

It has been recognized for more than 30 years (Shear 1944) that administration of bacterial endotoxin causes hemorrhagic necrosis in certain mammalian tumors. Carswell and coworkers (1975) found that if mice are pretreated with agents that cause hyperplasia of the reticuloendothelial system, such as BCG, then within approximately 2 hours of endotoxin treatment, a substance called tumor necrosis factor (TNF) is released into their serum. Readily detected in serum containing only a small part of the amount of endotoxin necessary to necrotize tumors directly, TNF has been characterized both in rabbit (Matthews and Watkins 1978) and mouse (Mannel et al. 1980a) as a glycoprotein of 60,000 molecular weight, with an isoelectric point of 4.8. Tumor necrosis factor is separable from a 15,000 molecular weight helper activity for the antisheep red cell plaque-forming cell response of nude mouse spleen cells. This non-TNF–helper factor in serum has been identified as macrophage-derived LAF. Yang and Nowotny (1974) showed that TNF-like tumor resistance can be adoptively transferred with spleen cells activated with LPS either *in vivo* or *in vitro*.

More recently, *in vivo* studies have shown that the transfer of bone marrow cells derived from endotoxin-sensitive animals, into lethally irradiated, LPS-resistant C3H/ HeJ mice will restore to these mice the ability to necrotize tumors in response to LPS. These data strongly suggest a role for LPS-sensitive lymphoreticular cells in the production of TNF. Furthermore, it is now known that a macrophage cell line produces, in response to endotoxin, a factor capable of killing tumor cells *in vitro*, suggesting that the macrophage may be the cell responsible for TNF production (Mannel et al. 1980b).

Glucocorticoid Antagonizing Factor

Berry (1977) and colleagues have identified another mediator that appears in serum about 2 hours after endotoxin administration. This substance, glucocorticoid antagonizing factor (GAF), blocks induction of the hepatic enzymes tryptophan oxygenase, PEPCK, and glycogen synthetase. It appears that macrophages are the source of GAF, as well as TNF; however, GAF may be distinct from TNF in that it is not detected in nude mice (Mannel et al. 1980b). The properties of GAF as they are presently understood have been reviewed by Moore and co-workers (1978). Glucocorticoid antagonizing factor is a trypsin-sensitive macromolecule maximally produced approximately 2 hours after endotoxin treatment. When fractionated in conditions of low ionic strength, GAF has a molecular weight of 100,000 to 200,000. However, TNF also was observed to have a molecular weight of 150,000 and to be dissociated to a species of 60,000 molecular weight by gel filtration in high ionic strength buffer (Green et al. 1976).

Serum Amyloid A Inducer

Another early reactant derived from phagocytic cells is serum amyloid A (SAA) inducer. This as-yet chemically undefined substance was identified recently during studies of the mechanism of induction of the acute phase reactant SAA. The origin and properties of SAA will be discussed in the section Late Acute Phase Reactants.

The profile of acute phase SAA response of mice to endotoxin is characteristic of an inducible protein. Although LPS is a potent SAA inducer in all strains of mice except for endotoxin-resistant sublines, such as C3H/HeJ mice that can produce a normal SAA response to other inflammatory stimuli (McAdam and Sipe 1976), attempts to induce SAA production by treating various cultured cell lines with LPS were unsuccessful. It is now apparent that two cell populations (Rosenstreich and McAdam 1979) are involved in the acute phase SAA response. During the 2 or 3 hour lag following administration of an inflammatory agent, RNA synthesis is required for the subsequent elevation of SAA concentration (Sipe 1978). It was found that serum obtained from LPS-treated C3H/HeN mice during this lag period could induce an SAA response in C3H/HeJ mice although they are resistant to LPS. An SAA-inducing mediator was also detected in the culture supernatant of LPS-treated C3H/HeN macrophages but not C3H/HeJ macrophages or C3H/HeN spleen cells. It seems likely that a macrophage response to LPS gives rise to the mediator, which stimulates SAA synthesis (Sipe 1978) by hepatocytes. It is quite possible that SAA inducer is closely related to EP/LEM/LAF and perhaps to the mediator of CRP synthesis as well.

Role of Early Reactants in Acute Phase Response

The early acute phase reactants that have been discussed are produced by phagocytic cells in response to trauma or toxemia. Substances such as EP and SAA inducer, which appear within 3 hours of an inflammatory stimulus, are part of a signal to the entire body that cell injury and death has occurred and that systemic changes are necessary for recovery.

Much less is known about the chemistry of the early reactants EP, LEM, TNF, GAF, and SAA inducer than about the late acute phase reactants such as fibrinogen and C-reactive protein. In view of the recent work of Murphy and coworkers (Murphy et al. 1980) suggesting heterogeneity of rabbit EP, it is possible there may be a family of phagocytic cell products produced in response to inflammation, depending on the nature of the initial insult to the body. It appears that the early reactants are produced in hormonal concentrations, unlike the late reactants for which concentrations approaching milligrams per milliliter are observed during an acute phase response. These early acute phase reactants of macrophage origin are part of an immediate, and initially local, host response to injury (Figure 1). They serve to initiate the systemic part of the host response by acting on the brain to produce fever, by acting on connective tissue to remove dead and injured cells, and by mobilizing liver to reduce synthesis of "normal" serum proteins to allow synthesis of acute phase reactants such as CRP and SAA, which are important in recovery from injury. One can speculate that the early reactants also initiate the "shut-off" mechanism for the acute phase response; thus a rapid return to pre-acute phase conditions can be achieved by the host.

Late Acute Phase Reactants

C-Reactive Protein

The structure and function of C-reactive protein, and its characterization as an acute phase reactant have been thoroughly reviewed by Kushner (1980). More than 50 years ago, Tillet and Francis observed that a precipitate formed when serum from patients with acute infectious diseases was mixed with the C-polysaccharide of pneumococcus. The precipitated serum protein, C-reactive protein, was initially thought to be absent from normal serum; however, it is now recognized to be a trace constituent of serum, the concentration of which becomes greatly elevated after tissue injury.

C-reactive protein was purified originally in 1941 by Abernethy, McLeod, and Avery, using a calcium-dependent salt fractionation. They noted that the presence of lipid was required for calcium and sodium sulfate to precipitate CRP from serum. McCarty crystallized CRP in 1947, and Gotschlich and Edelman determined in 1965

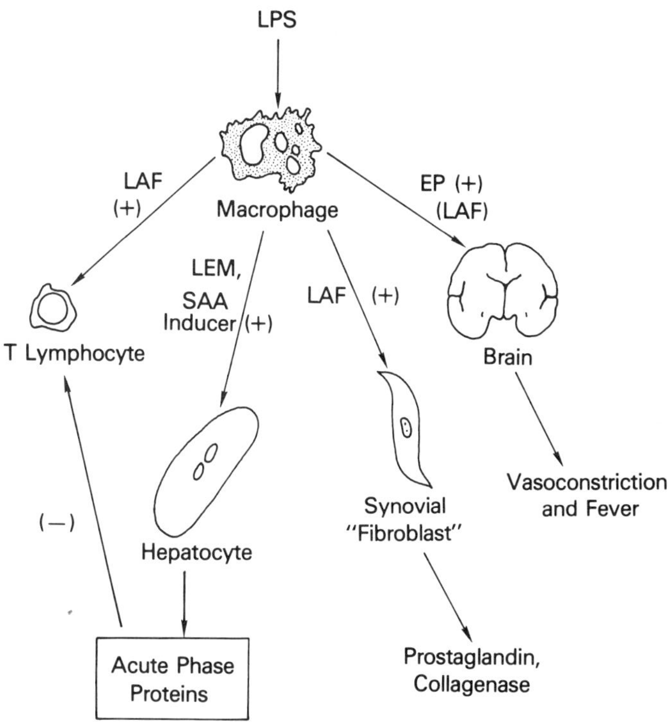

Figure 1. Endotoxin-induced inflammation: early and late acute phase reactants.

that CRP consists of five or six probably identical subunits of approximately 23,000 molecular weight, noncovalently associated as an oligomer of 120,000 to 140,000 molecular weight. Recent morphologic studies by Osmand and colleagues (1977) have shown that purified CRP consists of five subunits identical in appearance and arranged in cyclic symmetry. Primary structure analysis indicates that the molecular weight of pentameric CRP is 105,000. The higher values determined for CRP in serum may be caused by its interaction with serum lipids or perhaps by the association of additional subunits with CRP pentamers.

Although the basic physiological role of CRP in the host response to injury is not clearly understood, it would seem to be important; CRP-like proteins have been preserved throughout the course of evolution and have been shown to exist in fish and chickens as well as mammals (Pepys et al. 1978). The promptness of the return of the acute-phase CRP response to baseline concentration suggests CRP may be consumed during fulfillment of its biological function. Model studies have demonstrated that CRP has a specific affinity for phosphoryl choline residues. In the presence of calcium, CRP activates the classical complement pathway on binding to either C-polysaccharide or phosphocholine-containing liposomes (Richards et al. 1977). Twenty years ago Kushner and Kaplan observed that CRP is deposited on the surfaces of necrotic cells at inflammatory sites. It has been suggested that CRP combines with the phospholipid constituents of necrotic cell membranes at sites of inflammation to activate the complement system. This would eventually promote the removal of damaged tissue by complement-mediated phagocytosis.

CRP exhibits considerable amino acid sequence homology with the CH3-domain of human IgG, and it has been postulated that CRP may have originated as a primitive effector in the recognition of nonself. The complement-fixing site of IgG lies within the CH2 rather than the CH3 domain; however, it has not been possible to assign a conservative sequence within immunoglobulins to which the function of complement fixation can be attributed.

Evidence suggests that CRP may serve as an immunoregulator during acute inflammation, and CRP has been shown to promote the complement-dependent reactions of adherence and phagocytosis and to suppress platelet aggregation. C-reactive protein binds to T-lymphocyte subpopulations with resultant inhibition of the mixed lymphocyte reaction, antigen-induced lymphocyte proliferation, and lymphokine production. Recently, purified CRP has been shown to suppress antibody induction to T-dependent antigens by interacting with T-cells and generating a suppressive T-cell population (Mortenson 1979).

A radioimmunoassay survey has found that normal adult individuals have CRP concentrations that are frequently less than 1 µg/ml, and concentrations less than 0.1 µg/ml are found in umbilical cord sera. Within a few hours following an inflammatory stimulus, serum CRP concentration begins to increase significantly. Kushner and his collaborators (1978) studied a group of patients with acute myocardial infarction and found that the mean serum doubling time of CRP was 8.2 hours. The maximum CRP concentration was a reflection of the extent of tissue injury; values approaching 300 µg/ml were measured in patients with severe acute inflammatory states. These large CRP concentrations are as much as 1000 times baseline concentration in normal individuals (Kushner 1980). Human CRP concentrations generally reach maximum values between 1 and 3 days after an acute phase stimulus, then fall rapidly to half maximal concentrations within a day. However, persistently elevated CRP concentrations occur frequently in chronic inflammatory states associated with tuberculosis, rheumatoid arthritis, and malignancy. It is now clear that only serial determinations showing an elevation of CRP concentration are likely to be of significance in diagnosis, in view of the rapid changes that occur from day to day. Hurliman, et al. (1966) established that CRP synthesis takes place in the liver. More recently, Kushner and Feldman (1978) studied the CRP response to turpentine-induced local inflammation in rabbits, using an immunoenzymatic technique to visualize CRP in livers from rabbits at intervals after intramuscular injection of turpentine. C-reactive protein was detected only in hepatocytes; 8 hours after turpentine injection, CRP was found in occasional periportal hepatocytes. At the time of its maximum concentration in serum, CRP was found in most hepatocytes. These morphologic studies suggest that a mediator, acting initially in portal zones, is responsible for triggering cellular CRP synthesis. More recently, Kushner and colleagues (Schultz et al. 1980) have achieved *in vitro* synthesis of CRP with cultured hepatocytes obtained from inflamed rabbits.

Serum Amyloid P

There is substantial amino acid sequence homology between CRP and the serum glycoprotein, serum amyloid P (SAP). Amyloid P (AP) is found as a constituent of virtually all amyloid deposits, and its pentameric morphology led to its being named P-component (Bladen et al. 1966). Like CRP, SAP binds to ligands in the presence of calcium, and this property was a major factor in its being copurified with the first component of complement (Painter 1977).

It is of interest to note that SAP has been identified in three independent studies as C1t (Painter 1977), P-component of amyloid (SAP) (Skinner et al. 1974), and the 9.5S serum alpha glycoprotein (Haupt et al. 1972). These three independently identified protein preparations are virtually identical by antigenic and chemical criteria (Binnette et al. 1974). Serum amyloid P consists of approximately 12% carbohydrate and is a normal trace constituent of serum.

The morphology of SAP is very similar to that of CRP; when examined by electron microscopy it appears to be two pentameric discs interacting face to face. Considerable homology is present in the amino acid sequences of CRP and SAP. Biochemical studies have shown that SAP has a molecular weight of 230,000 to 290,000 (Pepys et al. 1978) and consists of subunits ranging from 19,000 to 27,000 in molecular weight. It behaves as an acute phase reactant in the mouse (Pepys et al. 1979), although SAPs fluctuation during the acute phase response in human beings is much less pronounced. While CRP does fluctuate as an acute phase reactant in mouse, the range of its concentration span is nanograms per milliliter rather than micrograms per milliliter as in the case of man (Siboo and Kulisek 1978).

C-reactive protein has not been implicated either actively or passively in amyloidosis. Its synthesis as an acute phase reactant is similar to that of serum amyloid A, to be discussed in the following section. On the other hand, SAP is a glycoprotein and normally is present in serum at a much higher concentration than CRP. The concentration increase of SAP during the acute phase is much more similar to that of other serum glycoproteins, as discussed in the following sections. The fact that SAP/AP is found in amyloid deposits containing fibrils other than amyloid A, raises intriguing questions as to what role it plays in the pathogenesis of amyloidosis. Clearly its calcium-dependent binding to fibrils and other constituents of amyloid deposits would contribute to organ dysfunction.

Serum Amyloid A

The historical, pathogenetic, chemical, and immunological aspects of amyloidosis have been reviewed thoroughly by Cohen and Cathcart (1980), by Glenner (1980), and by Franklin (1980). Amyloidosis is characterized by the accumulation in tissues of insoluble protein fibrils having the beta pleated sheet conformation. Amyloid deposits are identified morphologically and histologically by the presence of non-branching fibrils of indefinite length. They are approximately 80 to 100 Å in width, are stained by the planar hydrophobic dye Congo red, and exhibit green polarization color after Congo red staining.

Since 1970, biochemical investigations have revealed several distinct polypeptides that can accumulate extracellularly as insoluble amyloid fibrils. It was determined initially that the fibril protein associated with primary amyloidosis is the immunoglobulin light chain and its fragments. Shortly thereafter a fibril protein distinct from immunoglobulin was found to be associated with secondary amyloidosis, which is seen in various chronic conditions such as arthritis, infections, and familial Mediterranean fever. This fibril has since been designated amyloid A and the fibril protein is called AA. Both immunoglobulin amyloid and amyloid A are normally associated with systemic forms of the disease. There are also localized amyloid deposits associated with cells of neuroendocrine origin thought to be composed of peptide hormones or their precursors. An amyloid fibril protein larger than calcitonin, but sharing a common amino acid sequence with it, has been isolated from amyloid deposits of medullary carcinoma of the thyroid. Also, an amyloid fibril protein isolated from human senile hearts appears to be another unique protein, distinct from immunoglobulin or protein AA. Prealbumin has been recently identified as a constituent of the amyloid fibrils associated with the heredofamilial Portuguese amyloidosis (Costa et al. 1978). As shown in Table 1, prealbumin is one of the serum proteins for which concentration is decreased during pathologic conditions.

The amyloid deposits that accumulate in most of the spontaneous and induced amyloidoses in experimental animals contain amyloid A fibrils. The mouse model has proved exceedingly useful for studies of the origin and structure of SAA and its role in the pathogenesis of experimental amyloidosis. Amyloidosis can be induced in mice by chronic or repeated inflammatory stimulation. Repeated daily injections of casein is the time-honored method, but endotoxin, Freund's adjuvant, turpentine oil, and numerous other antigenic and inflammatory agents also have been employed successfully.

Amyloid fibrils can be isolated from tissue homogenates and separated from soluble tissue proteins by differential centrifugation. The fibrils can be solubilized by chemicals that dissociate noncovalent bonds of proteins. The solubilized AA proteins thus obtained range in molecular weight from 5,000 to 10,000.

Serum amyloid A (SAA) was identified by its cross-reaction with antibodies to denatured amyloid A fibrils. Both human and murine SAA proteins differ from the antigenically related tissue AA in that they are soluble and have a higher molecular weight, 160,000 to 180,000. It was noted previously that SAA was discovered during immunochemical studies of the origin of the secondary amyloid fibril protein AA. Since the biological function of SAA is unknown, immunoassay remains its only means of detection. Upon gel filtration of SAA in dissociating agents, a 12,000 to 13,000 molecular weight species bearing antigenic determinants common to AA is obtained. The structure of native SAA is not yet well defined; however, human (Benditt and Ericksen 1977), rabbit (Skogen et al. 1979), and murine SAA (Sipe et al. 1979) are isolated with high-density lipoprotein fraction of serum on ultracentrifugation. The dissociation of the low molecular weight SAA species from the high molecular weight species is accompanied by an unfolding of some determinants common to AA, and the dissociated SAA frequently is not soluble on return to physiological conditions.

It is not known to what extent, if any, amyloid A fibrils found in spleen, liver, kidney, and other tissues are derived from soluble circulating SAA. Amino acid sequence studies reveal considerable, but not absolute, sequence homology between the amino terminal amino acid sequences of AA and SAA. Thus, amyloid fibrils appear to be derived from SAA or a closely related polypeptide by proteolysis primarily at the carboxyl terminus. Elucidation of the cell of origin of the AA fibril protein is a crucial point in understanding the pathogenesis of amyloidosis. Evidence suggests that SAA/AA proteins are made by different types of cells. However, all of the available evidence to date indicates that the liver is the major source of acute phase serum SAA (Benson and Kliner 1979; Selinger et al. 1980).

Although the molecular aspects of the pathogenesis of secondary amyloidosis are not known, one can speculate that the cellular and molecular aspects of synthesis of the amyloid A polypeptide are similar to SAA biosynthesis. The induction of the acute-phase SAA response and the induction of experimental amyloidosis by intraperitoneal injection of complete Freund's adjuvant with added *Mycobacterium butyricum* are both two-phase processes. However, the time frame of the acute phase response is hours, whereas the time frame of amyloidogenesis is days. Perhaps, when the mechanism of clearance of circulating SAA is understood the molecular aspects of amyloid A fibril deposition will be clearer.

Benson and Aldo-Benson (1979) have reported recently that SAA exerts an immunosuppressive effect early in the antibody response of mice to sheep red blood cells. They found that extensive washing of spleen cells reversed the effect, suggesting that SAA could be acting at the level of the cell membrane. This work is the first hint of what physiological role SAA might be playing in the normal host inflammatory response to cell and tissue injury. Lavie and coworkers (1978) found that isolated and denatured SAA is degraded by an enzyme of human blood monocytes located on the cell surface. They presented evidence that an aberrant proteolysis of SAA may play a role in the pathogenesis of amyloidosis. Thus it may be that SAA, like CRP, is consumed in the execution of its normal biological function.

Other Serum Proteins (Glycoproteins)

The two inducible acute phase reactants described as late acute phase reactants, CRP and SAA, do not apparently contain carbohydrate. However, it is likely that an acute inflammatory response probably affects the synthesis of most, if not all, serum proteins synthesized by hepatocytes (Table 1). It is well documented that the concentrations of fibrinogen, haptoglobin, SAP, seromucoid, ceruloplasmin, alpha-1-acid glycoprotein, and other glycoproteins, as well, are increased during the acute phase response. On the other hand, plasma concentrations of albumin, transferrin, and prealbumin apparently decrease.

The decrease in transferrin concentration apparently is due to excess catabolism over synthesis (Gordon 1976). Rupp and Fuller (1979) have recently shown that a leukocytic factor, perhaps LEM or EP, stimulates fibrinogen synthesis in cultured hepatocytes and depresses albumin synthesis in a reciprocal fashion. It is known (Kwan and Fuller 1977) that the increased acute phase fibrinogen synthesis results from an increased quantity of fibrinogen messenger RNA rather than from an increased rate of translation of fibrinogen messenger RNA. Rupp and Fuller have also observed that cortisol is required for the *in vitro* effects of the leukocytic factor on hepatic protein synthesis.

Arachidonic acid metabolites, which are precursors of prostaglandins, have been implicated in the synthesis of ceruloplasmin and haptoglobin. Voelkel and co-workers (1978) observed that in rabbits bearing the VX2 carcinoma direct correlation existed between the amount of prostaglandin E2 in the serum and the amount of circulating ceruloplasmin and haptoglobin. Increased production of ceruloplasmin and haptoglobin was blocked by indomethacin. Therefore, prostaglandin E2 or its metabolites, or both, appear to be positive regulators of ceruloplasmin and haptoglobin levels.

Haptoglobin is an alpha-2-glycoprotein which combines with hemoglobin to form a complex that exhibits peroxidase activity. Haptoglobin is the major factor regulating the renal threshold for hemoglobin. Formation of the high molecular weight haptoglobin–hemoglobin complex prevents excessive loss of iron through urinary excretion. The haptoglobin–hemoglobin complex is rapidly taken up and catabolized by the reticuloendothelial system.

Ceruloplasmin is a blue copper-containing alpha-2-glycoprotein that can catalyze the four electron reduction of dioxygen to two molecules of water with concomitant substrate oxidation. The physiologic substrate of ceruloplasmin is thought to be ferrous iron, which when oxidized is incorporated into transferrin. Ceruloplasmin is also of importance to copper transport and to the maintenance of copper homeostasis in tissues.

Thus, a picture is emerging of a general alteration of liver protein synthesis during inflammation. It appears that the early acute phase response to inflammation effects a decrease in albumin and probably prealbumin synthesis in order to allow for rapid and increased synthesis of SAA and CRP and numerous glycoproteins that play an essential role in host defense. Increased fibrinogen synthesis will promote recovery from tissue damage by being readily available for blood clotting. The early part of the host defense response to trauma and infection involves formation of highly reactive free radicals that may themselves trigger further inflammation. Some of the acute phase serum proteins such as ceruloplasmin, haptoglobin, and transferrin may contribute to the antioxidant function of plasma during host defense. C-reactive protein may be involved in the removal of damaged tissue and cells, and may have an immunosuppressive role in order to minimize the triggering of a host autoimmune reaction during trauma. Serum amyloid A may be involved in catabolism of membrane constituents from dead and injured cells and in promoting membrane formation in new cells. Like CRP, SAA could also serve to suppress autoimmune responses of the host to its own injured tissues.

The molecular mechanisms by which phagocytic cell products such as SAA inducer and EP quickly induce and then stop acute phase reactant synthesis by liver, have barely began to be elucidated. Future studies of the function of the late acute phase reactants will have a close relationship to studies of cell–cell interaction.

Serum Factors that Modulate the Immune Response

There are a number of other serum substances, often associated with inflammatory and neoplastic conditions that suppress the immune response. This subject has been well reviewed by Tomasi (1976) and by Cooperband and colleagues (1976), taking into account the ever-present problems of nonspecificity and toxicity. Some of the serum-suppressive factors that have been studied include immunoregulatory alpha-globulin (IRA), low-density lipoproteins (LDL), alpha-fetoprotein (AFP), and normal serum.

Immunoregulatory alpha-globulin (IRA) is considered possibly to be part of the natural immunoregulatory system. Studies carried out for the past 20 years or so have shown that the alpha globulin fraction of serum from normal individuals and various patients' sera can suppress antibody formation, prolong skin grafts, suppress PHA and antigen-induced mitogenesis, the MLR and MIF production, and inhibit E-rosette formation. Because IRA does not inhibit plaque formation in response to thymus-independent antigens, and because its suppressive activity can be absorbed by T-lymphocytes and removed easily by washing, it is believed that IRA exerts its immunosuppressive activity by acting toward the T-cell. Immunoregulatory alpha-globulin activity adsorbed by T-lymphocytes can be removed by washing. A noncytotoxic alphaglobulin fraction has been purified by ion exchange chromatography. The bulk of suppressor activity can be separated from the alpha globulin fraction by acid dialysis, in the form of 500 to 5000 molecular weight trypsin-sensitive peptides. These peptides have a half-life of 4 days in mice.

Metcalfe and Tavill (1975) have hypothesized that rat alpha-1 macroglobulin and its homolog in man, alpha-2-macroglobulin, play a role in initiating the acute-phase biosynthetic response by liver. They believe the alpha globulin may bind and inactivate proteinases released in response to tissue injury, necrosis, and inflammation, and the macroglobulin–proteinase complex may stimulate liver to synthesize acute-phase reactants. Ford and coworkers (1973) separated a peptide suppressor factor from the macroglobulin fraction of patients' serum. It seems possible that IRA peptides may arise from the interaction of macroglobulin and proteinases, although Cooperband and colleagues (1976) suggest that the peptide may be noncovalently bound to a larger alpha globulin. Human alpha-2-macroglobulin is known to alter the activity of proteases toward large substrates as compared to substrates of low molecular weight (Twumasi et al. 1977).

Two types of immunosuppressive low-density lipoprotein (LDL) have been found. The first, rosette inhibitory factor (RIF), is found in the serum of some patients with acute viral hepatitis. Rosette inhibitory factor disappears on recovery and may represent a hybrid LDL that is a unique molecular complex resulting from hepatocellular injury. The second type of LDL, which is distinct from and may play a more important role than RIF, has recently been found in normal human serum and is a discrete subpopulation of LDL called L inhibitor, or LDL-In, that interferes with early events in T cell activation. It inhibits mitogenesis and the MLR, decreases *in vivo* antibody production, but does not affect E rosette formation.

Alpha fetoprotein (AFP) is a major serum protein found in embryos. It has been found to have significant amino acid sequence homology with albumin and may be an embryonic analog of albumin, similar to the fetal and adult hemoglobin system. *In vitro* studies of AFP synthesis by hepatocytes have shown it is synthesized in the G-1 and S phases, while albumin is made in the mid-S phase and G-2 phase. Many laboratories have found AFP to have immunoregulatory properties, since it inhibits antibody formation and mitogenesis, GVH reaction, and the MLR. It has been shown that AFP is distinct from fetuin, which is actually stimulatory. Receptors for AFP have been demonstrated on a subpopulation of murine T-lymphocytes, and lymphocytes from cancer patients and their families have been shown to have surface AFP. The variability in AFP suppression observed in different laboratories may be because of the lability of the suppressive activity or to variability in AFP preparations with respect to glycosylation and ligand binding. Alternatively, some investigators have suggested that the immunosuppressive effects of AFP may be caused by contamination with IRA.

Normal serum from mice has long been known to suppress growth of cultured cells at concentrations greater than 1%. Our recent recognition of the existence of SAA and its immunosuppressive properties, as well as the other serum substances already discussed, provide possible explanations for these phenomena and make it important for investigators to be aware of these potential variables in experimental mice.

Fibronectins

Fibronectin is a high molecular weight protein that exists in both a soluble form in plasma and in an insoluble form in connective tissue and at the surface of fibroblasts (as reviewed by Yamada and Olden 1978). The fibronectins are dimeric glycoproteins with subunit molecular weights of 200,000 to 250,000. Cell surface or cellular fibronectin has also been studied as large, external, transformation-sensitive (LETS) protein and as cell-surface protein (CSP). Plasma fibronectin is very similar, although probably not identical, to cell-surface fibronectin; the plasma glycoprotein is also known as cold insoluble globulin. Purified plasma fibronectin becomes cold insoluble when it is complexed with fibrin and fibrinogen. Opsonic protein and antigelatin factor, both of which were identified as affecting phagocytosis, are now recognized to be identical to plasma fibronectin (Molnar et al. 1979).

Fibronectins are implicated in many cellular activities, apparently because they serve as adhesive proteins that bind cells to other cells or to substrata. The concentration of cell-surface fibronectin varies with the cell cycle and is depleted in malignant transformation. Cellular fibronectin appears to be attached loosely to the cell membrane and is continuously secreted by or sloughed from cells. Hepatocytes, although the site of synthesis of most plasma proteins, do not contain fibronectin.

The concentration of fibronectin in blood is decreased after severe trauma and after reticuloendothelial system blockade following injection of colloids into experimental animals. It is thought that plasma fibronectin may play an important role after injury in the removal of collagenous debris from blood via the sessile macrophages of the reticuloendothelial system. Opsonic protein and antigelatin factor, both of which were identified as affecting phagocytosis, are now recognized to be identical to plasma fibronectin. It seems highly probable that future studies will reveal a wider role for fibronectin in host defense.

Conclusion

The chemical composition of plasma can be considered to be the mirror of the metabolic activities of the body. There are numerous fluctuations in those serum constituents that carry out antiinflammatory functions to aid the body in its recovery from injury and trauma. The acute phase changes in concentration of plasma constituents may be accomplished by varying the rate of protein synthesis, the concentration of messenger RNA, or the rate of protein catabolism. These three mechanisms for varying protein concentration, together with the variety of phagocytic cell products that modulate production of late acute phase reactants, could allow for a kind of fine tuning of the body's response to the many different kinds of inflammatory stimuli.

Thus a picture of host defense and restitution is emerging that involves a carefully synchronized sequence of events involving lymphoid cells, parenchymal cells, brain, synovial cells, and probably numerous other sites of the body. A model for the involvement of serum factors associated with inflammation in the body's antiinflammatory response is shown in Figure 1. The inflammatory stimulus, such as endotoxin, interacts with circulating or fixed tissue macrophages that then give rise to a cadre of mediators for the host inflammatory response. Thus, the brain is stimulated to produce fever and the liver is stimulated to alter its protein-synthesizing activity so that proteins that aid in restitution can be made. Proteins such as fibrinogen, CRP, and SAA may suppress autoimmune reactions and serve to clear damaged tissue and cells from the body as part of the normal host defense. In situations of prolonged inflammation, increased concentrations of late acute phase reactants may give rise to pathological conditions such as amyloidosis.

The similarities and differences in the early acute phase reactants need to be resolved in terms of assessing the diversity of the acute phase response. Studies of the mechanism of action of these phagocytic cell products will be difficult, because they are minor constituents of plasma. However, such studies may be expected to augment our understanding of eukaryotic protein synthesis and metabolism in addition to advancing our knowledge of the biochemistry of inflammation. Since the function of many serum constituents is poorly understood at this time, it is likely that inquiries into the function of the late acute phase reactants will aid our understanding of the complex milieu, serum.

The authors wish to express their gratitude to Esther Vetula for her very capable assistance in the preparation of this manuscript and to J. J. Oppenheim for his critical review of this manuscript.

References

Atkins, E., Bodel, P. (1972) Fever. Physiol. Med. 286,27–34.

Atkins, E., Wood, W.B. (1955) Studies on the pathogenesis of fever. II. Identification of an endogenous pyrogen in the blood stream following the injection of thyphoid vaccine. J. Exp. Med. 102,499–516.

Badwey, J.A., Curnutte, J.T., Karnovsky, M.L. (1979) The enzyme of granulocytes that produces superoxide and peroxide. N. Engl. J. Med. 300,1157–1160.

Beeson, P.B. (1948) Temperature-elevating effect of a substance obtained from polymorphonuclear leukocytes. J. Clin. Invest. 27,524A.

Benditt, E.P., Eriksen, N. (1977) Amyloid protein SAA is associated with high density lipoprotein from human serum. Proc. Natl. Acad. Sci. U.S.A. 74,4025–4028.

Benson, M.D., Aldo-Benson, M. (1979) Effect of purified protein SAA on immune response in vitro: mechanisms of suppression. J. Immunol. 122,2077–2082.

Benson, M.D., Kliner, E. (1979) Synthesis of serum amyloid protein A (SAA) by hepatocytes in mice treated with casein. J. Immunol. 124,495–498.

Berry, L.J. (1977) Bacterial toxins. C.R.C. Crit. Rev. Toxicol. 5,239–318.

Binette, P., Binette, M., Calkins, E. (1974) The isolation and identification of the P-component of normal human plasma proteins. Biochem. J. 143,253–254.

Bladen, H.A., Nylen, M., Glenner, G.G. (1966) The ultrastructure of human amyloid as revealed by the negative staining technique. J. Ultrastructure Res. 14,449–459.

Carswell, E.A., Old, L.J., Kassel, R.L., Green, S., Fiore, N., Williamson, B. (1975) An endotoxin-induced serum factor that causes necrosis of tumors. Proc. Natl. Acad. Sci. U.S.A. 72,3666–3670.

Cohen, A.S., Cathcart, E.S. (1980) Amyloidosis. In: *Contemporary Hematology–Oncology,* vol. 1. LaBue, J., Gordin, A.S., Filber, R., Muggia, S.M. (eds.) New York, Plenum Press, pp. 243–246.

Cooperband, S.R., Badger, A.M., Mannick, J.A. (1976) Nonhormonal serum suppressive factors. In: *Mitogens in Immunobiology.* Oppenheim, J.J., Rosenstreich, D.L. (eds.) New York, Academic Press, pp. 5550.

Costa, P.P., Figueira, A.S., Bravo, F.R. (1978) Amyloid fibril protein related to prealbumin in familial amyloidotic polyneuropathy. Proc. Natl. Acad. Sci. U.S.A. 75,4499–4503.

Dinarello, C.A., Wolff, S.M. (1979) Mechanisms in the production of fever in humans. Sem. Infect. Dis. 6,173–192.

Dinarello, C.A., Wolff, S.M. (1978) Pathogenesis of fever in man. N. Engl. J. Med. 298,607–612.

Dormandy, T.L. (1978) Free-radical oxidation and antioxidants. Lancet 1,647–650.

Ford, W.H., Caspary, E.A., Shenton, B. (1973) Purification and properties of a lymphocyte inhibition factor from human serum. Clin. Exp. Immunol. 1,169–179.

Franklin, E.C. (1980) Immunopathology of the amyloid disease. Hospital Practice. Sept. 70–77.

Fridovich, I. (1978) The biology of oxygen radicals. Science 201,875–880.

Glenner, G.G., Page, D.L. (1976) Amyloid deposits and amyloidosis. N. Engl. J. Med. 302,1283–1292; 1333–1342.

Gordon, A.H. (1976) The acute phase plasma proteins. In: *Plasma Protein Turnover.* Bianchi, R., Mariani, G., McFarlane, A.S. (eds.) Baltimore, University Park Press, pp. 381–394.

Green, S.A., Dobrjansky, E.A., Carswell, R.L., Kassel, R.L., Old, L.J., Fiore, N., Schwartz, M.K. (1976) Partial purification of a serum factor that causes necrosis of tumors. Proc. Natl. Acad. Sci. U.S.A. 73,381–385.

Haupt, H., Heimburger, N., Kranz, T., Baudner, S. (1972) Human serum protein mit hoer afinitat zu carboxymethyl cellulose. Z. Physiol. Chem. 353,1841–1849.

Hurliman, J., Thorbecke, G.J., Hochwald, G.M. (1966) The liver as the site of C-reactive protein formation. J. Exp. Med. 123,365–378.

Kampschmidt, R.F. (1974) Effects of leukocytic endogenous mediator on metabolism and infection. Ann. Okla. Acad. Sci. 4,62–67.

Koj, A. (1974) Acute phase reactants. In: *Structure and Function of Plasma Proteins,* vol. 1, Allison, A.C. (ed.) New York, Plenum Press, pp. 73–131.

Kushner, I. (1980) The acute phase reactants and the erythrocyte sedimentation rate. In: *Textbook of Rheumatology,* Kelley, W., Ruddy, S., Sledge, C., Harris, E., (eds.) Philadelphia, Saunders, 669–676.

Kushner, I., Broder, M.L., Karp, D. (1978) Control of the acute phase response. Serum C-reactive protein kinetics after acute myocardial infarction. J. Clin. Invest. 61,235–239.

Kushner, I., Feldman, G. (1978) Control of the acute phase response. Demonstration of C-Reactive protein synthesis and secretion by hepatocytes during acute inflammation in the rabbit. J. Exp. Med. 148,466–477.

Kwan, S.W., Fuller, G.M. (1977) Immunochemical characterization of fibrinogen induction in rat liver. Biochem. Biophys. Acta 475,659–668.

Lavie, G., Zucker-Franklin, D., Franklin, E.C. (1978) Degradation of serum amyloid A protein by surface-associated enzymes of human blood monocytes. J. Exp. Med. 148, 1020–1031.

McAdam, K. P. W. J., Elin, R. J., Sipe, J. D., Wolff, S. M. (1978) Changes in human serum amyloid A and C-reactive protein after etiocholanolone-induced inflammation. J. Clin. Invest. 61,390–394

McAdam, K.P.W.J., Sipe, J.D. (1976) Murine model for human secondary amyloidosis: Genetic variability of the acute-phase serum protein SAA response to endotoxins and casein. J. Exp. Med. 144,1121–1127.

Mannel, D.N., Ferrar, J.J., Mergenhagen, S.E. (1980a) Separation of a serum-derived tumoricidal factor from a plaque-forming cell helper mediator. J. Immunol. 124,1106–1110.

Mannel, D.N., Meltzer, M., Mergenhagen, S.E. (1980b) Generation and characterization of a lipo-polysaccharide-induced and serum-derived cytotoxic factor for tumor cells. Infect. Immunol. 28,-204–211 (in press).

Matthews, N., Watkins, J.F. (1978) Tumor-necrosis factor from the rabbit I. Mode of action, specificity and physiocochemical properties. Br. J. Cancer 38,302–309.

Merriman, C.R., Pulliam, L.A., Kampschmidt, R.F. (1977) Comparison of leukocytic pyrogen and leukocytic endogenous mediator. Proc. Soc. Exp. Biol. Med. 154,224–227.

Metcalfe, J., Tavill, A.S. (1975) A proposed role for alpha-1-acute phase globulin synthesis by the perfused rat liver. Br. J. Exp. Pathol. 56,570–578.

Molnar, J., Gelden, F.B., Lai, M.Z., Siefring, G.E., Credo, R.B., Lovandi, L. (1979) Purification of opsinically active human and rat cold-insoluble globulin (plasma fibronectin). Biochemistry 18, 3909–3916.

Moore, R.N., Goodrum, K.J., Berry, L.J. (1978) Mediation of endotoxin effects by macrophages. J. Reticuloend. Soc. 19,187–197.

Mortenson, R.F. (1979) C-reactive protein (CRP)-mediated inhibition of the induction of *in vitro* antibody formation. Cell. Immunol. 44,270–282.

Murphy, P.A., Hanson, D.A., Simon, P.L., Willoughby, W.E., Windle, B. (1980) Properties of two distinct endogenous pyrogens secreted by rabbit macrophages. In: *Microbiology 1980.* Schlessinger, D. (ed.), ASM Publications, Washington, D.C.

Osmand, A.P., Friedenson, B., Gewurz, H., Painter, R.H., Hofmann, T., Sheltin, E. (1977) Proc. Natl. Acad. Sci. U.S.A. 74,739–744.

Painter, R.H. (1977) Evidence that C1t (amyloid P-component) is not a subcomponent of the first component of complement (CI). J. Immunol. 119,2203–2205.

Pepys, M.B., Baltz, M., Gomer, K., Davies, A.J.S., Doenhoff, M. (1979) Serum amyloid P-component is an acute-phase reactant in the mouse. Nature 278,259–261.

Pepys, M.B., Dash, A.C., Fletcher, T.C., Richardson, N., Munn, E.A., Feinstein, A. (1978) Analogues in other mammals and in fish of human plasma proteins, C-reactive protein and amyloid P component. Nature 273,168–170.

Richards, R.L., Gewurz, H., Osmand, A.P., Alving, C.R. (1977) Interactions of C-reactive protein and complement with liposomes. Proc. Natl. Acad. Sci. U.S.A. 74,5672–5676.

Rosenstreich, D.L., McAdam, K.P.W.J. (1979) Lymphoid cells in endotoxin induced production of the amyloid-related serum protein SAA. Infect. Immunol. 23,181–183.

Rosenwasser, L.J., Dinarello, C.A., Rosenthal, A.S. (1979) Adherent cell function in murine T-lymphocyte antigen recognition. IV. Enhancement of murine T-cell antigen recognition by human leukocytic pyrogen. J. Exp. Med. 150,709–714.

Rupp, R.G., Fuller, G.M. (1979) Comparison of albumin and fibrinogen biosynthesis in stimulated rats and cultured fetal rat hepatocytes. Biochem. Biophys. Res. Comm. 88,327–334.

Schultz, D., Macintyre, S., Kushner, I. (1980) Studies of induction of CRP synthesis by primary rabbit hepatocytes. Ann. N.Y. Acad. Sci., 349,387–389.

Selinger, M.J., McAdam, K.P.W.J., Kaplan, M.M., Sipe, J.D., Vogel, S.N., Rosenstreich, D.L. (1980) Monokine induced synthesis of serum amyloid A protein by hepatocytes. Nature 285,498–500.

Shear, M.J. (1944) Chemical treatment of tumors. IX. Reactions of mice with primary subcutaneous tumors to injection with hemorrhage-producing bacterial lipopolysaccharide. J. Natl. Can. Inst. 4,461–476.

Siboo, R., Kulisek, E. (1977) A fluorescent immunoassay for the quantification of C-reactive protein. J. Immunol. Meth. 23,59–67.

Sipe, J.D. (1978) Induction of the acute phase serum protein SAA requires both RNA and protein synthesis. Br. J. Exp. Pathol. 59,305–310.

Sipe, J.D., Vogel, S.N., Ryan, J.L., McAdam, K.P.W.J., Rosenstreich, D.L. (1979) Detection of a mediator derived from endotoxin-stimulated macrophages that induces the acute phase SAA response in mice. J. Exp. Med. 150,597–606.

Skinner, M., Cohen, A.S., Shirahama, T., Cathcart, E.S. (1974) P-component (pentagonal unit) of amyloid. Isolation, characterization and sequence analysis. J. Lab. Clin. Med. 84,604–614.

Skogen, B., Borresen, A.C., Natvig, J.B., Berg, K., Michaelsen, T.E. (1979) High density lipoprotein as carrier for amyloid-related protein SAA in rabbit serum. Scand. J. Immunol. 10,39–45.

Tomasi, T. (1976) Serum factors which suppress the immune response. In: *Regulatory Mechanisms in Lymphocyte Activation.* Lucas, D.O. (ed.) Academic Press pp. 219–248.

Twumasi, D.Y., Liener, I.E., Galdston, M., Levytska, V. (1977) Activation of human leukocyte elastase by human alpha 2-macroglobulin. Nature 267,61–63.

Voelkel, E.F., Levine, L., Apler, C.A., Tashijan, A.H. (1978) Acute phase reactants ceruloplasmin and haptoglobin and their relationship to plasma prostaglandins in rabbits bearing the VX2 carcinoma. J. Exp. Med. 148,1078–1087.

Yang, C., Nowotny, A. (1974) Effect of endotoxin on tumor resistance in mice. Infect. Immunol. 9,95–100.

Yamada, K.M., Olden, K., (1978) Fibronectins-adhesive glycoproteins of cell surface and blood. Nature 275,179–184.

The Effect of Glucocorticoids and Other Hormones on Inflammatory and Immune Responses

H. C. Stevenson, M.D. and A. S. Fauci, M.D.

It is known that hormones exert a significant modulatory influence on inflammatory and immune responses of animals and human beings. These hormonal effects are best understood with regard to the glucocorticoids, but a substantial amount of information concerning the immunoregulatory effects of other hormones has been accumulated in recent years. This chapter reviews some of the current information concerning the impact of hormones on inflammatory and immune responses. Distinctions between *in vivo* and *in vitro* systems and between human and animal studies are emphasized, concentrating on those phenomena that most likely reflect human *in vivo* mechanisms.

The chapter is divided into sections focusing on certain of the divergent aspects of the subject: (1) receptor theories of hormone action, (2) relevant differences between animal versus human and *in vivo* versus *in vitro* assays of hormone function, (3) effects of glucocorticoids on leukocyte traffic, (4) effects of glucocorticoids on leukocyte function, (5) effects of glucocorticoids on the humoral factors involved in immune and inflammatory responses, and (6) significant advances in the study of the effects of other hormones on inflammatory and immunologic reactions.

Cell Receptors and Hormone Action

Current theories concerning the mechanisms of action of hormones on target tissues generally divide hormones into two major groups based on their chemical structure: (1) catecholamines and the peptide hormones and (2) the steroid hormones (Baxter and Harris 1975; Kahn et al. 1977; O'Malley and Buller 1977). The general model for catecholamines and peptide hormone function centers around a plasma membrane-bound hormone-specific receptor on the target cell (Kahn et al. 1977). Once a receptor hormone complex is formed at the cell surface, a "second messenger" substance is stimulated in the cytoplasm, which in turn either stimulates protein synthesis, causes various membrane changes, or modulates intracellular enzyme activity as the means whereby the specific effect of the hormone on the target tissue is effected. The second messenger has been identified for some hormone systems, the best understood one

From the Clinical Physiology Section, Laboratory of Clinical Investigation, National Institute of Allergy and Infectious Diseases, National Institutes of Health, Bethesda, Maryland.

being cyclic AMP (Kahn et al. 1977). However, for other peptide hormones, the exact second messenger substance and the mechanisms whereby that second signal affects the target cell physiology have not been entirely elucidated.

On the other hand, the general model for steroidal hormone activity centers around a receptor that is situated in the target cell cytoplasm (see Figure 1) (Baxter and Harris 1975; O'Malley and Buller 1977). The steroid hormone diffuses through the plasma membrane and binds to the specific steroid-binding protein receptor. Once binding occurs, a steroid-receptor complex is formed that diffuses into the nucleus where it stimulates the nuclear DNA to transcribe specific RNA. This RNA directs the production of new proteins, which presumably are the final effectors of the hormone's action (Baxter and Harris 1975). Certain aspects of receptor physiology have been worked out in greater detail particularly with regard to the sex steroids (O'Malley and Buller 1977). With progesterone, for example, the target-cell cytoplasm contains a divalent receptor composed of similar but slightly different proteins, the A and B subunits. When the A and B subunits each bind a hormone molecule, the entire complex migrates to the nucleus, where the B subunit attaches to a specific nonhistone protein on the DNA. The A subunit then stimulates the transcription of DNA in this region into RNA molecules, which code for the desired effector proteins (O'Malley and Buller 1977). In the rat immune system, thymocytes that are lysed or inhibited by glucocorticoids have been shown to contain steroid receptors (Munck and Brinck-Johnson 1968). Events known to be associated with thymocyte cell death, such as glucose transport inhibition, have been shown to be associated with RNA and protein synthesis following hormone-receptor triggering (Hallahan et al. 1973). Other animal systems have demonstrated that the biologic response of target tissues correlates well with the binding of steroid hormones to their receptors (Baxter and Harris 1975).

The importance of steroid receptors to glucocorticoid effects on the human immune response has been more difficult to delineate. As will be discussed below, human lymphocytes are not lysed by even suprapharmacologic concentrations of glucocorticoids (Claman 1971, 1972; Fauci et al. 1976). Thus, more subtle biologic endpoints must be used to measure the physiologic effects of glucocorticoids on the human immune response, such as changes in cell trafficking or functional capabilities (to be discussed below). Although both normal (Neifeld et al. 1977) and malignant (Lippman et al. 1973) human lymphocytes express steroid receptors, steroid effect in normal human lymphocytes has not correlated with the numbers of steroid receptors present in the target-cell population (Lippman and Barr 1977). This discrepancy may be related to a current lack of sensitivity in our hormone-receptor and target-cell functional assays in humans.

It thus appears that in the final analysis both types of hormones alter the physiology of target tissues by stimulating protein synthesis, modulating existing enzymes, or altering cell membrane characteristics, or a combination of the three (Baxter and Harris 1975). Although a precise chemical definition of the events resulting from a hormone's interaction with its target tissue receptors has not been reached, a great deal of interest currently exists in this area.

Differences Between Man and Animal Immune Responsiveness to Hormones and the Contrast Between *in vitro* and *in vivo* Studies

Studies of the effects of *in vitro* hormones on various aspects of the immune response, as well as studies of the effects of *in vivo* hormonal manipulations in various animal models, greatly have enhanced our knowledge of the mechanism of action of these

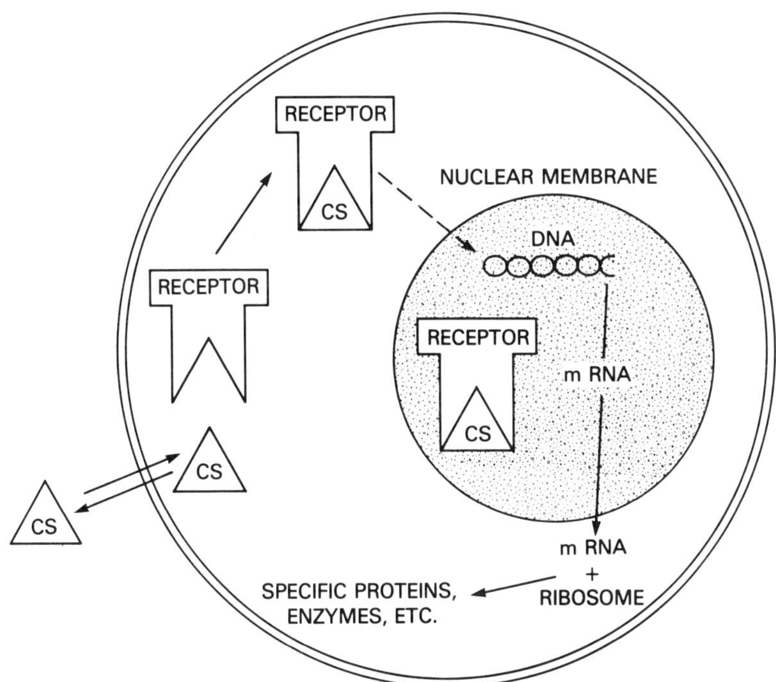

Figure 1. Model of corticosteroid receptor mechanisms. Corticosteroid molecule first binds to its receptor in the target cell cytoplasm. This hormone-receptor complex then migrates to nucleus and stimulates transcription of mRNA from specific segments of nuclear DNA. This mRNA codes for specific proteins that presumably are final mediators of hormone's effect on target cell.

agents. However, certain cautions are warranted when one attempts to extrapolate data derived in these systems to the *in vivo* mechanisms of action of these hormones in man. Although these caveats are indicated clearly for glucocorticoids, they also merit consideration when addressing investigation in other hormonal systems as well. In most *in vitro* studies of glucocorticoids, the concentrations of steroids used generally to depress various *in vitro* cell functions are suprapharmacologic, do not replicate the hormonal *in vivo* diurnal fluctuations, and are usually not attainable *in vivo* for any appreciable period of time (Fauci et al. 1976). Animal studies in which glucocorticoids are administered *in vivo* often use enormous drug concentrations; furthermore, when leukocytes from treated animals are placed in *in vitro* assay conditions, they usually are washed free of glucocorticoids. With *in vivo* administration of hormone in man, studies are almost invariably performed on leukocytes that remain in the circulation following drug administration and obviously not on those cells which have been depleted from the intravascular compartment. Secondly, it has been emphasized that animal models used in the study of glucocorticoid action on the immune response can differ greatly in their responses to these agents. Mice, rats, and rabbits are considered corticosteroid-sensitive, with a marked tendency to lympholysis in the presence of glucocorticoids (Claman et al. 1971). Table 1 summarizes some of the specific distinctive responses of the rodent immune system to glucocorticosteroids. Man, monkeys, and guinea pigs are considered to be corticosteroid-resistant species, so many studies showing profound effects on immunological parameters in a corticosteroid-sensitive animal do not apply to man. Thus, unless otherwise indicated, the next four

Table 1. Distinctive Responses of the Murine (Corticosteroid-Sensitive Species) Immune System to Glucocorticosteroids

Peripheral blood, spleen, and lymph-node lymphocytes are lysed by glucocorticoids
Glucocorticoids lyse murine immature (cortical) thymocytes; mature medullary thymocytes are steroid resistant
Serum-antibody levels and splenic PFC are depressed following *in vivo* glucocorticoid administration
Corticosteroids block the effector limb of murine cytotoxic lymphocytes in response to xenogeneic and allogeneic targets
Murine uncommitted B-lymphocytes are more susceptible to steroid depletion than T-lymphocytes

sections will consider predominantly those effects of corticosteroids shown to be operative in humans.

Effect of Glucocorticoids on Leukocyte Movement

Lymphocytes

As mentioned above, in corticosteroid-sensitive species, glucocorticoids cause a profound circulating lymphopenia predominantly by causing lympholysis and cell death. A commensurate lymphocyte depletion can also be seen in other lymphocyte-containing organs, specifically the thymus, spleen, and lymph nodes (Branceni and Arnason 1966; Claman 1972; Ingle 1938). In man and other corticosteroid-resistant species, normal lymphocytes exposed to suprapharmacologic concentrations of glucocorticoids are not lysed (Claman 1971, 1972). Despite this fact, a profound but transient circulating blood lymphopenia is seen 4 to 6 hours after a single parenteral or oral dose of glucocorticoids administered to man, with a return to baseline levels within 24 hours (Fauci and Dale 1974; Fauci et al. 1976). The same pattern is seen in those patients who receive a single dose of steroids on a daily basis (Cooper et al. 1977) and in those persons who receive chronic alternate-day glucocorticoid therapy (Fauci and Dale 1975a). (See Figure 2.)

Shortly after documenting that lymphocytes in corticosteroid-resistant species are not killed by glucocorticoids (Claman 1972), it was found that the glucocorticoid-induced blood lymphopenia seen in human beings was due to retrafficking of lymphocytes into other lymphoid compartments (Fauci 1975; Fauci and Dale 1974; Fauci and Dale 1975a,b). Glucocorticoid administration affects the recirculating lymphocyte (found in the peripheral blood but capable of circulating into all the lymphoid organs) by causing its migration out of the intravascular space. On the contrary, the nonrecirculating lymphocyte (one confined essentially to the intravascular space) is relatively unaffected in its traffic pattern by glucocorticoids (Fauci 1975; Fauci and Dale 1975b). Of those recirculating lymphocytes that migrate out of the intravascular space secondary to a glucocorticoid effect, a selectively greater depletion of T-cells is seen than of any other lymphocyte subpopulations (Fauci and Dale 1974; Yu et al. 1974). Of interest is the fact that those T-cells with Fc receptors for IgM (T_M) are depleted preferentially from the circulation by glucocorticoids, as opposed to those T-cells with receptors for IgG (T_G) (Haynes and Fauci 1978). Since T_M cells have been shown to confer help to the immunoglobulin production of mitogen-stimulated B-cells, and T_G cells are known to be suppressors of the same system (Moretta et al. 1977), the T_M lymphocyte depletion observed *in vivo* may be one mechanism by which glucocorticoids are able to modulate the immune response selectively. The relative effects of glucocorticoids directly on the functional capabilities of these T-cell subsets will be discussed below.

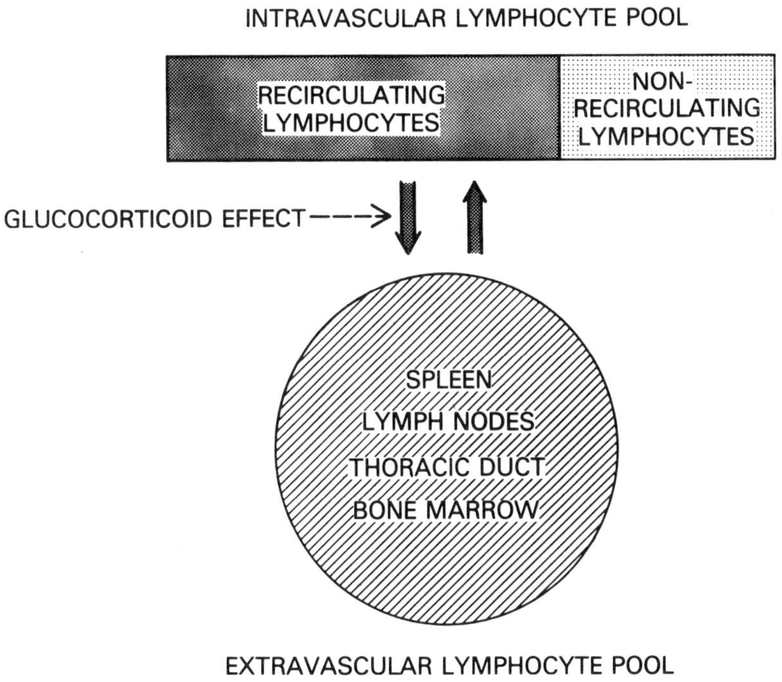

INTRAVASCULAR LYMPHOCYTE POOL

RECIRCULATING LYMPHOCYTES

NON-RECIRCULATING LYMPHOCYTES

GLUCOCORTICOID EFFECT ——→

SPLEEN
LYMPH NODES
THORACIC DUCT
BONE MARROW

EXTRAVASCULAR LYMPHOCYTE POOL

Figure 2. Effects of glucocorticoids on circulatory kinetics of the lymphocyte. Administration of glucocorticoids causes recirculating lymphocytes to migrate out of the intravascular space. T-cells appear to be selectively more depleted than other lymphocyte subpopulations.

Monocytes

Human studies have demonstrated a profound peripheral blood monocytopenia following glucocorticoids administration that is similar in magnitude and time course as that seen with the lymphocyte populations (Fauci and Dale 1974; Fauci and Dale 1975a). In mice, this glucocorticoid-induced monocytopenia may be only partially attributed to a decrease in the ability of the bone marrow to release monocytes into the peripheral blood as previously described (Thompson and van Furth 1970). It has been postulated that monocytes may be redistributed out of the intravascular space following glucocorticoid administration in a fashion similar to that seen with lymphocytes, but this remains unproved because of technical constraints of such studies. In fact it has been shown in mice that though liver Kupffer cells (fixed-tissue macrophages) are normally generated from circulating monocytes, there is no migration of monocytes to the liver in hydrocortisone-treated animals (Crofton et al. 1978).

It has been demonstrated that monocytes are even more sensitive to the antiinflammatory effects of glucocorticoids than neutrophils, since not only are they blocked in their migration to an inflammatory focus by a single dose of glucocorticoid (Boggs et al. 1964), but reaccumulation of cells is slower for monocytes than for neutrophils (see below) (Dale et al. 1974).

Neutrophils

The *in vivo* administration of pharmacologic doses of glucocorticoids in man produces three dramatic changes in the kinetics of neutrophils: (1) mature neutrophils are released at an accelerated rate from the bone marrow (Thompson and van Furth 1970),

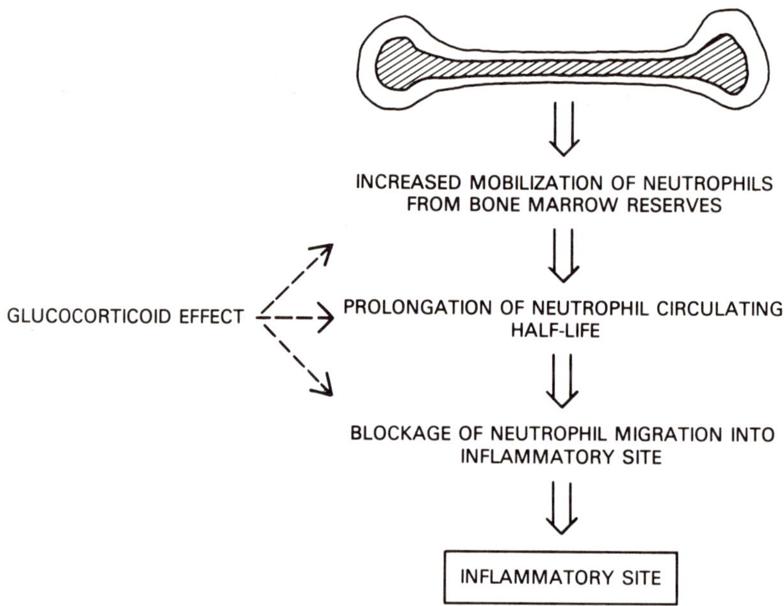

Figure 3. Manner in which glucocorticoids alter normal neutrophil traffic pattern. Normal release rate of neutrophils from bone marrow is accelerated in presence of glucocorticoids. In addition, a decrease in rate of neutrophil egress from intravascular space is seen. In particular, glucocorticoids inhibit migration of circulating neutrophils into inflammatory loci.

(2) the neutrophil circulating half-life increases (Athens et al. 1961), and (3) egress of neutrophils from the intravascular space decreases (Dale et al. 1974). This third effect may be directly related to the well-documented observation that glucocorticoids are able to decrease the numbers of both neutrophils and monocytes that accumulate at an inflammatory locus, a phenomenon that is evident within 2 hours of hormone administration in pharmacologic doses (Boggs et al. 1964). The decrease seen in the ability of neutrophils to leave the intravascular space following glucocorticoid administration may be partially caused by a decrease in the cell's adherence capabilities seen in the presence of glucocorticoids. Glucocorticoids block the adherence of neutrophils to the vascular endothelium of an inflammatory locus (Ebert and Barclay 1952) and when administered *in vivo* to human beings, they block the *in vitro* adherence of neutrophils to nylon wool columns (MacGregor et al. 1974). This decrease in the ability of neutrophils to leave the intravascular space and migrate into inflammatory loci is probably the major mechanism of the neutrophil-related antiinflammatory effects of glucocorticoids and is likely the major cause for the host defense defect seen in patients receiving daily corticosteroid therapy (Fauci et al. 1976). (See Figure 3.)

Eosinophils and Basophils

The administration of glucocorticoids is known to produce a profound decrease in the number of circulating eosinophils in man (Thorn 1973). Since the migration of eosinophils into immediate hypersensitivity skin test sites is blocked by glucocorticoid administration (Zweiman et al. 1976), and since the eosinophil appears to normally recirculate between body compartments in a somewhat more limited fashion than the lymphocyte (Dale et al. 1976), it is likely that glucocorticoid-induced eosinopenia is due to a redistribution of eosinophils out of the intravascular space into other body compartments (Anderson 1969). As with the other human leukocytes, eosinophils are not believed to be lysed on exposure to glucocorticoids.

Because of the technical constraints in obtaining purified suspensions of human basophils, very little is known about the effects of glucocorticoids on the kinetics of human basophil circulation.

The precise mechanism whereby glucocorticoids alter the traffic patterns of human leukocytes is not known. However, for each cell type to enter or leave the intravascular compartment, it must traverse the endothelial lining of the microvasculature. It has been postulated that movement through this barrier requires a complementary inter-action between the leukocyte and the endothelial-lining cell and that glucocorticoids somehow modulate this interaction (Fauci et al. 1976). Since glucocorticoids have been shown to be capable of altering cell-membrane function (Baxter and Forsham 1972; Thompson and Lippman 1975), such changes at the cell membrane level of the leukocyte or the endothelial cell may be responsible for the dramatic changes in normal leukocyte trafficking seen following glucocorticoid administration.

Effect of Glucocorticoids on Leukocyte Movement

Lymphocytes

The plethora of studies regarding lymphocyte activation, proliferation, differentiation, cytotoxic capabilities, receptor function, surface markers, mediator release, response to mediators, and other functions modulated by glucocorticoids precludes an in-depth review in this chapter; the reader is referred to other reviews on these subjects (Claman 1971, 1972; Baxter and Harris 1975; Claman 1972; Fauci et al. 1976; Gabrielson and Good 1967; Zurier and Weissman 1973; Fauci et al. 1976). However, some of the most recent observations particularly relevant to human immunobiology will be discussed below.

In vitro immunoglobulin production by B-lymphocytes appears to be enhanced by glucocorticoids in physiologic ranges (Sherman et al. 1973) and can be increased further by increasing the concentration of glucocorticoids to pharmacologic levels (Fauci et al. 1978). It has been suggested that glucocorticoids may enhance the immunoglobulin secreting function of B-cells by promoting cell maturation through the complex interaction of a variety of signals, including antigen presentation, T-cell factors, and hormone-receptor complexes on B-cells (Rusu and Cooper 1975). How-ever, as discussed previously, *in vivo* glucocorticoids are known to deplete helper-cell numbers (T_M) more than suppressor-cell numbers (T_G) (Haynes and Fauci 1978), whereas *in vitro* glucocorticoid suppresses functionally T_G-cells while having little effect on the function of T_M-cells (Gupta and Good 1977). Similar studies have shown glucocorticoids capable of inhibiting suppressor cells in some patients with common variable hypogammaglobulinemia (Waldmann et al. 1976). *In vivo* studies have demonstrated that antibody production is unchanged or may be enhanced following glucocorticoid administration in animals (Rusu and Cooper 1975) and human beings (Tuchinda et al. 1972). The complex interplay between the effects of glucocorticoids on B-cells themselves and on the various immunoregulatory lymphocyte subpopula-tions modulating B-cell antibody production is an area of great current interest.

Subpopulations of T lymphocytes other than T_G and T_M are also affected by glucocorticoids. Cutaneous delayed hypersensitivity is impaired during daily gluco-corticoid administration (Gabrielson and Good 1967), but this has been shown to be secondary to a defect in the ability to recruit macrophages to the skin site, rather than an active suppression of sensitized T-lymphocytes (Seebohm et al. 1954; Weston et al. 1973a). The suppressive nature of corticosteroids on cell-mediated immunity appears to be mediated mainly via the effects on production of and response to lymphokines (see Effects of Glucocorticosteroids).

Glucocorticoid administration will suppress lymphoproliferative responses to an-tigen stimulation at levels approaching the physiologic range, whereas much higher doses of drug are required to suppress mitogen-stimulated lymphocyte proliferation (Balow et al. 1975; Fauci and Dale 1974; Vischer 1972). The possibility that this *in vitro* phenomenon may in fact form the basis for an *in vivo* regulatory mechanism of T-cell autoreactivity is reflected in those studies that demonstrate suppression of the autologous mixed lymphocyte reaction (MLR) by physiologic doses of glucocorticoids without suppression of the allogeneic MLR (Ilfeld et al. 1977). However, glucocor-ticoids have been shown capable of inhibiting a cell in the human spleen that has a suppressor function in the allogeneic MLR (Sampson et al. 1975).

Only high concentrations of *in vitro* glucocorticoids are capable of inhibiting lymphocyte-mediated cytotoxicity. Cytotoxic capability of lymphocytes from rubella-immune patients against rubella-infected xenogeneic targets was suppressed by mod-erate concentrations of hydrocortisone (Thong et al. 1975); only high concentrations of glucocorticoids could suppress the cytotoxicity of lymphocytes from skin graft recipients against donor fibroblasts (Lundgren 1970), and only suprapharmacologic concentrations of hydrocortisone could suppress MLR-induced cytotoxicity against allogeneic targets (Balow et al. 1977). Lymphocytes exposed to hydrocortisone *in vivo* demonstrated decreased cytotoxicity induced by the MLR, which was related directly to the depletion of the numbers of T-lymphocytes from the circulation, implying that individual cell function remained unimpaired and that the decreased effector function was related to a quantitative deficiency of a certain cell type (Balow et al. 1976). Human antibody-dependent cell-mediated cytotoxicity (ADCC) is relatively resistant to modulation by *in vitro* glucocorticoids (Parrillo and Fauci 1978a). *In vivo* glucocor-ticoid administration actually increases ADCC, because the glucocorticoid-induced lymphocyte redistribution phenomenon described earlier actually selectively enriches for the ADCC effector cell population in the peripheral blood (Parrillo and Fauci 1978a). The only lymphocyte cytotoxic function suppressed easily by glucocorticoids is spontaneous cytotoxicity or natural killing. Though the effector cell for this function is felt to be identical to the K-cell that mediates ADCC (Cerottini and Brunner 1974), spontaneous cytotoxicity is depressed profoundly by both *in vivo* and *in vitro* hydro-cortisone (Parrillo and Fauci 1978a).

Monocytes and Macrophages

Although macrophages may be unable to participate in cutaneous delayed hypersen-sitivity reactions in the presence of steroids because their ability to respond to lymphokines such as migratory inhibition factor (MIF) is suppressed (see below), many of the other potential effects of glucocorticoids on monocyte–macrophage func-tion are not agreed on entirely. One that is generally accepted is the finding that doses of glucocorticoids that do not suppress *in vivo* and *in vitro* neutrophil function do significantly suppress human monocyte microbicidal activity (Rhinehart et al. 1974,

1975). Another clear-cut observation is that exposure to glucocorticoids *in vitro* exerts a profound depressive influence on human monocyte chemotaxis (Rhinehart et al. 1974), whereas *in vitro* chemotaxis was unaffected by *in vivo* steroid administration (Rhinehart et al. 1975). Animal studies of the *in vivo* and *in vitro* effects of glucocorticoids on phagocytosis have shown conflicting results in both types of experiments (Fleer et al. 1978; Hsu 1969; van Furth and Jones 1975; vanZwet et al. 1975; Wiener et al. 1972). Animal alveolar macrophages have decreased cytotoxic capabilities in an ADCC system in the presence of high concentrations of steroids (Hunninghake and Fauci 1977a,b). However, this was attributed to an inhibition by glucocorticoids of macrophage Fc-receptor binding of antibody-coated targets. In this regard, other studies have been unable to demonstrate an inhibitory role of glucocorticoids on the binding kinetics of Fc receptors of monocyte–macrophages in either animals or human beings (Abramson et al. 1970; Fleer et al. 1978). Cytotoxic function of human monocytes was depressed following *in vivo* glucocorticoid administration, but this was attributed to the profound steroid-induced monocytopenia rather than to a direct suppressive effect on the cytotoxic capabilities of the effector cells themselves (Parrillo and Fauci 1978b). Studies of the effects of *in vitro* steroids on human monocytes are few in number because of the current lack of a method for easily purifying large quantities of human monocytes. Since monocytes are among the predominant cell types involved in the evolution of granuloma formation (Epstein 1967), and since glucocorticoids are the mainstays of therapy of several of the idiopathic human granulomatous diseases (Fauci et al. 1976), further work in the area of steroid-induced modulation of monocyte kinetics and functional capabilities obviously is needed.

Neutrophils

The predominant effect of glucocorticoids on the neutrophil is reflected much more in the altered traffic pattern of this cell (as discussed previously) than in a direct effect on functional capability, since alteration in neutrophil function at pharmocologic doses of steroids is not demonstrated consistently. Although there are reports of suppression of neutrophil phagocytosis and killing with very high concentrations of *in vitro* glucocorticoids (Dale and Petersdorf 1973; Mandell et al. 1970), assessments of human neutrophil phagocytosis following *in vivo* administration of as much as 1000 mg of methylprednisolone have shown unimpaired phagocytosis and bactericidal capabilities (Allison and Adcock 1965; Glasser et al. 1977). *In vitro* chemotaxis of neutrophils has been shown to be inhibited only by very high concentrations of glucocorticoids (Rhinehart et al. 1974; Ward 1966).

Whether or not glucocorticoids can inhibit lysosomal enzyme release of neutrophils is also debated. Since it has been long known that high-dose *in vitro* glucocorticoids can stabilize lysosomal membranes in liver (Weissman and Thomas 1973), and since the contents of neutrophil lysosomal granules are known to be important mediators of inflammation (Weissman 1967), many laboratories have attempted to determine if glucocorticoids cause neutrophil lysosomal membrane stabilization. Although some studies show no stabilization of neutrophil lysosomes even by very high concentrations of steroids (Persellin and Ku 1974), other groups have demonstrated that high doses of glucocorticoids *in vitro* can inhibit both the phagocytic (Wright and Malawister 1973) and nonphagocytic (Ignarro 1977) release of lysosomal enzymes by neutrophils. Whether these high *in vitro* concentration ranges are applicable to the pharmacologic antiinflammatory *in vivo* effects of glucocorticoids is uncertain. Of interest is the

Table 2. Summary of the Alterations in Leukocyte Function Induced by Glucocorticoids

Lymphocytes
    Delayed hypersensitivity skin testing suppressed by inhibition of recruitment of macrophages
    Lymphocyte proliferation to antigens suppressed more easily than proliferation to mitogens
    Mixed leukocyte reaction proliferation suppressed
    High *in vitro* concentrations suppress T-lymphocyte–mediated cytoxicity
    ADCC (antibody-dependent cell-mediated cytoxicity) not depressed
    NK (natural cytotoxicity) suppressed
    Regulatory effects on helper and suppressor cell populations

Monocyte–macrophages
    Cutaneous delayed hypersensitivity suppressed by inhibition of lymphokine effect on the
        macrophage in certain systems and suppression of lymphokine production in others
    Possible blockade of Fc receptor binding and function
    Depressed bactericidal activity
    Possible decrease in monocyte chemotaxis

Neutrophil
    Probably no effect on phagocytic and bactericidal capability
    ADCC (antibody-dependent cellular cytotoxicity) increased
    Probably decreased lysosomal release but little effect on lysosomal membrane stabilization at
        pharmacologic concentrations
    Chemotaxis inhibited only by suprapharmacologic concentrations

Adapted from Parrillo and Fauci, 1979

finding that human neutrophil-mediated ADCC is actually increased following *in vivo* administration of glucocorticoids (Parrillo and Fauci 1978b).

Because of profound circulating eosinopenia seen following steroid administration, little is known about the direct effects of *in vivo* glucocorticoids on human eosinophil functional capabilities.

Table 2 summarizes the modulatory effects of glucocorticoids on the various leukocyte subpopulations.

Effect of Glucocorticosteroids on Humoral Factors Involved in the Immune Response

Antibody Production

Glucocorticoid-induced suppression of antibody production has been demonstrated in corticosteroid-sensitive species and is possibly secondary to the primary lympholytic action of glucocorticoids in these animals (Claman 1972). However, total human serum immunoglobulin levels fall only slightly when high enough doses of glucocorticoids are given. This is believed to be caused by decreased protein synthesis and increased protein catabolism (Butler and Rossen 1973). Specific antibody production is not depressed by glucocorticoid administration (Claman 1972, 1975; Fauci et al. 1976; Tuchinda et al. 1972). As a result of effects on the suppressor regulatory mechanisms of B-lymphocyte function described previously, *in vivo* and *in vitro* antibody production may actually be increased by glucocorticoids (Fauci et al. 1978; Sherman et al. 1973; Tuchinda et al. 1972).

Complement

Although high doses of glucocorticoids have been reported to produce up to a 60% reduction in some of the complement components of guinea pigs (Atkinson et al. 1973), the physiologic significance of this decrease is unclear. Detectable alteration in complement metabolism does not occur as a result of glucocorticoid administration in man (Claman 1975).

Kinins and Plasminogen Activator

Although the vasoactive peptides known as kinins have a theoretical impact in some human inflammatory responses (Kellermeyer and Graham 1968), early reports of glucocorticoid-induced inhibition of the *in vitro* and *in vivo* activation (Cline and Melmon 1966) of kinins have not been confirmed (Lefer and Inge 1974). However, glucocorticoids are capable of blocking the production of plasminogen activator by macrophages (Vassalli et al. 1976; Werb 1978) and neutrophils (Granelli-Peperno et al. 1977). Since this enzyme appears to be correlated with the ability of cells to migrate into inflammatory loci, the regulation of this activity may be an important antiinflammatory effect of steroids (Granelli-Piperno et al. 1977).

Lymphokines

Corticosteroids may suppress some aspects of cell-mediated immunity by interfering with the function of lymphokines. Antigen-induced *in vitro* generation of macrophage MIF in guinea pig lymphocytes is not affected by hydrocortisone (Balow and Rosenthal 1973), whereas mitogen-induced generation of MIF is decreased by *in vitro* methylprednisolone (Bendtzen 1975), and chronic *in vivo* hydrocortisone administration produces a marked decrease in antigen-induced MIF production (Balow et al. 1975). Some studies show that macrophage responses to MIF are impaired in the presence of glucocorticoids (Balow and Rosenthal 1973), whereas others show they are unaffected (Bendtzen 1975). Studies of other lymphokines have shown that production of chemotactic lymphokine was decreased in both man (Ruhl et al. 1974) and guinea pigs (Wahl 1975) exposed to hydrocortisone *in vitro*. Lymphotoxin generation was also depressed in humans by *in vitro* hydrocortisone (Williams and Granger 1969), whereas the production of macrophage-activating factor (MAF) in guinea pigs was not affected by *in vitro* steroids (Weston et al. 1973b). Complete understanding of the effects of corticosteroids on lymphokines will elucidate some of the immunologic effects of glucocorticoids.

Immediate Hypersensitivity Reactions

Although corticosteroids have been used clinically for years in the effective treatment of immediate hypersensitivity disease states (Type I reactions), very little objective laboratory evidence for the mechanism of their action is known. Although it is clear that steroids cause eosinopenia, vasoconstriction, and potentiate catecholamines in man, corticosteroids seem not to affect specific or total IgE levels and are not capable of blocking anaphylaxis or histamine shock in animals (Claman 1975).

Table 3. Experimental Basis for Clinical Observations Regarding Altered Immune Status in the Presence of Glucocorticoids

Suppression of acute inflammatory reactions
    Marked reduction in circulating monocytes and eosinophils
    Inhibition of the migration of granulocytes into inflammatory foci
    Inhibition of humoral factors such as plasminogen activator and collagenase

Impaired delayed-type hypersensitivity
    Impaired ability to recruit macrophages and lymphocytes to inflammatory site
    Possible inability of macrophages to respond to lymphokines

Defects in reticuloendothelial function
    Decreased clearance of antibody-coated targets, possibly secondary to Fc receptor impairment
    Impaired macrophage microbicidal activity for certain infectious agents

Table 3 summarizes the experimental basis for our current knowledge about the clinically observed antiinflammatory and immunosuppressive properties of glucocorticoids.

Effect of Hormones Other than Glucocorticoids on Inflammation and the Immune Response

Recent studies, primarily with animal models, have revealed real and potential interactions between hormones other than glucocorticoids and immunologic function.

Dwarf Mice

Hypopituitary dwarf mice belong to a strain that inherits, as an autosomal recessive characteristic, a hypofunctioning anterior pituitary gland and thus from birth produces no growth hormone and diminished quantities of thyroid stimulating hormone, adrenocorticotropic hormone, and prolactin (Bartke 1964). After the first 2 weeks of life, dwarf mice demonstrate a progressive deterioration of its immune system, predominantly of the thymic-dependent functions (see Table 4). Blastogenic responses to T-cell mitogens are depressed (Duquesnoy et al. 1969), as is the ability to reject skin grafts (Fabris et al. 1971a) and B-cell responses to thymus-dependent antigens (Fabris et al. 1971a). Since dwarf mice appear to have normal hematopoietic stem-cell potential (Duquesnoy 1972), the argument has been made that the observed immunodeficiency is secondary to a defect in the thymic microenvironment (that is at least partially under hormonal control) which does not allow for proper differentiation of lymphoid stem cells into mature T-lymphocytes (Fabris et al. 1971b). This argument is strengthened by the observation that the immunodeficiency of dwarf mice can be prevented completely by the administration of bovine growth hormone and thyroxine within the first months of life, provided the thymus is intact (Fabris et al. 1971b, 1972). Thus, although dwarf mice have more than a single hormone defect, it appears that they provide an animal model of the hormonal control (particularly growth hormone) of the thymic-dependent immune system. There are no reports, however, of immunodeficiency associated with congenitally absent growth hormone in human beings. Paradoxically, although dwarf mice were studied initially as an animal model for the role of immunodeficiencies in the development of malignancies, it now appears

Table 4. Comparison between the Hormonal Aberrancies and the Altered Immune Function of the Hypopituitary Dwarf Mouse

Endocrine system dysfunction
    Nearly absent growth hormone production
    Reduced secretion by anterior pituitary of prolactin, thyroid-stimulating hormone, and
        adrenocorticotropic hormone
    Reduced adrenocortical hormone activity
    Poorly developed thyroid
    Underdeveloped gonads
    Altered secretion by pancreatic islets

Immune system characteristics
    Hypoplasia of thymus and thymus-dependent areas
    Lymphopenia
    Low blastogenic responses to T-cell mitogens
    Normal serum immunoglobulin levels
    Impaired antibody responses to thymic-dependent antigens
    Impaired cutaneous delayed hypersensitivity
    Depressed ability to reject allografts
    Normal hemopoietic stem cell potential
    Decreased autoimmune reactivity

that dwarf mice have a decreased incidence of both malignancies (Piantanelli and Fabris 1978) and autoimmune phenomena (Pelletier et al. 1976) when compared to littermate controls.

Androgrens

Orchiectomy in rats has been shown to increase the size of the thymus and lymph nodes (Castro 1974a). *In vivo* androgens have been shown to accelerate the rate but not the total amount of antibody formation in animals (Tolentino 1975), to deplete lymphocytes from the thymic cortex (Sobhon and Jirasattham 1974), and to prolong allograft survival in mice (Castro 1974b). *In vitro* testosterone caused a mild decrease in blastogenic responses of phytohemagglutinin (PHA) stimulated human lymphocytes at hormone concentrations of 0.1 μg/ml and completely ablated the response at 80 μg/ml (Wyle and Kent 1976). Testosterone administered *in vivo* to NZB mice prolonged their survival and decreased the degree of autoimmune phenomena observed when given to castrated male or female animals (Roubinian et al. 1978). Whether these effects are of physiologic significance in man is not known. (See Table 5.)

Estrogen, Progesterone, and Birth Control Pills

Estrogen administration in large doses to rats has also been noted to cause lymphoid tissue atrophy (Sobhon and Jirasattham 1974; vonHaam and Rosenfeld 1942). Skin allograft survival in animals is prolonged following estrogen and progesterone therapy (Waltman et al. 1971). In contrast to testosterone, *in vivo* estradiol administered to guinea pigs reduces not only the rate but also the maximal titer of antibody production in penicillin–hapten-stimulated animals (Feigen et al. 1978). PHA-induced blastogenic responses in human lymphocytes following *in vitro* estradiol administration show a similar inhibitory curve to that seen for androgens (Wyle and Kent 1976). Chronic *in vivo* estradiol causes a reversible reduction in natural killer cell activity of castrated

Table 5. Effects of the Nonglucocorticoid Hormones on the Immune Response

Growth hormone	Appears essential for the maturation and function of thymic-dependent immune responses in the dwarf mouse.
Thyroxine	Stimulatory to PHA blastogenic responses and antisheep erythrocyte PFC production in mature mice. Also appears to affect the maturation of thymic-dependent immune functions in immature mice.
Insulin	Physiologic doses are stimulatory to thymic-derived immune responses in rats.
Androgens	Prolong allograft survival in mice and decrease the PHA blastogenic responses in human lymphocytes *in vitro*.
Estrogens and oral contraceptives	*In vitro* and *in vivo* administration inhibits human lymphocyte PHA blastogenic responses. Also, decreases natural killer-cell activity and prolongs allograft survival in animals *in vivo*.

male and female mice (Seaman et al. 1978). *In vitro* addition of diethylstilbestrol diminished the PHA blastogenic responses of both bovine (Chrisman and Lasley 1975) and human lymphocytes (Ablin et al. 1974a). Lymphocyte blastogenic responses induced by PHA were similarly depressed in men with prostatic carcinoma undergoing diethylstilbestrol treatment (Ablin et al. 1974b) and in women taking several different kinds of birth control pills (Barnes et al. 1972). However, other thymic-dependent functions shown to be enhanced by birth control pills include the numbers of active sheep erythrocyte rosette-forming cells (Sathoh et al. 1977) and dinitrochlorobenzene skin-sensitization responses (Gerrestsen et al. 1975) in women taking a variety of oral contraceptives.

Pregnancy

The mechanisms of alteration of the maternal immune response during pregnancy are incompletely understood at this time. The mixed lymphocyte reaction is depressed in mice during midpregnancy (Fabris et al. 1977), and depression of other *in vivo* and *in vitro* cell-mediated immune functions have been reported (Purtilo et al. 1972; Thong et al. 1973). Although the levels of serum immunoglobulin do not change with pregnancy (Horne et al. 1970), anti-sheep plaque-forming cells (PFC) are increased in pregnant mice (Fabris 1973a), suggesting either a direct effect of the pregnant state on B-lymphocytes or on their regulatory cell populations. Theories to explain why the maternal immune response is not activated against histoincompatible fetal antigens include (1) the trophoblast is somehow able to mask its foreign antigens from the maternal immune system with a mucoprotein barrier (Currie 1968), (2) either fetal antigen-overload or circulating serum-blocking factors could lead to a state of maternal tolerance (Hellstrom et al. 1969), (3) fetal suppressor cells actively inhibit the maternal immune response (Olding and Oldstone 1976), and (4) fetal hormones could suppress actively the maternal immune response (Adcock et al. 1973).

Commercial preparations of human chorionic gonadotropin (HCG), a trophoblastic hormone, have been shown to reversibly depress PHA blastogenic responses of human lymphocytes at *in vitro* concentrations of 100 IU/ml (which is within the maternal serum range during the second semester of pregnancy) (Adcock et al. 1973). In addition, HCG levels of 10,000 IU/ml (conceivable local concentrations at the trophoblast) inhibit lymphocyte blastogenic responses completely (Adcock et al. 1973). Commercially obtained HCG when given *in vivo* to nonpregnant mice was able to depress the MLR responses of those animals and increase their antisheep erythrocyte PFC responses (Fabris et al. 1977). However, other studies have suggested that HCG is not the active inhibitor in the commercial preparations used and that some contaminant of these preparations (likely another placental hormone) may be responsible for the *in vitro* suppression of lymphocyte activity observed (Muchmore and Blaese 1976).

Thyroxine

When adult rats are thyroidectomized, depressions of their lymphocyte PHA blastogenic responses, specific antibody production, and circulating lymphocyte counts have been observed (Piantanelli and Fabris 1978). When thyroid ablation is performed on neonatal rats, these depressions are much more dramatic, and the animals also exhibit greatly delayed allograft rejection (Fabris 1973b). It has been observed in some strains of mice that both serum thyroxine levels and anti-sheep erythrocyte PFC capacity decline with age. Administration of daily thyroxine to these old mice produces a fivefold increase in PFC responses (to the level of young mice) (Fabris 1973b). Similar data in humans with congenital or acquired hypothyroidism are not available.

Insulin

Insulin is believed to stimulate mainly thymic-dependent functions, since alloxan-treated diabetic rats lose most of their PHA blastogenic responses and their MLR capabilities. However, their specific antibody production remains normal (Piantanelli and Fabris 1978). Rats treated with physiologic doses of insulin have enhanced lymphocyte-mediated cytotoxic responses, an effect that may be mediated by intracellular cyclic GMP (Strom et al. 1975).

Conclusion

The purpose of this chapter has been to review the modulating effects of hormones on inflammatory and immune responses. Although it is clear that the majority of the effects of steroid hormones are inhibitory to animal inflammatory and immune responses, the mechanisms of these actions for the glucocorticoids appear to be different for the corticosteroid-resistant and the corticosteroid-sensitive species. Although the sex steroids also inhibit the immune response predominantly, details about the mechanisms of this action or about species differences are still forthcoming. Nonsteroidal hormones seem predominantly to play a stimulatory role in regulating the development and function of the immune response in animals, but data are relatively sparse in this area of human immunobiology.

We have warned the reader about the dangers inherent in the interpretation of animal data as they relate to possible *in vivo* mechanisms in human beings, and we have stressed that the use of suprapharmacologic concentrations of hormones in many *in vitro* studies makes their relevance to *in vivo* mechanisms questionable at best. We must also state that since change in one hormone level can signal change in another hormone level (e.g., the interplay between insulin and growth hormone levels in relation to the serum glucose), and since change in one immune function can signal change in another immune function (e.g., factors that depress cell-mediated immune functions may enhance B-lymphocyte function because of decreased regulatory cell control), the total matrix of how the endocrine system interacts with animal inflammatory processes and immune responses would appear to be quite complex indeed. Since the ability to effect alterations in immunoregulation with just one type of steroid hormone (the glucocorticoids) has had profound effects on the therapy of human inflammatory and immune-mediated disease states (Fauci et al. 1976), we have reason to expect that further understanding of the mechanisms of action of these and other hormones will provide additional insight into the regulatory mechanisms of the immune system and their potential for treating human immunologically mediated diseases.

References

Ablin, R.J., Bruns, G.R., Guinan, P.D., Bush, I.M. (1974a) Diethylstilbestrol exposure and lymphocytic impairment. J. Am. Med. Assoc. 229,1863.

Albin, R.J., Bruns, G.R., Guinan, P., Sadough, N., Bush, I.M. (1974b) Immunosuppressive effect of estrogen on thymic dependent lymphocytic blastogenesis. Urol. Res. 2,69.

Abramson, N., LoBuglio, A.F., Landly, J.H., Cotran, R.S. (1970) The interaction between human monocytes and red cells. Binding characteristics. J. Exp. Med. 132,1191.

Adcock, E.W., Teasdale, F., August, A.S., Cox, S., Meschia, G., Battaglia, F.C., Naughton, M.A. (1973) Human chorionic gonadotropin: its possible role in maternal lymphocyte suppression. Science 181,845.

Allison, F., Adcock, M.H. (1965) Failure of pretreatment with glucocorticoids to modify the phagocytic and bactericidal capacity of human leukocytes for encapsulated type I pneumococcus. J. Bacteriol. 89,1256.

Anderson, V. (1969) Autoradiographic studies of eosinophil kinetics: effects of cortisol. Cell Tissue Kinet. 2,139.

Athens, J.W., Haab, D.P., Raab, S.O., Mavez, A.M., Ashenbucker, H., Cartwright, G.E., Wintrobe, M.M. (1961) Leukokinetic studies. IV. The total blood circulating and marginal granulocyte pools and the granulocyte turnover rate in normal subjects. J. Clin. Invest. 40,989.

Atkinson, J.P., Schreiber, A.D., Frank, M.M. (1973) Effect of cortisone therapy on serum complement components. J. Immunol. 111,1061.

Balow, J.E., Rosenthal, A.S. (1973) Glucocorticoid suppression of macrophage migration inhibitory factor. J. Exp. Med. 137,1031.

Balow, J.E., Hurley, D.L., Fauci, A.S. (1975) Immunosuppressive effects of glucocorticosteroids: Differential effects of acute versus chronic administration on cell mediated immunity. J. Immunol. 114,1072.

Balow, J.E., Parillo, J.E., Hunninghake, G.W., Fauci, A.S. (1976) Mechanisms of corticosteroid suppression of cytotoxic responses to alloantigens. Proc. Dialysis Transplant. Forum 6,27.

Balow, J.E., Hunninghake, G.W., Fauci, A.S. (1977) Corticosteroids in human lymphocyte-mediated cytotoxic reactions. Effects on the kinetics of sensitization and on the cytolytic capacity of effector lymphocytes in vitro. Transplantation 23,322.

Barnes, E.W., MacCuish, A.C., Loudon, N.B., Jordan, J., Irvine, W.J. (1972) Depressed lymphocyte response to PHA in women taking oral contraceptives. Lancet 1,1185.

Bartke, A. (1964) Histology of the anterior hypophysis, thyroid, and gonads of two types of dwarf mice. Anat. Rec. 149,225.

Baxter, J.D., Forsham, P.H. (1972) Tissue effects of glucocorticoids. Am. J. Med. 53,573.

Baxter, J.D., Harris, A.W. (1975) Mechanism of glucocorticoid action: general features, with references to steroid mediated immunosuppression. Transplant. Proc. 7,55.

Bendtzen, K. (1975) Drug effects on human leucocyte migration and migration inhibitory activity from lymphocytes stimulated with concanavalin A. Acta Pathol. Microbiol. Scand. (Sec. C.) 14,267.

Boggs, D.R., Athens, J.W., Cartwright, G.E., Wintrobe, M.M. (1964) The effect of adrenal glucocorticosteroids upon cellular composition of inflammatory exudates. Am. J. Pathol. 44,763.

Branceni, D., Arnason, B.G. (1966) Thymic involution and recovery: Immune responsiveness and immunoglobulins after neonatal prednisolone in rats. Immunology 10,35.

Butler, W.T., Rossen, R.D. (1973) Effects of corticosteroids on immunity in man. I. Decreased serum IgG concentration caused by 3 or 5 days of high doses of methylprednisolone. J. Clin. Invest. 52,2629.

Castro, J.E. (1974a) Orchiectomy and the immune response. I. Effect of orchiectomy on lymphoid tissues in mice. Proc. R. Soc. Lond. (Biol.) 185,425.

Castro, J.E. (1974b) Orchiectomy and the Immune response. II. Response of orchiectomized mice to antigens. Proc. R. Soc. Lond. (Biol.) 185,437.

Cerottini, J.C., Brunner, K.T. (1974) Cell-mediated cytotoxocity, allograft rejection, and tumor immunity. Adv. Immunol. 18,67.

Chrisman, C.L., Lasley, J.F. (1975) Effect of diethylstilbesterol diphosphate on mitotic activity in bovine lymphocyte cultures. Cytologia 40,817.

Claman, H.N. (1972) Corticosteroids and lymphoid cells. N. Engl. J. Med. 287,388.

Claman, H.N. (1975) How corticosteroids work. J. Allergy Clin. Immunol. 55,145.

Claman, H.N., Moorehead, J.W., Benner, W.H. (1971) Corticosteroids and lymphoid cells in vitro. I. Hydrocortisone lysis of human, guinea pig, and mouse thymus cells. J. Lab. Clin. Med. 18,499.

Cline, M.J., Melmon, K.L. (1966) Plasma kinins and cortisol: a possible explanation of the anti-inflammatory action of cortisol. Science 153,1135.

Cooper, D.A., Petts, V., Luckhurst, E., Penny, R. (1977) The effect of acute and prolonged administration of prednisolone and ACTH on lymphocyte subpopulations. Clin. Exp. Immunol. 28,267.

Crofton, R.W., Diesselhoff-Den Dulk, M.M.L., vanFurth, R. (1978) The origin, kinetics, and characteristics of the Kupffer cells in the normal steady state. J. Exp. Med. 148,1.

Currie, G.A. (1968) The immunology of pregnancy: the foetal-maternal barrier. Proc. R. Soc. Med. 61,1206.

Dale, D.C., Petersdorf, R.G. (1973) Corticosteroids and infectious diseases. Med. Clin. North Am. 57,1277.

Dale, D.C., Fauci, A.S., Wolff, S.M. (1974) Alternate day prednisone leukocyte kinetics and susceptibility to infections. N. Engl. J. Med. 291,1154.

Dale, D.C., Hubert, R.T., Fauci, A.S. (1976) Eosinophil kinetics in the hypereosinophilic syndrome. J. Lab. Clin. Med. 87,487.

Duquesnoy, R.J. (1972) Immunodeficiency of the thymus-dependent system of the Ames dwarf mouse. J. Immunol. 108,1578.

Duquesnoy, R.J., Rodey, G.E., Holmes, B.H., Good, R.A. (1969) Studies of the relation of the thymus-dependent lymphoid system and the pituitary gland. Fed. Proc. 28,376.

Ebert, R.H., Barclay, W.R. (1952) Changes in connective tissue reaction induced by cortisone. Ann. Intern. Med. 37,506.

Epstein, W.L. (1967) Granulomatous hypersensitivity. Prog. Allergy 11,36.

Fabris, N. (1973a) Immunological reactivity during pregnancy in the mouse. Experientia 29,610.

Fabris, N. (1973b) Immunodepression in thyroid deprived animals. Clin. Exp. Immunol. 15,601.

Fabris, N., Pierpaoli, W., Sorkin, E. (1971a) Hormones and the immunological capacity. III. The immunodeficiency diseases of the hypopituitary Snell-Bragg dwarf mouse. Clin. Exp. Immunol. 9,209.

Fabris, N., Pierpaoli, W., Sorkin, E. (1971b) Hormones and the immunologic capacity. IV. Restorative effects of developmental hormones or of lymphocytes on the immunodeficiency syndrome of the dwarf mouse. Clin. Exp. Immunol. 9, 227.

Fabris, N., Pierpaoli, W., Sorkin, E. (1972) Lymphocytes, hormones and aging. Nature 240,557.

Fabris, N., Piantanelli, C., Muzzioli, M. (1977) Differential effect of pregnancy or gestagens on humoral and cell-mediated immunity. Clin. Exp. Immunol. 28,306.

Fauci, A.S. (1975) Corticosteroids and circulating lymphocytes. Transplant. Proc. 7,37.

Fauci, A.S., Dale, D.C. (1974) The effect of *in vivo* hydrocortisone on subpopulations of human lymphocytes. J. Clin. Invest. 53,240.

Fauci, A.S., Dale, D.C. (1975a) Alternate-day prednisone therapy and human lymphocyte subpopulations. J. Clin. Invest. 55,22.

Fauci, A.S., Dale, D.C. (1975b) Effect of hydrocortisone on the kinetics of normal human lymphocytes. Blood 46,235.

Fauci, A.S., Dale, D.C., Balow, J.E. (1976) Glucocorticosteroid therapy: mechanisms of action and clinical considerations. Ann. Intern. Med. 84,304.

Fauci, A.S., Pratt, K.R., Whalen, G. (1978) Activation of human B lymphocytes. IV. Regulatory effects of corticosteroids on the triggering signal in the PFC response of human peripheral blood B lymphocytes to polyclonal activation. J. Immunol. 119,598.

Feigen, G.A., Fraser, R.C., Peterson, N.S. (1978) Sex hormones and the immune response. II. Perturbation of antibody production by estradiol 17B. Int. Archs. Allergy Appl. Immunol. 57,488.

Fleer, A., Van Schaik, M.K.T., Von Dem Borne, A.E.G., Englefriet, C.P. (1978) Destruction of sensitized erythrocytes by human monocytes *in vitro* effects of cytochalasin B, hydrocortisone and colchicine. Scand. J. Immunol. 8,515.

Gabrielson, A.E., Good, R.A. (1967) Chemical suppression of adaptive immunity. Adv. Immunol. 6,91.

Gerretsen, G., Bleumink, E., Kremer, J., Nater, J.P. (1975) Dinitrochlorobenzene sensitization test in women on hormonal contraceptives. Lancet 1,347.

Glasser, L., Heustis, D.W., Jones, J.F. (1977) Functional capabilities of steroid-recruited neutrophils harvested for clinical transfusion. N. Engl. J. Med. 297,1033.

Granelli-Piperno, A., Vassalli, J.D., Reich, E. (1977) Secretion of plasminogen activator by human polymorphonuclear leukocytes. Modulation by glucocorticoids and other effectors. J. Exp. Med. 146,1693.

Gupta, S., Good, R.A. (1977) Subpopulations of human T lymphocytes. II. Effects of thymopoietin, corticosteroids, and irradiation. Cell Immunol. 34,10.

Hallahan, C., Young, D.A., Munck, A. (1973) Time course of early events in the action of glucocorticoids on rat thymus cells *in vitro*. J. Biol. Chem. 248,2922.

Haynes, B.F., Fauci, A.S. (1978) The differential effect of *in vivo* hydrocortisone on the kinetics of subpopulations of human peripheral blood T lymphocytes. J. Clin. Invest. 61,703.

Hellstrom, K.E., Hellstrom, K., Brown, J. (1969) Abrogation of cellular immunity to antigenically foreign mouse embryonic cells by a serum factor. Nature 224,914.

Horne, C.H., Howie, P.W., Wein, R.J., Goidie, R.B. (1970) Effect of combined estrogen-progestogen oral contraceptives on serum levels of 2-macroglobulin, transferrin albumin and IgG. Lancet 1,49.

Hsu, H.S. (1969) Cellular basis of cortisone-induced host susceptibility to tuberculosis. Am. Respir. Dis. 100,677.

Hunninghake, G.W., Fauci, A.S. (1977a) Immunologic reactivity of the lung. III. Effects of corticosteroids on alveolar macrophage cytotoxic effector cell function. J. Immunol. 118,146.

Hunninghake, G.W., Fauci, A.S. (1977b) Immunologic reactivity of the lung. VII. Effect of corticosteroids and cyclophosphamide on the Fc receptor function of alveolar macrophages. Cell. Immunol. 32,228.

Ignarro, L.J. (1977) Glucocorticoid inhibition of nonphagocytic discharge of lysosomal enzymes from human neutrophils. Arthritis Rheum. 20,73.

Ilfeld, D.N., Krakauer, R.S., Blaese, R.M. (1977) Suppression of the autologous mixed lymphocyte reaction by physiologic doses of hydrocortisone. J. Immunol. 119,428.

Ingle, D.J. (1938) Atrophy of the thymus in normal and hypophysectomized rats following the administration of cortin. Proc. Soc. Exp. Biol. Med. 38,443.

Kahn, C.R., Mesyesi, K., Bar, R.S., Eastman, R.C., Flier, J.S. (1977) Receptors for peptide hormones. New insight into the pathophysiology of disease states in man. Ann. Intern. Med. 86,205.

Kellermeyer, R.W., Graham, R.C. (1968) Kinins: possible pathophysiologic and pathologic roles in man. N. Engl. J. Med. 279,859.

Lefer, A.M., Inge, T.F. (1974) Lack of interaction between glucocorticoids and the kallikrein-kinin system. Proc. Soc. Exp. Biol. Med. 145,658.

Lippman, M.E., Barr, R. (1977) Glucocorticoid receptors in purified subpopulations of human peripheral blood lymphocytes. J. Immunol. 118,1971.

Lippman, M.E., Halterner, R.H., Leventhal, B.G., Perry, S., Thompson, E.B. (1973) Glucocorticoid-binding proteins in human acute lymphoblastic leukemia blast cells. J. Clin. Invest. 52,1715.

Lundgren, G. (1970) *In vitro* cytotoxicity by human lymphocytes from individuals immunized against histocompatibility antigens. I. Kinetics and specificity of the reactions. Influence of metabolic inhibitors and anti-lymphocyte serum. Clin. Exp. Immunol. 6,664.

MacGregor, R.R., Spagnuolo, P.J., Lentnek, A.L. (1974) Inhibition of granulocyte adherence by ethanol, prednisone, and aspirin measured with an assay system. N. Engl. J. Med. 291,642.

Mandell, G.L., Rubin W., Hook, E.W. (1970) The effect of an NADA oxidase inhibitor (hydrocortisone) on polymorphonuclear leukocyte bactericidal activity. J. Clin. Invest. 49,1381.

Moretta, L., Webb, S.R., Grossi, L.E., Lydyand, P.W., Cooper, M.D. (1977) Functional analysis of two human T cell subpopulations: help and suppression of B cell responses by T cells bearing receptors for IgM or IgG. J. Exp. Med. 146,184.

Muchmore, A.V., Blaese, R.M. (1976) Immunoregulatory properties of fractions from human pregnancy urine: Evidence that human chorionic gonadotropin is not responsible. J. Immunol. 116,881.

Munck, A., Brinck-Johnson, T. (1968) Specific and non-specific physiochemical interactions of glucocorticoids and related steroids with rat thymus cells *in vitro*. J. Biol. Chem. 243,5556.

Neifeld, J.P., Lippman, M.E., Tormey, D.C. (1977) Steroid hormone receptors in normal human lymphocytes. Induction of glucocorticoid receptor activity by phytohemagglutinin stimulation. J. Biol. Chem. 252,2922.

Olding, L.B., Oldstone, M.B.A. (1976) Thymus-derived peripheral lymphocytes from human newborns inhibit division of their mothers lymphocytes. J. Immunol. 116,682.

O'Malley, B.W., Buller, R.E. (1977) Mechanisms of steroid hormone action. J. Invest. Dermatol. 68,1.

Parrillo, J.E., Fauci, A.S. (1978a) Comparison of the effector cells in human spontaneous cellular cytotoxicity and antibody-dependent cellular cytotoxicity: Differential sensitivity of the effector cells to *in vivo* and *in vitro* corticosteroids. Scand. J. Immunol. 8,99.

Parrillo, J.E., Fauci, A.S. (1978b) Mechanisms of corticosteroid action on lymphocyte subpopulations. III. Differential effects of dexamethasone administration on subpopulations of effector cells mediating cellular cytotoxicity in man. Clin. Exp. Immunol. 31,116.

Parrillo, J.E., Fauci, A.S. (1979) Mechanisms of glucocorticoid action on immune processes. Annu. Rev. Pharmacol. Toxicol. 19,179.

Pelletier, M., Montplaisiros, Dardenne, M., Back, J.F. (1976) Thymic hormone activity and spontaneous autoimmunity in dwarf mice and their litter mates. Immunology 30,783.

Persellin, R.H., Ku, L.C. (1974) Effects of steroid hormones on human polymorphonuclear leukocyte lysosomes. J. Clin. Invest. 54,919.

Piantanelli, L., Fabris, N. (1978) Hypopituitary dwarf and athymic nude mice and the study of the relationships among thymus, hormones and aging. Birth Defects 14,315.

Purtilo, D.T., Hallgren, H.M., Yunis, E.J. (1972) Depressed maternal lymphocyte response to phytohemagglutin in humans. Lancet 1,769.

Rhinehart, J.J., Balcerzak, S.P., Sagone, A.L., LoBuglio, A.F. (1974) Effects of corticosteroids on human monocyte function. J. Clin. Invest. 54,1337.

Rhinehart, J.J., Sagone, A.L., Balcerzak, S.P., Ackerman, G.A., LoBuglio, A.F. (1975) Effects of corticosteroid therapy on human monocyte function. N. Engl. J. Med. 292,235.

Roubinian, J.R., Talal, N., Greespan, J.S., Goodman, J.R., Siiteri, P.K. (1978) Effect of castration and sex hormone treatment on survival, antinucleic acid antibodies and glomerulonephritis in NZB/NZW F_1 mice. J. Exp. Med. 147,1568.

Ruhl, H., Vogt, W., Bochert, G., Schmidt, S., Moelle, R., Schaova, H. (1974) Effect of L-asparaginase and hydrocortisone on human lymphocyte transformation and production of a mononuclear leukocyte chemotactic factor *in vitro*. Immunology 26,989.

Rusu, V.M., Cooper, M.D. (1975) *In vivo* effects of cortisone on the B cell line in chickens. J. Immunol. 115,1370.

Sampson, D., Groteluescher, C., Kauffjan, H.M. (1975) The human splenic suppressor cell. Transplantation 20,362.

Satoh, P.S., Fleming, W.E., Johnston, K.A., Ozmun, J.M. (1977) Active E-rosette formation in women taking oral contraceptives. N. Engl. J. Med. 296,54.

Seaman, W.E., Blackman, M.A., Gindlart, T.D., Roubinian, J.R., Loeb, J.M., Talal, N. (1978) B-estradiol reduces natural killer cells in mice. J. Immunol. 121,2193.

Seebohm, P.M., Tremaine, M.M., Jeter, W.S. (1954) The effect of cortisone and adrenocorticotrophic hormone on passively transferred delayed hypersensitivity to 2, 4 dinitrochlorobenzene in guinea pigs. J. Immunol. 121,2193.

Sherman, N.A., Smith, R.S., Middleton, E., Jr. (1973) Effect of adrenergic compounds, aminophylline and hydrocortisone on *in vitro* immunoglobulin synthesis by normal human peripheral lymphocytes. J. Allergy Clin. Immunol. 52,13.

Sobhon, P., Jirasattham, C. (1974) Effect of sex hormones on the thymus and lymphoid tissues of ovariectomized rats. Acta Anat. 89,211.

Strom, T.B., Bear, R.A., Carpenter, C.B. (1975) Insulin-induced augmentation of lymphocyte-mediated cytotoxicity. Science 187,1206.

Thompson, J., van Furth, R. (1970) The effect of glucocorticosteroids on the kinetics of mononuclear phagocytes. J. Exp. Med. 131,429.

Thong, Y.H., Steele, R.W., Vincent, M.M., Hensen, S.A., Bellanti, J.A. (1973) Impaired *in vitro* cell-mediated immunity to rubella virus during pregnancy. N. Engl. J. Med. 289,604.

Thong, Y.H., Hersen, S.A., Vincent, M.M., Rola-Pleszcyzynski, M., Walser, J.B., Bellanti, J.A. (1975) Effect of hydrocortisone on *in vitro* cellular immunity to viruses in man. Clin. Immunol. Immunopathol. 34,363.

Thorn, G.W. (1973) The adrenal cortex: reflections, progress and speculations. Trans. Assoc. Am. Physicians 86,65.

Tolentino, P. (1975) Androgens and antibody formation. Pharmacol. Ther. 1,209.

Tuchinda, M., Newcomb, R.W., DeVald, B.L. (1972) Effect of prednisone treatment on the human immune response to keyhole limpet hemocyanin. Int. Arch. Allergy Appl. Immunol. 42,533.

van Furth, R., Jones, T.C. (1975) Effect of glucocorticosteroids on phagosome–lysosome interaction. Infect. Immun. 12,888.

vanZwet, T.L., Thompson, J., van Furth, R. (1975) Effect of glucocorticosteroids on the phagocytosis and intracellular killing by peritoneal macrophages. Infect. Immun. 12,699.

Vassalli, J.D., Hamilton, J., Reich, E. (1976) Macrophage plasminogen activator: Modulation of enzyme production by anti-inflammatory steroids, mitotic inhibitors, and cyclic nucleotides. Cell 8,271.

Vischer, T.L. (1972) Effect of hydrocortisone on the reactivity of thymus and spleen cells of mice to *in vitro* stimulation. Immunology 23,777.

vonHaam, E., Rosenfeld, I. (1942) The effect of estrone on antibody production. J. Immunol. 43,109.

Wahl, S.M. (1975) Corticosteroid inhibition of chemotactic lymphokine production by T and B lymphocytes. Ann. N.Y. Acad. Sci. 256,375.

Waldmann, T.A., Broder, S., Krakauer, R., MacDermott, R.P., Durm, M., Goldman, C., Mead, B. (1976) The role of suppressor cells in the pathogenesis of common variable hypogammaglobulinemia and the immunodeficiency associated with myeloma. Fed. Proc. 35,2067.

Waltman, S.R., Burde, R.M., Berrios, J. (1971) Prevention of corneal homograft rejection by estrogens. Transplantation 11,194.

Ward, P.A. (1966) The chemosuppression of chemotaxis. J. Exp. Med. 124,209.

Weissman, G. (1967) The role of lysosomes in inflammation and disease. Annu. Rev. Med. 18,97.

Weissman, G., Thomas, L. (1973) Studies on lysosomes. II. The effect of cortisone on the release of acid hydrolases from a large granule fraction of rabbit liver induced by an excess of Vitamin A. J. Clin. Invest. 42, 661.

Werb, Z. (1978) Biochemical actions of glucocorticoids on macrophages in culture. J. Exp. Med. 147,1695.

Weston, W.L., Mandel, M.J., Yeckley, J.A., Krueger, G.G., Claman, H.N. (1973a) Mechanism of cortisol inhibition of adoptive transfer of tuberculin sensitivity. J. Lab. Clin. Med. 82,366.

Weston, W.L., Claman, H.N., Krueger, C.G. (1973b) Sites of action of cortisol in cellular immunity. J. Immunol. 110,880.

Wiener, E., Marmary, Y., Curelaru, Z. (1972) The *in vitro* effect of hydrocortisone on the uptake and intracellular digestion of particulate matter by macrophages in culture. Lab. Invest. 26,220.

Williams, T.W., Granger, G.A. (1969) Lymphocyte *in vitro* cytotoxicity: correlation of derepression with release of lymphotoxin from human lymphocytes. J. Immunol. 103,170.

Wright, D.G., Malawister, S.E. (1973) Mobilization and extracellular release of granular enzymes from human leukocytes during phagocytosis: inhibition by colchicine and cortisol but not by salicylate. Arthritis Rheum. 16,749.

Wyle, F.A., Kent, J.R. (1976) Immunosuppression by sex steroid hormones. I. The effect upon PHA and PPD stimulated lymphocytes. Clin. Exp. Immunol. 27,407.

Yu, D.T.Y., Clements, P.J., Paulus, H.E., Peter, J.B., Levy, J., Barnett, E.V. (1974) Human lymphocyte subpopulations. Effects of corticosteroids. J. Clin. Invest. 53,565.

Zurier, R.B., Weissman, G. (1973) Anti-immunologic and anti-inflammatory effects of steroid therapy. Med. Clin. North. Am. 57,1295.

Zweiman, B., Slott, R.I., Atkins, P.C. (1976) Histologic studies of human skin test responses to ragweed and compound 48/80. III. Effects of alternate-day steroid therapy. J. Allergy Clin. Immunol. 58,657.

Inflammation and Wound Healing

Sharon M. Wahl, Ph.D.

Wound healing involves the participation of several different kinds of cells that arrive sequentially in the tissue to remove foreign material and debris, eliminate bacteria, and repair injury (Bryant 1977; Menaker 1975; Ross 1968). Since people do not have the ability to regenerate a lost limb or duplicate injured tissue, as do some lower vertebrates, we must depend on the process of wound healing. It is this process of epithelial, endothelial, and fibroblast proliferation with the concomitant production of collagen that substitutes for our inability to regenerate injured tissue. Wound repair, which does not vary significantly from one type of tissue to another, is also similar whether it is induced mechanically or secondary to an inflammatory response. The injury may be the consequence of bacteria, foreign agents, trauma, excess heat, cold, chemical stimuli, or other agents, but the inflammation and tissue repair proceed along similar pathways. Most of the cellular and physiologic mechanisms involved in wound healing are identical to those of other inflammatory responses, immune responses, neoplasia, or rejection of transplanted tissues. For example, the insult in an infection is bacterial, whereas wound healing occurs generally in response to the mechanical trauma of injury; yet the host responses to both are remarkably similar. Bacterial contamination is also a frequent complication of wounds and leads to a cellular and vascular response to the bacterial-induced injury with repair following the destruction and removal of bacteria and necrotic tissue. A comparable cellular and vascular response occurs in wound repair. Since inflammatory and immunological processes are involved intimately in wound healing it provides a useful model for delineating their contribution to restoring host integrity.

Whenever tissues or cells are injured, soluble chemical mediators are released which initiate the inflammatory process. First (within minutes), arteriolar dilatation occurs, followed by an increased rate of blood flow through the microvasculature. The increased blood flow is responsible for bringing large numbers of leukocytes, plasma proteins, oxygen, and nutrients to the affected area. Coincidentally, the increased blood flow can carry away larger amounts of toxic products so they will not accumulate as readily

From the Laboratory of Microbiology and Immunology, National Institute of Dental Research, National Institutes of Health, Bethesda, Maryland.

in the tissue. As capillary dilatation occurs, vascular permeability increases, resulting in an exudation of fluid (edema) from the vessels to the tissue. The vasodilatation is responsible for the redness (rubor) of inflammation. Escaping plasma fluid and proteins (albumin, fibrinogen, antibodies, etc.) accumulate in the tissues, concentrating the red cells remaining in the capillaries. This packing of red cells slows the blood flow, and the white cells begin to adhere to the endothelial cells of the vessel wall and commence migration from the vessels to the inflammatory locus. Neutrophils are the first to emigrate, followed by the monocytes and, later, the lymphocytes. These inflammatory cells are then responsible for removing foreign material and debris. Enzymes released by the leukocytes digest foreign protein substances and dead cells. The inflammatory cells may also mediate subsequent repair of the injured tissue. In an acute inflammatory response, lymphocytes may play a minor role as they emigrate slowly and may not appear at the site until after the acute stage has passed. However, if the inflammation is antigen induced or chronic in nature, the lymphoid cells play a prominent role. Thus, the identification of an acute inflammatory reaction requires the demonstration of vascular dilatation, fluid exudation, and leukocyte accumulation. From this stage, the inflammation may progress or become resolved, depending on the nature of the injurious agent, the presence or absence of bacteria, and other host conditions. Following injury, the host attempts to protect itself and the inflammatory process provides the defensive elements.

While the phases of wound healing are interrelated and interdependent, it is expedient to categorize the healing process into four stages: inflammation, epithelialization, vascular proliferation, and fibroplasia and repair.

Inflammation

Early Transudative Phase

The skin wound provides a simple model for studying the reparative process (Figure 1) and does not differ qualitatively from the healing of a defect caused by trauma, excision, or a burn. Initially, blood flows into the opening created by the cut and clots with the formation of fibrin. This extravascular fibrin unites the edges of the wound and provides an early seal that serves as a framework for the migration of inflammatory cells. As the clot loses fluid, it forms the scab of dried fibrin and cell fragments that protect the wound. Hemorrhage is followed by an immediate capillary constriction, then by capillary dilatation and increased permeability, the consequences of the release of vasoactive agents such as histamine, bradykinin, serotonin, and prostaglandins (see Chapters 12, 13, and 15). The increased permeability leads to transudation of plasma proteins, complement, and biologically active polypeptides, some of which may play a role in chemotactic attraction of inflammatory cells. Furthermore, this transudate provides all the essential nutrients for cell growth, as well as for the growth of any invading bacteria.

Cellular Phase

Neutrophils. The process of emigration of leukocytes from blood vessels was described as early as 1843 (Harris 1960). A localized slowing of the rate of blood flow in the dilated vessels occurs with the margination of the leukocytes to the vascular

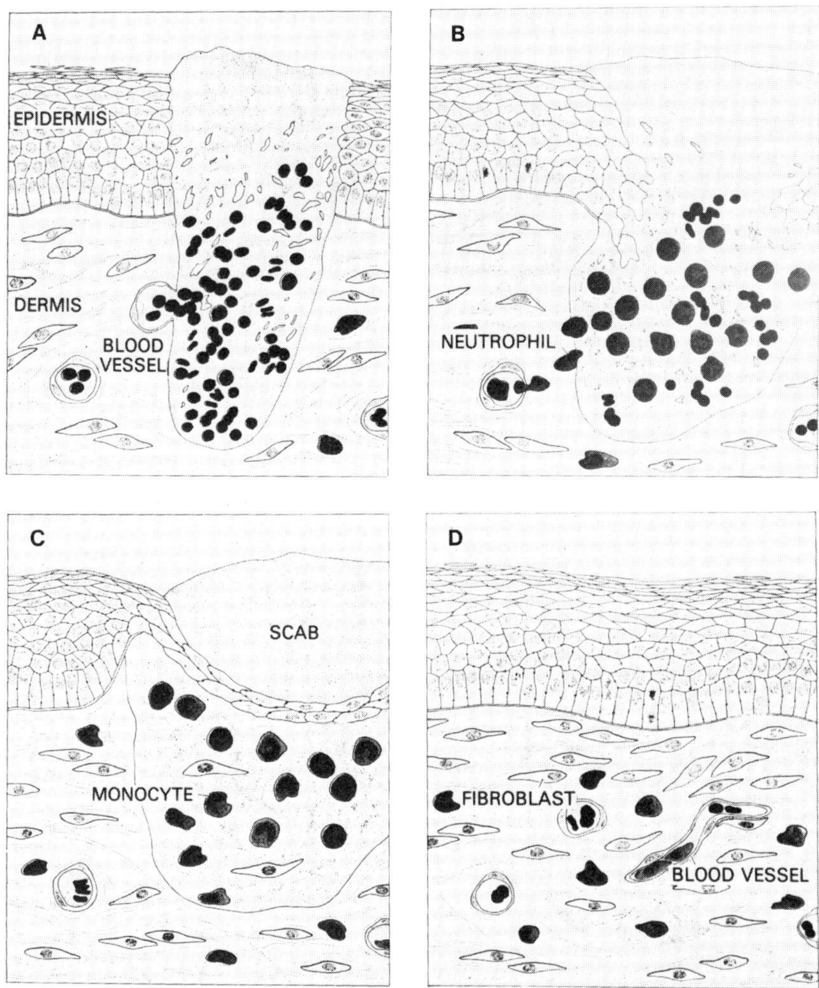

Figure 1. (A) At the time of wounding gap fills with blood from broken vessels and cellular debris from the surface (epidermal) and subsurface (dermal) layers. Blood contains red cells and fibrinogen molecules, which form into strands that unite the wound's edges. Cells of the dermis called fibroblasts surround the wound. (B) One day later neutrophils enter the wound from surrounding blood vessels and begin to ingest bacteria and debris. The epidermal cells, held together by desmosomes, have also intruded. Other epidermal cells begin to reproduce *(note dividing nuclei)* so that those that migrate into the wound will be replaced. (C) Two days later wedges of epidermis have met under the scab, and the next generation of inflammatory cells, called monocytes, have entered the wound to complete the scavenging process. Meanwhile, the fibroblasts that make the collagen and other connective-tissue proteins of the dermis have migrated into the defect. (D) Seven days later the scab has sloughed from the restored epidermis. A few monocytes and neutrophils remain in the wound. At this stage the fibroblasts actively make collagen and other dermal-tissue proteins. The new connective tissue is denser than the original, unwounded dermis and is permeated by blood vessels. (From Ross 1968 with permission.)

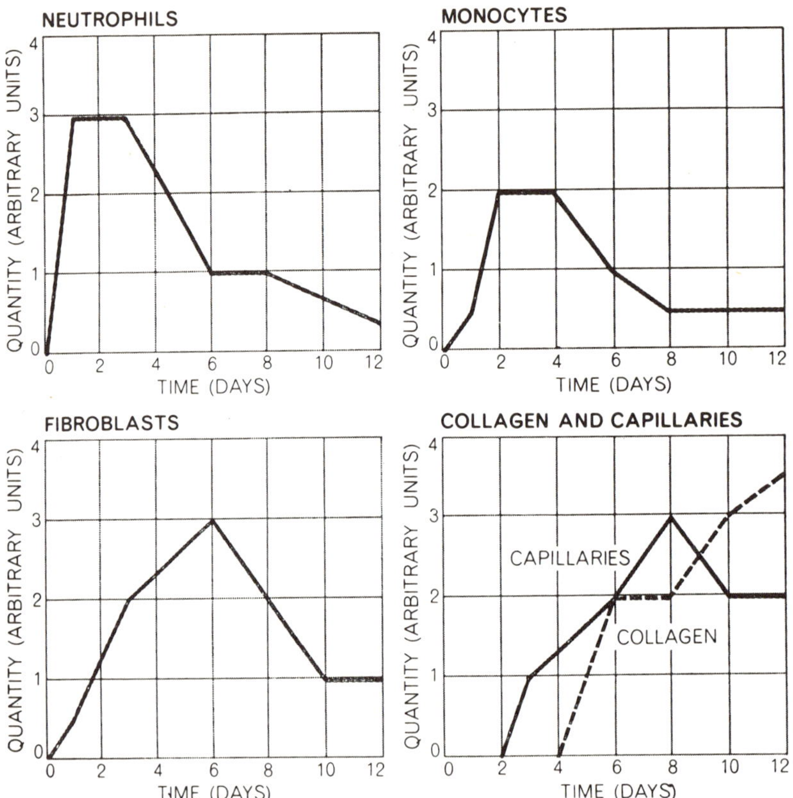

Figure 2. Kinetics of appearance of the major cell populations that occupy the wound. The stages overlap one another and the arrival of each cell population may be dependent upon the activities of the preceding cells. From Ross (1968).

endothelium, primarily in the postcapillary venules. The leukocytes insert pseudopodia between the endothelial cells and force their way through the extracellular space, penetrating the extravascular space through the basement membrane. The mechanism of this traversal through the endothelium remains somewhat of a mystery, because passage occurs rapidly (within 2 to 8 minutes) without any visible openings in the endothelium. Cochrane and Aikin (Cochrane and Aikin 1966; Cochrane 1968) have studied neutrophil migration from the vasculature in immune-complex–mediated injury and found that passage of neutrophils through the basement membranes followed preliminary destruction of this connective tissue barrier by extracellularly secreted neutrophil enzymes. Collagenase, elastase, cathepsin D and E, and neutral proteases are leukocyte enzymes effective in degrading the proteins found in the basement membrane.

The neutrophils, a migrating phagocytic population, are the first inflammatory cells to enter the area of injury (Figures 1 and 2). A number of agents appear capable of initiating this cellular egress from the vasculature. The mechanisms of this neutrophil movement have been well documented and reviews are available (Wilkinson 1974; Gallin and Quie 1978). Numerous chemotactic molecules (i.e., capable of initiating directed cellular migration) have been identified primarily in *in vitro* assays. Hageman-factor activation, which occurs during the coagulation process, leads to the formation of kallikrein and plasminogen activator, both of which stimulate chemotaxis of leukocytes (Kaplan et al. 1972, 1973). Activation of the complement system (see Chapter 14) by inflammatory agents such as immune complexes, bacterial-derived endotoxins, and aggregating platelets, as well as proteases, results in the cleavage of certain complement components with the release of low molecular weight peptides, including C5a, chemotactic for leukocytes.

Within 4 to 6 hours significant numbers of neutrophils are present in the wound site (Figure 2). These cells are important in the phagocytosis (Stossel 1974), and elimination of bacteria that are present in nearly every wound. In addition, the release of stored enzymes is important in the digestion of extracellular debris and also may potentiate the inflammatory response (see Chapter 3). However, healing can occur in the absence of neutrophils, since the normal sequence of inflammatory events is not altered when an experimental animal is made neutropenic with antineutrophil serum and then wounded (Simpson and Ross 1972). Thus, these cells do not appear to be obligatory precursors of the remainder of the cellular inflammatory response.

Macrophages. After the neutrophils appear, macrophages begin to arrive about 12 hours after wounding, and maximal numbers appear at 48 to 72 hours. These cells are subject to many of the same chemotactic stimuli as the neutrophils (Snyderman and Mergenhagen 1976), and these stimuli presumably cause monocytes to leave the circulation and migrate into the tissues. Additional stimuli for monocyte migration from the vessels into the surrounding tissue may include collagen and its cleavage products, which are generated by neutrophil digestion of the basement membrane and damaged collagen in the injured tissue (Postlethwaite and Kang 1976). Lymphokines with chemotactic properties (see Chapter 10) produced by lymphocytes exposed to foreign material or antigenically altered damaged tissue components in the wound may also influence the migration of these mononuclear cells. The macrophages are phagocytic cells with intracellular digestive properties that promote wound healing by eliminating bacteria, tissue and neutrophil debris, and fibrin deposits. In contrast to the neutrophils, however, the outcome of inflammatory lesions depends to a considerable extent on the participation of macrophages. Leibovich and Ross (1975) demonstrated delayed wound healing in animals depleted of macrophages by anti-macrophage serum and corticosteroids. Not only was wound debridement inhibited, but subsequent fibrosis was also delayed.

Once macrophages arrive at the injured site, they must recognize what to attack. For example, these cells will ingest aged or damaged erythrocytes but do not phagocytize normal erythrocytes. Recognition is antibody dependent and is mediated through the macrophage receptors for Fc components of immunoglobulins (Huber and Fudenberg 1968) and for the complement component, C3b (Lay and Nussenzweig 1968). Particles opsonized by antibody or complement or both are more readily recognized and become available for ingestion at the macrophage surface.

Macrophages become activated by numerous agents normally present in the area of a wound, including bacterial endotoxins and lymphokines, as well as by phagocytosis. Activation leads to an increase in adhesiveness, phagocytic activity, and in the macrophages' content of hydrolytic enzymes (Table 1), which are essential in the digestion and transport of cellular and organic debris from the wound site. Fibrin and

Table 1. Enhanced Functions of Activated Macrophages

Adherence and spreading	Secretion of neutral proteases
Transport and metabolism of glucose	Plasminogen activator
	Elastase
Respiratory activity	Collagenase
Phagocytosis	
Phagolysosomes	Monokine synthesis
	Lymphocyte-activating factor
Prostaglandin synthesis	Fibroblast chemotactic factor
Acid hydrolase activity	Fibroblast-activating factor
Cathepsins	Colony-stimulating factor
Amylase	Interferon
Lipase	
β-glucuronidase	
Peroxidase	
Ribonuclease	
Deoxy-ribonuclease	
Acid phosphatase	
Lysozyme	

dead cells are the inevitable result of wounding and must be digested and absorbed by cellular enzymatic and lysosomal action. Once particles have been ingested they become envacuolated into phagosomes, where they are digested by the plethora of hydrolytic enzymes identified within these cells (Table 1). Furthermore, the macrophages have the capacity to synthesize new enzymes and new surface membranes containing receptors so that phagocytosis can resume. Activated macrophages also produce large amounts of plasminogen activator (Unkeless et al. 1974), which cleaves plasminogen enzymatically to form plasmin. Plasmin, in turn, degrades the large quantities of fibrin in the area of tissue damage and also acts on other substrates, including Hageman factor and complement components, to generate other inflammatory mediators.

One of the important leukocyte enzymes in injured tissue repair is collagenase. Since 30% of the body is composed of collagen it is highly likely that damaged collagen will occur in most injuries. Removal of damaged connective tissue is essential before the laying down of new collagen in the repair of the wound. The enzyme collagenase is the only known enzyme that can make the initial attack on the helical portion of the collagen molecule under physiological conditions. Thus, collagenase is believed to play an essential role in the degradation of this protein. Following this break, the molecule becomes susceptible to further degradation by nonspecific proteases. Increased levels of collagenase have been found in inflamed tissues, including healing wounds in which connective tissue has been destroyed. Neutrophils, which predominate early in acute inflammatory lesions, produce collagenase. However, at later stages when relatively few neutrophils are present and when collagen breakdown is most evident, macrophages are present in large numbers; these cells have been shown to synthesize collagenase. The mechanism of induction and regulation of collagenase production by activated macrophages has been investigated by Wahl and colleagues (1974, 1975, 1977). Macrophages must be activated to produce and release this enzyme, since nonstimulated cells neither store the enzyme nor release it into culture medium. As illustrated in Figure 3, the macrophages can be activated to synthesize collagenase by bacterial endotoxin (LPS), macrophage-activating factor (MAF) produced by activated lymphocytes, or a variety of other stimulating agents.

Macrophage production of collagenase is regulated by prostaglandins. This is based on *in vitro* observations that production of collagenase by activated macrophages was inhibited significantly by indomethacin, a prostaglandin synthetase inhibitor, and it could be restored by the addition of exogenous prostaglandins to these cells. Macrophages secrete significantly elevated levels of prostaglandins when activated. Elevated prostaglandin levels have been identified in many inflammatory lesions and play a prominent role in perpetuation of the response (see Lewis 1977; Kuehl 1977; Chapters 13 and 15). The regulation of macrophage activity may be one of their many functions. The pathway of prostaglandin influence on the macrophages is through the modulation of intracellular cyclic nucleotides (McCarthy et al. 1978). Elevations in $3'-5'$ cyclic AMP initiated by the prostaglandins trigger the synthesis and release of the enzyme collagenase. Once the collagen molecules have been cleaved by collagenase and further degraded by nonspecific proteases, repair of the injury proceeds with the production of new connective tissue.

Figure 3. Activation of macrophages by the lymphokine, macrophage activating factor (MAF), or by lipopolysaccharide (LPS) causes these cells to produce prostaglandins (PGE), which increase intracellular levels of $3'-5'$cAMP to trigger collagenase synthesis. Collagenase cleaves the collagen molecule which can then be further degraded by other proteolytic enzymes.

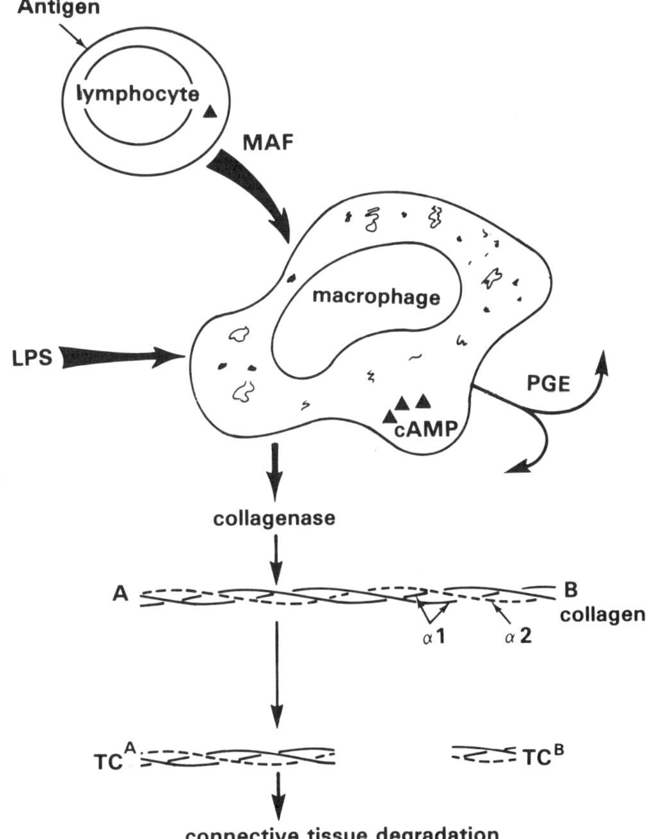

In addition to their enhanced enzymatic activities, stimulated macrophages also synthesize several molecules involved in the perpetuation of the inflammatory process, such as complement proteins and interferon (Table 1). Other molecules produced by activated macrophages include mediators such as lymphocyte activating factor (LAF) (Gery and Waksman 1972), fibroblast activating factor (FAF) (Wahl et al. 1979), fibroblast chemotactic factor (Tsukamoto et al. 1981), and colony stimulating factor (CSF) (Eaves and Bruce 1974), which regulate the functions of other cells.

Lymphoid Cells. In addition to the neutrophils and macrophages, lymphoid cells may be involved in the inflammation associated with tissue damage or may initiate the tissue damage in response to antigenic stimuli. The types of cells involved in the reaction depend on the nature of the stimulus. Although the role of lymphoid cells in antigen-induced inflammation is clear, their role in nonspecifically induced inflammation, such as in a wound caused by mechanical trauma, is less obvious. Plasma cells may produce antibodies to bacterial and other foreign body antigens present and persisting in the wound site. Trentham et al. (1978) have also demonstrated cell-mediated immune responses to collagen. Whatever the stimulus, once activated, the lymphocytes release lymphokines including chemotactic factor, and migration inhibition factor which may promulgate the inflammatory responses (see Chapter 10).

Epithelialization

As inflammation occurs, the epidermal cells close the surface of the wound (Odland and Ross 1968). This regeneration of the epithelium is necessary in the restoration of the integrity of the tissue and may become essential for the survival of the organism. The normal epithelium is regenerating constantly, with the new cells created by mitosis in the basal layer continually replacing those in the more peripheral layers of the epidermis. However, in contrast to the normal situation, mitosis following wounding occurs in those cells close to the edge of the wound and in the outer layers. Skin epithelialization proceeds in four stages (Montandon et al. 1977). The first involves the mobilization of basal cells, which appear to lose their attachment to the underlying dermis. In the second stage, these epithelial cells migrate along the remaining basement membrane or the fibrin deposits, which form a pathway along which the cells can advance. By this mechanism of contact guidance, the cells appear to roll or slide over one another at about 0.2 mm/day. As these epithelial cells migrate, they are actively phagocytic and ingest and degrade serum proteins, fibrin, and other obstacles within their path.

The third stage of epithelialization is a proliferative phase that commences when the epithelial cells are close to the margins of the wound. The basal cells proliferate first, and migrating cells also divide, replenishing the supply of epithelial cells. Simple interruption of the epithelial surface may enable the cells to migrate and replicate; these processes may be inhibited in normal circumstances because of the compression effect of the surrounding cells. Some evidence suggests that the epidermis itself must be damaged directly for the mitotic response to occur. Bullough and Laurence (1964) suggest that increased proliferation is due to the local loss of an inhibitor present and extractable in the normal epidermis. Whatever the mechanism, there is at this time no direct evidence for a role of the inflammatory leukocytes in controlling the functioning of these epithelial cells.

When the migrating cells come into contact with each other they are contact inhibited and cease migrating. These epithelial cells then begin the final phase of epithelialization as they fill the area between the wound margins. This process of differentiation in the skin involves production of keratin and establishment of characteristic intracellular junctions. By the time the leading edges meet and form a continuous layer beneath the scab, the epithelial cells regain their normal appearance and function. Epithelial regenerative processes must generally precede the formation of new connective tissue, and a wound is not considered healed until its epithelial surface is covered. However, epithelialization alone is not sufficient for restoration of functional integrity, since it is only a cellular repair and has little tensile strength.

Vascular Proliferation

Microvascular proliferation is an essential component of many biological processes including wound healing, tumor growth, and chronic inflammation. In this process, large numbers of small blood vessels form throughout the wound. The development of new vessels in granulation tissue takes place through division of endothelial cells lining small blood vessels. Capillaries originate as budlike structures on nearby vessels, penetrate the wound, become canalized, and ramify throughout the wound by the division of their cells. As the capillaries from different sites migrate through the wound they intersect to form an interconnecting network of vessels.

Several possible mechanisms for the initiation of this vascular growth have been proposed. It has been suggested that an oxygen gradient may be partially responsible for the branching of new vessels into the region. Endothelial growth may also represent a nonspecific reparative response to injury caused by vasoactive mediators, cytotoxic factors, or by direct trauma. Alternatively, it may represent proliferation mediated in some way by the inflammatory cells. Because of the localization of inflammatory cell infiltrates in such close proximity to areas of endothelial cell replication, an association between this angiogenesis and mononuclear cells has been postulated. Lymphocytes have been implicated, since capillary proliferation occurs in local graft-versus-host reactions (Sidky and Auerbach 1975) as well as in other immunologically induced host responses (Graham and Shannon 1972; Anderson et al. 1975), but direct evidence is lacking. Recent studies by Polverini and co-workers (1977a, 1977b) demonstrated that macrophages may play a role in controlling vascular proliferation. Neovascularization was assayed in the normally avascular cornea of the guinea pig eye by injecting macrophages or macrophage culture medium into the stroma and observing the growth of vascular sprouts into the cornea. Obvious directional ingrowth of small blood vessels toward the macrophage injection site occurred. The proliferating capillaries underwent extensive branching in response to activated, but not control, macrophage supernatants, suggesting it was necessary for the macrophages to be activated for them to influence vascular proliferation. Neutrophils and activated lymphocytes were ineffective. These studies suggest that included among its diverse functions in an inflammatory response, the macrophage may mediate microvascular proliferation through the secretion of an active molecule.

Once the continuity of the connective tissue has been reestablished, many of the new capillaries regress. The wound changes from a tissue rich in blood vessels and actively dividing cells into one with a much simpler structure with the final scar being poorly vascularized. However, this initial ingrowth of capillaries in the granulation tissue precedes the appearance of significant numbers of fibroblasts. The functioning unit of repair in granulation and scar tissue has been described as a capillary loop with surrounding fibroblasts and mononuclear cells. Capillaries provide oxygen and nutrients, while the fibroblasts synthesize collagen, which is invaded by macrophages and more fibroblasts. New capillaries branch into the developing scar as the process repeats itself to complete repair.

Fibroplasia and Repair

Toward the end of the inflammatory response, the area of the wound containing the newly formed capillary buds is gradually invaded by fibroblasts (Figure 1), forming the granulation tissue. The fibroblasts repair the injury by secreting the collagen and the protein polysaccharides that form the scar tissue. These cells develop from the division of local fibroblasts, and there is evidence that they are capable of directed migration in response to appropriate stimuli. While fibroblasts appear in the wound by the end of the first day, appreciable numbers do not become evident until the fourth or fifth day after injury (Figure 2) with the ingrowth of the granulation tissue capillaries.

Fibroblast Migration

The mechanisms that induce fibroblasts to enter the wound area, proliferate, and increase protein synthesis are currently under study in several laboratories. It has been shown recently that fibroblasts are capable of directed migration toward agents or molecules that might be found in an area of damaged tissue and inflammation. Collagen (Types I, II, and III) and peptides derived by collagenase cleavage of collagen molecules or by direct tissue damage are chemotactic stimuli for fibroblasts (Postlethwaite et al. 1978). A soluble mediator produced by activated macrophages causes fibroblasts to migrate toward the highest concentration of this molecule in an *in vitro* assay (Tsukamoto et al. 1981). Furthermore, these cells appear to respond chemotactically to a factor produced by antigen- or mitogen-stimulated lymphocytes, which appears to be distinct from the factor that attracts macrophages (Postlethwaite et al. 1976).

Fibroblast Proliferation

In addition to the migrating population of fibroblasts that invades the damaged area, the increased number of fibroblasts can, in part, be accounted for by increased proliferation of these cells. The fluid transudate from the vasculature may provide many of the requirements for cell growth. Recent studies may help clarify possible mechanisms for this enhanced fibroblast division. During platelet aggregation one or several factors are released into the serum that stimulate proliferation of fibroblasts (Rutherford and Ross 1976). In addition, macrophages in intimate contact with fibroblasts in the granulation tissue produce a soluble factor (monokine) that activates fibroblasts (Wahl et al. 1979) or additionally, macrophages can activate a serum component which stimulates fibroblast division (Leibovich 1978). The monokine that modulates fibroblast function is a 40,000 to 60,000 molecular weight molecule released early (1 to 4 hours) after macrophage activation, with maximum production at 24 to 48 hours *in vitro*. By release of such a mediator, the macrophages may play an important role in modulating the repair of the wounded area (Figure 4).

Not only have macrophages been shown to influence fibroblast division, but lymphocytes as well can cause resting fibroblasts to divide (Figure 4). Specific antigen-

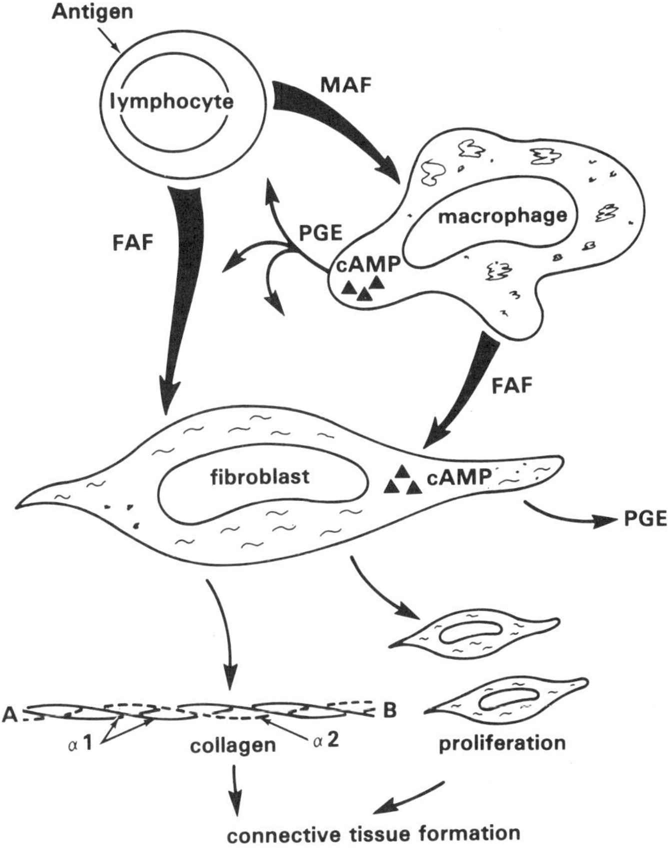

Figure 4. Stimulation of sensitized lymphocytes by specific antigen initiates production of the lymphokine, fibroblast activation factor (FAF), which stimulates fibroblasts to divide and to increase their output of collagen and other proteins. These cells also produce prostaglandins (PGE). Another lymphokine, macrophage activation factor (MAF) stimulates macrophages, which also produce FAF to enhance fibroblast functions including proliferation and collagen synthesis resulting in connective tissue formation.

triggered lymphocytes from sensitized animals produce a soluble lymphokine, fibroblast activating factor (FAF), that stimulates division and other fibroblast functions (Wahl et al. 1978). Thus, it appears that either by specific or nonspecific mechanisms, inflammatory cells can regulate fibroplasia and repair.

Fibroblast Protein Synthesis

The expanded populations of fibroblasts must synthesize the macromolecules of the intercellular matrix to repair the injury. First to be synthesized are the polysaccharide components, including the glycosaminoglycans and the proteoglycans, then collagen synthesis occurs. Lymphocytes (Wahl et al. 1978; Johnson and Ziff 1976) and macrophages (Wahl et al. 1979) appear to play a role in initiating this protein synthesis by fibroblasts that take on the characteristics of actively synthesizing cells, including large nuclei with one or more nucleoli, prominent rough endoplasmic reticulum, and Golgi apparatus. The collagen molecule being synthesized by these cells is a three-stranded helix made up by repeated polymerization of proline, lysine, glycine, and other amino acid residues. The collagen fibrils are formed by aggregation of collagen molecules that organize into fibers forming the connective tissue. Without collagen there is no strength or permanence in the healing wound. Although collagen is often considered inert, it is produced by fibroblasts and other cells and is lysed and absorbed in response to the body's functional needs. Scar tissue begins to form as the small collagen molecules aggregate into larger and larger bundles called fibers. As the synthesis of connective tissue proteins begins to decrease, the process of remodeling takes place, giving the scar tissue its great strength. During remodeling many of the randomly oriented smaller collagen fiberrs formed earlier are broken down and reassembled into thick bundles. As the bundles form they orient themselves along the lines of stress in the healing wound. The final character of the scar is determined by the opposing play of collagen synthesis and destruction.

A readily resolved acute inflammatory response may result in minimal scarring. However, if the injurious agent persists for weeks or longer, the inflammation becomes chronic. Chronic inflammation is characterized by a predominantly mononuclear cell infiltrate and fibroblast proliferation. When this inflammation persists for prolonged periods, considerable scarring may occur, causing tissue deformities such as fibrous replacement of functional parenchymal cells in the liver or kidney, narrowing of the bowels, other organs or vessels, adhesions between serosal surfaces, and keloid formation. Chronic inflammation usually causes permanent damage to the involved tissues. Another type of chronic response is represented by granuloma formation which is an organized cellular reaction to a persistent noxious agent. Granulomatous diseases include tuberculosis, histoplasmosis, shistosomiasis, silicosis, and berylliosis and may be associated with extensive fibrosis. Whereas fibrosis is a normal repair process following tissue injury, in chronic inflammation, such as that found in granulomatous disease, the irreversible fibrosis may lead to organ damage with a potentially fatal outcome. Thus, the host response to an irritant or antigen resulting in fibroplasia and collagen secretion may have deleterious consequences.

Conclusion

Resolution of an inflammatory reaction as exemplified by the response of injured skin is dependent on the participation of several inflammatory cells which infiltrate the wounded area and perform their individual functions. These inflammatory cells, including neutrophils and macrophages—the major scavengers of the body—eliminate dead cells and other debris, such as denatured proteins, bacteria, and other foreign materials. These cells are equipped with the necessary machinery to engulf and degrade this debris at the inflammatory site. Following phagocytosis, the vacuoles that are formed fuse with primary lysosomes containing the hydrolytic enzymes for digestion. In addition, some of these enzymes may be released from the cells so that both intracellular and extracellular destruction of debris may occur. During the later stages of the inflammatory process, macrophages, as well as lymphoid cells, may modulate noninflammatory cell functions in bringing about repair of the wounded tissue. Evidence now exists that macrophages can influence vascular proliferation through the release of a soluble mediator. Furthermore, macrophages and lymphocytes can secrete mediators that stimulate fibroblasts to divide and to produce increased amounts of collagen, which is essential to wound repair. Thus, lymphokines and monokines, produced locally at the inflammatory site may contribute to the final repair and restoration of functional integrity of an area of injury.

References

Anderson, N.D., Anderson, A.O., Wyllie, R.G. (1975) Microvascular changes in lymph nodes draining skin allografts. Am. J. Pathol. 81,131.

Bryant, W.M. (1977) Wound Healing. Clinical Symposia 29,1.

Bullough, W.S., Laurence, E.B. (1964) Mitotic control by internal secretion: The role of the chalone-adrenaline complex. Exp. Cell. Res. 36,192.

Cochrane, C.G. (1968) Immunologic tissue injury mediated by neutrophilic leukocytes. Adv. Immunol. 9,97.

Cochrane, C.G., Aikin, B.S. (1966) Polymorphonuclear leukocytes in immunologic reactions. The destruction of vascular basement membrane *in vivo* and *in vitro*. J. Exp. Med. 124,733.

Eaves, A.C., Bruce, W.R. (1974) *In vitro* production of colony-stimulating activity I. Exposure of mouse peritoneal cells to endotoxin. Cell Tissue Kinet. 7,19.

Gallin, J.I., Quie, P.G. (eds.) (1978) *Leukocyte Chemotaxis: Methods Physiology, and Clinical Implications.* New York, Raven Press.

Gery, I., Waksman, B.H. (1972) Potentiation of the T-lymphocyte response to mitogens. II. The cellular source of potentiating mediators. J. Exp. Med. 136,143.

Graham, R.C., Shannon, S. (1972) Peroxidase arthritis. II. Lymphoid cell-endothelial interactions during developing immunologic inflammatory responses. Am. J. Pathol. 69,7.

Harris, H. (1960) Mobilization of defensive cells in inflammatory tissue. Bacteriol. Rev. 24,3.

Huber, H., Fudenberg, H.H. (1968) Receptor sites of human monocytes for IgG. Int. Arch. Allergy 34,18.

Johnson, R.L., Ziff, M. (1976) Lymphokine stimulation of collagen accumulation. J. Clin. Invest. 58,240.

Kuehl, F.A., Jr. (1977) Prostaglandins in the regulation of immune and inflammatory responses. In: *Immunopharmacology.* Hadden, J.W., Coffey, R.G., Spreafico, F. (eds.) New York, Plenum Press, p. 145.

Kaplan, A.P., Guetzl, E.J., Austen, K.F. (1973) The fibrinolytic pathway of human plasma II. The generation of chemotactic activity by activation of plasminogen proactivator. J. Clin. Invest. 52,2591.

Kaplan, A.P., Kay, A.B., Austen, K.F. (1972) A prealbumin activator of prekallikrein: III. Appearance of chemotactic activity for human neutrophils by the conversion of human prekallikrein to kallikrein. J. Exp. Med. 135,81.

Lay, W.H., Nussenzweig, V. (1968) Receptors for complement on leukocytes. J. Exp. Med. 128,991.

Leibovich, S.J. (1978) Production of macrophage-dependent fibroblast-stimulating activity (M-FSA) by murine macrophages. Effects on BALBc 3T3 fibroblasts. Exp. Cell. Res. 113,47.

Leibovich, S.J., Ross, R. (1975) The role of the macrophage in wound repair. A study with hydrocortisone and antimacrophages serum. Am. J. Pathol. 78,71.

Lewis, G.P. (1977) Prostaglandins in inflammation. J. Reticul. Soc. 22,389.

McCarthy, J.B., Wahl, S.M., Rees, J., Olsen, C.E., Sandberg, A.L., Wahl, L.M. (1980) Mediation of macrophage collagenase production by 3'5' cyclic adenosine monophosphate. J. Immunol. 124,2405.

Menaker, L. (1975) *Biologic Basis of Wound Healing.* Hagerstown, Md., Harper and Row.

Montandon, D., D'Andiran, J., Gabbiari, G. (1977) The mechanism of wound contraction and epithelialization. Clinics in Plastic Surgery 4,325.

Odland, G., Ross, R. (1968) Human wound repair. I. Epidermal regeneration. J. Cell. Biol. 39,135.

Polverini, P.J., Cotran, R.S., Gimbrone, M.A., Jr., Unanue, E.R. (1977a) Activated macrophages induce vascular proliferation. Nature 269,804.

Polverini, P.J., Cotran, R.S., Sholley, M.M. (1977b) Endothelial proliferation in the delayed hypersensitivity reaction: An autoradiographic study. J. Immunol. 118,529.

Postlethwaite, A.E., Kang, A.H. (1976) Collagen- and collagen peptide-induced chemotaxis of human blood monocytes. J. Exp. Med. 143,1299.

Postlethwaite, A.E., Seyer, J.M., Kang, A.H. (1978) Chemotactic attraction of human fibroblasts to type I, II, and III collagens and collagen-derived peptides. Proc. Natl. Acad. Sci. U.S.A. 75,871.

Postlethwaite, A.E., Snyderman, R., Kang, A.H. (1976) The chemotactic attraction of human fibroblasts to a lymphocyte-derived factor. J. Exp. Med. 144,1188.

Ross, R. (1968) Wound healing. Sci. Am. 220,40.

Rutherford, R.B., Ross, R. (1976) Platelet factors stimulate fibroblasts and smooth muscle cells quiescent in plasma serum to proliferate. J. Cell. Biol. 69,196.

Sidky, Y.A., Auerbach, R. (1975) Lymphocyte induced angiogenesis. A quantitative and sensitive assay of the graft-vs-host reaction. J. Exp. Med. 141,1084.

Simpson, D.M., Ross, R. (1972) The neutrophilic leukocyte in wound repair. A study with antineutrophil serum. J. Clin. Invest. 51,2009.

Snyderman, R., Mergenhagen, S.E. (1976) Chemotaxis of macrophages In: *Immunobiology of the Macrophage* Nelson, D.S. (ed.) New York, Academic Press, p. 323.

Stossel, T.P. (1974) Phagocytosis. New England J. Med. 290,717.

Trentham, D.E., Townes, A.S., Kang, A.H., David, J.R. (1978) Humoral and cellular sensitivity to collagen in Type II collagen-induced arthritis in rats. J. Clin. Invest. 61,89.

Tsukamoto, Y., Helsel, W.E., Wahl, S.M. (1981) Macrophage production of fibronectin, a chemoattractant for fibroblasts (submitted for publication).

Unkeless, J.C., Gordon, S., Reich, E. (1974) Secretion of plasminogen activator by stimulated macrophages. J. Exp. Med. 139,834.

Wahl, L.M., Olsen, C.E., Sandberg, A.L., Mergenhagen, S.E. (1977) Prostaglandin regulation of macrophage collagenase production. Proc. Nat. Acad. Sci. U.S.A. 74,4955.

Wahl, L.M., Wahl, S.M., Mergenhagen, S.E., Martin, G.R. (1974) Collagenase production by endotoxin-activated macrophages. Proc. Nat. Acad. Sci. U.S.A. 71,3598.

Wahl, L.M., Wahl, S.M., Mergenhagen, S.E., Martin, G.R. (1975) Collagenase production by lymphokine-activated macrophages. Science 187,261.

Wahl, S.M., Wahl, L.M., McCarthy, J.B. (1978) Lymphocyte-mediated activation of fibroblast proliferation and collagen production. J. Immunol. 121,942.

Wahl, S.M., Wahl, L.M., McCarthy, J.B., Chedid, L., Mergenhagen, S.E. (1979) Macrophage activation by mycobacterium water soluble compounds and synthetic muramyl depeptide. J. Immunol. 122,2226.

Wilkinson, P.C. (1974) *Chemotaxis and Inflammation*. London, Churchill Livingstone.

Index